ELEMENTARY BIOCHEMISTRY

(For Undergraduate Students)

J.L. JAIN
MSc, PhD, FIES, FLS (London)
Former Reader in Botany Department
MS College, Saharanpur (UP), INDIA

SUNJAY JAIN
BSc, MBBS, MS (Ophth.), FCLI
Affiliate, SEE Intl. (USA)

NITIN JAIN
BSc, PGDBA, ERP (USA)
OCP (USA)

S. CHAND
PUBLISHING

S. Chand And Company Limited

(ISO 9001 Certified Company)

S. Chand And Company Limited
(ISO 9001 Certified Company)
Head Office: D-92, Sector–2, Noida – 201301, U.P. (India), Ph. 91-120-4682700
Registered Office: A-27, 2nd Floor, Mohan Co-operative Industrial Estate,
New Delhi – 110 044, Phone: 011-49731800
www.**schandpublishing.com**; e-mail: **info@schandpublishing.com**

Marketing Offices:

Chennai	:	Ph: 23632120; chennai@schandpublishing.com
Guwahati	:	Ph: 2738811, 2735640; guwahati@schandpublishing.com
Hyderabad	:	Ph: 40186018; hyderabad@schandpublishing.com
Jalandhar	:	Ph: 4645630; jalandhar@schandpublishing.com
Kolkata	:	Ph: 23357458, 23353914; kolkata@schandpublishing.com
Lucknow	:	Ph: 4003633; lucknow@schandpublishing.com
Mumbai	:	Ph: 25000297; mumbai@schandpublishing.com
Patna	:	Ph: 2260011; patna@schandpublishing.com

First Edition 1983
Second Revised Edition 1990
Third Revised Multicolour Edition 2007, Reprint 2014
Single Colour Edition 2018; Reprint 2020, 2022

Reprint 2023

ISBN : 978-81-219-2816-8 **Product Code :** H6BCH68BCHM10ENAC07O

PRINTED IN INDIA

By Vikas Publishing House Private Limited, Plot 20/4, Site-IV, Industrial Area Sahibabad, Ghaziabad – 201 010 and Published by S. Chand And Company Limited, A-27, 2nd Floor, Mohan Co-operative Industrial Estate, New Delhi – 110 044.

Dedicated to
my reverend mother

LATE SMT. KAPOORI DEVI JAIN

(20-06-1903 — 28-05-1987)

Humanitarian, Compassionate, Religious and a True Mother

– J.L. Jain

"M" is for the million things she gave me,
"O" means only that she's growing old,
"T" is for the tears she shed to save me,
"H" is for her heart of purest gold ;
"E" is for her eyes, with love-light shining,
"R" means right, and right she'll always be,
Put them all together, they spell "MOTHER",
A word that means the world to me."

— *Howard Johnson*

"We must trust in nothing but facts which are presented to us by nature and cannot deceive."

—Antoine Laurent Lavoisier (LT, 1743-1794)

ABOUT THE AUTHOR

Dr. J.L. Jain

Dr. J.L. Jain (born, 20.12.1938) has about 35 years of teaching experience of undergraduate and postgraduate classes. He did M.Sc. (Botany), with specialization in Cytogenetics and Plant Breeding, from Agra University, Agra and was conferred Ph.D. (Virology) from Meerut University, Meerut. He has published about a dozen research papers in various journals of international repute. He has also contributed articles of varied interest in many Magazines and Periodicals.

Dr. Jain has authored many text books pertaining to Botany and Biochemistry, catering to the needs of High School to postgraduate and research students and are being used in India and abroad. He has been an active member of the Leadership Project in Botany, sponsored by the University Grants Commission, New Delhi. He is presently associated with many universities, and prestigious institutions such as Banasthali Vidyapeeth, Jaipur and Madhya Pradesh Bhoj (Open) University, Bhopal, as an expert and counsellor.

Dr. Jain is the recipient of **'Sahitya Bhooshan'** degree from Lucknow, in 1953. He has the rare distinction of being selected fellow of the coveted **'The Linnean Society of London'**, the oldest extant society in the world (founded in 1788) devoted to Life Sciences and patronized by Her Most Excellent Majesty Queen Elizabeth II and was awarded **F.L.S.** in 1994, in appreciation of his everlasting contributions in Life Sciences.

PREFACE TO THE THIRD EDITION

At the outset, the senior author of the book welcomes his two sons, Dr. Sunjay Jain and Er. Nitin Jain who have joined him as coauthors of this text, a credit which would have been given earlier to them as they were helping in a latent way in the evolution of the book for the past many years.

Thirty four years have elapsed since one of us (JLJ) embarked on the intellectual journey of writing a textbook of Biochemistry. As Biochemistry is growing at a dazzling pace, each edition has been demanding in a different way. In this latest 3rd edition, the book has been thoroughly revised, enlarged and updated with not even a single chapter left untouched. Besides, **four new chapters,** *viz.*, **Water : The Solvent of Life, pH and Buffers, The Atom and Chemical Bonds,** and **Biochemistry Experiments** have been added to enhance the scope and utility of the book. Addition of **six new appendices** is an added charm of the book, including `**The Nobel Prizes**', the world's most venerated awards. Above and over all, this time the book has been presented in **multicolour edition** with profuse colour illustrations so as to increase its clarity, understandability and legibility, especially of the diagrams.

It is hoped that the present book, in its revised and enlarged multicolour form, would serve in a still better way, the authors are keenly desirous of.

Gratitude is expressed to the students and teachers, both from India and abroad, who have sent in their valuable suggestions which have been given due consideration. We are sincerely thankful to our publishers, specially to Shri Ravindra Kumar Gupta, the Managing Director of the company, for his wholehearted cooperation and goodwill gesture accorded to.The entire staff of the company deserves appreciation for their unflinching support during the gestation period of this book, specifically Mr. Shishir Bhatnagar, the Editor and Mr. Benny K. Abraham for design and layout of this venture. We are also deeply indebted to our own Mrs. Mithlesh K. Jain M.A., for her sustained support of this endeavour from its inception; her wisdom has made all the difference. In fact, we are grateful to our whole family for their patience, buoyancy and encouragement of this venture which was more arduous than anticipated.

Healthy criticism and suggestions for further improvement of the book are solicited.

4/228, Taunk Building
Kacheri Ghat
AGRA - 282 004
Phone: 0562-3253621

J.L. Jain
Sunjay Jain
Nitin Jain

"The preface of a book is kept first, written last and read least."
— *Alfred Lotke*

PREFACE TO THE FIRST EDITION

Biochemistry is a new branch of science which aims at answering, in chemical language, questions such as 'What is the life made of ?' and 'How does it work ?' Whereas the eye works at the gross level of visible objects, the microscope reaches down to the cellular level, exposing details of the various cell organelles, including nuclei and other particles. Biochemistry, however, works at a still finer level that cannot be accessed by the ultra-modern optical or phase-contrast microscopes. In other words, it operates at a molecular level and thus brings to light the hidden secrets of life. The rapid development and enormous expansion of every phase of biochemistry has not only markedly enriched our knowledge about the nature of life but has also made biochemistry the very language of life itself.

Realising the paramount importance of this discipline of science, degree courses in biochemistry are now offered in a good number of colleges and universities. But the students offering this subject at undergraduate level come as raw hand, as biochemistry is not taught to them at the school level. It is, in fact, to meet the requirements of such students that this book has been written. Besides, the book may also serve an useful purpose for higher studies.

The book contains the basic concept of biochemistry written in a manner suited to the broad spectrum of the college students. The matter has been lucidly presented, orderly arranged and profusely illustrated. References have been included at the end of each chapter in order to guide the reader to the classical and current literature. A few appendices are given at the end. These may, however, prove useful to the reader. Some illustrations have been adapted from standard texts, journals and research papers. The sources of all such figures have been duly acknowledged in their legends. The author feels greatly beholden to their authors and publishers.

I am also thankful to my publishers, specially to Sri Shyam Lal Gupta and Sr. T.N. Goel for taking keen interest and in bringing out the book in its present form. Finally, I am grateful to my wife Mrs. Mithlesh K. Jain M.A., for inspiring me in proceeding from thoughts to print.

Helpful suggestions for improvement are welcome.

Saharanpur **J.L. JAIN**

LIST OF CHAPTERS

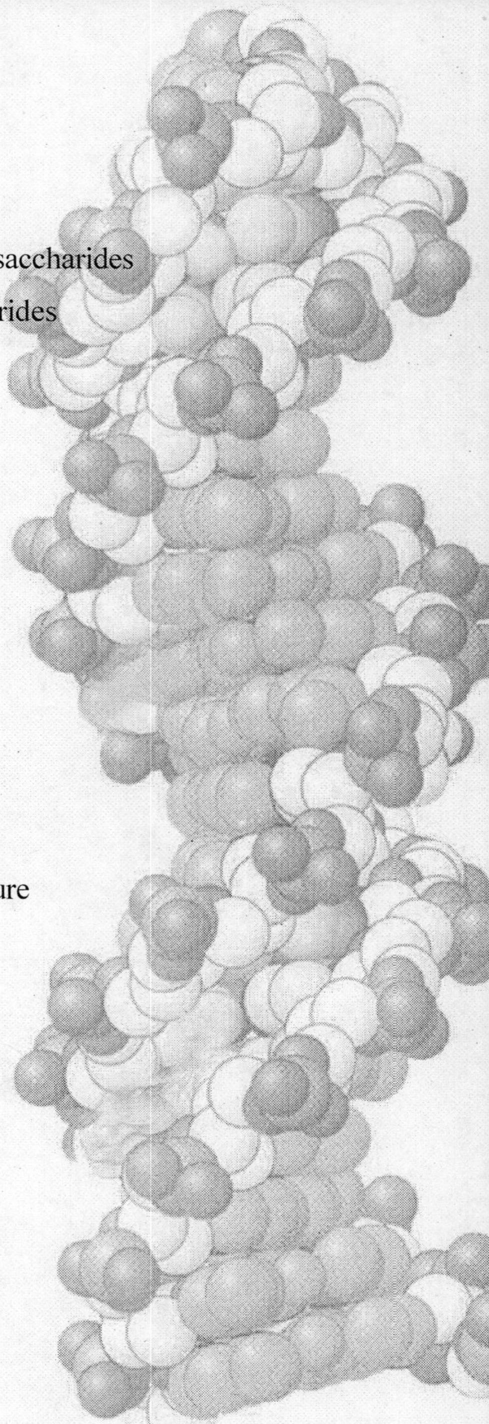

CONTENTS

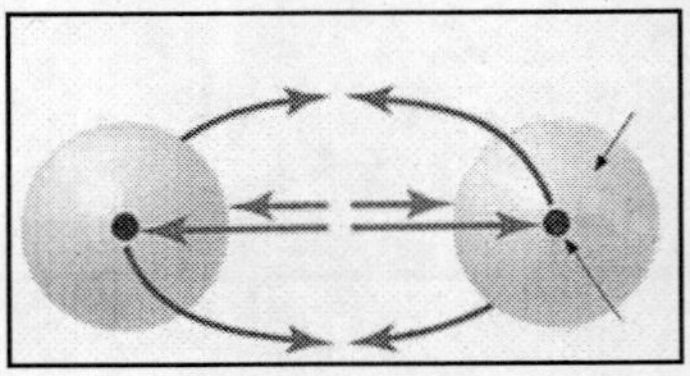

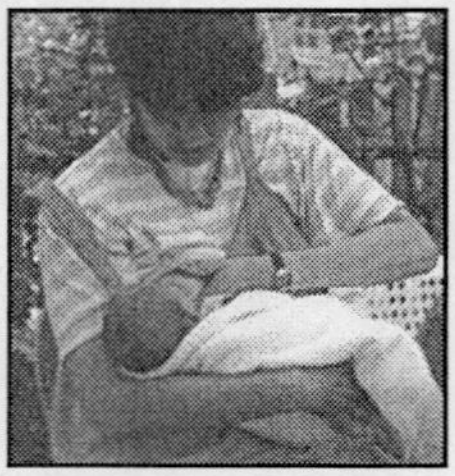

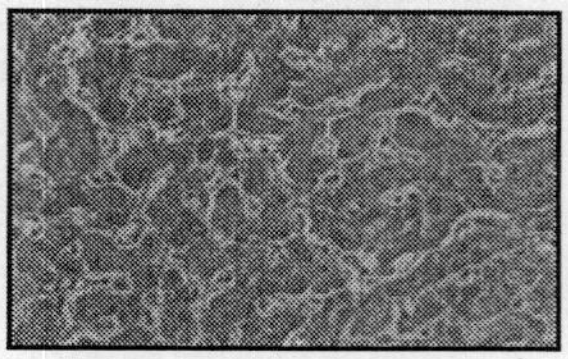

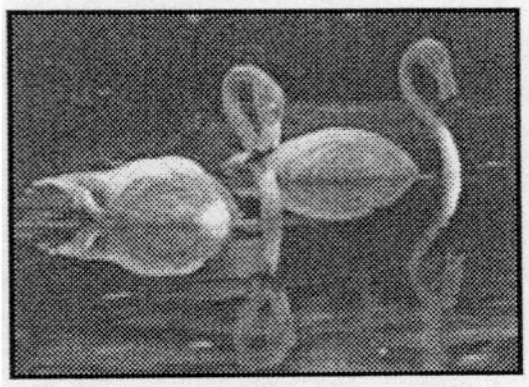

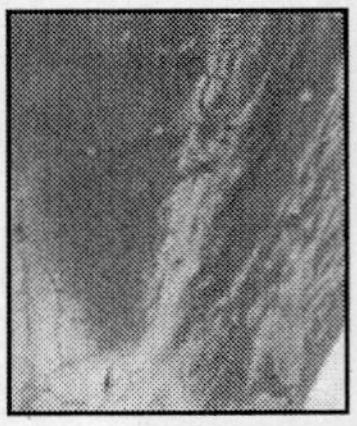

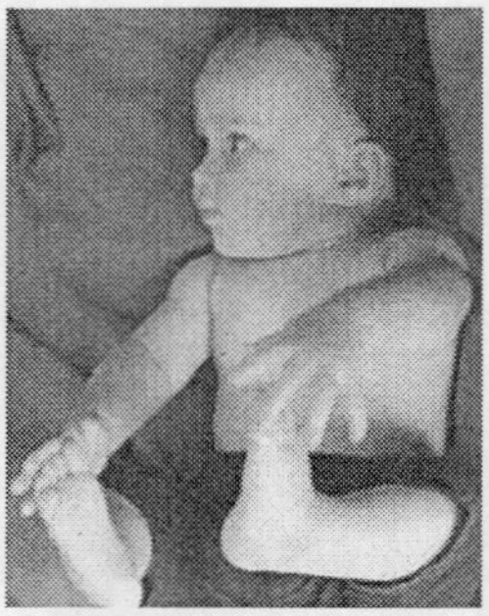

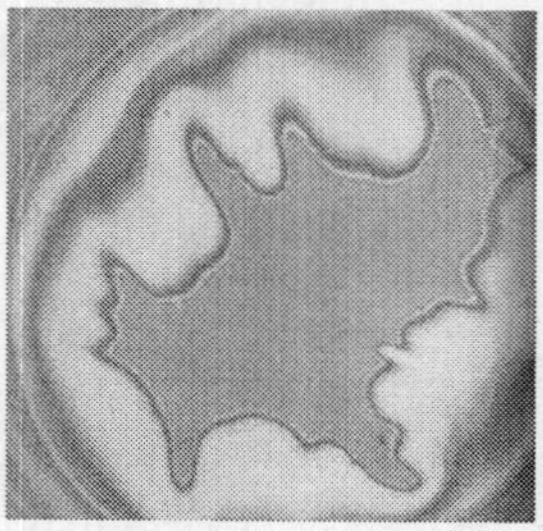

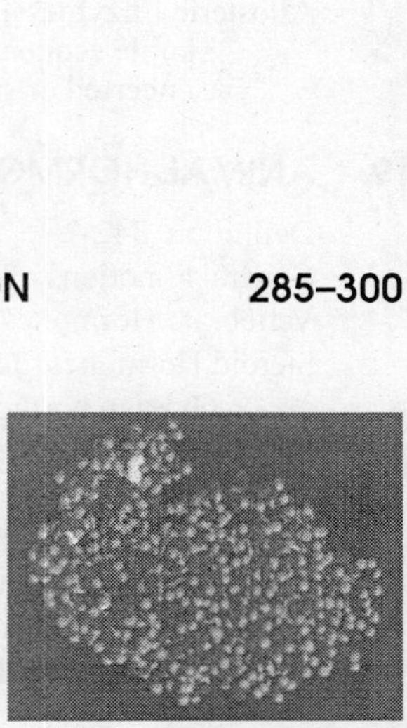

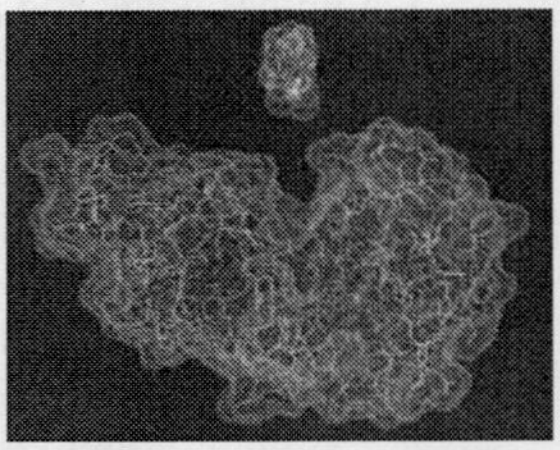

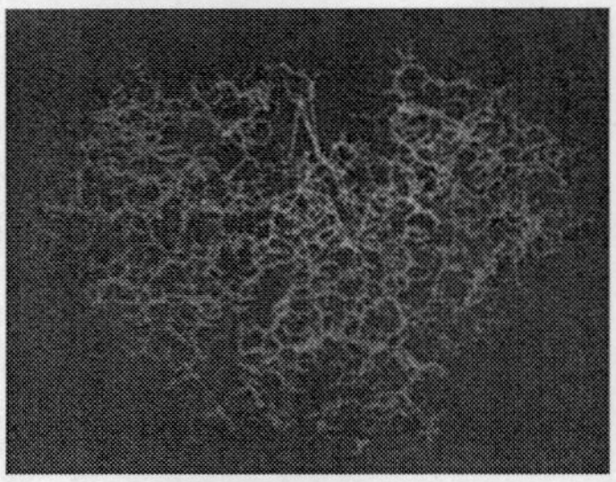

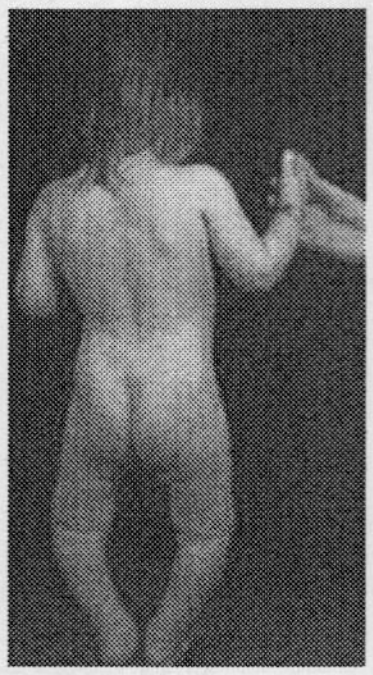

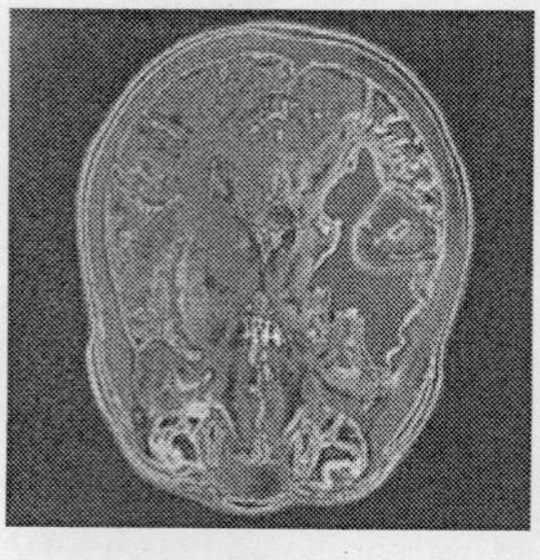

CHAPTER

1 Introduction

INSIDE

- Definition
- Historical Resume
- Biochemistry as Molecular Logic of Living Organisms
- Nature
- Major Organic Compounds of Animate Objects
- The Scientific Method of Approach
- Importance

Ubiquitous nature of the basic chemicals
All the plants and animals in this photograph are made from the same basic types of chemicals, *i.e.*, carbohydrates, proteins, lipids and nucleic acids.

DEFINITION

The term **Biochemistry*** (*bios*^{G} = life) was first introduced by a German chemist **Carl Neuberg** in 1903. Biochemistry may be defined as a science concerned with the chemical nature and chemical behaviour of the living matter. It takes into account the studies related to the nature of the chemical constituents of living matter, their transformations in biological systems and the energy changes associated with these transformations. Such studies have been conducted in the plant and animal tissues both. Broadly speaking, biochemistry may thus be treated as a discipline in which biological phenomena are analyzed in terms of chemistry. The branch of biochemistry, for the same reason, has been variously named as **Biological Chemistry** or **Chemical Biology**.

In fact, biochemistry originated as an offshoot from human physiology when it was realized that the chemical analysis of urine, blood and other natural fluids can assist in the diagnosis of a particular disease. Hence in its infancy, biochemistry was accordingly known as **Chemical Physiology**. But physiology now covers the study of normal functions and phenomena of living beings. And biochemistry is concerned particularly with the chemical aspects of these functions

*For a detailed treatise of the subject Biochemistry as a whole, consult same 3 authors another multicolour text, **FUNDAMENTALS OF BIOCHEMISTRY,** 6th edition (2005), published by **S. Chand & Company Ltd., New Delhi**.

and phenomena. In other words, biochemistry is but one of the many ways of studying physiology. The two may be compared by watching the monkeys in a zoo which means studying the physiology of behaviour. But if the behaviour of animal molecules is studied, rather than the whole animals, it would form the study of biochemistry.

Modern biochemistry has two branches, descriptive biochemistry and dynamic biochemistry. **Descriptive biochemistry** is concerned with the qualitative and quantitative characterization of the various cell components and the **dynamic biochemistry** deals with the elucidation of the nature and the mechanism of the reactions involving these cell components. While the former branch is more a concern of the organic chemist, the latter branch has now become the language of modern biochemistry.

However, as the knowledge of biochemistry is growing speedily, newer disciplines are emerging from the parent biochemistry. Some of the disciplines are **enzymology** (science of the study of enzymes), **endocrinology** (science dealing with the endocrine secretions or the hormones), **clinical biochemistry, molecular biochemistry** etc. Along with these branches certain link specialities have also come up such as agricultural biochemistry, pharmacological biochemistry etc.

HISTORICAL RESUME

In terms of history, biochemistry is a young science. It started largely as an offshoot of organic chemistry and later incorporated ideas obtained from physical chemistry. In fact, the science of biochemistry may be regarded to have begun with the writings of **Theophrastus Bombastus von Hohenheim,** better known as **Philippus Aureolus Paracelsus** (Lifetime or LT, 1493–1541), a Swedish physician and alchemist who laid the foundation of chemotherapy as a method of treating diseases and also the promoter of '*doctrine of signatures*.' He first acquired the knowledge of chemistry of his time and then entered the field of medicine to apply his knowledge of chemistry. He proclaimed, "Life processes are essentially of chemical nature and diseases can be cured by medicines". Later, his followers notably **Jan Baptist van Helmont** (LT, 1577–1644) amalgamated the science of chemistry with medicine which emerged under the name of 'medical chemistry' (or Iatrochemistry).

The basis of biochemistry was, in fact, laid down by chemists like Scheele and Lavoisier. **Karl Wilhelm Scheele** (LT, 1742–1786), a swedish pharmacist, discovered the chemical composition of various drugs and the plant and animal materials. He also isolated a number of substances such as citric acid from lime juice, lactic acid from sour milk, malic acid from apple and uric acid from urine. Scheele, thus, laid the foundation of descriptive biochemistry. Similarly, **Antoine Lavoisier** (LT, 1743–1794), a French chemist, who studied the composition of air and propounded the theory of conservation of matter, put dynamic biochemistry on firm standfootings. He developed the concept of oxidation and also clarified the nature of animal respiration. He had concluded that respiration could be equated with combustion and that it was slower but not essentially different from the combustion of charcoal. Lavoisier is often spoken of as '*father of modern biochemistry*'.

Scheele can probably be called the unluckiest chemist in History, for although he discovered several elements such as barium, chlorine, manganese and many others, he does not receive undisputed credit for having discovered even a single one.

The earliest book relating to Biochemistry was *'Lectures in Animal Chemistry'* published by the famous Swedish chemist **Jöns Jacob Berzelius** (LT, 1779–1848), in 1806.

In 1828, **Friedrich Wöhler**, a German chemist, synthesized urea, a substance of biological origin, in the laboratory from the inorganic compound ammonium cyanate. This achievement was the unexpected result of attempts to prepare ammonium cyanates through the treatment of metal cyanates with ammonium salts. As Wöhler phrased it in a letter to a colleague, "I must tell you that I can prepare urea without requiring a kidney or an animal, either man or dog." This was a shocking statement in its time, for it breached the presumed barrier between the living and the nonliving. Consequently, this rendered the vitalistic theory of organic materials untenable. *Vitalistic theory* (= *the doctrine of vitalism*) maintained that organic compounds could be synthesized only through the agency of a vital force, supposed to be present only in living tissues.

Wöhler's work was followed by the synthesis of acetic acid from inorganic materials by another German chemist **Adolf Wilhelm Hermann Kolbe** (LT, 1818–1884), in 1845. However, the final blow to the theory of vital force was given by a French Chemist **Pierre Eugene Marcellin Berthellot** (LT, 1827–1907) who synthesized a host of organic compounds (such as methyl alcohol, ethyl alcohol, methane, benzene, acetylene) from inorganic compounds in 1850s. Vitalism was, thus, quietly laid to rest. Organic synthesis remains very much alive.

FRIEDRICH WÖHLER
(LT, 1800-1882)

Wöhler, a German chemist, is well-known for the historical synthesis of urea. He worked with Liebig on the benzoyl derivatives. Wöhler earned worldwide acclaim for his teaching during his 46 years at the University of Göttingen. Between 1845 and 1866, he lectured to some 8,250 students. In his lifetime, 13 and 15 editions of his Organic and Inorganic texts, respectively, were published. He is also credited with the discovery of calcium carbide and isolation of beryllium and yttrium.

Justus von Liebig, a German chemist and discoverer of chloroform and who is often termed as '*father of agricultural chemistry*', arrived at the conclusion that "the nutritive materials of all green plants are inorganic substances." He wrote many books which provided an impetus in the early development of biochemistry. Of special interest is his book '*Organic Chemistry in Its Applications to Physiology and Pathology*', published in 1842. **Michel Chevreul** (LT, 1786–1889) demonstrated through studies on saponification that fats were composed of glycerol and fatty acids, of which he isolated several. The excellent researches conducted by the great German biochemist **Hermann Emil Fischer** (LT, 1852–1919) may be regarded as landmark in the development of structural biochemistry. In the course of his studies, this remarkable man completely revolutionized research concerning the structures of carbohydrates, amino acids and fats. Although nucleic acids are the newest of the 4 great groups of biochemical materials, their discovery goes back to observations by **Friedrich Miescher** (LT, 1844–1895) in 1869. His discovery of nucleic acids in the nuclei of pus cells, obtained from discarded surgical bandages, led him to investigate the distribution and the properties of these compounds.

JUSTUS VON LIEBIG
(LT, 1803–1873)

Justus von Liebig, by age 36, headed the world's largest laboratory and school for training chemists. He was one of the early investigators of large-scale research. Today, his laboratory with its furnishings is preserved as a museum in Giessen, Germany. Liebig's oft-repeated quotation, for which he is famous, reads as :

"The secret of all those who make discoveries is to look upon nothing as impossible."

During the first half of the nineteenth century, the studies on heat conducted mainly by **V. Mayer** and **Ludwing von Helmholtz** (LT, 1821–1894) led to the formulation of the 'laws of thermodynamics' which are essential to the understanding of energy relations in biological systems.

Besides respiration, the other physiological process to attract the attention of biochemists was that of digestion. Main contributions towards this were made by **van Helmont, Abbé Lazaro Spallanzani** (LT, 1729–1799), **René Antoine de Réaumur, William Beaumont** and **Claude Bernard**.

Claude Bernard (LT, 1813–1878) of Paris was perhaps the greatest of these. His contributions included the discovery of liver glycogen and its relation to blood sugar in health and disease. He noted the digestive properties of pancreatic juice and began research in muscle and nerve physiology.

The process of fermentation is probably the single most important process around which the interest of biochemists persisted for a considerable period. By about 1780, fermentation had been recognized by **Theodor Schwann** (LT, 1810–1882) as a biological process. He established that yeast was a plant capable of converting sugar to ethanol and carbon dioxide. However, many of the leading chemists of the day, including Berzelius, Wöhler and Leibig, considered yeast to be nonliving and fermentation to be caused solely by oxygen. Ridicule by Liebig and Wöhler delayed acceptance of Schwann's views until the illustrious French microbiologist **Louis Pasteur** (LT, 1822–1895) presented evidence overwhelming all objections. He founded the useful branch of Microbiology in 1857 and identified several organisms that carried out various fermentations, including that leading to butyric acid, a type performed by organisms that function without oxygen. He defined fermentation as "*la vie sans l'air*" (life without air). Pasteur, thus, introduced the concept of aerobic and anaerobic organisms and their associated fermentations. These conclusions again aroused the ire of Liebig who again took sharp exception but this time with less effect. Such studies on fermentations climaxed by the demonstration in 1887 of **Eduard Buchner** (LT, 1860–1917) that sugars could be fermented by cell-free extracts of yeast. This discovery represents the cornerstone of much of the enzymological and metabolic study of the twentieth century since it led to techniques of isolation and identification that ultimately permitted the study of enzymes and the individual reactions concerned. It also expunged, yet once and for all, any traces of vitalism (or vitalistic theory) still lingering.

Researches conducted by pioneers such as Arrhenius, van't Hoff and Ostwald on electrolytic dissociation and osmotic pressure led physical chemists to turn their attention to biological phenomena. **Soren Sörensen** (LT, 1868–1939), a Danish chemist, developed our concept on pH, **Jacques Loeb** (LT, 1859–1924) studied the colloidal behaviour of proteins and their effect on the cell, **Leonor Michaelis** placed the concept of chemical compound formation between enzyme and substrate on an experimental basis and **Wendell Stanley,** a biochemist working at the Rockfeller Institute, New York, showed that viruses are nucleoproteins. For his fundamental research on TMV, Stanley was awarded a share of the 1946 Nobel Prize in Chemisty. Later, many instruments were invented such as Van Slyke blood gas apparatus, the ultracentrifuge of **Theodore Svedberg** (LT, 1884-1971) and the electrophoresis apparatus of **Arne W.K. Tiselius**. The use of isotopes in biochemical research by **Urey** and **Schoenheimer** and the application of chromatography, first developed by **Martin** and **Synge,** opened a new chapter in modern biochemistry.

With the advent of twentieth century, biochemistry burst into full bloom. Important developments took place rapidly on several fronts including nutrition. The significance of unknown food factors was clearly recognized by **Frederick Gowland Hopkins** at Cambridge University and his associates, who developed the *concept of deficiency diseases*. Extensive series of feeding experiments utilizing synthetic diets were conducted mainly by Babcook McCollum, Osborne, Mendel and Sherman. As a result, many deficiency diseases such as scurvy, rickets, beriberi and pellagra were recognized and their curative agents, which were called vitamins by a Polish biochemist **Casimer Funk,** were isolated and subsequently characterized.

Buchner's work on cell-free fermentation of sugars was actively extended in many laboratories, including those of **Harden** and **Young, Embden** and **Meyerhof,** with the result the complete biochemical pathway known as *Embden-Meyerhof-Parnas pathway* (or *glycolysis*) was elucidated. The researches conducted by **Warburg, Heinrich Wieland, Keilin** and **Theorell** led to the discovery of enzymes and cofactors involved in cellular oxidation. Later, **Fritz Albert Lipmann** (LT, 1899–1986) and **Kurt Henseleit** made notable observations on the significance of the terminal pyrophosphate linkages of ATP as an energy storage reservoir. **Albert Szent-Györgyi** and **Hans Adolf Krebs** (LT, 1900–1981) of England studied the fate of lactate (or pyruvate) during aerobic oxidation. This led to the development of a sequence of reactions known as *Krebs cycle* (or *citric acid cycle*). Later studies revealed that the fatty acids and amino acids, upon oxidation, also yield intermediates that are identical with those in the Krebs cycle, thus providing a common mechanism for the liberation of energy from all foodstuffs. **Frederick Sanger** established the complete amino acid sequence of the protein hormone insulin and **Vincent du Vigneaud** proved the structure of the nonapeptide hormones of posterior pituitary by direct synthesis.

The brilliant studies by **Linus Carl Pauling** (LT, 1901–1994) and **Robert Corey** (LT, 1897–1971) led to the concept of a secondary structure of protein molecules in the form of an α-helix. A similar kind of structure for the nucleic acids was also elucidated. **James Dewey Watson** and **Francis Harry Compton Crick,** in 1953, proposed that a double-stranded DNA molecule could be made by binding bases on adjacent strands to each other by hydrogen bonding. This *base-pairing hypothesis* was confirmed by the quantitative data of an Austrian refugee biochemist, **Erwin Chargaff** and was soon followed by the enzymatic synthesis of DNA by **Arthur Kornberg.** These synthetic macromolecules have properties that suit the Watson-Crick hypothesis. Soon after, the base sequence of transfer RNA molecules specific for different amino acids was determined by many workers such as **Robert W. Holley, Medison, Zachan** and others. A final and accurate list of the base sequences in messenger RNA that code for each of the amino acids was made available as a result of the brilliant researches by **Marshall Nirenberg**, who used synthetic nucleotides as messenger molecules to identify the coded base sequences for each of the amino acids.

By 1835, **Jönes Jacob Berzelius,** a Swedish chemist, had clearly recognized the importance of catalysis in controlling the rates of chemical processes. He suggested that the formation of biological materials was controlled by catalytic actions and cited an enzyme from potato (*i.e., potato diastase*) as an example of a biological catalyst effective in the hydrolysis of starch. He reasoned that all materials of living tissues are formed under the influence of catalytic action, a conclusion thoroughly established by subsequent work. Later, many catalysts were isolated, purified to some extent and the associated reactions investigated kinetically. However, the chemical nature of these so-called biocatalysts or enzymes remained unknown until 1926, when **James B. Sumner** (LT, 1887–1955) at Cornell University, for the first time, crystallized the enzyme *urease* from the extracts of Jack bean and demonstrated its protein nature. For many years, Sumner's discovery was greeted with skepticism and derision, most particularly by the renowned German biochemist **Richard Willstätter**, an authoritative figure, who insisted that enzymes are low-molecular-weight compounds and that the protein found in urease crystals was merely a contaminant. Sumner was a tenacious man, armed with a body of convincing evidence and did not surrender to authority and even successfully repeated his experiments in Willstätter's laboratory. Subsequently, **Johan Howard Northrop** and **Kunitz** crystallized a series of pancreatic and gastric enzymes. This work and subsequent isolation studies clearly confirmed that enzymes are proteins and established Sumner as the '*father of modern enzymology*'. Invaluable work in this field has been accomplished by **Stanford Stein**, **William Moore, Max Ferdinand Perutz, John C. Kendrew** and **David C. Phillips.**

Perhaps the greatest impetus in biochemical researches is nowadays being given to the mechanisms for the regulation of the synthesis of cellular compounds. The phenomenon of feedback inhibition of enzyme activity by the end product of a reaction sequence illustrates a self-regulating mechanism. Enzyme induction and repression—the acceleration or inhibition of synthesis of an enzyme—have been described. These discoveries have led to a hypothesis proposed in 1961 by two Frenchmen, **Francois Jacob** and **Jacques Monod,** suggesting that the DNA molecules consist of areas in which genes are maintained in an inactive state (by repressors) until they need to be activated for the production of messenger RNA molecules. In 1963, Jacob and Monod, with **Jean-Pierre Changeux**, also

> Both **Jacob** and **Monod** shared the 1965 Nobel Prize for their work on the discovery of a class of genes which regulate the activities of other genes, along with **A. Lwoff**, their compatriot. Jacob, whose medical studies were interrupted by World War II, was seriously wounded while serving in the Free French forces and was a decorated veteran. After the war, he finished his medical training, but his physical disabilities prevented him from fulfilling his original desire to practice surgery.

proposed a theory to explain the molecular aspects of regulation of the catalytic activity of enzymes. These two findings introduced scientific concepts about the control of genetic andmetabolic functions of organisms which soon established a new field of scientific enquiry — the study of biological regulation.

At about this point, the strands of scientific development (shown in Fig. 1–1) –biochemistry, cell biology, and genetics–became inextricably woven, and the new science of **molecular biology** emerged. The distinction between molecular biology and biochemistry is not always clear, because both disciplines take as their ultimate aim the complete definition of life in molecular terms. The term

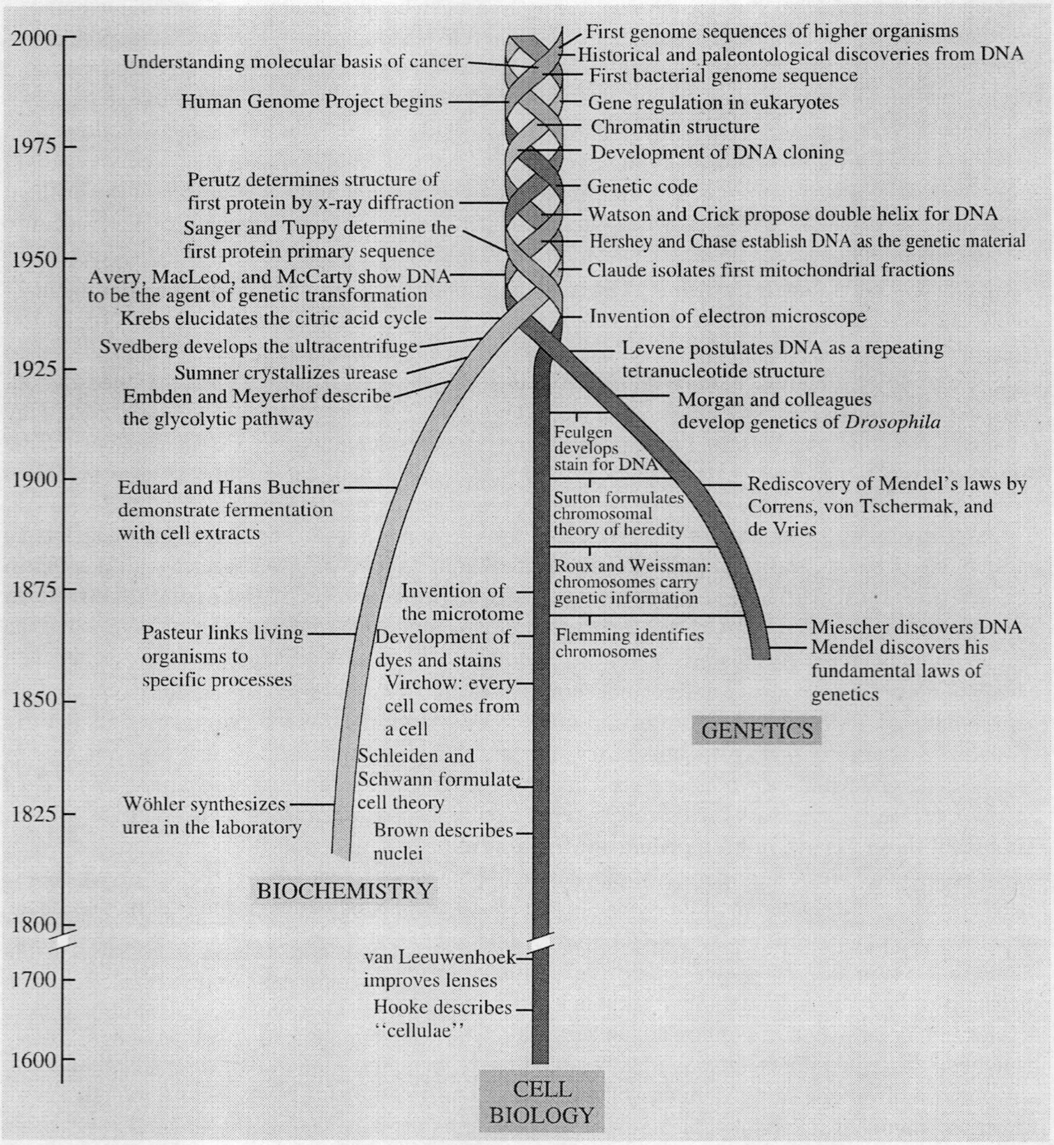

Fig. 1–1. Intertwining of historical traditions of biochemistry, cell biology and genetics

These 3 disciplines of science, originally considered quite distinct among each other, have become interweaved to yield a true molecular biology, which is the subject matter of present-day biochemistry.

(*Adaped from Becker WM, Kleinsmith LJ and Hardin J, 2000*)

molecular biology is often used in a narrower sense, to denote the study of nucleic acid structure and function and the genetic aspects of biochemistry, an area which one might more accurately call *molecular genetics.* Molecular biology and biochemistry are perhaps differentiated more easily by the orientations of their users than by the research problems being addressed. It may be said that while a biochemist thinks like chemist, a molecular biologist thinks like biologist. Even this distinction is somewhat artificial, since scientists in either field must use the approaches of all relevant disciplines, including chemistry, biology, and physics. In fact, 3 of the most powerful research techniques used by biochemists were developed by physicists : **electron microscopy (EM)**, which has revealed the minutest details of cellular structure, and **x-ray diffraction** and **nuclear magnetic resonance (NMR),** which have revealed the precise 3-dimensional structure of huge biological molecules (=biomolecules).

In the 1970s, molecular biology reached a new zenith with the introduction of recombinant DNA technology, which has brought about a revolution in biological research and has been instrumental in the emergence of a new branch of science called biotechnology. Use of this technique in science has allowed researchers to isolate eucaryotic genes to study their structure, regulation and expression.

In this era of biotechnology, biochemistry will continue to occupy a position of central importance. As a basic science, it is in the foreranks of many scientific endeavours that stand to make a reality of the statement that "the twentieth century belongs to the biologists." It was about 125 years ago that a German **Ernst Hoppe–Seyler,** one of the foremost chemists (and possibly the first biochemist) of the period, established a journal for those studies in physiological chemistry he believed should be recognized as the new discipline of biochemistry. If alive today, he would be justifiably proud of the science whose niche he helped define.

BIOCHEMISTRY AS MOLECULAR LOGIC OF LIVING ORGANISMS

Biochemistry operates at a molecular level and brings to light the hidden secrets of life. Therefore, in considering the molecules of various biological compounds, it is necessary to have an idea of scale (Fig. 1–2). The angström (Å) unit (1 Å = 10^{-10} meter or 10^{-8} centimeter or 0.1 nanometer) is customarily used as the measure of length at atomic level. Small biomolecules such as amino acids, sugars etc., are many angströms long whereas biological marcromolecules are much larger: for example, hemoglobin, an oxygen-carrying protein in red blood cells, has a diameter of 65 Å. Ribosomes, the protein-synthesizing organelles of the cell, have diameters of about 300 Å. Most viruses fall within a range of 100 Å (= 10 nm) to 1,000 Å(= 100 nm). Cells are mostly a hundred times as large, in the range of micrometers (μm). For example, a red blood cell is 7 μm (= 7 × 10^4 Å) long. As the limit of resolution of the light microscope is about 0.2 μm (= 2,000Å), most of the studies of biological structures in the range between 1 Å (= 0.1 nm) and 10^4Å (= 1μm) have been conducted with the help of electron microscope and *x*-ray diffraction.

> The term **angström** is named after **Anders J. Ångström** (LT, 1814–1874), a spectroscopist. Å is used for lengths shorter than 100 Å, whereas nm or μm is used for longer dimensions.

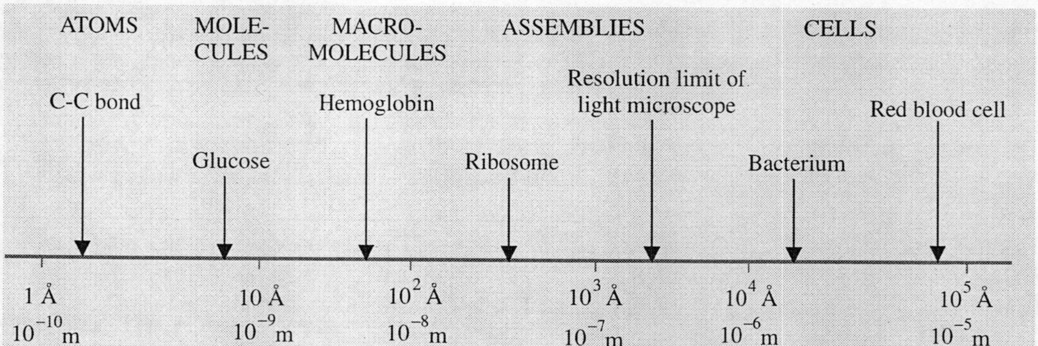

Fig. 1–2. Dimensions of biomolecules, assemblies and cells

The approximate dimensions of the components of the hierarchy of organization in cells is as follows :

Atoms	Å
Micromolecules (Amino acids)	1 nm
Macromolecules (Proteins)	5 to 500 nm
Organelles	nm to μm
Cells	μm to cm

The biomolecules are in a state of flux (Fig. 1–3). The enzymes change their substrate into product in milliseconds (1 m sec = 10^{-3} second). Some enzymes are even more efficient and catalyze their substrate in even few microseconds (1 μ sec = 10^{-6} sec). The unwinding of the DNA double

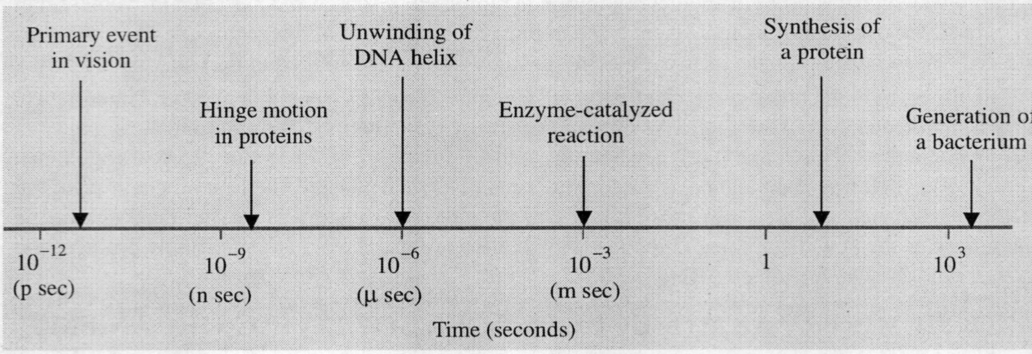

Fig. 1–3. Rates of some biological processes

helix, which is essential for its replication, is completed in a microsecond. The rotation of one sphere of a protein with respect to another takes place in nanoseconds (1 n sec = 10^{-9} sec). It is remarkable to note that the primary event in vision — a change in structure of the light-absorbing group — occurs within a few picoseconds (1 p sec = 10^{-12} sec) after the absorption of a photon.

The molecular events are associated with energy changes (Fig. 1–4). The ultimate source of energy for the living objects is the sun. The energy of visible light, say green, is 57 kilocalories per mole (kcal/mol). ATP, which is the universal currency of energy, has a usable energy content of about 12 kcal/mol. The amount of energy needed for the cleavage of a covalent C—C bond is 83 kcal/mol. Hence, the covalent skeleton of biomolecules is stable in the absence of enzymes and inputs of energy. On the contrary, thermal energy is enough to make and break noncovalent bonds in living systems, which typically have energy contents of only a few kilocalories per mole.

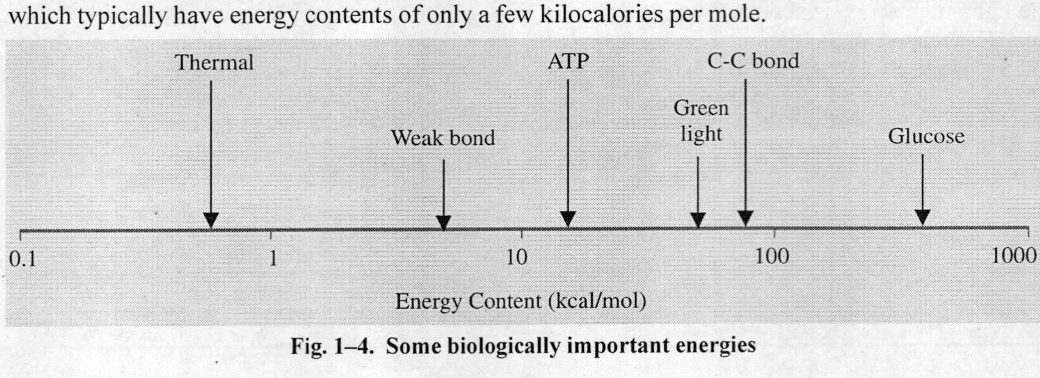

Fig. 1–4. Some biologically important energies

NATURE

Biochemistry has been defined as **"the chemistry of living things"**. The living things are composed of lifeless molecules which, when examined individually, exhibit all the physical and chemical laws that are characteristic of inanimate bodies. Yet living organisms possess extraordinary attributes, not shown by inanimate molecules. At this stage, henceforth, it shall be useful to distinguish 'living' from 'nonliving'. Although the word 'life' is difficult to define but it can be associated with certain characteristics which are common to the living objects and are usually not found in the nonliving or lifeless objects. All **living objects** are characterized by their capacity of movement, power of growth, respiratory activity, sense of irritability and above all the power of reproduction.

Certain identifying characteristics of the living matter or the '*signs of life*', as they are also called, are enumerated below :

1. Living organisms are *highly complicated and organized structures* and contain a very large number of different organic molecules. For example, a single bacterial cell of *Escherichia coli* contains 5,000 different kinds of organic compounds, including as many as 3,000 different kinds of proteins and 1,000 kinds of nucleic acids. In contrast, the inanimate objects such as clay, sand, rock, sea water etc., consist of random mixtures of simpler chemical molecules.

2. Each component unit of a living object appears to have a *specific purpose or function*, whether it be a macroscopic structure (heart, lungs, brain) or a microscopic intracellular structure (nucleus). Even the individual chemical compounds in cells (carbohydrates, proteins, lipids) have specific functions to perform. But it is meaningless or futile to ask about the function of various chemical compounds present in inanimate bodies. They just happen to be there.

3. The living organisms have the *ability to extract, transform and use energy* from their environment, either in the form of organic nutrients or the radiant energy of sunlight. Living organisms are never at equilibrium within themselves or with their surroundings. On the other hand, the inanimate bodies do not use energy to maintain its sturcture or else to do work. Rather when left to itself, it tends to come to equilibrium with its surroundings.

4. The most remarkable attribute of the living organisms is their *capacity for self-replication*, a characteristic which can be regarded as the very quintessence of the living state. Lifeless objects, on the contrary, do not grow and reproduce in forms identical in mass, shape and internal structure.

It should, however, be realized that the boundary between the living and nonliving objects is not always well demarcated. For example, certain diseases are caused by extremely small filterable substances called viruses. These can also reproduce when introduced into the environment of living cells and a few of them have been isolated in a purified cyrystalline form. Chemically, the crystalline viruses such as tobacco mosaic virus (TMV) are nucleoproteins and have no apparent features of living things. Yet TMV, when inoculated into a healthy leaf of a tobacco plant, multiplies rapidly and causes the onset of a disease termed tobacco mosaic disease. Thus, an inanimate crystalline substance, on inoculation, starts behaving as if it were living.

MAJOR ORGANIC COMPOUNDS OF ANIMATE OBJECTS

Living beings contain a wide variety of organic compounds, besides the ubiquitous water and other inorganic compounds. Major organic molecules present in the living beings are: carbohydrates, proteins (of course, including enzymes), lipids and nucleic acids. Table 1–1 lists some details of these compounds.

TABLE 1–1. Major compounds of living beings

Organic compound	*Building block*	*Some major functions*	*Examples*
Carbohydrate :			
Monosaccharide	—	Energy storage; physical structure	Glucose, fructose, galactose
Disaccharide	Monosaccharides	Energy storage; physical structure	Lactose, maltose, sucrose

Polysaccharide	Monosaccharides	Energy storage; physical structure	Starch, cellulose, chitin, inulin, pectin
Protein	Amino acids	Enzymes; toxins; physical structures	Antibodies; viral surface; flagella; pili
Lipid :			
Triglycerides	Fatty acids and glycerol	Energy storage; thermal insulation; shock absorption	Fat, oil
Phospholipids	Fatty acids, glycerol, phosphate, and an R group*	Foundation for cell membranes	Plasma (cell membranes)
Steroids	Four-ringed structure†	Membrane stability	Cholesterol
Nucleic acid	Ribonucleotides; Deoxyri\bonucleotides	Inheritance; instructions for protein synthesis	DNA, RNA

* R group = a variable portion of a molecule.

† Technically, steroids are neither polymers nor macromolecules.

THE SCIENTIFIC METHOD OF APPROACH

Although biologists employ a variety of approaches in conducting research, the experimentally -oriented biologists such as biochemists and microbiologists often use the general approach known as the **scientific method.** They first gather **observations** of the process to be studied and then develop a tentative hypothesis—an educated guess—to explain the observations (Fig. 1-5). Thus, the **hypothesis** is simply a tentative explanation to account for observed phenomena. This step often is inductive and creative because there is no detailed, automatic technique for generating hypotheses. Next, they decide what information is required to test the hypothesis and collect this information through observation or carefully designed experiments. Then, they decide whether the hypothesis has been supported or falsified. If it has failed to pass the test, the hypothesis is rejected, and a new explanation or hypothesis is constructed. If the hypothesis passes the test, it is subjected to more severe testing. The procedure often is made more efficient by constructing and testing alternative hypotheses and then refining the hypothesis that survives testing. This general approach is often called the **hypothetico-deductive method.** One deduces predictions from the currently-accepted hypothesis and tests them. In *deduction,* the conclusion about specific cases follows logically from a general premise ("if . . ., then . . .," reasoning).

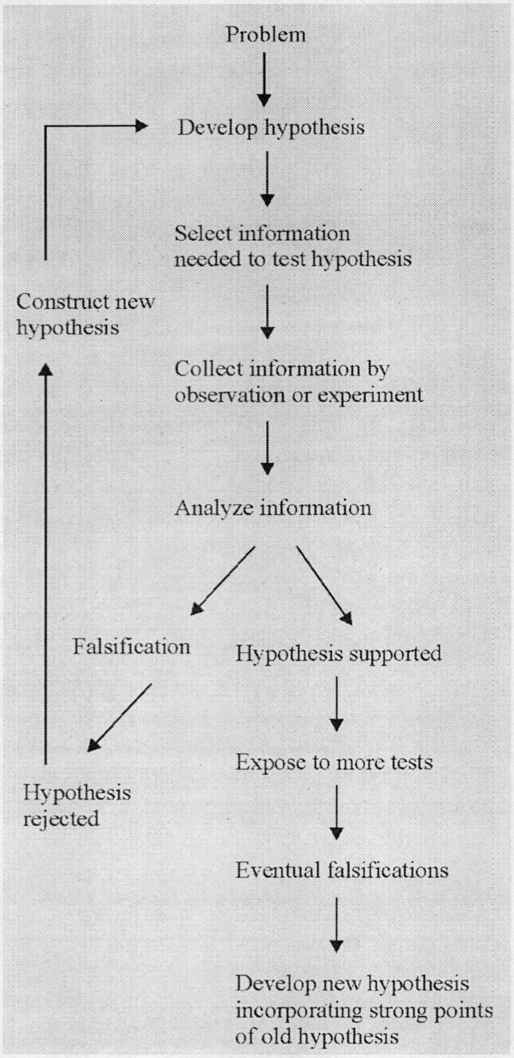

Fig. 1-5. The hypothetico-deductive method
This approach is most often used in scientific research.

Induction is the opposite. A general conclusion is reached after considering many specific examples. Both types of reasoning are used by scientists.

When carrying out an experiment, it is essential to use a control group as well as an experimental group. The *control group* is treated precisely the same as the *experimental group* except that the experimental manipulation is not performed on it. In this way, one can be sure that any changes in the experimental group are due to the experimental manipulation rather than to some other factor not taken into account.

If a hypothesis continues to survive testing, it may be accepted as a valid theory. The term **theory** is applied to a hypothesis that has been extensively tested and that ties together and arranges the results of a number of observations and experiments. It provides a reliable, systematic, and rigorous account of an aspect of nature. It is important to note that *hypotheses and theories are never absolutely proven.* Scientists simply gain more and more confidence in their accuracy as they continue to survive testing, fit with new observations and experiments, and satisfactorily explain the observed phenomena.

IMPORTANCE

Modern biochemistry is relatively a new branch and much of the work has been conducted during the present century. The enormous growth of biochemical literature has permitted in many cases the correlation of physical events in biological systems with the help of chemical processes. For a biochemist, therefore, a physiological process does not merely mean to elucidate the nature of the chemical substances involved in it but also to find out the physical relations among these substances and of these substances to the environment. Thus, the principal objective of biochemistry is to fill the wide gap between the highly integrated functions of the living cell and the various properties of its individual chemical constituents. A biochemist, therefore, has to perform an important arduous task of carrying the research work with utmost sincerity, patience and honesty. **Prof. Hopkins** (1931) has rightly remarked :

> "He (biochemist) should be bold in experiment but cautious in his claims. His may not be the last word in the description of life, but without his help the last word will never be said."

The *American Society of Biological Chemists* (1965) has, indeed, worked out a definition of a *biochemist* as a guideline for eligibility for membership in that society. The definition reads as follows :

> "A biochemist is an investigator who utilizes chemical, physical or biological technics to study the chemical nature and behaviour of living matter."

Ernest Baldwin (1937) entrusts the biochemist with the task of, "the study of physicochemical processes associated with the manifestations of what we call life—not the life of some particular animal or group of animals, but life in its most general sense."

PROBLEMS

It is very difficult to learn something well without somehow participating in it. The chaper-end problems are therefore an asset of any textbook. This book contains few problems in each chapter which are of regurgitory type. Rather they are designed to make one think and to offer insights not discussed in the text. The difficulties in solving these problems range from those that require only a few moments reflection to those that might take an hour or more of concentrated effort to work out.

1. What concerns biochemistry ?
2. Prove that biochemistry is an amalgamation mainly of chemistry, biology, physics and bioenergetics.
3. What type of scientific approach is normally used in conducting biochemical researches ?
4. Which theory gave way to the theory of organic synthesis ?
5. Who coined the word 'biochemistry' ?
6. Who is considered the '*father of modern biochemistry*'?

7. Which is the oldest available literature in biochemistry ?
8. Who is a biochemist ?

ANSWERS TO PROBLEMS

1. Biochemistry may be defined as a science concerned with the chemical nature and chemical behaviour of the living matter. It takes into account the studies related to the nature of the chemical constituents of living matter, their transformations in biological systems and the energy changes associated with these transformations. Such studies have been conducted in plant and animal tissues both. Broadly speaking, biochemistry may thus be treated as a discipline in which biological phenomena are analyzed in terms of chemistry. The branch of biochemistry, for the same reason, has been variously named as **Biological Chemistry** or **Chemical Biology**.
2. The body of living organisms contains a number of molecules (biomolecules), most of which are of organic type. To study their behaviour within the organisms, it is essential first to known about their structure and properties which is usually studied under **chemistry**, esp., organic chemistry. The transformations of these molecules to other molecules involves the energy which is either taken in or given out; such studies are a part of another fascinating branch called **thermodynamics** or **bioenergetics**. The various transformations and the physiological processes (respiration, digestion, excretion, synthesis and degradation of biomolecules, photosynthesis etc) that go on in the living body are all governed by definite physical principles, which are integral part of the study of **physics**. For all this to study, a thorough knowledge of the external and internal structure of the organism, its behaviour and response to various stimuli, its genetic make-up and its niche are all required to be studied, which comes under the purview of **biology**. Thus, truly biochemistry is an amalgamation of chemistry, biology, physics and bioenergetics.
3. **Hypothetico-deductive type** of scientific approach.
4. **Vitalistic theory** (= the doctrine of vitalism).
5. A German chemist **Carl Neuberg,** in 1903.
6. A French chemist **Antoine Laurent Lavoisier** (LT, 1743-1794).
7. '*Lectures in Animal Chemistry*', written by a Swedish chemist Jöns Jacob Berzelius (LT, 1779-1848) and published in 1806.
8. A biochemist is an investigator who utilizes chemical, physical or biological techniques to study the chemical nature and behaviour of living matter. He should be bold in experiment but cautious in his claims. His may not be the last word in the description of life, but without his help the last word will never be said.

CHAPTER

2 Water : The Solvent of Life

INSIDE

- General Considerations
- Physical Properties of Water
- Structure of Water Molecule
- Weak Interactions in Aqueous Solutions
- Role of Noncovalent Interactions
- Role of Water in Life

The Source of Life
Water, most of it in the oceans, covers three-fourths of Earth's surface. The oceans teem with life.

GENERAL CONSIDERATIONS

Water is the mother liquor of all forms of life. It is the vital essence, miracle of Nature, and the great sustainer of life. The essentiality of water for living systems is quite evident as without water, there is no life. The essentiality of water is a continuous reminder of the aqueous origin of life. It was in the solvent water that the chemical reactions of biological processes evolved. All aspects of cell structure and function are adapted to the physical and chemical properties of water. The strong attractive forces between water molecules result in water's solvent properties. A meagre tendency of water to ionize is of utmost importance to the structure and function of biomolecules. The water molecule and its ionization products (H^+ and OH^-) greatly influence the structure, assemblage and properties of all the cellular components, including enzymes and other proteins, nucleic acids and lipids. Water is needed not only for biochemical reactions, but also for transporting substances across membranes, maintaining body temperature, dissolving waste products for excretion and producing digestive fluids.

Water is the medium in which the first cells arose, and the solvent in which most biochemical transformations take place. The properties of water have shaped the course of evolution and exert a decisive influence on the structure of biomolecules in aqueous solution. Many of the weak interactions within and between biomolecules are strongly influenced by the solvent properties of water. Even water-insoluble components, such as membrane lipids, interact with each other in ways dictated by the polar properties of water.

No other substance on earth is as abundant as water. It is almost everywhere in air, clouds, oceans, lakes, rivers, springs or glaciers. In the 5 km layer below the sea level on Earth, water is nearly 6 times as abundant as all other substances put together. And, none other occurs in 3 states—solid, liquid and gaseous—at the same time. It is water that had conditioned our climate. The water in oceans, seas and the atmosphere (vapour) acts as an accumulator of heat. In hot weather, it absorbs heat and in cold, it gives up heat; thus, it keeps the planet warm. Without water, our planet would be cooled to subzero temperatures long ago and all forms of life would have perished. It takes 10 litres of water to produce 1 litre of petroleum, 100 litres for 1 kg of paper, 4,500 litres for 1 kg of rice, 4,500 litres for l ton of cement and 20,000 litres to produce 1 ton of steel. This roughly gives us an idea of the extent of water usage.Water has deluded us into believing that it is abundant and inexhaustible. But it is merely an illusion rather than a reality. The volume of water on the earth is about 1,36,00,00,000 cubic kilometres which covers nearly 70% of the earth's surface.The global scenario is that 97% of water is in the sea, 2% locked up in the Arctic and Antarctic oceans, and 1% is fresh water. Of this 1% fresh water, 0.22% is in the underground acquifer and the remaining 0.78% of the world's water fills the rivers, streams, lakes and ponds. No doubt, water is a renewable source but it is finite too. So it has to be used judiciously. The daily water requirement of an average person is as follows :

15 litres – for cooking and drinking needs

35-40 litres – for toilet needs

65 litres – for bathing, washing, and other sundry needs.

Water accounts for ca 70% or more of the weight of most organisms. In human adults also, total body water accounts for about 70% of the lean body mass. Variations observed are mainly due to differences in fat contents. In obese males, water constitutes a lower percentage of body weight (45–60%) than in lean individuals (55 – 70%). Adult lean females have a low water content (45 – 60%) and the value in infants can be in the range 65 – 75%. Our normal body temperature of 37°C is maintained mainly because water is expelled by lungs and skin. However, a loss of 10% of water in our body is serious; a loss of 20% fatal. In humans, water is not evenly distributed between the two major compartments (intracellular and extracellular) of the body (Table 2–1). In an adult male of 70 kg body weight, 70% of water (*i.e.*, about 30 litres) is found in intracellular fluids. Of this, about 4 litres are found in bones which does not readily participate in fluid exchange. Nearly 30% of water (*i.e.*, about 12 litres) is found in extracellular fluid. Plasma (3 litres), interstitial fluid (8.5 litres) and transcellular fluid (l litre) come under this category.

Table 2–1. Water distribution in an adult male of 70 kg body weight

Compartment	*Total body water (in litres)*	
Intracellular fluids	30	
Extracellular fluids		
Plasma	3	12.5
Interstitial and lymph	8.5	
Transcellular fluid	1	

The maintenance of **water balance**, the equilibrium between water intake and output, is a critical aspect of metabolism. An adult in water balance generally takes in and loses about 2.8 litres of water per day in temperate climate and between 3.3 and 7.3 litres per day in tropical climate (Table 2–2). Besides the water obtained from foods and liquids, **metabolic water** is also made available through the oxidation of food in the body. Oxidation of 100 g each of fat, carbohydrate and protein yields 107,

55 and 41g of water, respectively. Water losses occur by evaporation (water vapour in exhaled air and perspiration) and by excretion of urine and feces.

Table 2–2. Daily water intake and output (or loss) from the human body

Intake/output	*Temperate climate*		*Tropical climate*	
Intake				
As liquids	1.5	2.8	2 – 5	3.3 – 7.3
In foods	1.0		1 – 2	
Oxidation of foods	0.3		0.3	
Output (or Loss)				
Urine	1.5	2.8	1.0 – 1.5	3.3 – 7.3
Feces	0.1		0.1 – 0.2	
Evaporation through skin	0.8		1.8 – 5.2	
Evaporation through lung	0.4		0.4	

If the output of water significantly exceeds its intake, **dehydration** occurs due to diarrhea, vomiting, fever and high environmental temperatures. Dehydration is especially serious in young children because their body pool of water is small and hence can be readily depleted. If water accumulates in tissues, then **edema** occurs. Edema is often observed in children suffering from **kwashiorkor,** a nutritional disease caused by an inadequate intake of protein. In the Carribeans, children with kwashiorkor are called *sugar babies* because of their plump appearance (a deceiving indication of good health).

CLINICAL IMPLICATIONS

Kwashiorkor is the Bantu word meaning displaced child. This refers to the fact that kwashiorkor appears in infants when they are no longer nursed by their mothers, *i.e.*, are displaced from the breast by the birth of a new baby. Kwashiorkor is a protein malnourishment syndrome that results from a diet adequate in calories but deficient in protein. It is characterized by degeneration of the liver, severe anemia, and inflammation of the skin. Kwashiorkor is especially common among children in industrially-underdeveloped countries where the diet consists primary of a single plant material, as in Indonesia, where rice is the main food, and in parts of Africa and Latin America, where corn (maize) is the principal staple. Worldwide, corn is the second or third largest crop and it is a staple for nearly half the world's malnourished people (For this reason, intense efforts have been directed at breeding stations for producing corn with a higher quality protein. The results have been marvellous; the quality protein maize (QPM) has a protein quality twice that of normal corn and almost equal to milk. If found fit on field trials, it will be good news for the impoverished populations in Mexico and Central America where maize constitutes about 85% of the grain consumed).

Early clinical evidence of protein malnutrition is vague but does include lethargy, apathy or irritability. When well advanced, it results in retarded growth, lack of stamina, loss of muscular tissue, a secondary immunodeficiency, increased susceptibility to infections, and edema of dependent parts. The child may develop impaired appetite, anorexia, flabbiness of subcutaneous tissue and loss of muscle tone and may show little interest in play or in toys. The liver may enlarge early or late; fatty infiltration is common. Edema usually develops early; failure to gain weight may be masked by edema, which is often present in internal organs before it can be recognized externally in the face and limbs. The affected children are said to have a characteristic "*sugar baby*" *appearance* due to their bloated body. Renal plasma flow, glomerular filtration rate, and renal tubular function are decreased. The heart may be small in early stages, but is usually enlarged later.

PHYSICAL PROPERTIES OF WATER

The physical properties of water differ markedly from those of other solvents. For example, water as a hydride of oxygen (H_2O) has a higher melting point, boiling point, heat of vapourization and surface tension than do the comparable hydrides of sulfur (H_2S) and nitrogen (NH_3) and most other common liquids (Table 2–3). These unusual properties are a consequence of strong attractions between adjacent water molecules, which give liquid water great internal cohesion.

Table 2–3. Some physical properties of water and other common liquids

Substance	*Melting point, °C*	*Boiling point, °C*	*Heat of vapourization, cal/g*	*Heat capacity, cal/g*	*Heat of fusion, cal/g*	*Surface tension*	*Dielectric constant*
Water	0	100	540	1,000	79.7	72.8	80
Methanol	– 98	65	263	0.600	22	–	33
Ethanol	– 117	78	204	0.581	24.9		
Propanol	– 127	97					
Butanol	– 90	117					
Acetone	– 95	56	125	0.528	23	23.7	21.4
Butane	– 135	0.5					
Hexane	– 98	69					
Chloroform	– 63	61	59	0.226		27.1	5.1
Benzene	6	80	94	0.500	30	28.9	2.3
Hydrogen sulfide	– 85	– 60	132	–	16.7		
Ammonia	– 78	– 33	327	1.120	84		
Hydrofluoric acid	– 92	19	360	–	54.7		

In humans, about 60% of red blood corpuscles (RBCs) and 92% of blood plasma are water. About 75% of most other tissues comprises of water; the only exceptions are such relatively inert tissues as hair, nails and the solid portion of bones. Water is also a principal constituent of the environment in which organisms thrive. Water has some unusual properties of physiological importance. These are described below :

1. Expansion on freezing. Most substances decrease in volume (and hence increase in density) as their temperature decreases. But in case of water, there is a temperature at which its density exceeds that at higher or lower temperatures. This temperature is 4 °C. In fact, water just above the freezing point is heavier than water at the freezing point. Therefore, it moves towards the bottom, freezing begins at the surface and the bottom is last to freeze. Organisms living at the bottoms of fresh-water lakes are, hence, protected from freezing. It may, however, be concluded that :

(*a*) While almost all substances contract on cooling, water expands. If water could contract on cooling, it would have become heavier and would have sunk. Thus, all water on earth would gradually have become ice.

(*b*) When temperature of water is raised above 0 °C, its volume decreases upto 4 °C and thereafter increases. Water thus has the minimum volume and hence the maximum density of 1.00 at

4 °C. The volume of water at 4 °C increases either on heating or cooling it.

(*c*) While all substances increase in volume when they are melted, the volume of ice decreases when melted : volume of 1g of ice is 1.09 cc but when it melts into water, it occupies only 1cc.

(*d*) Frozen water is less dense than liquid water.

2. Uniquely high surface tension. Like a stretched membrane, the surface of a liquid tends to contract as much as possible. This phenomenon is called *surface tension*. Water has the highest surface tension (of 72.8) of any known liquid. And it is the reason why water rises to unusually high levels in narrow capillary tubes. This has great significance in physiology.

3. Uniquely high heat capacity. There occurs a smaller temperature rise in water as compared to most other substances, when a given amount of heat is applied. Thus, water acts as a temperature buffer. It maintains its temperature more successfully than most other substances. We may, thus, say that has a high heat capacity (1,000 cal/g).

4. High solvent power. Water is a solvent for a great number of molecules which form ionized solutions in water. It may, thus, be called a universal solvent which facilitates chemical reactions both outside of and within biological systems.

Table 2–4 lists some of the unique features of water.

Table 2–4. Water and its features

	Property	*Chemistry*	*Result*
1.	Universal solvent	Polarity	Facilitates chemical reactions
2.	Adheres and is cohesive	Polarity; hydrogen bonding	Serves as transport medium
3.	Resists changes in temperature	Hydrogen bonding	Helps keep body temperature constant
4.	Resists change of state (from liquid to ice and from liquid to steam)	Hydrogen bonding	Moderates earth's temperature
5.	Less dense as ice than as liquid water	Hydrogen bonding changes	Ice floats on water

(*a*) (*b*)

Fig. 2-1. Cohesion among water molecules

(*a*) Cohesion among water molecules allows water striders to skate across the surface of still waters.

(*b*) In giant redwoods, cohesion holds water molecules together in continuous strands from the roots to the topmost leaves even 100 meters above the ground.

It is due to the cohesive property of water molecules that allows water striders to skate across the surface of still waters and empowers giant redwoods to raise water about 100 metres above the ground (Fig. 2-1).

STRUCTURE OF WATER MOLECULE

In a water molecule (Figs. 2–2 and 2–3*a*, 3*b*), each hydrogen atom shares an electron pair with the oxygen atom. The geometry of the water molecule is dictated by the shapes of the outer electron orbitals of the oxygen atom, which are similar to the bonding orbitals of carbon. These orbitals describe a rough tetrahedron, with a hydrogen atom at each of the two corners and unshared electrons at the other two. The H—O—H bond angle is 104.5°, 5° less than the bond angle of a perfect tetrahedron which is 109.5°; the nonbonding orbitals of the oxygen atom slightly compress the orbitals shared by hydrogen.

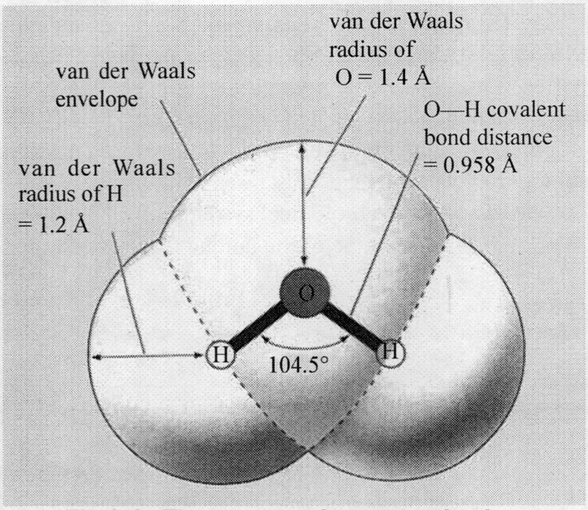

Fig. 2–2. The structure of a water molecule

The outline represents the van der Waals envelope of the molecule (where the attractive components of the van der Waals interactions balance the repulsive components). The skeletal model of the molecule indicates its covalent bonds. The H – O – H bond angle is 104.5°. Both hydrogen atoms carry a partial positive charge and the oxygen a partial negative charge, creating a dipole.

The oxygen nucleus attracts electrons more strongly than does the hydrogen nucleus (*i.e.*, the proton); oxygen is more electronegative. The sharing of electrons between H and O is therefore unequal; the electrons are more often in the vicinity of the oxygen atom than of the hydrogen. This unequal electron sharing creates two electric dipoles in the water molecule, one along each of the H—O bonds. The oxygen atom bears a partial negative charge (δ^-), and each hydrogen a partial positive charge (δ^+). The resulting electrostatic attraction between the oxygen atom of one water molecule and the hydrogen of another water molecule (Fig. 2–3*c*) constitutes a **hydrogen bond.**

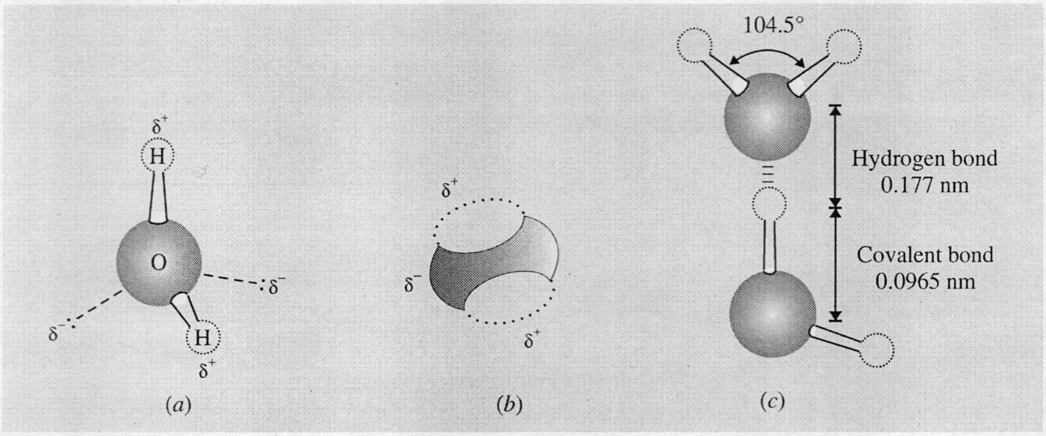

Fig. 2–3. The dipolar nature of the water molecule

A. Ball-and-stick model (dashed lines represent the nonbonding orbitals). There is nearly tetrahedral arrangement of the outer shell electron pairs around the oxygen atom; the two hydrogen atoms have localized partial positive charges and the oxygen atom has two localized partial negative charges.

B. Space-filling model.

C. Two water molecules joined by a hydrogen bond (designated by 3 thick horizontal lines) between the oxygen atom of the upper molecule and a hydrogen atom of the lower one. Hydrogen bonds are longer and weaker than covalent O — H bonds.

Polarity of water. Knowing the electronegativity of 2 atoms allows one to predict whether a bond between them will be covalent or ionic. The larger the difference in electronegativities of 2 atoms, the more likely they are to form an ionic rather than a covalent bond. Sodium and chlorine, for example, have a large difference in electronegativities and hence form ionic bonds. Carbon and nitrogen, on the contrary, have similar, moderate electronegativities and they usually form covalent bonds.

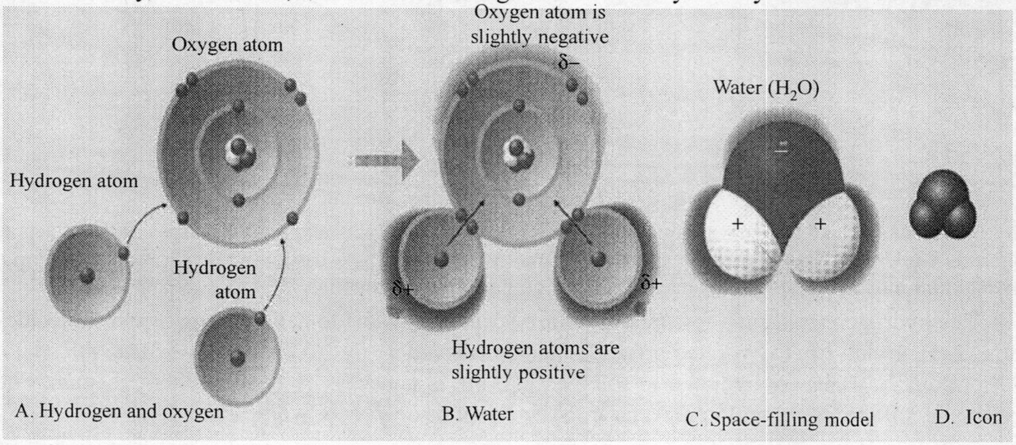

Fig. 2-4. The polarity of water

A. Two atoms of hydrogen and one atom of oxygen share electrons in two covalent bonds to form a molecule of water. **B.** Because the oxygen atom is more electronegative than the hydrogen atoms, the electrons spend more time hovering around the oxygen end of the water molecule. As a result, the oxygen end has a slightly negative charge, while the hydrogens have a slight positive charge. Such a partial charge, less than one full electron, is symbolized by δ. **C.** Space-filling model of water molecule. **D.** Icon for water.

Molecules that have uneven distributions of electrical charge are said to be **polar**, since they have positive and negative poles in the same way that a magnet has 2 poles (Fig. 2-4). When a polar molecule, such as water, comes close to an ion or to another polar molecule, its negative pole points toward the other molecule's positive pole, and its positive pole toward a neighbouring negative pole. Molecules with approximately uniform charge distributions are said to be **nonpolar.**

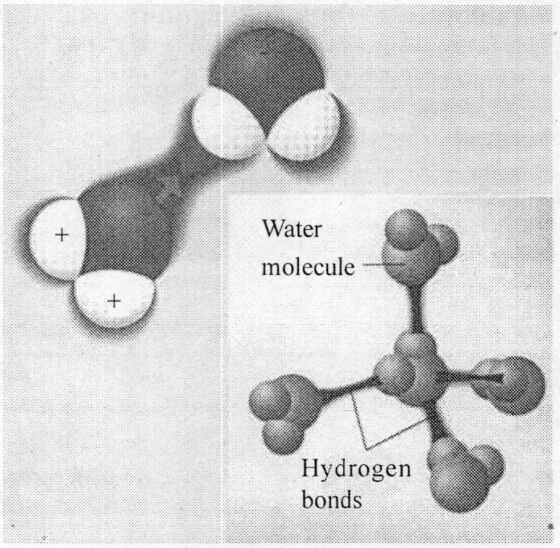

Fig. 2-5. Hydrogen bonds betweeen water molecules

The hydrogen atoms of one water molecule are attracted to the oxygen atom of another water molecule.

Hydrogen bonds between water molecules. When a hydrogen atom attaches to a highly electronegative atom such as oxygen or nitrogen, the resulting covalent bond is polar. In this case, the hydrogen atom acquires a slight positive charge. Such a hydrogen atom can then participate in a hydrogen bond–a weak interaction to a negatively-charged atom in another molecule (Fig. 2-5). The most common hydrogen bonds are those between water molecules, but other hydrogen bonds also play a critical role in the structure of proteins and DNA.

Hydrogen bonds are weaker than covalent bonds. The hydrogen bonds in liquid water have a **bond energy** (the energy required to break a bond) of only about 20 kJ/mol, as compared to 460 kJ/mol for the covalent O—H bond. At room temperature, the thermal energy of an aqueous solution (*i.e.*, the kinetic energy resulting from the motion of individual atoms and molecules) is of the same order as that required to break hydrogen bonds. When water is heated, the resulting temperature increase causes the faster motion of individual water molecules. Although at any given time, most of the molecules in liquid water are hydrogen-bonded, the lifetime of each hydrogen bond is less than 1×10^{-9} s. Nevertheless, the very large number of hydrogen bonds between molecules confers great internal cohesion on liquid water.

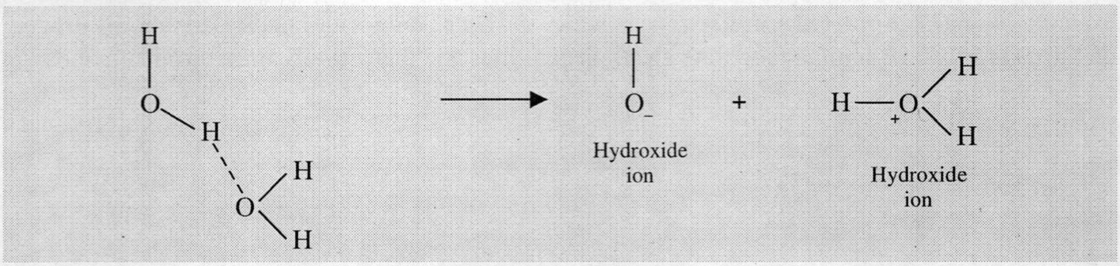

Fig. 2–6. Formation of hydronium ion

Water itself has a slight tendency to ionize and can act both as a weak acid and as a weak base. When it acts as an acid, it releases a proton to form a **hydroxyl ion.** When it acts as a base, it accepts a proton to form a **hydronium ion.** Most protons in aqueous solutions exist as hydronium ions.

The hydrogen atoms in a few molecules are occasionally lost to neighbouring water molecules (Fig. 2–6), giving rise to hydrated proton, called a *hydronium ion,* H_3O^+ and a *hydroxide ion,* OH^-. All protons in water are hydrated to some extent, but is unnecessary to write them as such. The symbol H^+ refers to any proton in water irrespective of its degree of hydration.

WEAK INTERACTIONS IN AQUEOUS SOLUTIONS

A. Hydrogen Bonding Between Water Molecules

The nearly tetrahedral arrangement of the oxygen electrons (bond angle 104.5°) allows each water molecule to form hydrogen bonds with 4 neighbouring water molecules. At any moment in liquid water at room temperature, each water molecule forms hydrogen bonds with an average of 3.4 other water molecules (Fig. 2–7). The water molecules are in continuous motion in the liquid state, hence hydrogen bonds are constantly and swiftly being broken and formed.

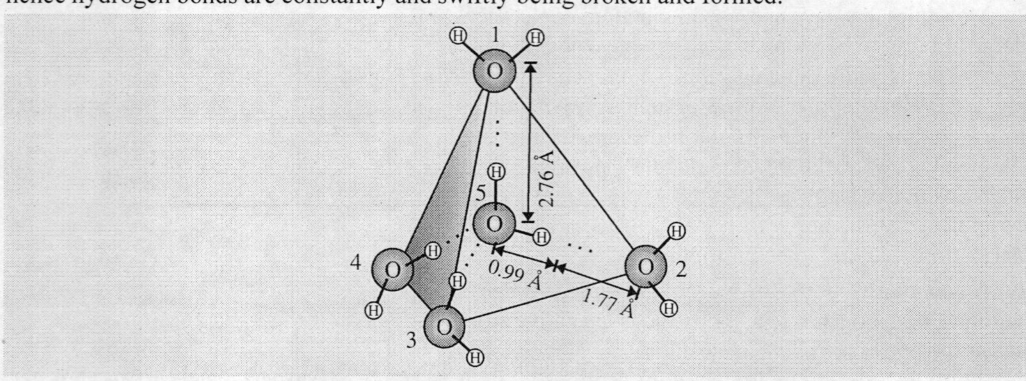

Fig. 2–7. Tetrahedral hydrogen bonding of a water molecule, in ice

Molecules 1, 2 and 5 are in the plane of the paper with molecule 3 above and molecule 4 below the plane. Molecules 1, 2, 3 and 4 are positioned at the corners of a regular tetrahedron.

In ice, however, each water molecule is fixed in space and forms hydrogen bonds with 4 other water molecules to produce a regular lattice structure (Fig. 2–8). Much thermal energy is needed to break the large number of hydrogen bonds in such a lattice and this is the reason for a relatively high melting point of water. When ice melts or water evaporates, heat is taken up by the system :

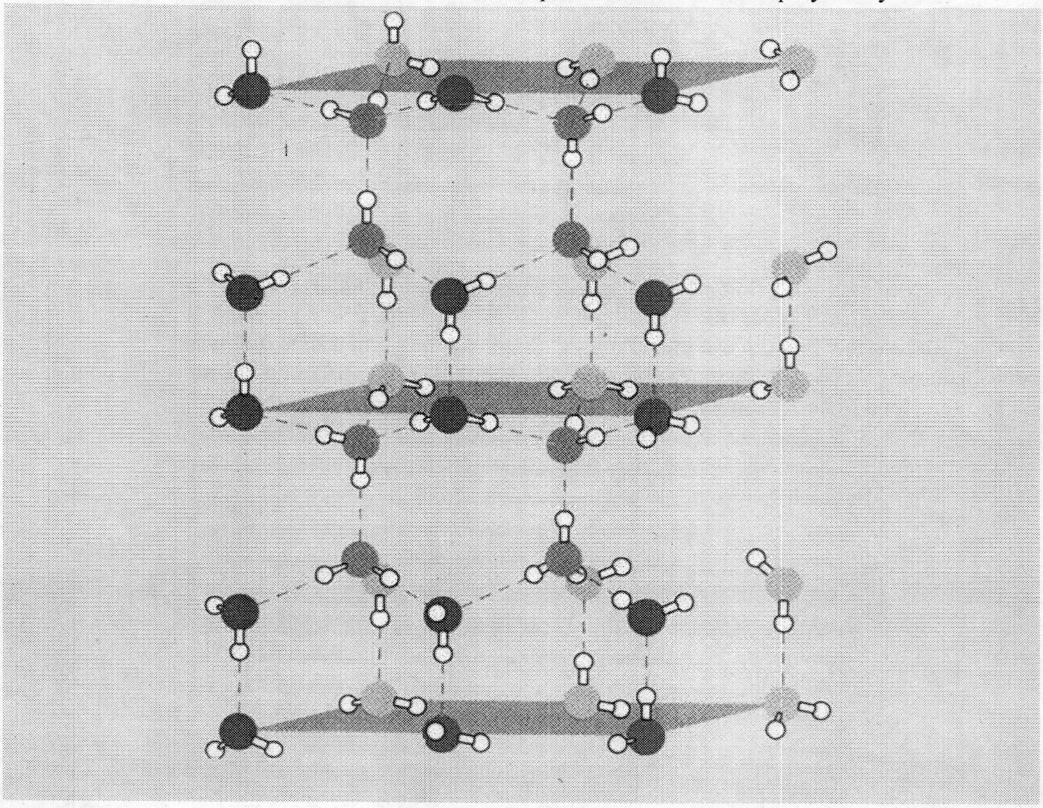

Fig. 2–8. The structure of ice

The tetrahedral arrangement of the water molecules is a consequence of the roughly tetrahedral disposition of each oxygen atom's sp^3 - hybridized bonding and lone pair orbitals. Oxygen and hydrogen atoms are represented, respectively, by *red* and *white* spheres, and hydrogen bonds are indicated by *dashed lines*. Note the open structure that gives ice its low density relative to liquid water. (*After Pauling LC, 1960*)

$$H_2O\ (s) \longrightarrow H_2O\ (l) \qquad \Delta H = +5.9\ kJ/mol$$

$$H_2O\ (l) \longrightarrow H_2O\ (g) \qquad \Delta H = +44.0\ kJ/mol$$

B. Hydrogen Bonding Between Water and Solute Molecules

Hydrogen bonding is not unique to water. They readily form between an electronegative atom (usually oxygen or nitrogen) and a hydrogen atom covalently bonded to another electronegative atom in the same or another molecule (Fig. 2–9). However, hydrogen atoms covalently bonded to carbon atoms (which are not electronegative), do not participate in hydrogen bonding. The distinction explains why butanol (CH_3 CH_2 CH_2 CH_2.OH) has a relatively high boiling point of 117°C in contrast to butane (CH_3 CH_2 CH_2 CH_3) which has a boiling point of only –0.5°C. Butanol has a polar hydroxyl group and, hence, can form hydrogen bonds with other butanol molecules.

Uncharged but polar biomolecules such as sugars dissolve readily in water because of the stabilizing effect of the many hydrogen bonds that form between the hydroxyl groups or the carbonyl oxygen of the sugar and the polar water molecules. Alcohols, aldehydes and ketones all form hydrogen

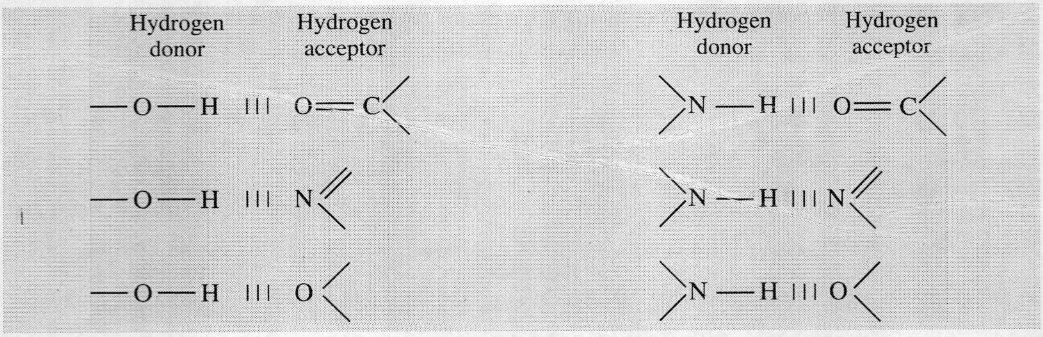

Fig. 2–9. Common types of hydrogen bonds

Note that in biological systems, the electronegative atom (*i.e.*, the hydrogen acceptor) is usually oxygen or nitrogen. The distance between two hydrogen-bonded atoms varies from 0.26 to 0.31 nm.

bonds with water, as do compounds containing N—H bonds (Fig. 2–10), and the molecules containing such groups tend to be soluble in water.

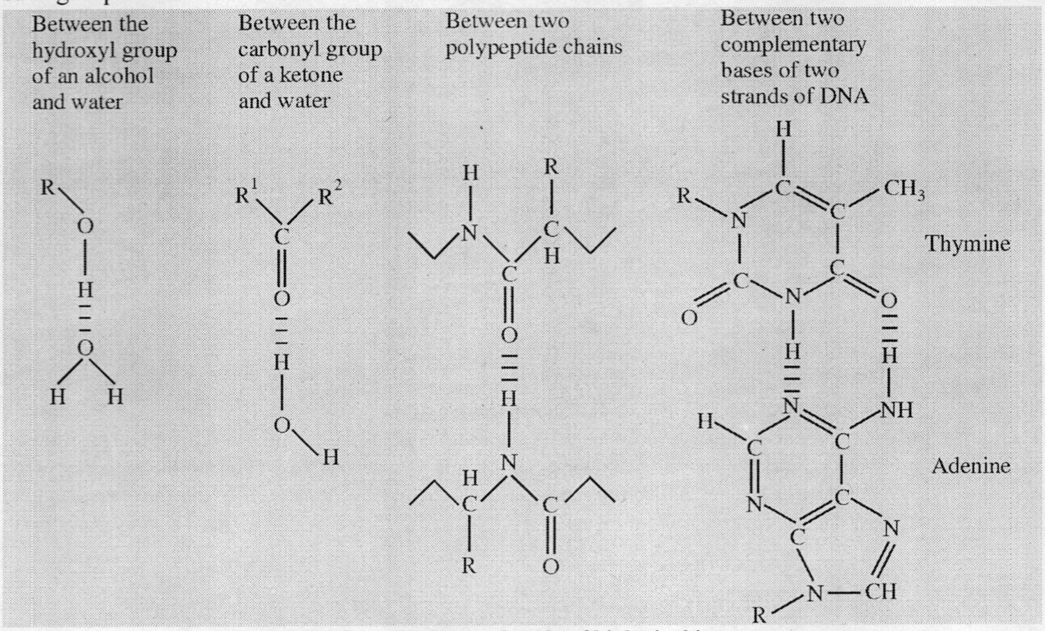

Fig. 2–10. Some hydrogen bonds of biological importance

Hydrogen bonds are strongest when the bonded molecules are oriented to maximize electrostatic interaction. This happens when the hydrogen atom and the two atoms that share it are in a straight line (Fig. 2–11). *Hydrogen bonds are, thus, highly directional* and are capable of holding two hydrogen-bonded molecules or groups in a specific geometric arrangement.

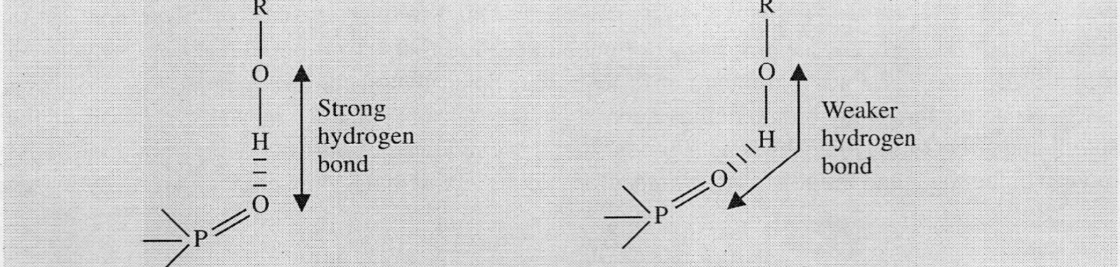

Fig. 2–11. Directionality of the hydrogen bond

The attraction between the partial electric charges is greatest when the three atoms involved (in this case, O, H and O) lie in a straight line.

C. Interaction Between Water and Charged Solutes

Water is a polar solvent. It readily dissolves most biomolecules, which are generally charged or polar compounds (Table 2–5). Compounds that dissolve readily in water are **hydrophilic** (*hudor*^{G} = water; *philic*^{G} = loving). In contrast, nonpolar solvents (such as chloroform and benzene) are poor solvents for polar biomolecules, but readily dissolve nonpolar biomolecules such as lipids and waxes

Table 2–5. Some examples of polar, nonpolar and amphipathic biomolecules

Biomolecule	*Ionic form at pH 7*
Polar	
Glucose	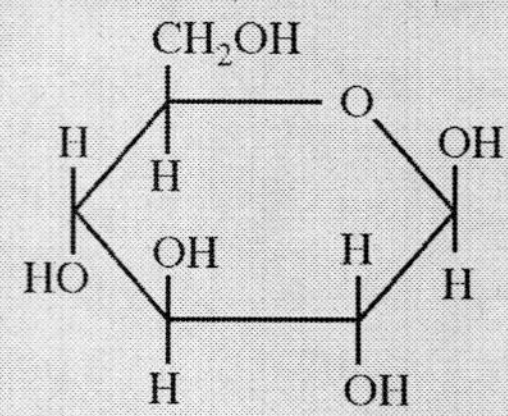
Glycine	$^{+}NH_3-CH_2-COO^{-}$
Aspartic acid	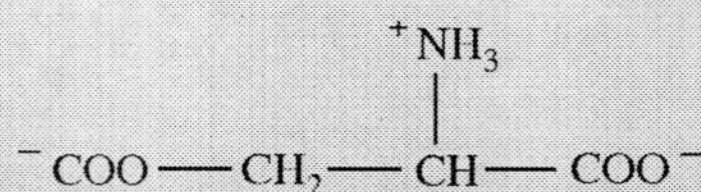
Lactic acid	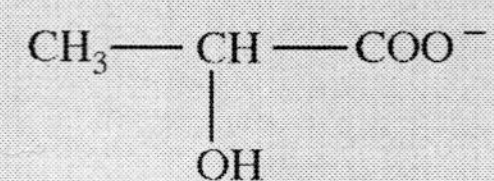
Glycerol	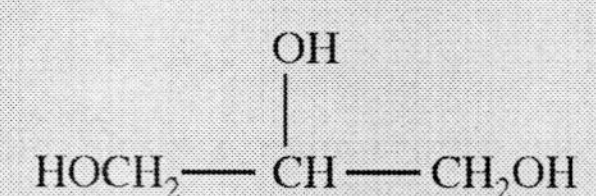
Nonpolar	
Typical wax	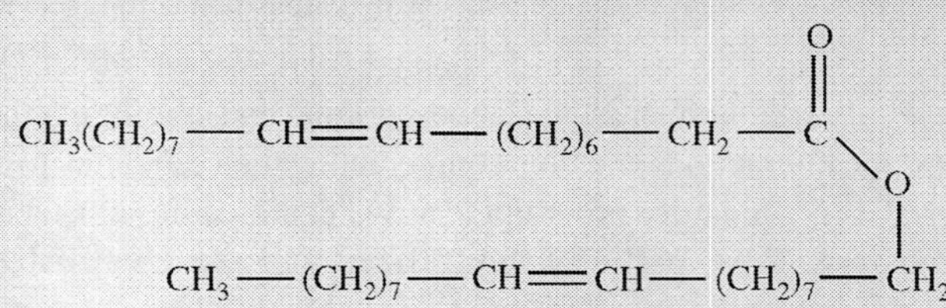
Amphipathic	
Phenylalanine	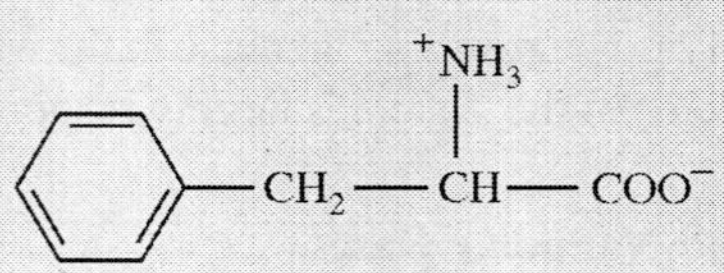
Phosphatidylcholine	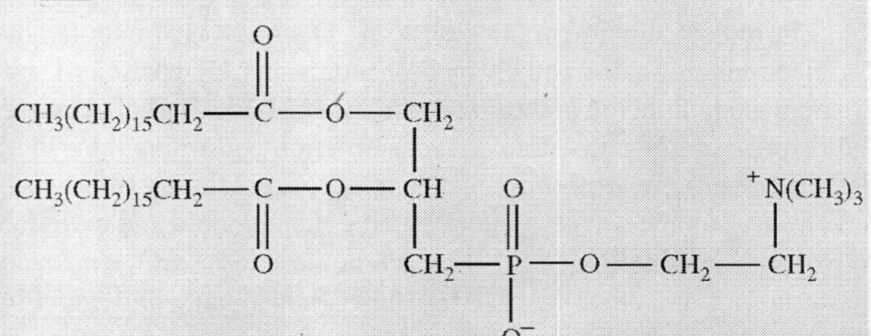

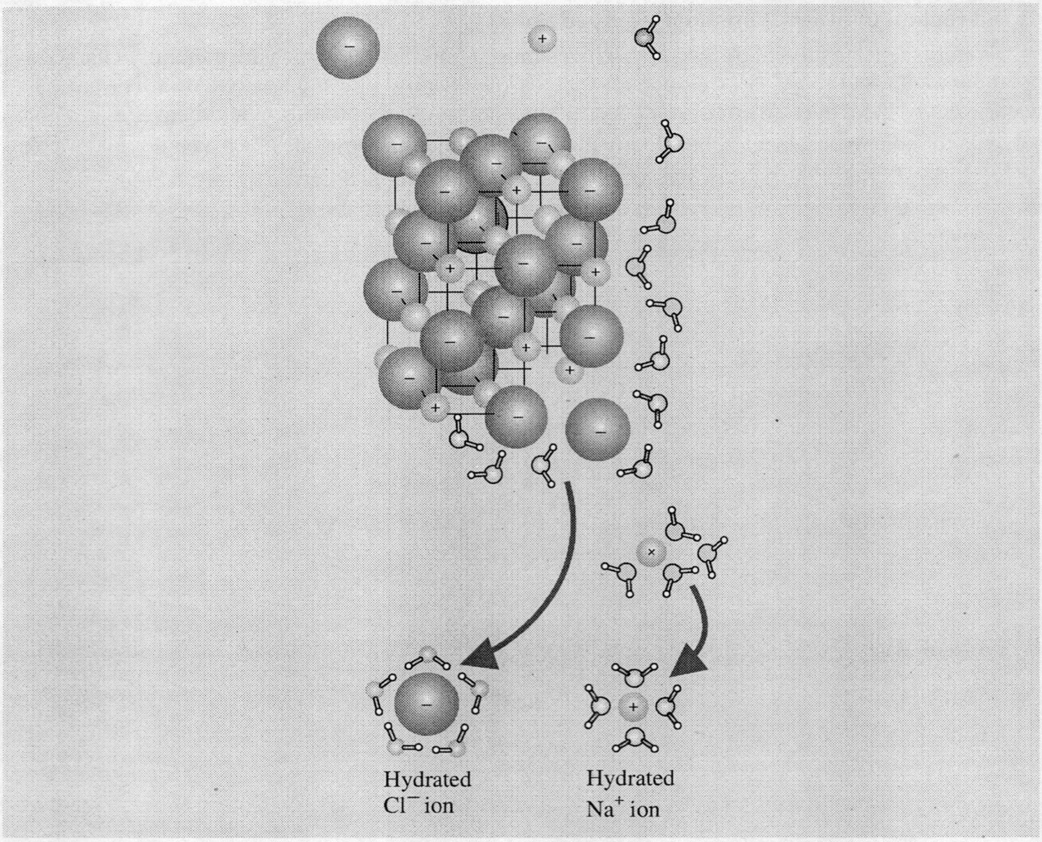

Fig. 2–12. Interaction between water and crystalline salts such as NaCl

Note that water dissolves many crystalline salts by hydrating their component ions (in this case, Na^+ and Cl^- ions).

(Adapted from *AL Lehninger, DL Nelson and MM Cox, 1993*)

Water dissolves salts such as NaCl by hydrating and stabilizing the Na^+ and Cl^- ions, weakening their electrostatic interactions and thus counteracting their tendency to associate in a crystalline lattice. In fact, the NaCl crystal lattice is disrupted as water molecules cluster about the Cl^- and Na^+ ions. The ionic charges are, thus, partially neutralized and the electrostatic attractions necessary for lattice formation are weakened. As a result, the Na^+ and Cl^- ions leaving the crystal lattice acquire far greater freedom of motion (Fig. 2–12). The resulting increase in the entropy (a measure of the degree of randomness or disorder of a system) of the system is largely responsible for the ease of dissolving the salt in water.

D. Interaction Between Water and Nonpolar Gases

The biologically important gases CO_2, O_2 and N_2 are nonpolar. In the diatomic molecules O_2 and N_2, electrons are shared equally by both atoms. In CO_2, each C = O bond is polar, but the two dipoles are oppositely directed and cancel each other (Table 2–6). The movement of these molecules from the disordered gas phase into aqueous solution constrains their motion and therefore represents a decrease is entropy. As a consequence, these gases are very poorly soluble in water. Some organisms have water-soluble carrier proteins (such as hemoglobin and myoglobin) that facilitate the transport of oxygen. Carbon dioxide gas forms carbonic acid (H_2CO_3) in aqueous solution, and is transported in that form. Two other gases, NH_3 and H_2S, also have biological roles in some organisms. These gases are polar and dissolve easily in water.

Table 2–6. Solubilities of some gases in water

Gas	*Structure**	*Polarity*	*Solubility in water (g/L)*	*Temperature (°C)*
Nitrogen	N≡N	Nonpolar	0.018	40
Oxygen	O=O	Nonpolar	0.035	50
Carbon dioxide	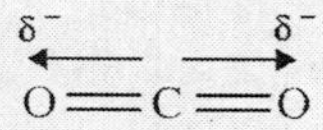	Nonpolar	0.970	45
Ammonia	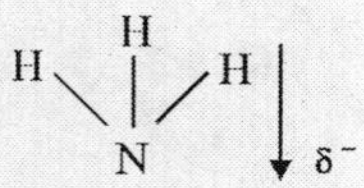	Polar	900	10
Hydrogen sulfide	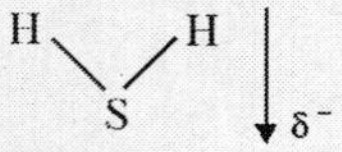	Polar	1,860	40

* The arrows represent electric dipoles; there is a partial negative charge (δ^-) at the head of the arrow, a partial positive (δ^+; not shown here) at the tail.

E. Interaction Between Water and Nonpolar Compounds

When water is mixed with a hydrocarbon such as benzene or hexane, two phases form: neither liquid is soluble in the other. Shorter hydrocarbons such as ethane have small but measurable solubility in water. **Nonpolar compounds** such as benzene, hexane and ethane are *hydrophobic* (*hudor*[G] = water; *phobic*[G] = hatred) as they are unable to undergo energetically favourable interactions with water molecules. In fact, they interfere with the hydrogen bonding among water molecules. All solute molecules or ions dissolved in water interfere with the hydrogen bonding of some water molecules in their immediate vicinity, but polar or charged solutes (such as NaCl) partially compensate for the lost hydrogen bonds by forming new solute-water interactions. The net change in enthalpy (the heat content of the reacting system) for dissolving these solutes is usually small. Hydrophobic solutes offer no such compensation. Addition of water to such solutes may hence result in a small gain of enthalpy; the breaking of hydrogen bonds requires the addition of energy to the system. Furthermore, dissolving hydrophobic solutes in water results in a measurable decrease in entropy.

Amphipathic compounds contain regions that are polar (or charged) and regions that are nonpolar (Table 2–5). When amphipathic compounds are mixed with water, the two regions of the solute molecule experience conflicting tendencies: the polar or charged, hydrophilic region interacts favourably with the solvent and tends to dissolve whereas the nonpolar hydrophobic region has the opposite tendency, to avoid contact with the water. The nonpolar regions of the molecules cluster together to present the smallest area to the solvent, and the polar regions are arranged to maximize their interaction with the aqueous solvent. These stable structures of amphipathic compounds in water, called **micelles**, may contain hundreds or thousands of molecules. The forces that hold the nonpolar regions of the molecules together are called **hydrophobic interactions.**

Many biomolecules are amphipathic (Table 2–5): proteins, pigments, the sterols and phospholipids of membranes and certain vitamins all have polar and nonpolar surface regions. Structures composed of these molecules are stabilized by hydrophobic interactions among the nonpolar regions. Hydrophobic interactions among lipids and between lipids and proteins, are the most important determinants of structure in biological membranes; also hydrophobic interactions between nonpolar amino acids stabilize the 3–'D' folding patterns of proteins.

Hydrogen bonding between water and polar solutes also causes some ordering of water molecules but the effect is less significant than with nonpolar solutes.

F. Van der Waals Interactions

van der Waals interactions (named after **Johannes Diederik van der Waals**) are weak, nonspecific, interatomic attractions and come into play when any two uncharged atoms are 3 to 4 Å apart. Though weaker and less specific than electrostatic and hydrogen bonds, van der Waals interactions are no less important in biological systems. The basis of a van der Waals bond is that *the distribution of electronic charge around an atom changes with time.*

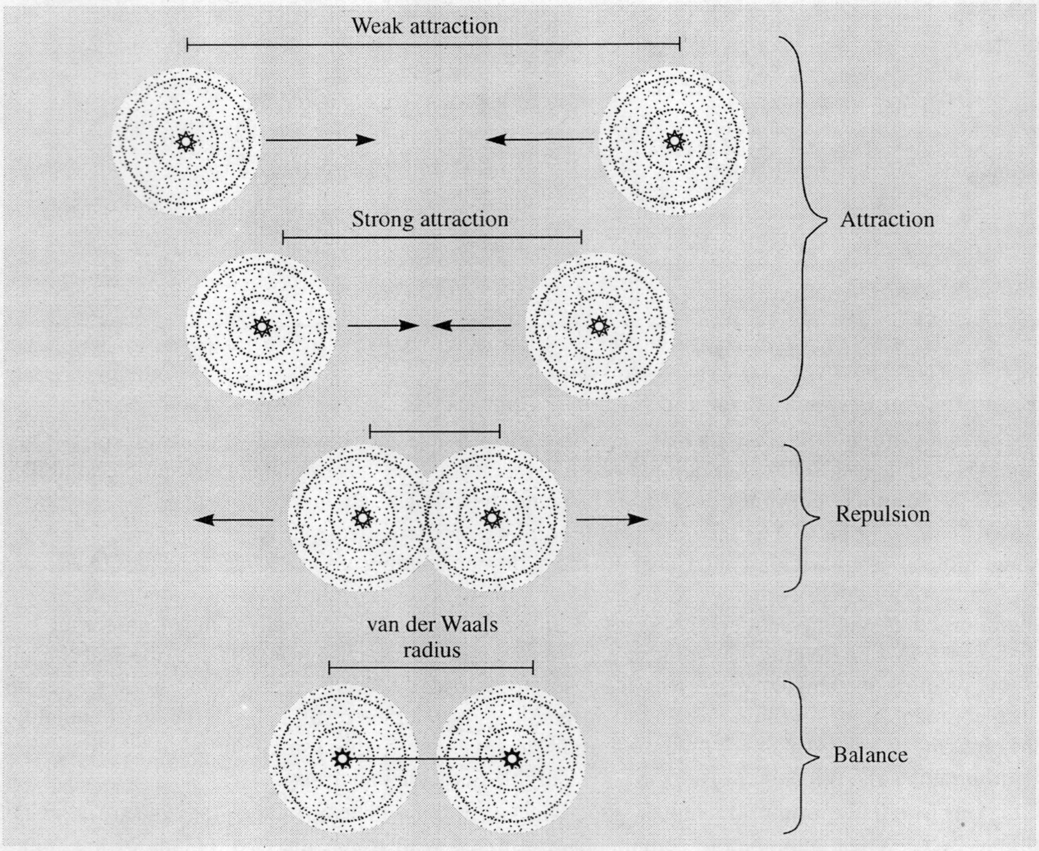

Fig. 2–13. Schematic of van der Waals forces of attraction, repulsion and balance between two atoms

All types of molecules exhibit van der Waals forces which arise from the attraction of the bound electrons of one atom for the nucleus of another. When two atoms are far apart, there is a very weak attraction which becomes stronger as the atoms move closer together (Fig. 2–13). However, if the atoms move close enough for their outer electron shells to overlap, then a force of repulsion ocurs. At a certain distance, defined as the *van der Waals contact radius*, there is a balance between the forces of attraction and those of repulsion. Each type of atom has a specific van der Waals contact radius (Table 2–7). At this point of balance, the two atoms are separated by the *van der Waals contact distance* (Fig. 2–14). The contact distance between an oxygen and carbon atom, for example, is 3.4 Å which is obtained by adding 1.4 and 2.0 Å, the contact radii of the O and C atoms, respectively.

Table 2–7. van der Waals contact radii of atoms and group

Atom/Group	*Radius (in Å)*
H	1.2
O	1.4
N	1.5
S	1.85
P	1.9
C	2.0
CH_3	2.0

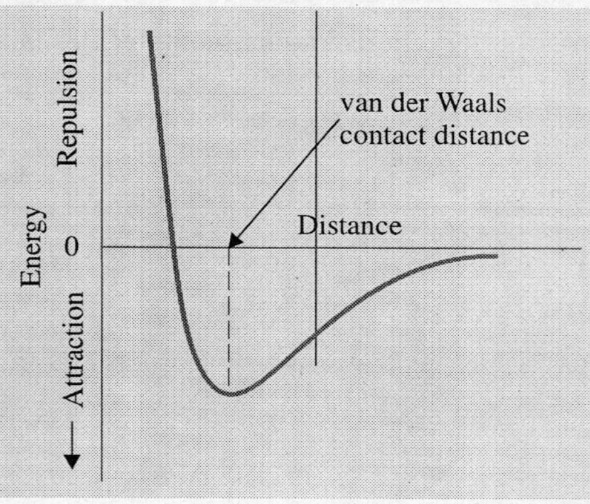

Fig. 2–14. Energy of a van der Waals interaction as a function of the distance between two atoms

The van der Waals bonding energy between two average atoms is very small, *i.e.*, about 1 kcal/mol, which is only slightly greater than the average thermal energy (0.6 kcal/mol) of molecules at room temperature. It is considerably weaker than a hydrogen or electrostatic bond, which is in the range of 3 to 7 kcal/mol. The Energy of a most favourable or the van der waals contact distance. The energy rises rapidly owing to electron-electron repulsion as the atoms move closer together than this distance.

Table 2–8. Strengths of bonds common in biomolecules

Type of bond	*Bond dissociation energy (kJ/mol)*	*Type of bond*	*Bond dissociation energy (kJ/mol).*
Single bonds		**Double bonds**	
O—H	461	C = O	712
H—H	435	C = N	615
P—O	419	C = C	611
C—H	414	P = O	502
N—H	389	**Triple bonds**	
C—O	352	N ≡ N	930
C—C	348	C ≡ C	816
S—H	339	**Noncovalent bonds or interactions**	
C—N	293	Hydrogen bonds	4–20
C—S	260	van der Waals forces	4–20
N—O	222	Hydrophobic bonds	4–20
S—S	214	Ionic interactions	4–20

Since van der Waals interactions are very week, it is customary to call them as a force rather than a bond. The crystalline compounds possessing these forces are very soft and have low melting points. In fact, the more a symmetrical the molecule, the greater are the *van der Waals forces*. These forces are greater for compounds than for atoms and molecules of elements.

ROLE OF NONCOVALENT INTERACTIONS

The noncovalent interactions described above (hydrogen bonds and ionic interactions among charged groups and hydrophobic interactions and van der Waals interactions among nonpolar groups) are much weaker than covalent bonds (Table 2–8). The input of about 350 kJ of energy is required to break a mole (6×10^{23}) of C—C single bonds, and of about 410 kJ to break a mole of C—H bonds, but only 4-8 kJ is sufficient to disrupt a mole of typical van der Waals interactions (Table 2–9).

Table 2–9. Some weak interactions among biomolecules in aqueous solvent

Weak interaction		*Stabilization energy (kJ/mol)*
Hydrogen bonds		
Between neutral groups	>C=O···H—O—	8 – 21
Between peptide bonds	>C=O···H—N<	8 – 21
Ionic interactions		
Attraction	—$^+NH_3$ → ← ^-O—C(=O)—	42
Repulsion	—$^+NH_3$ ⟷ H_3N^+—	≈ – 21
Hydrophobic interactions	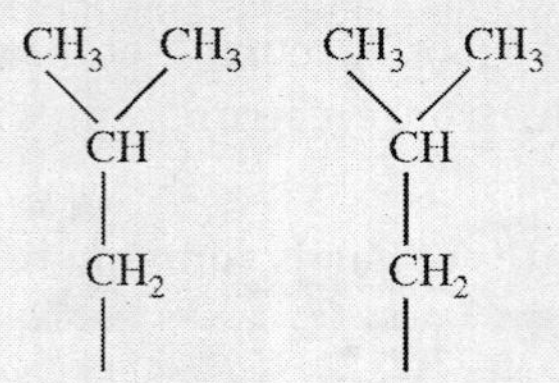	4 – 8
van der Waals interactions	Any two atoms in close proximity	4

Hydrophobic interactions are similarly weak, and ionic interactions and hydrogen bonds are only a little stronger; a typical hydrogen bond can be broken by the input of about 20 kJ/mol. In aqueous solvent at 25°C, the available thermal energy is of the same order as the strength of these weak interactions. Consequently, hydrogen bonds and ionic, hydrophobic and van der Waals interactions are continuously formed and broken.

Although these 4 types of interactions are individually weak relative to covalent bonds, the cumulative effect of many such interactions in a protein or nucleic acid can be very significant. For example, the noncovalent binding of an enzyme to its substrate may involve several hydrogen bonds and one or more ionic interactions, as well as hydrophobic and van der Waals interactions. The formation of each of these weak bonds contributes to a net decrease is free energy; this binding free energy is released as bond formation stabilizes the system.

The most stable (native) structure of most macromolecules is that in which weak-bonding possibilities are maximized. The folding of a single polypeptide or polynucleotide chain into its 3-dimensional shape is determined by this principle. The binding of an antigen to a specific antibody depends on the cumulative effects of many weak interactions. The energy released when an enzyme binds noncovalently to its substrate is the main source of catalytic power for the enzyme. The binding of a hormone or a neurotransmitter to its cellular receptor protein is the result of weak interactions.

ROLE OF WATER IN LIFE

A saying goes, 'there is no life without water'. Life and water are inextricably connected (Fig. 2-15). Life probably originated in water. Wherever life is found, there is water; and wherever liquid water is found, there is life. Water makes up more than 70% of the material of living organisms themselves and covers more than 75% of Earth's surface. It is the medium in which most cells are constantly bathed and the major component of cells themselves. Not only do most biochemical reactions occur in water, but water itself participates in many biochemical reactions.

Fig. 2-15. Essentiality of water to organisms

(*Courtesy : Francois Gohier*)

Killer whale (*Orcinus orca*) spend their entire lives in the ocean.

PROBLEMS

1. Why can we blow a bubble with soap water but not with plain water ?
2. Why does ice melt when salt is sprinkled on it ?
3. Why is salt not soluble in oil but soluble in water ?
4. When a piece of paper is wet with water, it tears off easily. But the same paper if soaked in oil doesn't lose strength. Why ?
5. Why does the water film move away when we touch a wet surface ?
6. Why does water wet the glass while mercury does not ?
7. Why does ice melt but wax harden when subjected to pressure ?
8. Why are ice cubes preferred to water at 0°C to cool a soft drink ?
9. When glucose is added and stirred, the water feels cooler. Why ?
10. How can hydrogen, a combustible, and oxygen, a supporter of combustion when combined in the form of water extinguish fire ?
11. Why is snow white although water of which it is made up of is transparent ?

12. Why is water liquid but hydrogen sulfide a gas ?
13. Why is sea water salty ?
14. How do fishes survive in frozen lakes ?
15. Why do we feel thirsty ?
16. What is the difference between ice and snow ?

ANSWERS TO PROBLEMS

1. Surface tension of water is so strong that it does not allow water to form stable bubbles when blown through a pipe. Soaps and detergents contain a basic cleaning agent called surfactant or surface active agent. *Surfactants lower the surface tension of water sufficiently to enable a film of soap solution to be blown as bubbles.* So soapy water is used by children to blow bubbles.

2. A pure liquid such as water freezes completely at a constant temperature which is known as the freezing point for that liquid. *But the addition of any soluble impurity lowers the freezing point of the mixture so formed.* When salt is sprinkled on ice, a part of it dissolves in the water film that always covers the ice surface. As the freezing point of the solution so formed is much lower than the freezing point of water, it remains in a liquid state and continues dissolving more salt and melting more ice. *This phenomenon is used to clear snow from roads and side-walks in cold countries during winter.*

3. Water is a polar solvent, that is, its molecules behave like dipoles with two oppositely-charged ends. Common salt is a crystalline substance made up of sodium (Na^+) and chloride (Cl^-) ions, held together by strong electrostatic forces. When salt is put in water, the water molecules get attached to the sodium and chloride ions and neutralize the force of attraction between them. The negatively-charged chloride ions get surrounded by the positive ends of the water molecules, while the positively-charged sodium ions get surrounded by the negative ends of the water molecules. As a result, water molecules tend to pull the ions away from the solid crystal lattice, making salt dissolve in water. Oil, on the other hand, is not a polar solvent and its molecules are not able to neutralize the electrostatic force of attraction between ions. Salt, therefore, does not dissolve in oil.

4. Paper is made up of thin sheets of cellulosic fibres and other ingredients mixed together. It is sized to make it suitable for writing. The cellulosic fibres are held together by weak forces called the van der Waal's forces. When paper is dipped in water, the sizing material used for refining the paper being water-soluble becomes soft. Water being a polar solvent, weakens the van der Waal's forces that hold the fibres together. As a result, the fibres come off easily when pulled and so a wet paper tears off easily. On the other hand, oil being a nonpolar solvent does not affect the water-soluble sizing material. Nor does it weaken the forces holding the cellulosic fibres together. So paper soaked in oil does not tear off as easily as paper soaked in water does.

5. The surface of water behaves like any stretched membrane. In other words, when it is ruptured at any point it shrinks away. This property is known as surface tension. Some substances such as oil, grease or detergents have the property of reducing the surface tension when added to water. *When we touch a wet surface with our fingertips, the oil on our fingertips which is always present on the unwashed finger reduces the surface tension of water at that point and the water film moves away.* The area of the surface from which the water film moves away depends on how oily the finger is.

6. Wetting is a phenomenon which depends to a large extent on the difference between the adhesive force between a liquid and the surface in contact and the cohesive force between the molecules of the liquid which is also a measure of its surface tension. Mercury does not wet the surface of glass because of its higher surface tension which is six times that of water. In case of mercury, the cohesive force is much stronger than the adhesive force, that is the force of attraction between molecules of mercury and of glass. As a result, mercury neither spreads out nor wets the surface of glass. On the other hand, in case of water the adhesive force between water and glass is stronger than the cohesive force between water molecules which makes water to spread out and cling to the surface of glass wetting it.

7. Ice is nothing but solidified water in which H_2O molecules are bound together by the weak hydrogen bonds and form an open cage-like structure. When ice is subjected to pressure, the hydrogen bonds being the weaker ones break easily. Consequently, the H_2O molecules get separated and ice melts into water. Whereas in the case of paraffin wax, which is hydrocarbon petroleum product, carbon atoms are held together by covalent bond. Also, since it is plastic solid, it usually deforms under pressure without the application of heat. When pressure is applied on soft wax, the molecular spaces decrease and molecules tend to come closer, thereby causing hardening of wax.

8. Ice cubes are preferred to water at 0°C to cool a soft drink because when ice melts, it absorbs a certain amount of heat known as latent heat until it is melted. *The heat absorbed does not raise the temperature of the drink but goes entirely into changing the ice into water. Thus ice cubes added to a drink chill it by absorbing heat from it.* On the other hand, chilled water at 0°C simply mixes up with the drink and the temperature attained by the mix will only be an average of the temperatures of the two. Hence, ice cubes are preferred over chilled water for cooling a soft drink.

9. Whenever a solute is dissolved in water, some energy change takes place. In most cases, the amount of this change is very small and is hardly noticeable. But in some cases, it is quite large and leads to large changes in temperature – a rise or fall depending upon whether the heat of solution is positive or negative. *Glucose has a negative heat of solution, i.e., heat is absorbed when glucose is dissolved in water. As a result, the temperature drops and the solution feels cool.*

10. Although hydrogen is a combustible gas and oxygen a supporter of combustion, when combined to form water they lose their individual properties. Water is not a mixture but a compound formed by the chemical reaction between hydrogen and oxygen in a definite proportion. Since the properties of a compound are entirely different from the properties of its constituent elements, water does not burn or support combustion. Rather because of its high heat capacity, it is an excellent coolant and fire-extinguisher.

11. Both snow and foam look white due to the same reason. While foam is a collection of large number of bubbles, snow is made up of large number of ice-crystals. In both snow and foam, the gaps between the bubbles or the crystals is filled by air. Light rays falling on such a collection of crystals get bent and reflected by the crystals. However, all the crystals do not reflect or refract the light in the same direction. Thus, light gets scattered haphazardly. The presence of air adds to this scattering. The light reaching an observer looking at the snow, therefore, is just the scattered light and so the snow looks white. In case of water, the situation is different. Here since the water surface is smooth, very little light gets scattered although a part of it is reflected. Almost all of the light falling on a clear water surface passes through it to the other side making water look transparent.

12. Oxygen and sulphur belong to the same group in the periodic table, so theoretically the hydrides of these elements should have very similar properties. But in reality, there is a large difference in the physical properties of these hydrides. For example, water is a liquid and hydrogen sulphide is a gas. This unusual behaviour of water arises from its unique property of forming weak bonds called hydrogen bonds. In water, large number of H_2O molecules are held together by these bonds which form between O and H, thus effectively forming a giant molecule. As is the rule, with the increase in the molecular weight, the boiling point of water rises, compared to what it should be, had there been no hydrogen bonding, and it remains a liquid at room temperature. This does not happen with hydrogen sulphide and so it remains a gas.

13. Seawater is salty because it contains many dissolved salts, mostly sodium chloride with small proportions of potassium, magnesium, calcium and carbonates. The salt comes mainly with river water that flows through rocks eroded by frost and rain. The gradual wearing away of mountains releases chemicals which are carried down by rivers to the ocean as dissolved salts. Some salt also enters seawater from the rocks beneath the seabed. The river water carrying the salts to the sea does not taste salty because the salts are present in extremely small concentration. Whereas in the case of the oceans, water is continuously evaporating even as more and more salt continues to be added with river water. This, over millions of years has led to the high concentration of dissolved salts in seawater.

14. Lakes that appear frozen from the top never freeze completely. Only the top layer freezes and a thick layer of ice floats on water which remains liquid below and in which fishes and other aquatic animals survive. With the onset of winter, water starts cooling. As it cools, it shrinks in volume and becomes denser, but only until it reaches 4°C. Below that temperature, water expands and its density drops. This comparatively lighter water moves up and finally at 0°C it turns into ice. Since the density of ice is less than that of water, the ice floats. Also, ice being a very poor conductor of heat, further cooling of the water below it stops. Thus, the temperature of water below the ice layer remains at 4°C and fishes survive in it.

15. About 60 per cent of our body weight is water. Water is constantly lost from our body through exhalation, perspiration and urine. At the same time, we obtain a good amount of water from our food, but it is not enough to keep the balance of water our body needs. That is why we need to drink water from time to time. When the water balance is disturbed, either due to excessive loss by perspiration or due to insufficient intake, the salt content of the blood changes. This change activates the thirst centre in our brain which makes us feel thirsty.

16. Though ice and snow are both made up of water, there is a difference between the two. Snow is nothing but the frozen atmospheric vapour which falls in winters on earth as light flakes whereas ice is simply frozen water. Another difference is that snow fall can be seen only in winters at areas of high altitudes and places near the polar regions whereas ice can be seen in any season of the year in our refrigerator.

CHAPTER

3 pH and Buffers

INSIDE

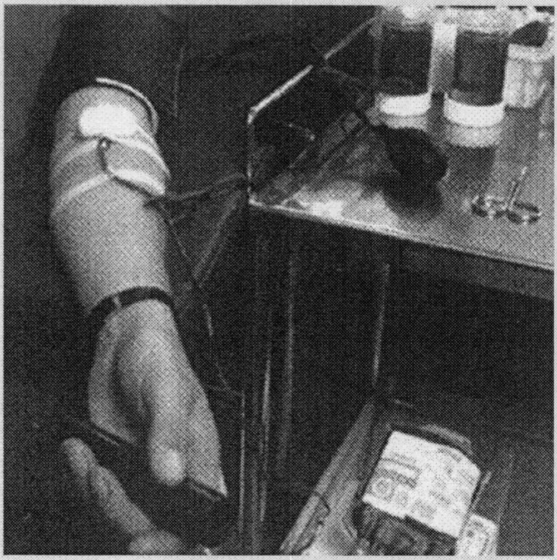

The buffer solution
The blood being taken from this donor's arm is a buffer solution with a pH very close to 7.4.

IONIZATION OF WATER, WEAK ACIDS AND WEAK BASES

Although many of the solvent properties of water can be explained in terms of the uncharged H_2O molecules,the small degree of ionization of water to hydrogen ions (H^+) and hydroxide ions (OH^-) must also be taken into account. *Like all reversible reactions, the ionization of water can be described by an equilibrium constant.* When weak acids or weak bases are dissolved in water, they can contribute H^+ by ionizing (if acids) or consume H^+ by being protonated (if bases). These processes are also governed by equilibrium constants. The total hydrogen ion concentration from all sources, which is experimentally measurable, is expressed as the pH of the solution.

Water molecules have a slight tendency to undergo reversible ionization to yield a hydrogen ion and a hydroxide ion :

$$H_2O \rightleftharpoons H^+ + OH^- \qquad ...(3.1)$$

This reversible ionization is crucial to the role of water in cellular functions. Henceforth, we must have a means of expressing the extent of ionization of water in quantitative terms. This is discussed below.

The position of equilibrium of any chemical reaction is given by its **equilibrium constant.** For the general reaction,

$$A + B \rightleftharpoons C + D \qquad ...(3.2)$$

an equilibrium constant can be defined in terms of concentrations of reactants (A and B) and products (C and D) present at equilibrium :

$$K_{eq} = \frac{[C][D]}{[A][B]} \qquad ...(3.3)$$

The equilibrium constant is fixed and characteristic for any given chemical reaction at a specific temperature. It defines the composition of the final equilibrium mixture of that reaction, irrespective of the starting amounts of reactants and products. Conversely, one can calculate the equilibrium constant for a given reaction at a given temperature, if the equilibrium concentrations of all its reactants and products are known.

> Strictly speaking, the **bracketed terms** represent molar activity, rather than molar concentration.

As mentioned, the degree of ionization of water at equilibrium (Eqn. 3.1) is small; at 25°C, only about one of every 10^7 molecules in pure water is ionized at any instant. The equilibrium constant for the reversible ionization of water (Eqn. 3.1) is :

$$K_{eq} = \frac{[H^+][OH^-]}{[H_2O]} \qquad ...(3.4)$$

In pure water at 25°C, the concentration of water is 55.5 M (*i.e.*, grams of H_2O in 1 litre divided by gram molecular weight or 1000/18 M = 55.5 M). This value is essentially constant in relation to the very low concentrations of H^+ and OH^-, namely 1×10^7 M. Accordingly, on substituting 55.5 M in the equilibrium constant expression (*i.e.*, Eqn. 3.4), we get :

$$K_{eq} = \frac{[H^+][OH^-]}{55.5\ M} \qquad ...(3.5)$$

which, on rearranging, becomes :

$$(55.5\ M)\,(K_{eq}) = [H^+]\,[OH^-] = K_w \qquad ...(3.6)$$

where K_w designates the product (55.5 M) (K_{eq}), the **ion product of water** at 25°C.

The value for K_{eq} is 1.8×10^{-16} M at 25°C as calculated from electrical conductivity measurements. Substituting this value for K_{eq} in Eqn. 3.6 gives :

$$(55.5\ M)\,(1.8 \times 10^{-16}\ M) = [H^+]\,[OH^-]$$
$$99.9 \times 10^{-16}\ M^2 = [H^+]\,[OH^-]$$
$$1.0 \times 10^{-14}\ M^2 = [H^+]\,[OH^-] = K_w$$

Thus, the product $[H^+]\,[OH^-]$ in aqueous solutions at 25°C always equals $1 \times 10^{-14}\ M^2$. When there are exactly equal concentrations of both H^+ and OH^-, as in pure water, the solution is said to be at **neutral pH**. At this pH, the concentration of H^+ and OH^- can be calculated from the ion product of water as follows :

$$K_w = [H^+]\,[OH^-] = [H^+]^2$$

Solving for $[H^+]$ gives :

$$[H^+] = \sqrt{K_w} = \sqrt{1 \times 10^{-14} M^2}$$
$$[H^+] = [OH^-] = 10^{-7}\ M.$$

As the ion product of water is constant, whenever the concentration of H^+ ions is greater than 1×10^{-7} M, the concentration of OH^- must become less than 1×10^{-7} M, and vice versa. When the concentration of H^+ is very high, as in a solution of hydrochloric acid, the OH^- concentration must be very low.

THE pH SCALE

The ion product of water, K_w, is the basis for the **pH scale** (Table 3–1). It is a convenient means of designating the actual concentration of H^+ (and thus of OH^-) in any aqueous solution in the range

Table 3–1. The pH scale

[H⁺] *(M)*	*pH*	*[OH⁻]* *(M)*	*pOH**
10^0(1)	0	10^{-14}	14
10^{-1}	1	10^{-13}	13
10^{-2}	2	10^{-12}	12
10^{-3}	3	10^{-11}	11
10^{-4}	4	10^{-10}	10
10^{-5}	5	10^{-9}	9
10^{-6}	6	10^{-8}	8
10^{-7}	7	10^{-7}	7
10^{-8}	8	10^{-6}	6
10^{-9}	9	10^{-5}	5
10^{-10}	10	10^{-4}	4
10^{-11}	11	10^{-3}	3
10^{-12}	12	10^{-2}	2
10^{-13}	13	10^{-1}	1
10^{-14}	14	10^0(1)	0

* The expression of pOH is sometimes used to describe the basicity, or OH^- concentration of a solution. pOH is defined by the expression pOH = – log [OH^-], which is analogous to the expression for pH. *Note that for all cases, pH + pOH = 14.*

between 1.0 M H^+ and 1.0 M OH^-. Biochemical reactions are often defined in terms of hydrogen ion (H^+) concentrations. In 1909, Soren Sörensen, a Danish biochemist, used a logarithmic scale for expressing the H^+ concentration. This scale was called pH, where *p* stands for power and H for hydrogen ion concentration. He defined pH of a solution as the negative logarithm of the concentration (in moles/litre) of hydrogen ions. Thus,

$$\text{pH} = \log \frac{1}{[H^+]} = -\log [H^+]$$

The symbol *p* denotes "negative logarithm of".

For a precisely neutral solution at 25°C, in which the concentration of hydrogen ions is 1.0 × 10^{-7} M, the pH can be calculated as follows:

$$\text{pH} = \log \frac{1}{1 \times 10^{-7}} = \log (1 \times 10^7)$$

$$= \log 1.0 + \log 10^7$$

$$= 0 + 7.0$$

$$= 7.0$$

The value of 7.0 for the pH of a precisely neutral solution is not an arbitrarily chosen figure. *It is derived from the absolute value of the ion product of water at 25 °C, which by*

In fact, the term **pH** was introduced in 1909 by **Sörensen** who defined it as:

"Für die Zahl *p* schlage ich den Namen Wasserstoffionenexponent und die Schreibweise pH vor. Unter dem Wasserstoffionexponenten (pH) einer Lösung wird dann der Briggsche Logarithmus des reziproken Wertes des auf Wasserstoffionen bezagenen Normalitäts faktors de Lösung verstanden."

The translation into English is :

"For the sign *p* I propose the name 'hydrogen ion exponent' and the symbol pH. Then, for the hydrogen ion exponent (pH) of a solution, the negative value of the Briggsian logarithm of the related hydrogen ion normality factor is to be understood."

pH (short for "potential of hydrogen") is a symbol and denotes the relative concentration of hydrogen ions in a solution. pH values extend from 0 to 14; the lower the value, the higher the acidity or the more hydrogen ions the solution contains. Water at 25 °C has a concentration of H ion of 10^{-7}; the pH, therefore, is 7.

convenient coincidence is a round number.

To calculate the pH of a solution :

1. Calculate the hydrogen ion concentration, $[H^+]$.
2. Calculate the base 10 logarithm of $[H^+]$.
3. pH is the negative of the value found in step 2.

For example, for pure water at 25°C :

$$pH = -\log [H^+] = -\log 10^{-7} = -(-7) = 7.0$$

The number of hydrogen ions present in a solution is a measure of the **acidity of the solution.** All acids do not ionize completely when dissolved in water, *i.e.*, all the molecules of acid do not ionize and exist in the solution as electrically-charged particles. The hydrogen ion concentration is a measure, therefore, of the amount of *dissociated acid* rather than of the anmount of acid present. Strong acids dissociate more freely than weak acids; hydrochloric acid, for example, dissociates freely into H^+ and Cl^- whereas carbonic acid, a weak acid, dissociates much less freely into H^+ and CO_3^-. The number of free hydrogen ions is a measure of its acidity rather than an indication of the type of molecule from which the hydrogen ions originated. The **alkalinity of a solution** is dependent upon the number of hydroxyl ions present. Water is a neutral solution because each molecule contains one H^+ and one OH^-. For each molecule of water dissociated, there is one H^+ and one OH^-, each one neutralizing the other.

The pH scale was developed taking water as the standard. It is an experimental fact that only 1 mole in 5,50,000,000 moles of water ionizes into a H^+ and OH^-. This is the same proportion as 1 gram hydrogen ion in 10,000,000 litres of water. Hence, 1 litre of water contains 1/10,000,000 (or $1/10^7 = 10^{-7}$) of a gram of H^+. Later, for every day use, only the 'power' figure was used and the symbol pH placed before it.

A neutral solution, such as water, where the number of hydrogen ions is balanced by the same number of hydroxyl ions, has a pH of 7.0. The range of the pH scale is from 0 to 14. If the pH is 0, it

Fig. 3-1. pH of washing soda indicates it is basic in nature

would mean that 1 litre of water contained 1/1 = 1 gram hydrogen ion; or, at the other end of the scale, if there were no hydrogen ion present, it would be written $1/10^{14}$ or 10^{-14} or pH 14. The pH scale, thus, runs from 1 to 14; neutrality being at pH 7.0. Solutions having a pH lesser than 7 are acidic, *i.e.*, the concentration of H^+ is greater than that of OH^-. Conversely, solutions having a pH more than 7 are basic or alkaline, *i.e.*, denote an excess of OH^- over H^+ (Table 3–1). Washing soda for example, has a pH of ca 11 (Fig. 3-1).

Note that *the pH scale is logarithmic, not arithmetic*. Thus, when the pH of a solution decreases one unit from 5 to 4, the H^+ concentration has increased tenfold from 10^{-5} to 10^{-4} M, since decimal logarithms are used for the pH scale. Similarly, when the pH has increased three-tenth of a unit from 6.0 to 6.3, the H^+ concentration has decreased from 10^{-6} M to 5×10^{-7} M. To say that two solutions differ in pH by 1 pH unit means that one solution has ten times the H^+ concentration of the other, but it does not tell us the absolute magnitude of the difference. Figure 3–2 gives the pH of some common aqueous fluids. A coca cola drink (pH 3.0) or red wine (pH 3.7) has an H^+ concentration approximately 10,000 times greater than that of blood (pH 7.4). If we now apply the term of pH to the ion product expression for pure water, we obtain another useful expression :

$$[H^+] \times [OH^-] = 1.0 \times 10^{-14}$$

On taking logarithmic of this equation :

$$\log [H^+] + \log [OH^-] = \log (1.0 \times 10^{-14}) = -14$$

and then multiplying by −1, we get :

$$-\log [H^+] - \log [OH^-] = 14$$

If we now define $-\log [OH^-]$as pOH, a definition similar to that of pH, we have an expression relating the pH and pOH in any aqueous solution :

$$pH + pOH = 14$$

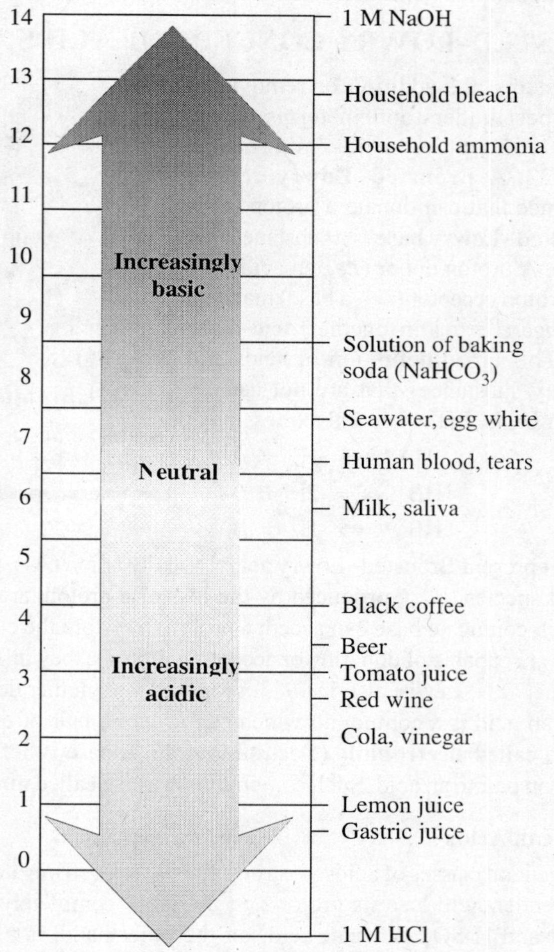

Fig. 3–2. The pH of some aqueous fluids

Measurement of pH is one of the most important and frequently used procedures in Biochemistry (Fig. 3-3). The pH affects the structure and activity of biological macromolecules; for example, the catalytic activity of enzymes. Measurements of the pH of the blood and urine are commonly used in

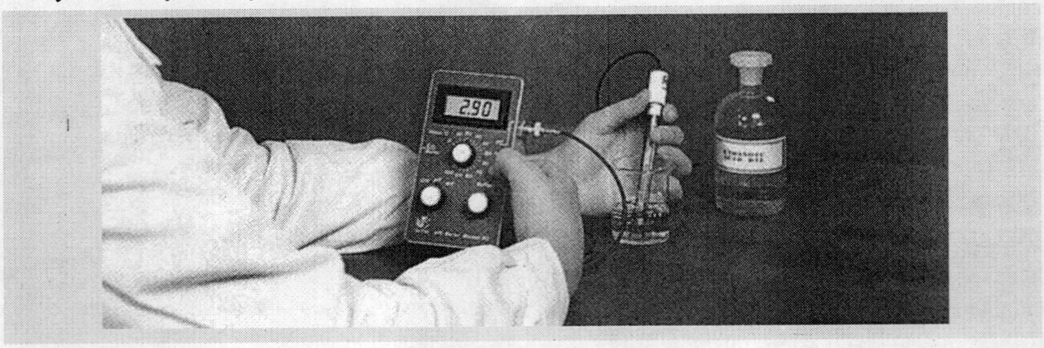

Fig. 3-3. Testing pH by pH meter

diagnosing disease. The pH of the blood plasma of severely diabetic people, for example, is often lower than the normal value of 7.4; this condition is called **acidosis.** In certain other disease states, the pH of the blood is higher than normal, a condition called **alkalosis.**

BRÖNSTED–LOWRY CONCEPT OF ACIDS AND BASES

In aqueous systems, the addition or removal of hydrogen ions is best understood in terms of the Brönsted–Lowry concept of acids and bases, propounded in 1923. A **Brönsted–Lowry acid** is defined as a substance that can donate a proton (H^+); conversely, a **Brönsted–Lowry base** is a substance that can accept a proton. A proton donor (*i.e.*, an acid) and its corresponding proton acceptor (*i.e.*, a base) make up a **conjugate** (*coniungere*^{L} = to join together) **acid–base pair** (Table 3–2). This broad definition of acids and bases includes many substances that are not usually considered acidic or basic. In the general expressions :

J.N. Brönsted **M. Lowry**

Both famous for developing the familiar concept of acids and bases.

$$HA \rightleftharpoons H^+ + A^-$$
$$HB^+ \rightleftharpoons H^+ + B:$$
$$HB^- \rightleftharpoons H^+ + B^{2-}$$

HA, HB^+ and HB^- represent Brönsted–Lowry acids, and the anion (A^-), neutral species (B:) and the negatively–charged species (B^{2-}), produced by the loss of a proton, are their respective conjugate bases. Note that each conjugate base can accept a proton to restore the corresponding acid. Because many biomolecules are capable of donating or accepting protons, they influence the H^+ concentration in biological systems. **G.N. Lewis**, also in 1923, proposed yet another definition of acids and bases. According to him, an acid is a compound which can accept a pair of electrons from a base. Such compounds are also called **electrophile** (electron-loving). A base is defined as a compound which can donate an electron pair to an acid. Such compounds are also called **nucleophile** (nucleus-loving).

A. Strong and Weak Acids

There are two general classes of acids — strong and weak. A **strong acid** is defined as a substance that has a greater tendency to lose its proton and therefore completely dissociates (or ionizes) in water, such as HCl and H_2SO_4. A **weak acid**, on the other hand, is a molecule that has a lesser

Table 3–2. Some Brönsted–Lowry acids and their conjugate bases*

Acid	*Undissociated acid (HA)*	*Conjugate base (A^-)*
Acetic Acid	CH_3COOH	CH_3COO^-
Propionic acid	CH_3CH_2COOH	$CH_3CH_2COO^-$
Pyruvic acid	$CH_3COCOOH$	CH_3COCOO^-
Phosphoric acid	H_3PO_4	$H_2PO_4^-$
Dihydrogen phosphate	$H_2PO_4^-$	HPO_4^{2-}
Monohydrogen phosphate	HPO_4^{2-}	PO_4^{3-}
Ammonium ion	NH_4^+	NH_3
Carbonic acid	H_2CO_3	HCO_3^-
Bicarbonate ion	HCO_3^-	CO_3^{2-}
Water	HOH	OH^-
Glycine	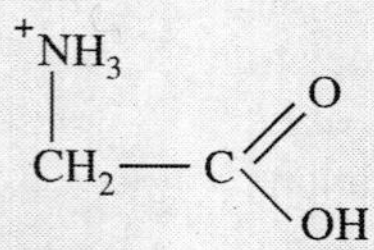 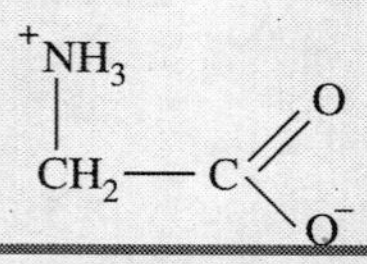	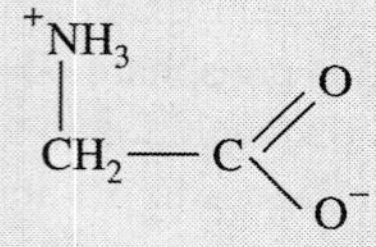 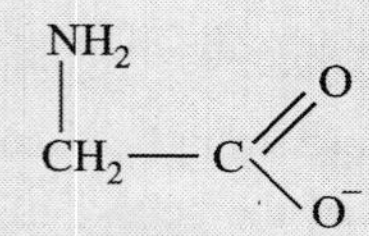

* Some compounds, such as acetic acid, are **monoprotic**, as they can give up only one proton. Others are **diprotic** (carbonic acid and glycine) or **triprotic** (phosphoric acid).

tendency to lose its proton (or, in other words, displays a high affinity for its proton) and, therefore, does not readily dissociate in water, such as CH_3COOH.

B. Ionization of Weak Acids

The selective dissociation of weak acids in water is a characteristic. It is of great importance to Biochemistry because of its role in influencing the H^+ concentration of an aqueous environment. The dissociation of the weak organic compound, acetic acid, is written as :

$$CH_3COOH \rightleftharpoons H^+ + CH_3COO^-$$

At a given temperature, the extent of ionization at equilibrium can be calculated by the following equation :

$$K_a' = [H^+]\,[CH_3\ COO^-]\,/\,[CH_3\ COOH]$$

The expression is comparable to that used to describe the dissociation of water, except that the symbol K_a' (apparent ionization constant) is substituted for K'_{eq} (equilibrium constant). The change recognizes the reaction as an ionization. The prime (′) symbol on the K_a' is used to identify the ionization constant as *apparent* (value based on the concentrations of reactants and products) rather than as a *true* value (K_a' corrected for deviation of the system from ideal behaviour).

The slight amount of ionization that a 1 M solution of acetic acid undergoes ($K_a' = 1.8 \times 10^{-5}$ M at 25°C) can be readily calculated by leting $x = [H^+]$ and $[CH_3COO^-]$ and not by correcting [CH_3 COOH], which is a relatively large concentration, for the small amount of x produced.

$$x^2 = 1.8 \times 10^{-5}\ \text{M}$$

$$x = 4.2 \times 10^{-3}\ \text{M or } 0.0042\ \text{M}$$

Thus, only slightly over 0.4 per cent of a 1 M solution of acetic acid is ionized at 25°C. The pH of the solution is 2.38 (*i.e.*, – log 0.0042).

C. Titration of a Weak Acid by a Strong Base

Titration is used to determine the amount of an acid in a given solution. In this procedure (Fig. 3-4), a measured volume of the acid is titrated with a solution of a strong base (usually NaOH) of known concentration. The NaOH is added in small increments until the acid is consumed (or neutralized), as determined with an indicator dye or with a pH meter. The concentration of the acid in the original solution can be calculated from the volume and concentration of NaOH added. A plot of the pH against the amount of NaOH added (a **titration curve**) reveals the pK_a of the weak acid.

Fig. 3-4. A pH titration in action

Equilibrium constants for ionization reactions are more usually called *ionization or dissociation constants* and are often designated K_a. Table 3–3 lists dissociation constants of some weak acids. Stronger acids, such as formic and lactic acids, have higher dissociation constants; weaker acids, such as dihydrogen phosphate ($H_2PO_4^-$), have lower dissociation constants. Also included in the Table 3–3 are values of pK_a, which is analogous to pH and is defined by the equation :

$$pK_a = \log \frac{1}{K_a} = -\log K_a$$

The more strongly dissociated the acid, the lower is its pK_a.

A plot of the course of the titration reveals the property of Brönsted–Lowry acids and their conjugate bases that makes them useful in Biology. When acetic acid (CH_3COOH) is titrated with NaOH (Fig. 3–5), the greatest changes in pH take place at the beginning and end of the titration. The region of least change occurs at the midpoint of titration, when exactly 0.5 equivalents of base have been added. At this point, the concentration of undissociated acid (CH_3 COOH or HA) is equal to that of its anion (CH_3COO^- or A^-). At these particular concentrations of HA and A^-, the pH (4.76) is equal to the pK_a' value.

Table 3–3. Dissociation constant and pKa of some common weak acids (proton donors) at 25°C

Acid	*Formula*	K_a *(M)*	pK_a
Formic acid	$HCOOH$	1.78×10^{-4}	3.75
Acetic acid	CH_3COOH	1.74×10^{-5}	4.76
Propionic acid	CH_3CH_2COOH	1.35×10^{-5}	4.87
Lactic acid	$CH_3CH(OH)COOH$	1.38×10^{-4}	3.86
Phosphoric acid	H_3PO_4	7.25×10^{-3}	2.14
Dihydrogen phosphate	$H_2PO_4^-$	1.38×10^{-7}	6.86
Monohydrogen phosphate	HPO_4^{2-}	3.98×10^{-13}	12.40
Carbonic acid	H_2CO_3	1.70×10^{-4}	3.77
Biocarbonate	HCO_3^-	6.31×10^{-11}	10.20
Ammonium	NH_4^+	5.62×10^{-10}	9.25

The most important point about the titration curve of a weak acid is that it shows graphically that a weak acid and its anion — a conjugate acid-base pair—can act as a buffer.

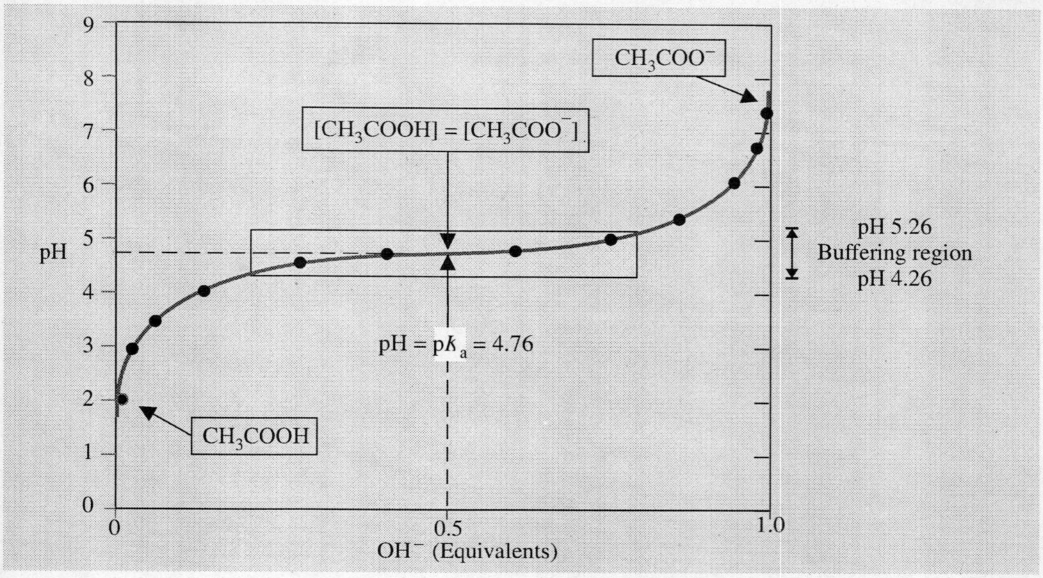

Fig. 3–5. The titration curve of acetic acid

After the addition of each increment of NaOH to the acetic acid solution, the pH of the mixture is measured. This value is plotted against the fraction of the total amount of NaOH required to neutralize the acetic acid (*i.e.*, to bring it to pH ≈ 7). The points so obtained yield the titration curve. Shown in the boxes are the predominant ionic forms at the points designated. At the midpoint of the titration, the concentration of the proton donor and proton acceptor are equal. The pH at this point is numerically equal to the pK_a of acetic acid. The screened zone is the useful region of buffering power.

Fig. 3–6 compares the titration curves of 3 weak acids with very different dissociation constants: acetic acid (pK_a = 4.76), dihydrogen phosphate (pK_a = 6.86) and ammonium ion (pK_a = 9.25). Although the titration curves of these acids have the same shape, they are displaced along the pH axis because these acids have different strengths. Acetic acid is the strongest and loses its proton most readily, since its K_a is highest (pK_a lowest) of the three. Acetic acid is already half dissociated at pH 4.76. $H_2PO_4^-$ loses a proton less readily, being half dissociated at pH 6.86. NH_4^+ is the weakest acid of the three and becomes half dissociated only at pH 9.25.

BUFFERS

A **buffer solution** is one that resists a change in pH on the addition of acid (H^+) or base (OH^-), more effectively than an equal volume of water. Most commonly, the buffer solution consists of a mixture of a weak Brönsted acid and its conjugate base; for example, mixtures of acetic acid and sodium acetate or of ammonium hydroxide and ammonium chloride are buffer solutions. A buffer system consists of a weak acid (the proton donor) and its conjugate base (the proton acceptor). As an example, a mixture of equal concentrations of acetic acid and acetate ion, found at the midpoint of the titration curve in Fig. 3–5, is a buffer system. The titration curve of acetic acid has a relatively flat zone extending about 0.5 pH units on either side of its midpoint pH of 4.76. In this zone, there is only a small change in pH when increments of either H^+ or OH^- are added to the system. This relatively flat zone is the buffering region of the acetic acid-acetate buffer pair. At the midpoint of the buffering region, where the concentration of the proton donor (acetic acid) exacts equals that of the proton acceptor (acetate), the buffering power of the system is maximal, *i.e.*, its pH changes least on addition of an increment of H^+ or OH^-. The pH at this point in the titration curve of acetic acid is equal to its pK_a. The pH of the acetate buffer system does change slightly when a small amount of H^+ or OH^- is

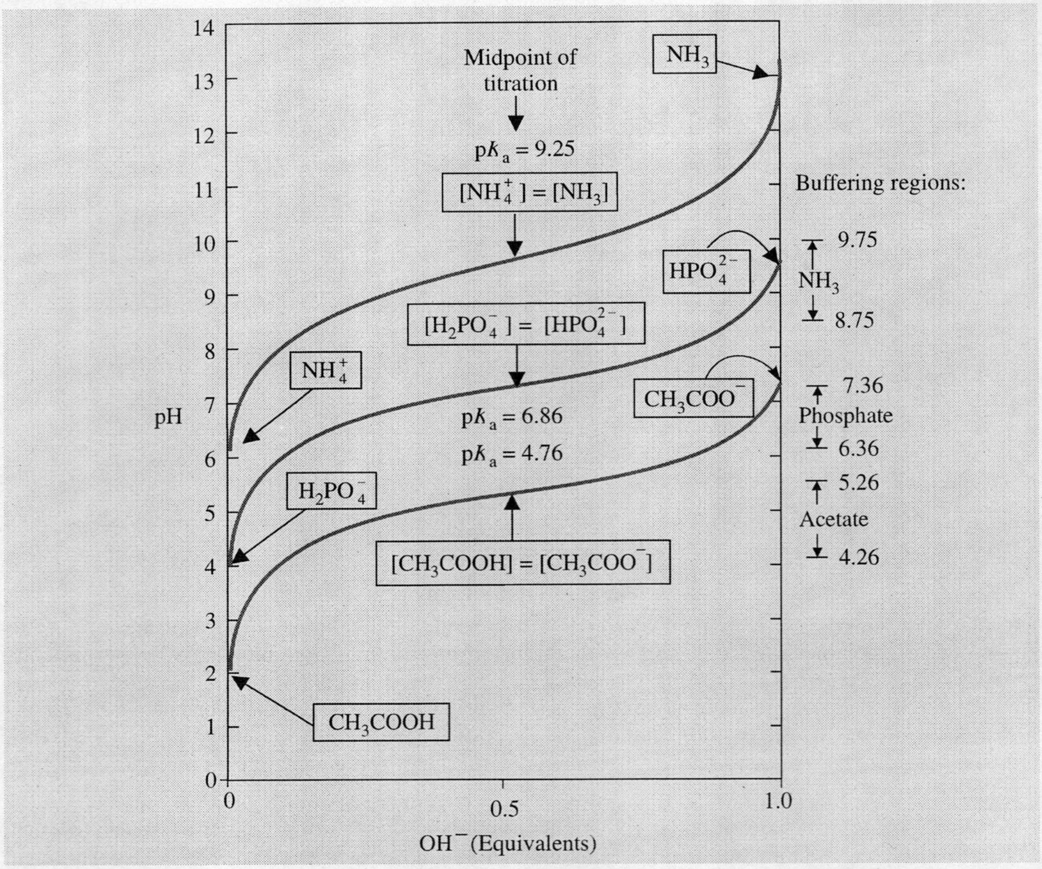

Fig. 3–6. Comparison of the titration curves of 3 weak acids, CH_3COOH, $H_2PO_4^-$ and NH_4^+

The predominant ionic forms at designated points in the titration are given in boxes. The regions of buffering capacity are indicated at the right. Conjugate acid-base pairs are effective buffers between approximately 25 and 75% neutralization of the proton-donor species.

added, but this change is very small compared with the pH change that would result if the same amount of H^+ (or OH^-) were added to pure water or to a solution of the salt of a strong acid and strong base, such as NaCl, which have no buffering power.

Each conjugate-acid-base pair has a characteristic pH zone in which it is an effective buffer (Fig. 3–3). The $H_2PO_4^- / HPO_4^{2-}$ pair has a pK_a of 6.86 and thus can serve as a buffer system near pH 6.86; the NH_4^+/NH_3 pair, with a pK_a of 9.25, can act as a buffer near pH 9.25.

Let us consider the mechanism by which a buffer solution exerts control over large pH changes, by taking example of a buffering system consisting of acetic acid and sodium acetate.

$$CH_3COOH + CH_3COONa \longrightarrow Na^+ + H^+ + 2CH_3COO^-$$

If alkali (NaOH) is added to this system, it will form salt and no free H^+ or OH^- will be available :

$$CH_3COOH + CH_3COONa + NaOH \longrightarrow 2CH_3COONa + H_2O$$

If, however, acid (HCl) is added to this system, it will also form salt and no free H^+ or OH^- will be available :

$$CH_3COOH + CH_3COONa + HCl \longrightarrow NaCl + 2CH_3COOH$$

In either cases, there is no change in H^+ concentration. The buffer acts almost as if it were 'absorbing' the added free hydrogen or hydroxyl ions.

In discussing the quantitative aspects of buffer action, we must note that two factors determine the effectiveness or *capacity* of buffer solution :

(*a*) *Molar concentration of the buffer components.* The buffer capacity is directly proportional to the concentration of the buffer components. The concentration of a buffer refers to the sum of the concentration of the weak acid and its conjugate base. Thus, a 0.1 M acetate buffer could contain 0.05 mole of acetic acid and 0.05 mole of sodium acetate in 1 litre of water. It could also contain 0.065 mole of acetic acid and 0.035 mole of sodium acetate in 1 litre of water.

(*b*) *Relative concentrations of the conjugate base and the weak acid.* Quantitatively, it should seem evident that the most effective buffer would be one with equal concentrations of acidic and basic components, since such a mixture could furnish equal quantities of acidic and basic components to react, respectively with alkali or acid. An inspection of the titration curve for acetic acid (Fig. 3–3) shows that the minimum change in pH resulting from the addition of a unit of base (or acid) occurs at the pK_a for acetic acid. At this pH, the ratio CH_3COO^- to CH_3COOH is 1. On the contrary, at values of pH far removed from the pK_a (and therefore at ratios of conjugate base to acid greatly differing from unity), the change in pH for unit of acid or alkali added is much larger.

The hydrogen ion concentration of most body fluids and secretions is on the alkaline side. Urine may be acid, and gastric juice may be very acid, but these are exceptions. Many influences tend to change this alkalinity, but the buffers present prevent marked fluctuations in hydrogen ion concentrations. The pH of blood, for instance, stays within the limits 7.3 to 7.5 in health. When these limits are exceeded, acidosis or alkalosis with alarming symptoms results, frequently with dire consequences. Table 3.4 gives some of the pH values for various human fluids.

Table 3–4. pH values of human body fluids and secretions

Body fluid/Secretion	*pH*	*Body fluid/Secretion*	*pH*
Gastric juice (parietal secretion)	0.87	Blood	7.4
		Cerebrospinal fluid	7.4
Urine	6.0	Intestinal juice	7.7
Milk	6.6 – 6.9	Bile	7.8
Saliva	7.2	Pancreatic Juice	8.0
Aqueous humor of eye	7.2		

It is quite difficult to keep a solution at constant pH if no buffer is present because of the influence of the CO_2 of the air or the alkali of the glass container or because of other influences. Consequently, buffers are frequently required. Various mixtures, consisting of definite amounts of the acid or base and its respective salt have been prepared. Since such buffer sets maintain their pH indefinitely, they are used in the indicator method of determining pH. It should be noted that buffers can be prepared not only by titration of a weak acid with OH^- (or titration of a solution of A^- with H^+), but also by addition of A^-, the salt of HA, to a solution of HA.

A. Henderson–Hasselbalch Equation

The quantitative relationship among pH, buffering action of a mixture of weak acid with its conjugate base, and the pK_a of the weak acid is given by a simple expression called **Henderson-Hasselbalch Equation.** The titration curves of acetic acid, $H_2PO_4^-$ and NH_4^+ (Fig. 3–6) have nearly identical shapes, suggesting that they all point towards a fundamental law or relationship. This is actually the case. The shape of the titration curve of any weak acid is expressed by Henderson–Hasselbalch equation. This equation is simply a useful way of restating the expression for the dissociation constant of an acid. For the dissociation of a weak acid HA into H^+ and A^-, the Henderson–Hasselbalch equation can be derived as follows :

$$K_a = \frac{[H^+][A^-]}{[HA]}$$

1. Rearrange the K_a equation to solve for (H^+) :

$$[H^+] = K_a \frac{[HA]}{[A^-]}$$

2. Convert to logarithmic functions :

$$\log [H^+] = \log K_a + \log \frac{[HA]}{[A^-]}$$

3. Make the expression negative (or multiply by –1) :

$$-\log [H^+] = -\log K_a - \log \frac{[HA]}{[A^-]}$$

4. Substitute pH for – log $[H^+]$ and pK_a for $-\log K_a$

$$\text{pH} = pK_a - \log \frac{[HA]}{[A^-]}$$

5. Now, to remove the minus sign, invert the last term, *i.e.*, $-\log \frac{[HA]}{[A^-]}$ to obtain Henderson-Hasselbalch equation :

$$\text{pH} = pK_a + \log \frac{[A^-]}{[HA]}$$

The equation is expressed more generally as :

$$\text{pH} = pK_a + \log \frac{[\text{proton acceptor}]}{[\text{proton donor}]}$$

This equation fits the titration curve of all weak acids and enables one to deduce a number of important quantitative relationships. Henderson-Hasselbalch equation is of great predictive value in protonic equilibria as illustrated below :

A. When [A–] = [HA] or when an acid is exactly half neutralized : Under these conditions,

$$\text{pH} = pK_a + \log \frac{[A^-]}{[HA]} = pK_a + \log \frac{1}{1} = pK_a + 0 = pK_a.$$

Therefore, at half neutralization, pH = pK_a. The equation, thus, shows why the pK_a of a weak acid is equal to the pH of the solution at the midpoint of its titration.

B. When the ratio [A–]/[HA] = 100 to 1 :

$$\text{pH} = pK_a + \log \frac{[A^-]}{[HA]} = pK_a + \log \frac{100}{1} = pK_a + 2$$

C. When the ratio [A–]/[HA] = 1 to 10 :

$$\text{pH} = pK_a + \log \frac{[A^-]}{[HA]} = pK_a + \log \frac{1}{10} = pK_a + (-1)$$

If the equation is evaluated at several ratios of $[A^-]$ / [HA] between the limits 10^3 and 10^{-3}, and the calculated pH values plotted, the result obtained describes the titration curve for a weak acid as shown in Fig. 3–7.

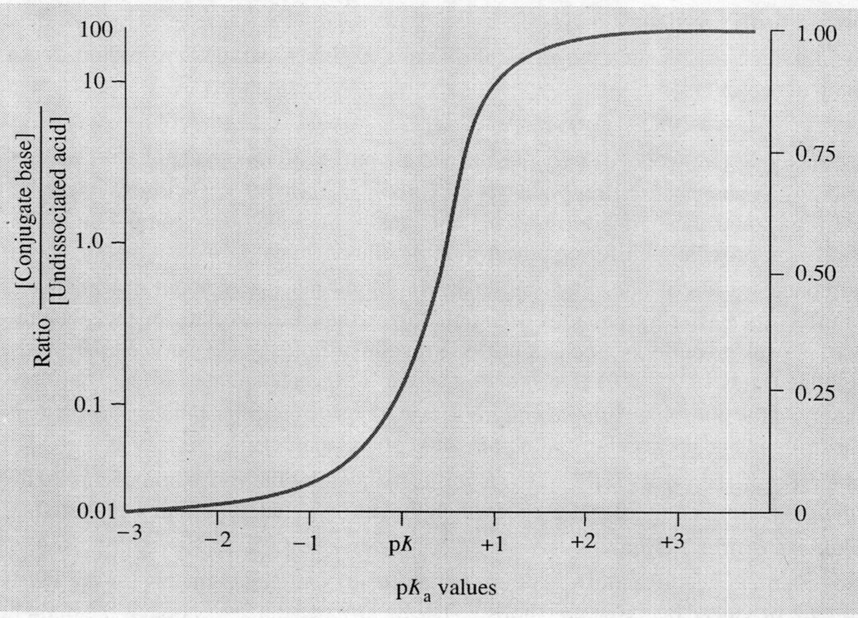

Fig. 3–7. General form of a titration curve calculated from the Henderson-Hasselbalch equation

BIOLOGICAL BUFFER SYSTEMS

Almost every biological process is pH-dependent; a small change in pH produces a large change in the rate of the process. This is true not only for the many reactions in which the H^+ ion is a direct participant, but also for those in which there is no apparent role for H^+ ions. The enzymes and many of the molecules on which they act, contain ionizable groups with characteristic pK_a values. The protonated amino ($-NH_3^+$) and carboxylic groups of amino acids and the phosphate groups of nucleotides, for example, function as weak acids; their ionic state depends upon the pH of the solution in which they are dissolved.

Cells and organisms maintain a specific and constant cytosolic pH, keeping biomolecules in their optimal ionic state, usually near pH 7. In multicelled organisms, the pH of the extracellular fluids (blood, for example) is also tightly regulated. Constancy of pH is achieved primarily by **biological buffers** : mixtures of weak acids and their conjugate bases. Table 3–5 lists some important buffering systems of body fluids which help maintaining pH. A certain amount of many of these is usually present in the body and cellular fluids, and so the maintenance of a constant pH depends on a complex system.

Table 3–5. Body fluids and their principal buffers

Body fluids	*Principal buffers*
Extracellular fluids	Biocarbonate buffer Protein buffer
Intracellular fluids	Phosphate buffer Protein
Erythrocytes	Hemoglobin buffer

A description of various buffering systems follows :

1. The Phosphate Buffer System

This system, which acts in the cytoplasm of all cells, consists of $H_2PO_4^-$ as proton donor and HPO_4^{2-} as proton acceptor :

$$H_2PO_4^- \rightleftharpoons H^+ + HPO_4^{2-}$$

The phosphate buffer system works exactly like the acetate buffer system, except for the pH range in which it functions. The phosphate buffer system is maximally effective at a pH close to its pK_a of 6.86 (see Table and Fig.), and thus tends to resist pH changes in the range between 6.4 and 7.4. It is, therefore, effective in providing buffering power in intracellular fluids.

Since the concentration of phosphate buffer in the blood plasma is about 8% of that of the bicarbonate buffer, its buffering capacity is much lower than bicarbonate in the plasma. The concentration of phosphate buffer is much higher in intracellular fluid than in extracellular fluids. The pH of intracellular fluids (6.0 – 6.9) is nearer to the pK_a of the phosphate buffer. Therefore, the buffering capacity of the phosphate buffer is highly elevated inside the cells and the phosphate is also effective in the urine inside the renal distal tubules and collecting ducts.

In case the ratio of $[HPO_4^{2-}]$ / $[HPO_4^-]$ tends to be changed by the formation of more $H_2PO_4^-$, there occurs the renal elimination of $H_2PO_4^-$ for which the ratio ultimately remains unaltered.

2. The Bicarbonate Buffer System

This is the main extracellular buffer system which (also) provides a means for the necessary removal of the CO_2 produced by tissue metabolism. The bicarbonate buffer system is the main buffer in blood plasma and consists of carbonic acid as proton donor and bicarbonate as proton acceptor :

$$H_2CO_3 \rightleftharpoons H^+ + HCO_3^-$$

This system has an equilibrium constant

$$K_1 = \frac{[H^+][HCO_3^-]}{[H_2CO_3]}$$

and functions as a buffer in the same way as other conjugate acid-base pairs. It is unique, however, in that one of its components, carbonic acid, is formed from dissolved (*d*) carbon dioxide and water, according to the reversible reaction :

$$CO_2(d) + H_2O \rightleftharpoons H_2CO_3$$

which has an equilibrium constant given by the expression :

$$K_2 = \frac{[H_2CO_3]}{[CO_2(d)][H_2O]}$$

Carbon dioxide is a gas under natural conditions, and the concentration of dissolved CO_2 is the result of equilibration with CO_2 of the gas phase (*g*) :

$$CO_2(g) \rightleftharpoons CO_2(d)$$

This process has an equilibrium constant given by :

$$K_3 = \frac{[CO_2(d)]}{[CO_2(g)]}$$

The pH of a bicarbonate buffer system depends on the concentration of H_2CO_3 and HCO_3^-, the proton donor and acceptor components. The concentration of H_2CO_3, in turn, depends on the concentration of dissolved CO_2, which, in turn, depends on the concentration or partial pressure of CO_2 in the gas phase.

With respect to the bicarbonate system, a $[HCO_3^-]$ / $[H_2CO_3]$ ratio of 20 to 1 is required for the pH of blood plasma to remain 7.40. The concentration of dissolved CO_2 is included in the $[H_2CO_3]$ value, *i.e.*,

$$[H_2CO_3] = [H_2CO_3] + [CO_2\ (\text{dissolved})]$$

If there is a change in the ratio in favour of H_2CO_3, **acidosis** results. This change can result from a decrease in $[HCO_3^-]$ or from an increase in $[H_2CO_3]$. Most common forms of acidosis are metabolic or respiratory. *Metabolic acidosis* is caused by a decrease in $[HCO_3^-]$ and occurs, for example, in uncontrolled diabetes with ketosis or as a result of starvation. *Respiratory acidosis* is brought about when there is an obstruction to respiration (euphysema, asthma or pneumonia) or depression of respiration (toxic doses of morphine or other respiratory depressants). If acidosis is not treated promptly, the patient may go into a comma.

Alkalosis results when $[HCO_3^-]$ becomes favoured in the bicarbonate/carbonic acid ratio. *Metabolic alkalosis* occurs when the HCO_3^- fraction increases with little or no concomitant change in H_2CO_3. Severe vomiting (loss of H^+ as HCl) or ingestion of excessive amounts of sodium bicarbonate (bicarbonate of soda) can produce this condition. *Respiratory alkalosis* is induced by hyperventilation because an excessive removal of CO_2 from the blood results in a decrease in $[H_2CO_3]$. Hyperventilation can result in anxiety, hysteria, prolonged hot baths or lack of O_2 at high altitudes. Alkalosis can produce convulsive seizures in children and tetany, hysteria, prolonged hot baths or lack of O_2 as high altitudes. Alkalosis can produce convulsive seizures in children and tetany in adults (characterized by sharp flexion of the wrist and ankle joints, muscle twitchings, and cramps).

The pH of blood is maintained at 7.4 when the buffer ratio $[HCO_3^-]/[H_2CO_3]$ becomes 20. If the bicarbonate neutralizes any acid or base, there may be the change of buffer ratio and the blood pH value. But the buffer ratio remains by the respiratory elimination of H_2CO_3 as CO_2 or the urinary elimination of HCO_3^-.

Since cells contain much lower amounts of HCO_3^-, the importance of bicarbonate buffer inside the cell is negligible.

3. The Protein Buffer Systems

The protein buffers are very important in the plasma and the intracellular fluids but their concentration is very low in cerebrospinal fluid, lymph and interstitial fluids. The proteins exist as anions serving as conjugate bases (Pr^-) at the blood pH 7.4 and form conjugate acids (HPr) accepting H^+. They have the capacity to buffer some H_2CO_3 in the blood.

$$H_2CO_3 + Pr^- \rightleftharpoons HCO_3^- + HPr$$

4. The amino acids buffer system

This system also operates in humans. Amino acids contain in their molecule both an acidic (–COOH) and a basic (– NH_2) group. They can be visualized as existing in the form of a neutral zwitterion in which a hydrogen atom can pass between the carboxyl and amino groups. The glycine may, thus, be represented as :

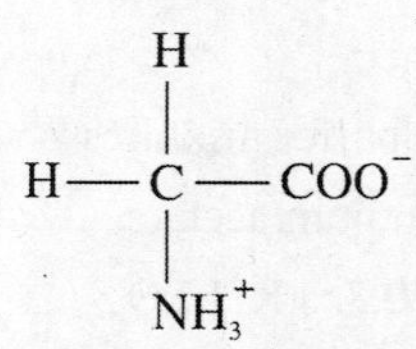

By the addition or subtraction of a hydrogen ion to or from the zwitterion, either the cation or anion form will be produced :

$$^{+}H_3N{-}CH_2{-}COOH \xleftarrow{+H^+} {}^{+}H_3N{-}CH_2{-}COO^- \rightleftharpoons H_2N{-}CH_2{-}COO^- + H^+$$

Cation form — Zwitterion — Anion form

Thus, when OH^- ions are added to the solution of amino acid, they take up H^+ from it to form water, and the anion is produced. If H^+ ions are added, they are taken up by the zwitterion to produce the cation form. In practice, if NaOH is added, the salt $H_2N{-}CH_2{-}COONa$ would be formed and the addition of HCl would result in the formation of amino acid hydrochloride, ClH —H_3N—CH_2—COOH, but these substances would ionize in solution to some extent to form their corresponding ions. Hemoglobin and plasma proteins act as buffers in a similar way.

Amino acids differ in the degree to which they will produce the cation or anion form. In other words, a solution of an amino acid is not neutral but is either predominantly acidic or basic, depending on which form is present in greater quantity. For this reason, different amino acids may be used as buffers for different pH values, and a mixture of them possesses a wide buffer range.

5. The Hemoglobin Buffer Systems

These buffer systems are involved in buffering CO_2 inside erythrocytes. The buffering capacity of hemoglobin depends on its oxygenation and deoxygenation. Inside the erythrocytes, CO_2 combines with H_2O to form carbonic acid (H_2CO_3) under the action of carbonic anhydrase. At the blood pH 7.4, H_2CO_3 dissociates into H^+ and HCO_3^- and needs immediate buffering. Oxyhemoglobin (HbO_2^-), on the other side, loses O_2 to form deoxyhemoglobin (Hb^-) which remains undissociated (HHb) by accepting H^+ from the ionization of H_2CO_3. Thus, Hb^- buffers H_2CO_3 in erythrocytes :

$$HbO_2^- \rightleftharpoons Hb^- + O_2$$

$$Hb^- + H_2CO_3 \rightleftharpoons HHb + HCO_3^-$$

Some of the HCO_3^- diffuse out into the plasma to maintain the balance between intracellular and plasma bicarbonates. This causes influx of some Cl^- into erythrocytes along the electrical gradient produced by the HCO_3^- outflow (*chloride shift*).

$HHbO_2$, produced in lungs by oxygenation of HHb, immediately ionizes into H^+ and HbO_2^-. The released hydrogen ions (H^+) are buffered by HCO_3^- inside erythrocyte to form H_2CO_3 which is dissociated into H_2O and CO_2 by carbonic anhydrase. CO_2 diffuses out of erythrocytes and escapes in the alveolar air. Some HCO_3^- return from the plasma to erythrocytes in exchange of Cl^- and are changed to CO_2.

$$HHb + O_2 \rightleftharpoons HHbO_2 \rightleftharpoons HbO_2^- + H^+$$

$$HCO_3^- + H^+ \rightleftharpoons H_2CO_3 \rightleftharpoons H_2O + CO_2$$

PROBLEMS

1. What is the pH of each of the following solutions ?
   (*a*) 0.35 M hydrochloric acid
   (*b*) 0.35 M acetic acid
   (*c*) 0.035 M acetic acid
2. What is the pH of the following buffer mixtures ?
   (*a*) 1 M acetic acid plus 0.5 M sodium acetate
   (*b*) 0.3 M phosphoric acid plus 0.8 M KH_2PO_4
3. You need to make a buffer whose pH is 7.0, and you can choose from the weak acids shown in Table 3-3. Briefly explain your choice.
4. Why does pure water have no taste or colour ?

ANSWERS TO PROBLEMS

1. (*a*) pH = 0.456.
   (*b*) pH = 2.608 by approximation, 2.609 exact.
   (*c*) pH = 3.108 by approximation, 3.113 exact.
2. (*a*) pH = 4.46.
   (*b*) pH = 2.57.
3. The best choice would be a mixture of $H_2PO_4^-$ and HPO_4^{2-}, which has $pK_a = 6.86$.
4. Physical properties of a compound, such as colour, taste, touch etc, depend on its pH factor. The pH further depends upon the concentration of H^+ and OH^- ions in that compound. For example, compounds having higher concentration of H^+ ions are sour like acids and compounds having higher OH^- ion concentration are bitter like bases. Water has an equal concentration of H^+ and OH^- ions. Thus, it has neither sour nor a bitter taste and hence it is tasteless. Similarly, it is colourless and does not have an odour.

CHAPTER

4 The Atom and Chemical Bonds

INSIDE

- Elemental Composition of Living Matter
- Structure of An Atom
- Ionization Potential
- Nature of Chemical Bonding
- Types of Chemical Bonding
  - Electrovalent bond
  - Covalent bond
  - Coordinate bond
  - Noncovalent bond

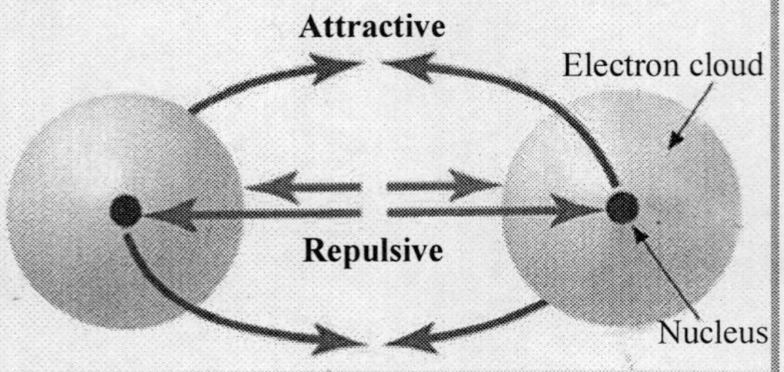

The chemical bond
A covalent H–H bond is the net result of attractive and repulsive electrostatic forces. The nucleus–electron attractions (blue arrows) are greater than the nucleus–nucleus and electron–electron repulsions (red arrows), resulting in a net attractive force that holds the atoms together to form an H_2 molecule.

Living organisms are composed of lifeless molecules. When these molecules are isolated and examined individually, they conform to all the physical and chemical laws that describe the behaviour of inanimate matter. Yet living organisms possess extraordinary attributes not shown by any random collection of molecules. The composition of living matter is strikingly different from that of the inanimate world. **Antoine Laurent Lavoisier** noted the relative chemical simplicity of the "mineral world" and contrasted it with the complexity of "plant and animal worlds"; the later he knew were composed of compounds rich in the elements carbon, oxygen, nitrogen, and phosphorus.

ELEMENTAL COMPOSITION OF LIVING MATTER

With the exception of certain metal ions, Biochemistry is, for the most part, related to the chemistry of the elements of the second and third periods of the Periodic Table. On earth, all cells, regardless of their origin (animal, plant, or microbial), contain the same elements in approximately the same proportions (Table 4–1). Thus, of the 107 known elements, only about 20 are essential for terrestrial life. Perhaps there is some logical chemical explanation for their selection.

Table 4–1. The elemental composition of living cells

Element	*Composition by weight (%)*	*Element*	*Composition by weight (%)*
O	65		
C	18	Cu, Zn, Se, Mo, F, Cl, I, Mn, Co, Fe	0.70
H	10		
N	3		
Ca	1.5		
P	1.0		
K	0.35		
S	0.25	Li, Sr, Al, Si, Pb, V, As, Br	Traces*
Na	0.15		
Mg	0.05		

* Variable occurrences in cells. No known function in most cases.

Six nonmetals (O, C, H, N, P, S), which contribute almost 98% of the total mass of cells, provide the structural elements of protoplasm. From them, the functional components of cells (such as walls, membranes, genes, enzymes etc) are formed. These 6 elements all occur in the first three periods of the Periodic Table (refer Table 4–2).

Table 4–2. The structural elements of protoplasm*

Period	*Group*							
	I	*II*	*III*	*IV*	*V*	*VI*	*VII*	*VIII*
1	(H)							He
2	Li	Be	B	(C)	(N)	(O)	F	Ne
3	Na	Mg	Al	Si	(P)	(S)	Cl	Ar

*The six common nonmetals have been shown encircled.

The relative abundance of these 6 lighter elements in the seas, crust and atmosphere of earth does not by itself explain their utilization for life. Aluminium, for example, is more abundant than carbon, but it performs no known function essential to life. By contrast, the intrinsic chemical properties of these elements suggest their unique suitability as building blocks for life. Desirable features of structural elements apparently are as follows (**Rodwell, Victor W**, 1979) :

(*a*) Small atomic radius

(*b*) The versatility conferred by the ability to form 1–, 2–, 3– and 4– electron bonds

(*c*) The ability to form multiple bonds.

Carbohydrates and lipids contain C, H and O while proteins and nucleic acids contain in addition N and S, and N and P respectively. Na, K, Mg, Ca, and Cl are usually found in ionic forms whose concentrations must remain within narrow limits. The presence of Bo, Si, V, Mn, Fe, Co, Cu, Zn and Mb, though in ultratrace amounts, is essential for the functioning of enzymes. These elements are used as cofactors in the catalytic function of enzymes.

Small atoms form the tightest and most stable bonds — a distinct advantage for structural elements. **H, O, N and C are the smallest atoms capable of forming 1– , 2–, 3– and 4– electron bonds respectively.** Utilization of all possible types of electron bonds permits maximum versatility in molecular design. So also does the ability to form multiple bonds, a property confined almost entirely to P, S, and the elements of period 2. Advantages of carbon-based versus silicon-based life include the following :

1. Greater chemical stability of C—C bonds than the Si—Si bonds.
2. The ability of carbon, but not of silicon, to form multiple bonds; for example, the oxides of carbon are diffusible monomeric gases, whereas the oxides of silicon is a viscous polymer.
3. The stability of C—C bonds, but not of Si—Si bonds, to rupture by nucleophilic reagents (electron-rich elements or compounds) such as O_2, H_2O or NH_3.

Nucleophilic attack is the attack of an electron-rich centre upon an electron-deficient centre.

Similar factors uniquely qualify P and S for utilization in energy transfer reactions. **Energy transfer is facilitated by bonds susceptible to nucleophilic attack.** A most common example is the nucleophilic attack of the 6-OH of glucose on the terminal P—O—P bond of ATP, forming ADP plus glucose-6-phosphate. P and S resemble Si in that P—O—P or S—O—S bonds, like Si—Si bonds, are susceptible to nucleophilic rupture by virtue of their unoccupied third orbitals. However, unlike Si, P and S form multiple bonds (more versatile), a consequence of their smaller atomic diameters. Most energy transfer reactions in Biochemistry may be visualized as resulting from attack of a nucleophil (N) on the unoccupied third orbital of a phosphorus atom :

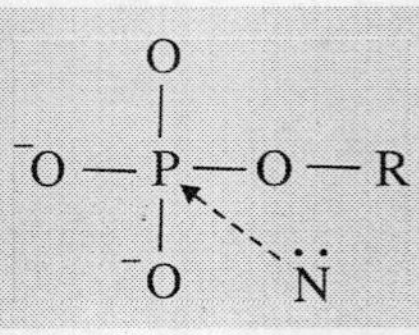

The characteristic chemical and physical properties of the chemical elements of life are the same throughout the known universe. It, thus, seems probable that if life exists elsewhere, the same elements are employed for the same or similar reasons.

STRUCTURE OF AN ATOM

Each matter is composed of very small particles called **atoms**, which cannot be created, destroyed or subdivided. Atoms of the same element are similar to one another and equal in weight. Atoms of different elements have different properties and weight. Although, at one time, the atoms were conceived to be the smallest particles, subatomic particles were later recognized in due course of time. The 3 fundamental subatomic particles are: proton, neutron and electron. Besides these fundamental particles, about 35 other atomic particles are also known to exist. Many of them are, however, extremely unstable and they merely represent a bundle of energy. Some of the stable particles other than fundamental particles, are positron, photon, neutrino, graviton and antiproton. These particles are, however, of little importance in the study of Biochemistry because their

ERNEST RUTHERFORD
(LT, 1871–1937)

Ernest Rutherford grew up on a farm in New Zealand before entering Cambridge University to study under J.J. Thomson. Although he was a physicist, for his pioneering work in atomic structure Rutherford was awarded the **Nobel Prize in Chemistry in 1908.**

existence is rarely encountered in biological systems.

Regarding the arrangement of fundamental particles inside an atom, a British chemist **Ernest Rutherford** (1911) proposed the most satisfactory model which is accepted even today with some modifications (Fig. 4–1). Accordingly, an atom is made up of a central *nucleus* containing positively-charged protons and neutral neutrons, surrounded by negatively-charged electrons which move around it (the nucleus) in discrete, successive, concentric volumes in space known as *orbits* or *shells*. The model is analogous to the sun's planetary system but differs from it in having the subatomic particles, protons and electrons as charged.

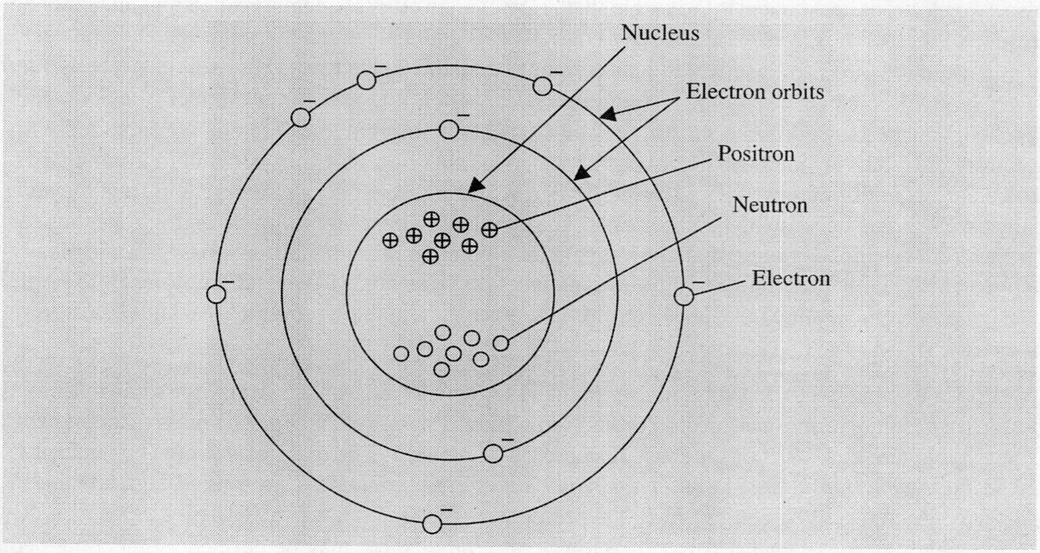

Fig. 4–1. Structure of an atom

The electron shells are numbered (from within) as 1, 2, 3, 4, 5, 6 and 7 and are designated by the letters *K, L, M, N, O, P, Q* respectively (Table 4–3). Each shell has a certain number of electrons and the maximum number of electrons for each shell is fixed. The maximum number is given by $2n^2$, where n is the serial number of the shell. Thus, the maximum number of electrons in *K, L, M, N, O, P, Q* shells will be 2, 8, 18, 32, 50, 72, 98 respectively. The maximum number of electrons in the outermost shell is 8 and in the penultimate shell is 18.

Table 4–3. Distribution of electrons in shells and subshells

Shell number (n)	1	2L		3			4 etc			
Shell name	K	L		M			N			
Subshell number	0	0	1	0	1	2	0	1	2	3
Subshell name	*s*	*s*	*p*	*s*	*p*	*d*	*s*	*p*	*d*	*f*
Number of electrons in subshells	2	2	6	2	6	10	2	6	10	14
Total number of electrons in a subshell		2			18			32		

The shells are subdivided into subshells. The number of subshells in a shell is equal to the number of the shell from within. K shell has one subshell called *s*; second L shell has two subshells *s* and *p*; the third M shell has three subshells *s*, *p* and *d* and fourth N subshell has four subshells *s, p, d* and *f*.

The subshells *s, p, d* and *f* can have a maximum of 2, 6, 10 and 14 electrons, respectively.

The position of electrons in the various shells and subshells are represented as follows. Major shells in which the electrons exist are indicated by the numbers 1, 2, 3 etc and the subshells designated by *s, p, d, f* etc. The superscript on *s, p, d* and *f* gives the number of electrons in the subshell. Thus, $1s^2$ indicates the presence of two electrons in the s subshell of the first major shell (*K*). Similarly, $4f^8$ indicates the presence of 8 electrons in the *f* subshell of fourth major shell (*N*).

Hydrogen atom is the simplest atom which consists of one proton, one neutron and one electron. In this atom, the lone electron is situated in the lone orbit or shell around the nucleus. In helium atom also, two electrons are situated in the single shell. However, in other elements the electrons are arranged in several shells. Thus, a **neon atom** has two shells of 2 and 8 electrons (a total of 10 electrons) and **argon atom** has three shells of 2, 8 and 8 electrons (a total of 18 electrons). These orbits or shells may never have more electrons than a certain maximum. When this maximum number of electrons is reached, the shell is said to be *saturated*. The elements whose outermost shells are saturated with electrons are relatively inert and do not participate in the chemical reactions under normal conditions. The elements in this category (He, Ne, Ar, Kr, Xe, Rd) are gases at normal temperature and they are called *noble gases* because of their inertness. The elements, whose atoms have electrons one more or one less (or even higher values) than the inert gas configuration in their outermost shell, are chemically active. This chemical activity may be interpreted as a tendency of those atoms to acquire noble configuration by accepting or losing electron(s). Acquiring the stable configuration results in lowering of energy.

The mass of an atom depends entirely upon its nucleus. A neutron has nearly the same mass as a proton. In absolute terms, each proton or neutron weighs 1.66043×10^{-24} g. The mass of an electron is negligible, about 1/1823 of the mass of a proton or neutron. The mass of a proton or neutron is called the *atomic mass unit* (*amu*). Thus, if a carbon atom contains 6 protons, 6 neutrons and 6 electrons, its amu will be equivalent to the proton + neutron, *i.e.*, 6 + 6 = 12. This amu is also the atomic weight of the element. Thus, *atomic weight* of an element may be defined as the combined weight of protons and neutrons, each taken as a unit weight. The number of protons on an atom is called the *atomic number* of the atom. Thus, the atomic number of the carbon is 6. Although the number of neutrons for a particular atom is fixed, it may vary, sometime giving rise to different species of the same atom. The different atomic species, having the same atomic number (as they have same proton number) but different atomic weight (as they have different neutron numbers) are called *isotopes*. Thus, a hydrogen atom with two neutrons (named as deuterium) is an isotope of normal hydrogen atom which contains only one neutron. The number of protons and electrons in both species of the hydrogen is same.

IONIZATION POTENTIAL

The amount of energy required the remove the most loosely-held electron from a normal (uncharged) gaseous atom is called ionization potential. It may be expresed as :

$$A(g) \xrightarrow{\text{EP}} A^{+}(g) + e^{-}$$

where, g denotes gaseous form of the atom A, and EP denotes the amount of energy required to remove one electron (*i.e.*, ionization potential).

The removal of electron generates a positive charge on the atom. The magnitude of ionization potential depends upon several factors. If the **radius of the atom** is large, the orbiting electrons are farther from the positively-charged nucleus and it is easier to remove them. Therefore, the ionization potential of the atoms with larger radii is smaller. The **magnitude of the positive charge on the nucleus** also affects the ionization potential. If the charge (positive) on the nucleus is increased, it becomes more difficult to remove electrons and therefore the ionization potential increases. The potential also depends upon the **number of electron orbits.** The inner shells or orbits of the electrons act as a shield or screen between the nucleus and the outer shells. Therefore, each inner shell existing between the nucleus and the outermost shell produces a shielding effect, which decreases the force of attraction between the nucleus and the electrons of outer shells. Hence, *higher the number of inner shells, lesser the value of ionization potential.*

An atom can lose more than one electron during ionization. The amount of energy required to remove one electron from an uncharged atom is commonly called *first ionization potential*. The energy required for the removal of one electron from a charged ion, *i.e.*, a second electron from the original atom is called *second ionization potential.*

$$A^{+}(g) \xrightarrow{\text{II EP}} A^{2+}(g) + e^{-}$$

Similarly, *third ionization potential* represents the energy required to remove one electron from a 2^{+} ion, *i.e.*, third electron from the original atom.

$$A^{2+}(g) \xrightarrow{\text{III EP}} A^{3+}(g) + e^{-}$$

NATURE OF CHEMICAL BONDING

A perusal of the periodic table shows that besides the inert gases (He, Ne, Ar, Kr, Xe, Rd–all of them have 8 electrons in their outermost shells except He where outermost shell is also the first shell (*K*) and cannot have more than 2 electrons), there are 3 types of elements :

(*a*) those having 1 to 3 electrons in the outermost shell,

(*b*) those having 5 to 7 electrons in the outermost shell, and

(*c*) those having 4 electrons in the outermost shell.

Since the driving force for chemical combination is the tendency to acquire inert gas stable configuration, **first types of elements** will try to stabilize themselves by losing 1, 2 or 3 electrons because it is difficult to gain 7, 6 or 5 electrons than to lose 1, 2 or 3 electrons for obtaining stability. Similarly, **elements of the second type** will try to stabilize themselves by gaining 3, 2 or 1 electrons as losing 5, 6, or 7 electrons will be more difficult. Thus, the elements of the first and second type usually combine by either losing or gaining electrons.

The **elements of the third type,** of which *carbon* is representative, have to lose 4 electrons or gain 4 electrons in order to obtain inert gas configuration and if the element has balanced elecropositive and electronegative characters then either process (of losing or gaining electrons) becomes still more difficult. In such cases, the element achieves the stable configuration by sharing of these electrons with other atoms. Besides carbon, one more element is in a similar position and that is *hydrogen* which may attain stable configuration by either losing or gaining one electron. It can also do so by sharing of electrons with other elements.

TYPES OF CHEMICAL BONDING

Several atoms are assembled and held together to form thousands of molecules which participate in the building and function of physical and biological systems. A **bond** is any force which holds two atoms together. The formation of bond between two atoms is due to some redistribution or regrouping of electrons to form a more stable configuration. The regrouping of electrons in the combining atoms may take place in either of the 3 ways :

(*a*) by a *transfer* of one or more electrons from one atom to another— **electrovalent bonding**

(*b*) by a *sharing* of one or more pairs of electrons between the combining atoms— **covalent bonding**

(*c*) by a combination of the two processes of *transfer and sharing*— **coordinate bonding.**

Electrovalent or Polar or Ionic Bond

Ionic bond formation takes place between atoms of strongly electropositive and strongly electronegative elements. An element preceding an inert gas in the periodic table is strongly electronegative and the element immediately following the inert gas is strongly electropositive. For example, chlorine is electronegative while sodium is electropositive. According to **W. Kossel** (1916), a transfer of electron(s) takes place from the outermost shell of the electropositive atom to the outermost shell of the electronegative atom, resulting in the formation of stable positive and negative ions respectively which are held together by electrostatic forces of attraction to form a *molecule* or more precisely an *ion pair*. Thus,

$$A^{e} + B \longrightarrow A^{+} + B^{-e} \quad \text{or} \quad A^{+} \approx B^{-e}$$

The atoms involved are electrically neutral before combining. The element A which has lost its electrons is known as electropositive whereas element B which has gained the electrons is termed electronegative element. The compound formed by electron transfer is termed as *electrovalent* by **Langmuir** (1919) because the resulting compound is electrolyte. It is also called *polar* since the molecule develops a positive and a negative pole. The electrovalent compounds always exist in ionic form, are hard and nonvolatile, have high melting and boiling points because of stronger nature of the bond and are soluble in polar solvents (such as water and alcohol) and because of the presence of ions conduct electricity in solution or in the fused state. The electrovalent compounds having identical electronic configuration exhibit the phenomenon of isomorphism. For example, NaF (2, 8), magnesium oxide (2, 8), calcium chloride (2, 8, 8) and potassium sulfide (2, 8, 8) are isomorphous.

Electrovalent linkage is common in inorganic compounds like NaCl (Fig. 4-2), Na_2O, CaO etc. In the case of NaCl,

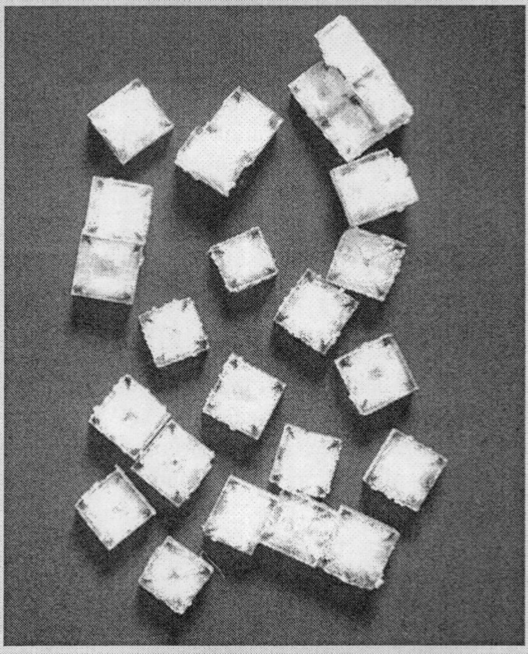

Fig. 4-2. Cubic crystals of sodium chloride
The bond involved is ionic bond.

$$\underset{\cdot}{Na} + \cdot\ddot{\underset{\cdot\cdot}{Cl}}: \longrightarrow Na^{+} + :\ddot{\underset{\cdot\cdot}{Cl}}:^{-}$$

The electronic configuraion of sodium atom is $1s^2\ 2s^2\ 2p^6\ 3s^1$ and that of chlorine is $1s^2\ 2s^2\ 2p^6\ 3s^2\ 3p^5$. After transfer of one electron from sodium to chlorine, the sodium atom contains a configuration ($1s^2\ 2s^2\ 2p^6$) similar to neon, and chlorine attains a configuration ($1s^2\ 2s^2\ 2p^6\ 3s^2\ 3p^6$) which is similar to that of argon. The Na^+ can be said to be isoelectronic with neon while Cl^- is isoelectronic with argon. The bond between the ions results from the attraction between oppositely-charged ions. This linking of the ionized atoms is the formation of ionic bonds. The strength of the bond is roughly proportional to the product of the charges on ions and inversely proportional to the square of the distance between the effective centres of ions.

Ionic bonds may be formed by transfer of more than one electrons also. For example, in the case of Na_2O, one oxygen atom can form an ionic bond by accepting one electron from each of the two sodium atoms :

$$\begin{matrix}\underset{\cdot}{Na}\\ \underset{\cdot}{Na}\end{matrix} + \ddot{\underset{\cdot\cdot}{O}}: \longrightarrow 2Na^{+} + :\ddot{\underset{\cdot\cdot}{O}}:^{2-} \longrightarrow Na_2O$$

Oxygen can also form ionic bonds by accepting two electrons from a single atom of the element which has a low second ionization potential. For example, in the case of CaO, one calcium atom can link to one oxygen atom by transferring its two electrons :

$$Ca: + \ddot{\underset{\cdot\cdot}{O}}: \longrightarrow Ca^{2+} + :\ddot{\underset{\cdot\cdot}{O}}:^{2-} \longrightarrow CaO$$

The number of the charge (+ or –) or the number of electrons transferred during an ionic bond formation is called the *electrovalence number* or the *electrovalency* of the participating atoms. Thus, sodium has an electrovalency of 1 while calcium or oxygen has 2. The sign (+ or –) of this number depends upon whether the electron is donated or received. The element donating electron has a positive (+) electrovalency while that receiving electron has a negative (–) sign. Since the number of electrons transferred from an electropositive element to an elecronegative element during an ionic bond formation is equal to the number of electrons present in its outermost orbit, the electrovalency for such elements is the number of electrons present in outermost orbit. Similarly, the electrovalency of an electronegative element is the number of electrons which it can accept to achieve an inert gas configuration.

Covalent or Nonpolar Bond

Covalent bond formation, first suggested by **G.N. Lewis (1916)**, consists in sharing or holding a pair of electrons in partnership between two combining atoms, so that the pair counts towards the electronic grouping of both atoms. By this mechanism also, the stability akin to the inert gas is attained by each atom. For each pair of electrons to be shared between two atoms, each of the constituent atom contributes one electron :

$$A\cdot + B\cdot \longrightarrow A:B$$

This type of linkage which is the result of equal contribution and equal sharing of electrons is known as covalent bond. The compound formed

GILBERT NEWTON LEWIS

(LT, 1875–1946)

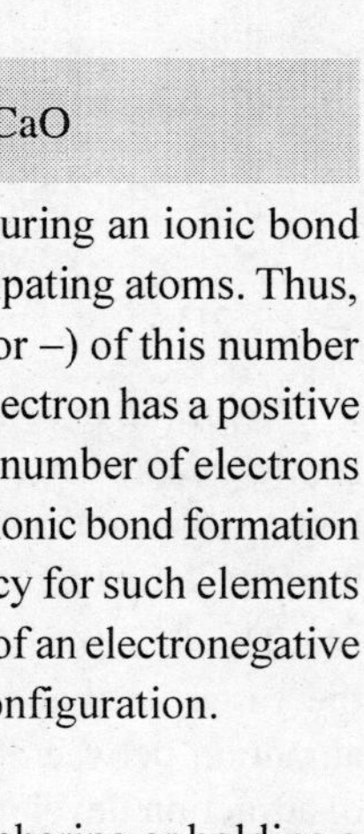

Gilbert Newton Lewis was one of the foremost American chemists of the first half of the twentieth century. In addition to his pioneer work in describing chemical bonding through the symbolism named after him, Lewis was a driving force in the introduction of thermodynamics into the mainstream of chemistry, and he made important contributions to acid-base theory.

by electron sharing is termed as *covalent* or *nonpolar* by **Langmuir** (1919). The covalent compounds always exist in molecular form, are nonelectrolytes or nonionizable, soluble in organic solvents (such as benzene, ether, pyridine etc) and have low melting and boiling points because of weaker nature of the bond. They are usually liquids or gases and are generally soft, easily-fusible and volatile. They are nonconducting in the fused state or in solution. The covalent bond is rigid and directional and as such there is a possibility of position isomerism and stereoisomerism amongst these compounds.

Covalent linkage is common in organic compounds, although inorganic compounds also have it. In covalent compounds, one pair of shared electrons corresponds to a single bond, two pairs to double bond, three pairs of electrons or six shared electrons to triple bond. Some common examples from 3 categories are :

(*a*) One pair of electrons : single bond

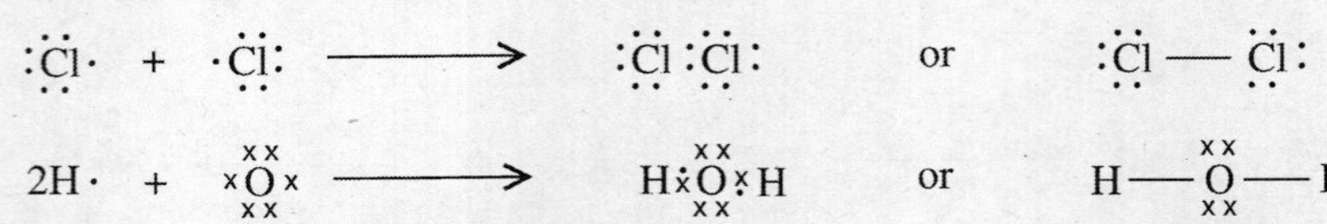

(*b*) Two pairs of electrons : double bond

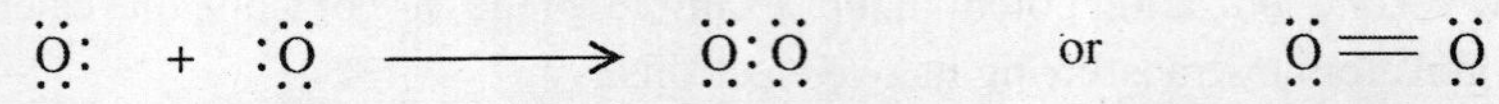

(*c*) Three pairs of electrons : triple bond

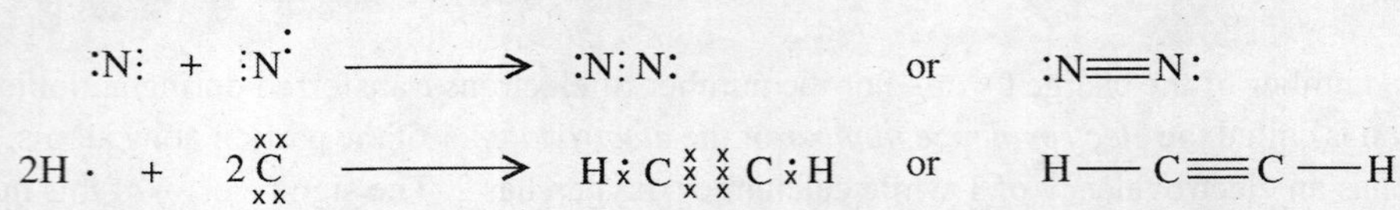

With the exception of hydrogen atom which by sharing one electron easily attains the helium configuration, it may be observed that only negative atoms can form covalent bonds by sharing electrons between them. Positive atoms contain an excess of electrons, which they have to lose to attain a stable inert gas structure; and therefore they form electrovalent bonds.

The force of bond formation in covalent bonds is the same as that in ionic bonds, *i.e.*, electrostatic attraction between the two atoms, although this force of attraction develops in a different manner. When two atoms, destined to link through a covalent bond, come within a definite range, the wave function of the electrons of two atoms overlap each other (Fig. 4–3). This overlapping causes the accumulation of the negative charge between the two atomic nuclei. The accumulated negative charge, in turn, attracts the nuclei (with positive charges) of the two atoms and, thus, they are held together.

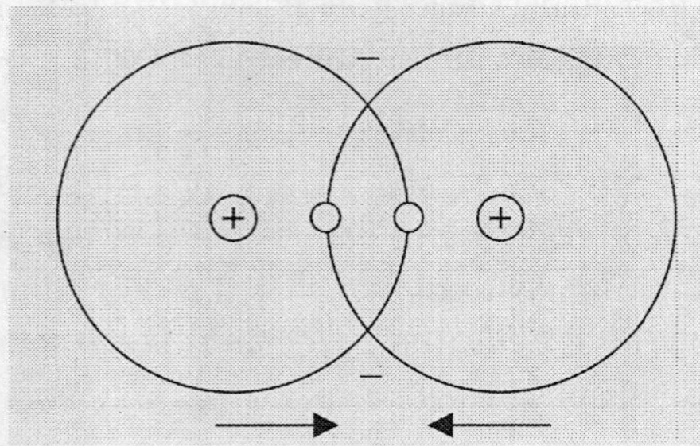

Fig. 4–3. Covalent sharing and bond formation between two atoms

A. Electron affinity or electronegativity

Electronegativity of an atom is a measure of its power to attract electrons that it is sharing in a covalent bond. Atoms of different elements have different electronegativities. Atoms with higher *atomic number* (magnitude of the positive charge on the nucleus) and less *atomic radius* (the distance of the valence electrons from the nucleus) have higher electronegativities. Electronegativity values of some elements commonly encountered are given in Table 4–4. When two elements, that differ in

Table 4–4. The electronegativities of some elements

Element	*Electronegativity**	*Element*	*Electronegativity**
F	4.0	Cu	1.9
O	3.5	Fe	1.8
Cl	3.0	Co	1.8
N	3.0	Ni	1.8
Br	2.8	Mo	1.8
S	2.5	Zn	1.6
C	2.5	Mn	1.5
I	2.5	Mg	1.2
Se	2.4	Ca	1.0
P	2.1	Li	1.0
H	2.1	Na	0.9
		K	0.8

*The higher the number, the more electronegative is the element.

electronegativity, form a covalent bond (for example, C and O), that bond is polarized; the shared electrons are more likely to be in the region of the more electronegative atom (O) than of the less electronegative (C). In the extreme cases of two atoms of very different electronegativity (Na and Cl, for example), one of the atoms actually gives up the electron(s) to the other atom, resulting in the formation of ions and ionic interactions such as those in solid NaCl.

B. Polar and nonpolar covalent bonds

Polar bond. Polar covalent bonds may be characterized as transition state between ionic and covalent bonds. In this case, there is neither the complete transfer of electrons from one atom to the other nor equal sharing. When a covalent bond is formed between two atoms with *different electronegativities*, the electrons involved in the bond are not shared equally. The atom with higher electronegativity pulls the bonding electrons closer to it. In other words, the electron density in the molecular orbital would be greater around the atom with higher electronegativity. The result of this displacement of the molecular orbital toward the more electronegative atom will acquire a small negative charge and the less electronegative atom will acquire a small positive charge. For example, in the case of **C—Cl bond**, chlorine is more electronegative than carbon. As such, the electron density in the molecular orbital would be higher around chlorine atom than around the carbon atom. Thus, the chlorine atom acquires a small negative charge and the carbon atom acquires a small positive charge. This may be indicated as in Fig. 4–4. Such a bond which appears to have a positive end and a negative end is said to be a **polar bond**.

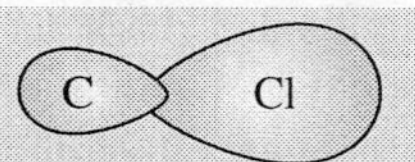

or $C^{\delta+}$ — $Cl^{\delta-}$

Fig. 4–4. C—Cl bond is a polar bond.

(The Greek letter delta, δ stands for *small.* The symbol δ^+ means a small positive charge and δ^- means a small negative charge.)

Water molecule is another good example of a polar bond. In this molecule, sharing of electrons between one oxygen and two hydrogen atoms is not equal. The electrons of hydrogen atoms are attracted more towards the nucleus of the oxygen atom. Consequently, oxygen develops a δ^- while hydrogen develops a δ^+ charge. Yet another common example of a polar bond is H—Cl, which may be depicted as $H^{\delta+}$—$Cl^{\delta-}$ (Fig. 4—5). A polar molecule like that of HCl having two centres of charge is called a **dipole** and the molecule becomes **dipolar**. Although the ends of molecule may have partial charges, the molecule itself is neutral because of the equal number of protons and electrons.

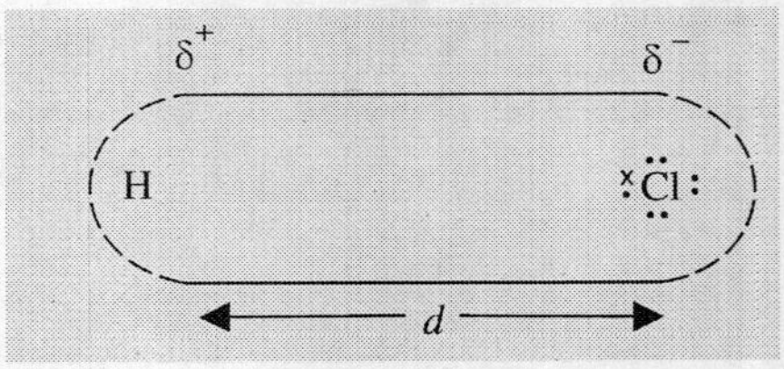

Fig. 4–5. A dipole for HCl molecule *d* is the distance between the positive and negative centres and is called bond distance or bond length.

The degree of a polarity of a polar molecule (*i.e.*, a molecule having a polar covalent bond) is expressed in terms of its **dipole moment** (μ) which is equal to the product of electric charge (*e*) in **e.s.u.** and the distance (*d*) in Å between the positive and negative centres. Thus,

$$\mu = e \times d$$

As *e* is of the order of 10^{-10} e.s.u. and *d* of the order of 10^{-8} cm, the dipole moment (μ) is, therefore, of the order of 10^{-18} and this unit is known as **Debye** (D). Thus, 1D = 10^{-18} e.s.u. cm = 10^{-10} e.s.u. Å. The μ is a vector quantity and is often indicated by an arrow parallel to the line joining the points of charge and pointing towards the negative end. Thus, $H^{\delta+}$—$Cl^{\delta-}$ can also be represented as H—Cl. Greater the value of the dipole moment of a molecule, greater is the polarity of the bond between the atoms.

As expected, polar compounds have properties intermediate between ionic and covalent compounds. Because of the dipole attraction between the atoms, polar compounds have higher melting and boiling points than nonpolar covalent compounds. These compounds are hard but not as hard as ionic compounds.

Nonpolar bond. When a covalent bond is formed between two atoms with *same electronegativities* (C—C, H—H and F—F), the electrons involved in the bond are shared equally. Electron density in the molecular orbitals bonding the two atoms together is same around each atom. There is no positive and negative end. Such a bond is said to be **nonpolar bond** (Fig. 4–6).

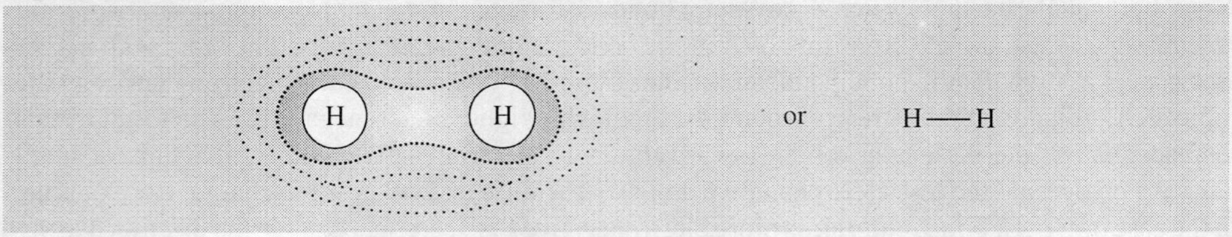

Fig. 4–6. H—H bond is a nonpolar bond.

C. Bond length or bond distance

The distance between two atomic nuclei in a covalent molecule is called the **bond distance** or **bond length.** The distance is such that the repulsion between atomic nuclei is balanced by the stability acquired through the overlapping of orbitals. The bond lengths are characteristic properties of a molecule and give information about its structure, properties etc. Bond lengths are measured by x-ray crystallography and by microwave spectroscopy. Bond lengths are generally affected by hybridization, electronegativity, steric condition etc. Some typical bond lengths are given in Table 4–5.

D. Bond angle

The 3 *p* orbitals of an atom are at the angle of 90° to each other. Hence, bonds formed by these should have an angle of 90°. Similarly, sp^3, sp^2 and *sp* orbitals in hybridized carbon atoms have an angle of 109.5°, 120° and 180° respectively. For example, in methane, where the four groups attached to sp^3 orbitals are identical, the angle is 109.5°. But if one of these groups is different then usually the angle is distorted. Thus, in bromopropane, C—C—Br bond angle is 114.2°.

Table 4–5. Bond lengths* of some common covalent bonds

Bond	*Bond length (in Å)*	*Typical compound*
C—C	1.54	Ethane, Propane
	1.48	Butadiene
	1.38	Butadiyne
C=C	1.34	Ethylene
	1.28	Butatriene
C≡C	1.20	Acetylene
C—H	1.11	Methane
	1.10	Benzene
	1.08	Acetylene
C—O	1.41	Ethanol
	1.34	Formic acid
C=O	1.20	Formaldehyde
	1.16	Carbon dioxide
C—N	1.47	Methylamine
	1.36	Formamide
C—Cl	1.78	Methyl chloride
C—Br	1.94	Methyl bromide
C—I	2.14	Methyl iodide
O—H	0.96	Methanol
N—H	1.01	Methyl amine

*Note that for most bonds, the value of bond length lies between 1 and 2 Å.

H—O—H bond angle in water and H—S—H bond angle in H_2S should be similar or 90° but the measured angle is 104.5° and 92.1°, respectively. Usually in strained molecules, the bond angles may be greatly distorted. An understanding of bond angle is essential for the study of the shape and geometry of the molecule. Bond angles in some molecules have been shown in Fig. 4–7.

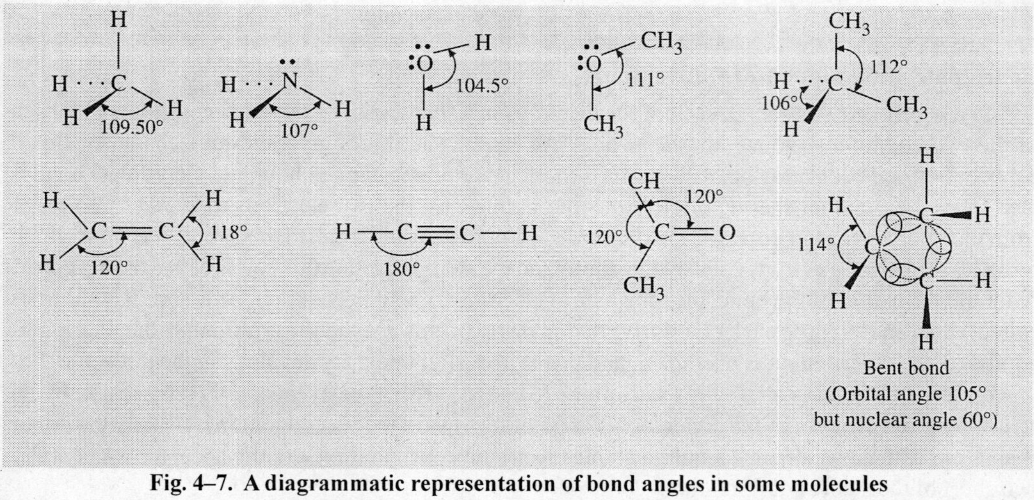

Fig. 4–7. A diagrammatic representation of bond angles in some molecules

E. Bond strength and bond energy

The energy required to break a bond into its constituents is referred to as bond energy and is a measure of **bond strength**. The greater the energy required to break a bond, the greater will be its strength. Thus, bond energies provide a useful picture of the strengths of various bonds. Bond energy depends upon the relative electronegativities of the elements involved. The number of electrons shared also influences bond strength: double bonds are stronger than single bonds, and triple bonds are stronger yet. Bond strengths are usually determined by quantitative measurements of heats of chemical reactions (calorimetry) and by spectroscopic methods. The strength of a bond is expressed as bond energy, in joules. Bond energy can be thought of as either the amount of energy required to break a bond or the amount of energy gained by the surroundings when two atoms form the bond. Some of the important bond energies are given in Table 4–6.

> In biochemistry, calories have often been used as units of energy– bond energy and free energy. The **joule** is the unit of energy in the International System of Units. For conversions, 1 cal is equal to 4.18 J.

Table 4–6. Strength of bonds common in biomolecules

Type of bond	*Bond dissociation energy (kJ/mol)*	*Type of bond*	*Bond dissociation energy (kJ/mol)*
Single bonds		**Double bonds**	
O—H	461	C = O	712
H—H	435	C = N	615
P—O	419	C = C	611
C—H	414	P = O	502
N—H	389	**Triple bonds**	
C—O	352	N ≡ N	930
C—C	348	C ≡ N	816
S—H	339		
C—N	293	**Noncovalent bonds or interactions**	
C—S	260	Hydrogen bonds van der Waals interactions Hydrophobic interactions Ionic interactions	4–0
N—O	222		
S—S	214		

Coordinate or Semipolar or Dative Bond

In a purely covalent bond, each of the electrons in a shared pair or duplet is contributed by each of the two combining atoms; in other words, we have a sharing of give-and-take character. Equal contribution and equal sharing of electrons is thus characteristic of polar bonding. Coordinate bond is also formed by mutual sharing of electrons but in this case the two electrons that are shared come from the same atom (**Perkins**, 1921). The shared pair of electrons is called *lone pair*. The atom which provides the pair of electrons is called the **donor** and the atom accepting this pair is called the **acceptor**. After the formation of the bond, the lone pair of electrons is held in common. This sort of bonding is called **coordinate** (**Sidgwick**) or **dative** (**Menzies**). In this mechanism, although the sharing is equitable, the contribution is one-sided, and therefore a slight polarity develops in the molecule. For this reason, this bond is also called **semipolar** (**Sugden**). This type of linkage is represented by an arrow, pointing away from the donor atom (or pointing toward the acceptor atom). Usually the donor is an atom which has already acquired stable electronic configuration and the acceptor is generally two short of the stable configuration.

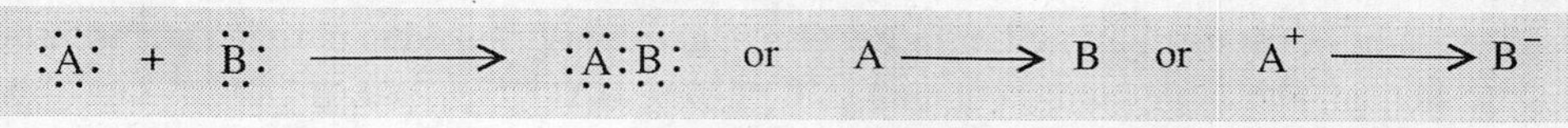

Because of slight increase in electron density near about the acceptor atom, the acceptor acquires a small negative charge (δ^-) and simultaneously the donor atom acquires a small positive charge (δ^+) because of slight decrease in electron density around it. Formation of O_3, SO_2, SO_3, boron hydride–ammonium complex and hydronium ion involves the coordinate bond.

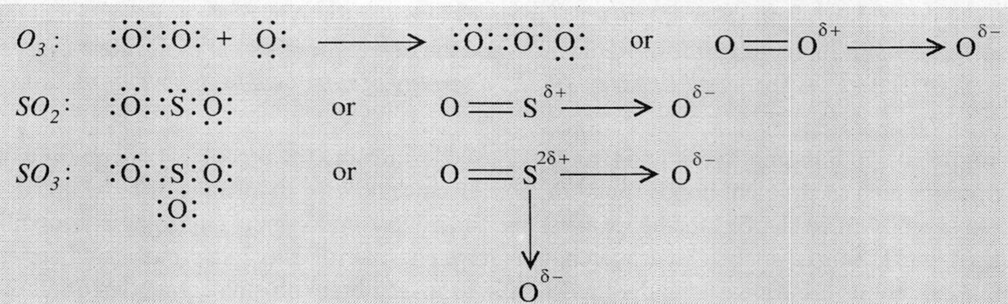

Boron hydride-ammonium complex :

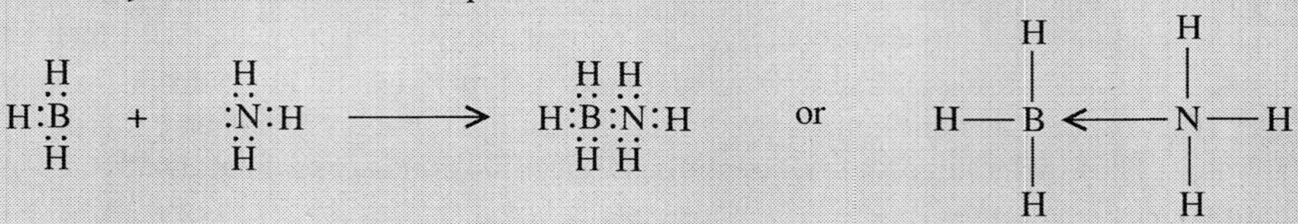

Hydronium ion :

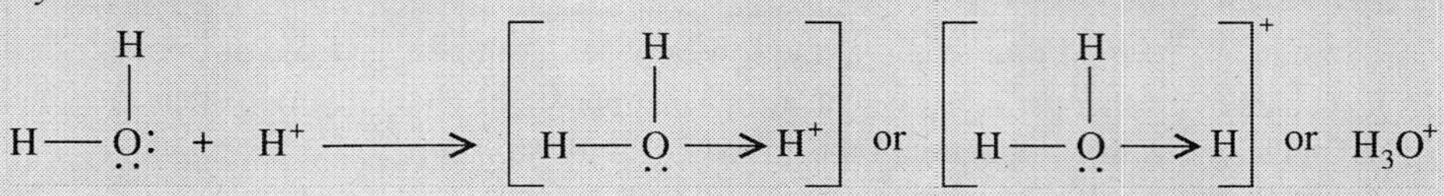

Coordinate compounds exhibit characteristics similar to covalent compounds :

1. They do not ionize in water and are poor conductors of electricity.
2. They are very sparingly soluble in water but dissolve readily in organic slovents.
3. Since a coordinate linkage is semipolar, the coordinate compounds possess melting and boiling points which are higher than those of purely covalent compounds but lower than those of ionic compounds.
4. The coordinate linkage is also rigid and directional. Therefore, such compounds exhibit space isomerism.
5. The coordinate linkage is easily broken when donor and acceptor are molecules which are capable of independent existence.

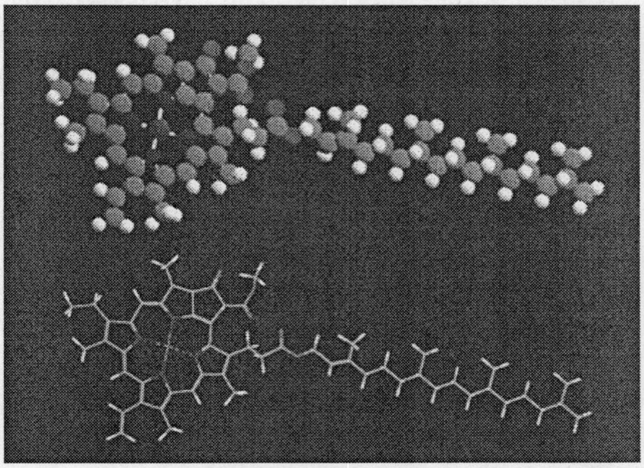

Fig. 4-8. Model of chlorophyll which contains coordinate covalent bond

Coordinate bonds are less common in organic compounds. However, *they are commonly found in complex compounds*. The two important biological molecules containing coordinate covalent bonds are **chlorophyll** (Fig. 4-8) and **heme** in which either Mg or Fe atom is linked coordinately to nitrogen atoms, respectively. In **vitamin B_{12}** also, a Co atom is linked to nitrogen atoms through coordinate bonds. The compound containing coordinate covalent bonds are called **coordinate complexes.**

Noncovalent Bonds or Interactions

In addition to covalent bonding, there are weaker forces of interaction that profoundly influence conformation of biomolecules and their function. These **noncovalent forces**, as they are called, play key roles in the faithful replication of DNA, the folding of proteins into intricate 3-dimensional forms, the specific recognition of substrates by enzymes, and the detection of signal molecules. *Indeed, all biological structures and processes depend on the interplay of noncovalent interactions as well as covalent ones.*

With respect to bonding, weak and strong are used to indicate the amount of energy in a bond. **Strong bonds** such as covalent bonds found in biomolecules require an average of 100 kilocalories/mole or kcal mol^{-1}, to be cleaved and hence are stable and seldom break under physiological conditions. In contrast, **weak bonds,** such as hydrogen bonds, have energies of 2 to 7 kcal mol^{-1} and are easily broken. Weak bonds are transient; individually they form and break in small fractions of a second. The transient nature of noncovalent interactions confers flexibility on macromolecules, such as proteins and nucleic acids, that is critical to their function. Furthermore, the large number of noncovalent interactions in a single macromolecule makes it unlikely that at any given moment, all the interactions will be broken; thus macromolecular structures are stable over time. The four fundamental noncovalent bonds are : electrostatic (or ionic) bonds, hydrogen bonds, hydrophobic bonds and van der Waals forces or bonds. They differ in geometry, strength and specificity and are profoundly affected by the presence of water.

A. Electrostatic or ionic bonds

Ionic bonds are formed due to the attraction between atoms or groups, of opposite charges (+ and –). A charged group on a substrate can attract an oppositely-charged group on an enzyme. The force (F) of such an electrostatic attraction is given by Coulomb's law :

$$F = \frac{q_1 q_2}{r^2 D}$$

where, q_1 and q_2 = charges of the two atoms or groups,
r = distance between the two atoms or groups, and
D = dielectric constant of the medium.

The attraction is strongest in a vacuum where D is 1 and is weakest in a medium such as water where D is 80. The distance between oppositely-charged atoms in an optimal electrostatic attraction is about 2.8 Å. The average bond energy of ionic bonds in aqueous solution is about 5 kcal mol^{-1}. This kind of attraction is also called *saline bond, salt linkage, salt bridge* or *ion pair*.

Ionic bonding occurs in crystals and salts that are ionized when dissolved in water. For example, NaCl is a salt composed of Na^+ (cation) and Cl^- (anion). Common examples of the salts of biomolecules are sodium acetate, potassium pyruvate and ethanolamine chloride (Fig. 4–9). This type of interaction even permits the bonding between two different molecules in heteroproteins (for example, in nucleoproteins, between the negatively-charged nucleic acid and the positively-charged basic proteins, esp., the histones).

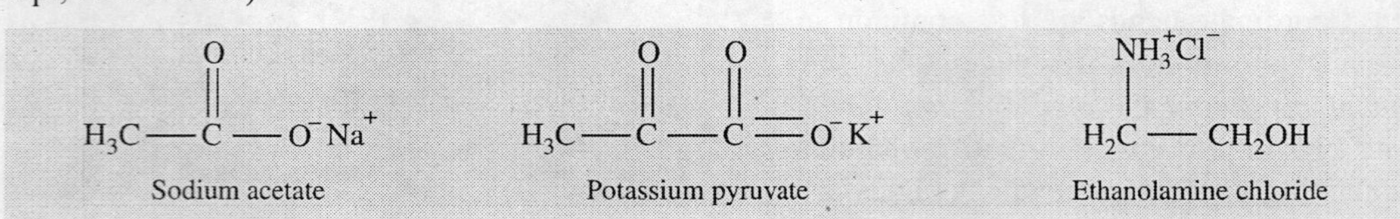

Fig. 4–9. Some salts of biomolecules

B. Hydrogen bonds

Hydrogen bonds can be formed between uncharged molecules as well as charged ones. *In a hydrogen bond, a hydrogen atom is shared by two other atoms* (Fig. 4–10). The atom to which the hydrogen is more tightly linked is called the *hydrogen donor*, whereas the other atom is called as the *hydrogen acceptor*. The acceptor has a partial negative charge that attracts the hydrogen atom. In fact, a hydrogen bond can be considered as an intermediate in the transfer of a proton from an acid to a base.

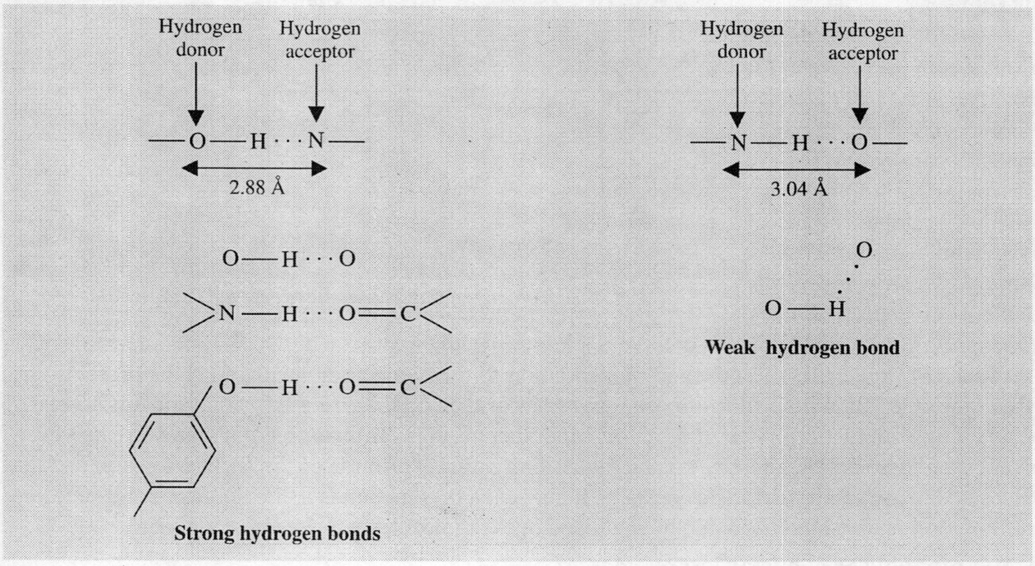

Fig. 4–10. Structure and types of hydrogen bonds

The donor in a hydrogen bond in biological systems is an oxygen or nitrogen atom that has a covalently-attached hydrogen atom. The acceptor is either oxygen or nitrogen. The bond lengths of different types of hydrogen bonds along with their bond energies are given in Table 4–7. A perusal of the table reveals that the length of a hydrogen bond is intermediate between that of a covalent bond and a van der Waals bond. The bond energies of hydrogen bonds range between 2 and 7 kcal/mol.

Table 4–7. Bond lengths and bond energies of some hydrogen bonds

Bond	*Length (Å)*	*Energy (kcal mol^{-1})*
—O—H···N<	2.88	7
—O—H···O=	2.70	6
—O—H···O<		6
>N—H···N<	3.10	2 – 4
>N—H···O=	3.04	2 – 3

Hydrogen bonds are stronger than the van der Waals but much weaker (about 1/10th) than covalent bonds. The strongest hydrogen bonds are those in which the donor, hydrogen and acceptor atoms are colinear.

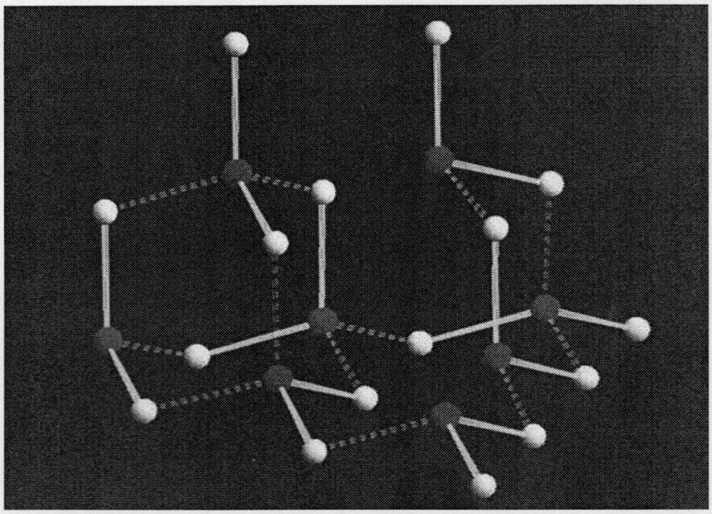

Fig. 4-11. Hydrogen bonds seen between water molecules

Hydrogen bonds in biomolecules (water, for example; Fig. 4-11) are also more specific than other weak bonds because they require particular complementary groups that donate or accept hydrogen. *An important feature of hydrogen bonds is that they are highly directional*. Hydrogen bonding is of 2 types : intramolecular (within a molecule) and intermolecular (between two molecules). Both types are common to many macromolecules such as proteins and DNA molecule. **Intramolecular bonding** gives rise to chelation, *i.e.*, ring formation and this normally occurs only with the formation of 5-, 6- or 7-membered rings. **Intermolecular bonding,** however, gives rise to association, thereby raising the boiling point; it also raises the surface tension and the viscosity, but lowers the dielectric constant. It may exist in compounds in the liquid or solid state and its formation is very much affected by the shape of the molecules, *i.e.*, by the steric factor.

C. Hydrophobic or nonpolar interactions

The essentiality of hydrophilic (water-loving) properties of biomolecules is obvious. That hydrophobic (water-fearing) characteristics can be valuable may not be as readily apparent . As in the formation of micelles, hydrophobic groups of macromolecules, if in proper spatial relation, will interact (*not bond*) to the exclusion of solvent molecules (water) and thereby reside in a hydrophobic environment. On the contrary, hydrophilic groups usually remain exposed to the aqueous environment where they interact with water molecules. Hydrophobic interactions are a major driving force in the folding of macromolecules, the binding of substrates to enzymes and the formation of membranes that define the boundaries of cells and their internal compartments.

In macromolecules such as proteins, the acceptance or rejection by the aqueous environment of the hydrophilic and hydrophobic moieties respectively exerts a dominant influence on their final conformation. Here the nonpolar side chains of neutral amino acids tend to be closely associated with one another. *The relationship is nonstoichiometric*; hence no true bond may be said to exist. *This clustering together of nonpolar molecules or groups in water is called hydrophobic interaction*. The familiar sight of dispersed oil droplets coming together in water to form a single large oil drop is an analogous process.

To understand the basis of hydrophilic attractions, let us take an example wherein a single nonpolar molecule, such as hexane, is introduced into some water (Fig. 4–12). A cavity in the water is created, which temporarily disrupts some hydrogen bonds between water molecules. The dispersed water molecules then reorient themselves to form a maximum number of hydrogen bonds. The water molecules around the hexane molecule are much more ordered than elsewhere in the solution. Now consider the arrangement of two hexane molecules in water. The two possibilities are : either they sit in two small cavities (Fig. 4–12A) or in a single larger one (Fig. 4–12B). The experimental fact is that the two hexane molecules come together and occupy a single large cavity. This association releases some of the more ordered water molecules around the separated hexanes. In fact, the basis of a hydrophilic attraction is this enhanced freedom of released water molecules. *Nonpolar solute molecules are driven together in water not primarily because they have a high affinity for each other but because water bonds strongly to itself.*

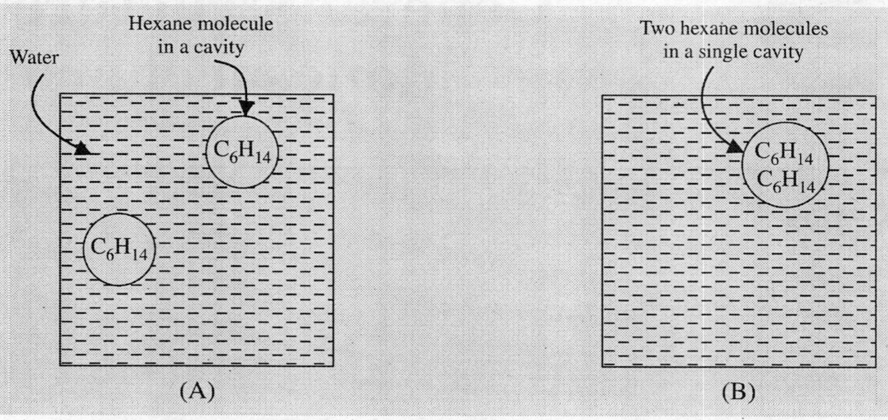

Fig. 4–12. Schematic of two molecules of hexane in a small volume of water
The hexane molecules occupy different cavities in the water structure (A), or they occupy the same cavity, which is energetically more favoured (B).

D. van der Waals interactions

These have already been dealt with in detail in Chapter 2 (Water : The Solvent of Life) on pages 26 and 28.

PROBLEMS

1. Rank the following in terms of expected dipole moment, and explain your choice :

$$H_2S, CCl_4, H_3\overset{+}{N}—CH_2—COO^-, H_3\overset{+}{N}—CH_2—CH_2—CH_2—COO^-$$

2. Why has the element carbon maximum number of organic compounds ?

ANSWERS TO PROBLEMS

1. $CCl_4, H_2S, H_3\overset{+}{N}—CH_2—COO^-, —H_3\overset{+}{N}—CH_2—CH_2—CH_2—COO^-$. CCl_4 is symmetrical, $\mu = 0$; H_2S will be comparable to H_2O; the latter two involve separation of whole charges, and in the last, separation is greatest.
2. Carbon is an element which is most abundant in nature. It is found in 5 allotropic forms : graphite, diamond, fullerenes, nanotubes and the recently found nanofoam. Carbon atom is tetravalent, *i.e.*, it can combine with 4 atoms. The 4 bonds of carbon atom are directed towards the apices of a regular tetrahedron. Consequently, the compounds formed by carbon are very stable.

   The carbon-carbon skeleton forms a long chain or a ring or both. Such compounds are found in nature in a large number and are the basis of life. They constitute a separate branch of chemistry– *Organic Chemistry*.

   Sometimes, the carbon-carbon structure also takes up atoms of another elements such as hydrogen, oxygen, nitrogen, sulfur etc. Compounds of such types are also in large number and are dealt with under *Inorganic Chemistry*.

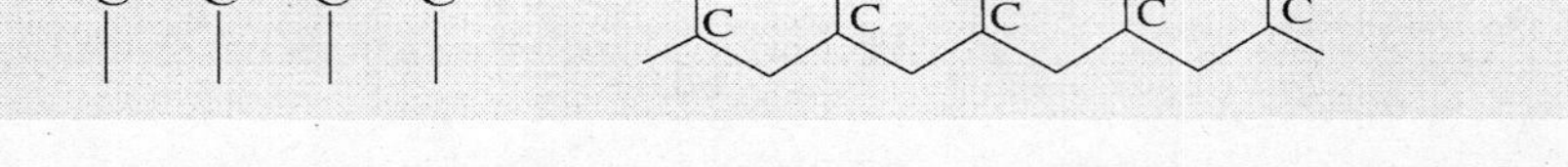

CHAPTER

5 Carbohydrates-I
General Structure of Monosaccharides

INSIDE

- Importance
- Nomenclature and Definition
- Classification
- Asymmetry
- Isomerism
- Kiliani Cyanohydrian Synthesis
- Optical Isomerism
- Formulation of Monosaccharides
  - Formula of Glucose
    - Linear form
    - Ring form
  - Haworth Perspective formula

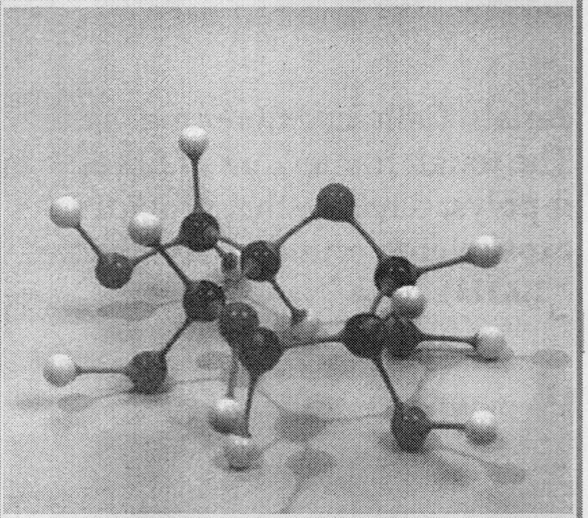

Ball-and-stick model of a glucose molecule

IMPORTANCE

The carbohydrates, often termed as sugars, are the '*staff of life*' for most organisms. On the basis of mass, they are the most abundant class of biomolecules in nature. Carbohydrates are also known as *saccharides* (*sakcharon*^{G} = sugar or sweetness) since many of those of relatively small molecular weight have a sweet taste, although this is not true of those with large molecules. They are widely distributed molecules (*moles*^{L} = mass) in both plant and animal tissues. They are indispensable for living organisms, serving as skeletal structures in plants and also in insects and crustaceans. They also occur as food reserves in the storage organs of plants and in the liver and muscles of animals. In addition, they are an important source of energy required for the various metabolic activities of the living organisms; the energy being derived as a result of their oxidation. They also serve to lubricate skeletal joints, to provide adhesion between cells and to confer biological specificity on the surface of animal calls.

Plants are considerably richer in carbohydrates in comparison to the animals. In fact, animal and plant tissues differ widely in the relative abundance of the various major classes of constituent chemicals (Table 5–1).

Table 5–1. Percentage composition of various major classes of constituent chemicals in plants and animals

Organism	*Water*	*Carbohydrates*	*Proteins*	*Lipids*	*Ash*
Animal	60	1	20	15	4
Plant	60	30	5	1	4

NOMENCLATURE AND DEFINITION

The term 'carbohydrate' was originally coined for this class of compounds as most of them were '**hydrates of carbon**' or could be represented by the general formula, $C_x(H_2O)_y$. Later, it was found that some of them, such as deoxyribose ($C_5H_{10}O_4$) and rhamnose ($C_6H_{12}O_5$) do not have the required ratio of hydrogen to oxygen. In addition, certain other carbohydrates are now known to possess nitrogen (*e.g.*, glucosamine, $C_6H_{13}O_5N$), phosphorus or sulfur also and obviously do not coincide with the above general formula. Furthermore, formaldehyde (H.CHO or CH_2O), acetic acid (CH_3.COOH or $C_2H_4O_2$) and lactic acid (CH_3.CHOH.COOH or $C_3H_6O_3$) which have C, H and O and the ratio of H : O is also the same as in water, but are not a carbohydrates. Hence, *the continued usage of the term 'carbohydrate' is for convenience rather than exactness.*

To accommodate a wide variety of compounds, the carbohydrates are nowadays broadly defined as **polyhydroxy aldehydes or ketones and their derivatives** or as substances that yield one of these compounds on hydrolysis.

CLASSIFICATION

Carbohydrates are usually classified in 3 groups :

A. **Monosaccharides** or **Monosaccharoses** (*mono*[G] = one; *sakcharon*[G] = sugar). The monosaccharides, often called *simple sugars*, are compounds which possess a free aldehyde (—CHO) or ketone (= CO) group and 2 or more hydroxyl (—OH) groups. They are, in fact, the simplest sugars and cannot be hydrolyzed into smaller units. Their **general formula** is $C_n(H_2O)_n$ or $C_nH_{2n}O_n$.

The monosaccharides may be subdivided into trioses, tetroses, pentoses, hexoses, heptoses etc., depending upon the number of carbon atoms they possess; and as aldoses or ketoses, depending upon whether they contain aldehyde or ketone group. Some important examples are :

Name	**Formula**	**Aldoses** (*Aldo sugars*)	**Ketoses** (*Keto sugars*)
Trioses	$C_3H_6O_3$	Glycerose	Dihydroxyacetone
Tetroses	$C_4H_8O_4$	Erythrose	Erythrulose
Pentoses	$C_5H_{10}O_5$	Ribose	Ribulose
Hexoses	$C_6H_{12}O_6$	Glucose	Fructose
Heptoses	$C_7H_{14}O_7$	Glucoheptose	Sodoheptulose

Both these characters (*i.e.*, the number of carbon atoms and the nature of functional group present) may also be combined into one. Thus, for example, glycerose (= glyceraldehyde) is an aldotriose; ribulose, a ketopentose and glucose, an aldohexose. It is noteworthy that, *except fructose, ketoses are not as common as aldoses.* The most abundant monosaccharide in nature is the 6-carbon sugar, D-glucose.

Sometimes, a distinction in naming between aldoses and ketoses is also maintained. The suffix

-*oses* is kept reserved for the aldoses and the suffix -*uloses* is used for ketoses. Thus, glucose is a hexose and fructose, a hexulose. However, a few ketoses are named otherwise, such as fructose (*fructus*[L] = fruit) as fruits are a good source of this sugar.

> **Anselme Payen,** codiscoverer of diastase, also isolated a compound common to cell walls of higher plants, which he named 'cellulose'. His naming of this polysaccharide, in fact, introduced the **-ose** suffix into the nomenclature of carbohydrates.

B. **Oligosaccharides** or **Oligosaccharoses** (*oligo*[G] = few). These are *compound sugars* that yield 2 to 10 molecules of the same or different monosaccharides on hydrolysis. Accordingly, an oligosaccharide yielding 2 molecules of monosaccharide on hydrolysis is designated as a dissaccharide, and the one yielding 3 molecules of monosaccharide as a trisaccharide and so on. The **general formula** of disaccharides is $C_n(H_2O)_{n-1}$ and that of trisaccharides is $C_n(H_2O)_{n-2}$ and so on. A few examples are :

Disaccharides – Sucrose, Lactose, Maltose, Cellobiose, Trehalose, Gentiobiose, Melibiose

Trisaccharides – Rhamninose, Gentianose, Raffinose (= Melitose), Rabinose, Melezitose

Tetrasaccharides – Stachyose, Scorodose

Pentasaccharide – Verbascose

C. **Polysaccharides** or **Polysaccharoses** (*poly*[G] = many). These are also compound sugars and yield more than 10 molecules of monosaccharides on hydrolysis. These may be further classified depending on whether the monosaccharide molecules produced as a result of the hydrolysis of polysaccharides are of the same type (homopolysaccharides) or of different types (heteropolysaccharides). Their *general formula* is $(C_6H_{10}O_5)_x$. Some common examples are :

Homopolysaccharides – Starch, Glycogen, Inulin, Cellulose, Pectin, Chitin

Heteropolysaccharides – "Specific soluble sugar" of pneumococcus type III, Hyaluronic acid, Chondrotin

ASYMMETRY

Jacobus H. van't Hoff, the first Nobel Laureate in Chemistry (1901) and **Joseph A. Le Bel,** in 1894, introduced the **concept of tetrahedral carbon atom**. It is now recognized that the carbon atom has the shape of a tetrahedron in which the carbon nucleus resides in the centre of this tetrahedron and the 4 covalent bonds extend out to the corners of it. The angle between any two covalent bonds is 109°28′ (Fig. 5–1).

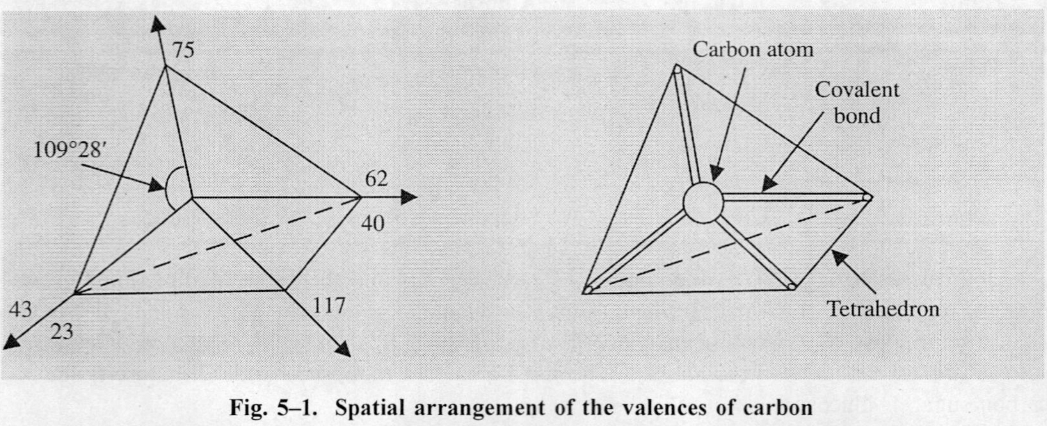

Fig. 5–1. Spatial arrangement of the valences of carbon

In the formula given on page 71 for glucose, it may be observed that a different group is attached to each of the 4 valences of carbon atoms 2 to 5. For example, the 4 groups attached at carbon atom 2 are :

JACOBUS HENRICUS VAN'T HOFF
(LT, 1852-1911)

A Dutch chemist. Best known for his hypothesis about the structure of carbon atom. Made notable contributions in the field of Physical Chemistry, particularly on the theory of dilute solutions. Won the first-ever **Noble Prize in Chemistry in 1901** for his work on rates of reactions, chemical equilibrium and osmotic pressure.

JOSEPH A. LE BEL
(LT, 1847-1930)

A French chemist. Best known for his stereochemical theory of carbon and its compounds. Co-discoverer of the cause of optical activity with van't Hoff. Synthesized the first optically-active compound with asymmetric nitrogen atom.

— CHO, — H, — OH and HO — C — H

Thus, a carbon atom to which 4 different atoms or groups of atoms are attached is said to be **asymmetric** or **chiral** (*cheir*[G] = hand).

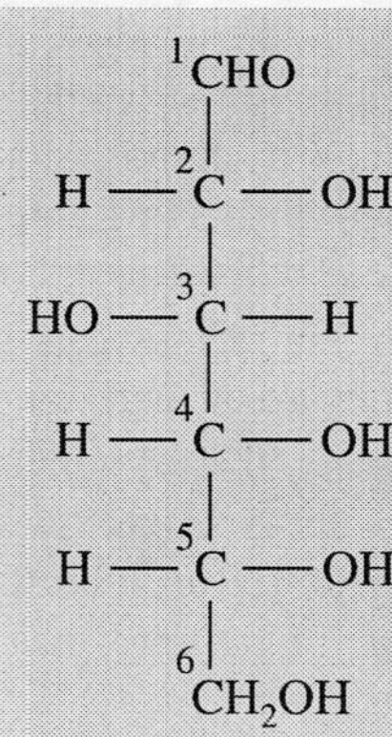

Glucose

[The carbon atoms of a sugar are numbered from the end closest to the aldehyde or ketone group.]

ISOMERISM

The term **isomer** (*isos*[G] = equal ; *meros*[G] = part) was originally applied by **Jönes Jacob Berzelius**, in 1827, to different compounds with the same molecular formula, and the phenomenon was called **isomerism**. The presence of asymmetric carbon atoms in the carbohydrates makes possible the formation of isomers in them. It shall, hence, be beneficial at this stage to discuss in brief about isomers.

The isomers are of 2 types : structural isomers and stereoisomers. **Structural isomers** have the same molecular formula but possess different structures. The difference in structure may be exhibited either in the length of the carbon chain (**chain isomers**) or in the position of a substituent group (**positional isomers**) or in possessing different functional groups (**functional-group isomers**).

Stereoisomers, on the other hand, have the same molecular formula and the same structure but differ only in spatial configuration. Stereoisomers are of 2 types : geometrical and optical. **Geometrical** or **cis-trans** (*cis*[L] = same side; *trans*[L] = across) **isomers** arise from peculiar geometry of compounds having a double bond within the carbon chain. It may be illustrated by *cis-trans* pair, maleic and fumaric acids.

Maleic acid
(*cis* form)

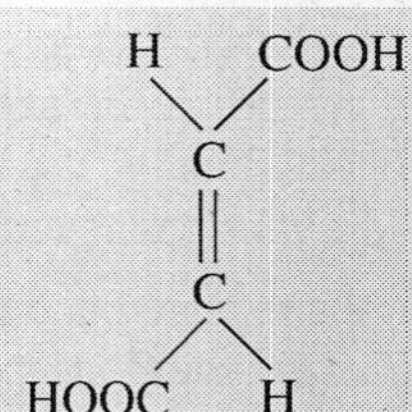

Fumaric acid
(*trans* form)

Optical isomers (= enantiomers) differ from each other in the disposition of the various atoms or groups of atoms in space around the asymmetric carbon atom. These are, in fact, the mirror image of each other. These may also be likened to left- and right-handed gloves. For example, the glyceraldehyde has only one asymmetric carbon atom (numbered as 2) and it can, therefore, exist in 2 isomeric forms :

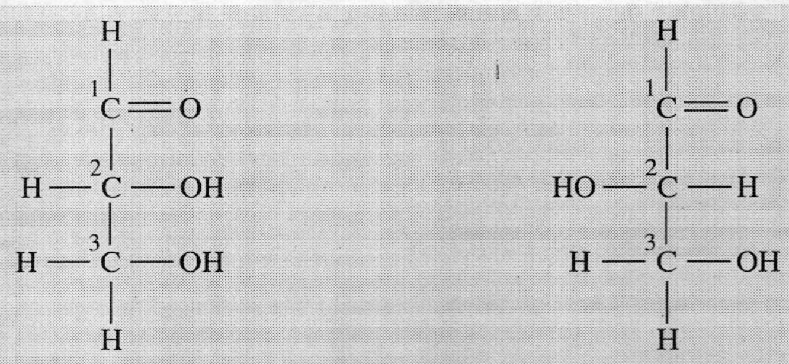

D-glyceraldehyde L-glyceraldehyde

One form in which H atom at carbon 2 is projected to the left side and OH group to the right is designated as **D-form** and the other form where H atom is projected to the right side and OH group to the left is called as **L-form** (note the use of small capital letters D and L) Similarly, a compound having 2 asymmetric carbon atoms (*e.g.*, a tetrose) might exist in 4 optical isomeric froms.

The number of possible isomers of any given compound, thus, depends upon the number of asymmetric carbon atoms present in its molecule. According to the **rule of n** (also called as **Le Bel-van't Hoff rule**), the total number of optical isomers of a compound will be equal to 2^n, where n represents the number of asymmetric carbon atoms present in the molecule.

In fact, in sugars with 2 or more asymmetric carbon atoms, the designation of D- and L- forms depends upon the orientation of the H and OH groups around the lowermost asymmetric carbon atom (*i.e.*, the asymmetric carbon furthest from the aldehyde or keto group). The distribution of the H and OH groups on the other carbon atoms in the molecule is of no importance in this connection. It is interesting to note that *majority of the monosaccharides found in the human body are of D type*. However, some sugars do occur naturally in their L form such as L-arabinose, a pentaaldose.

KILIANI CYANOHYDRIN SYNTHESIS

A method for the synthesis of monosaccharides was first proposed by Heinrich Kiliani in 1886. It is, in fact, a method by which the chain length of a carbohydrate may be increased. The application of Kiliani synthesis to D-glyceraldeyde resulting in the production of 2 tetroses, D-erythrose and D-threose is shown in Fig. 5–2.

The process is based upon the addition of HCN to the carbonyl group of aldehydes (or ketones) of the sugars forming cyanohydrin (Reaction-1). This reaction creates a new asymmetric carbon atom, marked by an asterisk in the figure. Thus, two compounds differing in conformation about the newly-formed asymmetric carbon atom are formed. These two cyanohydrins are, then, hydrolyzed to produce carboxylic acids (Reaction-2) which are later converted to γ-lactones or inner esters (Reaction-3). Finally, the lactones are reduced to the corresponding aldoses, containing one carbon atom more than their parent sugar (Reaction-4).

The process can be repeated and the 4 isomeric D-pentoses may be produced; and from these the 8 isomeric D-hexoses would also result.

The structural relationships of the monosaccharides (the aldoses, for example) may be visualized by the formulae given in Fig. 5–3. It is easy to remember the configurations of the aldoses by

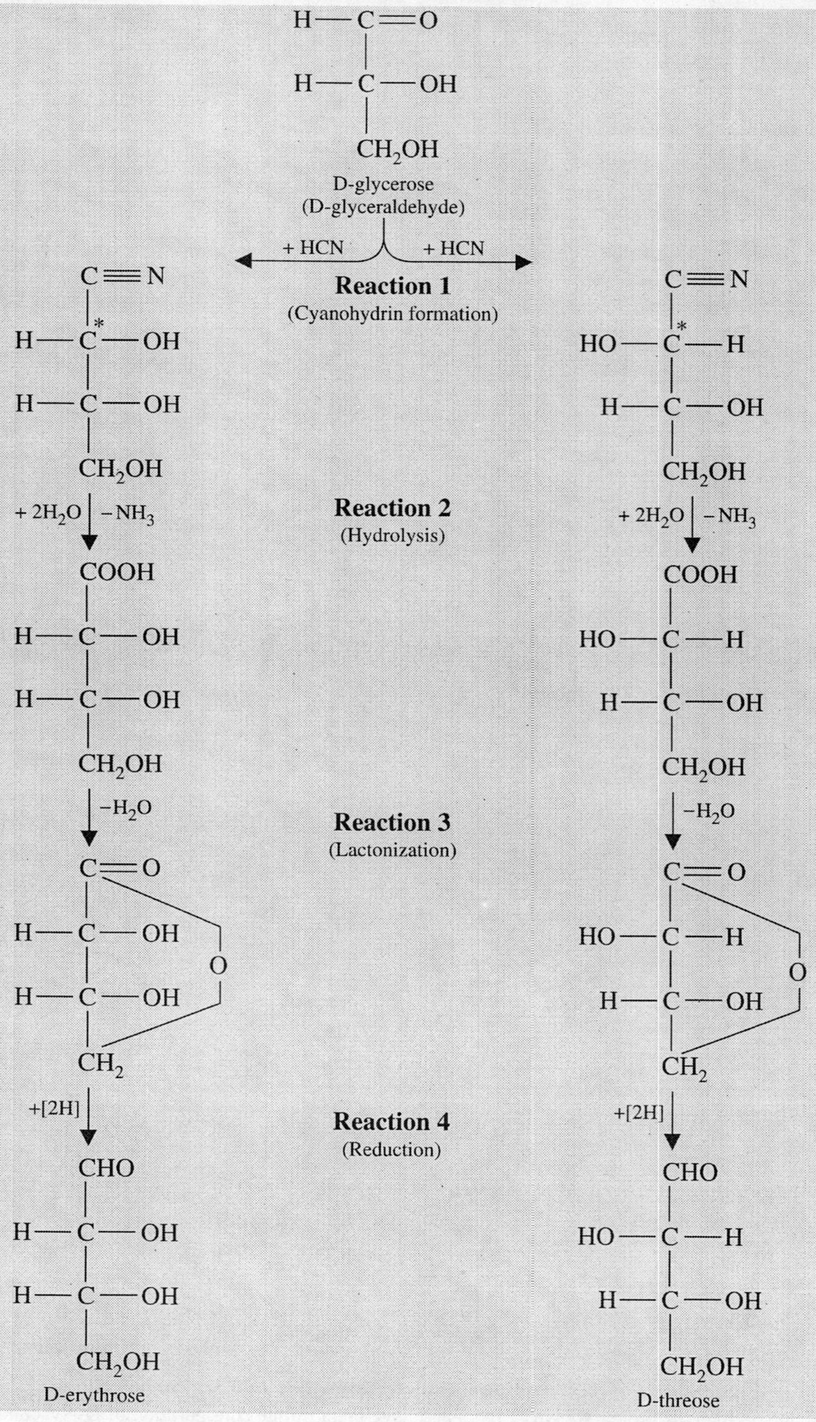

Fig. 5–2. Application of the cyanohydrin synthesis to D-glyceraldehyde

recalling the *Rosanoff scheme*, drawn out on the chart and mneumonic—ET; RAXL; AAGMGIGT.

Any two sugars which differ from each other only in the configuration around a single

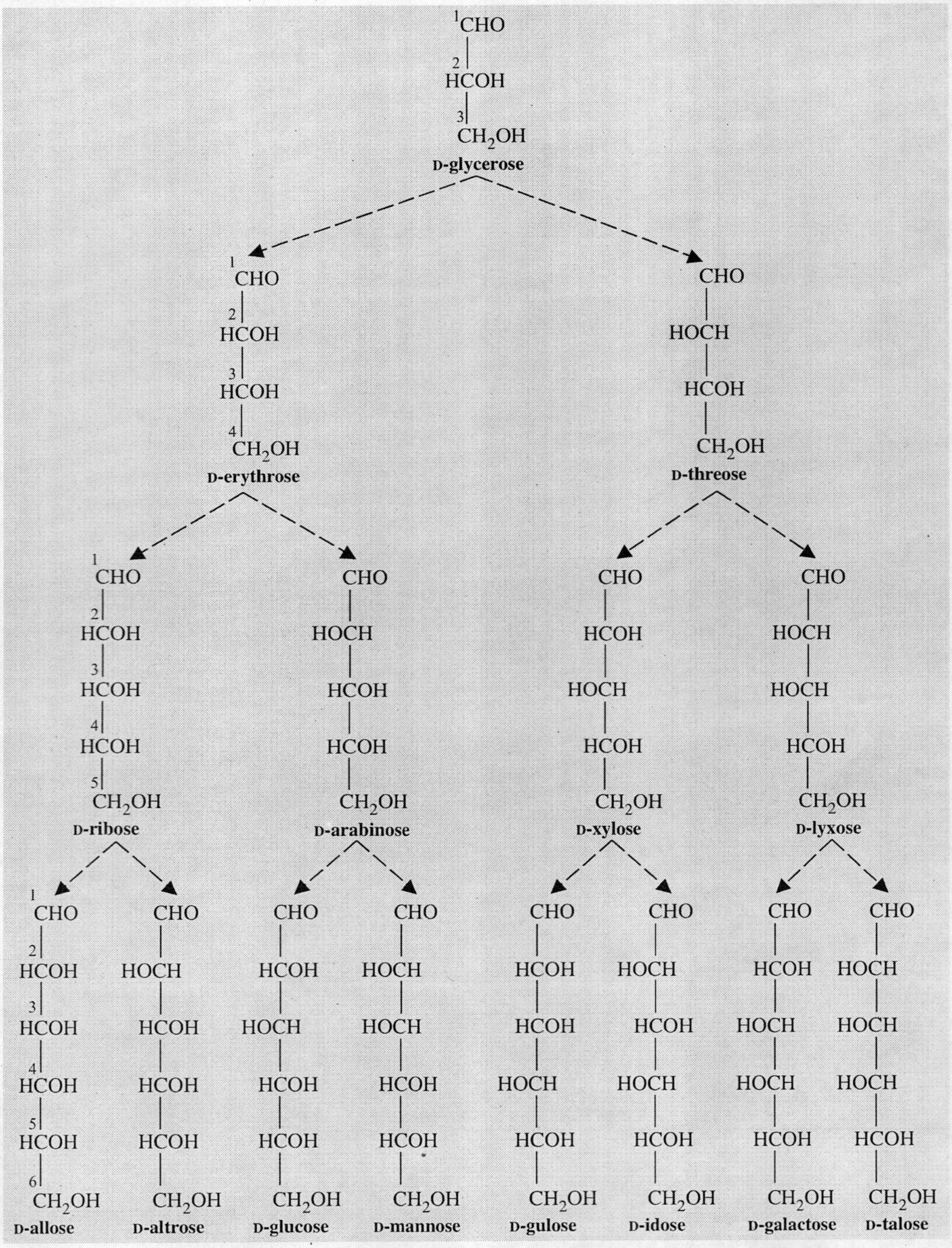

Fig. 5–3. D-aldoses containing three, four, five and six carbon atoms

Aldoses contain an aldehyde group (shown in *blue*) and have the absolute configuration of D-glyceraldehyde at the asymmetric centre (shown in *red*) farthest from the aldehyde group. The numbers indicate the standard designations for each carbon atom. The formulae of L-aldoses are in each case the mirror images of these structures.

asymmetric carbon atom other than the carbonyl carbon atom are called **epimers.** Glucose and galactose, for example, form an epimeric pair as they differ with respect to carbon 4 only. Similarly, glucose is also epimeric with mannose (differing in C_2 configuration) and allose with altrose (differing also in C_2 configuration).

The two D- and L-forms of a compound constitute a pair of **enantiomers** (*enantios*[G] = opposite) or **enantiomorphs**. Thus, D-erythrose and L-erythrose are enantiomers; also D-glucose and L-glucose. The enantiomers (also called optical isomers or double image isomers) are nonsuperimposable mirror images of each other but are chemically identical in their reactions. They agree in their melting points, solubility etc., but differ in their ability to rotate the plane of polarized light in a polarimeter; a solution of one of the two enantiomers rotates the plane to the right, and a solution of the other to the left.

The conversion of D-glyceraldehyde into an aldotetrose yields D-erythrose and D-threose. These are called **diastereoisomers** *i.e.*, isomers but not mirror images of each other. In fact, diastereoisomers are different forms of a compound with two asymmetric centres. It may, hence, be inferred that if two optical isomers are not enantiomers, they are related as diastereoisomers. They differ from each other in their melting points, solubilities and the chemical properties, in general.

> The term **'conformation'** was first introduced by **Walter Norman Haworth** (1929), an English chemist. In its broadest sense, conformation has been used to describe different spatial arrangements of a molecule which are not superimposable. This means that, in effect, the terms *conformation* and *configuration* are equivalent. In the classical sense, the definition of conformation does not include the internal forces acting on the molecule. The term conformation, however, is the spatial arrangement of the molecule when all the internal forces acting on the molecule are taken into account. In more restricted sense, the term conformation is used to denote different spatial arrangements arising due to twisting or rotation of bonds of a 'given' configuration (used in the classical sense). The terms **rotational isomers** and **constellations** have also been used in the same sense as conformations.

OPTICAL ISOMERISM

The presence of asymmetric carbon atoms in the compound also confers optical activity to it. When a beam of polarized light is passed through an optically active solution, it may be rotated either to the right or to the left, depending upon the type of optical isomer present. A compound rotating the plane of polarized light to the right is called as *dextrorotatory* (*dexter*[L] = right, on the right side; *rotatus*[L] = pp of *rotare*[L], to turn) or clockwise and is designated as *d* or (+) type. If the compound causes rotation of polarized light to the left, it is said to be *levorotatory* (*laevus*[L] = left, on the left side) or anticlockwise and is denoted as *l* or (–) type. The direction in which the light is rotated (or in other words, the optical rotation) is a specific property of the molecule. It should, however, be emphasized that *the optical rotation is not at all related with the two D and L forms of a compound.* Thus, D-glucose is dextrorotatory and D-fructose is levorotatory. If it is desired to indicate the direction of rotation, the two names may be written as D(+)-glucose and D(–)-fructose. Similarly, D-erythrose is levorotatory and D-threose, dextrorotatory.

When equal amounts of dextrorotatory and levorotatory isomers are present, the resulting mixture becomes optically inactive because the optical activities of each isomer cancel each other. Such a mixture is called a **racemic** or **dl-mixture** or **(±)-conglomerate** and this process of converting an optically active compound into the racemic modification is known of **racemisation** (*racemus*[L] = grape). Compounds produced synthetically are invariably racemic in nature because equal chances exist for the formation of two types of optical isomers or the *d* and *l* antipodes.

It may be concluded that all monosaccharides are optically active (except aldobiose and ketotriose) since they contain one or more asymmetric carbon atoms in their molecules.

Mutarotation

When a monosaccharide is dissolved in water, the optical rotatory power of the solution gradually changes until it reaches a constant value (**Dubrunfaut**, 1846). A freshly-prepared aqueous solution of α-D-glucose, for instance, has a specific rotation, $[\alpha]_D^{20}$ of +112.2°. And when this solution is allowed to stand, the rotation falls to +52.7° and remains constant at this value. The final stage can be attained more quickly either by heating the solution or by adding some catalyst which may be an acid or an alkali. This gradual change in specific rotation is known as **mutarotation** or changing rotation. In fact, this term reflects the discovery of the phenomenon by way of changes in the optical rotation of certain carbohydrates. The terms **multirotation** and **birotation** have sometimes also been used for mutarotation. The value of mutarotation for α-D-glucose is (+112.2°) – (+52.7°) or +59.5°. A fresh solution of β-D-glucose, on the other hand, has a rotation value of +18.7°; on standing, it also changes to the same value, +52.7°. *All reducing sugars (except a few ketoses) undergo mutarotation.* The specific rotations for the anomers of glucose and for the aqueous equilibrium mixture are :

$$\underset{\alpha\text{-D-glucose}}{[\alpha]_D^{20} = +112.2^\circ} \longrightarrow \underset{\text{Mutarotational equilibrium mixture}}{[\alpha]_D^{20} = +52.7^\circ} \longleftarrow \underset{\beta\text{-D-glucose}}{[\alpha]_D^{20} = +18.7^\circ}$$

These substances (open and ring forms) interconvert due to a dynamic equilibrium. Although the ring form predominates, each molecule spends some time in the open form, when asymmetry is lost. When the open-chain form passes back to the ring form, it is possible for it to give rise to either anomer. Thus, a pure sample of an anomer only remains so when in the solid state. In solution, particularly at alkaline pH, a pure sample of one anomer will change to an equilibrium mixture of the two forms. There need not be equal quantities of the two forms at equilibrium because there are other centres for asymmetry and so anomers (like *allo* and *threo* pairs of amino acids) do not have identical physical properties, including ΔG° of formation. Equilibrium between the anomers is called mutarotation and it can be followed by observing optical rotation (Fig. 5–4).

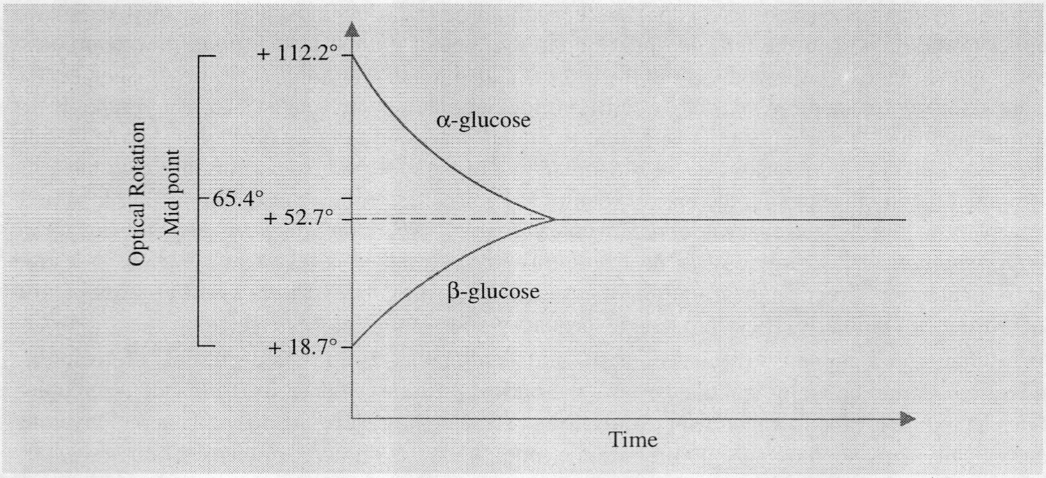

Fig. 5–4. The experimental detection of mutarotation

FORMULATION OF MONOSACCHARIDES

Formula of Glucose—Linear Form

An organic chemist intends to incorporate as many characteristics as possible in writing the formulae (Fig. 5–5). When the molecular formula of glucose, $C_6H_{12}O_6$ is written in a form as propounded by Fittig and Baeyer, it tells us the presence of an aldehyde and 5 hydroxyl groups. Replacement of this formula by Fischer projection formula informs the reader about some more details. For example, the presence of 4 asymmetric carbon atoms in the molecule indicates towards the existence of 2^4 or 16 stereoisomers.

Ring Form

However, even Fischer formula fails to describe certain other properties of glucose. Hence, another method of representing the structure was searched for. The aldehyde or ketone group of a sugar can react with hydroxyl groups of alcohols forming **hemiacetals** or **hemiketals**, respectively. For the larger sugars (n = >4), this happens *within the same molecule* to form a 5- or 6-membered ring.

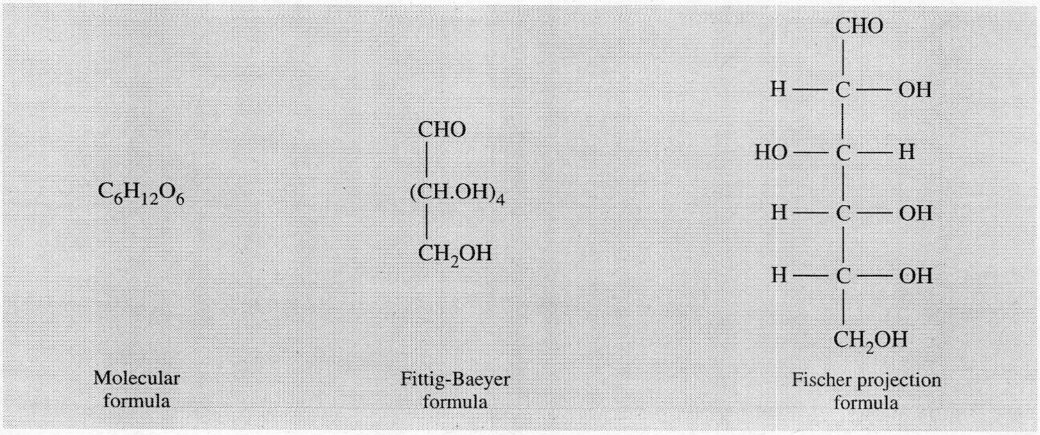

Fig. 5–5. Formulation of glucose

In fact, the angles of the tetrahedral carbon atom tend to bend the glucose molecule forming a ring. The aldehyde group and the alcohol group of carbon 5 or carbon 4 readily approach each

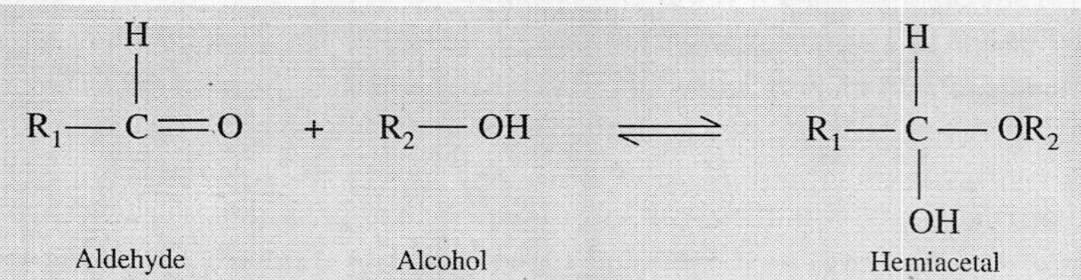

Formation of a hemiacetal

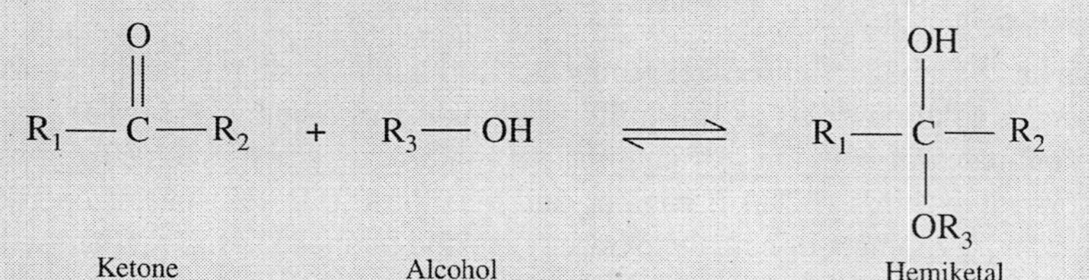

Formation of a hemiketal

other forming intramolecular *hemiacetals*. Similarly, the keto group of a ketose can also approach the alcohol group of either carbon 5 or carbon 4 forming intramolecular *hemiketals*. This results in the formation of either a 6-membered ring (**pyranose form**) or a 5-membered ring (**furanose form**). A 7-membered ring, however, becomes too strained to allow participation of the OH group of carbon 6 of aldohexoses with the CHO group in ring formation. The pyranose form has a lower $\Delta G°$ of formation than the furanose form.

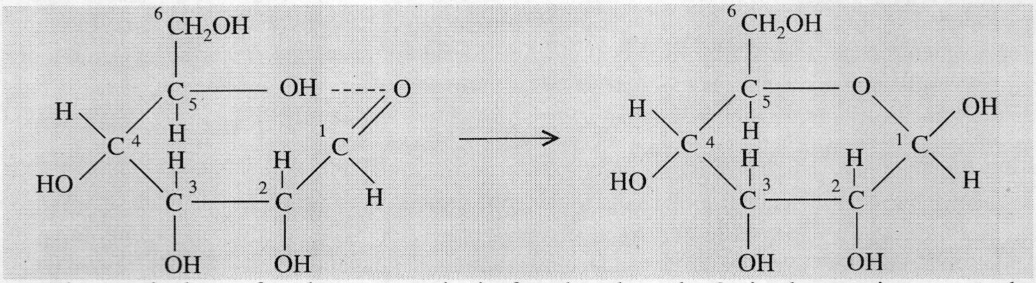

The terminology of such structures is, in fact, based on the 2 simple organic compounds, exhibiting a similar ring structure. These are **pyran** and **furan**. *The pyranose forms of the sugars are more stable than the furanose forms in solution.*

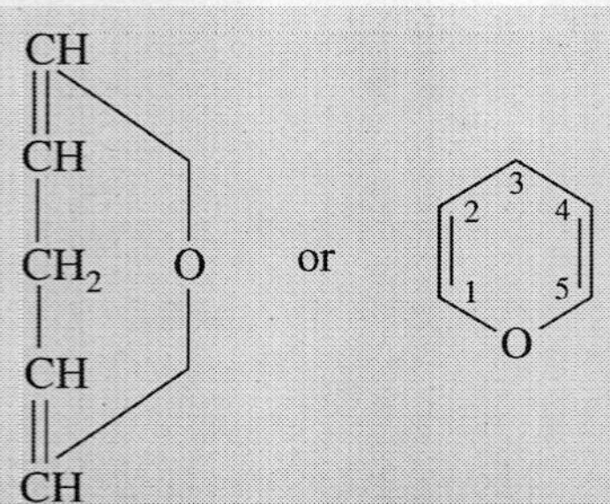

Pyran
(the parent compound of pyranoses)

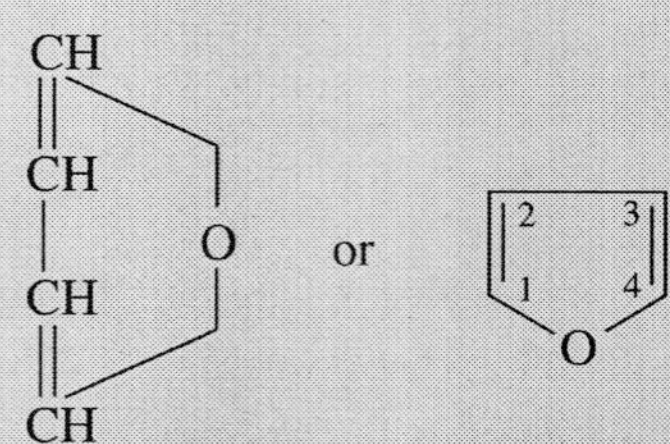

Furan
(the parent compound of furanoses)

The formation of a ring structure in the glucose molecule creates a new centre of asymmetry, *i.e.*, the carbon 1 (carbon 2 in ketoses). One may, therefore, find that the ring form of glucose can exist as either of the 2 isomeric forms, termed as α and β isomers. These two isomers are, in fact, diastereoisomers rather than enantiomers because the α-form of D-glucose is not the mirror image of the β-form. They are also known as **anomers** (*ano*[G] = upper) as they differ in configuration only around the hemiacetal or anomeric carbon atom (*i.e.*, carbon 1). The designation of α- and β-forms is based on the suggestion of Freudenberg (1933) who stated that those anomers having the same configuration at both the anomeric and penultimate carbon atoms should be called as α-form while in β-form the configuration would be different at both these carbon atoms. In other words, the configuration of H atoms at both these carbons (anomeric and penultimate) is **erythro** (written on the same side of the structure) in α-form and **threo** (written on opposite sides) in β-form. The α- and β-forms of D-glucose in both pyranose and furanose rings along with the open chain formula are shown in Fig. 5–6.

As shown in the above figure, the hemiacetal formation is of reversible nature. Thus, D-glucose in aqueous solution exists as an equilibrium mixture of 5 compounds. The pyranose form, however, predominates in solution. An equilibrium mixture of glucose contains about one-third (37%) α anomer, two-thirds (62%) β anomer and very little (<< 1%) of the open-chain form. Likewise, the α and β anomers of both the pyranose and the furanose forms of fructose interconvert through the open-chain form. In fact, under physiologic conditions in solution, monosaccharides with 5 or more carbons are typically more than 99% in the ring forms. The distribution between pyranose and furanose forms depends on : (*a*) the particular sugar structure, (*b*) the pH, (*c*) the solvent composition, and (*d*) the temperature.

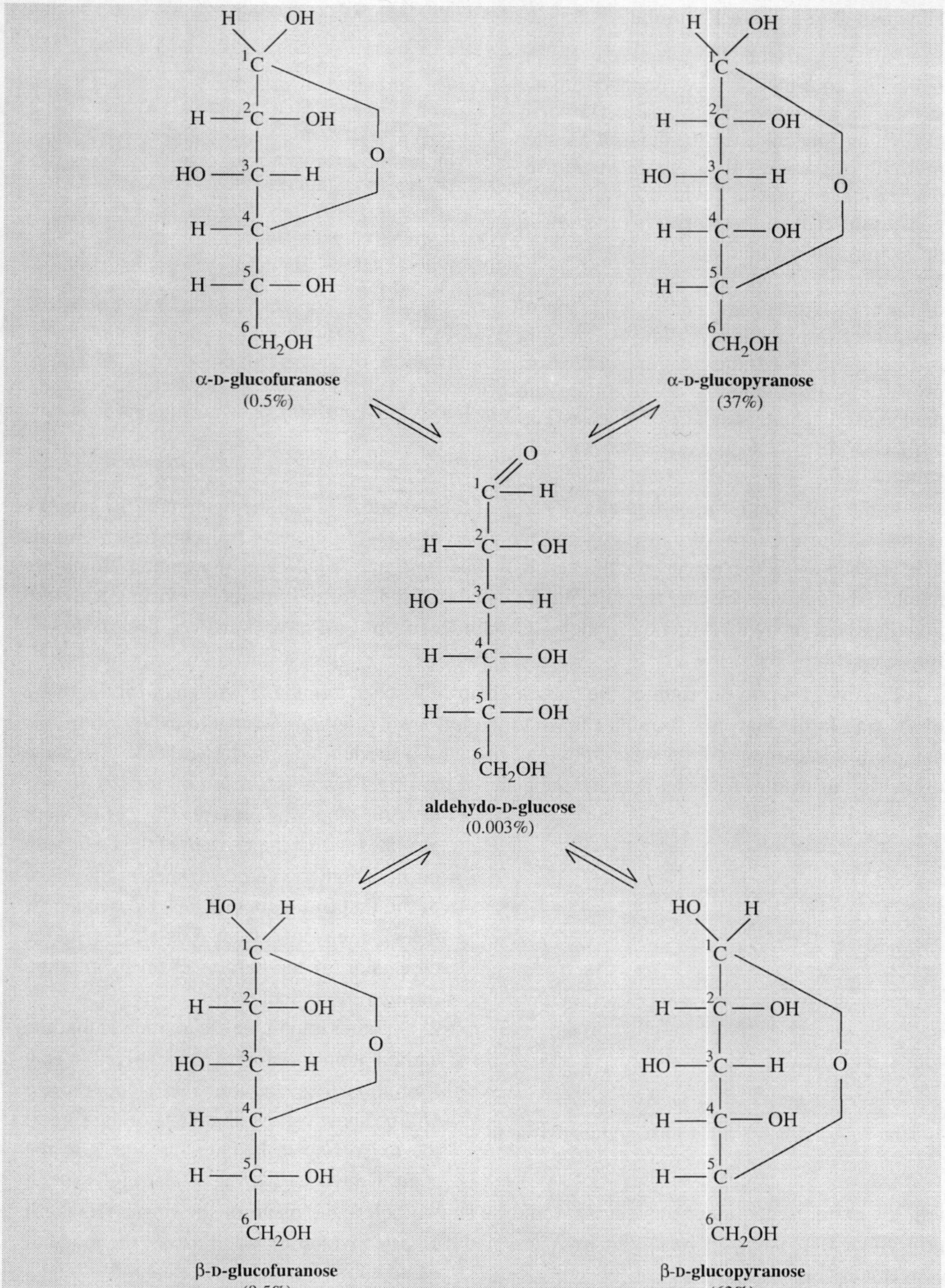

Fig. 5–6. Various forms of D-glucose present in an aqueous solution

[Note that in α-form, the OH group at anomeric carbon 1 is also on the right side like that at the penultimate carbon atom 5 ; whereas in β-form, the OH group at carbon 1 is on the left and that on carbon 5 is on the right side. Figures in parentheses indicate the molecular percentage present in an equilibrium mixture at pH 7.]

Haworth Perspective Formula

Although Fischer formulae are useful in indicating configurational differences among sugars, they are tedious to write. **Haworth** (1929), an English chemist, devised another method of representing their structure, resembling a hexagon (pyranose form) or a pentagon (furanose form). In such a projection, the carbon atoms in the ring are not explicitely shown. The approximate plane of the ring is perpendicular to the plane of the paper, with the heavy line on the ring projecting toward the reader. Haworth projections more clearly approximate the predominant "chair" structure of glucose (Fig. 5–7) and other hexose sugars that exist in solutions.

WALTER NORMAN HAWORTH

(LT, 1883–1950)

Walter Norman Haworth, a Briton, had been a student of **Otto Wallach** and **W.H. Perkin** at the Universities of Manchester and Göttingen. He was awarded **1937 Nobel Prize in Chemistry** for his work on the constitution of carbohydrates and vitamin C, along with Paul Karrer of Switzerland. Haworth's contributions include : methylation of sugars, pyranose and furanose structures of monosaccharides, structure of starch, cellulose and amygdalin, and structure and synthesis of ascorbic acid.

In the pyranose form, carbon atoms 1 to 5 of the aldohexoses and the O atom of the ring are represented in the form of a hexagon and the plane of this ring projects from the paper. Bonds that are nearer to the reader are thickened in this type of formula and the carbon atoms are numbered clockwise. Fischer formula may be converted to Haworth formula by twisting it around imaginatively or by constructing models. Certain '**assumptions**' have, however, been followed for conversion. These are :

1. Any group to the right of the carbon chain in Fischer formula is written *down* the plane of the ring in the Haworth formula ; those to the left are written *up*. A notable exception to this assumption is carbon atom 5 where H atom, present on the left side, is shown below the plane of the ring instead of showing it above the plane of the ring. This is as a result of torsion needed to bring about ring closure. The carbon atom 5 rotates so that the O atom of the OH group on the anomeric carbon comes in the plane of the first five carbon atoms. Consequently, the H atom on carbon 5 is shifted to the other side of the carbon chain by rotating through more than 90°.

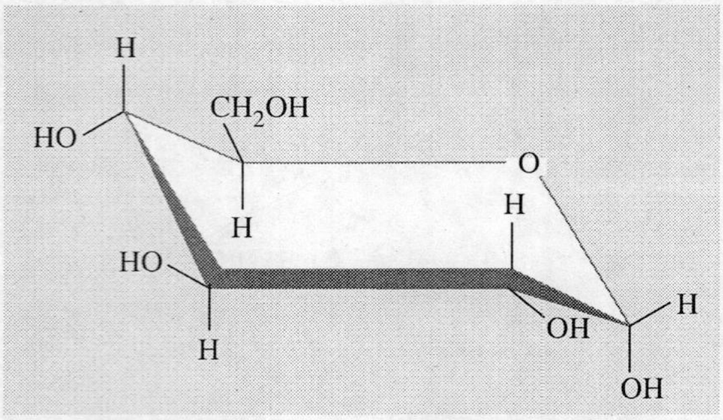

Fig. 5–7. "Chair" form of α-D-glucopyranose

2. Another important assumption is that the carbon atom (or atoms) not involved in ring formation (*e.g.*, carbon 6 in aldopyranses) will be shown above the plane of the ring, if the oxide ring in Fischer formula is to the right. Conversely, if the oxide ring is to the left, the extra carbon atom (or atoms) would be shown below the plane of the ring in Haworth projection. Two examples (α-D-glucopyranose and β-L-glucopyranose) will illustrate the mode of conversion (Fig. 5–8).

For the sake of simplicity, while writing Haworth formulae, the H and OH groups are not written and the bond indicating the single hydrogen atom is not shown. Thickened line in lower half of the hexagon is also omitted. The extra carbon atoms are, however, written as such. Thus, the above two compounds may be simply represented as in Fig. 5–9.

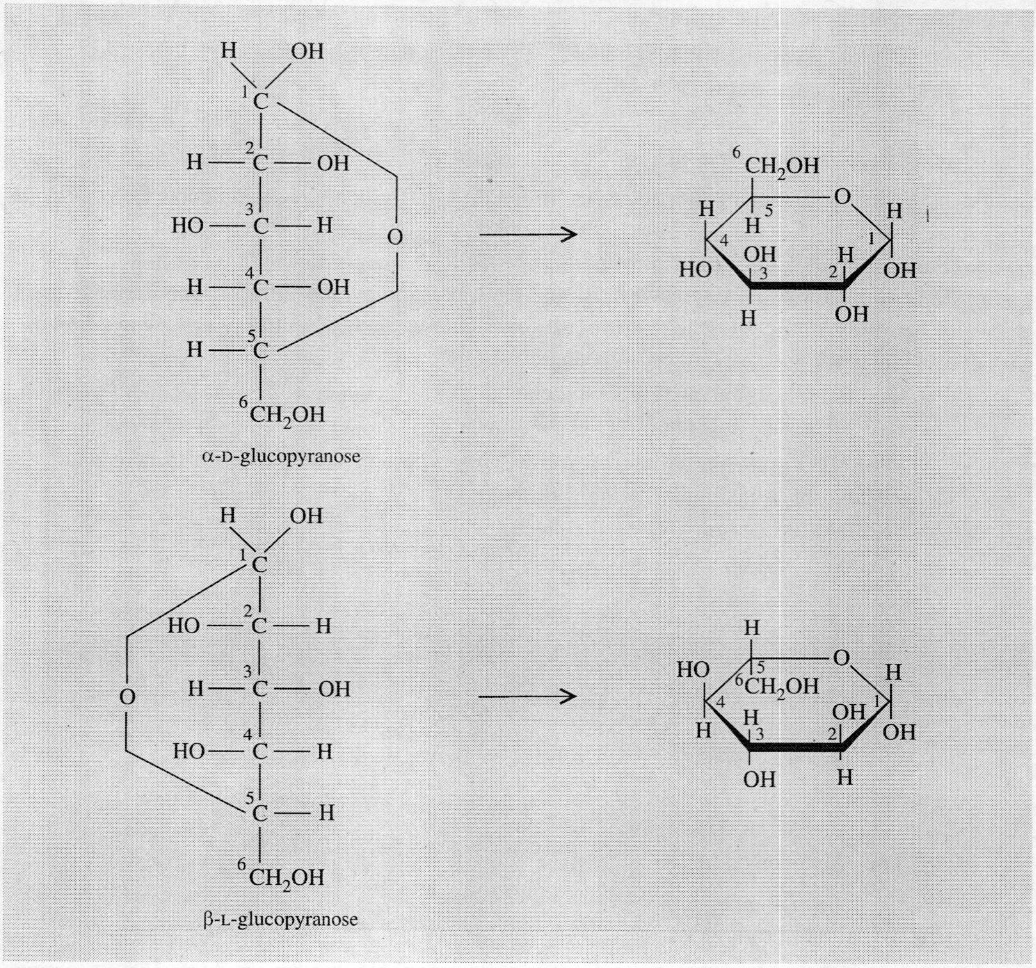

Fig. 5–8. Mode of conversion of two hexose sugars from Fischer projection formula to Haworth perspective formula

Fig. 5–9. Simplified Haworth formulae for two hexose sugars

PROBLEMS

1. Account for the origin of the term *carbohydrate*.
2. Indicate whether each of the following pairs of sugars consists of anomers, epimers, or an aldose-ketose pair :

   (a) D-glyceraldehyde and dihydroxyacetone

(*b*) D-glucose and D-mannose
(*c*) D-glucose and D-fructose
(*d*) α-D-glucose and β-D-glucose
(*e*) D-ribose and D-ribulose
(*f*) D-galactose and D-glucose

3. Does the oxygen atom attached to C-1 in methyl α-D-glucopyranoside come from glucose or methanol ?

4. Fructose in its β-D-pyranose form accounts for the powerful sweetness of honey. The β-D-furanose form, although sweet, is not as sweet as the pyranose form. The furanose form is the more stable form. Draw the two forms and explain why it may not always be wise to cook with honey.

5. (*a*) Compare the number of reducing ends to nonreducing ends in a molecule of glcogen.
(*b*) Glycogen is an important fuel storage form that is rapidly mobilized. At which end—the reducing or nonreducing—would you expect most metabolism to take place ?

6. Draw Haworth projections for the following :
(*a*) in α-furanose form. Name the sugar.

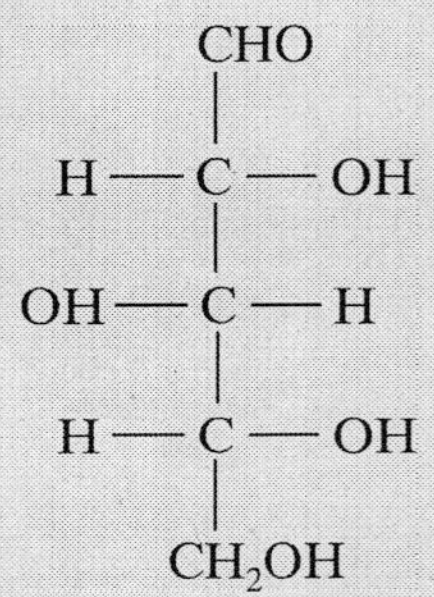

(*b*) The L isomer of (*a*)
(*c*) α-D-GlcNAc
(*d*) α-D-fructofuranose

7. α-D-galactopyranose rotates the plane of polarized light, but the product of its reduction with sodium borohydride (galactitol) does not. Explain the difference.

ANSWER TO PROBLEMS

1. Carbohydrates were originally regarded as *hydrates* of *carbon* because the empirical formula of many of them is $(CH_2O)_n$.

2. (*a*) aldose-ketose; (*b*) epimers; (*c*) aldose-ketose; (*d*) anomers; (*e*) aldose-ketose; (*f*) epimers.

3. From methanol.

4. Heating converts the very sweet pyranose form to the more stable but less sweet furanose form. Consequently, it is difficult to accurately control the sweetness of the preparation, which also accounts for why honey loses sweetness with time.

5. (***a***) Each glycogen molecule has one reducing end, whereas the number of nonreducing ends is determined by the number of branches, or α-1,6 linkages.

(***b***) Because the number of nonreducing ends greatly exceeds the number of reducing ends in a collection of glycogen molecules, all of the degradation and synthesis of glycogen takes place at the nonreducing ends, thus maximizing the rate of degradation and synthesis.

6. (*a*)

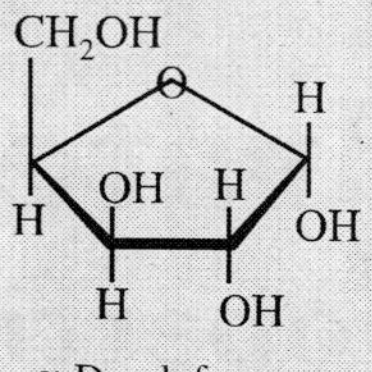

α-D-xylofuranose

(*b*)

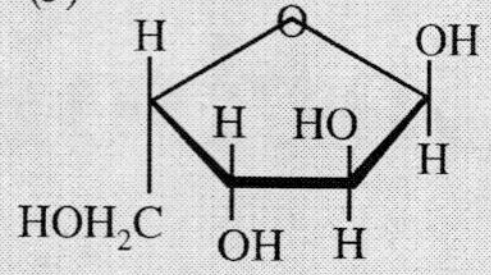

(*c*)

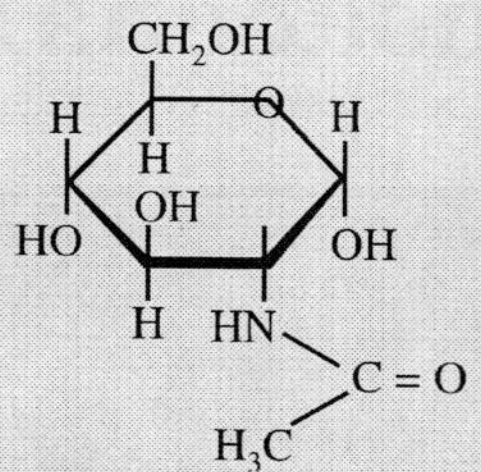

(*d*)

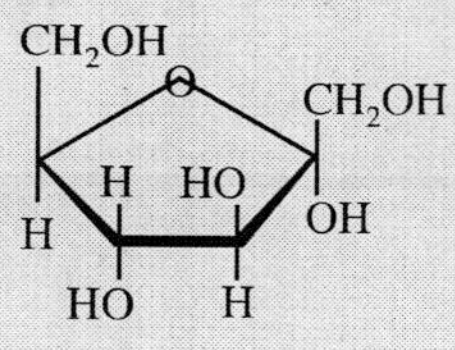

7. Galactitol has the following structure:

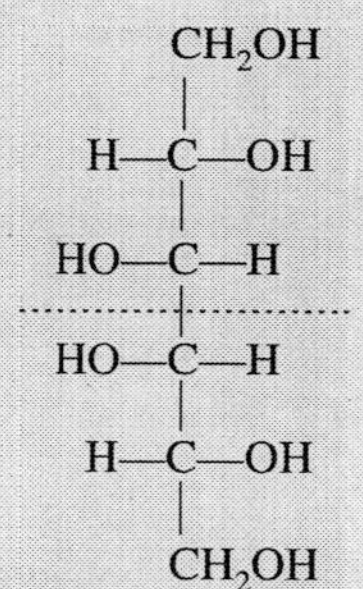

Since it has a plane of symmetry between C-3 and C-4, it is optically inactive. Such compounds, which contain asymmetric carbons but have no net optical activity, are called *meso* forms.

CHAPTER

6 Carbohydrates-II
Properties of Monosaccharides

INSIDE

- Reaction of Glycosidic OH Group
- Reaction Involving Both Glycosidic and Alcoholic OH Groups
- Reaction of Alcoholic OH Group
- Reaction Involving Both Alcoholic OH and CHO/CO Groups
- Reactions of CHO/CO Group

Carbohydrates in food as important sources of energy Starch, found in plant-derived food such as pasta, consists of chains of linked glucose molecules. These chains are broken down into individual glucose molecules for eventual use in generation of ATP and building blocks for other molecules.

Carbohydrates possess active groups which are responsible for their chemical behaviour. These groups are glycosidic OH, alcoholic OH and CHO (or CO). Most of the reactions of mutarotating sugars take place in aqueous solution so that all the 3 forms become available for the reaction.

$$\alpha\text{-form} \rightleftharpoons \text{aldehydo/keto form} \rightleftharpoons \beta\text{-form}$$

The above equilibrium is shifted in favour of the group which reacts. Some reagents, however, react with more than one group.

REACTION OF GLYCOSIDIC OH GROUP

Reaction with alcohol. The glycosidic OH group of mutarotating sugars react with alcohols to form α- and β-**glycosides** or **acetals**. Thus, glucose forms glucosides and fructose forms fructosides.

The glucosides, or glycosides in general, do not exhibit mutarotation as the aldehyde group in them is converted to the acetal group. A number of glycosides occur in nature, *e.g.*, *phlorhizin* (glucose + phloretin) in rose bark ; *digitonin* (4 galactose + xylose + digitogenin) in foxglove leaves; *amygdalin* (2 glucose + 2 mandelonitril) from bitter almonds and *saponin* (sugar + sapogenin) from soapwort. They are useful as medicaments.

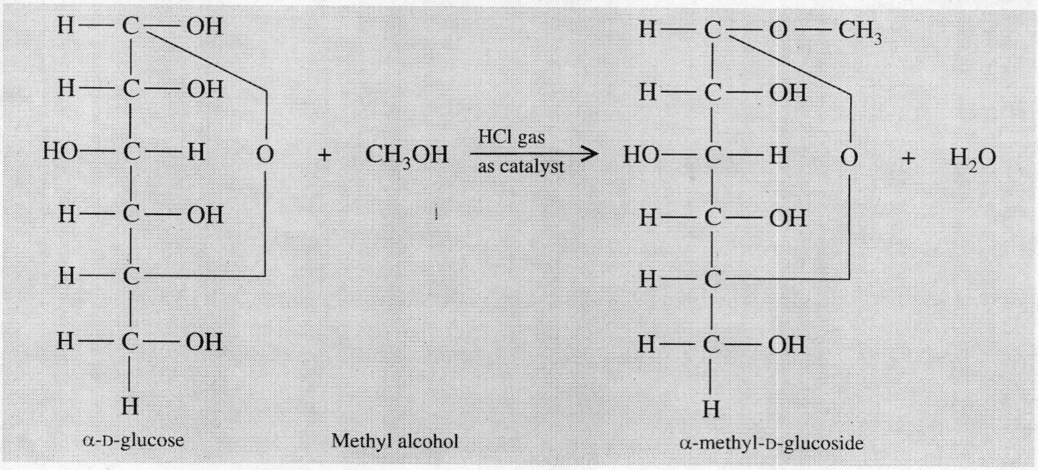

REACTION INVOLVING BOTH GLYCOSIDIC AND ALCOHOLIC OH GROUPS

Reaction with acetic anhydride (Esterification). The glycosidic and alcoholic OH groups of monosaccharides and disaccharides react with acetylating agents (like acetic anhydride in pyridine) to form acetate derivatives called **esters.**

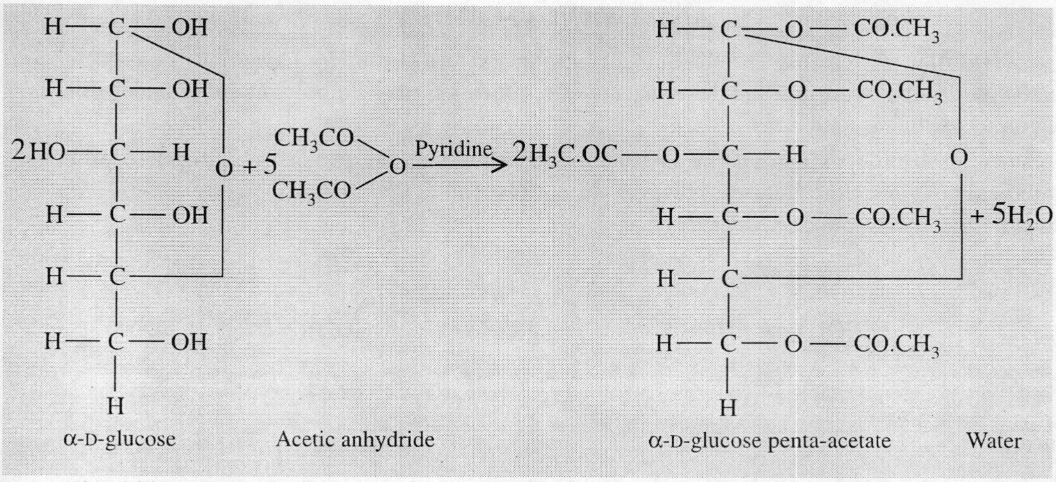

The ability of sugars to form esters indicates the presence of alcohol groups in their molecule. As glucose upon acetylation yields a penta-acetate derivative, it obviously contains 5 OH groups.

REACTION OF ALCOHOLIC OH GROUP

Reaction with methyl iodide (Etherification). The alcoholic OH groups of monosaccharides and disaccharides are converted to ether groups upon treatment with methylating agents.

[The glycosidic OH group of the sugars is, however, first protected with an R group by converting these sugars to glycosides which are, then, allowed to react with the methylating agent.

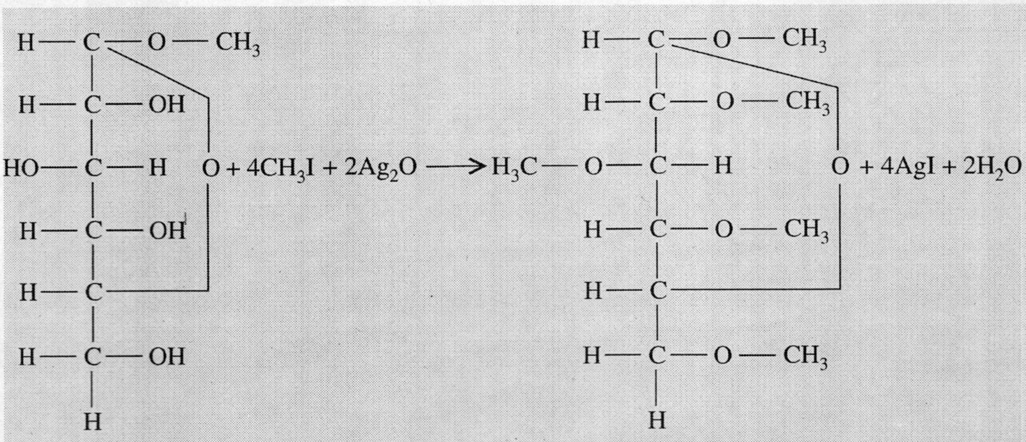

α-methyl glucose — Methyl iodide — Silver oxide — Tetramethyl ether of α-methyl glucoside — Silver iodide — Water

This reaction has been extensively used to ascertain which OH groups in a sugar are free and available for reaction. Thus, α-methyl glucoside is methylated at carbon 2, 3, 4 and 6 but not at carbon 5 which is involved in hemiacetal link.

REACTION INVOLVING BOTH ALCOHOLIC OH AND CHO/CO GROUPS

Oxidation with acids. The alcoholic OH group and CHO group (or CO group) are oxidized to carboxyl groups by certain oxidizing agents. The oxidation may be brought about under mild or vigorous oxidizing conditions.

(*a*) With mild oxidants (like HOBr) – Only the aldehyde group is oxidized to produce **monocarboxylic acids.** Ketoses, however, do not respond to this reaction. Hence, *this reaction is used to distinguish aldoses from ketoses.*

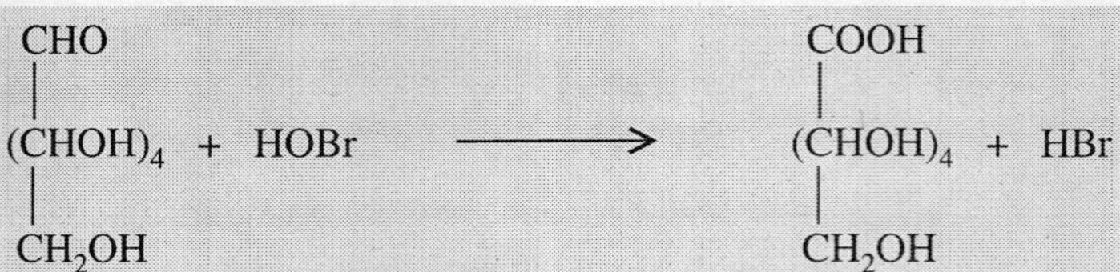

Glucose — Gluconic acid (= Aldonic acid)

(*b*) With strong oxidants (like conc. HNO_3) – Both the aldehyde group (or ketone group) and the primary alcohol group are oxidized to yield **dicarboxylic acids.** With aldoses, acids with same number of carbon atoms are obtained whereas ketoses react to produce acids with fewer number of carbon atoms (reaction on the next page).

REACTIONS OF CHO/CO GROUP

1. Oxidation with metal hydroxides. Metal hydroxides like $Cu(OH)_2$, AgOH and Bi $(OH)_3$ oxidize the free aldehyde (or ketone) group of mutarotating mono- and di-saccharides and are, at the same time, themselves reduced to the lower oxides or to the free metals. Thus, $Cu(OH)_2$ is reduced to Cu_2O and AgOH and $Bi(OH)_3$ are reduced to the free metal, Ag and Bi. The reaction with cupric hydroxide is given below :

$$\underset{}{\text{Reducing sugar}} + \underset{\text{Blue}}{2\,Cu^{++}} \longrightarrow \text{Oxidized sugar} + 2\,Cu^{+}$$

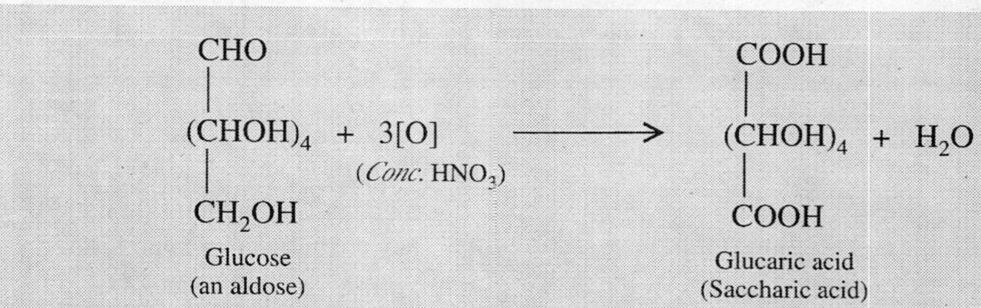

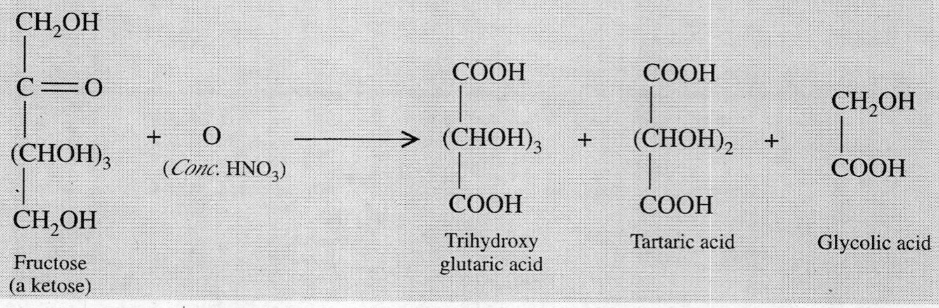

$$2\,Cu^{+} + 2\,OH^{-} \longrightarrow \underset{\text{Yellow}}{2\,Cu.OH} \longrightarrow \underset{\text{Red}}{Cu_2O} + H_2O$$

Cupric ion (Cu^{++}) is, in fact, the most common oxidizing agent of these. It is the active ingredient in Fehling's, Benedict's and Barfoed's reagents.

Fehling's reagent consists of 2 solutions : *solution A* contains 7% $Cu.SO_4$ and *solution B* contains 25% KOH and 35% sodium potassium tartarate. When the two solutions are mixed in equal amounts, a clear blue solution results for the tartarate forms a soluble complex with the copper hydroxide produced. This solution is widely used as an oxidizing agent preferably in quantitative sugar determinations.

With Fehling's reagent, the overall reaction occurs as follows :

$$\underset{\text{Glucose}}{C_6H_{12}O_6} + \underset{\text{Fehling's solution}}{2\,Cu(OH)_2} \longrightarrow \underset{\text{Gluconic acid}}{C_6H_{12}O_7} + \underset{\text{Cuprous oxide}}{Cu_2O} + 2H_2O$$

In clinical laboratories, however, a modification of Fehling's reagent is used for the qualitative determination of glucose in urine. This is known as **Benedict's reagent** and consists of a single solution containing $CuSO_4$, $Na_2.CO_3$ (in place of KOH) and sodium citrate (in place of sodium potassium tartarate). Sodium carbonate reduces alkalinity to such an extent that, unlike the Fehling's reagent, the two solutions ($Cu.SO_4$ and sodium citrate) here are kept mixed indefinitely as one solution.

Barfoed's reagent utilizes a weakly acid solution for oxidation. It contains 7% cupric acetate and 1% acetic acid. This reagent can be used to distinguish monosaccharides from disaccharides as the former give a red cuprous oxide precipitate in dilute acid more quickly than the latter. *Monosaccharides are, thus, more active reducing agents than the disaccharides.*

Carbohydrates with free carbonyl groups or in hemiacetal form give positive tests to these reagents without having been hydrolyzed first and are referred to as **'reducing'** sugars ; others (*i.e.*, the acetal types) are then **'nonreducing'** sugars (Table 6–1).

Table 6–1. Differences between reducing and nonreducing sugars

Reducing sugar	*Nonreducing sugar*
1. Carbohydrates with a free aldehyde (at C-1) or a free ketone (at C-2) group.	1. Aldehyde or ketone group is not free but instead utilized in bond formation.
2. They are in hemiacetal or hemiketal form.	2. They are in acetal or ketal form.
3. Do exhibit mutarotation.	3. Do not exhiblit mutarotation.
4. Do form osazones with phenyl hydrazine.	4. Do not form osazones.
5. Do form oximes with hydroxylamine.	5. Do not form oximes.
Examples – Glucose, Fructose, Lactose, Maltose, Cellobiose	*Examples* – Sucrose, Glycogen, Inulin

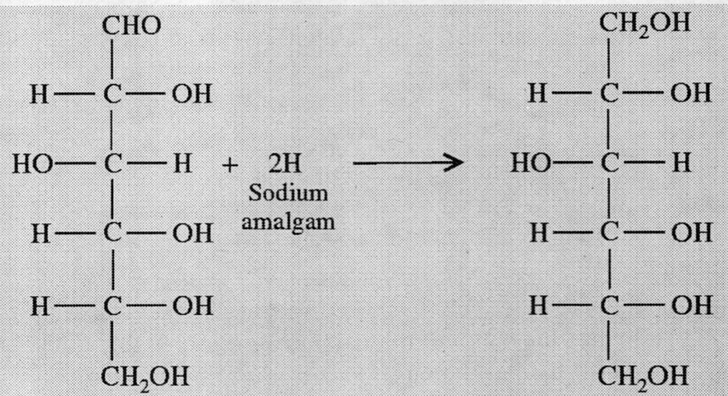

Glucose → Sorbitol (= Glucitol)

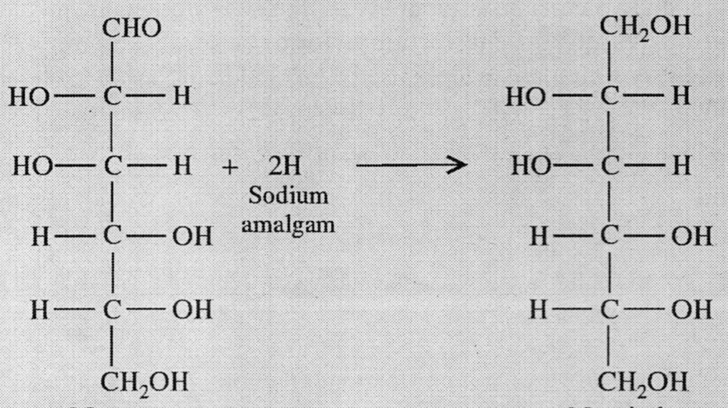

Mannose → Mannitol

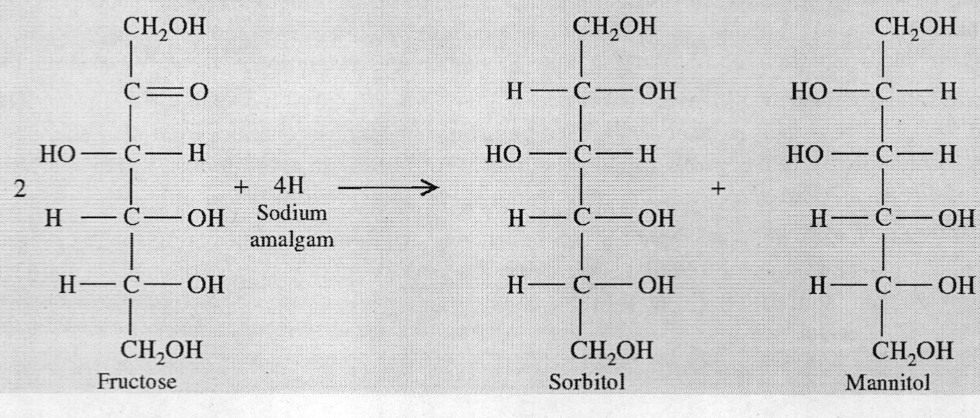

Fructose → Sorbitol + Mannitol

The **Molisch test** employs sulfuric acid for dehydration and α-naphthol as the phenol. The acid also hydrolyzes acetal groups present and thus makes this test probably the most general of all the tests for carbohydrates.

Since the alkali present in some of the above tests causes fragmentation of the sugar molecule, a mixture of hydroxy acids of different chain length is obtained upon oxidation of carbohydrates. Hence, these tests, although widely employed in analytical methods for sugars, are of little value in preparative chemistry.

2. **Reduction.** The sugars may be reduced in various ways depending upon the type of reducing agent used.

(*a*) With sodium amalgam – The monosaccharides are reduced to their corresponding **alcohols** by treating them with reducing agents like Na-amalgam. Thus, glucose yields sorbitol (= glucitol), mannose yields mannitol, galactose yields dulcitol, fructose yields a mixture of sorbitol and mannitol, and glyceraldehyde yields glycerol.

Sorbitol and mannitol both are important and are used in the manufacture of surface-active agents and explosives, respectively.

(*b*) With strong mineral acids – With *hexoses,* the reduction leads to the formation of 5-hydroxymethylfurfural, which on further heating is transformed to levulinic acid.

This reaction forms the basis of some colour tests (*e.g.*, Molisch test and Bial's orcinol test) for sugars as the aldehyde products of these reactions condense with certain organic phenols to give characteristic coloured products.

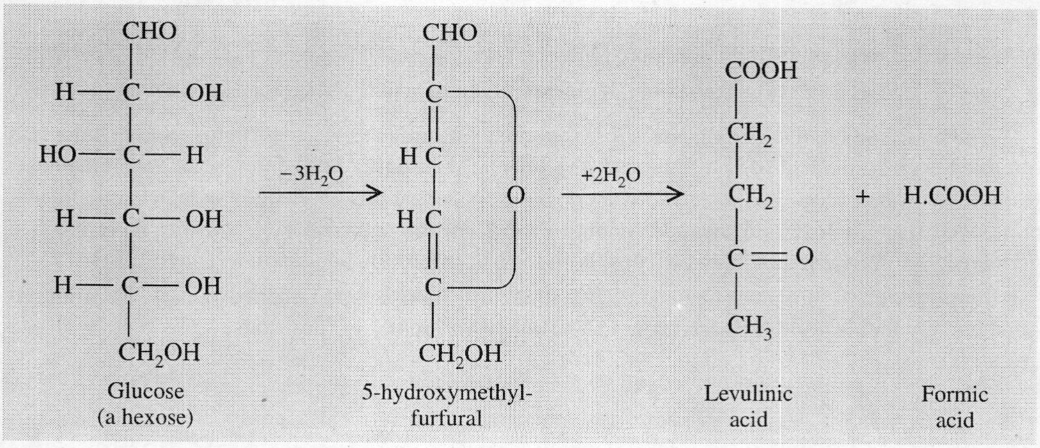

With *pentoses*, a product **furfural** is obtained on heating.

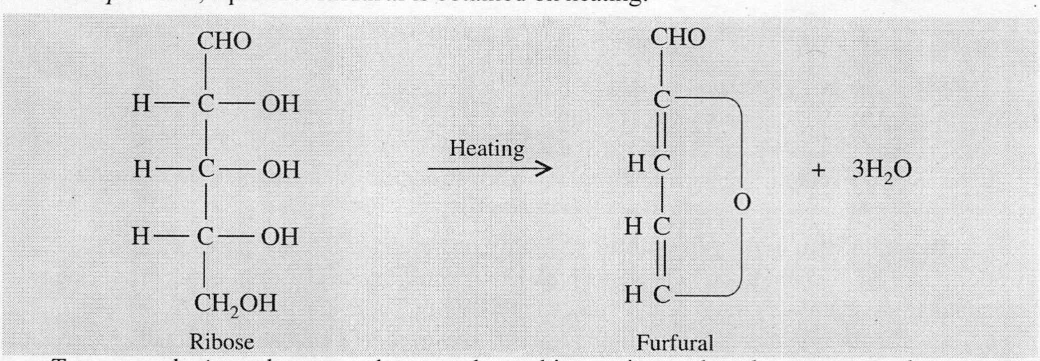

Tetroses and *trioses*, however, do not undergo this reaction as they do not possess the necessary minimum 5 carbon atoms for furfural formation.

(*c*) With dilute alkalies – Glucose, fructose and mannose are interconvertible in weak alkaline solutions such as $Ca(OH)_2$ and $Ba(OH)_2$ at low temperatures. This reaction is known as **Lobry de**

Bruyn–Alberda van Ekenstein transformation. This reaction, proposed in 1890, is of special significance as a similar reaction is supposed to take place in the human body. The mechanism of this reaction involves **enolization**, the migration of a proton from a carbon atom onto the oxygen of an adjacent carbonyl group, resulting in the formation of an unsaturated alcohol called **enol** (*ene* = unsaturation ; *ol* = alcohol). The interconversion between glucose, fructose and mannose may, in fact, be visualized as occurring through a common enediol form (Fig. 6–1).

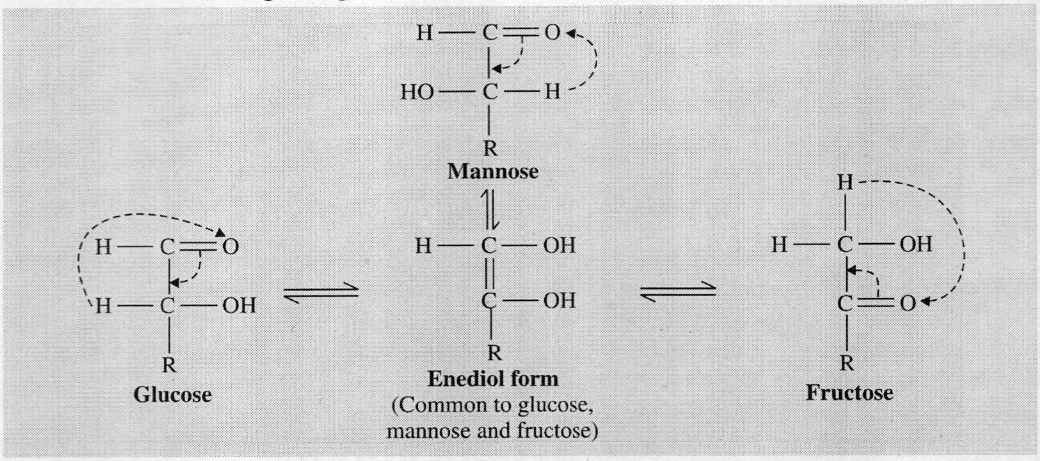

Fig. 6–1. Interconversion of glucose, mannose and fructose (Enolization)

The above mechanism emphasizes that the same mixture (of glucose, fructose and mannose) is obtained even if, instead of D(+)-glucose, the starting material is D(–)-fructose or D(+)-mannose. Using deuterium oxide, **Topper** et al, in 1951, have supported this enolization mechanism but have concluded that fructose and mannose cannot arise from the same enediol and that there are two geometric isomeric enediol intermediates, both forms capable of changing into fructose.

If the mixture is, however, heated to 37°C, the acidity increases and a series of enols are produced in which the double bonds migrate from O—C link to various C—C positions.

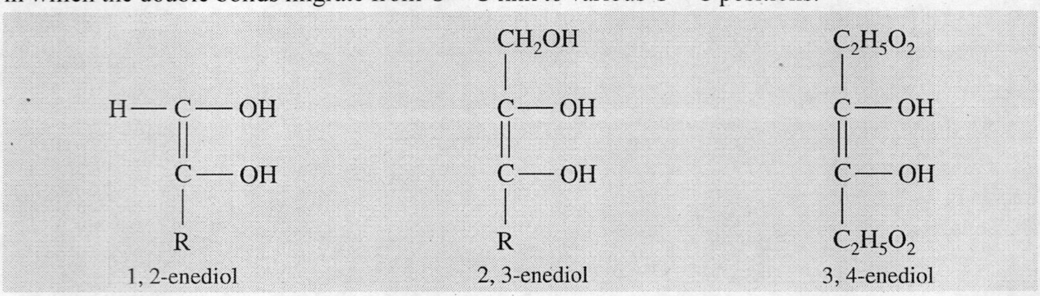

In concentrated alkali, sugar caramelizes and many decomposition products are formed. Certain coloured pigments develop and salts may also be formed.

3. **Reaction with hydrogen cyanide (Kiliani synthesis).** As already discussed (see page 72), the addition of HCN to mutarotating sugars creates a new asymmetric carbon atom, producing two new compounds namely, cyanohydrin A and cyanohydrin B. These are, then hydrolyzed with dilute mineral acids to form hydroxy acids which are later converted to γ-lactones and finally reduced to two new sugars containing one more carbon atom than the parent sugar.

4. **Reaction with alanine.** The aldehyde group of carbohydrates condenses with the amino group of alanine to form Schiff's base.

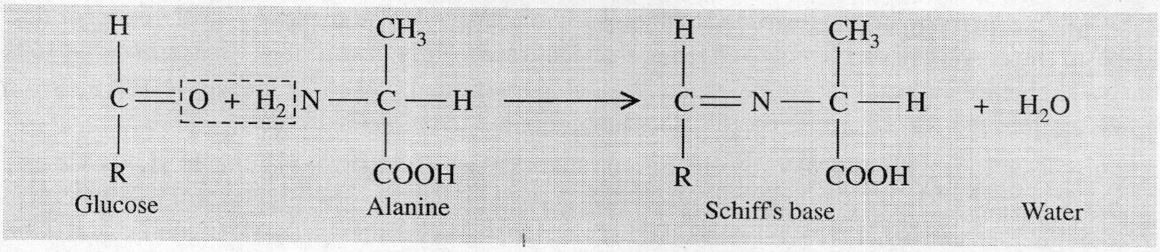

The browning reaction (Maillard's reaction) occurring during the baking of bread and other mixtures of carbohydrates and proteins is believed to be due to the formation of Schiff's base between the amino groups of proteins and the aldehyde groups of carbohydrates (**Mertz**, 1960).

5. Reaction with phenyl hydrazine. Reaction with phenyl hydrazine involves only 2 carbon atoms viz., the carbonyl (i.e., the aldehyde or ketone group) and the adjacent one. One mole of phenyl hydrazine reacts with one mole of aldose (or ketose) to form a mole of hydrazone. With a second mole of phenyl hydrazine, the hydrazone is oxidized to **aldohydrazone** and the phenyl hydrazine itself is reduced to aniline and ammonia. Finally, a third mole of phenyl hydrazine reacts with the aldohydrazone to produce **osazone.** The hydrazone may, in fact, be regarded as a special type of Schiff's base and the osazone as a double Schiff's base (Fig. 6–2).

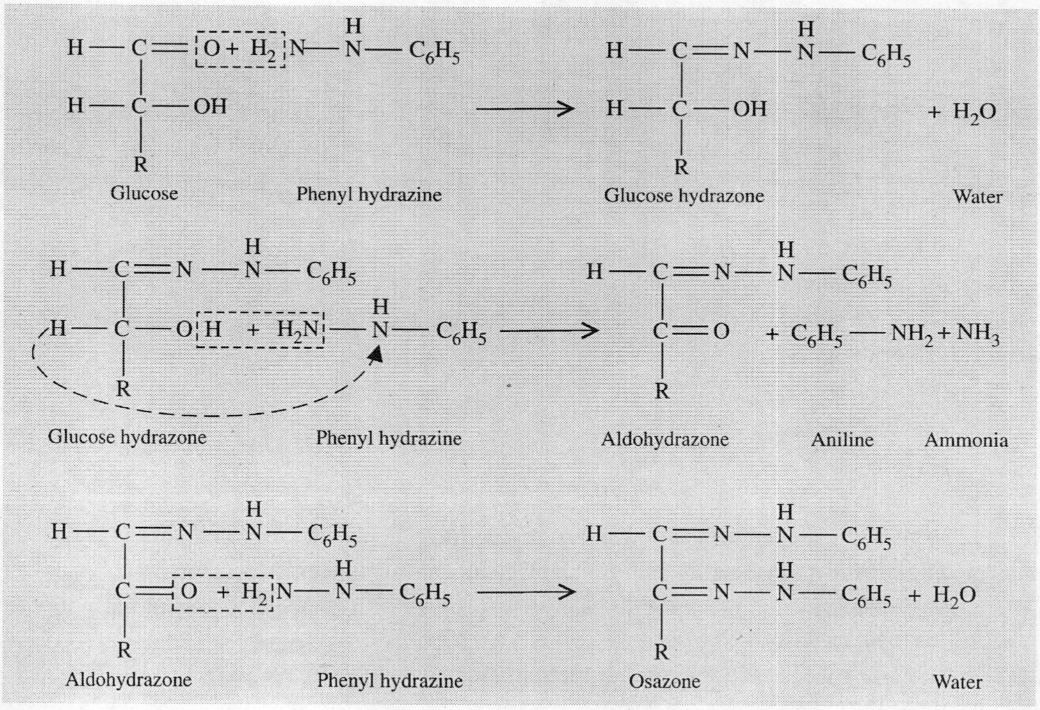

Fig. 6–2. Osazone formation

The reaction with ketose will also take place in a likewise manner.

The osazones of reducing (mutarotating) sugars are crystalline, yellow and usually insoluble compounds and hence may also be recovered. These have characteristic crystal structure and melting points and are, therefore, frequently used in the identification of sugars. It may, however, be noted that glucose, fructose, and mannose would yield the same osazone owing to their similar structure in the unaffected part of the molecule (from C_3 to C_6). Moreover, the asymmetry at C_2 is destroyed during the reaction. Obviously, galactose would, then, form a different osazone as it has a different configuration below C_2.

Osazone formation is not restricted to monosaccharides but is also shown by most disaccharides. Lactosazone crystals have a typical hedgehog shape and are readily identifiable. Maltosazone crystals have a characteristic petal-like appearance and a cluster of them looks like the sunflower. Sucrose, however, does not form osazone since it does not contain a functional carbonyl group.

6. **Reaction with hydroxylamine.** Simple sugars react with hydroxylamine to yield **oximes.**

$$\begin{array}{lcl} CH\fbox{O + H_2}N{-}OH & & CH{=}N{-}OH \\ \quad | & & \quad | \\ (CHOH)_4 & \longrightarrow & (CHOH)_4 \quad + \quad H_2O \\ \quad | & & \quad | \\ CH_2OH & & CH_2OH \end{array}$$

Glucose Hydroxylamine Aldoxime Water

7. **Fermentation.** Monosaccharides such as glucose, fructose and mannose are readily fermented by yeast. The process of yeast fermentation is very complex. During this process, sugar phosphate and sugar acid phosphate are formed. Ordinarily, this process results in the formation of alcohol and carbon dioxide.

$$\underset{\text{Monosaccharide}}{C_6H_{12}O_6} \xrightarrow[\text{(yeast)}]{\text{Fermentation}} \underset{\text{Alcohol}}{2C_2H_5OH} + 2CO_2$$

PROBLEMS

1. Sucrose is commonly used to preserve fruits. Why is glucose not well suited for preserving fruits ?
2. Which of the following carbohydrates is most rapidly fermented by brewer's yeast (*Saccharomyces cerevisiae*) :
   (*a*) glycogen
   (*b*) maltose
   (*c*) lactose
   (*d*) glucose
   (*e*) starch
3. Which is the most general of all the tests for carbohydrates ?
4. Why is glucose not well suited for preserving fruits ?
5. Explain the mechanism of enolization by chemical equation only.
6. Why the reduction of cupric to cuprous is profuse with Benedict's reagent but not with Barfoed's reagent ?
7. How can glucose and fructose be differentiated by their reactions ?
8. What are glycosides ?

ANSWERS TO PROBLEMS

1. Glucose is reactive because its open-chain form contains an aldehyde group.
2. (*d*) glucose
3. Molisch test
4. Glucose in its open-chain form contains an aldehyde group which is reactive and is, hence, unsuitable for preservation.

5.

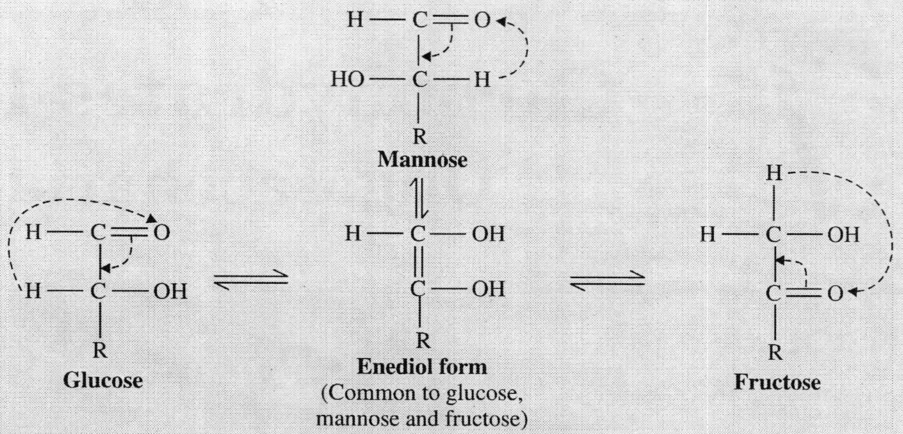

6. Benedict's reagent contains sodium carbonate (Na_2CO_3) which with its alkalinity produces enediol form. This is not possible with Barfoed's reagent which is a weakly acid solution of glacial acetic acid (CH_3COOH). As enediol is formed with Benedict's reagent, reduction is profuse.

7. Glucose and fructose can be differentiated by the following 2 tests :
   1. Seliwanoff's test. When fructose is boiled with Seliwanoff's reagent (resorcinol in HCl), a cherry-red colour develops. However, glucose does not respond to this test.
   2. Foulger's test. On boiling with Foulger's reagent (urea in H_2SO_4 with stannous chloride), fructose produces a blue colour while glucose gives a green colour.

8. Glycosides are plant products though they can be synthesized in the laboratory. In a glycoside, a sugar molecule is linked to a nonsugar with an oxygen bridge. The nonsugar is called aglycone. Some of the glycosides have profound biochemical effects. For example, cardiac glycosides like the derivative of *Digitalis* have effect on the heart. **Streptomycin**, the well-known antibiotic, **digitonin** which is used to precipitate cholesterol, and **quabain**, an inhibitor of sodium potassium ATPase, are all glycosides.

CHAPTER

7 Carbohydrates-III
Oligosaccharides

INSIDE

- Conformations of Pyranose and Furanose Rings
- Sucrose or Saccharose
- The Artificial or Synthetic Sweeteners
- Lactose
- Maltose
- Cellobiose
- Isomaltose
- Trehalose

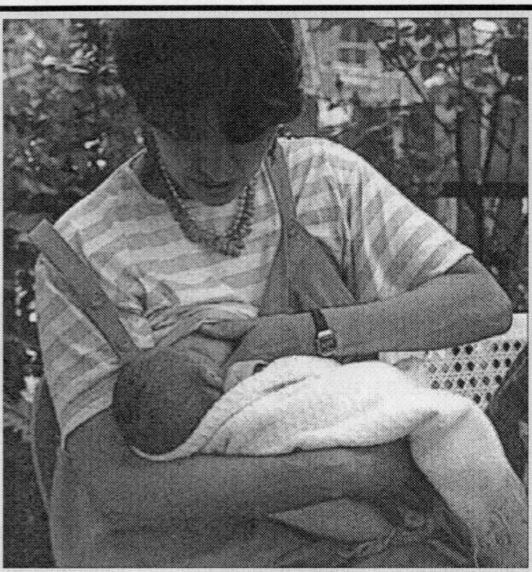

Lactose in milk

Milk contains all of the food chemicals that a new baby needs. Lactose is the main carbohydrate present in milk.

The oligosaccharides yield 2 to 10 monosaccharide molecules on hydrolysis. *Disaccharides are the most common oligosaccharides found in nature*. Most of the naturally-occurring representatives occur in plant rather than in animal sources. The three most important disaccharides found free in appreciable quantities are sucrose, lactose and maltose ; a few others are cellobiose and trehalose.

CONFORMATIONS OF PYRANOSE AND FURANOSE RINGS

To name disaccharides, certain rules are followed. First, the compound is written with its nonreducing end to the left. An *O* precedes the name of the first (left) monosaccharide unit, as a reminder that the sugar-sugar linkage is through an oxygen atom. The configuration at the anomeric carbon joining the first (left) monosaccharide unit to the second is then given (α or β). To distinguish 5- and 6-membered ring structures, "furanosyl" or "pyranosyl" is inserted in the name of each monosaccharide unit. The two carbon atoms joined by the glycosidic bond are then shown in parentheses, with an arrow connecting the two numbers; for example, $(1 \rightarrow 4)$ shows that C-1 of first sugar residue is joined to C-4 of the second. If there is a third residue, the second glycosidic bond is described next, by the same conventions. To shorten the description of a complex **polysaccharide**, 3-letter abbreviations for each monosaccharide are often used.

Following this convention for naming oligosaccharides, sucrose is therefore named *O*–α–D–glucopyranosyl–(1→2)–β–D–fructofuranoside. Because most sugars are the D-enantiomers, and the pyranose form predominates, a shorthand version of the formal name of such compounds is used, giving the configuration of the anomeric carbon and naming the carbons joined by the glycosidic bond. In this abbreviated nomenclature, sucrose is Glc (α1→2) Fru.

Based on the type of linkage present in their molecule, the disaccharides may by classified as follows :

Disaccharides

- Nonreducing (= Glycosylglycosides)
  - $C_1 \rightarrow C_1$ *glycosidic linkage*: Trehalose
  - $C_1 \rightarrow C_2$ *glycosidic linkage*: Sucrose
- Reducing (= Glycosylaldoses or glycosylketoses)
  - $C_1 \rightarrow C_4$ *glycosidic linkage*: Lactose, Maltose, Cellobiose
  - $C_1 \rightarrow C_6$ *glycosidic linkage*: Gentiobiose, Melibiose, Isomaltose

SUCROSE OR SACCHAROSE
(Table sugar, Cane sugar, Beet sugar)

Occurrence. Sucrose is the common sugar of commerce and kitchen (hence, also called `household' sugar) and is widely distributed in all photosynthetic plants. It is the chief constituent of sugarcane (*Saccharum officinarum*), beet (*Beta vulgaris*) and maple (*Acer saccharina*) and is also present in pineapple (*Ananas sativus*) and carrot (*Daucas carota*). Sucrose occurs in varying amounts in different plant organs such as fruits, seeds, flowers and roots. Nectar of flowers is particularly rich in sucrose. It is the raw material for honey. *Sucrose is probably the only foodstuff used in the crystalline form.*

Sucrose is something of a riddle in plant biochemistry. Although D-glucose is the major building block of both starch and cellulose, sucrose is a major intermediate product of photosynthesis. It is the predominant form in which sugar is transported in most plants from the leaves to other organs of plants *via* their vascular system.

Chemistry. It is derived commercially from either sugarcane or beet. Refinement removes the yellow-brown pigments of unrefined sugar to produce the white crystal form of table sugar, while leaving a syrup which does not crystallize. This syrup, known as **molasses,** is also a commercial product and is the least refined form of sucrose. In plants, sucrose is formed by elimination of a molecule of water from glycosidic hydroxyl groups (marked by an asterisk) or α-D-glucose and β-D-fructose (Fig. 7–1.)

It is noteworthy that fructose possesses the furanose ring structure in the sucrose molecule, although the pyranose ring is the dominant and more stable form in the free ketohexose. The linkage between the two disaccharide moieties (= units) of sucrose is a glycosidic one between C_1 of glucose and C_2 of fructose.

Properties. Sucrose is a white crystalline solid, soluble in water and with a melting point 180°C. When heated above its melting point, it forms a brown substance known as **caramel**. Concentrated sulfuric acid chars sucrose, the product being almost pure carbon. It is dextrorotatory and has a specific rotation of + 66.7°. It is by far the sweetest of the 3 common disaccharides (sucrose, lactose, maltose). It is also sweeter than glucose (refer Table 7–1). It crystallizes in colourless crystals.

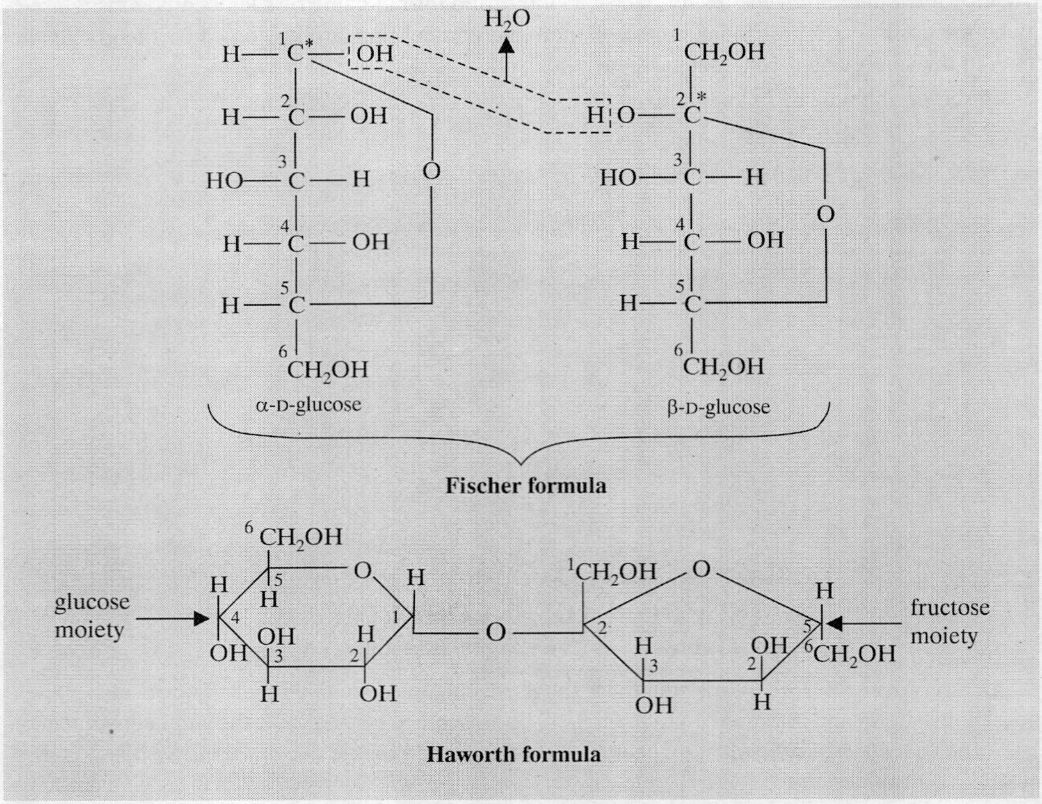

Fig. 7–1. Structure of sucrose

O-α-D-glucopyranosyl-(1 → 2)- β- D-fructofuranoside or Glc (α1 → 2) Fru

It can be seen from the formula of sucrose that both the carbonyl groups (marked by an asterisk) are involved in the formation of glycosidic bond. Consequently, sucrose contains no active group and, therefore, does not exhibit those properties which depend upon the presence of this group. Obviously, sucrose does not exhibit mutarotation and is *not* a reducing sugar. It also does not form an osazone or an oxime.

Hydrolysis. Upon hydrolysis, sucrose yields equimolar mixture of glucose and fructose which is often called **invert sugar.** This mixture readily reduces Fehling's solution and other alike reagents. The name 'invert sugar' is given to this mixture because the levorotatory fructose, thus produced, changes (or inverts) the previous dextrorotatory action of the sucrose. In fact, sucrose (which is dextrorotatory with a specific rotation, $[\alpha]_D^{20}$ of + 66.7°), upon hydrolysis, gives a mixture of equimolar quantities of D(+) glucose (dextrorotatory ; $[\alpha]_D^{20}$ of + 52.7°) and D(–) fructose (levorotatory; $[\alpha]_D^{20}$ of – 92.0°). And as the levorotation of fructose is greater than the dextrorotation of glucose, the mixture so obtained is levorotatory, contrary to the initial dextrorotatory sucrose

Table 7–1. Sweetness index* of some sugars and noncaloric sweeteners

Sugar / Sweetener	*Relative sweetness*
Sugars	
Lactose	0.16
Raffinose	0.20
D-galactose	0.32
Maltose	0.32
D-xylose	0.40
Glycol	0.50
D-glucose	0.74
Sucrose	*1.00*
Invert sugar	1.23
D-fructose	1.73
Sweeteners	
Sucaryl sodium	30
Aspartame	180
Acesulfame-K	200
Stevia extract	300
Saccharin	400
Neohespiridine dihydrochalcone	1,000
Monellin	2,000

* Sweetness index is ascertained with respect to sucrose whose sweetness level is arbitrarily taken as one.

$$\underset{\substack{\textbf{Sucrose}\\ [\alpha]_D^{20} = +66.7^\circ\\ \text{dextrorotatory}}}{C_{12}H_{22}O_{11}} + \underset{\textbf{Water}}{H_2O} \xrightarrow{\text{Invertase}} \underset{\substack{\textbf{Glucose}\\ [\alpha]_D^{20} = +52.7^\circ\\ \text{dextrorotatory}}}{C_6H_{12}O_6} + \underset{\substack{\textbf{Fructose}\\ [\alpha]_D^{20} = -92.0^\circ\\ \text{levorotatory}}}{C_6H_{12}O_6}$$

Mixture of glucose and fructose is levorotatory with $[\alpha]_D^{20}$ value of (92.0 – 52.7)= – 39.3°.

This reaction which is called "inversion of sucrose" is catalyzed by the enzyme invertase and also by H^+ ions. Animals cannot absorb sucrose as such, but it is made available for absorption by the enzyme sucrase present in the intestinal mucosal cells. As already pointed out, this enzyme catalyzes the hydrolysis of sucrose to D-glucose and D-fructose, which are readily absorbed in the bloodstream.

> **Invertase** is also called *sucrase, saccharase* or *6-D-fructosidase* the first name being the most preferred one. This enzyme has been found to occur in green leaves, fruits, grains, stems, potato tubers, some roots, pollen and such lower plants as fungi and bacteria. It is especially abundant in yeast. Salts of heavy metals (Ag, Cu, Hg) inhibit its action. Maximum activity is obtained with low concentrations of sucrose (5 –10%). Sucrase also hydrolyzes the sugars gentianose, raffinose, stachyose and, to some extent, inulin.

It is interesting to note that the invert sugar is more sweeter than the sucrose itself owing to the presence of fructose in invert sugar which is sweetest of all the sugars. This also explains why honey which contains a large proportion of invert sugar is sweeter than sucrose. The invert sugar is used in candy both because of its sweetness and also because the monosaccharides do not crystallize.

THE ARTIFICIAL OR SYNTHETIC SWEETENERS

Sweeteners are customarily grouped into two categories, based on the difference in the amount of energy provided by them. These two groupings are :

A. Nutritive or **Caloric Sweeteners.** These contain calories, hence have nutritive value. These include sugar sweeteners (*e.g.*, refined sugars, high fructose corn syrup, crystalline fructose, glucose, dextrose, corn sweeteners, honey, lactose, maltose, invert sugars, concentrated fruit juice) and sugar alcohols.

B. Nonnutritive or **Noncaloric Sweeteners.** These contain no calories, hence have no nutritive value. These have been developed artificially, hence also called as **artificial sweeteners.** As they sweeten with little volume, they may also be referred to as **high-intensity sweeteners.** They have been developed especially for obese or diabetics, for whom sugar consumption is harmful. Artificial sweeteners stimulate the taste buds of the tongue that are stimulated by sugars but, unlike sugars, they are of no food value and may have harmful effects, if used to excess. Some sweeteners are considered 'Generally Recognized As safe' (GRAS) ingredients and others are considered food additives. The safety limit of food additives or conditions of use is expressed as the acceptable daily intake (ADI), that is, the estimated amount per kilogram body weight that a person can safely consume every day over a lifetime without risk. ADI is a conservative level – it usually reflects an amount 100 times less than the maximum level at which no observed adverse effects occur in animals.

A description of some important artificial sweeteners (Fig. 7–2) follows :

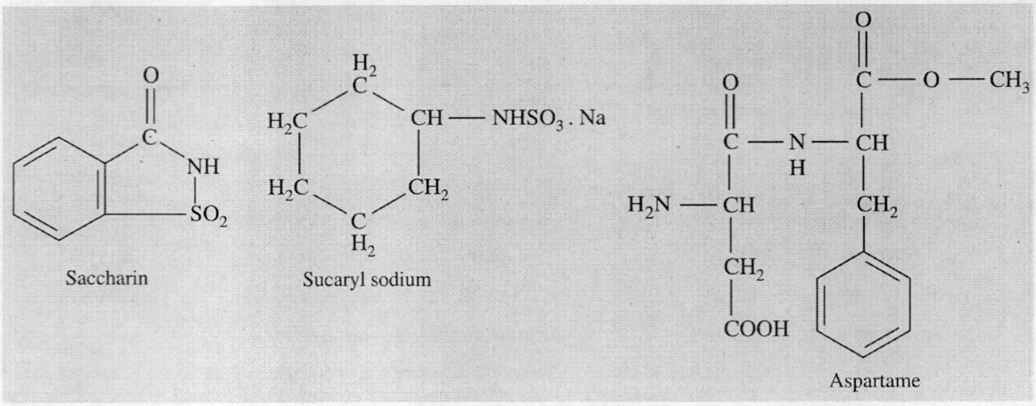

Fig. 7–2. Some artificial sweeteners

1. Saccharin. Saccharin is the first artificial sweetener which was discovered accidentally by **Constantine Fahlberg** in 1879. *It is the most-widely used artificial sweetener*. In fact, saccharin was not used extensively as a sweetener until the advent of World War I when the use of sugar was strictly rationed. When World War II hit, sugar rationing started again, leading to another significant rise in saccharin use. Saccharin is sold under the tradename Sweetex™. It has a molecular formula $C_6H_4SO_2$ CONH, and its chemical structure is different from that of a carbohydrate. It possesses no OH group but rather a sulfonamide. It is a noncaloric sweetener and is manufactured from toluene, $C_6H_5CH_3$. Saccharin is a white, crystalline, sparingly soluble solid with a m.p. 227 °C. It is heat-stable and is 400 times sweeter than sucrose by weight in aqueous solutions (Table 7-1). It is advised that pregnant women avoid saccharin.

While working on new food preservatives, **Fahlberg** accidentally spilled a compound he had synthesized. When he ate his dinner that night, he noticed that the intense residual sweetness of the chemical still lingered on his hands. He named the compound **saccharin** after the Latin ***saccharum*** which means sugar, referring sweetness in this context.

2. **Sucaryl sodium.** It is also a sulfonamide and is 30 times sweeter than sucrose.

3. **Monellin.** This protein sweetener is present in the sap of serendipity berries, fruit of a western African plant, *Dioscoreophyllum cumminsii*. This sweet protein is composed of two noncovalently-associated polypeptide chains (one having 42 amino acid residues and the other 50 residues) and is 2,000 times as sweet as sucrose. Monellin's sweetness requires the undissociated protein since neither protomer has a sweet taste. Presently, monellin is being studied as a nonfattening, nontoxic food sweetener for human use. In case of the African plant *Katemfe*, its intense sweetness is attributed to two proteins, which together are 1,600 times sweeter than sucrose.

4. **Aspartame.** This is an essentially noncaloric, sugar-free sweetener being commercially used (Technically, aspartame is nutritive as it contains 4 calories per gram, but since it is added in such tiny doses that it really makes little, if any, contribution to the total calorific intake). Aspartame is a synthesized dipeptide, containing aspartic acid (Asp) and the methyl ester of phenylalanine (Phe). Aspartame is marketed under the tradenames Nutrasweet™ (when included in food) and Equal™ (when sold as a powder). It is about 180 times sweeter than sucrose by weight (*i.e.*, about half as sweet than saccharin) and *does not have the bitter aftertaste, often associated with saccharin.* It, thus, has a better taste profile than saccharin. Interestingly, (S–S)-aspartame is a sweetener whereas its enantiomer, (S–R)-aspartame is bitter in taste. However, aspartame has a shelf life of about 6 months, after which it breaks down into its components and loses its sweetening power. Also, it splits at high temperatures (which means that it cannot be used in hot or baked items). Intestinal enzymes hydrolyze aspartame to phenylalanine, aspartic acid and methanol in the ratio 5 : 4 : 1.

It is now widely used in the drink and food industry, *e.g.*, soft drinks, cold cereals, beverages, gelatin desserts, puddings, toppings and fillings for baked goods, and cookies. Aspartame has an additional advantage as it does not promote tooth decay, hence also used in gum and candies. Some users have reported suffered from adverse reactions to aspartame such as headache, dizziness, seizure and nausea. Such persons must abstain from using it. Because aspartame is made from phenylalanine, people with phenylketonuria (who cannot metabolize Phe) should avoid aspartame. Also, the pregnant women are advised to refrain from aspartame. Aspartame is approved for use in more than 100 nations.

LACTOSE
(Milk sugar)

Occurrence. Lactose is solely of animal origin and is found in the milk of mammals. Human milk contains about 6.8 % (and 0.3% of higher oligosaccharides) ; cows' milk, about 4.8% (refer Table 7-2); hence, mother's milk is about 1.5 times sweeter than cow's milk. It is an unique product of mammary glands and is not found in other parts of the animal body. A minute quantity of lactose is also present in the milk as neuramin lactose. However, during pregnancy, it may be found in urine. Interestingly and curiously enough, there is no lactose in the milk of seals and other close relatives.

> **Kuhn** (1949) has, however, found **lactose** to be present in the pollen of *Forsythia* plant.

Table 7–2. Percentage composition of milk

Contents	*Human milk*	*Cow's milk*
Water	88.5	87.0
Fat	3.3	3.5
Lactose	6.8	4.8
Casein	0.9	2.7
Lactalbumin + other proteins	0.4	0.7
Ash (containing minerals)	0.2	0.7

Chemistry. Lactose is prepared commercially from whey by evaporation to crystallization. Whey is obtained as a by-product in the manufacture of cheese. Lactose may be formed by elimination of a molecule of water from the glycosidic OH group of β-D-galactose and the alcoholic OH group on carbon atom 4 of D-glucose (Fig. 7-3). As D-glucose exists in 3 forms (α, *aldehydo* and β), it is obvious to have 3 forms of lactose *viz*., α form, *aldehydo* form and β-form. The aldehydo form is present in traces only. The formula shown in Fig. 7–3 is that of α-lactose.

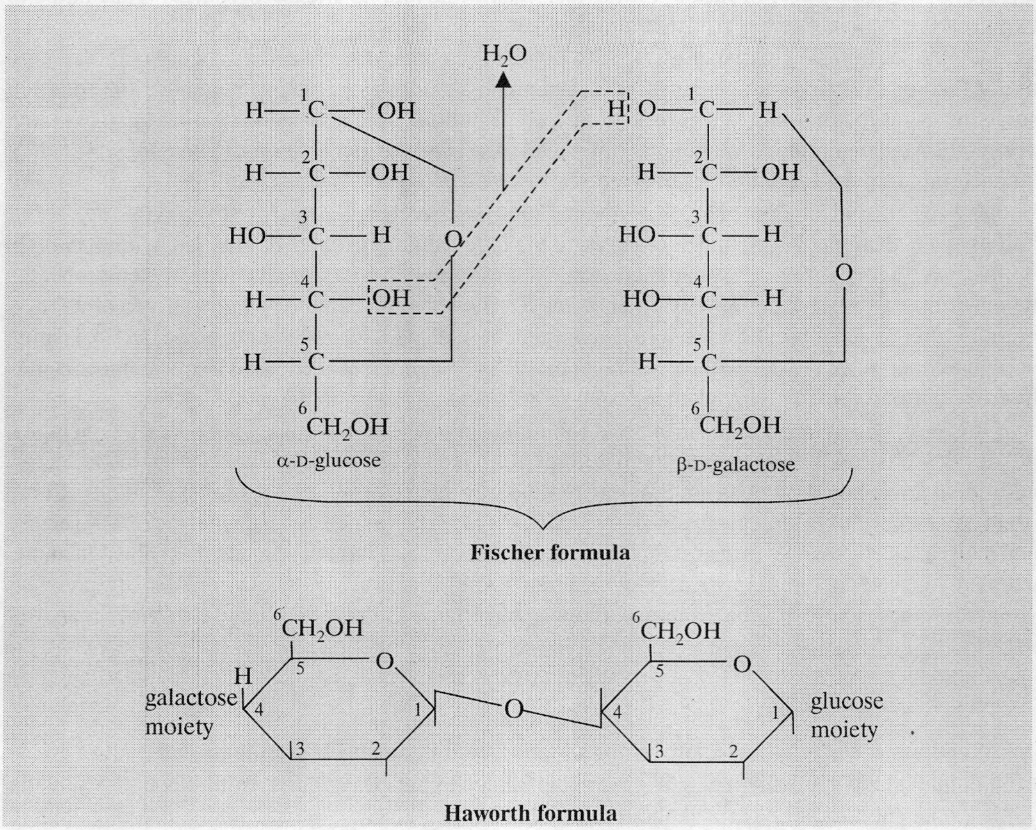

Fig. 7–3. Structure of α-lactose

Lactose, in solution, consists of an equilibrium mixture of all the 3 forms. Using the simplified scheme of representing the Haworth formula, the 3 forms may be shown as below :

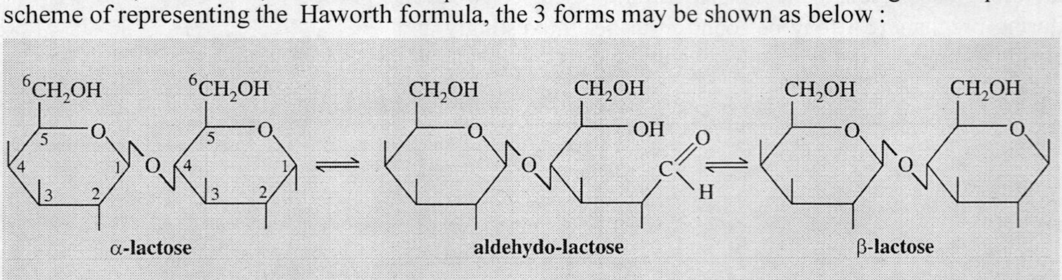

It may be emphasized that the α- and β-forms of lactose differ from each other only in hemiacetal configuration ; the linkage in both the forms being β-1, 4-galactoside.

Properties. Lactose is a white, crystalline solid with a melting point 203°C (with decomposition) and is also dextrorotatory. The α- and β-forms have a specific rotation of + 90° and +35° respectively. The equilibrium mixture has a specific rotation +52.5°. It is less soluble in water and much less sweet than sucrose. The α-diastereoisomer is less soluble than the β-form. Sometimes, the α-isomer crystallizes in ice-cream, making the product seem sandy in texture. The more soluble β isomer is used in diets for infants. Since lactose has a free carbonyl group on carbon atom 1 of glucose unit, it is a reducing sugar. Hence, it exhibits mutarotation, reduces Cu^{++} ions to Cu^{+} ions and forms osazone and oxime. It crystallizes in rhombic prisms with a mole of water.

Hydrolysis. Upon hydrolysis with the enzyme, *lactase*, it yields an equimolar mixture of glucose and galactose. It should be noted that lactase is a β-glycosidase, *i.e.*, splits β-glycosides (hence, it is identical with emulsin). Lactose itself cannot be absorbed from the intestine into the bloodstream unless it is first hydrolyzed into monosaccharide units. The enzyme lactase is actively secreted by the intestinal mucosal cells in suckling infants and therefore the lactose, in the form of milk suckled by them, is easily hydrolyzed to the component monosaccharide units, thus effecting absorption by the intestinal tract. Although milk is the universal food of newborn mammals and one of the most complete human foods, many adult humans cannot digest milk because they are deficient in intestinal lactase, the enzyme that hydrolyzes the milk sugar, lactose. Such individuals may, therefore, show **lactose intolerance,** which is genetically determined. It is characterized by abdominal bloating, cramps, flatulence, colic pains, abnormal intestinal flow, nausea and watery diarrhea. All these symptoms appear within 30 to 90 minutes after ingesting milk or its unfermented by-products (fermented dairy products such as yogurt and cheese create no intolerance problems). North Europeans and their descendants, which include the majority of North American whites, are most tolerant of milk. Many other ethnic groups are generally intolerant to milk, including the Japanese, Chinese, Jews in Israel, Eskimos, South American Indians and most African blacks. Only about 30% of North American blacks are tolerant; those who are tolerant are mostly descendants of slaves brought from east and central Africa, where dairying is traditional and tolerance to lactose is high. Lactase deficiency appears to be inherited as an autosomal recessive trait and is usually first expressed in adolescene or young adulthood. The prevalence of lactase deficiency in human populations varies greatly. For example, 3% of Danes are deficient in lactase, compared with 97% of Thais.

> **Lactose intolerance** should not be confused with the genetic galactosemia.

MALTOSE
(Malt sugar)

Occurrence. Maltose does not occur abundantly in nature. However, its occurrence has been occasionally reported. It is the major product of enzymic hydrolysis of starch. Sprouting cereal grains are rich in *amylases* which split the starch present to dextrins and maltose. Malt, prepared from sprouting barely, is an excellent source of maltose.

Chemistry. Maltose is the simplest of the disaccharides and is produced by the action of malt (which contains the enzyme *diastase*) on starch :

$$(C_6H_{10}O_5)_n \quad + \quad \frac{n}{2}\, H_2O \xrightarrow{\text{Diastase}} \frac{n}{2}\, C_{12}H_{22}O_{11}$$

Maltose may be considered as originating by splitting out a molecule of water from the glycosidic OH group of α-D-glucose and the alcoholic OH group on carbon atom 4 of D-glucose (Fig. 7–4).

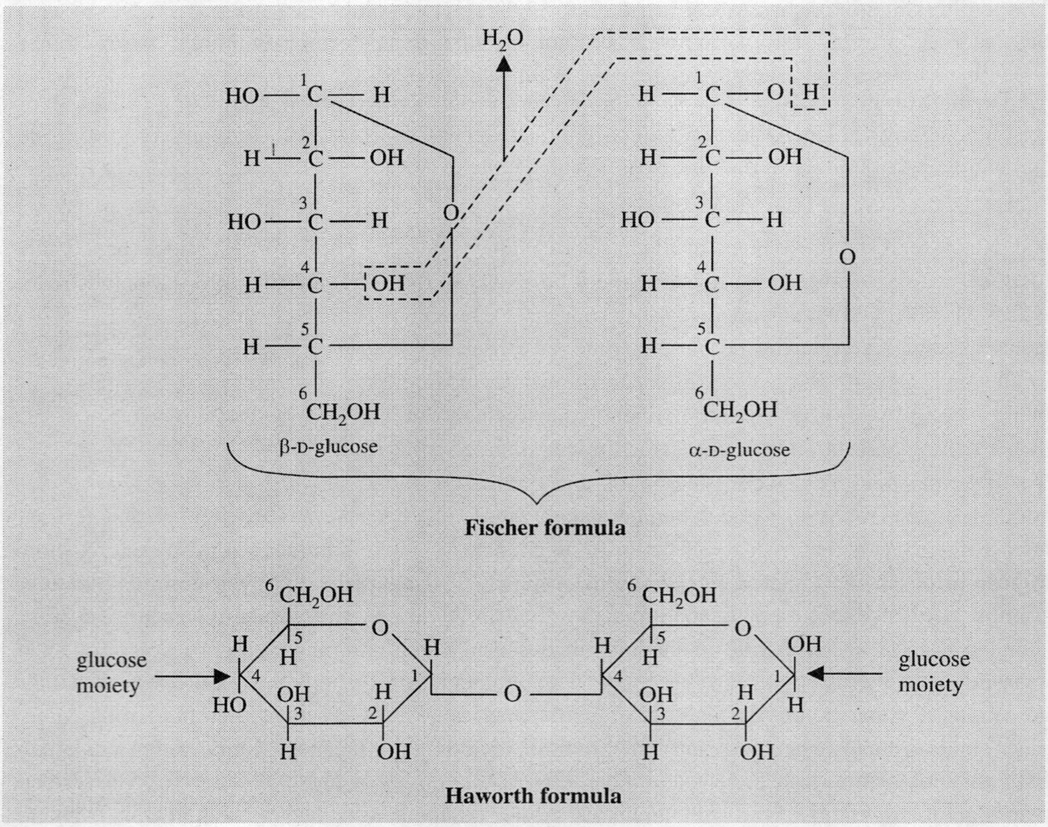

Fig. 7–4. Structure of β-maltose
***O*-α-D-glucopyranosyl-(1→ 4) -β–D-glucopyranose or Glc (α1 → β4)Glc**

Maltose, like lactose, has one free hemiacetal group. Consequently, it too exists in 3 forms ; α, β and *aldehydo*. Maltose is usually found in β-form. Maltose, in solution, also exists as an equilibrium mixture of all the 3 forms. It may be noted that the linkage in both α- and β-forms of maltose is always an α-1, 4-glucoside.

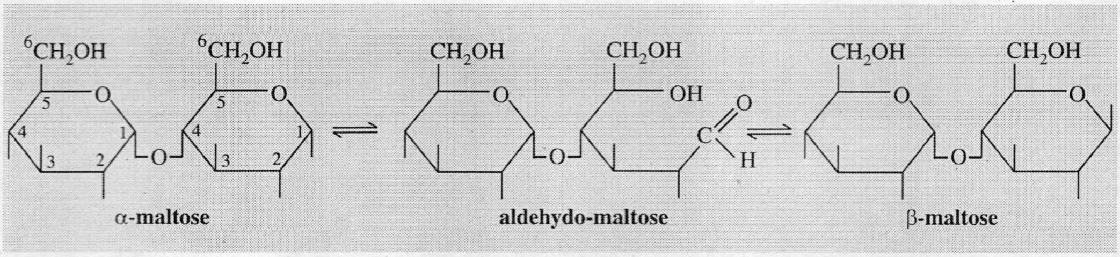

Properties. Maltose is a white crystalline solid, with a melting point 160–165 °C. It is soluble in water and is dextrorotatory. Because of the free aldehyde group, maltose is also a reducing sugar and forms osazone with phenyl hydrazine. It is also capable of exhibiting mutarotation.

Hydrolysis. Maltose is easily hydrolyzed into 2 identical units of glucose by dilute acids or by the enzyme, *maltase*, found in the intestine. It should also be noted that the enzyme maltase hydrolyzes or splits only α-glycoside linkages and has no action upon β-glycosides. It has, hence, been used to ascertain the presence of this bond in disaccharides and polysaccharides. Hydrolysis may also be brought about by the enzyme, *diastase* (= *amylase*), found in sprouting barley.

CELLOBIOSE

Occurrence. It is probably present in only traces in nature. However, it is apparently released during the digestion of a polysaccharide, cellulose by the *cellulases* of microorganisms.

Chemistry. Cellobiose is identical with maltose except that the former has a β-1, 4-glucosidic linkage (Fig. 7–5) in contrast to the α-1, 4-glucosidic of the latter. It also exists in an equilibrium mixture of 3 forms : α, β and *aldehydo*. On hydrolysis, cellobiose yields glucose units only.

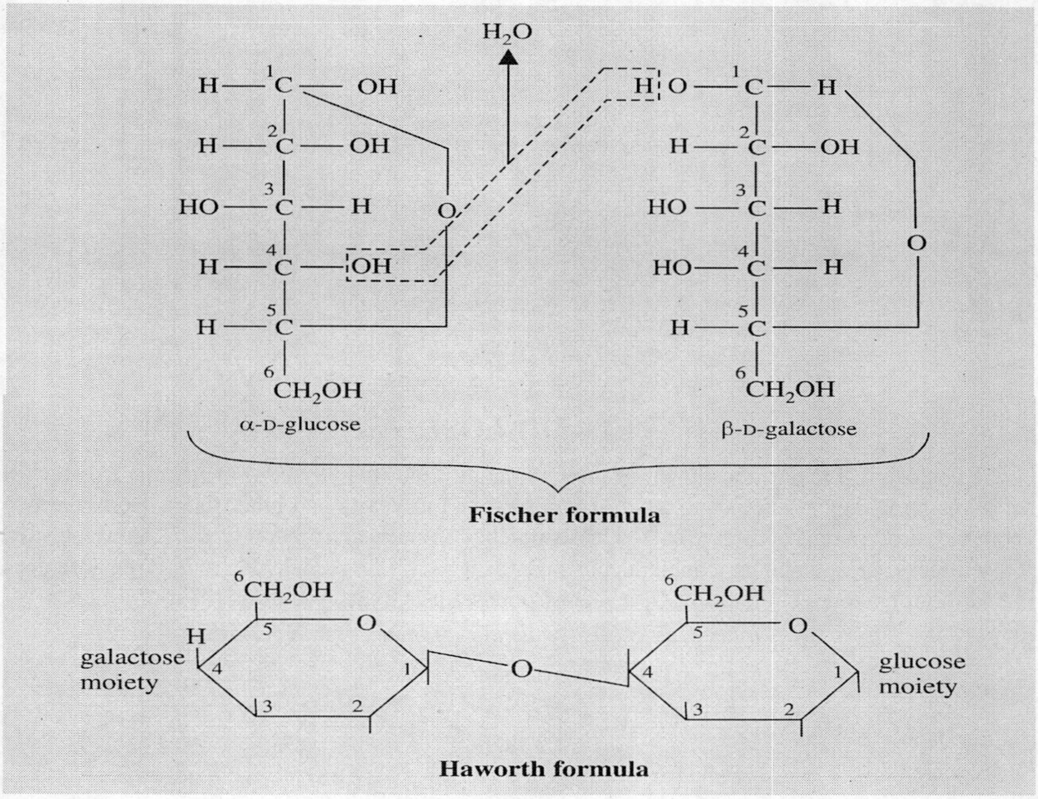

Fig. 7–5. Structure of α-cellobiose

O-β-D-glucopyranosyl-(1→ 4) -α-D-glucopyranose or Glc (β1 → α4)Glc

Properties. Cellobiose is a white crystalline solid with a melting point 225 °C. It is soluble in water and is dextrorotatory. Since cellobiose contains a free hemiacetal group, it is also a reducing sugar and undergoes mutarotation in aqueous solution and forms an oxime or osazone. In fact, all the disaccharides with a free hemiacetal group (lactose, maltose, cellobiose, etc.) are reducing sugars and as such reduce cupric copper to cuprous oxide in the Fehling test and the silver ion to metallic silver in the silver mirror reaction. However, they are not as powerful reducing agents as are the monosaccharides glucose, galactose and fructose, and do not readily reduce Barfoed's reagent to cuprous copper. The reason for this difference is that while C-1 of one of the two glucose residues is free to revert back to the carbonyl (= reducing) form, that on the second residue is stabilized in the cyclic form by the existence of the glycosidic linkage.

Hydrolysis. When hydrolyzed with dilute acids or by the enzyme *emulsin,* cellobiose yields identical units of D (+)-glucose. It may be noted that the enzyme emulsin splits only β-glycosidic linkages (*cf* maltase).

ISOMALTOSE

It is similar to maltose except that it has an α-1, 6-glucoside linkage (Fig. 7–6).

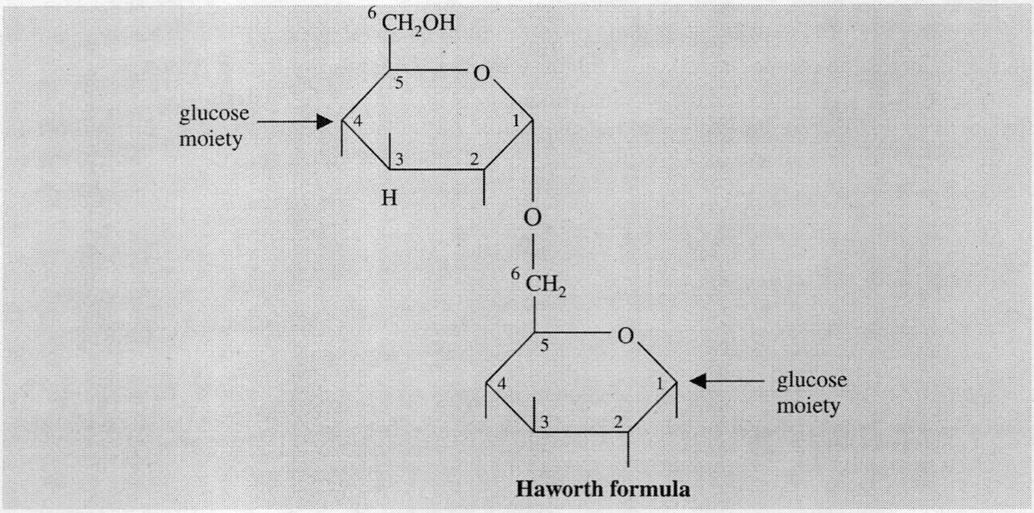

Fig. 7–6. Structure of isomaltose

O-α-D-glucopyranosyl-(1→ 6) -α–D-glucopyranose or Glc (α1 → α6)Glc

TREHALOSE

Trehalose is a major constituent of the circulating fluid (hemolymph) of insects, in which it serves as an energy storage compound. It is also found in yeasts and other fungi. Here, the two anomeric carbon atoms of the two α-D-glucose moieties connect each other (Fig. 7–7). Consequently, it resembles sucrose in being a nonreducing sugar as it has no free aldehyde group. Trehalose does not form an osazone. On hydrolysis, it yields glucose. It is hydrolyzed by HCl to 2 moles of D-glucose.

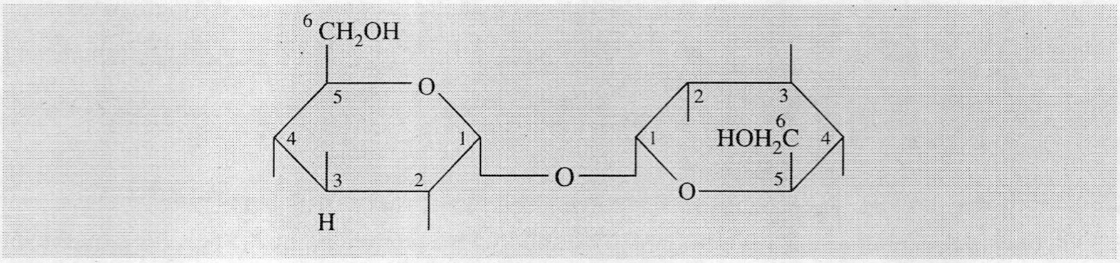

Fig. 7–7. Structure of trehalose

O-β-D-glucopyranosyl-(1→ 4) -α–D-glucopyranose or Glc (β1 → α4)Glc

Two isomeric forms of trehalose are recognized :

(*a*) *Isotrehalose* (β-D-glucopyranosyl-β-D-glucospyranoside)

(*b*) *Neotrehalose* (α-D-glucopyranosyl-β-D-glucopyranoside)

A summary chart of the important disaccharides (Table 7-3) is given on the next page.

Table 7-3. Summary of occurrence, structure and properties of some important disaccharides

Name	*Occurrence*	*Constituent monosaccharides**	*Oxide linkage involved*	*Groups involved in linkage*	*Properties*	*Hydrolysis*
Sucrose (Cane sugar)	Solely of plant origin : found in sugaracane, beet and maple	α-D-glucose and β-D-fructose	glycosidic linkage between C_1 of glucose and C_2 of fructose	glycosidic OH on C_1 glycosidic OH on C_2	Dextrorotatory; a non-reducing sugar; does not exhibit mutarotation; does not form osazone;	Yields an equimolar mixture of glucose and fructose on hydrolysis with invertase
Lactose (Milk sugar)	Solely of animal origin; found in milk of mammals	β-D-galactose and D-glucose	β-1, 4-galactoside	glycosidic OH on C_1 alcoholic OH on C_4	Dextrorotatory; a reducing sugar; exhibits mutarotation; forms osazone; nonfermentable.	Yields an equimolar mixture of galactose and glucose on hydrolysis with lactase
Maltose (Malt sugar)	Does not occur abundantly in nature; malt from sprouting barley an excellent source	α-D-glucose and D-glucose	α-1, 4-glucoside	glycosidic of OH on C_1 alcoholic OH on C_4	A reducing sugar; exhibits mutarotation; forms osazone; fermentable	Yields identical units of glucose on hydrolysis with maltase or diastase
Cellobiose	Probably present in only traces in nature	β-D-glucose and D-glucose	β-1, 4-glucoside	glycosidic OH on C_1 alcoholic OH on C_4	A reducing sugar; exhibits mutarotation	Yields identical units of glucose on hydrolysis with β-glycosides or with dilute acids
Trehalose	Major sugar of insect hemolymph; also occurs in fungi and yeasts	α-D-glucose and α-D-glucose	α-1, 1-glucoside	glycosidic OH on C_1 glycosidic OH on C_1	A nonreducing sugar; does not exhibit mutarotation; does not form osazone	Yields glucose units on hydrolysis

* All the constituent monosaccharides are present in pyranose ring form except β-D-fructose which has a furanose ring structure.

PROBLEMS

1. The manufacture of chocolates containing a liquid centre is an interesting application of enzyme engineering. The flavored liquid centre consists largely of an aqueous solution of sugars rich in fructose to provide sweetness. The technical dilemma is the following : the chocolate coating must be prepared by pouring hot melted chocolate over a solid (or almost solid) core, yet the final product must have a liquid, fructose-rich centre. Suggest a way to solve this problem. (**Hint :** The solubility of sucrose is much lower than the solubility of a mixture of glucose and fructose.)
2. Although lactose exists in two anomeric forms, no anomeric forms of sucrose have been reported. Why ?
3. A sample of disaccharide is either lactose or sucrose. No reddish precipitate forms in Fehling's reaction, unless the compound is first warmed in dilute acid. Is it lactose or sucrose ? Explain.
4. The carbohydrate portion of some glycoproteins may serve as a cellular recognition site. In order to perform this function, the oligosaccharide moiety of glycoproteins must have the potential to occur in a large variety of forms. Which can produce a larger variety of structures: oligopeptides composed of five different amino acid residues or oligosaccharides composed of five different monosaccharide residues ? Explain.
5. Sucrose is commonly used to preserve fruits. Why ?
6. Although the disaccharide lactose exists in two anomeric forms, no anomeric forms of the disaccharide sucrose have been reported. Why ?
7. In oriental and black adults, symptoms such as abdominal distention, cramps, pain and diarrhea develop on consumption of milk. Explain.
8. Why sucrose, which contains in its molecule glucose and fructose which are reducing sugars, is itself a nonreducing sugar ?
9. Does sugar contain any nutrients ?

ANSWERS TO PROBLEMS

1. Prepare a slurry of sucrose and water for the core; add a small amount of invertase; immediately coat with chocolate.
2. Sucrose is not a reducing sugar.
3. Lactose; in sucrose, the anomeric carbons of both glucose and fructose are involved in the glycosidic bond and are not available to reduce Fehling's reagent. Lactose is a reducing sugar that converts Fe^{3+} to Fe^{2+}, which precipitates as the red oxide.
4. Oligosaccharides; their subunits can be combined in more ways than the amino acid subunits of oligopeptides. Each of the several hydroxyl groups can participate in glycosidic bonds, and the configuration of each glycosidic bond may be either α or β. Furthermore, the polymer may be linear or branched.
5. In sucrose, both the carbonyl groups of its 2 moieties (*i.e.*, α-D-glucose and β-D-fructose) are involved in the formation of glycosidic bond. Consequently, sucrose contains no active group and is, hence, well suited for preservation.
6. Both the active carbonyl groups (present on C_1 of α-D-glucose and C_2 of β-D-glucose) are involved in the formation of glucosidic bond. Hence, no anomeric forms of sucrose are available. Thus, sucrose is not a reducing sugar because of the absence of the active group.
7. This is due to a substance, lactose. The ability to secrete the enzyme *lactase*, which hydrolyzes lactose, decreases with age. The unabsorbed lactose passes through the small intestine to the

large intenstine, where it is fermented by intestinal bacteria causing distention, pain and diarrhea.

8. In the formation of a sucrose molecule, the reducing aldehyde group of glucose and the ketonic group with adjacent hydroxyl group of fructose are both involved and result in the formation of an oxide linkage. Hence, sucrose is a nonreducing sugar. But the hydrolysis of sucrose breaks the glycosidic oxygen bond, releasing free glucose and fructose.
9. Refined white sugar is devoid of any nutrients. *Molasses* is a natural sweetener that is very nutritious. It contains B vitamins, calcium, iron, magnesium, potassium and chromium as well as other micronutrients. It may be tried as a substitute for sugary pancake syrups.

CHAPTER

8 Carbohydrates-IV Polysaccharides

INSIDE

- Homopolysaccharides
  - Starch
  - Glycogen
  - Inulin
  - Cellulose
  - Pectin
  - Chitin
  - Hemicelluloses
  - Xylan
- Heteropolysaccharides
  - Hyaluronic Acid
  - Chondroitin
  - Chondroitin Sulfates
  - Dermatan Sulfate
  - Keratosulfate
  - Heparin
  - Vegetable 'gums'
  - Agar-agar

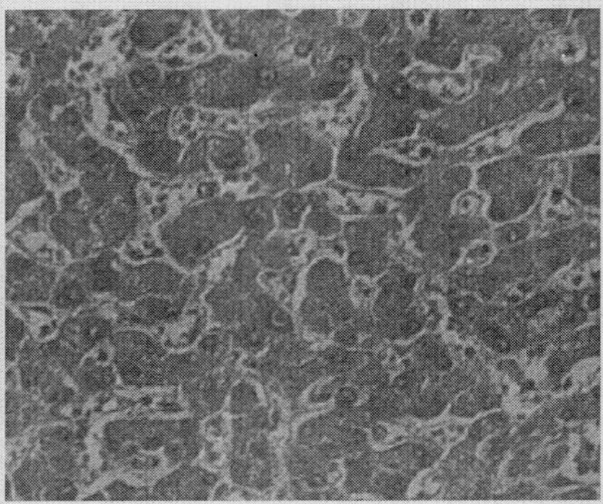

A thin section of liver containing glycogen (shown by *pink* colour)

Polysaccharides (variously called as *glycanes* or *polyholosides* or *polyosides*) are high molecular weight carbohydrates which, on hydrolysis, yield mainly monosaccharides or products related to monosaccharides. They may also be regarded as **polymeric anhydrides of simple sugars**. D-glucose is the commonest component of polysaccharides. However, D-and L-galactose, D-mannose, D-xylose, L-arabinose as well as D-glucuronic, D-galacturonic, D-mannuronic acids, D-glucosamine, D-galactosamine and amino uronic acids also occur as constituents of polysaccharides. The various polysaccharides differ from one another not only in the composition of the constituent monosaccharide but also in the molecular weight, in the nature of the chain (whether linear or branched), in the type of glycosidic bond (whether α or β) and in the type of linkage (whether 1→2, 1→3, 1→4, or 10→6) involved in the respective monosaccharide units. *A great majority of carbohydrates of nature occur as polysaccharides*. Chemically, the polysaccharides may be distinguished into **homopolysaccharides** (or **homoglycanes**), which yield, on hydrolysis, a single monosaccharide and **heteropolysaccharides** (or **heteroglycanes**), which produce a mixture of monosaccharides on hydrolysis.

> The **small arrow** shows the direction of the glycosidic bond which is always from the hemiacetal hydroxyl to some alcohol or hemiacetal group of the following sugar.

Based on their functional aspect, the polysaccharides may be grouped under two heads :

(*a*) **Nutrient** (or **digestible**) **polysaccharides.** These act as metabolic reserve of monosaccharides in plants and animals, *e.g.*, starch, glycogen and inulin.

(*b*) **Structural** (or **indigestible**) **polysaccharides.** These serve as rigid mechanical structures in plants and animals, *e.g.*,cellulose, pectin and chitin and also hyaluronic acid and chondroitin.

HOMOPOLYSACCHARIDES

These yield, on hydrolysis, a single monosaccharide. They serve as both storage (starch, glycogen, inulin) and structural (cellulose, pectin, chitin) polysaccharides.

STARCH

Occurrence. It is the most important reserve food material of the higher plants and is found in cereals, legumes, potatoes and other vegetables. *More than half the carbohydrate ingested by humans is starch*. Sago starch is obtained from sago palm, *Meteroxylon rumphii* ; the arrowroot from *Maranta arundinacea* and the tapioca, a starchy food, from *Manihot utillissima*. It is usually present inside the plant cells as compact insoluble granules which may be spherical, lens-shaped or ovoid, and which have a distinctly layered structure.

Chemistry. Natural starches consist of two components : *amylose* (15–20%), a long unbranched straight-chain component and *amylopectin* (80–85%), a branched chain polysaccharide. Starch from waxy corn is notable as it consists practically of amylopectin component, there being no amylose (Table 8.1).

α-amylose or simply **amylose** (Fig. 8–1) has a molecular weight range of 10,000 to 50,000. It may be formed in plant cells by elimination of a molecule of water from glycosidic OH group of one α-D-glucose molecule and alcoholic OH group on carbon 4 of the adjacent α-D-glucose molecule. The linkage in amylose is, thus, an α-1, 4-glucoside, like that in maltose. Enzymic hydrolysis of amylose with *amylase*, henceforth, yields maltose units mainly. Amylose may be considered as an anhydride of α-D-glucose units.

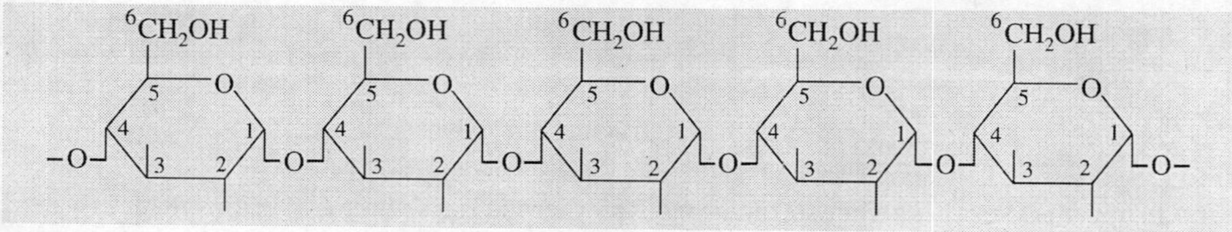

Fig. 8–1. Structure of amylose (= A-fraction)

Various types of starches differ in the amount of the amylose component present in them (Table 8–1).

β-amylose or **isoamylose** or **amylopectin** (Fig. 8–2) has a high molecular weight range of 50,000 to 1,000,000, thus indicating the presence of 300–5,500 glucose units per molecule. This possesses the same basic chain of α-1, 4-glucoside linkage like that of amylose but has, in addition, many side chains attached to the basic chain by α-1, 6-glucoside linkages, similar to those in isomaltose. It may, thus, be seen that the glucose unit, present at each point of branching, has substituents not only on carbon atoms 1 and 4 but also on carbon atom 6. In other words, these glucose units have 3 points of attachment to serve as branching points. The average chain length is about 24 glucose units. Amylopectin, upon incomplete hydrolysis, yields the disaccharide isomaltose.

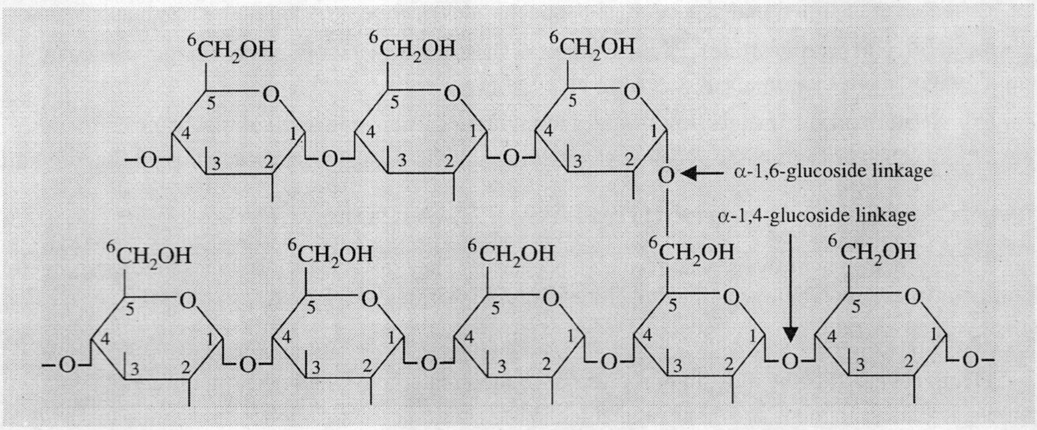

Fig. 8–2. Structure of amylopectin (= B-fraction)

The detailed structure of amylopectin (Fig. 8–3) is still speculative. The general consensus appears to be that amylopectin is composed of 3 types of chain, A, B and C ; each chain consisting of 24 glucose units. *A-chains* are linked to *B-chains* in 1, 6, or 1, 3 manner. These B-chains are further linked to other B-chains, the terminal one of which is linked to a single *C-chain*, which characteristically possesses a free reducing group. There are 2 different types of combinations of these chains which explain reasonably well the properties of amylopectin : the laminated structure as proposed by Haworth in 1937 and the randomly highly-branched structure as given by Meyer in 1940.

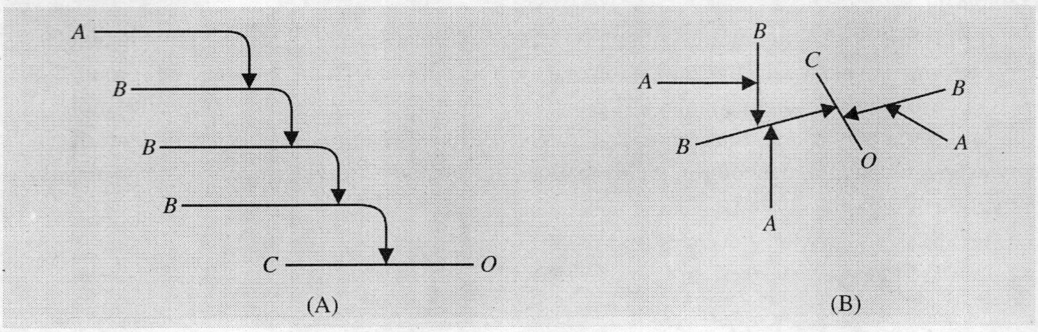

Fig. 8–3. Detailed structure of amylopectin

(A) Laminated structure (B) Randomly branched structure

Properties. Starch is a white soft amorphous powder and lacks sweetness. It is insoluble in water, alcohol and ether at ordinary temperature. The specific rotation of starch, $[a]_D^{20}$ is + 196°. The microscopic form of the starch grains is characteristic of the source of starch. Starch, on partial hydrolysis by boiling with water under pressure at about 250°C, breaks down into large fragments called **dextrins.** The resulting dextrins then confer stiffness to clothes that have been starched and ironed. Starch molecule is highly hydrated since it contains many exposed hydroxyl groups. With the result, starch, when extracted from granules with hot water, forms turbid colloidal solutions. In common with other polysaccharides, starch is a nonreducing carbohydrate since carbonyl groups of all units (except one of the two terminal ones) participate in the glycosidic linkages.

The two components of starch differ in following points :

I. Amylose has a simpler structure and hence more soluble in water than amylopectin. Because of this difference in solubility, the two components may be separated partially by keeping starch in water for a prolonged period. When potatoes are boiled, amylose is extracted by the hot water turning it milky and opalescent. The amylopectin left behind accounts for most of the starch in boiled potatoes.

II. Amylose is soluble without swelling in hot water whereas amylopectin swells in water.

III. Amylose is readily dispersed in water but does not form the characteristic gel or starch paste. However, when dispersed in sufficient concentrations, the amylopectin forms the typical starch gel.

IV. Amylose produces a typical blue colour with iodine whereas amylopectin gives a purplish colour with iodine. In fact, amylose is an open, helical molecule. The inside diameter of the helix is just large enough to accommodate an iodine molecule and it is the consequent change in the light-absorbing properties of the halogen that is responsible for the blue colour given in the starch-iodine test. By contrast, most other polysaccharides, including amylopectin, give only dull, reddish-brown colours.

Hydrolysis. One enzyme, *α-amylase*, found in the digestive tract of animals (in saliva and the pancreatic juice) hydrolyzes the linear **amylose** chain by attacking α (1→ 4) linkages at random throughout the chain to produce a mixture of maltose and glucose. β-*amylase*, an enzyme found in plants, attacks the nonreducing end of amylose to yield successive units of maltose (The prefixes α and β used with the amylases do not refer to glycosidic linkage, but simply designate these two enzymes). **Amylopectin** can also be attacked by α- and β-amylases, but the α (1 → 4) glycosidic bonds near the branching point in amylopectin and the α (1 → 6) bond itself are not hydrolyzed by these enzymes. A separate "debranching" enzyme, an *α (1→ 6) glucosidase* can hydrolyze the bond at the branch point. Therefore, the combined action of α-amylase and the α (1 → 6) glucosidase will hydrolyze amylopectin ultimately to a mixture of glucose and maltose.

> The suffix **-an** is used for designating polymers; for example, starch is a *glucosan* as it yields only glucose molecules on hydrolysis; similarly, inulin is a *fructosan* as it yields only fructose molecules on hydrolysis.

Starch is readily hydrolyzed by mineral acid with the final production of glucose. The course of hydrolysis may be followed by the gradual change in colour when treated with iodine (blue ⟶ purple ⟶ red ⟶ none) and also by the increase in concentration of reducing sugar. A compound which yields only glucose molecules on hydrolysis is called a **glucosan.** Hence, starch is a glucosan.

$$\underset{\substack{\text{Starch or Glycogen}\\\text{(insoluble)}}}{(C_6H_{10}O_5)_n} + \underset{\substack{(n-1)\text{ water}\\\text{molecules}}}{(n-1)\,H_2O} \xrightarrow{\text{Hydrolysis}} \underset{\substack{n\text{ glucose molecules}\\\text{(soluble)}}}{n(C_6H_{12}O_6)}$$

If hydrolysis of starch is interrupted early, dextrins are formed. The dextrins are white powders, and are used for making adhesives and confectionery, for sizing paper, etc. These materials possess adhesive properties and are used as adhesives on paper products.

GLYCOGEN

Occurrence. It is the major reserve food in animals and is often called **'animal starch'.** Glycogen is stored in the liver and muscles of animals. It is especially abundant in the liver where it may attain up to 7% of the wet weight. It occurs in animal cells as particles much smaller than the starch grains.

> Although of animal origin, a **glycogen-like polysaccharide** has also been described in maize.

Chemistry. Glycogen is a branched-chain polysaccharide and resembles amylopectin very much in structure, rather than amylose (for which there is no 'animal' equivalent), but has somewhat more glucose residues per molecule and about one-and-a-half times as many branching points. Also the chains are shorter (10–20 glucose units), and hence the molecule is even more highly branched and more compact. These differences, however, do not alter the functional behaviour of the molecule to any significant extent. Its molecular weight is $1 - 2 \times 10^7$.

Properties. It is a white powder and is more soluble in water than amylopectin. Hence, it readily forms suspension even in cold water. Glycogen is precipitated from aqueous solutions by addition of ethyl alcohol and is fairly stable in hot alkali. It is a nonreducing sugar and gives red colour with iodine. The red colour, however, disappears on boiling and reappears on cooling.

Hydrolysis. On incomplete hydrolysis with α-1,4-*glucanmaltohydrolase*, glycogen yields maltose. However, the acids completely hydrolyze it into glucose.

INULIN

Occurrence. It is the storage form of carbohydrate in the members of the family—Compositae, such as dahlias (*Dahlia sp.*), artichokes (*Helianthus tuberosus*) and dandelions (*Taraxacum officinale*). Inulin is stored in the tubers of the dahlia and artichoke and in the roots of dandelion. It is also found in onion and garlic. Animal lack it.

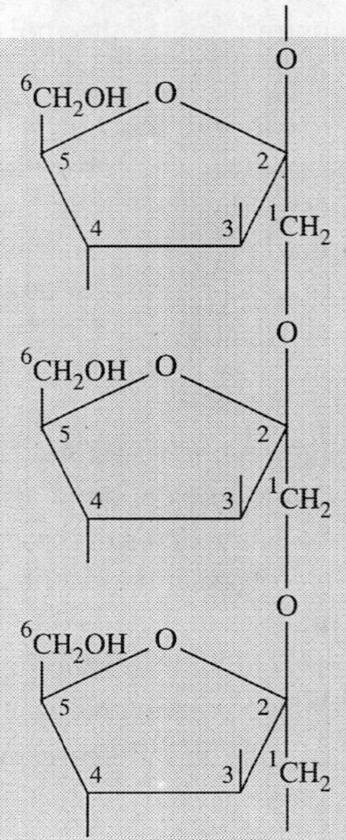

Fig. 8–4. Structure of inulin

[The two glucose units have not been shown for the sake of simplification.]

Chemistry. Inulin (Fig. 8–4) has a molecular weight of about 5,000 and consists of about 30–35 fructose units per molecule. It is formed in the plants by eliminating a molecule of water from the glycosidic OH group on carbon atom 2 of one β-D-fructose unit and the alcoholic OH group on carbon atom 1 of the adjacent β-D-fructose unit. In inulin, the fructose residues (in furanose form) are, thus, joined together in a straight chain by β-2, 1-fructosidic linkages. Inulin is, therefore, 2 → 1-β-polyfructose.

On hydrolysis, however, inulin also yields a small amount of glucose besides fructose molecules. It is now thought that there are 2 glucose units in the inulin molecule, one somewhere in the centre and the other at the reducing end of the chain.

Properties. It is a white tasteless powder and is insoluble in cold water. In warm water, it easily dissolves forming a colloidal solution which does not form a gel on cooling. Inulin solution does not give any colour with iodine. It is a nonreducing sugar.

Hydrolysis. Inulin, on complete hydrolysis by dilute acids, yields D-fructose and hence it is a **fructosan** or **levan**.

CELLULOSE

Occurrence. Cellulose is not only the most abundant extracellular structural polysaccharide of the plant world but is also *undoubtedly the most abundant of all biomolecules in the biosphere*. It is present in all land plants, but is completely lacking in meat, egg, fish and milk. It is, however, not metabolized by the human system. Some 10^{15} kg of cellulose is synthesized and degraded on earth each year! It shares as much as 50% of all the carbon in vegetation. It is the most widely distributed carbohydrate of the plants, although certain tunicates also possess it. Cellulose occurs in the cell walls of plants where it contributes in a major way to the structure of the organism. Lacking a skeleton of bone onto which organs and tissues may be organized, the higher plant relies on its cell walls to bear its own weight, whether it is the smallest *Wolffia* flower or the tallest *Sequoia* tree. Cellulose is often found associated with other structural substances such as lignin. Plant residues in soil consist of 40-70% cellulose. This high cellulose content emphasizes the importance of cellulolytic organisms in the mineralization process and in the carbon cycle. It occurs in nearly pure form in cotton (94%), flax (90%) and in the wood of many plants (60%). It is not usually found in bacterial cell walls, but it is the substance which keeps the cells of *Sarcina ventriculi* together in large packets or bunches.

Chemistry. The molecular weight of cellulose (Fig. 8–5) ranges between 200,000 and 2,000,000, thus corresponding to 1,250–12,500 glucose residues per molecule. It may be formed by taking out a molecule of water from the glycosidic OH group on carbon atom 1 of one β-D-glucose molecule and the alcoholic OH group on carbon atom 4 of the adjacent β-D-glucose molecule. It, thus, resembles in structure with amylose except that the glucose units are linked together by β-1, 4-glucoside linkages. Cellulose may, henceforth, be regarded as an anhydride of β-D-glucose units.

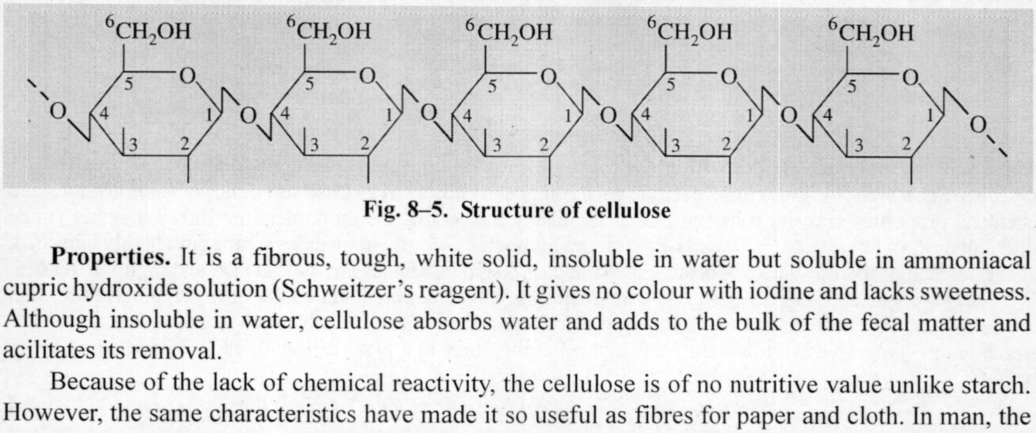

Fig. 8–5. Structure of cellulose

Properties. It is a fibrous, tough, white solid, insoluble in water but soluble in ammoniacal cupric hydroxide solution (Schweitzer's reagent). It gives no colour with iodine and lacks sweetness. Although insoluble in water, cellulose absorbs water and adds to the bulk of the fecal matter and acilitates its removal.

Because of the lack of chemical reactivity, the cellulose is of no nutritive value unlike starch. However, the same characteristics have made it so useful as fibres for paper and cloth. In man, the cellulose is not digested since it does not possess an enzyme capable of catalyzing the cleavage of β-glucoside bonds. Hence, it serves as an important source of "bulk" in the diet. However, the ruminants (cattles, sheep, goats, camels, giraffes) and certain wood-eating insects are able to digest it because the microorganisms, present in their digestive tract, do possess such digesting enzymes. Also certain wood-eating insects such as termites readily digest cellulose because their intestinal tract harbours a parasitic protozoan, *Trichonympha*. The latter secretes *cellulase*, a cellulose-hydrolyzing enzyme which enables termites to digest wood. Wood-rot fungi and bacteria also produce cellulase.

Hydrolysis. Cellulose is a relatively inert material and is completely degraded only under the most drastic conditions. For example, it may be hydrolyzed to glucose when treated with conc. H_2SO_4 or HCl or with conc. NaOH.

Biological importance. Huge quantities of cellulose are made annually by the plants of the globe. It is estimated that every day some 50 kg of cellulose is synthesized by the plant world for each human being on the earth. Cellulose is also a very useful substance. Wood, cotton, paper and cardboard are all largely cellulose. Dried wood contains about 65% cellulose, 30% lignin and 2% pentosans. In the manufacutre of **paper**, wood is cooked with water, lime and SO_2 to remove lignin leaving behind the cellulose. These fibres are then spread into mats and rolled into sheets which may be coated to provide a smooth writing surface. When heavy paper is treated with H_2SO_4, the translucent parchment paper is obtained.

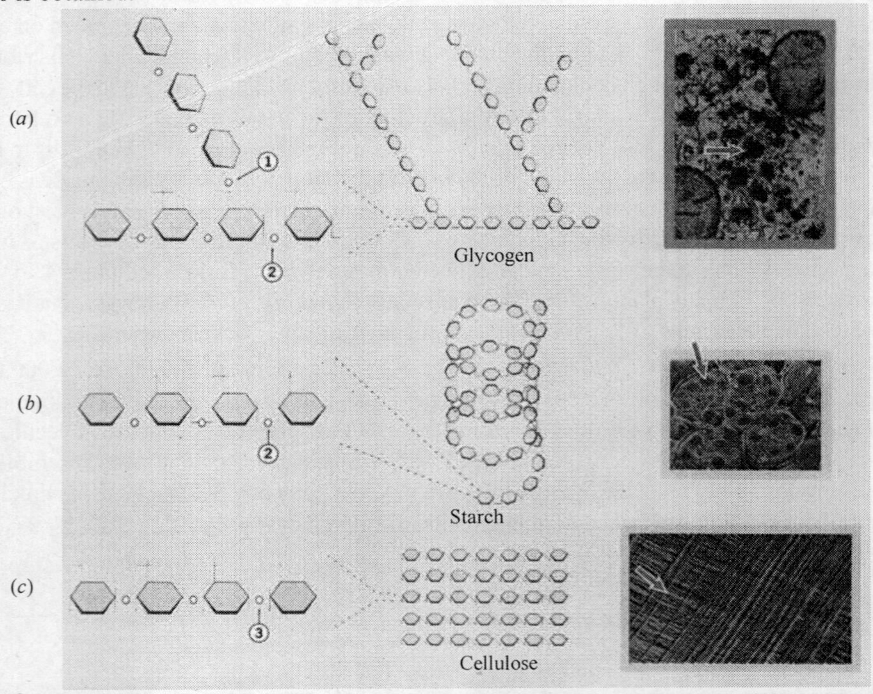

Fig. 8.6. Three polysaccharides with identical sugar monomers but dramatically different properties Gycogen (*a*), starch (*b*), and cellulose (*c*) are each composed entirely of glucose subunits, yet their physical and chemical properties are very different due to the distinct ways that the monomers are linked together (three different types of linkages are indicated by the circled numbers). *Glycogen* molecules are the most highly branched, *starch* molecules assume a helical arrangement, and *cellulose* molecules are unbranched and highly extended. Whereas glycogen and starch are energy stores, cellulose molecules are bundled together into tough fibres that are suited for their structural role. Colourized electron micrographs show glycogen granules in a liver cell, starch grains (amyloplasts) in a plant seed, and cellulose fibres in a plant cell wall; each is a indicated by an *arrow*.

[*Photo insets* : (*Top*) *Don Fawcett*; (*Centre*) *Jeremy Burgess*; (*Bottom*) *Cabisco*]

Cotton treated with alkali becomes somewhat translucent, acquires a silky lustre, and is called **mercerized cotton**. Rayon is made from cellulose by dissolving the material in NaOH and CS_2. The viscous material is forced through small jets into an acid solution of sodium bisulfite, regenerating the cellulose in small threads.

Cellulose nitrates are used in explosives, lacquers, celluloid and collodion. **Cellulose acetate** is widely used in photographic film and packing materials.

It may, thus, be easily visualized that the way the constituent monomeric units of various polysaccharides (glycogen, starch and cellulose, for example) are linked together has a profound bearing on the physical and chemical properties of these molecules, although the polysaccharides are made up of identical glucose units (Fig. 8-6).

PECTIN

Occurrence. Pectins are found as intercellular substances in the tissues of young plants and are especially abundant in ripe fruits such as guava, apples and pears. They are components of the middle lamella found between the cell wall and adjacent cells.

Chemistry. Pectin is a polysaccharide of α-D-galacturonic acid where some of the free carboxyl groups are, either partly or completely, esterified with methyl alcohol and others are combined with calcium or magnesium ions. Chemically, they are called polygalacturonides (Fig. 8–7) and have high molecular weights.

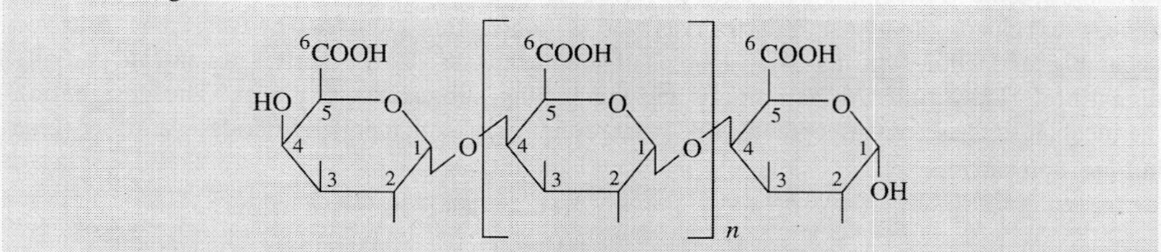

Fig. 8–7. Structure of α-polygalacturonic acid

Biological properties. The importance of pectin rests not so much on their quantitative role but rather on the functional part they play in plant stability and solidity. *The characteristic property of pectins is the ability of their solutions to gelate, i.e., form jellies.* Pectins, along with sugar, furnish the gelling characteristics of jellies and preserves made from fruit.

CHITIN

Occurrence. *Chitin is probably the most abundant polysaccharide of nature after cellulose.* It is also one of the most abundant biopolymer on the earth. It is found in fungi but principally among the arthropods (crabs and insects). The armour of crabs and the exoskeleton (= cuticula) of insects consist mostly of chitin and some protein. The excellent mechanical properties of insect skeleton are due to chitin. The chitin framework of lobster and crabs shells is impregnated and hardened with calcium carbonate.

Chemistry. Chitin (Fig. 8–8) is closely related to cellulose. Here the alcoholic OH group on carbon atom 2 of β-D-glucose units is replaced by an N-acetylamino group. It is, thus, a linear polymer of N-acetyl-D-glucosamine units joined together by β-1, 4-glucosidic linkages. Chitin's pronounced stability is based on the hydrogen bonding of the *N*-acetyl side chains. Like cellulose, chitin is indigestible by vertebrate animals.

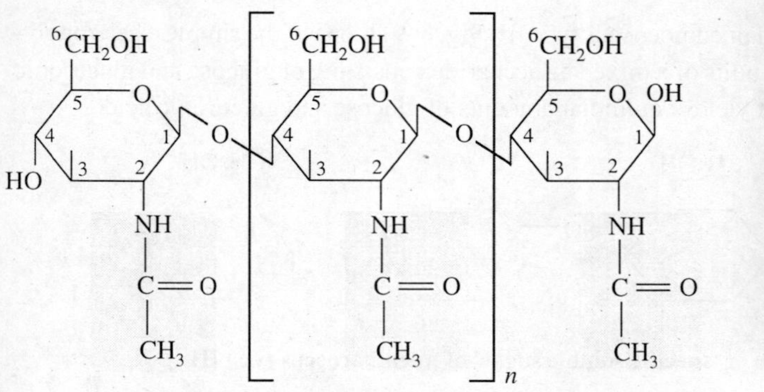

Fig. 8–9. Structure of chitin, a unique polysaccharide

Chitin has the same "alternating upside down" bonding of glucose molecules as cellulose has, but in chitin the glucose subunits are modified by the replacement of one of the hydroxyl groups with a nitrogen-containing functional group (*yellow*). Tough, slightly flexible chitin supports the otherwise soft bodies of arthropods (insects, spiders, and their relatives) and fungi.

Hydrolysis. On hydrolysis with mineral acids, chitin yields 2 final end products, namely, glucosamine and acetic acid. Glucosamine is an important component of some glycoproteins (=mucoproteins) such as mucin of saliva. *Chitinases* (from the gastric juice of snails or from bacteria), however, decompose the chitin to *N*-acetylglucosamine.

HEMICELLULOSES

Xylan and other related substances are collectively referred to as hemicelluloses. They are not structurally related to cellulose, nor do they contain the same building blocks but they are, at least partially, soluble in water or alkali. The hemicelluloses consist of either pentoses (xylose, arabinose) or hexoses (glucose, mannose, galactose) as well as uronic acids. Unrefined cereals, some fruits and vegetables and whole wheat are rich sources of hemicelluloses. They adsorb water and are partially digestible. They function as storage and supporting substances in plants. The designation 'hemicelluloses' has been discontinued since a large number of similar polysaccharides were discovered in fungi and bacteria.

Xylan is representative of the group and is, therefore, being discussed below.

XYLAN

Occurrence. *Xylan is the next most abundant and widely distributed carbohydrate after cellulose.* Straw and bark consist of upto 30% xylan; residues of sugarcane also contain about 30% xylan. It also occurs in conifer wood (7–12%) and deciduous wood (20–25%).

Chemistry. Xylan is a linear homopolymer of D-xylose (an aldopentose) in β-1→4 linkage, bearing side chains of 4–*O*–methylglucuronic acid and/or arabinose. Xylan can be derived from a cellulose chain by substituting hydrogen atoms for the —CH_2OH groups, but its polymer size (number of units per polymer) is considerably lower (30–100).

Properties. Xylan is more rapidly degraded by a large number of microorganisms than cellulose. Many cellulose-degrading organisms (such as *Sporocytophaga myxococcoides*) also produce *xylanase*. The ability to utilize xylans is very common among fungi. Xylan is even an excellent substrate for the cultivation of mushrooms. In bacteria, xylanase is formed constitutively in some organisms (*e.g.*, Clostridia) and in others it is inducible by xylan.

HETEROPOLYSACCHARIDES

These yield, on hydrolysis, a mixture of monosaccharides. They are numerous in both plants and animals.

The 'specific soluble sugar' of pneumococcus type III (Fig. 8–9) is one of the simplest heteropolysaccharides. It contains repeating units of a mixed disaccharide consisting of glucose and glucuronic acid. Apparently, on hydrolysis, it yields equimolar amounts of glucose and glucuronic acid.

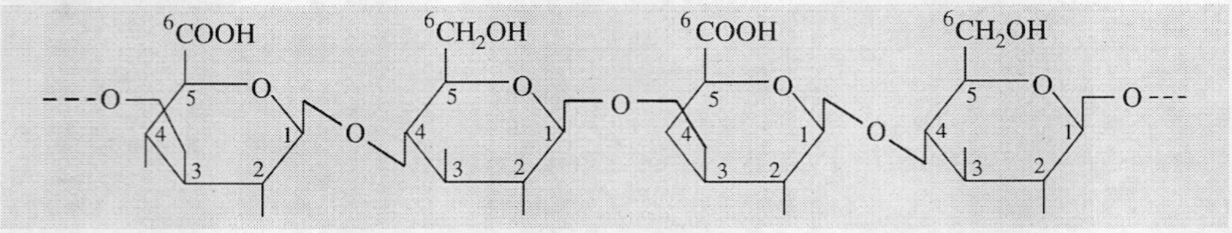

Fig. 8–9. Structure of 'specific soluble sugar' of pneumococcus type III

GLYCOSAMINOGLYCANS

Polysaccharides that are composed not only of a mixture of simple sugars but also of derivatives of sugars such as amino sugars and uronic sugars are called glycosaminoglycans (formerly called as *mucopolysaccharides*). They are gelatinous substances of high molecular weights (up to 5×10^6). Most of these act as structural support material for connective tissue or mucous substances of the body. They serve both as a lubricant and a cementing substance.

Structurally, they have a common feature. They consist of disaccharide units in which a uronic acid is bound by a glycosidic bond to the C_3 of an acetylated amino acid (1 → 3 linkage). These disaccharide residues are polymerized by 1 → 4 linkages to give a linear macromolecule. The uronic and sulfuric acid residues impart a strong acidic character to these substances. Table 8–1 lists the structural features of some common glycosaminoglycan.

Table 8–1. Some common glycosaminoglycans

Glycosaminoglycan	*Two components of the disaccharide units*	*Linkages**	
Hyaluronic acid	D-glucuronic acid + N-acetyl-D-glucosamine	β-1 → 3	β-1 → 4
Chondroitin sulfate A	D-glucuronic acid + N-acetyl-D-galactosamine-4-sulfate	β-1 → 3	β-1 → 4
Chondroitin sulfate C	D-glucuronic acid + N-acetyl-D-galactosamine-6-sulfate	β-1 → 3	β-1 → 4
Dermatan sulfate (Chondroitin sulfate B)	L-iduronic acid + N-acetyl-D-galactosamine-4-sulfate	α-1 → 3	β-1 → 4
Keratosulfate	D-galactose + N-acetyl-D-glucosamine-6-sulfate	β-1 → 4	β-1 → 3

* Linkage of the first column represents the linkage involved between the two monosaccharides of the disaccharide unit whereas the linkage of the second column is the one involved between the repeating disaccharide units.

HYALURONIC ACID

Occurrence. It is the most abundant member of glycosaminoglycans and is found in higher animals as a component of various tissues such as the vitreous body of the eye, the umbilical cord and the synovial fluid of joints. The high viscosity of the synovial fluid and its role as biological lubricant is largely due to the presence of its hyaluronic acid content (about 0.03%). Frequently, it is prepared from umbilical cord.

Chemistry. Hyaluronic acid (Fig. 8–10) has the least complicated structure among mucopolysaccharides. It is a straight-chain polymer of D-glucuronic acid and N-acetyl- D-glucosamine (NAG) alternating in the chain. Its molecular weight approaches approximately, 5,000,000. Here, apparently two linkages are invloved, β-1 → 3 and β-1 → 4. Hyaluronic acid is an acidic substance, because the carboxyl groups are largely ionized at cellular pH.

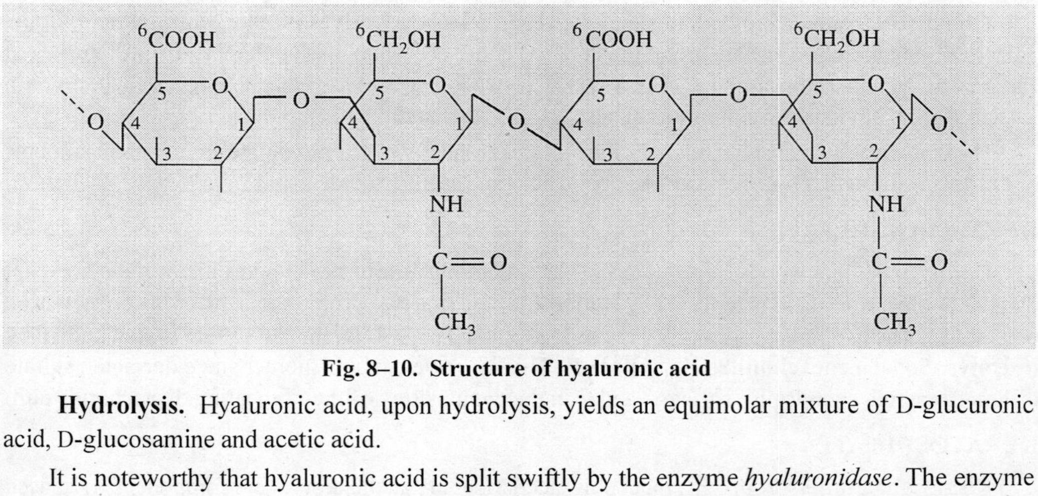

Fig. 8–10. Structure of hyaluronic acid

Hydrolysis. Hyaluronic acid, upon hydrolysis, yields an equimolar mixture of D-glucuronic acid, D-glucosamine and acetic acid.

It is noteworthy that hyaluronic acid is split swiftly by the enzyme *hyaluronidase*. The enzyme brings about depolymerization of hyaluronic acid (leading to a drop in its viscosity) and cleavage to smaller fragments. Hyaluronidase is a 'spreading factor' of skin and connective tissue. The depolymerization effect allows any foreign bodies (such as pigments, pathogenic bacteria) to penetrate the tissue, since the cementing substance is being dissolved. The enzyme also has a physiologic role in fertilization. The sperm is rich in hyaluronidase and hence can advance better in the cervical canal and finally fertilize the ovum.

CHONDROITIN

Occurrence. Chondroitin is of limited distributioin. It is found in cartilage and is also a component of cell coats. It is a parent substance for two more widely distributed mucopolysaccharides, chondroitin sulfate A and chondroitin sulfate B.

Chemistry. Chondroitin is similar in structure to hyaluronic acid except that it contains galactosamine rather than glucosamine. It is, thus, a polymer of β-D-glucuronido-1,3-N-acetyl-D-galactosamine joined by β-1 → 4 linkages.

Hydrolysis. On hydrolysis, chondroitin produces equimolar mixture of D-glucuronic acid, D-galactosamine and acetic acid.

CHONDROITIN SULFATES

Occurrence. The two chondroitin sulfate A and C are widely distributed and form major structural components of cartilage, tendons and bones. They are very often associated with collagen and probably with other proteins too.

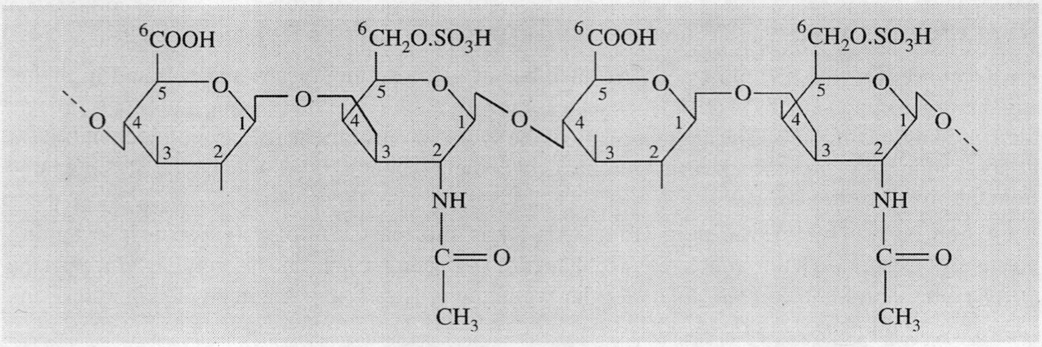

Fig. 8–11. Structure of chondroitin sulfate C

Chemistry. Chondroitin sulfates may be regarded as derivatives of chondroitin where, in the galactosamine moiety, a sulfate group is esterified either at carbon 4 as in chondroitin sulfate A or at carbon 6 as in chondroitin sulfate C (Fig. 8–11). The two linkages involved in both types of chondroitin sulfate would, obviously, be the same. These are β-1 → 3 and β-1 → 4.

Hydrolysis. On hydrolysis, chondroitin sulfates A and C yield approximately equivalent amounts of D-glucuronic acid, D-galactosamine, acetic acid and sulfuric acid.

DERMATAN SULFATE

Dermatan sulfate is a mucopolysaccharide structurally similar to chondroitin sulfate A except that the D-glucuronic acid is replaced by L-iduronic acid (the two uronic acids differ in configuration only at C_5). The two linkages involved are α-1 → 3 and β-1 → 4. Dermatan sulfate is also known by its conventional name, **chondroitin sulfate B**. This is, however, a misnomer since dermatan sulfate differs from both chondroitin sulfate A and C in the composition of their repeating disaccharide unit.

KERATOSULFATE

Keratosulfate differs from other mucopolysaccharides in that the uronic acid component is replaced by D-galactose. Here, the second acetylated amino sugar component (which is N-acetyl-D-glucosamine in this case) is esterified by a sulfate group at carbon 6. Although, the two alternating linkages involved are β-1 → 4 and β-1 → 3, in this case the linkage between the repeating disaccharide units is β-1 → 3 rather than β-1 → 4.

HEPARIN

Occurrence. Related to the sulfated mucopolysaccharides is heparin. It is present in liver, lung, arterial walls and, indeed, wherever mast cells are found, possibly for the purpose of neutralizing biogenic amines (*e.g.*, histamine).

Chemistry. Heparin (Fig. 8–12) is a heteropolysaccharide composed of D-glucuronic acid units, most of which (about 7 out of every 8) are esterified at C_2 and D-glucosamine-N-sulfate (= sulfonylaminoglucose) units with an additional O-sulfate group at C_6. Both the linkages of the polymer are alternating α-1 → 4. Thus, the sulfate content is very high and corresponds to about 5–6 molecules per tetrasaccharide repeating unit. The relative positions of the sulfate residues may also vary. Its molecular weight ranges between 17,000 and 20,000.

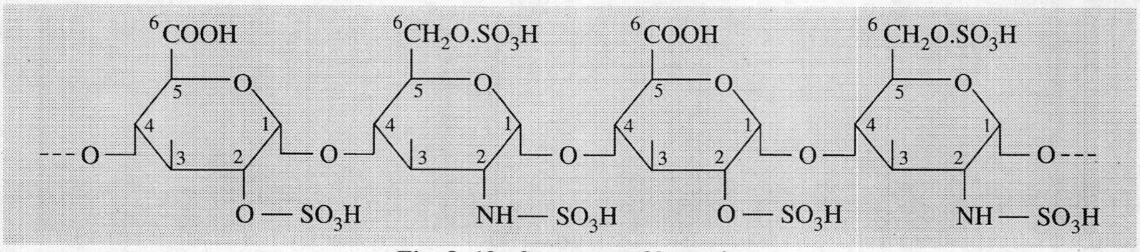

Fig. 8–12. Structure of heparin

Heparin acts as an anticoagulant. It prevents coagulation of blood by inhibiting the transformation of prothrombin to thrombin. This eliminates the effect of thrombin on fibrinogen.

The heteropolysaccharides described above contain two different sugars in their repeating units or monomers. However, there are many heteropolysaccharides which contain more than 2 carbohydrates in their repeating units. Vegetable 'gums' and agar-agar are two notable examples.

VEGETABLE 'GUMS'

These contain as many as 4 different monosaccharide units. Most common of these are D-glucuronic acid, D-mannose, D-xylose (the second most abundant sugar in the biosphere) and L-rhamnose.

AGAR-AGAR

It is a gelatinous polysaccharide produced by certain marine red algae (= rhodophycean members) such as species of *Gelidium, Gracilaria, Gigartina, Eucheuma, Campylaephora, Ahnfeltia, Hypnea,* etc. Japan is the major producer of agar-agar. It consists of D- and L-galactose, predominantly with 1 → 3 bonds and always contains some amount of sulfuric acid. On hydrolysis, it yields D- and L-galactose in a ratio of 9 : 1. It forms highly viscous gels and has melting point between 90 and 100°F. At lower temperature, it solidifies. It is insoluble in cold but soluble in hot water.

Agar-agar is of great value in the preparation of foodstuffs, particularly as articles of diet for invalids. It is used extensively in biological laboratories as the base material for the preparation of culture media, especially for bacteria and fungi. It is also used in the preparation of some medicines, and in cosmetics and leather industry. It is largely used as a solidifying agent in desserts, as a laxative and as a sizing material for textiles. It also finds application as an emulsifier in dairy products. Agar-agar is not utilized by man and hence adds to the bulk of the feces and helps in its propulsion.

A wide variety of heteropolysaccharides are found in animals and microbes. These, on hydrolysis, yield various types of monosaccharides. These more complex polysaccharides are found bound to proteins (**glycoproteins**) and lipids (**glycolipids**). Carbohydrates occur in them as an oligosaccharide moiety.

A summary chart of the important homopolysaccharides (Table 8-2) is given on the next page.

Table 8–2. Summary of occurrence, structure and properties of some important homopolysaccharides

Name	*Occurrence*	*Constituent monosaccharides**	*Oxide linkage(s) involved*	*Groups involved in linkage*	*Properties*	*Hydrolysis*
Starch	Major reserve food of higher plants; found in cereals, legumes, potatoes and other vegetables	α-D-glucose	(In amylose) α-1, 4-glucoside (In amylopectin) α-1, 4-glucoside and α-1, 6-glucoside	glycosidic OH on C_1 alcoholic OH on C_4 glycosidic OH on C_1 alcoholic OH on C_4 alcoholic OH on C_6	amylose gives blue colour with iodine; amylopectin gives purple colour with iodine; dextrorotatory	Yields glucose units only on hydrolysis, hence it is a glucosan
Glycogen	Major reserve food of animals; stored in liver and muscles of animals	α-D-glucose	α-1, 4-glucoside and α-1, 6-glucoside Chains are shorter and much more branched than amylopectin	glycosidic OH on C_1 alcoholic OH on C_4 alcoholic OH on C_6	a nonreducing sugar; gives red colour with iodine	Yields glucose units only on complete hydrolysis, hence a glucosan
Inulin	Found in tubers and roots of certain composite plants such as dahlias, artichokes and dandelions	β-D-fructose	β-2, 1-fructoside	glycosidic OH on C_2 alcoholic OH on C_1	a nonreducing sugar; gives no colour with iodine	Yields fructose units only on complete hydrolysis, hence a fructosan
Cellulose	Most abundant organic compound found in world; widely distributed in plants such as cotton and flax	β-D-glucose	β-1, 4-glucoside	glycosidic OH on C_1 alcoholic OH on C_4	gives no colour with iodine	Yields glucose units only on complete hydrolysis, hence a glucosan
Pectin	Fruits of many plants	α-D-galacturonic acid	α-1, 4-galactoside	glycosidic OH on C_1 alcoholic OH on C_4	—	—
Chitin	Most abundant in nature after cellulose; found in fungi and arthropods	N-acetyl-D-glucosamine	β-1, 4-glucoside	glycosidic OH on C_1 alcoholic OH on C_4	a nonreducing sugar	Yields N-acetyl-glucosamine on hydrolysis with chitinases

* All the constituent monosaccharides are present in pyranose ring form except β-D-fructose which has a furanose ring structure.

PROBLEMS

1. α-amylose is an unbranched glucose polymer. Why would this polymer not be as effective a storage form for glucose as glycogen ?
2. Both cellulose and α-amylose consist of (1→4)-linked D-glucose units and can be extensively hydrated. Despite this similarity, a person on a diet consisting predominantly of α-amylose (starch) will gain weight, whereas a person on a diet of cellulose (wood) will starve. Why ?
3. The practically pure cellulose obtained from the seed threads of the plant genus *Gossypium* (cotton) is tough, fibrous, and completely insoluble in water. In contrast, glycogen obtained from muscle or liver disperses readily in hot water to make a turbid solution. Although they have markedly different physical properties, both substances are composed of (1→4)-linked D-glucose polymers of comparable molecular weight. What features of their structures cause these two polysaccharides to differ in their physical properties ? Explain the biological advantages of their respective properties.
4. Why is it not possible to write on an oily paper ?
5. Why does an apple turn orangish when kept in the air ?
6. What is gun cotton ?
7. In what forms carbohydrates are stored in plants and animals ?
8. If one has diabetes, should he avoid potatoes because they raise one's blood pressure ?
9. Why starch forms turbid colloidal solution, when extracted from granules with hot water ?
10. Why do we need to fold a paper to tear it along a straight line ?
11. Why do the smooth pages of paper, after being dipped in water and dried, get wrinkled ?

ANSWERS TO PROBLEMS

1. As an unbranched polymer, α-amylose has only one nonreducing end. Therefore, only one glycogen phosphorylase molecule could degrade each α-amylose molecule. Because glycogen is highly branched, there are many nonreducing ends per molecule. Consequently, many phosphorylase molecules can release many glucose molecules per glycogen molecule.
2. The human intestinal enzyme that splits (1 →4) linkages between glucose residues is absolutely specific for the (α1→4) linkage; cellulose, with its (β1→4) linkages, cannot be digested.
3. Native cellulose consists of glucose units linked by (β1→4) glycosidic bonds. The β linkage forces the polymer chain into an extended conformation. A parallel series of these extended chains can form intermolecular hydrogen bonds, aggregating into long, tough, insoluble fibres. Glycogen consists of glucose units linked together by (α1→4) glycosidic bonds. The α linkage causes a bend in the chain and prevents the formation of long fibres. In addition, glycogen is highly branched. Because many of its hydroxyl groups are exposed to water, glycogen is highly hydrated, and can be extracted as a dispersion in hot water.

   The physical properties of these two polymers are well suited for their biological roles. Cellulose serves as a structural material in plants, consistent with its side-by-side aggregation into insoluble fibres. Glycogen serves as a storage fuel in animals. The highly hydrated glycogen granules with their abundance of nonreducing ends can be rapidly hydrolyzed by glycogen phosphorylase to release glucose-1-phosphate.
4. Writing with ink on a surface depends on both the ink and the surface. Ink contains a colouring substance (dye) dissolved in water or a volatile solvent. As paper is made of cellulose, when we write with a fountain pen, the colouring substance in the ink enters the fine capillaries through capillary action. When the solvent evaporates, dry ink remains in the capillaries, thus making the visible mark on the paper. If the surface of the paper is oily,

the oil repels the ink which as a result does not stick to the paper to produce a mark. So, it is not possible to write on oily paper with a fountain pen.

On the other hand, a ball-point pen does not write on oily paper for a different reason. When drawn over the oily surface, the ball on the tip of a ball-point pen slips and does not rotate as there is no friction. As a result, the thick ink does not flow out and does not write.

5. Apples contain tannins which act as *astringent agents i.e., they demoisturise the fruit*. Their concentration in the apple depends upon the season and the stage of maturity of the fruit. When the apple is green, the concentration of tannins is high and it decreases as the fruit matures. When the apple is sliced or cut, tannins get exposed to the air. In presence of some enzymes that apples contain, tannins combine with oxygen in the air to form oxides which impart an orangish hue to the sliced part.

6. Gun cotton is a highly inflammable, explosive chemical called cellulose nitrate. It is prepared by treating cotton, a natural almost pure (94%) cellulose, with concentrated nitric acid and hence the name gun cotton. Due to its inflammable nature and with the arrival of better plastic material, its use has been discontinued. However, it is still used for making certain kinds of explosives.

7. In the form of starch in plants and in the form of glycogen in animals.

8. No, any starch will increase one's blood sugar some. One should always eat a well-balanced diet that includes various starches.

9. Because starch molecule is highly hydrated since it contains many exposed hydroxyl groups.

10. Paper is made of cellulose fibres found in plant cell walls. The fibres are beaten into pulp and joined together by the action of water to form sheets of paper. It is these fibres which break when we try to tear a sheet of paper. As these fibres are randomly placed with respect to each other, they can break in any direction when one pulls a paper apart. However, when we fold a paper, the fibres get uniformly weakened along a straight line. So, when this folded paper is pulled apart to tear it, those weakened fibres give way easily and the paper gets torn along the straight line.

11. Paper is made of cellulose fibres and other ingredients mixed together. The smooth surface of finished paper results from the process of paper making itself. During manufacture, the wet film of pulp is drained, dried and finally calendered by pressing it between steam-heated rollers which set the fibres in place in a somewhat stretched condition. When paper is dipped in water, the sizing material gets loosened, the fibres soak water and get deformed. If this wet paper is allowed to dry, it gets wrinkled because the fibres dry in a deformed state.

CHAPTER

9 Proteins–I
General Structure

INSIDE

- Importance
- Amino Acids
- Peptides
- Chemical Bonds Involved in Protein Structure
- Protein Configuration
  - Primary structure
  - Secondary structure
  - Tertiary structure
  - Quaternary structure
- Prions
- Cleavage of a Protein
- Biological Roles of Proteins

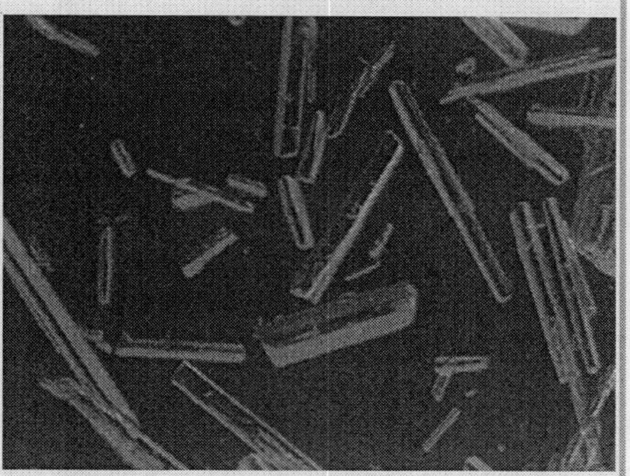

Crystals of glycine viewed under polarized light
Glycine has the simplest structure among all the amino acids which join together by peptide bonds to form proteins.

IMPORTANCE

The name protein (*proteios*[G] = pre-eminent or first) was first suggested, in 1838, by a Swedish chemist **Berzelius** to a Dutch chemist **Mulder**, who referred it to the complex organic nitrogenous substances found in the cells of the living beings. They are most abundant intracellular macromolecules and constitute over half the dry weight of most organisms. "Proteins occupy a central position in the architecture and functioning of living matter. They are intimately connected with all phases of chemical and physical activity, that constitute the life of the cell. Some proteins serve as important

Jöns Jacob Berzelius (LT, 1779-1848) was Professor of Chemistry at Stockholm, Sweden. He is famous for developing the modern system of chemical symbols. He isolated several elements and is also credited with the discovery of selenium (Se). He developed the concept of atomic weight. With Humphrey Davy, he laid the foundations of electrochemistry. He is also credited with publishing the earliest book on Biochemistry entitled 'Lectures in Animal Chemistry' in 1806.

Gerardus Johannes Mulder's (LT, 1802-1880) comment about **proteins** reads as follows:

> "There is present in plants and animals a substance which ... is without doubt the most important of the known substances in living matter, and without it, life would be impossible on our planet. This material has been named Protein."

structural elements of the body, for example, as hair, wool and collagen, an important constituent of connective tissue ; other proteins may be enzymes, hormones or oxygen- carriers. Still other proteins participate in muscular contraction, and some are associated with the genes, the hereditary factors."(**White, Handler** and **Smith**, 1964).

They are, therefore, essential to cell structure and cell function. The proteins with catalytic activity (enzymes) are largely responsible for determining the phenotype or properties of a cell in a particular environment. The total hereditary material of the cell or genotype dictates which type of protein the cell can produce. In fact, the proteins have built into their structure the information that instructs them in "what to do" (catalytic activity), "where to go" in the cell (intracellular organization) and "when and how to perform" (control of function through bacterial interactions with other proteins, small activators or inhibitors).

The constituent elements of proteins are carbon, hydrogen, oxygen, nitrogen and very rarely sulfur also. In certain complex proteins, phosphorus occurs as well. The elemental composition of proteins in plants and animals presents a great deal of variation (Table 9-1).

Table 9–1. Elemental composition of plant and animal proteins

Protein	*Carbon*	*hydrogen*	*Oxygen*	*Nitrogen*
Green plants	**54**	**7**	**38**	**0.003**
Mammals	**21**	**10**	**62**	**3.0**

Most animal proteins contain from 0.5 to 2.0 per cent sulfur. Insulin is, however, a notable exception to this in possessing about 3.4% sulfur.

AMINO ACIDS

Structure

All proteins are macromolecules because of their very high molecular weights. These are the polymers, *i.e.*, chain-like molecules produced by joining a number of small units of amino acids called monomers. The amino acids are, therefore, regarded as '**building blocks of proteins**'. The general formula of an amino acid along with its space-filling model is presented in Fig, 9-1.

The term **macromolecule** was coined by **Schaudinger** in 1920s.

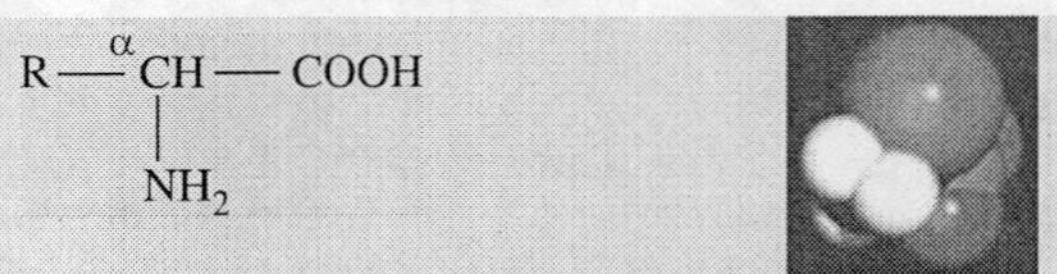

(a) General formula

(b) Space-filling model. The large green sphere represents the R group.

Fig. 9–1. General structure of an amino acid

Each amino acid is a nitrogenous compound having both an acidic carboxyl (— COOH) and a basic amino (—NH_2) group. R stands for the side chains that are different for each amino acid. R can be as simple as a hydrogen atom (H) as in glycine or a methyl group (—CH_3) as in alanine or a more complex structure. The first carbon is the part of the carboxyl group. The second carbon, to which is attached the amino group, is called the α-*carbon*. The α-carbon of most amino acids is joined by covalent bonds to 4 different groups. Thus, the α-carbon in all the amino acids is asymmetric except in glycine where the α-carbon is symmetric. Because of this asymmetry, the amino acids (of course, except glycine) exist in two optically active forms : those having — NH_2 group to the right are designated as D-forms and those having — NH_2 group to the left as L-forms (Fig. 9–2). However, the two amino acids, threonine and isoleucine have two asymmetric carbon atoms each and thus have $2^n = 2^2 = 4$ optical isomers. At pH 7.0, both the carboxyl and amino groups are ionized.

Fig. 9–2. Two optical isomers of amino acids

Specific Rotation

It is interesting to note that *the amino acids found in the proteins belong to the L-series*. Many of the naturally occurring L-amino acids rotate the plane of polarized light to the left (*i.e.*, they are levorotatory) while others rotate the plane of polarized light to the right (*i.e.*, they are dextrorotatory). Table 9–2. shows the specific rotation of some L-amino acids.

Table 9–2. Specific rotation of some amino acids isolated from proteins

Amino acid	*Specific rotation,* $[\alpha]_D^{25°}$
L-alanine	– 86.2
L-histidine	– 38.5
L-phenylalanine	– 34.5
L-threonine	– 28.5
L-serine	– 7.5
L-alanine	+ 1.8
L-glutamic acid	+ 12.0
L-isoleucine	+ 12.4
L-arginine	+ 12.5
L-lysine	+ 13.5

Thus, it is evident that the symbols D and L do not identify the property of light rotation, *i.e.*, D-isomers can be either dextrorotatory (*d*) or levorotatory (*l*); similarly, L-isomers can be either (*d*) or (*l*). However, to minimize confusion, the symbols *d* and *l* are usually not used nowadays.

Distribution in Proteins

The distribution of the 20 amino acids is not uniform in all proteins. Nearly 40% by weight of fibroin and 25% by weight of collagen are accounted for by glycine. Fibroin is all rich in alanine (30% by weight). Serine and threonine predominate in casein and phosvitin. Collagen (in connective tissue), gliadin (in wheat) and zein (in corn) are rich in proline. Human serum albumin with 585 amino acid residues has only one tryptophan moiety.

The **pulses** are notable as they lack S-containing amino acid, methionine (Met) but contain good amount of the basic amino acid, lysine (Lys); whereas **cereals** lack lysine but have sufficient quantity of methionine. When combined, these make good the deficiency of each other through mutual supplementation and are therefore better utilized in human body.

Location in Proteins

Amino acids with uncharged polar side chains are relatively hydrophilic and are usually on the outside of the proteins, while the side chains on nonpolar amino acids tend to cluster together on the inside. Amino acids with acidic or basic side chains are very polar, and they are nearly always found on the outside of the protein molecules.

Physical Properties

Regarding their physical characteristics, the amino acids are colourless crystalline substances. The crystal form may vary from slender needles (tyrosine) to thick hexagonal plates (cystine). They may be either tasteless (tyrosine), sweet (glycine and alanine) or bitter (arginine). It is interesting to note that (S)-phenylalanine is bitter in taste whereas its enantiomer (R)-phenylalanine is sweet. They have high melting points (above 200°C) and often result in decomposition. Amino acids are soluble in polar solvents such as water and ethanol but they are insoluble in nonpolar solvents such as benzene and ether. Some amino acids like tryptophan, tyrosine, histidine and phenylalanine absorb ultraviolet rays at 260–290 mμ. This property enables the identification of not only these amino acids but also the proteins which contain them.

The order of hydrophilicity and hydrophobicity of various amino acids is as follows :

Hydrophilic (= polar) amino acids :

Tyr > Ser > Asp > Glu > Asn > Gln > Arg

Hydrophobic (= apolar) amino acids :

Phe > Ala > Val > Gly > Leu > Cys

Although over 100 amino acids have been shown to be present in various plants and aminals, only 20 of them (L-isomers) are found as constituent of most proteins. These 20 amino acids of proteins are often referred to as **standard, primary** or **normal amino acids**, to distinguish them from others. The almost universal use of these 20 amino acids for the synthesis of various protein molecules is "one of nature's enigmatic rules." These have, therefore, rightly been called as the "**magic 20**", a phrase coined by **Francis Crick.**

Electrochemical Properties. All molecules possessing both acidic and basic groups might exist as uncharged molecules or in ionic form or as a mixture of both. Calculations have revealed that an aqueous solution of most amino acids, glycine for example, can have only one part of uncharged molecules in one lakh parts of the ionic form. Hence, it is more accurate to write the structure of glycine as

$$\underset{\displaystyle NH_3^+}{H-\underset{|}{CH}-COO^-} \quad \text{rather than} \quad \underset{\displaystyle NH_2}{H-\underset{|}{CH}-COOH}$$

In the ionic form, the proton migrates from the carboxyl to amino group, thus producing carboxylate (COO^-) and ammonium (NH_3^+) ions.

Amino acids react with both acids and bases. Hence, they are **amphoteric** in nature.

Thus, in acid solution, the COO^- ion acquires a proton and the amino acid becomes an ammonium salt of the acid. Conversely, in alkaline solution, the NH_3^+ ion loses a proton and the amino acid becomes the anion of a salt. These reactions are of reversible nature and depend on the pH of the medium. Henceforth, amino acids serve as buffers and tend to prevent pH change when an acid or a base is added.

$$\underset{}{R-CH(NH_3^+)-COO^-} + \underset{\text{(an acid)}}{HA} \rightleftharpoons R-CH(NH_3^+)-COOH + \underset{\text{(anion)}}{A^-}$$

$$R-CH(NH_3^+)-COO^- + \underset{\text{(a base)}}{BOH} \rightleftharpoons R-CH(NH_2)-COO^- + \underset{\text{(cation)}}{B^+} + H_2O$$

Classification

Three systems of classifying amino acids are in vogue.

A. On the basis of the composition of the side chain or R group. Threre are 20 different amino acids which regularly appear in proteins. These possess a side chain which is the only variable feature present in their molecules. The other features such as α-carbon, carboxyl group and amino group are common to all the amino acids. The common component of an amino acid appears in Fig. 9–3.

$$-\overset{\alpha}{C}H(NH_2)-COOH$$

Fig. 9–3. Common component of an amino acid

Based on the composition of the side chain, the twenty amino acids, whose structure is shown in Fig. 9–4, may be grouped into following 8 categories (**Fairley** and **Kilgour**, 1966) :

(*i*) *Simple amino acids.* These have no functional group in the side chain, *e.g.*, glycine, alanine, valine, leucine and isoleucine.

(*ii*) *Hydroxy amino acids.* These contain a hydroxyl group in their side chain, *e.g.*, serine and threonine.

(*iii*) *Sulfur-containing amino acids.* These possess a sulfur atom in the side chain, *e.g.*, cysteine and methionine.

(*iv*) *Acidic amino acids.* These have a carboxyl group in the side chain, *e.g.*, aspartic acid and glutamic acid.

(*v*) *Amino acid amides.* These are derivatives of acidic amino acids in which one of the carboxyl group has been transformed into an amide group (—$CO.NH_2$), *e.g.*, asparagine and glutamine.

(*vi*) *Basic amino acids.* These possess an amino group in the side chain, *e.g.*, lysine and arginine.

(*vii*) *Heterocyclic amino acids.* These amino acids have in their side chain a ring which possesses at least one atom other than the carbon, *e.g.*, tryptophan, histidine and proline.

(*viii*) *Aromatic amino acids.* These have a benzene ring in the side chain, *e.g.*, phenylalanine and tyrosine.

The classification given above is only a practical one and can conveniently be followed. It does not, however, strictly delimit the various categories. For example, tryptophan may also be included under aromatic amino acids and similarly, histidine under basic amino acids.

B. On the basis of the number of amino and carboxylic groups. McGilvery and **Goldstein** (1979) have classified various amino acids as follows :

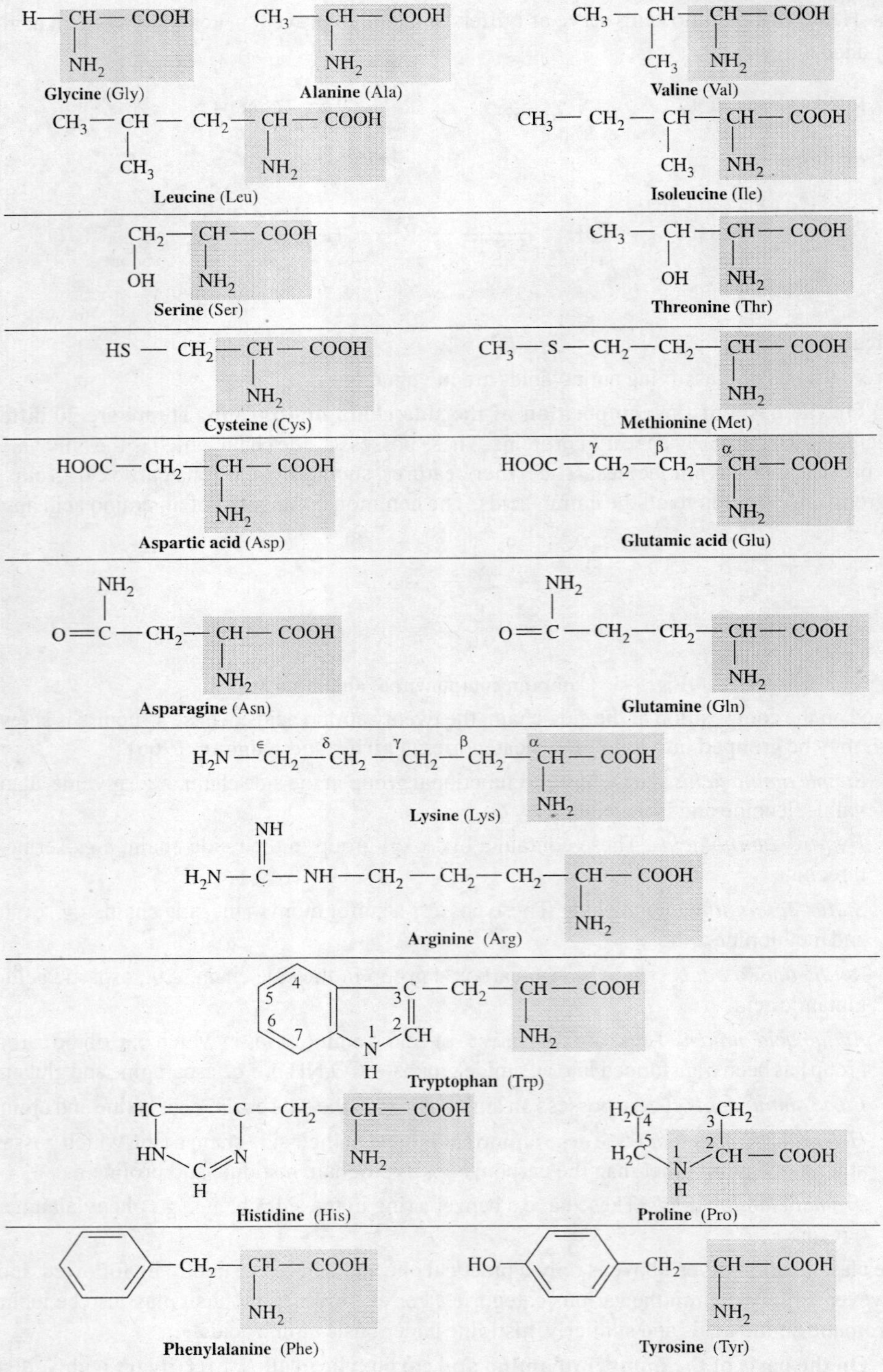

Fig. 9–4. The twenty amino acids ("Magic 20") found in proteins

The standard three-letter abbreviation for each amino acid is written in brackets.
The common component is shown in a *blue* enclosure.

I. Monoamino-monocarboxylic amino acids :

1.	Unsubstituted	—	Glycine Alanine, Valine, Leucine, Isoleucine
2.	Heterocyclic	—	Proline
3.	Aromatic	—	Phenylalanine, Tyrosine,Tryptophan
4.	Thioether	—	Methionine
5.	Hydroxy	—	Serine, Threonine
6.	Mercapto	—	Cysteine
7.	Carboxamide	—	Asparagine, Glutamine

II. Monoamino-dicarboxylic amnino acids : Aspartic acid, Glutamic acid

III. Diamino-monocaryboxylic amino acids : Lysine, Arginine, Histidine

C. On the basis of polarity of the side chain or R group. A more meaningful classification of amino acids is, however, based on the polarity of the R groups present in their molecules, *i.e.*, their tendency to interact with water at biological pH (near pH 7.0). The R groups of the amino acids vary widely with respect to their polarity from totally nonpolar or hydrophobic (water-hating) R groups to highly polar or hydrophilic (water-loving) R groups. This classification of amino acids emphasizes the possible functional roles which they perform in proteins. The system recognizes following 4 categories :

I. Amino acids with nonpolar R groups. The R groups in this category of amino acids are hydrocarbon in nature and thus hydrophobic. This group includes five amino acids with aliphatic R groups (alanine, valine, leucine isoleucine, proline), two with aromatic rings (phenylalanine, tryptophan) and one containing sulfur (methionine).

1. ***Alanine*** (α-aminopropionate). It was first isolated in 1888 from silk fibroin where it occurs in abundance, along with glycine and serine. *It is the parent substance of all the amino acids except glycine.* The various amino acids may be derived from alanine by replacement of one or two H atoms of the methyl group present on α-carbon atom. Alanine is the least hydrophobic of the 8 nonpolar (=hydrophobic) amino acids because of its small methyl side chain.

2. ***Valine*** (α-aminoisovalerate). It is widely distributed but rarely occurs in amounts exceeding 10%. It is a branched chain amino acid and can be derived from alanine by the introduction of two methyl groups in place of two H atoms of the methyl group present on α-carbon atom.

3. ***Leucine*** (α-aminoisocaproate). Its presence in proteins was shown, first of all the amino acids, by **Proust** in 1819. It was first isolated from cheese, but later was obtained in purer form from hydrolysates of wool. Leucine is generally more prevalent than valine or isoleucine in proteins. It is also a branched chain amino acid and is the next higher homologue of valine. As such, it has much in common with valine from the viewpoint of reactivity and function. It is also one of the few amino acids that are sparingly soluble in water.

4. ***Isoleucine*** (α-amino-β-methylvalerate). It is an isomer of leucine and is also a branched chain amino acid. It was discovered somewhat late (in 1904) by **Paul Ehrlich** (LT, 1854-1915) because it was difficult to separate from leucine owing to their similar chemical composition and properties. This amino acid is one of the 5 common ones that have more than one asymmetric carbon atoms. It has 2 asymmetric carbon atoms and thus occurs in 4 stereoisomeric forms, but the natural form obtained from protein hydrolysates is the L-*erythro* isomer, where erythro signifies that the orientation of the methyl and amino groups is similar to that found in erythrose, with the groups concerned on the same side of the carbon chain.

Valine, leucine and isoleucine are quite alike chemically and possess a branched carbon chain. As most animals cannot synthesize the branched carbon chain, these amino acids are, therefore, indispensable or essential in the diet.

5. ***Proline*** (2-pyrrolidinecarboxylate). *It is present in almost all proteins that have been studied*, but zein from corn and gelatin are relatively high in proline content. It is a cyclized derivative of

glutamic acid. Its α-amino group is not free but is substituted by a portion of its R group to yield a cyclic structure. The secondary amino group in proline is held in a rigid conformation. This reduces the structural flexibility of the protein because of the maximal steric hindrance of its side chain. *Unlike other amino acids, proline is very soluble in ethanol*. Moreover, it does not give many of the characteristic amino acid tests which are generally based on the presence of the unsubstituted α-amino group.

6. ***Phenylalanine*** (α-amino-β-phenylpropionate). It is one of the two common, clearly benzenoid amino acids. It is of widespread distribution and closely resembles tyrosine in structure. It also cannot be formed in animals because of its aromatic ring. Interestingly, (S)-phenylalanine is bitter whereas its stereoisomer, (R)-phenylalanine is sweet.

7. ***Tryptophan*** (α-amino-β-3-indole-propionate or β-indolylalanine). It was discovered in 1901 in the laboratory of **F.G. Hopkins**, one of the pioneers in biochemistry. *Tryptophan is the most complex amino acid found in proteins*. It is a heterocyclic amino acid and is a derivative of indole (with one N atom). Although widespread, it is usually limited in quantity. Tryptophan is the only amino acid of proteins which is nearly completely destroyed upon acid hydrolysis.

> Some authors use the term **"imino acids"** for proline and hydroxyproline which is not correct, because an imine is characterized by a $>CH = NH$ and not a $\substack{C \\ C}\!>NH$.

8. ***Methionine*** (α-amino-β-methylmercapto-butyrate). It is the only common amino acid possessing an ether linkage. Cereals have sufficient quantity of methionine whereas pulses lack in it. It is a methylation product of homocysteine. Apart from its role as a protein constituent and as an essential amino acid, methionine is also important as a donor of active methyl groups.

It may, however, be seen that **proline** (and also its 4-hydroxylated derivate, **hydroxyproline**) do not contain a primary amino group (—NH_2). The amino group in them is, in fact, utilized in ring formation. Both proline and hydroxyproline are actually the *substituted α-amino acids* as they have a secondary amino group $\left(\substack{C \\ C}\!>NH\right)$ in their molecule (By contrast, all other 19 protein amino acids are designated as *α-amino acids*). Since they are an integral part of the protein molecule and exhibit similar reactions, they are usually inadvertently referred to as amino acids.

II. Amino acids with polar but uncharged R groups. The R groups of these amino acids are more soluble in water *i.e.*, more hydrophilic than those of the nonpolar amino acids because they contain functional groups that form hydrogen bonds with water. This category includes 7 amino acids, *viz*., glycine, serine, threonine, tyrosine, cysteine, asparagine and glutamine. The polarity of these amino acids may be due to either a hydroxyl group (serine, threonine, tyrosine) or a sulfhydryl group (cysteine) or an amide group (asparagine, glutamine). The R group of glycine, a single hydrogen atom, is too small to influence the high degree of polarity of the α-amino and α-carboxyl groups.

9. ***Glycine*** or ***glycocoll*** (α-aminoacetate).*Glycine is the simplest amino acid and the only one lacking an asymmetric α-carbon atom*. Where it is present in a protein, the minimal steric hindrance of the glycine side chain allows much more structural flexibility than the other amino acids. Glycine represents the opposite structural extreme in comparison to proline. It is, therefore, unique in being optically inactive. It is one of the first to be isolated from proteins and has a characteristic sweet taste, hence so named (*glykos*^{G} = sweet). Glycine is present in abundance in scleroproteins. This occurs in especially large amounts in the protein gelatin (25% of the total amino acids) and in silk fibroin (40%). In addition, many nonprotein compounds also contain glycine as a component part, for instance glutathione (a common tripeptide), sarcosine (= N-methylglycine) and hippuric acid (= benzoylglycine). Methylation of glycine yields sarcosine which is found in many peptides. Further methylation produces betaine, a substance which exists exclusively in the zwitterionic or salt form (for the same reason, zwitterionic formulae are sometimes called as betaine structures).

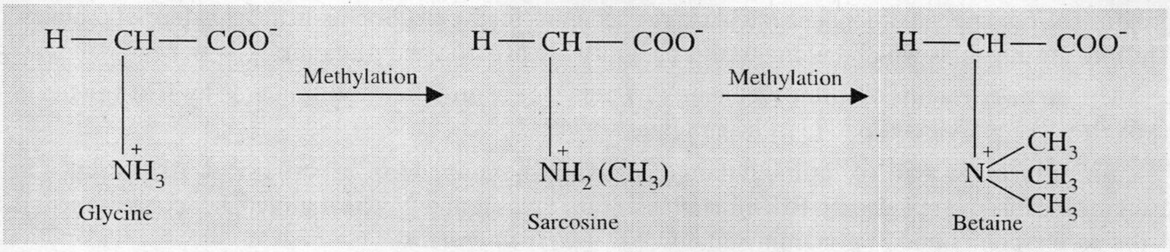

Glycine, the nonsymmetric amino acid is sometimes considered nonpolar. However, glycine's small R group (a hydrogen atom) exerts essentially no effect on the hydrophilicity of the molecule.

10. ***Serine*** (α-amino-β-hydroxypropionate; derived from the word, *serum*). It was first obtained from silk protein, sericin. It is found in all proteins but occurs in unusually high proportions in silk proteins, fibroin and sericin. This contains an alcoholic hydroxyl group which participates in ester formation. Esters with phosphoric acid have physiological significance as components of nucleotides and proteins. The unesterified serine residue appears to conduct special functions in many enzymes. The hydroxyl group on serine (and also on threonine) makes it much more hydrophilic and reactive than alanine and valine.

11. ***Threonine*** (α-amino-β-hydroxybutyrate). It was the last of the common amino acids to be discovered in proteins (**Meyer** and **Rose**, 1936). Its very name points towards its relationship with the sugar threose. Threonine is the next higher homologue to serine. It has two asymmetric carbon atoms and thus occurs in 4 stereoisomeric forms, namely D and L threonine and D and L allothreonine. The natural form is the L-*threo* isomer, hence its name. The *erythro* (unnatural) form of threonine is commonly referred to as allothreonine. Threonine is less abundant than serine in most proteins.

12. ***Tyrosine*** (α-amino β-(*p*-hydroxypyhenyl) propionate). It is the other aromatic amino acid normally found as a component of proteins. It was first isolated in 1857 from cheese and hence so named (*tyros*[G] = cheese). The phenolic group of tyrosine is weakly acidic and loses its proton above pH 9. It is sparingly soluble in water. It is destroyed during acid hydrolysis of proteins but may be isolated after enzymic hydrolysis. Some proteins (*e.g.*, protamines) contain almost no tyrosine.

13. ***Cysteine*** (α-amino-β-mercaptopropionate). It is the sulfur analogue of serine. *Cysteine is one of the most reactive amino acids found in proteins*. It contains sulfhydryl (SH) group which is quite reactive and esp., easily dehydrogenated. When it is dehydrogenated (*i.e.*, oxidized), two molecules join to form the amino acid cystine (see reaction on page 139). Fibrous proteins such as keratin from hair are especially rich in cystine (12%). *Cystine and its reduction product cysteine are together counted as one of the twenty amino acids.*

Cystine, from which the name 'cysteine' was later derived, was isolated in 1843 from urinary stones and was hence so named (*cystos*[G] = bladder).

14. ***Asparagine*** (β-amide of α-aminosuccinate). It was the first amino acid to be discovered in 1806 and was first isolated in 1813 from a plant, *Asparagus*, for which it was so named. This is the β-amide of aspartic acid and has been isolated from proteins after enzymic hydrolysis. Asparagine has long been known as a constituent of plant tissues.

15. ***Glutamine*** (γ-amide of α-aminoglutarate). It is the homologue of asparagine. This is the γ-amide of glutamic acid and has been isolated from proteins after enzymic hydrolysis. Free glutamine is found in many animal and plant tissues. It is even less stable toward hydrolysis than is asparagine. *Glutamine occurs widely in proteins.*

Glycine and cysteine have been included under the category of 'amino acids with nonpolar side chains' by **Alberts** *et al* (1994).

Amino acids of categories I and II are jointly referred to as *neutral amino acids* because each one of them contains one acidic and one basic group.

III. Amino acids with negatively charged (= acidic) R groups. These are monoamino-dicarboxylic acids. In other words, their side chain contains an extra carboxyl group with a dissociable proton. The resulting additional negative charge accounts for the electrochemical behaviour of proteins. The two amino acids which belong to this category are aspartic and glutamic.

16. **Aspartic acid** (α-aminosuccinate). Its presence in protein was discovered by **Ritthausen** in 1868. It is the parent compound of asparagine.

17. **Glutamic acid** (α-aminoglutarate). It is the homologue of aspartic acid. Its presence in protein was discovered also by **Ritthausen** in 1866. It was found in wheat gluten, hence so named. It is the parent compound of glutamine and *occurs widely in proteins*.

Both aspartic and glutamic acids in almost all proteins are linked to other amino acids only through their α-carboxyl and α-amino groups, not through their side chain carboxyl groups. Aspartic and glutamic acids are usually called aspartate and glutamate respectively to emphasize that their side chains are nearly always negatively charged at physiological pH.

IV. Amino acids with positively-charged (=basic) R groups. These are diamino-monocarboxylic acids. In other words, their side chain contains an extra amino group which imparts basic properties to them. Lysine, arginine and histidine belong to this category.

18. **Lysine** (α, ε-diaminocaproate). It has a second amino group at ε-position. Lysine is generally abundant in animal proteins but present in limited amounts in plant proteins, such as those of corn and wheat. Ionic charges are provided by protonation of the amino group of the e-carbon. It is nutritionally important since it is not synthesized by higher animals. Pulses contain good amount of lysine whereas cereals lack in it.

19. **Arginine** (α-amino-δ-guanidinovalerate). It is abundant in highly basic proteins of the cell nucleus (histones) and in sperm proteins, such as the protamines from salmon and herring sperm. In the latter, arginine may be as much as 80% of the total amino acids. It is found generally in all proteins, although in lesser amounts. Arginine is unique in possessing the guanidinium group, and due to which *it is more strongly basic than lysine*. Guanidinium group is very basic in character because its positive charge is stabilized by resonance. Protonation of the guanidinium group provides ionic charges.

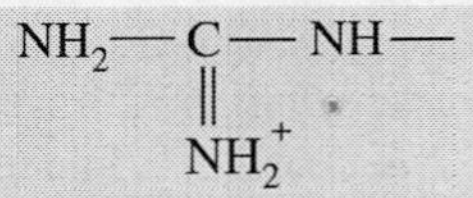

Guanidinium group (charged form) of arginine

20. **Histidine** (α-amino-β-imidazolopropionate). *Histidine is the last entry in the list of "magic 20".* It contains a weakly basic imidazolium R group with pK'_a value of 6.0, and, therefore, is less than 10% protonated at pH 7. In many enzyme proteins, it functions as a proton donor or acceptor. *Histidine is the only amino acid which has a proton that dissociates in the neutral pH range*. It is this property which allows certain histidine residues to play an important role in the catalytic activities of

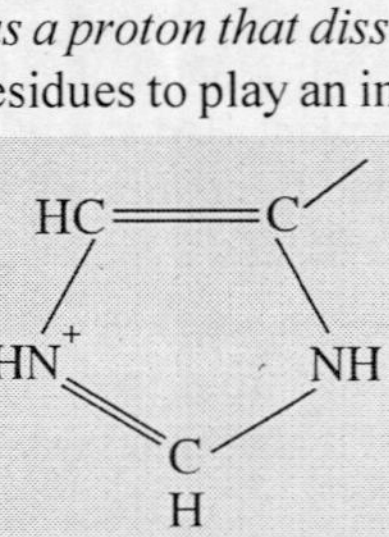

Imidazolium group (charged form) of histidine

some enzymes. Histidine occurs in limited quantities in most proteins, but hemoglobin, protamines and histones contain relatively large amounts. It is also a component in the simple peptides carnosine and anserine. Histidine's basic properties are clearly marginal. The two ring nitrogen atoms have a relatively weak affinity for an H^+ and are only partly positive at neutral pH.

Carnosine ($C_9H_{14}O_2N_4$) is an optically-active crystalline dipeptide ; m.p. 260°C ; found in muscle tissue.

Abbreviations and Symbols

Three-letter abbreviations (Table 9-3) are used in text books to designate amino acid residues in peptide sequences but *not to designate free amino acids*. **One-letter symbols** were devised for computer storage and retrieval of sequence date but are also being used for tabulation of long sequences.

Table 9-3. Abbreviations and symbols used for amino acids

Three-letter abbreviation †	*One-letter symbol* ‡	*Amino acid*
Ala	A	Alanine
Arg	R	Arginine
Asn, Asp-NH^*_2] Asx	N] B	Asparagine
Asp	D]	Aspartic acid
Cys, Cys-SH*	C	Cysteine
Gln, Glu-NH^*_2] Glx	Q] Z	Glutamine
Glu	E]	Glutamic acid
Gly	G	Glycine
His	H	Histidine
Ile	I	Isoleucine
Leu	L	Leucine
Lys	K	Lysine
Met	M	Methionine
Phe	F	Phenylalanine
Pro	P	Proline
Ser	S	Serine
Thr	T	Threonine
Trp	W	Tryptophan
Tyr	Y	Tyrosine
Val	V	Valine

† Note that the abbreviations for amino acids are the first letters of their names except tryptophan (Trp), isoleucine (IIe), asparagine (Asn) and glutamine (Gln).

‡ Also note that barring 3 letters (J, O, U), all the remaining 23 letters of English alphabet have been utilized in giving one-letter symbol to the standard amino acids. The symbols for the small amino acids are the first letters of their names (*e.g.*, A for alanine, G for glycine and L for leucine); the other symbols have been agreed upon by convention. However, the letter X is used for indicating an undetermined or nonstandard amino acid.

* Not preferred

The 3-letter abbreviations for the **derivatives of amino acids** are :

Cys-SO_3H	Cysteic acid
Cys-S-S-Cys	Cystine
fMet	N-formylmethionine
Hyl	Hydroxylysine
Hyp	Hydroxyproline
Pgl, pGlu, < Glu	Pyroglutamic acid

PEPTIDES

Peptide Bond

The amino acid units are linked together through the carboxyl and amino groups to produce the primary structure of the protein chain. The bond between two adjacent amino acids is a special type of amide bond, in which the hydrogen atom of amino (-NH_2) group is replaced by an R radical. Such a substituted amide bond is known as the **peptide bond**.

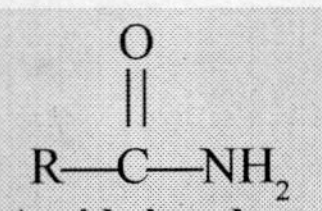

Amide bond
(an unsubstituted amide)

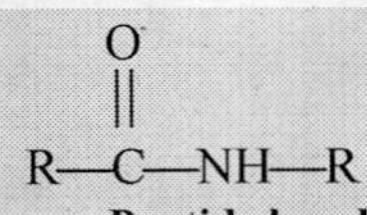

Peptide bond
(an *N*-substituted amide)

And the chain, thus formed by linking together of many amino acid units, is called a **peptide chain** (Fig. 9–5).

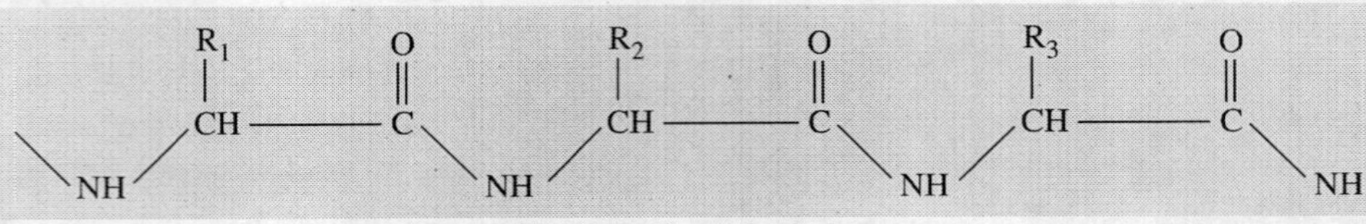

Fig. 9–5. Peptide chain

The characteristic structure of the peptide bond is shown in Fig. 9–6.

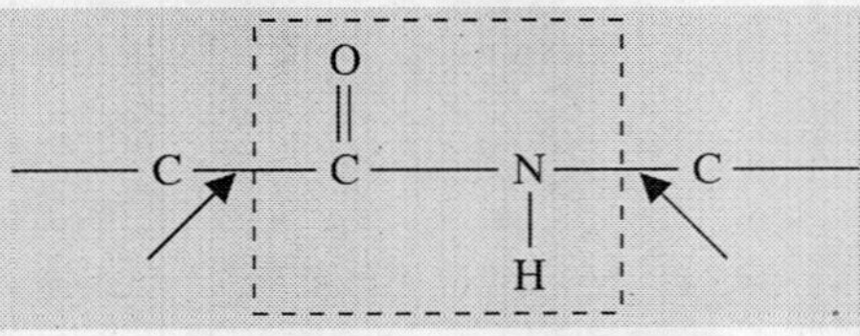

Fig. 9–6. Peptide bond

The peptide bond is shown enclosed in the dashed box. The four atoms (C, O, N, H) of the peptide bond form a rigid planar unit. There is no freedom of rotation about the C—N bond. On the contrary, the 2 single bonds (shown with arrows) on either side of the rigid peptide unit, exhibit a high degree of rotational freedom.

Each peptide chain is of considerable length and may possess from 50 to millions of amino acid units. Depending on the number of amino acid molecules composing a chain, the peptides may be termed as a **dipeptide** (containing 2 amino acid units), a **tripeptide** (containing 3 amino acid units) and so on. If a peptide is made up of not more than 10 amino acids, it is called an **oligopeptide ;** beyond that it is a **polypeptide**. Polypeptides when they are made up of over 100 amino acids are, sometimes, called as **macropeptides.** *Strictly speaking, the proteins are polypeptides with more than 100 amino acids.* All naturally-occurring important peptides, however, possess a shorter individual name, such as glutathione etc. Proteins differ widely in amino acid content. Various types of proteins in an organism may have varied amounts of a particular amino acid. Some amino acids are in abundance in one protein, may be in meagre amounts in others and may even be lacking in the rest. Tryptophan, for instance, lacks in certain proteins. However, *most of the proteins contain all the 20 amino acids.* As the number and manner in which the amino acids are grouped is highly variable, the number of proteins approaches almost to infinity.

Sometimes, the word **'proteinoids'** is used for short polypeptides containing up to 18 amino acids (**Fried GH**, 1990).

It is analogous to the indefinite number of words that can be formed with the 26 letters of English alphabet. While the words have to be restricted in length, there is no such restriction regarding the number of amino acids that may form a protein. According to an estimate given by **Erlene B.**

Cunningham (1978), if each protein molecule were to consist of only 250 amino acid residues, the utilization of all the 20 different monomers would permit the formation of 10325 different protein molecules ! In reality, protein molecules often contain more than 250 aminoacyl units, and hence there is possibility for an even greater number of different protein molecules.

N- and C-terminals

Each amino acid in the chain is termed a *residue*. The two ends of the peptide chain are named as amino terminal and carboxyl terminal or simply as an N-terminal and C-terminal respectively. These two terminal groups, one basic and another acidic, are the only ionizable groups of any peptide chain except those present in the side chain. The terminal amino acid with the free amino group is called as the *N-terminal amino acid* and the one with the free carboxyl group at the other end as *C-terminal amino acid.*

Representation of Peptide Chain

To fix the convention for representation of peptide structures in mind, it is helpful to imagine a rattlesnake moving from left to right across the page. The C-terminal residue forms its fangs and the N-terminal residue its rattle (Fig. 9–7).

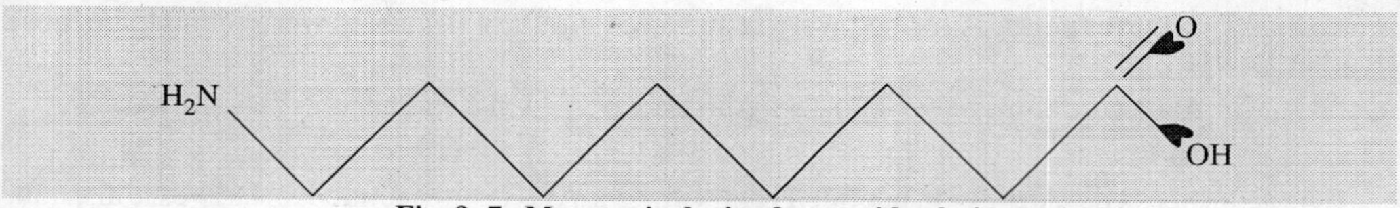

Fig. 9–7. Mnemonic device for peptide chain

Naming of Peptide Chain

In naming a polypeptide, the convention is that the N-terminal residue (which is shown at the left hand part of the structure) is written first and the C-terminal residue in the end. The names

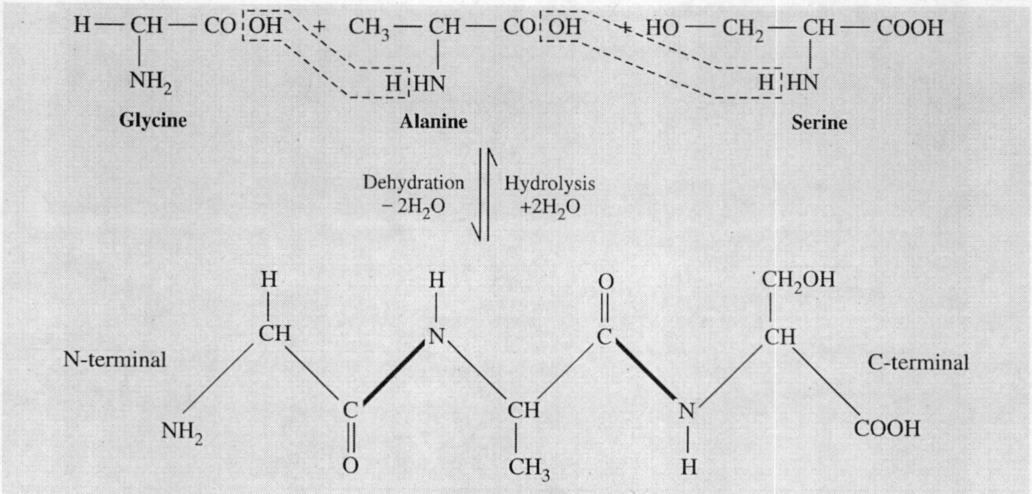

Glycyl-L-alanyl-L-serine

Fig. 9–8. Construction of a tripeptide chain from three different amino acids

The formation of each peptide bond (indicated by bold lines) involved splitting of a molecule of water. Thus, the process of synthesis of proteins may be deemed as essentially a dehydration synthesis.

of various intermediary amino acid residues are written in the same sequence as they are placed. Further, the names of all the amino acid residues, except the last one, are written by adding the suffix *-yl* because all these are the acyl groups. The name of the last amino acid, however, is written as such. For example, a tripeptide containing glycine, alanine and serine (structure shown in Fig. 9–8) is named as glycyl-L-alanyl-L-serine and abbreviated as Gly-Ala-Ser. If the sequence of amino acid in such a tripeptide is not known, the abbreviation would be (Gly, Ala, Ser), the parenthesis and commas indicating that only the compostiton of the tripeptide is known.

Similarly, glutathione or GSH (a tripeptide containing glutamic acid, cysteine and glycine) is named as γ-glutamyl-cysteyl-glycine and abbreviated as Glu-Cys-Gly. Glutathione is a naturally occurring and widely distributed polypeptide.

Determination of the Amino Acid Sequence of a Polypeptide

This can be explained by taking the example of a **dodecapeptide** whose composition was found to be Ala_2, Arg, Glu, Gly, Leu, Lys_2, Phe, Tyr_2, Val. It was determined that the N-terminal amino acid of the dodecapeptide was valine and the C-terminal amino acid, leucine. Hydrolysis of the dodecapeptide by trypsin yielded four peptides whose structures were determined and found to be those given in A to D.

Tyr-Glu-Lys	Phe-Gly-Arg	Val-Lys	Ala-Tyr-Ala-Leu
A	**B**	**C**	**D**

Since valine was the N-terminus and leucine the C-terminus of the dodecapeptide, it is apparent that peptide C must represent the amino acid sequence at the N-terminal end, and peptide D the amino acid sequence at the C-terminal end of the dodecapeptide. To establish the order of the A and B peptides in the interior of the dodecapeptide, another sample of the dodecapeptide was hydrolyzed by *chymotrypsin*, the four peptides formed were sequenced, and their structures were found to be those given in **E** to **H**.

Ala-Leu	Glu-Lys-Ala-Tyr	Val-Lys-Phe	Gly-Arg-Tyr
E	**F**	**G**	**H**

The sequences Gly-Arg-Tyr in peptide **H** and Glu-Lys-Ala-Tyr in peptide **F** clearly establish that peptide **B** must precede peptide **A** in the dodecapeptide. Hence, the structure of the dodecapeptide is unambiguously determined to be that as shown belew :

Val-Lys-Phe-Gly-Arg-Tyr-Glu-Lys-Ala-Tyr-Ala-Leu.

Stereochemistry of Peptide Chains

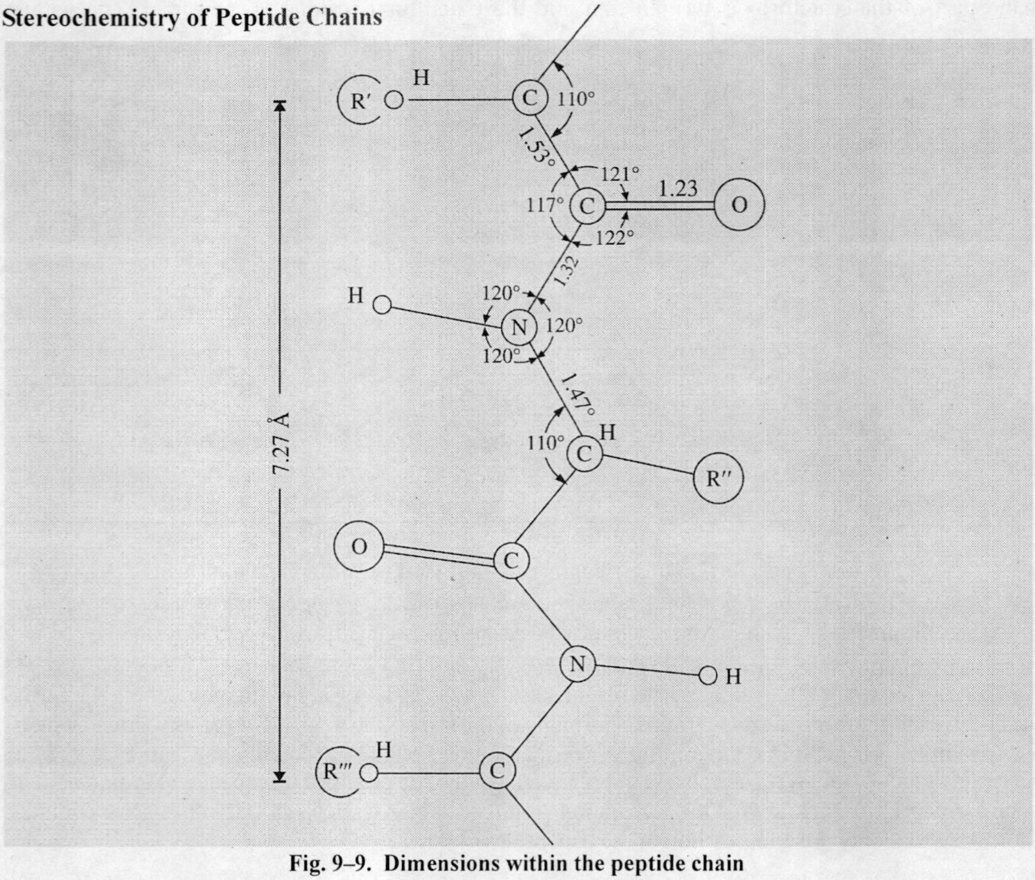

Fig. 9–9. Dimensions within the peptide chain

All proteins are made of amino acids of L-configuration. This fixes the steric arrangement at the α-carbon atom. The dimensions of the peptide chain are known exactly. These have been depicted in Fig. 9–9.

The peptide bond, which is an imide (substituted amide) bond, has a planar structure. The 6 atoms within the plane are related to each other by bond lengths and angles that vary little from amino acids residue to amino acid residue. Only 3 of these bonds are part of the peptide chain *per se* : the α-carbon to carbonyl carbon bond, the C–N bond, and the imide nitrogen to α-carbon bond. Since the double bond character of the C–N bond limits rotation about it, only the first and the last allow rotation. The rotation angles ϕ and ψ establish the relative positions of any 2 successive amide planes along the polypeptide chain. The α-carbon atoms can be thought of as swivel (a mechanism of two parts which can resolve the one on the other) centres for the adjacent amide planes.

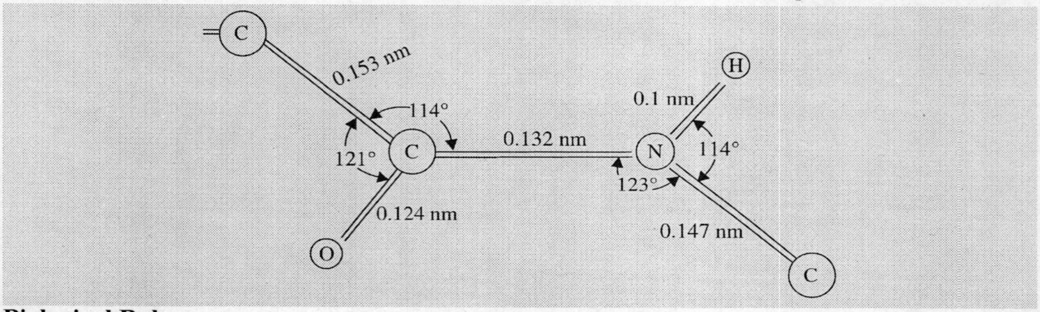

Biological Roles

Peptides participate in a number of biological activities.

1. They serve as intermediates in the formation of **proteins.**

2. They appear as constituents in a group of compounds called **alkaloids.** Majority of these have been isolated from fungi, although they are also found in higher plants. Ergotamine is a peptide alkaloid from rye ergot and has pronounced pharmacological properties. The four components of this alkaloid are lysergic acid, alanine, proline and phenylalanine.

3. Many of them possess **antibacterial activities** and are usually present in fungi and bacteria. Penicillin G with 3 components (valine, cysteine and phenylacetic acid) is a common antibiotics.

4. Certain other peptides serve as **growth factors.** Folic acid, a water-soluble vitamin, is a noteworthy example of it (see Chapter 22). Another group of peptides serving as growth factor for a variety of microorganisms is streptogenins.

5. Higher animals do synthesize certain peptides serving as **hormones** (see Chapter 19).

6. Certain peptides like glutathione participate in controlling the **oxidation-reduction potential** of the cell. This may also serve as a key intermediate in electron-transfer systems.

7. A direct correlation has been found to exist between the amount of peptides in the urine of the patients and their **mental state of disturbance.** A group of Norwegian doctors have found an excess of peptides in urine specimens from patients with psychiatric disturbances. The peptides have been shown to induce in animals some of the conditions for the development of psychiatric disorders which lead to mania, depression or schizophrenia. The urine tests can, thus, indicate if a person is suffering from mental illness. The cause of this hypersecretion of peptides is not yet well established. It may exist from birth in organic genetic derangements or may be induced in healthy people by environmental factors.

CHEMICAL BONDS INVOLVED IN PROTEIN STRUCTURE

Given a full assortment of amino acids, a cell can synthesize all of its protein components. *Protein synthesis is a multiple dehydration process* (refer Fig. 9–8). The net structure of a protein becomes possible as a result of linking together of various amino acid units. The union of these amino acids to each other forming a chain and also among various amino acid residues of different chains involves various types of chemical bonds (Fig. 9–10). These are described below.

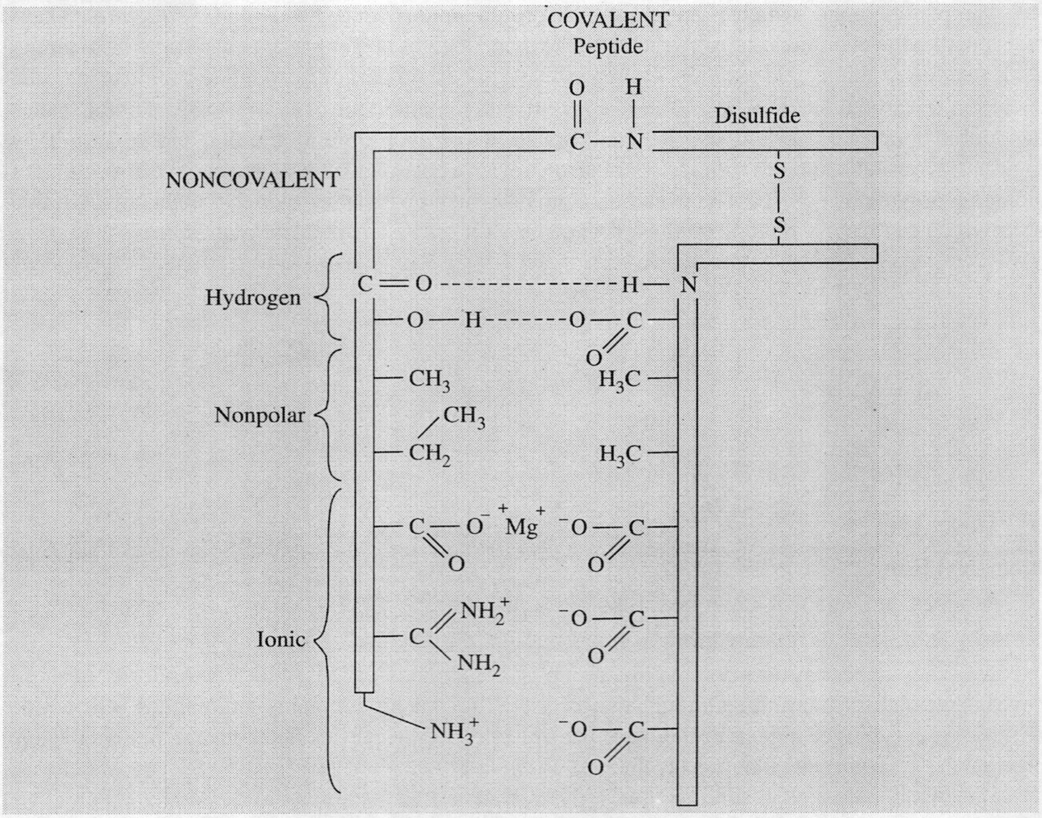

Fig. 9–10. Types of chemical bonds involved in protein structure

(*After Hartman PE and Suskind SR, 1969*)

A. Primary Bond

The principal linkage found in all proteins is the covalent **peptide bond, —CO—NH—** (**Fischer,** 1906). It is a specialized amide linkage where C atom of—COOH group of one amino acid is linked with the N atom of—NH_2 group of the adjacent amino acid. *Peptide bond is, in fact, the backbone of the protein chain.*

B. Secondary Bonds

Many of the properties of proteins, however, do not coincide with the linear chain structure, thus indicating that a variety of bonds other than the peptide exist in them. These secondary bonds, as they are called, hold the chain in its natural configuration. Some of the secondary bonds commonly found in proteins are listed below :

1. **Disulfide Bond (—S—S—).** In addition to the peptide bond, a second type of covalent bond found between amino acid residues in proteins and polypeptides is the disulfide bond, which is formed by the oxidation of the thiol or sulfhydryl (—SH) groups of two cysteine residues to yield a mole of **cystine**, an amino acid with a disulfide bridge (Fig. 9–11). In generalized form, the above reaction may be written as :

$$2R—SH + 1/2\ O_2 \rightleftharpoons R—S—S—R + H_2O$$

A disulfide bond is characterized by a bond strength of approximately 50 kilocalories per mole and a bond length of about 2Å between the two sulfur atoms. Hence, disulfide bond formation between 2 cysteine residues located some distance apart in the polypeptide chain requires that the polypeptide chain be folded back on itself to bring the sulfur groups close together. Although disulfide bridges are very strong, when compared to the strength of noncovalent bonds they are very short-range since, as with all covalent bonds, even a slight extension breaks them completely. They, therefore, only stabilize the tertiary structure when it has reached something approximating to its final form.

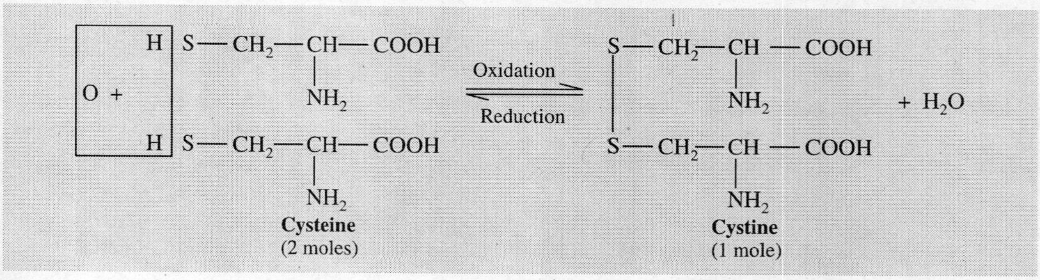

Fig. 9–11. Formation of cystine (= dicysteine)

Ocytocin (Fig. 9–12), a hormone stimulating the contraction of smooth muscles especially during childbirth, is an example where an internal disulfide bond is present between two cysteine units separated from each other in the peptide chain by 4 other amino acid units.

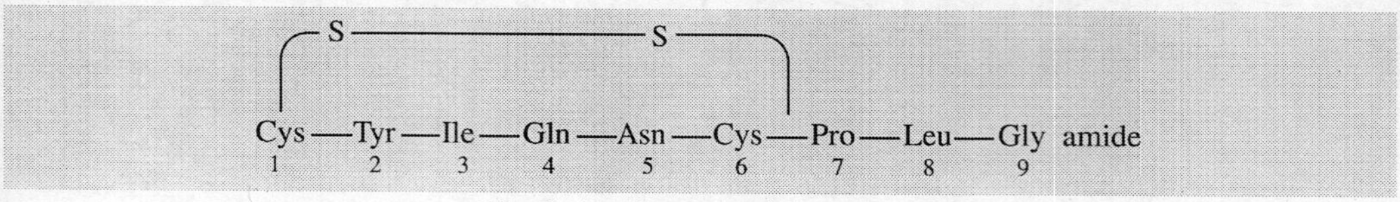

Fig. 9–12. Structure of ocytocin

Insulin (Fig. 9–13) is another excellent example where two peptide chains are linked together by 2 disulfide bonds. The presence of an internal disulfide bond in the glycyl (or A) chain between residues 6 and 11 is noteworthy.

These two chemical bonds, namely, peptide and disulfide, are relatively stable. Both these bonds, collectively or individually, maintain the linear form (or the *primary structure*) of the protein molecule.

2. Hydrogen Bond (>CO......HN<). When a group containing a hydrogen atom, that is covalently-bonded to an electronegative atom, such as oxygen or nitrogen, is in the vicinity of a second group containing an electronegative atom, an energetically favourable interaction occurs which is referred to as a hydrogen bond. The formation of a hydrogen bond is due to the tendency of hydrogen atom to share electrons with two neighbouring atoms, *esp.*, O and N. For example, the carbonyl oxygen of one peptide bond shares its electrons with the hydrogen atom of another peptide bond. Thus,

An interaction sets in between a C=O group and the proton of an NH or OH group if these groups come within a distance of about 2.8 Å. This secondary valence bond is symbolized by a dotted line, ... (The solid line, however, represents the normal covalent bond). The strength of the hydrogen bond is only 5 to 8 kilocalories per mole and is maximal when the bond is linear (Fig. 9–14). Hydrogen bonding between amides or peptides of the type depicted in Fig. 9–14 (*c*) plays an important role in stabilizing some conformations of the polypeptide chain.

Silk fibroin, composed mainly of glycine, alanine and serine units, is an example of the presence of hydrogen bonds involving the imide (>NH) and carbonyl (>C=O) groups of the peptide bonds. Here the hydrogen bonds link the vicinal peptide chains (refer Fig. 9–22).

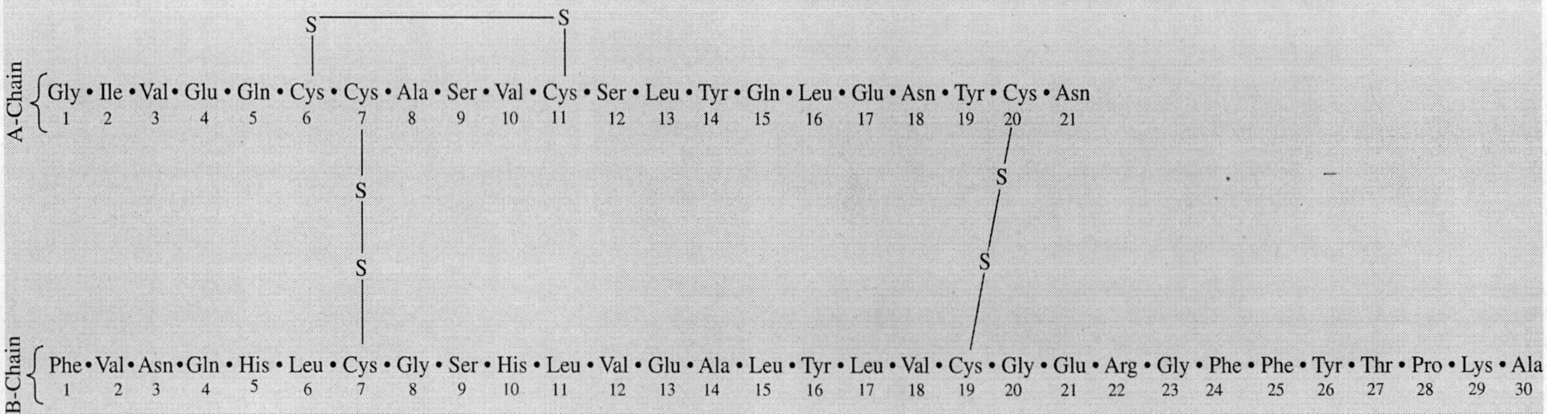

Fig. 9–13. Structure of bovine insulin

Numbers refer to specific amino acid residues. Glycine and phenylalanine are the N-terminal amino acids and the asparagine and alanine, the C-terminal amino acids. The A-chain of the insulin of man pig, dog, rabbit and sperm whale are identical. The B-chains of the cow, pig dog, goat and horse are identical. The amino acid replacements in the A-chain usually occur in positions 8, 9 and 10 (for details, refer Chapter 19 : Animal Hormones).

In other proteins like **keratin of wool**, however, the hydrogen bonds link the side chains so that a single peptide chain is held in a coiled or helical form (refer Fig. 9–21).

Since the binding energy of a hydrogen bond amounts to only 1/10th of that of a primary valence, the hydrogen bonds are relatively weak linkages but many such bonds collectively exert considerable force and help in maintaining the helical structure (or the *secondary structure*), characteristic of many proteins (**Mirsky** and **Pauling**, 1936).

3. Nonpolar or **Hydrophobic Bond.** Many amino acids (like alanine, valiné, leucine, isoleucine, methionine, tryptophan, phenylalanine and tyrosine) have the side chains or R groups which are essentially hydrophobic, *i.e.*, they have little attraction for water molecules in comparison to the strong hydrogen bonding between water molecules. Such R groups can unite among themselves with elimination of water to form linkages between various segments of a chain or between different chains. This is very much like the coalescence of oil droplets suspended in water.

The association of various R groups in this manner leads to a relatively strong bonding. It also serves to bring together groups that can form hydrogen bonds or ionic bonds in the absence of water. Each type linkage, thus, helps in the formation of the other; the hydrophobic bonds being most efficient in this aspect. The hydrophobic bonds also play important role in other protein interactions, for example, the formation of enzyme-substrate complexes and antibody- antigen interactions.

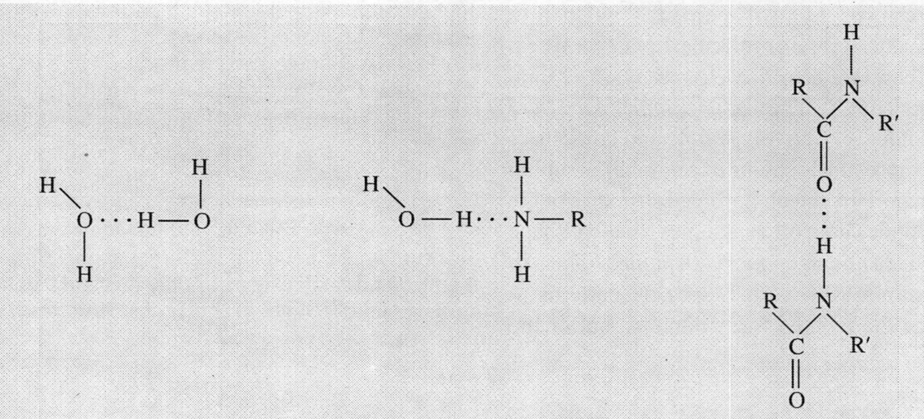

(*a*) (*b*) (*c*)

Fig. 9–14. Some common examples of hydrogen bonding

The hydrogen bonds (represented by a dotted line) have been shown between (*a*) two water molecules, (*b*) water and an amine, and (*c*) two amide groups.

4. Ionic or **Electrostatic Bond** or **Salt linkage** or **Salt bridge.** Ions possessing similar charge repel each other whereas the ions having dissimilar charge attract each other. For example, divalent cations like magnesium may form electrostatic bonds with 2 acidic side chains.

Another instance of ionic bonding may be the interaction between the acidic and basic groups of the constituent amino acids shown at the bottom of Fig. 9–10. The R groups of glutamic acid and aspartic acid contain negatively charged carboxylate groups, and the basic amino acids (arginine, histidine, lysine) contain positively charged amino groups in the physiological pH range. Thus, these amino acids contribute negatively charged and positively charged side chains to the polypeptide backbone. When two oppositely charged groups are brought close together, electrostatic interactions lead to a strong attraction, resulting in the formation of an electrostatic bond. In a long polypeptide chain containing a large number of charged side chains, there are many opportunities for electrostatic interaction. Intramolecular ionic bonds are rather infrequently used in the stabilization of protein structure but when they are so used, it is often with great effect. In fact, ionized groups are more frequently found stabilizing interactions between protein and other molecules. Thus, ionic bonds between positively charged groups (side chains of lysine, arginine and histidine) and negatively charged groups (COO^- group of side chain of aspartic and glutamic acids) do occur.

These ionic bonds, although weaker than the hydrogen bonds, are regarded as responsible for maintaining the folded structure (or the *tertiary structure*) of the globular proteins.

PROTEIN CONFIGURATION

To describe a complicated macromolecule like protein, the biochemists have, for convenience, recognized 4 **basic structural levels of organization** of proteins based on the degree of complexity of their molecule (Fig. 9–15). These structural levels were first defined by **Linderström–Lang** and are often referred to as primary, secondary, tertiary and quaternary. Three of these structural levels (primary, secondary and tertiary) can exist in molecules composed of a single polypeptide chain, whereas the fourth (*i.e.*, quaternary) involves interactions of polypeptides within a multichained protein molecule. In mathematical term, these are also depicted as 1°, 2°, 3° and 4°, respectively. The basic *primary structure* of a protein is relatively simple and consists of one or more linear chains of a number of amino acid units. This linear, unfolded structure or the polypeptide chain often assumes a helical shape to produce the *secondary structure*. This, in turn, may fold in certain specific patterns to produce the twisted three-dimensional or the

IRVING GEIS
(LT, 1908 – 1997)

Irving Geis is well known for his lucid visualizations of molecular structures, particularly proteins and nucleic acids. These have appeared in *Scientific American* for the past thirty five years and in major chemistry, biology, and biochemistry textbooks. He is a co-author with R.E. Dickerson, Director of the Molecular Biology Institute of UCLA, of 3 books 1. Chemistry. Matter and the Universe 2. The Structure and Action of Proteins, and 3. Hemoglobin: Structure, Function, Evolution and Pathology.

In addition to drawing, painting and writing, Irving Geis is a frequent lecturer at universities and medical schools on protein structure and function.

A recent Guggenheim fellowship made possible the assembly and cataloging of his drawings and paintings into **The Geis Archives of Molecular Structure.**

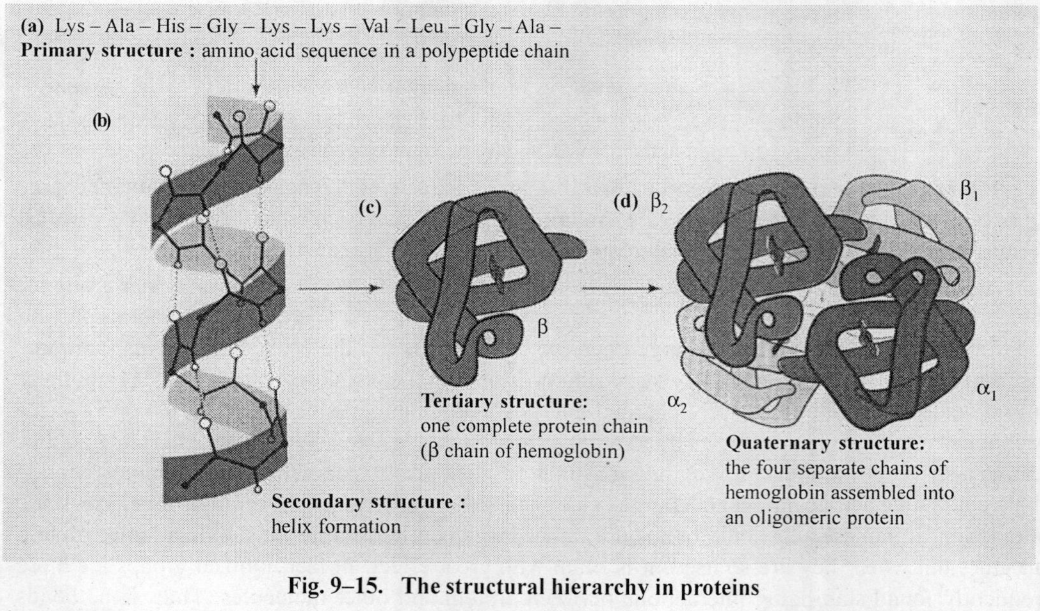

Fig. 9–15. The structural hierarchy in proteins

tertiary structure of the protein molecule. Finally, certain other proteins are made up of subunits of similar or dissimilar types of the polypeptide chains. These subunits interact with each other in a specific manner to give rise to the so-called *quaternary structure* of the protein. This, in fact, defines the degree of polymerization of a protein unit.

1. Primary Structure : Amino Acid Sequence

The *primary structure* of a protein refers to the number and sequence of amino acids, the constituent units of the polypeptide chain. The main mode of linkage of the amino acids in proteins is the peptide bond which links the α-carboxyl group of one amino acid residue to the α-amino group of the other. The proteins may consist either of one or of more peptide chains.

LINUS CARL PAULING
(LT, 1901-1994)

Linus Pauling (with his α-helix ball-and-stick model in the photograph) , the son of a German father and English mother, was an American physical chemist. Although, he had initially an undistinguished school education, Pauling obtained his doctorate degree in 1925 from California Institute of Technology (Caltech). He was one of the exceptional men of his times and carried the message of Quantum Mechanics to the New World. His pioneering application of principles of Quantum Physics to explain chemical properties, his investigations of molecular structures of proteins and, above all, his work on the nature of chemical bond rank among the outstanding pieces of chemical research of the century. To the graduate students of Chemistry in the 1940s and 50s, his book, *'The Nature of Chemical Bond'* (1939) became a Bible. At age 30, Pauling had become a full professor in 1931 and Chairman of Chemistry and Chemical Engineering some 6 years later. With Robert Corey (LT, 1867-1971), he investigated amino acids and polypeptides. Around 1940s, he and his colleagues conducted researches on antibody-antigen reactions. Pauling was the first one to devise a scale for comparing the electronegativity of different elements.

Pauling had his fair share of awards, which included, besides **Nobel Chemistry Prize (1954),** the prestigious Davy Medal of the Royal Society (1947), the Willard Gibbs Medal (1946) and the Presidential Award for Merit for his distinguished work during World War II in the Explosives Division of the Natural Research Commission (USA).

Soon after war, however, Pauling spoke bitterly against the nuclear arms race and advocated multilateral disarmament and an end to atomic testing. Following the publication of his famous book, *'No More War !'* (1958), he then sent, in Jan'58, a petition signed by 11,021 scientists to the United Nations, urging an end to the testing of nuclear weapons. For all this, he earned the **1962-Nobel Peace Prize,** which was presented to him on October 10, 1963, the day a US-Soviet partial nuclear test ban treaty came into force. He also won the **International Lenin Peace Prize**. His political activities took a heavy toll of his time and energy. The result was that he did not produce scientific work of significance after 1951. Nevertheless, during his career in science spanning more than 60 years, Pauling published several books and more than 1,000 scientific papers. He also established **Linus Pauling Institute of Science and Medicine** at Palo Alto.

However, Pauling's fame had been tarnished in his later years by his strong advocacy of vitamin C (ascorbic acid) as a 'wonder drug' and helpful in maintaining youthness for a longer period. He also published a book, *'Vitamin C and the Common Cold'* (1971) wherein he claimed that by ingesting 1,000 mg daily of vitamin C (instead of the 60 mg daily minimum as recommended by the U.S. National Research Council), a person would catch 45% fewer colds and suffer 60% fewer days of illness. His findings were bitterly criticized experimentally by Terene Anderson of the University of Toronto in Canada. But this should not minimize the magnitude of his earlier contributions. Till today, he holds the distinction of being **the only person to have won two Nobel Prizes on his own.** His life fell into 3 distinct phases : the first evoked *reverence*, the second *love* and the third *ridicule*. Truly, Pauling ranks among the most versatile scientists of more than one generation of 20th century.

2. Secondary Structure : Helix Formation or **Local Folding**

If the peptide bonds were the only type of linkage present in proteins, these molecules would have behaved as irregularly coiled peptide chains of considerable length. But the globular proteins, however, do show some regular characteristic properties, indicating the presence of a regular coiled structure in these molecules. This involves the folding of the chain which is mainly due to the presence of hydrogen bonds. Thus, folding and hydrogen bonding between neighbouring amino acids results in the formation of a rigid and tubular structure called a **helix**. This constitutes the *secondary structure* of proteins, which refers to the steric or spatial relationship of amino acids that are near to each other in the amino acid sequence.

Based on the nature of hydrogen bonding (whether intramolecular or intermolecular), **Pauling** and **Corey** (1951) identified two regular types of secondary structure in proteins : alpha helix (α-helix) and beta pleated sheet (β-pleated sheet).

α-Helix

Early x-ray diffraction studies conducted by **William Astbury** in 1930s, of fibrous proteins such as hair and wool, showed a major **periodicity** or repeat unit of 5.0 to 5.5 Å, indicating some regularity in the structure of these proteins. A minor repeat unit of 1.5 Å was also observed. With x-ray diffraction studies, **Pauling** and **Corey** (1951) found that a polypeptide chain with planar peptide bonds would form a right-handed helical structure by simple twists about the α-carbon-to-nitrogen and the

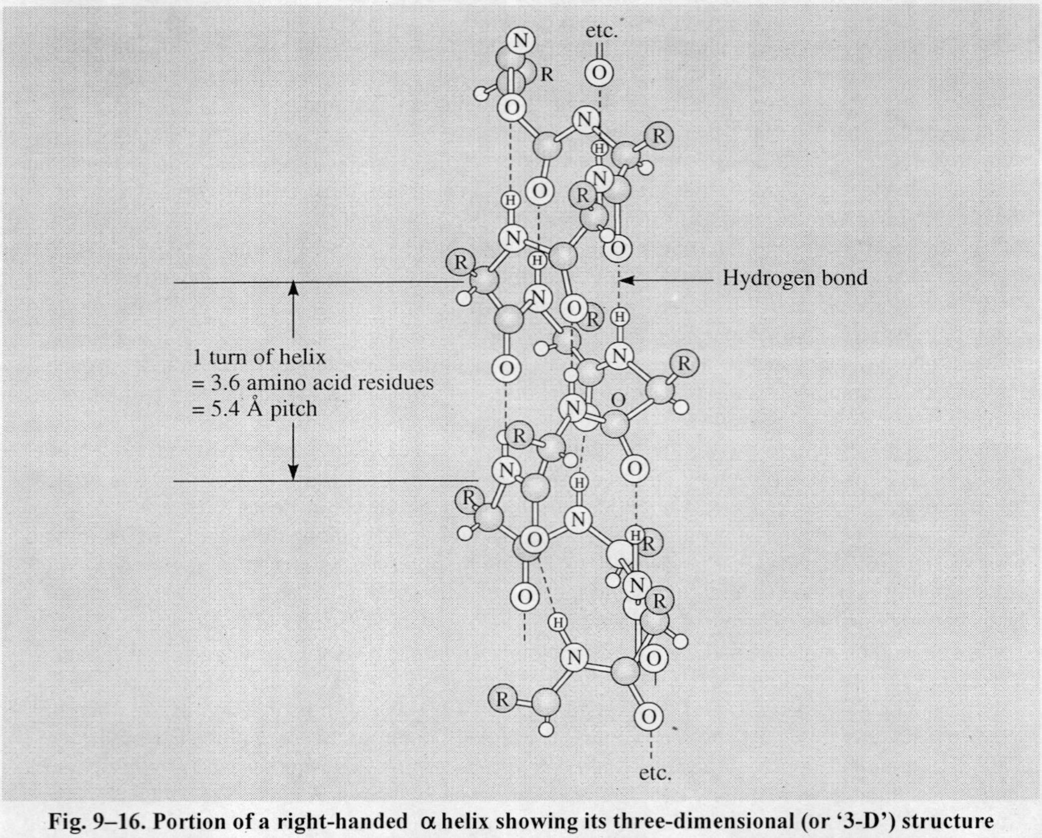

Fig. 9–16. Portion of a right-handed α helix showing its three-dimensional (or '3-D') structure

All the shaded balls represent carbon atoms or residues (R) of amino acids.

(*Redrawn from Linus Pauling and Robert Corey, 1955*)

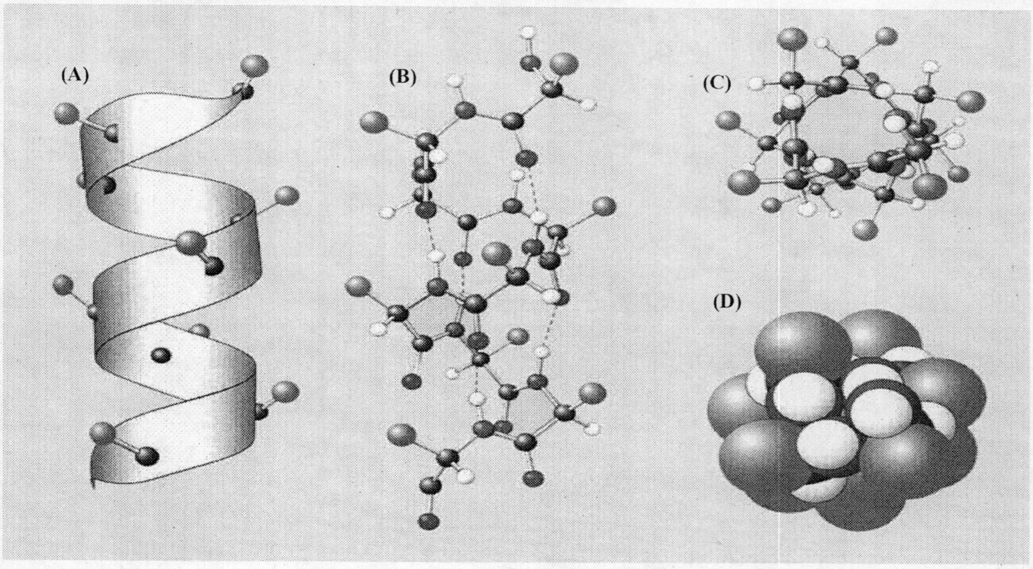

Fig. 9–17. Structure of the α helix

(A) A ribbon depiction with the α-carbon atoms and side chains (*green*) shown.

(B) A side view of a ball-and-stick version depicts the hydrogen bonds (*dashed lines*) between NH and CO groups.

(C) A cross-sectional view of the α-helix. The end view shows the coiled backbone as the inside of the helix and the side chains projecting outward. The van der Waal's radii of the atoms are larger than shown here for the sake of clarity. Hence, there is actually no free space inside the helix.

(D) A space-filling view of part (c) shows the tightly- packed interior core of the helix.

α-carbon-to-carboxyl carbon bonds. They called this helical structure as **α-helix.** The helix is so named because of the mobility of α-carbon atoms.

The α-helix is a rodlike structure. The tightly coiled polypeptide main chain forms the inner part of the rod, and the side chains extend outward in a helical array (Figs. 9–16 and 9–17). The α-helix is stabilized by hydrogen bonds between the NH and CO groups of the main chain. The CO group of each amino acid is hydrogen-bonded to the NH group of the amino acid that is situated four residues ahead in the linear sequence (Fig. 9–18). Thus, all the main chain CO and NH groups are hydrogen-bonded.

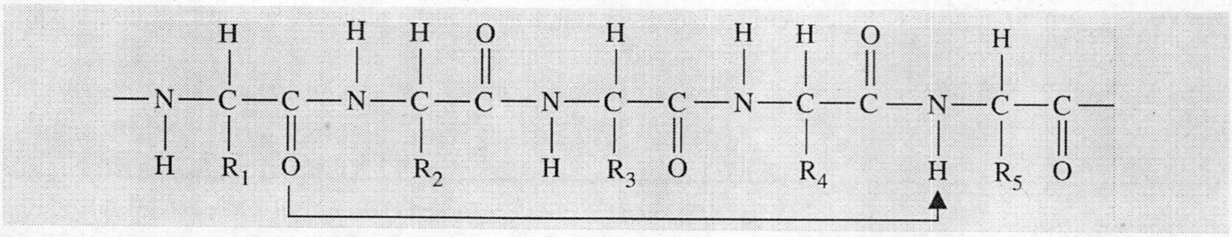

Fig. 9–18. Diagram showing that in the α-helix, the CO group of residue *n* is hydrogen-bonded to the NH group of residue (*n* + 4).

It is, thus, apparent that the α-helical structure depends on the intramolecular (= intrachain) hyrogen bonding between the NH and CO groups of peptide bonds. The hydrogen bonding occurs spontaneously and, as a result, a polypeptide can assume a rod-like structure with well-defined dimensions (Fig. 9–19). The α-helix (or α-conformation, as it is also called) has a pitch of 5.4 Å (= 0.54 nm) and contains 3.6 amino acids per turn of the helix, thereby giving a rise per residue of 5.4/ 3.6 = 1.5 Å (= 0.15 nm), which is the *identity* period of α-helix. The amino acid residues in an α-helix have conformations with $\phi = -60°$ and $\psi = -45°$ to $-50°$.

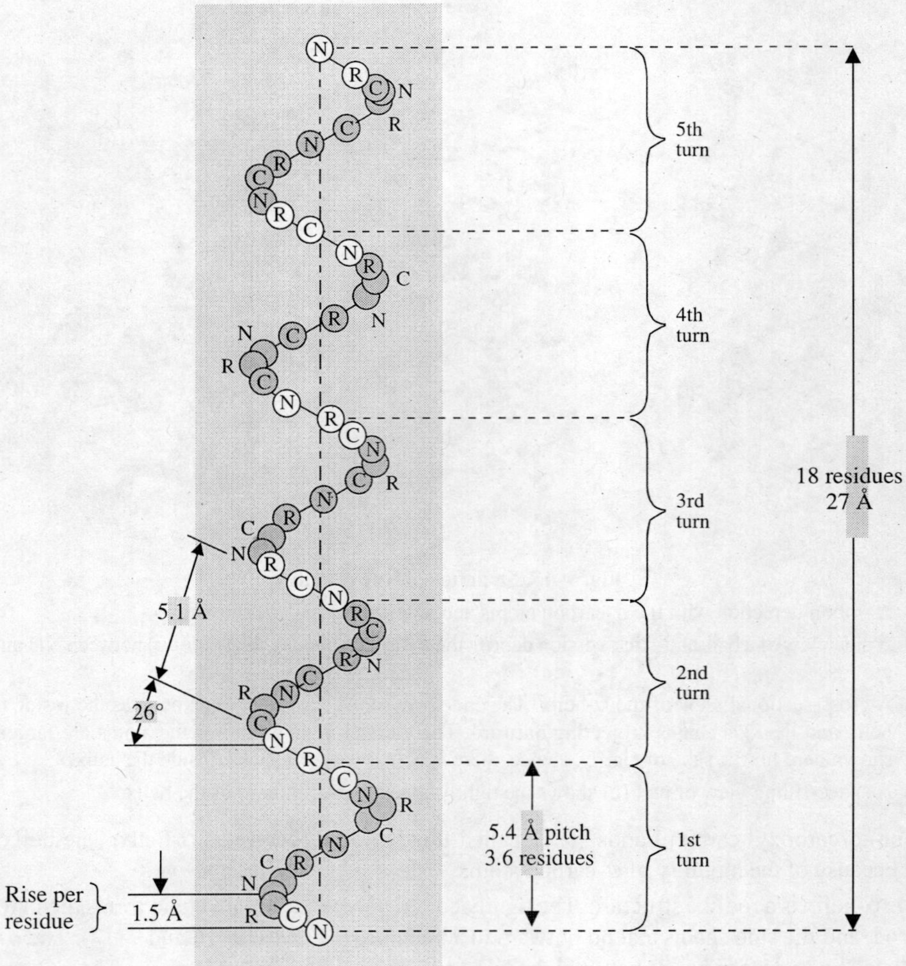

Fig. 9–19. Average dimensions of an α-helix

Note the average dimensions of an α-helix. The letters C and N represent the carbon and nitrogen atoms respectively in the peptide bonds. The letter R represents an α-carbon with a side chain attached. The shaded circles represent atoms back of the plane of the paper while the white circles indicate atoms above the plane.

(*Redrawn from Linus Pauling and Robert Corey, 1955*)

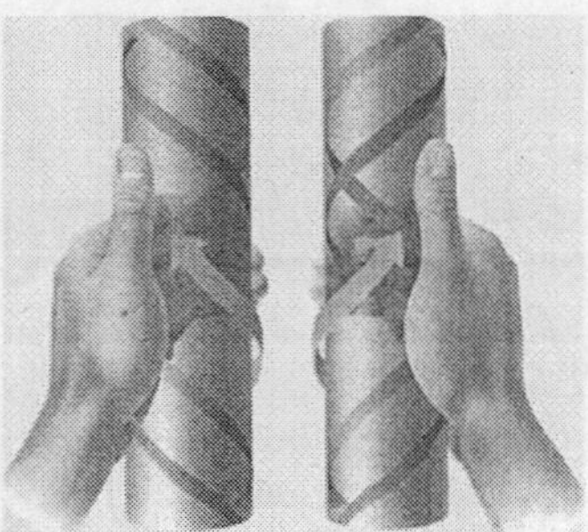

Fig. 9–20. Two types of helices

Right- and left-handed helices are related to each other in the same way as right and left hands.

A helix can be right-handed (clockwise) or left-handed (anticlockwise) (refer Fig 9–20) ; *α-helices of known polypeptides* (*i.e.*, *L-amino acids*) *are right-handed*. Biologically functional proteins do not usually exhibit cent per cent α-helical structure. Some have a high percentage of their residues in α-helical structures, *e.g.*, myoglobin and hemoglobin ; others have a low percentage, *e.g.*, chymotrypsin and cytochrome *C*. A very interesting feature of the α-helix, besides the periodicity, is the fact that the carbonyl group (>CO═O) of every peptide bond is in a position to form a hydrogen bond with the >N—H group of the peptide bond in the next turn of the helix, thereby contributing to the **stability** of the α-helix. Long polypeptides chains in the α-helical conformation, forming structures are reminiscent of a multistranded rope. The protofibril of hair, for example, is 20 Å in diameter and appears to be made up of 3 right-handed helices wound about each other to form a left-handed supercoil.

The model-building experiments have shown that an α-helix can form with either D- or L-amino acids. However, the residues must all be of one stereoisomer ; a d-amino acid will disrupt any regular structure consisting of L-amino acids, and vice versa. The preferential formation of such a helix over many other possible conformations in nature is due, in part, to the fact that it (α-helix) makes optimal use of internal hydrogen bonds.

Many globular proteins contain short regions of such α-helices, and those portions of a transmembrane protein that cross the lipid bilayer are usually α-helices because of the constraints imposed by the hydrophobic lipid environment. In aqueous environment, an isolated α-helix is usually not stable on its own. Two identical α-helices that have a repeating arrangement of nonpolar side chains, however, will twist around each other gradually to form a particularly stable structure known as a coiled coil. Long rod-like *coiled coils* are found in many fibrous proteins, such as intracellular α-keratin fibres that reinforce skin and its appendages.

α-helix occurs in the protein **α-keratin**, found in skin and its appendages such as hair, nails and feathers and constitutes almost the entire dry weight of hair, wool, feathers, nails, claws, quills, scales, horns, hooves, tortoise shell, and much of the outer layer of skin. The basic structural unit of α-keratin usually consists of 3 right-handed helical polypeptides in a left-handed coil that is stabilized by crosslinking disulfide bonds (Fig. 9–21).

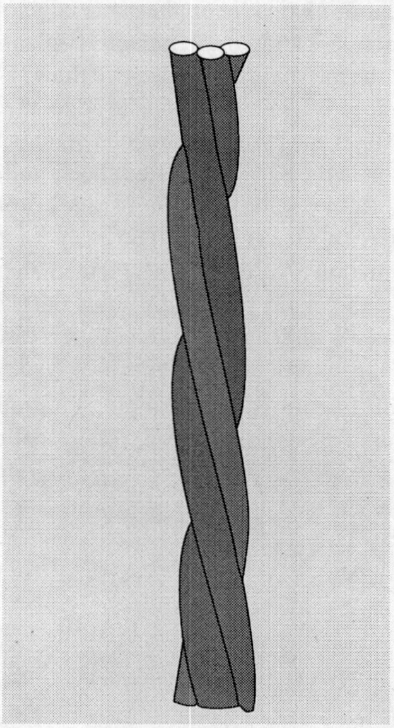

Fig. 9–21. Schematic of the 3 α-helical coils in keratin

β-Pleated Sheet

Pauling and Corey (1953) identified a second type of repetitive, minimum-energy or stable conformation, which they named **β-pleated sheet** (β because it was the second structure they elucidated, the α-heilx having been the first). The formation of β-pleated sheets depends on **intermolecular** (= interchain) hydrogen bonding, although intramolecular hydrogen bonds are also present. The pleated sheet structure is formed by the parallel alignment of a number of polypeptide chains in a plane, with hydrogen bonds between the >C = O and —N—H groups of adjacent chains. The R groups of the constituent amino acids in one polypeptide chain alternately project above and below the plane of the sheet, leading to a two-residue repeat unit. The β sheet structures are quite common in nature and are favoured by the presence of amino acids, glycine and alanine. Silk and certain synthetic fibres such as nylon and orlon are composed of β-structures.

The β-pleated sheet differs markedly from the rodlike α-helix :

1. A polypeptide chain in a β-pleated sheet, called a β-strand, has *fully extended conformation*, rather than being tightly coiled as in the α-helix.

2. The *axial distance* between adjacent amino acids in β-pleated sheets is 3.5 Å, in contrast with 1.5 Å for the α-helix.
3. β-sheet is stabilized by hydrogen bonds between NH and CO groups in *different* polypeptide strands, whereas in the α-helix, the hydrogen bonds are between NH and CO groups in the same strand.

There are two types of β-pleated sheet structures. If the N-terminal ends of all the participating polypeptide chains lie on the same edge of the sheet, with all C-terminal ends on the opposite edge, the structure is known as a **parallel β-pleated sheet.** In contrast, if the direction of the chains alternates so that the alternating chains have their N-terminal ends on the same side of the sheet, while their C-terminal ends lie on the opposite edge, the structure is known as the **antiparallel β-pleated sheet**. In other words, the hydrogen-bonded neighbouring polypeptides are aligned in the same N-to-C terminus direction in the parallel pleated sheets and in the opposite N-to-C direction in the antiparallel pleated sheets. Both parallel and antiparallel β-pleated sheets have similar structures, although the repeat period is shorter (6.5 Å 0.65 nm) for the parallel conformation in comparison to antiparallel conformation (7 Å or 0.7nm). Pleated-sheet structures can also be formed from a single polypeptide chain if the chain repeatedly folds back on itself. Although a β-pleated sheet structure is usually associated with structural proteins, it is also known to occur in the 3-dimensional structures of certain globular proteins, *e.g.*, the enzymes lysozyme and carboxypeptidase A.

Silk fibroin is one example of a protein that has the antiparallel pleated sheet structure. It is a member of a class of fibrillar proteins called α-keratins. Silk fibroin (Fig. 9–22) is composed mainly

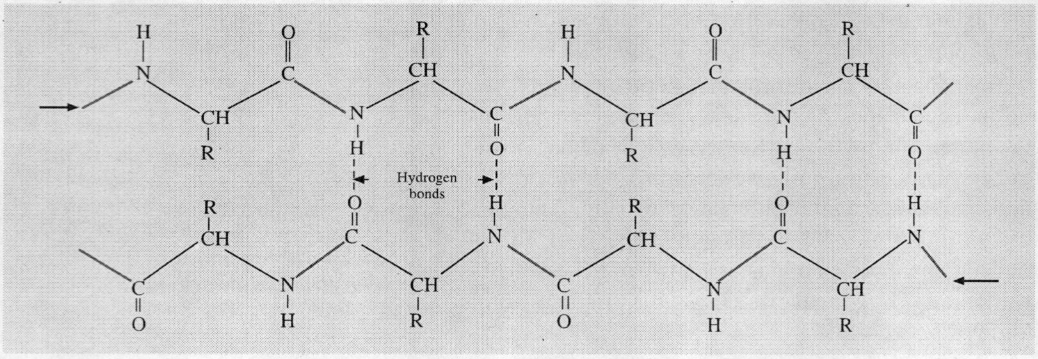

Fig. 9–22. A portion of two chains of silk fibroin

Note that the two chains runs in opposite direction. Peptide bonds are shown here by *red* thick lines.

of glycine, alanine and serine units linked together by the peptide bonds. Glycine constitutes approximately 45% of the total amino acid residues, and alanine plus serine compose another 42%. Hence, the R groups extending above and below the plane of the pleated sheet are small and allow the pleated sheets to stack. Here the two chains which run in opposite directions are linked by hydrogen bonds (shown by dotted lines). It may be noted that each of the two chains can also form hydrogen bonds with still other protein and so on. Very large protein aggregates are, thus, formed provided the R groups are relatively small. If the R groups are too large, the hydrogen bonding may not occur as the chains are held far apart for union. An interesting phenomenon occurs when hair or wool (α-keratins) is treated with moist heat and stretched because β-keratin structures with parallel β pleated sheets are produced.

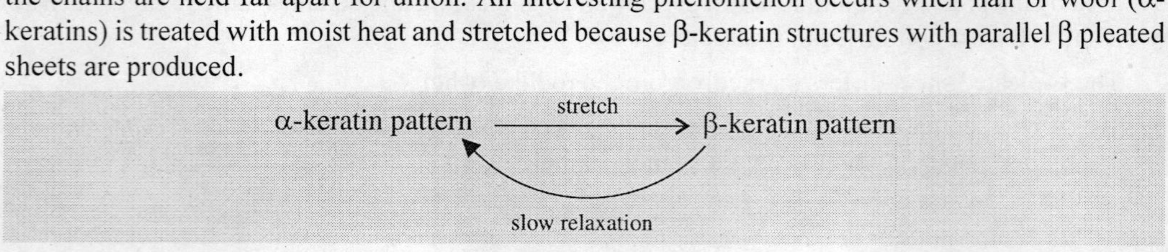

Random Coil

Besides the α-helix and β-pleated sheet structures of proteins, which were recognized by Pauling and Corey in 1950s, there also exist a third type of secondary structure in proteins called **random coil**. When a polypeptide contains adjacent bulky residues such as isoleucine or charged residues such as glutamic acid and aspartic acid, repulsion between these groups causes the polypeptide to assume a random coil configuration (Fig. 9–23).

Thus, we see that the R groups distributed along the polypeptide backbone determine the secondary structure adopted by different portions of the polypeptide (α-helix or β-pleated sheet), or the lack of a well-defined structure (random coil).

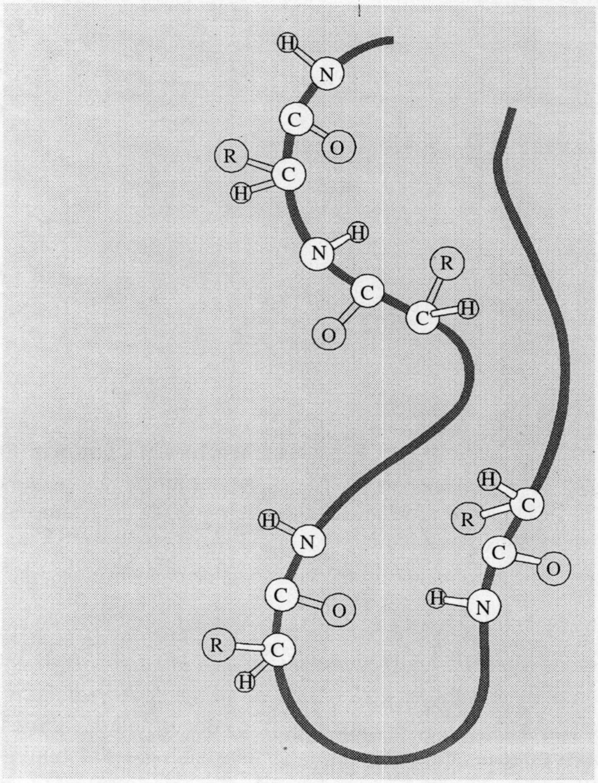

Fig. 9–23. A random coil

The backbone of the polypeptide is drawn, but only a few amino acid residues are shown.

(*Adapted from Francisco J. Ayala and John A. Kiger Jr., 1980*)

Other Secondary Structures

The α-helix and β-conformation are the major repetitive secondary structures easily recognized in a wide variety of proteins. The **γ-helix**, however, is an example of a highly hydrogen-bonded structure that is insufficiently stable to be used in proteins because of the lack of interatomic contacts. In addition to the α-helix and β-pleated sheet structures, other repetitive structures also exist in one or a few specialized proteins. Two such structures (β-turn and collagen triple helix) are described below :

β Turn or β Bend or Hairpin Bend

Most proteins have compact, globular shapes owing to reversals in the direction of their polypeptide chain. β turn (Fig. 9–24) is often found where a polypeptide chain abruptly reverses direction. β turns usually connect the ends of two adjacent segments of an antiparallel β pleated sheet, hence their name. The structure is a tight turn (~180°) involving 4 amino acids. The essence of this hairpin turn is that the CO group of residue *n* of a polypeptide is hydrogen-bonded to the NH group of residue *n* + 3. In other words, the hydrogen bond between the peptide groups of the first and fourth amino acid residues is involved in bend formation. Glycine and proline residues often occur in β turns ; the former because it is small and flexible and the latter because peptide bonds involving the imino nitrogen of proline readily assume the *cis* configuration, a form that is particularly amenable to a tight turn. β turns are often found near the surface of a protein. They are also called as **reverse turns.**

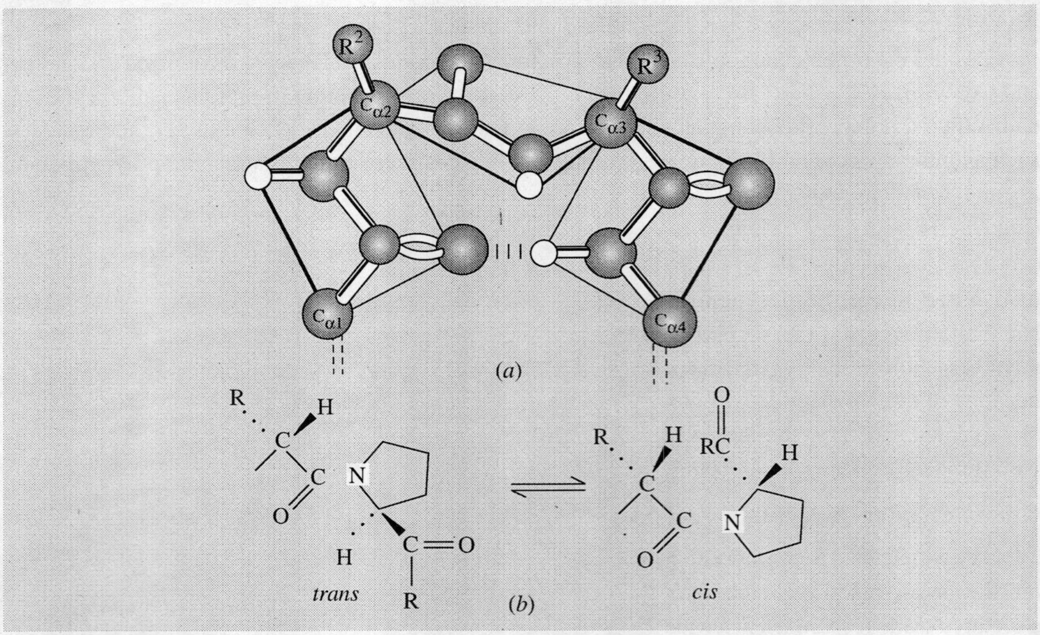

Fig. 9–24. Structure of a β turn or hairpin bend

(*a*) Hydrogen bonding in a tetrapeptide involving first and fourth amino acid residues. Here the NH and CO groups of residue 1 of the tetrapeptide are hydrogen-bonded, respectively, to the CO and NH groups of residue 4.

(*b*) The *trans* and *cis* isomers of a peptide bond involving the imino nitrogen of proline.

Collagen Triple Helix

Collagen is the most abundant protein of mammals. It is the principal structural element of the human body and makes up 25–33% of all the body protein. It is found in connective tissues such as tendons, cartilage, the organic matrix of bones and the cornea of the eye. This extracellular protein contains three helical polypeptide chains, each nearly 1,000 residues long. The amino acid sequence of collagen is remarkably regular : *nearly every third residue is glycine.* Proline is also present to a much greater extent than in most other proteins. Furthermore, collagen contains 4-hydroxyproline (Hyp), which is rarely found elsewhere. The percentage composition of predominant amino acids found in collagen is : Gly (35%), Ala (11%) and Pro + Hyp (25%). The amino acid sequence in collagen is generally a repeating tripeptide unit, Gly–X–Pro or Gly–X–Hyp, where X can be any amino acid. This repeating tripeptide sequence adopts a left-handed helical structure with 3 residues per turn.

Collagen is a rod-shaped molecule, about 3,000 Å (= 300 nm) long and only 15 Å (=1.5 nm) in diameter. The helical motif of its 3 chains is entirely different from that of the α helix. Hydrogen bonds within a strand (*i.e.*, intrachain hydrogen bonds) are absent. Instead, each of these 3 helices is stabilized by steric repulsion of the pyrrolidone rings of the proline and hydroxyproline residues. The 3 strands wind around each other in a cable fashion (Fig. 9–25), to form a **superhelix**. The superhelical

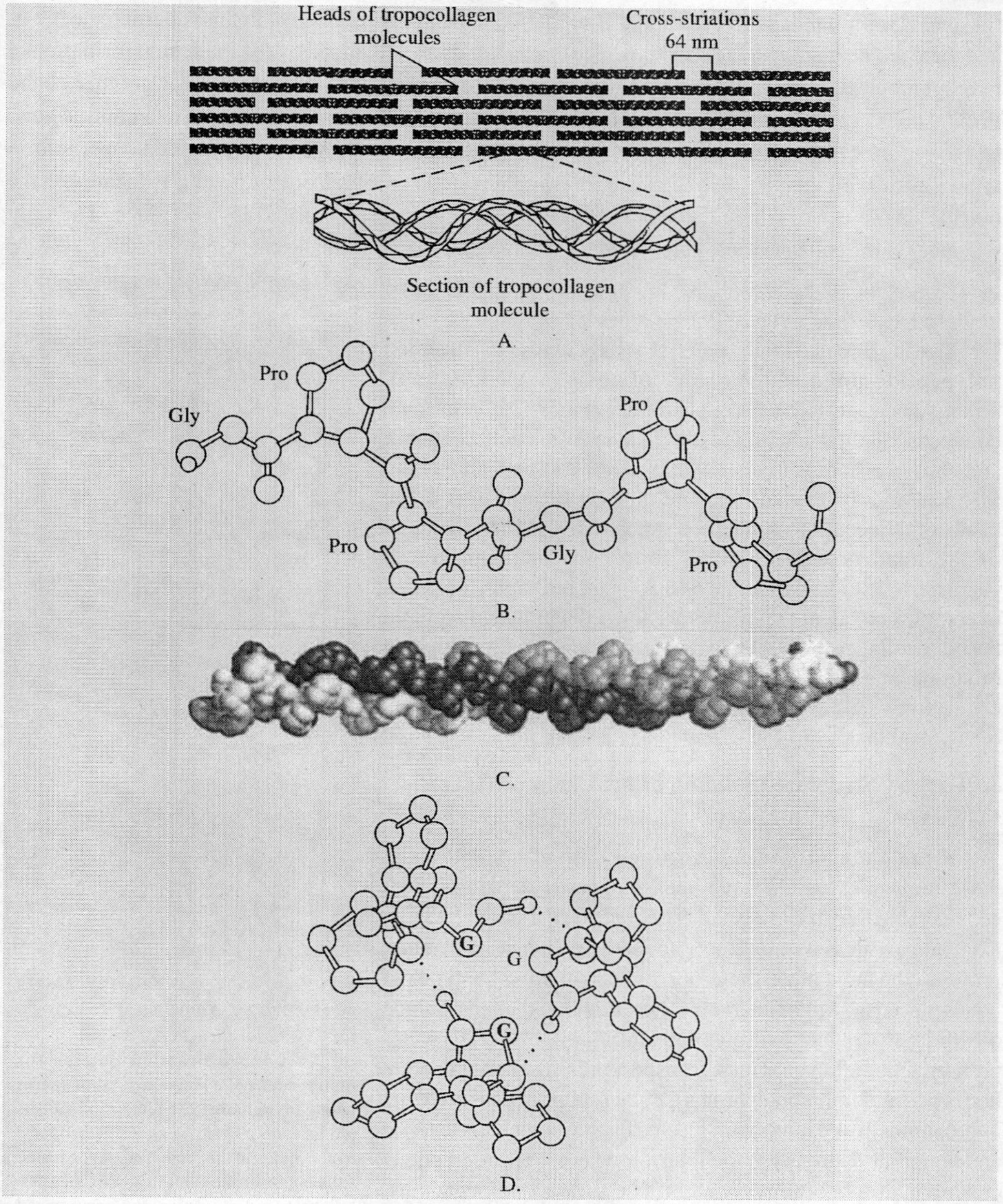

Fig. 9–25. The structure of collagen

A. **Collagen fibres and a section of tropocollagen molecule.** The basic collagen monomer is a triple helix, composed of 3 helical α chains. The collagen monomers before aligned in rows in which the molecules in one row are staggered relative to those in the neighbouring rows.

B. **Conformation of a single strand of the collagen triple helix.** The sequence shown here is Gly-Pro-Pro-Gly-Pro-Pro.

C. **Space-filling model of the collagen triple helix.** The three strands are shown in differnt shades.

D. **Cross section of a model of the triple-stranded helix of collagen.** Each strand is hydrogen-bonded to other two (... denotes a hydrogen bond). The α-carbon atom of a glycine residue in each strand is labelled G. Every third residue must be glycine because there is no space near the helix axis (centre) for a larger amino acid residue. Note that the pyrrolidone rings are on the outer side.

twisting is right-handed (cf α-keratin). The rise (axial distance) per residue in this superhelix is 2.9 Å, and there are nearly 3 residues per turn. The three strands are stabilized by the interchain formation of hydrogen bonds between the > C = O group of one chain and the >N—H group of another chain. Proline does not have a hydrogen atom attached to its nitrogen when participating in a peptide bond and therefore cannot participate in this interchain hydrogen-bonding. This 3-stranded superhelix is known as the **collagen triple helix**. The amino acid residue on either side of glycine is located on the outside of the cable, where there is room for the bulky rings of Pro and Hyp residues. The tight wrapping of the collagen triple helix provides great tensile strength with no capacity to stretch. *Collagen fibres can support up to 10,000 times their own weight and are said to have greater tensile strength than a steel wire of equal cross section.*

The collagen fibrils consist of recurring 3-stranded polypeptide units called **tropocollagen** (MW 3,00,000), arranged head-to-tail in parallel bundles (Fig. 9–25C). In some collagens, all three chains are identical in amino acid sequence and thus are homotrimers ; while in others, two chains are identical and the third differs and thus are heterotrimers. The heads of adjacent molecules are staggered, and the alignment of the head groups of every fourth molecule produces characteristic cross-striations 640 Å (= 64 nm) apart. A series of complex covalent cross-links are formed within and between the tropocollagen molecules in the fibril, leading to the formation of strong mature collagen. The rigid, brittle character of the connective tissue in older people is the result of an accumulation of covalent cross-links in collagen as they age.

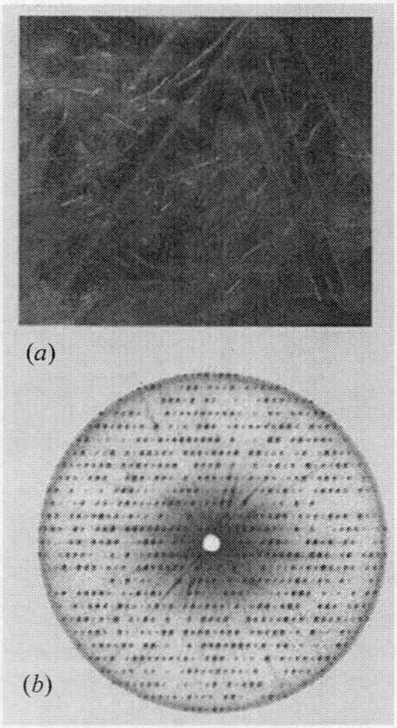

(*a*)

(*b*)

Fig. 9.26. Sperm whale myoglobin crystal and x-ray photograph

(*a*) **Crystal of myoglobin.**

(*b*) **An x-ray diffraction pattern (or photograph) of a single crystal of sperm whale myoglobin**. The pattern of spots is produced as a beam of x-rays is diffracted by the atoms in the protein crystal, causing the x rays to strike the film at specific sites. Information derived from the position and intensity (darkness) of the spots can be used to calculate the positions of the atoms in the protein that diffracted the beam. The intensity of each diffraction maximum (the darkness of each spot) is a function of the myoglobin crystal's electron density. The photograph contains only a small fraction of the total diffraction information available from a myoglobin crystal.

(*Courtesy : (a) Mel Pollinger*)

3. Tertiary Structure : Folding of the Chain or Overall Folding

If the globular proteins consisted only of a small helix, these molecules would have been elongated structures with considerable length and a small cross-sectional area (*i.e.*, a large axial ratio). But as we now know about the existence of globular proteins, the helix must, therefore, possess many other types of bonds placed at regular intervals. These additional bonds include disulfide, hydrogen, hydrophobic and ionic. In such globular proteins (including enzymes, transport proteins, some peptide hormones and immunoglobulins), polar groups because of their hydrophobicity are most often located on the molecule's exterior and nonpolar R groups in the interior, where their interactions create a hydrophobic environment. The *tertiary structure*, thus, involves the folding of the helices of globular proteins. It refers to the spatial arrangement of amino acids that are far apart in linear sequence and to the pattern of disulfide bonds. The dividing line between secondary and tertiary structure is, hence, a matter of taste. X-ray crystallographic studies have revealed the detailed 3-'D' structures of more than 300 proteins.

Myoglobin (Mb)

Myoglobin (*myo*[G] = muscle ; *globin*[G] = a type of protein) is a relatively small, oxygen-binding heme protein, found in muscle cells. It has the distinction of being the first globular protein to have its 3-'D' structure elucidated by x-ray diffraction studies. This was accomplished by **John C. Kendrew** at a resolution of 6 Å in 1957, 2 Å in 1959 and 1.4 Å in 1962 (Fig. 9–26). Myoglobin molecule (Fig. 9–27) contains a single polypeptide chain of 153 amino acid residues and a single prosthetic iron-porphyrin (or heme) group, identical with that of hemoglobin. The heme group is responsible for the deep red-brown colour of myoglobin (and also of hemoglobin). Myoglobin is especially abundant in the muscles of diving mammals such as the whale, seal and porpoise, whose muscles are so rich in this protein that they are brown. Storage of oxygen by

> Between 1912 and 1915, **Sir William Henry Bragg** and his son, **William Lawrence Bragg,** developed the technique of **x-ray diffraction** by determining the crystalline structure of NaCl. As the joint 1915 Nobel Prize recipients in physics, the Braggs became the only father-son combination to receive the award and W. Lawrence, who was 25 years old at the time, the youngest scientist so honoured.

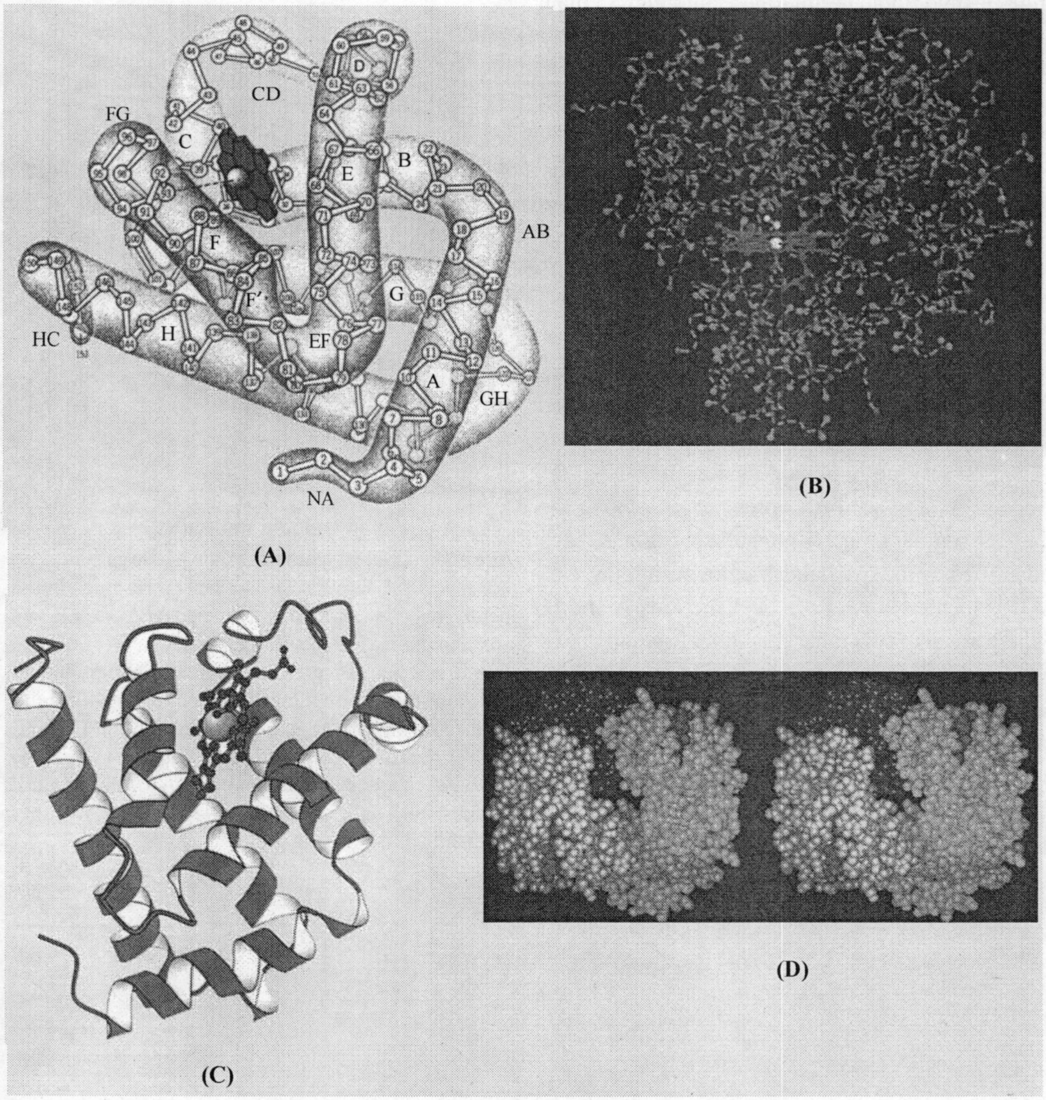

Fig. 9–27. The structure of sperm whale myoglobin

(A) The tertiary structure. Its 153 C_α positions are numbered from the N-terminus and its eight helices are sequentially labeled A through H. The last half of the EF corner is now regarded as a turn of helix and is therefore designated the F′ helix. The heme group is shown in *red*. Most of the amino acids are part of α helices. The nonhelical regions occur primarily as turns, where the polypeptide chain changes direction.
(B) The ball- and stick model. The three-dimensional structure of myoglobin shows the position of all of the molecule's atoms other than hydrogen and reveals many interactions between the amino acids.
(C) Ribbon model. A schematic view shows that the protein consists largely of α helices. The heme group is shown in black and the iron atom is shown as a purple sphere
(D) A stereo, space-filling model showing the quaternary structure of hemoglobin. The model shows the α_1, α_2, β_1, and β_2 subunits as colored yellow, green, light blue, and dark blue, respectively, Heme groups are red. The protein is viewed along its molecular twofold rotation axis which relates the $\alpha_1\beta_1$ protomer to the $\alpha_2\beta_2$ protomer.

(*Courtesy : (A) Irving Geis, (B) Ken Eward*)

muscle myoglobin permits these animals to remain submerged for long periods. The function of myoglobin is to bind oxygen in the muscles and to enhance its transport to the mitochondria, which consume oxygen during respiration.

Myoglobin (MW = 16,700) is an extremely compact macromolecule with oblate, spheroid shape and leaves little empty space in its interior. Its overall molecular dimensions are 45 × 35 × 25 Å. The sausagelike outline of the tertiary structure of myoglobin chain is folded into an odd, irregular form. The backbone of the molecule is made up of 8 almost-straight α-helical segments, designated from the N-terminus as A through H (Fig 9–28). The first residue in helix A is designated A_1, the second A_2 and so forth. Interspersed among the helical segments are 5 nonhelical regions, each identified by the two helical segments it joins, *e.g.*, AB is located between helical sections A and B. There are also two other nonhelical regions : two residues at the N-terminus (named NA_1 and NA_2) and five residues at the C-terminus (named HC_1 through HC_5). The longest helical segment has 23 amino acid residues and the shortest only 7. All the helical segments are of α-type and right-handed ; there being no β-structure (refer Table

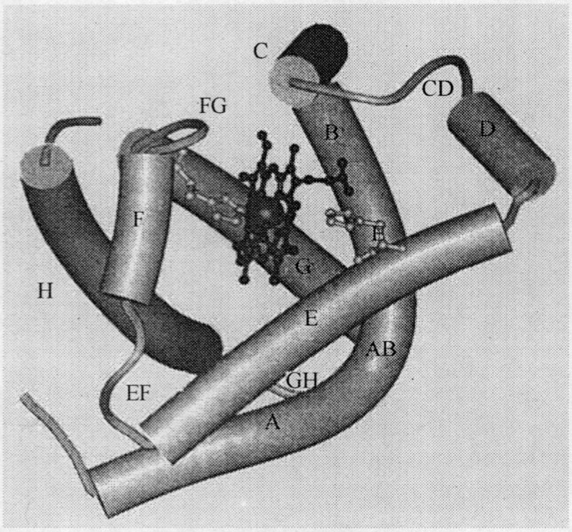

Fig. 9–28. The eight α helical segments of myoglobin molecule

The eight α helical segments (shown here as cylinders) are labeled A through H. Nonhelical residues in the bends that connect them are labeled AB, CD, EF, and so forth, indicating the segments they interconnect. A few bends, including BC and DE, are abrupt and do not contain any residues; these are not normally labeled. (The short segment visible between D and E is an artifact of the computer representation.) The heme is bound in a pocket made up largely of the E and F helices, although amino acid residues from other segments of the protein also participate.

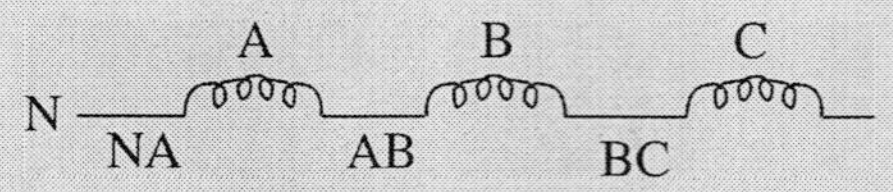

Table 9-4 Approximate amounts of α-helix and β-structure in some single-chain proteins*

Protein	*Total residues*	*Residues, %*	
		α-helix	*β-structure*
Cytochrome C	104	39	0
Ribonuclease	124	26	35
Lysozyme	129	40	12
Myoglobin	153	78	0
Chymotrypsin	247	14	45
Carboxypeptidase	307	38	17

* Portions of polypeptide chains, that are not accounted for by α-helix or β-structure, consist of bends, reverse turns and irregularly-coiled stretches.

(*Adapted from Cantor and Schimmel, 1980*)

9–4). Of the 153 amino acid residues, 121 (*i.e.*, 79%) are present on the helical regions and the remaining 32 amino acid residues are distributed over the nonhelical areas. The nonhelical areas possess various types of bonding such as hydrogen and nonpolar linkages. The flat heme group is tightly but noncovalently bound to the polypeptide chain. A notable feature of whale myoglobin is the absence of a disulfide bridge, since both cysteine and cystine residues are lacking. The electron density map of sperm whale myoglobin is presented in Fig. 9–29.

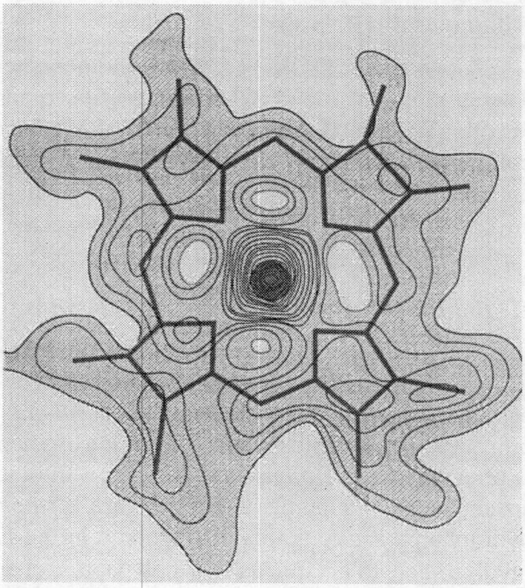

Fig. 9–29. A section through the 2.0-Å- resolution electron density map of sperm whale myoglobin

The heme group is represented by *red*. The large peak the centre of the map represents the electron-dense Fe atom.

(*After Kendrew JC, Dickerson RE, Strandberg BE, Hart RG, Davies DR, Phillips DC and Shore VC, 1960.*)

Other important features of the myoglobin molecule are listed below :

1. The molecule is very compact and leaves so little space in its interior as to accommodate only 4 water molecules.
2. All the polar R groups of the molecule, except two, are located on its outer surface and all of them are hydrated.
3. Most of the hydrophobic R groups are located in the interior of the molecule. Hydrophobic R groups of helices E and F form the sides of a pocket into which the hydrophobic heme group fits. The porphyrin ring of the heme is largely hydrophobic, except for the 2 propionic acid side chains which stick out of the pocket and into the aqueous environment.
4. Each of the 4 proline residues occurs at a turn. Other turns or bends contain serine, threonine and asparagine.
5. All the peptide bonds of the polypeptide chain are planar, with the carbonyl and amide groups being *trans* to each other.
6. The heme group is flat and rests in a crevice in the molecule. The iron atom in the centre of

the heme group has two coordination bonds that are perpendicular to the plane of the heme group. One bond is attached to the R group of histidine (93), whereas the other bond is the site to which an O_2 molecule is bound.

7. The inside and outside are well defined. There is little empty space inside. *The interior consists almost entirely of nonpolar residues* such as Leu, Val, Met, and Phe. On the contrary, Glu, Asp, Gln, Asn, Lys and Arg are absent from the interior of the protein. Residues with polar and nonpolar parts within them (Thr, Tyr, Trp) are so oriented that their nonpolar portions point inward. The only polar residues inside myoglobin are two histidines which have a critical function at the binding site. The exterior of the protein has both polar and nonpolar residues.
8. Myoglobin without its heme prosthetic group is designated *apomyoglobin*. The main function of apomyoglobin molecule is to provide a hydrophobic environment for the heme group and a properly oriented imidazole group to occupy the 5th coordination position of the iron. The presence of the hydrophobic environment and the proximal histidine enable the heme to combine reversibly with O_2 and prevent the oxidation of Fe^{2+} to Fe^{3+}.

Ribonuclease (RNase)

Ribonuclease (MW = 13,700) is another noteworthy example of a protein with tertiary structure. Two groups of workers (**Christian Anfinsen** and his associates, 1950, at the National Institute of Health ; **William H. Stein** and **Stanford Moore**, 1958, at Rockefeller Institute) have elucidated the complete structure of this pancreatic protein. Ribonuclease molecule (Fig. 9-30) consists of 124 amino acid units contained in a single polypeptide chain with 4 disulfide bridges. In establishing the amino acid sequence, the RNase was subjected to hydrolysis either by pepsin or by trypsin (for further details, refer Fig. 17–10 on page 300).

4. Quaternary Structure : Protein-Protein Interactions or Multichain Association

A fourth degree of complexity in protein structure has recently been recognized to be of great value in many proteins. Some globular proteins consist of 2 or more interacting peptide chains. Each peptide chain in such a protein is called a *subunit*. These chains may be identical or different in their primary structure. This specific association of a number of subunits into complex large-sized molecules is referred to as the *quaternary structure*. In other words, quaternary structure refers to the spatial arrangement of subunits and the nature of their contact. The same forces (disulfide, hydrogen, hydrophobic and ionic bonds) involved in the formation of tertiary structure of proteins are also involved here to link the various polypeptide chains.

Tobacco Mosaic Virus (TMV)

TMV, with 158 amino acid residues, is an instance of the protein-protein interactions. TMV particles are rod like, tubular structures measuring 3,000 Å length and 180 Å diameter. The tube is made up of a single RNA molecule (30,000 Å long) which is coiled round the central axis. The RNA molecule is a right-handed helical filament and is embedded in a protein matrix or the coat which is made up of 2,130 identical molecules of a single protein that interact to form a cylinder enclosing the RNA genome. The amino acid sequence of this coat protein (Fig. 9–31) was elucidated in 1962 independently by **H. Fraenkel-Conrat** and **H. Wittmann.** The coat protects the RNA molecule from disintegration by various environmental changes.

Hemoglobin (Hb)

Hemoglobin (*hemo*^{G} = blood ; *globin*^{G} = a protein, belonging to the myoglobin-hemoglobin family), the oxygen transporter in erythrocytes, constitutes about 90% of the protein of red blood cells. It is a tetrameric protein, *i.e.*, it contains 4 polypeptide chains. The four chains (of which two are of one kind and two of another) are held together by noncovalent interactions. Each chain contains a heme group and a single oxygen-binding site. *Hemoglobin A*, the principal hemoglobin in adults,

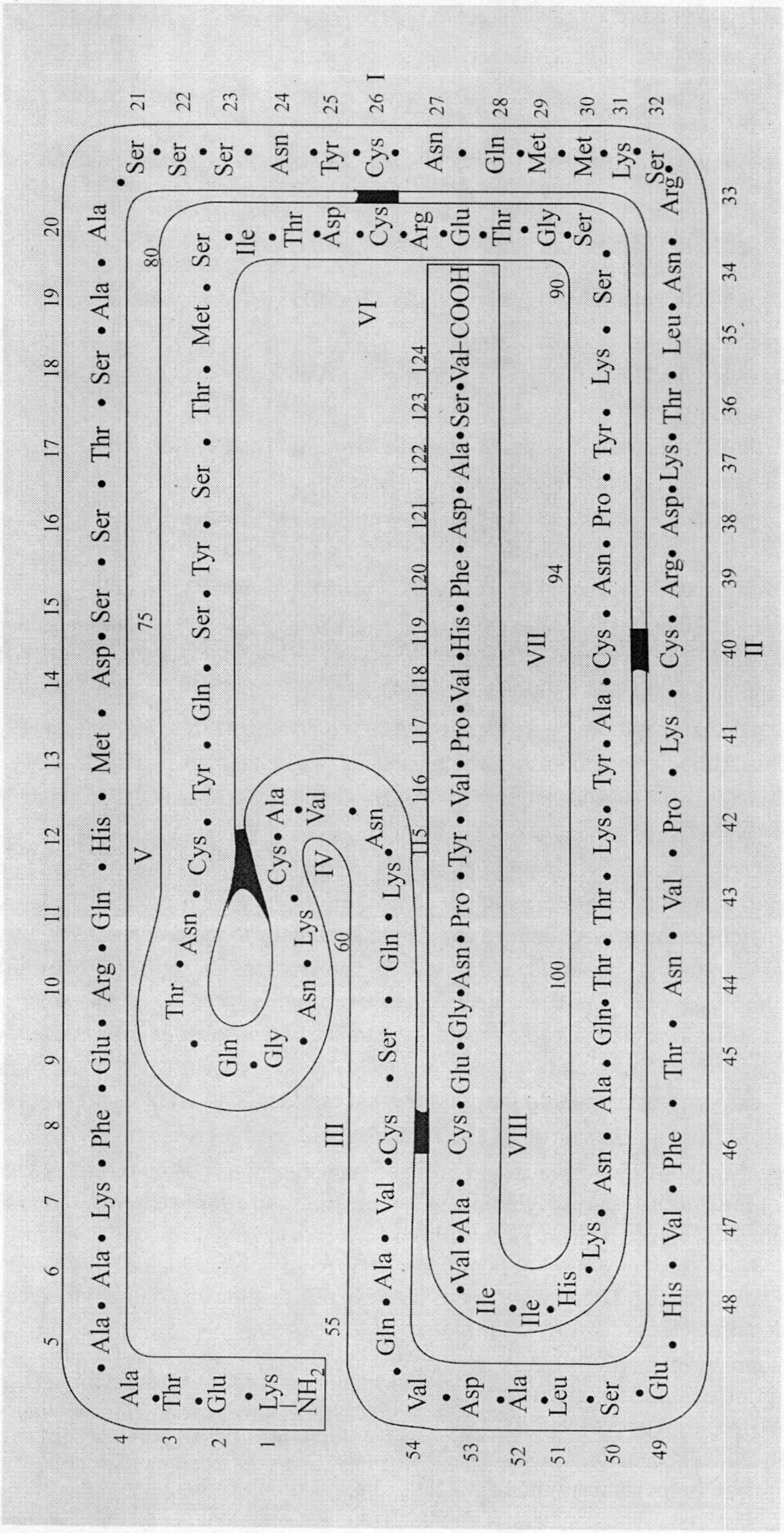

Fig. 9–30. The amino acid sequence of the enzyme ribonuclease from bovine

This enzyme molecule consists of 124 amino acid residues which are numbered. The 4 dark blocks between cysteines represent 4 disulfide bonds or bridges. Note that this illustration is diagrammatic; the polypeptide chain is actually folded to give a complex 3-dimensional configuration.

(After Smyth DG, Stein WH and Moore S, 1963)

AcSer-Tyr-Ser-Ile-Thr-Thr-Pro-Ser-Gln-Phe-Val-Phe-Leu-Ser-Ser-Ala-Trp-Ala-Asp-Pro-
1 5 10 15 20
Ile-Glu-Leu-Ile-Asn-Leu-Cys-Thr-Asn-Ala-Leu-Gly-Ans-Gln-Phe-Gln-Thr-Gln-Gln-Ala
25 30 35 40
Arg-Thr-Val-Val-Gln-Arg-Gln-Phe-Ser-Glu-Val-Trp-Lys-Pro-Ser-Pro-Gln-Val-Thr-Val-
45 50 55 60
Arg-Phe-Pro-Asp-Ser-Asp-Phe-Lys-Val-Tyr-Arg-Tyr-Asn-Ala-Val-Leu-Asp-Pro-Leu-Val-
65 70 75 80
Thr-Ala-Leu-Leu-Gly-Ala-Phe-Asp-Thr-Arg-Asn-Arg-Ile-Ile-Glu-Val-Glu-Asn-Gln-Ala-
85 90 95 100
Asn-Pro-Thr-Thr-Ala-Glu-Thr-Leu-Asp-Ala-Thr-Arg-Arg-Val-Asp-Asp-Ala-Thr-Val-Ala
105 110 115 120
Ile-Arg-Ser-Ala-Ile-Asn-Asn-Leu-Ile-Val-Glu-Leu-Ile-Arg-Gly-Thr-Gly-Ser-Tyr-Asn-
125 130 135 140
Arg-Ser-Ser-Phe-Glu-Ser-Ser-Ser-Gly-Leu-Val-Trp-Thr-Ser-Gly-Pro-Ala-Thr
145 150 155 158

Fig. 9–31. Amino acid sequence of the coat protein or tobacco mosaic virus (TMV)

Note that the serine in position 1 is acetylated (AcSer). Also note that there are no extended repetitions of an amino acid sequence, that there are no clusters of either polar or nonpolar amino acids, and that the frequencies of occurrence of different amino acids vary widely.

consists of two alpha (α) chains and two beta (β) chains. Adults also have a minor hemoglobin (~ 2% of the total hemoglobin) called hemoglobin A_2, which contains two delta (δ) chains in place of two β chains of hemoglobin A. Thus, the subunit composition of hemoglobin A is $\alpha_2\beta_2$, and that of hemoglobin A_2 is $\alpha_2\delta_2$. HbA and HbA_2 are the postnatal forms of hemoglobin.

Prior to birth, embryonic and fetal hemoglobins are used as oxygen carrier. The first hemoglobin to appear in embryonic development has the subunit composition $\zeta_2\,\varepsilon_2$: the two zeta (ζ) chains are analogous to the a chain and the two epsilon (ε) chains are analogous to the β chains. When, after about 6 weeks, ζ chain production ceases, the tetramer $\alpha_2\,\varepsilon_2$ appears, *i.e.*, ζ is replaced by α. A third embryonic hemoglobin, $\zeta_2\,\gamma_2$ has also been identified where ε is replaced by γ. These last two hemoglobins represent transition phases leading to the appearance of *fetal hemoglobin* (*HbF*), whose tetrameric composition is $\alpha_2\,\gamma_2$. The α and ζ chains contain 141 residues each. The β, γ and δ chains contain 146 residues each and have homologous (similar but not identical) amino acid sequences. The γ and δ chains differ from the β chain at 39 and 10 amino acid residues respectively.

Hemoglobin (MW = 64,500) provides an example of the interaction of unidentical protein subunits. Since hemoglobin is 4 times as large as myoglobin, much more time and effort were required to solve its 3- 'D' structure, finally achieved by **Max Perutz, John C. Kendrew** and their colleagues in 1959. They determined the 3- 'D' structure of adult hemoglobin (HbA) of horse, which is very similar to that of human hemoglobin (HbA). The x-ray analysis has revealed that the hemoglobin molecule is nearly spherical with a diameter of 55 Å (5 .5 nm).

Human hemoglobin protein (Fig. 9–32) consists of 4 polypeptide chains of two types, two α-chains and two β-chains. The polypeptide portion is collectively called as *globin*. The α-chain has valine at the N-terminal and arginine

The designations α and β, which are common labels in Biochemistry, are used simply to identify two different polypeptide chains and are not to be confused with α and β secondary structures or α and β amino acids.

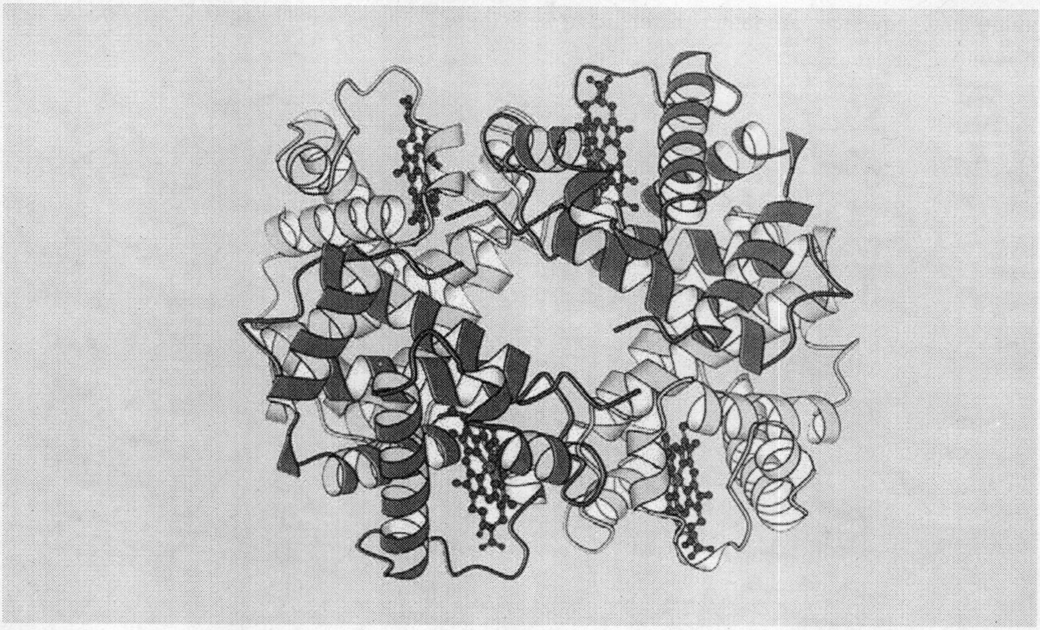

Fig. 9–32. Quaternary structure of hemoglobin molecule, showing interaction of 4 polypeptide chains Hemoglobin, which is composed of 2 α chains and 2 β chains, functions as a pair of αβ dimers. The structure of the two identical α subunits (*red*) is similar to but not identical with that of the two identical β subunits (*yellow*). The molecule contains four heme groups (*black* with the iron atom shown in *purple*).

at the C-terminal whereas in the β-chain, valine is situated at the N-terminal and histidine at the C-terminal. Each a chain is in contact with both β chains. In contrast, there are few interactions between the two α-chains or between the two β–chains. Each chain has a heme prosthetic group in a crevice near the exterior of the molecule. The heme groups are involved in the binding of oxygen. The α-chain has 141 residues and the β-chain, which is more acidic, has 146 residues. The protein, thus, has a total $(141 \times 2 + 146)$ 574 amino acid residues. Each of the 4 chains has a characteristic *tertiary structure*, in which the chain is folded. Like myoglobin, the α- and β-chains of hemoglobin contain several segments of α-helix, separated by bends. The α- and β-chains are held together as a pair by ionic and hydrogen bonds. The 2 pairs are then joined to each other by additional ionic bonds, hydrogen bonds and the hydrophobic forces. Thus, the 4 polypeptide chains fit together almost tetrahedrally to produce the characteristic *quaternary structure*. The hemes are 2.5 nm apart from each other and tilted at different angles. Each heme is partly burried in a pocket lined with hydrophobic R groups. It is bound to its polypeptide chain through a coordination bond of the iron atom to the R group of a histidine residue. The sixth coordination bond of the iron atom of each heme is available to bind a molecule of oxygen. Myoglobin and the α and β chains of hemoglobin have nearly the same tertiary structure. Both have well over 70% α-helical nature, both have similar lengths of α-helical segments and the bends have about the same angles.

Similarity in 3- 'D' Structure of Hemoglobin and Myoglobin

The 3-'D' structures of myoglobin and the α and β chains of human hemoglobin are strikingly similar (Fig. 9–33). The 8 helices in each chain of hemoglobin are virtually superposable on those of myoglobin. This close resemblance in the folding of their main chains was unexpected because their amino acid sequences are rather different. In fact, these 3 chains are identical at only 24 or 141 positions.

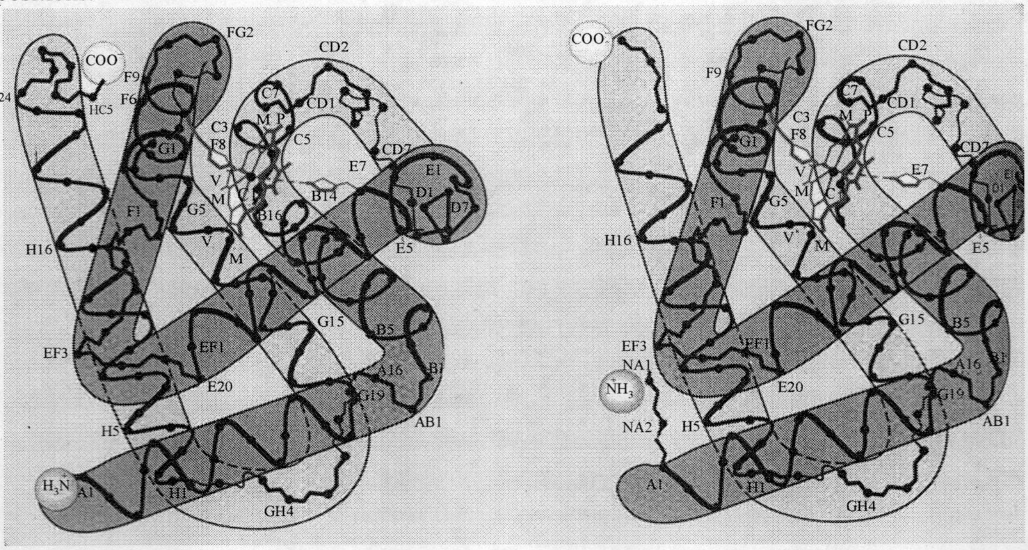

(*a*) (*b*)

Fig. 9–33. Comparison of the conformation of the main chain of myoglobin (*a*) and the β chain of hemoglobin A_1 (*b*)

Proximal His F8, distal His E7 and Val E11 side chains are shown. Other amino acids of polypeptide chain are represented by α-carbon positions only; the letters M, V and P. refer to the methyl, vinyl and propionate side chains of the heme. The overall structures are very similar, except at NH_2- terminal and COOH-terminal ends. (*Courtesy: Fersht A, 1977*)

The protein-protein interactions, thus, serve many important functions :

(*a*) protect nucleic acids of virus particles from destruction,

(*b*) conserve genetic information, and

(*c*) construct the various enzyme-complexes found in the cells.

The foregoing discussion and also the evidences presently available confirm the view that "the primary structure of the protein, which dictates the limitations of secondary and tertiary structure, also delineates the correct quaternary structure necessary for biological activity of the molecule." (**Hartman** and **Suskind**, 1969).

PRIONS

Until very recently, scientists believed that the diseases could be transmitted from one organism to another *via* viruses or microorganisms. After all, DNA and RNA were the carriers of genetic information. But now evidences have gathered to point out that some diseases are transmitted by a protein and nothing more. Table 9–5 lists some such diseases (called **prion diseases**) along with their host and geographic distribution.

In 1900, a fatal neurodegenerative disease called **Kuru** was reported only in a tribe of Papua New Guinea. **Carleton Gajdusek**, a virologist at the US National Institute of Health, discovered and understood the basis of the disease and the mode of its transmission. He pointed out that a peculiar ritual of the tribal people of eating the brain of dead relatives was the main mode of transmission of this disease. The frequency of Kuru gradually declined since the practice was abandoned. **Gajdusek** was awarded 1976 Nobel prize in medicine or physiology for this basic work, along with **Baruch Blumberg**, a virologist of the Institute of Cancer Research, Philadelphia.

Table 9–5. Prion disease in different species

Host	*Disease*	*Geographic distribution*	*First clinical observation*
Sheep	Scrapie	Cosmopolitan except Australia, New Zealand and some European countries	1730
Goat	Scrap	—	—
Man	Kuru ('Laughing death')	Papua New Guinea	1900
	Creutzfeldt-Jacob disease	Cosmopolitan	1920
	Gertsmann-Straussler–Scheinker syndrome	Cosmopolitan	1926
	Fatal familial insomnia	—	—
Mink	Transmissible mink encephalopathy	North America; Europe	1947
Mule deer	Chronic wasting disease	North America	1967
Cattle	Bovine spongiform encephalopahy, BSE (= Mad cow disease)	U.K., Ireland and some other European countries	1985

But **Stanley B. Prusiner**, an Americal biochemist, was not convinced about the nature of the infectious agent which was described by Gajdusek and others as 'unconventional slow virus'. However, **Prusiner** was against the so-called virus theory because of the following 3 reasons :

(*a*) The infective agent is extremely resistant to UV and ionizing radiations whereas all viruses are sensitive.

(*b*) The infectivity of the scrapie agent is not changed by treatment with nucleases (DNAase and RNAase enzymes). That means the infective agent does not carry DNA or RNA as genetic material which is the most essential component of a virus particle.

(*c*) Absence of any agent-specific antibody titre, strong tendency to aggregate and hydrophobicity go against the nature of a true virus.

Backed by these evidences, **Prusiner** and his colleagues in 1982, extracted the infectious material from hamster brains, and they gave the first blow to the virus theory. Prusiner suggested that the infectious agent, causing certain degenerative disorders of central nervous system (CNS) in animals and more rarely in humans, is a small proteinaceous infectious particle, which he called *prion* (pronounced as 'preeon') and the protein believed to be responsible for infection was called *prion-related protein* or *PrP.* Soon he found that there were 2 isoforms of PrP. The first one is the normal cellular **prion-related protein, PrP^c** (the superscript '*c*' denotes cellular), which is the nonpathological form; the PrP^c may act as an acetyl-choline receptor inducer and plays an important role in the transmission of nerve signal. The other isoform of PrP is infective and called **prion-related protein scrapie, PrP^{sc}** (the superscript 'sc' denotes scrapie, which is now used to refer to all infectious form of prion causing scrapie- like diseases in animals and humans). It is this form, in which the disordered N-terminal portion appears to fold into a β-sheet, that wrecks havoc with the nervous system. It is postulated that when PrP^{sc} is ingested into the body, it induces the conversion of PrP^c in the recipient to PrP^{sc}; thus the disease is transmitted. It is something like a bad guy, who converts a good guy into one of its 'bad' form. How this conversion is catalyzed is unknown, but it strongly suggests that PrP^c represents an especially stable "off-path" folding of the type hypothesized in the preceding section. Fig. 9–34 presents the computer-generated images of the two forms of human prion protein.

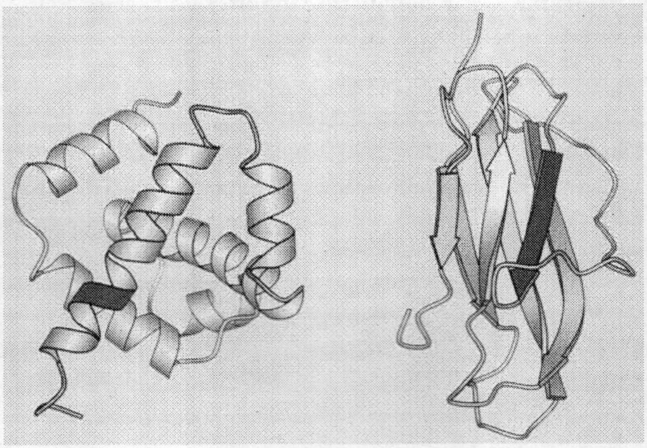

(a) Normal prion protein, PrPc

(b) Abnormal prion protein, Pr P^{sc}

Fig. 9–34. Computer-generated images of the two forms of human prion protein.

In 1988, **Prusiner** and his team reported that human prion diseases can certainly be inherited, *i.e.*, they could be heritable and communicable. Later studies on transmission of human prion to transgenic mice threw more light on prion diseases in 1994. When researchers tried to infect transgenic mice (carrying the human PrP gene) with human prion, there was no development of CNS dysfunction as expected. The mice could become susceptible to human prions only after removal of mouse PrP gene which is called **'gene knockout'** in genetics terminology.

Unfortunately, prion diseases have already played havoc among cattle population in some European countries. **Bovine spongiform encephalopathy (BSE),** or more commonly known as **mad cow disease,** took its toll in cattle population in Britain and elsewhere. Epidemiological studies revealed that the most probable cause of the epidemic was the inclusion of ruminant-derived meat and bone meal (which utilizes sheep brain, spinal cord etc.) in the cattle feed. Things became worse as reports of 12 persons afflicted with Creutzfeldt-Jakob disease came in. It was reported that these new cases occurred due to consumption of BSE-infected beef. These reports sent a shock wave throughout the world, apprehending the possible impact of prion diseases.

The recognition of the relationship of PrP to diseases won **Prusiner** the **Nobel prize in physiology or medicine in 1997**. His discovery of a new biological principle of infection goes by the name **'prion theory'.**

CLEAVAGE OF A PROTEIN

Certain proteolytic enzymes and chemical reagents cleave protein between specific amino acid residues (Table 9–6). The enzyme *trypsin*, for example, cuts on the carboxyl side of lysine or arginine, whereas the chemical *cyanogen bromide* splits peptide bonds next to methionine residues. Since these enzymes and chemicals cleave at relatively few sites in a protein molecule, they tend to create relatively large and relatively few peptides. If such a mixture of peptides is separated by biochemical procedures (chromatography, electrophoresis etc), a characteristic pattern or **peptide map** would be obtained which will be diagnostic of the protein from which the peptides were generated. This peptide map is, sometimes, referred to as the **protein's fingerprint.**

Table 9–6. Some reagents frequently used to cleave peptide bonds in proteins*

Reagent	*Amino acid 1*	*Amino acid 2*
Enzymes		
Trypsin	Lys or Arg	Any
Chymotrypsin	Phe, Trp or Tyr	Any
V8 protease	Glu	Any
Chemicals		
Cyanogen bromide	Met	Any
2-nitro-5-thiocyano-benzoate	Any	Cys

* The specificity for the amino acids on either side of the cleaved bond is indicated. The carboxyl group of amino acid 1 is released by the cleavage; this amino acid is to the left of the peptide bond as normally written.

BIOLOGICAL ROLES OF PROTEINS

Proteins are of utmost significance to biological systems. These are most critical to life and perform various functions. Some of their roles are given below.

1. Many proteins act as **catalysts**, thus usually enhancing the rate of chemical reactions to such extents as needed by the living cells.

2. The fibrous proteins serve as **components of the tissues** holding the skeletal elements together. Collagen is a structural unit of connective tissues.

3. The nucleoproteins serve as **carriers of genetic** characters and hence govern inheritance of traits.

4. Proteins also perform **transport functions.** Many compounds enter the cells and accumulate inside at much higher concentrations than expected from diffusion alone. These changes require the input of energy and are usually termed active transport. The mechanism of active transport involves proteins either as catalysts or as adsorbents or as both.

5. Various protein **hormones** are known. These regulate the growth of plants and animals, besides controlling many other physiological functions.

6. Under conditions of non-digestion and no chances for denaturation, the proteins accumulate inside the cells and produce **toxicity**. Venoms of snakes and insects are injected by biting into the blood. Certain of the foreign proteins present in venom are actually enzymes. These enzymes attack body tissue causing destruction of blood cells leading ultimately to death.

In other cases, the toxicity from foreign proteins results from responses by the affected animals. Allergic reactions exemplify this category. These reactions occur when an animal is exposed to a foreign protein to which it has been sensitized by prior exposure. The toxic disorders include skin blisters, swelling of limbs and respiratory congestion, leading sometimes to death.

7. Blood plasma, which is obtained after removal of the blood cells by centrifugal action, is essentially a solution of proteins in water. It is used for the **treatment of shock** produced by serious injuries and operations.

8. **Interferon (IF** or **IFN)** is a generic term which applies to a number of (about over 20) related low molecular weight, regulatory glycoproteins produced by many eukaryotic cells in response to numerous inducers : a virus infection, double-stranded RNA, endotoxins, antigenic stimuli, mitogenic agents, and many parasitic organisms capable of intracellular growth (*Listeria monocytogenes*, chlamydiae, rickettsias, protozoa). They are effective in treating viral diseases and cancer, and in eliminating its side effects. The most widely-studied property of interferons is their ability to 'interfere' (hence, their nomenclature) with the replication of viruses. *They are usually species-specific but*

virus-nonspecific. Interferon was discovered in 1957 in London by the British virologist **Alick Issacs** and a visiting scientist from Switzerland **Jean Lindenmann**, both of the National Institute for Medical Research, London. They found an agent responsible for such viral interference ; a protein released by cells exposed to a virus, that enables other cells to resist viral infection. They called it interferon (IF). Since then, several different classes of interferons have been identified. Human interferons have been classified into 3 classes, depending on which type of they are produced by :

(*a*) Leukocytes (L) or alpha interferons (IFN-α)

(*b*) Fibroblasts (F) or beta interferons (IFN-β)

(*c*) Immune type (I) or gamma interferons (IFN-γ)

In 1978, a Tokyo metropolitan medical team has cultivated this "wonder substance" using cells from the placenta taken at the time of birth (or parturition). Large-scale production of this protein is possible because the human cells reproduce in tissue culture. Because of their antiviral and antiproliferative activities, interferons have been considered important therapeutically in treating viral diseases such as hepatitis, encephalitis, cancer and even common cold. The IFs may not be a magic bullet like penicillin against all viruses, yet it may be a very useful addition to the armamentarium already available against viruses in general, and cancer in particular.

9. Peptides from humans called *defensins* have been found to be **antibiotic** in nature. Produced by the immune system, these cells smother and kill the invading pathogens. They are secreted by the epithelial cells lining the moist body surface of mammals and serve as the body's own disinfectants.

10. Another group of peptides called *endorphins* are found in the brain and are involved in the **suppression of pain**, creation of euphoric highs and feelings of joy. Candace Pert, a codiscoverer of the body's opiates, says "They reveal the harmony as well as the economy inherent in nature. Nothing is wasted". She adds "Once a molecule is found effective, it tends to be used over and over again in the ladder of life, in newer and ever more exciting configurations."

11. The frog's secretion *adenoregulin*, a 33-amino-acid-long peptide, works on the receptors in the brain which handle adenosine. Adenosine is a fundamental component in all human cell fuel. And the frog peptide seems to **enhance the binding of adenosine** by subtly altering the receptor in the brain.

12. The American scientists at the Chicago Medical School and the Harvard Medical School, in 1982, have isolated about 30 g of a glycopeptide, known as *S factor*, from 4.5 tones of urine from healthy males. The S factor is composed of alanine, glutamic acid, diaminopimelic acid, and muramic acid. The S factor acts **as a soporific,** *i.e.*, as a sleep-promoting agent and has little side effects. The researchers infused a very light concentration of the S factor into the brain of the rabbit and it was found that it induced a 50% increase in what is known as *slow wave sleep*— a deep dream-free sleep that occurs in animals and humans after sleep derivation and is normal as judged by various criteria.

PROBLEMS

1. Glycine is a highly conserved amino acid residue in the evolution of proteins. Why ?
2. The shape of hair is determined in part by the pattern of disulfide bonds in keratin, its major protein. How can curls be induced ?
3. Translate the following amino acid sequence into one-letter code : Leu-Glu-Ala-Arg-Asn-Ile-Asn-Gly-Ser-Cys-Ile-Glu-Cys-Glu-Ile-Ser-Gly-Arg-Glu-Ala-Thr.
4. The three-dimensional structure of biomolecules is more conserved evolutionarily than is sequence. Why is this the case ?
5. Would a homopolymer of alanine be more likely to form an α helix in water or in a hydrophobic medium ? Explain.

6. It has been postulated that the normal (noninfectious) form of prion differs from the infectious form only in secondary/tertiary structure.
   (*a*) How might you show that changes in secondary structure occur ?
   (*b*) How might you check for changes in quaternary structure ?
7. It is observed that chloride ion acts as a negative allosteric effector for hemoglobin. Suggest a possible explanation for why this should be so.
8. The structure of the amino acid isoleucine is :

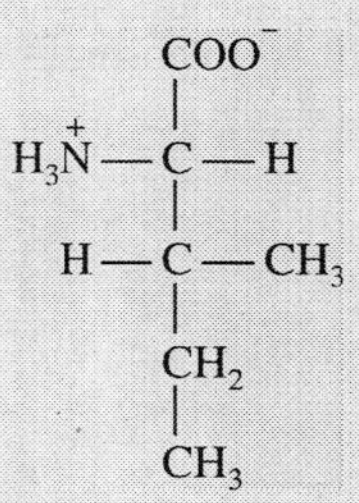

   (*a*) How many chiral centres does it have ?
   (*b*) How many optical isomers ?
   (*c*) Draw perspective formulas for all the optical isomers of isoleucine.
9. Lysine makes up 10.5% of the weight of ribonuclease. Calculate the minimum molecular weight of ribonuclease. The ribonuclease molecule contains ten lysine residues. Calculate the molecular weight of ribonuclease.
10. What is the approximate molecular weight of a protein containing 682 amino acids in a single polypeptide chain ?
11. Pepsin of gastric juice (pH $\approx$ 1.5) has a pI of about 1, much lower than that of other proteins. What functional groups must be present in relatively large numbers to give pepsin such a low pI ? What amino acids can contribute such groups ?
12. William Astbury discovered that the x-ray pattern of wool shows a repeating structural unit spaced about 0.54 nm along the direction of the wool fiber. When he steamed and stretched the wool, the x-ray pattern showed a new repeating structural unit at a spacing of 0.70 nm. Steaming and stretching the wool and then letting it shrink gave an x-ray pattern consistent with the original spacing of about 0.54 nm. Although these observations provided important clues to the molecular structure of wool, Astbury was unable to interpret them at the time. Given our current understanding of the structure of wool, interpret Astbury's observations.
13. A number of natural proteins are very rich in disulfide bonds, and their mechanical properties (tensile strength, viscosity, hardness, etc.) are correlated with the degree of disulfide bonding. For example, glutenin, a wheat protein rich in disulfide bonds, is responsible for the cohesive and elastic character of dough made from wheat flour. Similarly, the hard, tough nature of tortoise shell is due to the extensive disulfide bonding in its α-keratin. What is the molecular basis for the correlation between disulfide-bond content and mechanical properties of the protein ?
14. When wool sweaters or socks are washed in hot water and/or dried in an electric dryer, they shrink. From what you know of α-keratin structure, how can you account for this ? Silk, on the other hand, does not shrink under the same conditions. Explain.
15. In the following polypeptide, where might bends or turns occur ? Where might intrachain disulfide cross-linkages be formed ?

1 2 3 4 5 6 7 8 9 10 11 12 13 14 15 16 17 18
Ile–Ala–His–Thr–Tyr–Gly–Pro–Phe–Glu–Ala–Ala–Met–Cys–Lys–Trp–Glu–Ala–Gln–

19 20 21 22 23 24 25 26 27 28
Pro–Asp–Gly–Met–Glu–Cys–Ala–Phe–His–Arg–

16. How do depilatory creams remove hair ?

17. Which of the following characteristics are associated with myoglobin, hemoglobin, both of them or neither of them ?

(*a*) majority of structure in α-helical conformation

(*b*) oxygen carrier

(*c*) carbon dioxide carrier

(*d*) not an allosteric protein

(*e*) heme group(s) in a polar crevice(s)

(*f*) protoporphyrin IX

(*g*) sigmoidal oxygen dissociation curve

(*h*) tertiary structure

(*i*) quaternary structure

(*j*) blocked N-terminal residue

18. What is homocysteine ?

19. Why do some people have curly hair while in others hair grow straight ?

20. Why does our skin become wrinkled in old age ?

21. Why do some wounds leave scars while some do not ?

22. Why don't we feel pain when cutting hair / nails ?

23. What is the fraction of glycine (Gly) present as the completely uncharged form at the isoelectric point (pI) ?

24. Suppose you want to synthesize tripeptides using glycine, alanine and serine as building blocks.

(*a*) How many different tripeptides can you prepare using any of the three amino acids in any of the 3 positions, if you can use an amino acid more than once ?

(*b*) How many different tripeptides can you prepare using each amino acid only once ?

25. When wool sweaters or socks are washed in hot water and/or dried in an electric dryer, they shrink. Silk, on the other hand, does not shrink under the same conditions. Explain.

26. Proteins in a pea plant differ from those of a dog, yet they are built from similar amino acids. How can the difference be explained ?

27. Why amino acids are called ampholytes ?

28. What are the evidences for the existence of zwitterion structure of amino acids.

29. How many amino acids are involved in five complete turns of α-helix ?

30. Why L-phenylalanine is an essential amino acid whereas L-alanine is not ?

31. What is 'formol titration ?

ANSWERS TO PROBLEMS

1. Glycine has the smallest side chain of any amino acid. Its size often is critical in allowing polypeptide chains to make tight turns or to approach one another closely.

2. Disulfide bonds in hair are broken by adding a thiol and applying gentle heat. The hair is curled, and an oxidizing agent is added to re-form disulfide bonds to stabilize the desired shape.

3. LEARNINGSCIE(N)CEISGREAT.
4. Because three-dimensional structure is much more closely associated with function than is sequence, tertiary structure is more evolutionarily conserved than is primary structure. In other words, protein function is the most important characteristic, and protein function is determined by structure. Thus, the structure must be conserved, but not necessarily a specific amino acid sequence.
5. In a hydrophobic environment, the formation of intrachain hydrogen bonds would stabilize the amide hydrogen atom and carbonyl oxygen atoms of the polypeptide chain; so an α helix would form. In an aqueous environment, these groups would be stabilized by interaction with water, so there would be no energetic reason to form an α helix. Thus, the α helix would be most likely to form in an hydrophobic environment
6. Assuming that x-ray diffraction is not practical, we have:
   (*a*) Analysis by CD or by NMR.
   (*b*) Sedimentation equilibrium.
7. Chloride ions must interact with positively charged groups so as to stabilize the deoxy state. Thus, the higher the Cl^- concentration, the lower the O_2 affinity of hemoglobin will be.
8. (*a*) 2
   (*b*) 4
   (*c*)

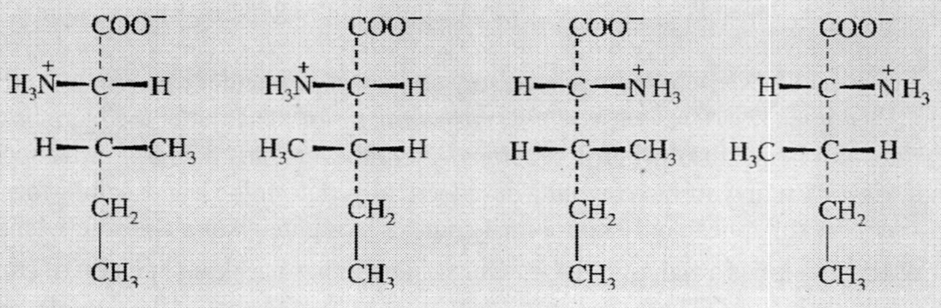

9. 1,220; 12,200
10. 75,020
11. Carboxyl groups; Asp and Glu
12. The principal structural units in the wool fibre polypeptides are successive turns of the α helix, which are spaced at 0.54 nm intervals. Steaming and stretching the fibre yields an extended polypeptide chain with the β conformation, in which the distances between the R groups are about 0.70 nm. On shrinking, the fibre reassumes an α-helical structure.
13. The disulfide bonds are covalent bonds, which are much stronger than the noncovalent bonds that stabilize most proteins. They serve to cross-link protein chains, increasing their stiffness, mechanical strength, and hardness.
14. Wool shrinks when polypeptide chains are converted from the extended conformation (β-pleated sheet) to the α-helical conformation in the presence of heat. The β-pleated sheets of silk, with their small, closely packed amino acid side chains, are more stable than those of wool.
15. At residues 7 and 19; prolines are often but not always found at bends. Between residues 13 and 24.
16. Methods of removal of hair are classified as *depilatory* (removal of hair from the surface of the skin) or *epilatory* (removal of intact hair with the root). Depilatory creams are substituted

mercaptans, such as thioglycollic acid, which reduce the disulfide bond in the peptide chains of keratin, a hair protein. As a result, the hair fibre swells and softens and can be washed from the skin.

17. (*a*) Both myoglobin and hemoglobin
(*b*) Both myoglobin and hemoglobin
(*c*) Hemoglobin
(*d*) Myoglobin
(*e*) Neither of the two
(*f*) Both myoglobin and hemoglobin
(*g*) Hemoglobin
(*h*) Both myoglobin and hemoglobin
(*i*) Hemoglobin
(*j*) Neither of the two

18. Homocysteine is a normal building block for protein, and too much can damage one's arteries. Vitamins B_6, B_9 and B_{12} can lower it.

19. Scalp hair is not the same among all the people in the world. The Negro has tightly curled hair while the American Indians have straight, lank and coarse hair. The Caucasians have hair that may be straight or wavy or curly. Australian natives have "frizzly" hair. So the type of hair one has depends on one's racial inheritance and is genetically determined. *The reason why someone's hair is curly while another's is not depends on the structure of the strand of hair.* If you cut a shaft of hair transversely and observe it under microscope, the difference would become clear. *The cut section of curly hair is flat.* The flatter the hair, the more easily it bends and hence the curlier it becomes. *Straight hair has a round cross-section.* Thus what kind of hair you have would be genetically determined and stable for a life-time. Cosmetic treatment can only temporarily curl or straighten hair as per our wishes.

20. Wrinkling of skin is a sign of old age. Our skin becomes wrinkled because of changes in the underlying structure. There are some connective tissues just under the outer layer of the skin which are made up of two types of protein fibres—collagen and elastin. The collagen provides the material for the tissue and the elastin fibres, which are smaller in number, give elasticity and suppleness to the skin. *With time, as one ages, the amount of elastin in the skin diminishes and the collagen fibres become disorganized. The cross-linking between collagen fibres also increases. Consequently, the tissue gradually loses its elasticity.* Thus, wrinkles appear on the face and other parts of the body. Another factor responsible for the development of wrinkles is sunlight. *The ultraviolet part of the sunlight accelerates the changes in the skin which cause wrinkles.*

21. A deep wound or cut usually leaves a scar whereas a superficial one seldom does so. A wound in which collagen fibres – situated deep in the dermis layer of the skin – are damaged leaves a scar. This is because the collagen fibres are arranged in the skin in a specific sequencial manner. To restore the same sequence when wound is healing is not possible. As a result of a change in the deposition pattern of collagen fibres, a scar is formed on the skin.

22. Pain is felt only when a nerve end is stimulated by a cut or injury. Since hair and nails do not have nerve endings, cutting them does not produce pain. However, cells at the base of a nail and hair are living cells and have nerve endings. So, if a nail is cut from the base or if a hair is plucked off, pain is felt.

23. 1 part in 10^7 (one ten-millionth)

24. (a) 27 (b) 6

25. Wood shrinks due to conversion of a polypeptide chain in the extended conformation (β-pleated sheet) to an α-helical conformation.

26. Proteins contain about 20 different amino acids. The number and manner in which the amino acids may be grouped is enormously high. Hence, the number of possible proteins is correspondingly large. From 20 different amino acids, it is theoretically possible to make 400 dipeptides and as many as 8,000 tripeptides.

27. Since amino acids contain at least one amino group and at least one carboxyl group, they are amphoteric. *In dry state*, amino acids form lower salts which are called dipolar ions or zwitterions. In the dipolar ion, the amino group remains as $-NH_3^+$ group and the carboxyl group remains as $-COO^-$ group. However, *in aqueous medium*, an equilibrium exists involving the dipolar ion, and the cationic and anionic forms of the amino acids :

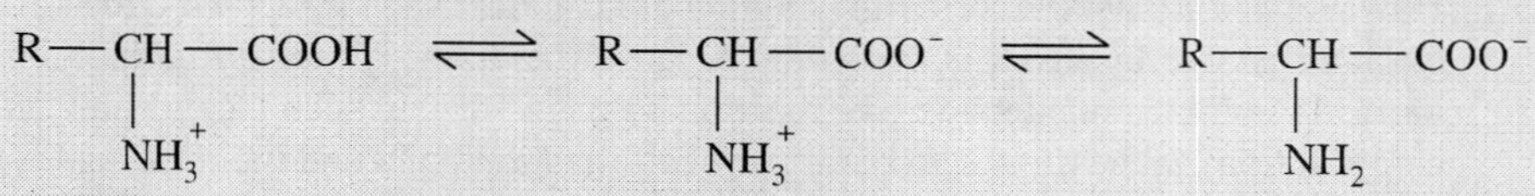

Cationic form (predominant at pH <2) — Zwitterionic form (predominant at isoelectric point) — Anionic form (predominant at pH > 11)

Hence, amino acids act as an ampholyte by donating a proton and also by accepting a proton.

28. Amino acids in solid crystalline form exist having inner salt structure, in which the $-NH_2$ group exists as $-NH_3^+$ and $-COOH$ group exists as $-COO^-$:

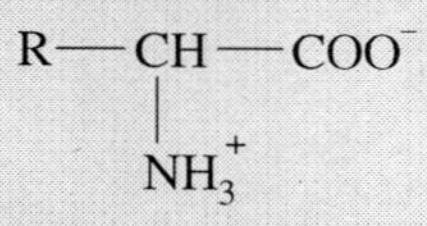

Following are the evidences for their existence :

(*a*) Amino acids are soluble in polar solvent like water.

(*b*) Amino acids, dissolved in water, have pH near the neutral region, which is in contrary to the carboxylic acids.

(*c*) Amino acids have high melting point (about 260°C) which is possible for ionic salts.

29. Per turn of helix = 3.6 amino acids

Five turns of helix = 3.6 × 5

= 18.0 amino acid

30. L-alanine can be formed within the body by metabolic activity of pyruvate. On the contrary, L-phenylalanine cannot by synthesized within the body and has to be supplied from outside in the diet.

$$\text{Glutamate} + \text{pyruvate} \rightleftharpoons \alpha\text{-ketoglutarate} + \text{alanine}$$

31. It is not possible to estimate the total acidity of an amino acid solution by direct titration with an alkali (such as NaOH) in view of the presence of $-NH_2$ group which will take up the H^+ released by the ionization of $-COOH$ group.

$$\underset{\displaystyle NH_2}{R-\underset{|}{CH}-COOH} \xrightleftharpoons{OH^-} \underset{\displaystyle NH_2}{R-\underset{|}{CH}-COO^-} + H^+$$

$$\underset{\displaystyle NH_2}{R-\underset{|}{CH}-COO^-} \xrightleftharpoons{H^+} \underset{\displaystyle NH_3^+}{R-\underset{|}{CH}-COO^-}$$

However, if 10% neutral formalin (solution of formaldehyde in water) is added in excess, the interference by $-NH_2$ is overcome because the amino group forms the dimethylol derivative with formaldehyde (see reaction on page 187). In such a condition, an amino acid solution can be titrated to complete neutralization point and the total acidity can be estimated. Such a titration of amino acids in presence of excess of formalin is called 'formol titration'.

CHAPTER

10 Proteins-II Classification

INSIDE

- Classification Based on the Source of Protein Molecule
- Classification Based on the Shape of Protein Molecule
  - Globular proteins
  - Fibrous proteins
- Classification Based on Composition and Solubility
  - Simple proteins
  - Conjugated proteins
  - Derived proteins
- Egg Proteins
- Milk Proteins

Bird feathers are made largely of the protein keratin.

The proteins are all remarkably similar in structure insofar as they contain amino acids. But as little is known so far regarding their structure, a classification based totally on this criterion is not possible. However, a few systems based on one or the other cirterion are given below.

CLASSIFICATION BASED ON THE SOURCE OF PROTEIN MOLECULE

Since long, the proteins have been traditionally divided into two well-defined groups: animal proteins and plant proteins. **Animal proteins** are the proteins derived from animal sources such as eggs, milk, meat and fish. They are usually called *higher-quality proteins* because they contain (and hence supply) adequate amounts of all the essential amino acids. On the other hand, **plant proteins** are called *lower-quality proteins* since they have a low content (limiting amount) of one or more of the essential amino acids. The *four most common limiting amino acids are methionine, lysine, threonine and tryptophan* (Table 10–1). Although plant proteins have limiting amounts of some (but not all) amino acids, it should not be construed that they are poor protein sources.

Table 10–1. Limiting amino acids in some plant proteins

Food	*Amino acid(s)*
Cereal grains and millets	Lysine, Threonine
Rice and soybeans	Methionine
Legumes (peas and beans)	Methionine, Tryptophan
Groundnuts	Methionine, Lysine, Threonine
Sunflower seeds	Lysine
Green leafy vegetables	Methionine

CLASSIFICATION BASED ON THE SHAPE OF PROTEIN MOLECULE

On the basis of the shape of protein molecule, the proteins have been grouped under two categories : globular and fibrous.

1. Globular or Corpuscular Proteins

These have an axial ratio (length : width) of less than 10 (usually not over 3 or 4) and, henceforth, possess a relatively spherical or ovoid shape. These are usually soluble in water or in aqueous media containing acids, bases, salts or alcohol, and diffuse readily. *As a class, globular proteins are more complex in conformation than fibrous proteins, have a far greater variety of biological functions and are dynamic rather than static in their activities.* Tertiary and quaternary structures are usually associated with this class of proteins. Nearly all enzymes are globular proteins, as are protein hormones, blood transport proteins, antibodies and nutrient storage proteins.

A simple functional classification of globular proteins is not possible because of 2 reasons :

(*a*) Firstly, these proteins perform a variety of different functions.

(*b*) Secondly, many widely-differing globular proteins perform almost similar functions.

However, **Conn** and **Stumpf** (1976) have classified globular proteins as follows:

Globular Proteins
- Cytochrome C
- Blood proteins
  - Serum albumin
  - Glycoproteins
  - Antibodies (= Immunoglobulins)
  - Hemoglobin
  - Hormones
- Enzymes
- Nutrient proteins

2. Fibrous or Fibrillar Proteins

These have axial ratios greater than 10 and, henceforth, resemble long ribbons or fibres in shape. *These are mainly of animal origin and are insoluble in all common solvents* such as water, dilute acids, alkalies and salts and also in organic solvents. Most fibrous proteins serve in a structural or protective role.

The fibrous proteins are extremely strong and possess two important properties which are characteristic of the elastomers. These are :

(*a*) They can *stretch* and later recoil to their original length.

(*b*) They have a tendency to *creep*, *i.e.*, if stretched for a long time, their basic length increases and equals the stretched length but, if the tension on the two ends of the fibril is relaxed, they creep to their shorter and shorter length. A large scar, for example, creeps to a smaller size if there is no tension on the scar. On the contrary, if the scar is in a region of high tension, the scar becomes larger and larger as happens in the skin of a person gradually becoming obese.

It is a heterogeneous group and includes the proteins of connective tissues, bones, blood vessels, skin, hair, nails, horns, hoofs, wool and silk. The important examples are:

I. **Collagens.** These are of mesenchymal origin and form the major proteins of white connective tissues (tendons, cartilage) and of bone. *More than half the total protein in mammalian body is collagen*; acted upon by boiling in water, dilute acids or alkalies to produce the soluble gelatins; *unique in containing high contents (12%) of hydroxyproline*; poor in sulfur since cysteine and cystine are lacking.

> ***Tendons*** connect muscle to bone, whereas ***ligaments*** attach one bone to another. Tendons are nonelastic but ligaments can be stretched, because they contain, in addition to collagen fibres, the protein elastin.

II. **Elastins.** Also of mesenchymal origin; form the major constituents of yellow elastic tissues (ligaments, blood vessels); differ from collagens in not being converted to soluble gelatins.

III. **Keratins.** These are of ectodermal origin; form the major constituents of epithelial tissues (skin, hair, feathers, horns, hoofs, nails); *usually contain large amounts of sulfur in the form of cystine*– human hair has about 14% cystine.

IV. **Fibroin.** It is the principal constituent of the fibres of silk; composed mainly of glycine, alanine and serine units.

CLASSIFICATION BASED ON COMPOSITION AND SOLUBILITY

This is nowadays the most accepted system of classification and is based on the proposals made by the committees of *British Physiological Society* (1907) and the *American Physiological Society* (1908). The system divides the proteins into 3 major groups, based on their composition *viz*., simple, conjugated and derived.

A. Simple Proteins or Holoproteins

These are of globular type except for scleroproteins which are fibrous in nature. This group includes proteins containing only amino acids, as structural components. On decomposition with acids, these liberate the constituent amino acids.

These are further classified mainly on their solubility basis as follows:

1. Protamines and histones. These are basic proteins and occur almost entirely in animals, mainly in sperm cells; possess simplest structure and lowest molecular weight (approximately 5,000); soluble in water; unlike most other proteins, not coagulated by heat; strongly basic in character owing to high content of basic amino acids (lysine, arginine); form salts with mineral acids and nucleic proteins. Protamines are virtually devoid of sulfur and aromatic amino acids. Histones are somewhat weaker bases and are, therefore, insoluble in NH_4OH solution, whereas the protamines are soluble.

e.g., protamines—*clupeine* from herring sperm, *salmine* from salmon sperm, *sturine* from sturgeon and *cyprinine* from carp,

histones—*nucleohistones* of nuclei; *globin* of hemoglobin.

2. Albumins. These are widely distributed in nature but more abundant in seeds; soluble in water and dilute solutions of acids, bases and salts; precipitated with a saturated solution of an acid salt like $(NH_4)_2.SO_4$ or a neutral salt like $Na_2.SO_4$; coagulated by heat.

e.g., leucosine in cereals, *legumeline* in legumes, *ovalbumin* from white of egg, *serum albumin* from blood plasma, *myosin* of muscles and *lactalbumin* of milk whey.

> **Pseudoglobulins** and **euglobulins** differentiated on the basis of solubility behaviour. A pseudoglobulin is soluble at very low ionic strength, whereas euglobulin remains sparingly soluble until the ionic strength is raised. Euglobulins remain soluble at the isoelectric point (pI), whereas pseudoglobulins do not.

3. Globulins. These are of two types—pseudoglobulins and euglobulins. Euglobulins are more widely distributed in nature than the pseudoglobulins; either soluble (pseudoglobulins) or insoluble (euglobulins) in water; precipitated with half saturated solution of $(NH_4)_2.SO_4$; coagulated by heat.

e.g., pseudoglobulins— *pseudoglobulin* of milk whey.

euglobulins— *serum globulin* from blood plasma, *ovoglobulin* from eggwhite; *myosinogen* from muscle; globulins of various plant seeds like hemp (*edestin*), soybeans (*glycinine*), peas (*legumine*), peach (*amandine*), oranges (*pomeline*); also potato (*tuberin*).

4. **Glutelins.** These have been isolated only from plant seeds; insoluble in water, dilute salt solutions and alcohol solutions but soluble in dilute acids and alkalies; coagulated by heat.

e.g., *glutenin* from wheat, *glutelin* from corn, *oryzenin* from rice, etc.

5. **Prolamines.** These have also been isolated only from plant seeds; insoluble in water and dilute salt solutions but soluble in dilute acids and alkalies and also in 60 – 80% alcohol solutions; not coagulated by heat

e.g., *gliadin* from wheat, *zein* from corn, *hordein* from oat, etc.

Some biochemists like **Karlson** (1968) are of the viewpoint that glutelins and prolamines should not be granted the status of exclusive classes since they are small groups of vegetable proteins occurring in grain kernels.

6. **Scleroproteins or Albuminoids.** These occur almost entirely in animals and are, therefore, commonly known as the '*animal skeleton proteins*'; insoluble in water, dilute solution of acids, bases and salts and also in 60–80% alcohol solutions; not attacked by enzymes.

e.g., *collagen* of bones, *elastin* in ligaments, *keratin* in hair and horry tissues and *fibroin* of silk.

B. Conjugated or Complex Proteins or Heteroproteins

These are also of globular type except for the pigment in chicken feathers which is probably of fibrous nature. These are the proteins linked with a separable nonprotein portion called *prosthetic group*. The prosthetic group may be either a metal or a compound. On decomposition with acids, these liberate the constituent amino acids as well as the prosthetic group.

Their further classification is based on the nature of the prosthetic group present. The various divisions are metalloproteins, chromoproteins, glycoproteins, phosphoproteins, lipoproteins and nucleoproteins. (Instead of metalloproteins, chromoproteins etc., the terms metalloproteids, chromoproteids etc., are sometimes used.)

1. **Metalloproteins.** These are the proteins linked with various metals. These may be of stable nature or may be more or less labile. Based on their reactivity with metal ions, the metalloproteins may be classified into 3 groups:

I. *Metals strongly bound by proteins*. Some heavy metals (Hg, Ag, Cu, Zn) become strongly binded with proteins like *collagen, albumin, casein* etc., through the —SH radicals of the side chains. Some other proteins have strong binding affinities for Fe (*siderophilin*) and Cu (*ceruloplasmin*). In these cases, the following pattern of binding may be present :

Siderophilin, also called as **transferrin**, is an important metalloprotein and constitutes about 30% of the total plasma protein. It has a molecular weight of about 90,000 and is capable of binding 2 atoms of iron per mole. It facilitates iron transport.

Ceruloplasmin is an important blue copper-binding protein in the blood of humans and other vertebrates. This protein contains about 90% of copper in serum. It has a molecular weight of about 150,000 and contains 8 atoms of copper per mole. Ceruloplasmin is only one of the many sialoglycoproteins whose removal from the bloodstream is triggered by the loss of sialic acid units. It probably functions by reversibly releasing and binding copper at various sites in the body, whereby regulating copper absorption. In its deficiency, the Wilson's disease develops in man which is characterized by hepatolenticular degeneration.

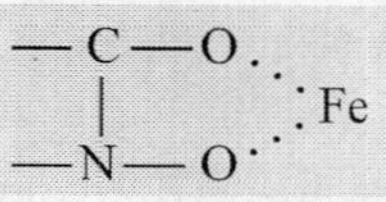

II. *Metals bound weakly by proteins.* Ca belongs to this category. Here the binding takes place with the help of radicals possessing the electron charge.

III. *Metals which do not couple with proteins.* Na and K belong to this group. These form compounds with nucleic acids where apparently electrostatic bonds are present.

2. **Chromoproteins.** These are proteins coupled with a coloured pigment. Such pigments have also been found among the enzymes like catalase, peroxidase and flavoenzymes. Similarly, chlorophyll is present in leaf cells in the form of a protein, the chloroplastin. The chloroplastin dissolves in water as a colloid and is readily denatured.

e.g., myoglobin, hemoglobin, hemocyanin, hemoerythrin, cytochromes, flavoproteins, catalase, etc.

3. **Glycoproteins and Mucoproteins.** These are the proteins containing carbohydrate as prosthetic group. Glycoproteins contain small amounts of carbohydrates (less than 4%), whereas mucoproteins contain comparatively higher amounts (more than 4%).

e.g., glycoproteins— *egg albumin*, *elastase*, certain *serum globulins* and also certain *serum albumins*.

mucoproteins— *ovomucoid* from egg white, *mucin* from saliva and *Dioscorea* tubers, *osseomucoid* from bone and *tendomucoid* from tendon.

4. **Phosphoproteins.** These are proteins linked with phosphoric acid; mainly acidic.

e.g., casein from milk and *ovovitellin* from egg yolk.

5. **Lipoproteins.** Proteins forming complexes with lipids (cephalin, lecithin, cholesterol) are called lipoproteins; soluble in water but insoluble in organic solvents.

e.g., lipovitellin and *lipovitellenin* from egg yolk; lipoproteins of blood.

6. **Nucleoproteins.** These are compounds containing nucleic acid and protein, esp., protamines and histones. These are usually the salt-like compounds of proteins since the two components have opposite charges and are bound to each other by electrostatic forces. They are present in nuclear substances as well as in the cytoplasm. These may be considered as the sites for the synthesis of proteins and enzymes.

e.g., nucleoproteins from yeast and thymus and also viruses which may be regarded as large molecules of nucleoproteins; *nucleohistones* from nuclei-rich material like glandular tissues; *nuclein*.

C. Derived Proteins

These are derivatives of proteins resulting from the action of heat, enzymes or chemical reagents. This group also includes the artificially-produced polypeptides.

I. ***Primary derived proteins.*** These are derivatives of proteins in which the size of protein molecule is not altered materially.

1. **Proteans.** Insoluble in water; appear as first product produced by the action of acids, enzymes or water on proteins.

e.g., edestan derived from edestin and *myosan* derived from myosin.

2. **Metaproteins or Infraproteins.** Insoluble in water but soluble in dilute acids or alkalies; produced by further action of acid or alkali on proteins at about 30–60°C.

e.g., acid and *alkali metaproteins.*

3. **Coagulated Proteins.** Insoluble in water; produced by the action of heat or alcohol on proteins. e.g., *coagulated eggwhite.*

II. ***Secondary derived proteins.*** These are derivatives of proteins in which the hydrolysis has certainly occurred. The molecules are, as a rule, smaller than the original proteins.

1. **Proteoses.** Soluble in water; coagulable by heat; produced when hydrolysis proceeds beyond the level of metaproteins; *primary proteoses* are salted out by half saturation with $(NH_4)_2.SO_4$ and precipitated by HNO_3 and picric acid; *secondary proteoses* are salted out only by complete saturation with $(NH_4)_2.SO_4$ but are not precipitated by HNO_3 or picric acid.

e.g., albumose from albumin; *globulose* from globulin.

2. **Peptones.** Soluble in water; noncoagulable by heat; produced by the action of dilute acids or enzymes when hydrolysis proceeds beyond proteoses; neither salted out by $(NH_4)_2SO_4$ nor precipitated by HNO_3 or picric acid.

3. **Polypeptides.** These are combinations of two or more amino acid units. In fact, *the proteins are essentially long chain polypeptides.*

Drawbacks. Although widely accepted, the system outlined above has certain discrepancies:

1. The classification is arbitrary.
2. The criterion of solubility is not well demarcated as some globulins (pseudoglobulins) are also soluble in water.
3. Protamines and histones should have been kept under derived proteins.
4. The group metaproteins is an artificial assemblage.

The two (second and third) systems of classification, described above, may be merged into one, as shown on page 177.

EGG PROTEINS

The eggs and milk are consumed as food by an appreciable portion of the global population. Hence, the egg and milk proteins deserve a special mention. Eggs contain 2 types of proteins : the egg white protein and egg yolk protein. The various components present in the egg white protein are listed in Table 10–2.

Table 10–2. Protein composition of egg white

Protein	*Percentage*
Total protein	10-11% (on wet basis) 82.8% (on dry basis)
Ovalbumin	70% of total proteins
Conalbumin	9%
Ovomucoid	13%
Globulins	
Lysozyme (G_1)	2.6%
(G_2) (G_3)	7%
Mucin	2%
Avidin	0.06%

(*Adapted from Fevold HL, 1951*)

PROTEINS

- Fibrous
  - **SIMPLE**
    (*Insoluble in water*)
    Scleroproteins
    Collagen
    Elastin
    Keratin
    Fibroin
  - **CONJUGATED**
    (*Insoluble in water*)
    Pigment in
    chicken feather ?
- Globular
  - **SIMPLE**
    - (*Soluble in water*)
      - **Protamines**
        clupeine, salmine,
        sturine, cyprinine
      - **Histones**
        nucleohistones of
        nuclei, globin
      - **Albumins**
        leucosine, legumeline,
        ovalbumin, lactalbumin
      - **Pseudoglobulins**
        pseudoglobulin from
        milk whey
    - (*Insoluble in water*)
      - **Euglobulins**
        ovoglobulin, myosinogen,
        glycinine, legumine
      - **Glutelins**
        glutenin, oryzenin
      - **Prolamines**
        gliadin, zein,
        hordein
  - **CONJUGATED**
    - **Metalloproteins**
      siderophilin
      ceruloplasmin
    - **Chromoproteins**
      hemoglobin
      hemocyanin
      cytochromes
      flavoproteins
    - **Glycoproteins**
      egg albumin
    - **Mucoproteins**
      ovomucoid, mucin
    - **Phosphoproteins**
      casein, ovovitellin
    - **Lipoproteins**
      lipovitellin
    - **Nucleoproteins**
      nucleohistones from
      nuclei, nuclein

When egg yolk is diluted with water, proteins precipitate. When the yolk is heated, the proteins undergo heat denaturation and precipitate. Egg yolk contains 2 phosphoproteins, lipovitellin and lipovitellenin, which differ from each other in the following characteristics :

	Lipovitellin	***Lipovitellenin***
Lipid concentration :	17-18%	36-41% (mainly lecithins)
Phosphorus concentration :	1%	0.29%
Protein present :	Vitellin	Vitellenin

Egg yolk also contains water-soluble protein that does not precipitate on dilution of the yolk. This fraction is called ***livetin***. The enzymatic activity of yolk lies in this fraction.

Whole egg is an excellent food because it is a very rich source not only of protein and lipid but also of most of the vitamins (except vitamin C) and most of the required minerals (except calcium).

MILK PROTEINS

Fig. 10-1. Milk
Milk, a source of nourishment for all mammals, is composed, in part, of a variety of proteins.

Milk (Fig. 10-1) contains about 0.6–0.7% protein which is not precipitated on acidification to pH 4.7. This represents about 20% of the protein contained in skim milk. These ***whey proteins*** are separated into 2 fractions : ***lactalbumin and lactoglobulin.***

The name casein is assigned to the fraction precipitated by acidifying milk to a pH of 4.7. It is present in cow's milk (3–3.5%) and human milk (0.3–0.6%). Casein may be further purified by redissolving and precipitating again. It is of 3 types : α, β and γ. These differ one another in their molecular weights, their rate of migration in an electric field and their phosphorus content.

PROBLEMS

1. Which amino acids are usually present in low amount in plant proteins ?
2. Which group of proteins is a heterogeneous group and has properties characteristic of the elastomers ?
3. How tendons are distinguished from ligaments ?
4. Why is a classification of globular proteins totally based on their structure not possible ?
5. Why are scleroproteins best suited for animal skeleton ?
6. What is casein ?
7. Following is the average composition of milk :

Water	87.30	Ash	0.73
Fat	3.67	Solids	12.69
Total protein	3.42	Solids not fat	8.77
Lactose	4.78		

(*After Jacobs MB, 1951*)

Some people maintain that milk is the perfect food and must be included in the diet of all people for proper nutrition. Is this statement correct ? Provide a biochemical basis for your answer.

ANSWERS TO PROBLEMS

1. Threonine (Thr), Methionine (Met), Lysine (Lys), Tryptophan (Trp).
2. Fibrous (or fibrillar) proteins

3. Tendons connect muscle to bone, whereas ligaments attach one bone to another. Tendons are nonelastic but ligaments can be stretched, because they contain, in addition to collagen fibres, the protein elastin.
4. A simple functional classification of globular proteins is not possible because of 2 reasons :
   (*a*) these proteins perform a variety of different functions.
   (b) many widely-differring globular proteins perform almost similar functions.
5. Scleroproteins (albuminoids) are simple fibrous proteins; they have the axial ratios greater than 10 and henceforth resemble long ribbons or fibres in shape. Moreover, these are insoluble in all common solvents such as water, dilute acids, alkalis and salts and also in organic solvents. Also, these are not attacked by enzymes. Hence, scleroproteins are best suited for animal skeleton and serve in a structural or protective role.
6. The name casein is assigned to the fraction precipitated by acidifying milk to a pH of 4.7. It is present in cow's milk (3-3.5%) and human milk (0.3 –0.6%). Casein may further be purified by redissolving and precipitating again. It is of 3 types : α, β, and γ. These differ one another in their molecular weights, their rate of migration in an electric field and their phosphorus content.
7. From a nutritional standpoint, people need a balanced diet to meet their biochemical requirements for energy, and to provide building blocks and cofactors for biosynthesis. These requirements can be met by dietary intake of a large variety of sources; there is no evidence that a specific food (e.g., milk) is required for normal nutrition.

CHAPTER

11 Proteins–III General Properties

INSIDE

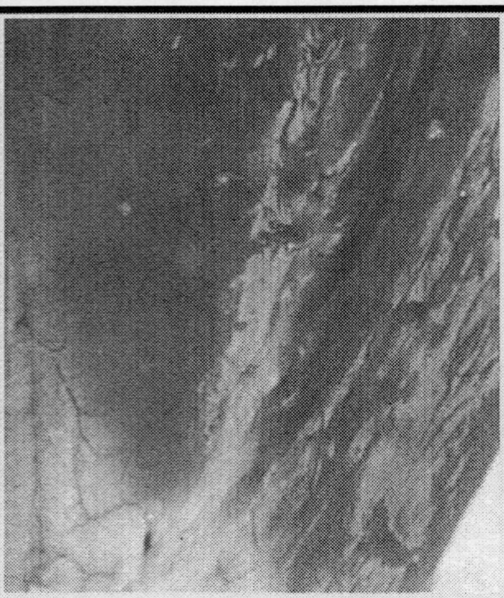

Tendon tissue (*pale*) linking muscle (*red*) to bone (left)

Tendons are made from fibrous connective tissue which contains many closely-packed collagen fibres. In some situations, such as in the legs of horses, the tendons must be able to withstand enormous forces.

The general properties of proteins are reminiscent of those of the amino acids.

PHYSICAL PROPERTIES

1. Colour and Taste. Proteins are colourless and usually tasteless. These are homogeneous and crystalline.

2. Shape and Size. As already discussed, the proteins range in *shape* from simple crystalloid spherical structures to long fibrillar structures. Two distinct patterns of shape have been recognized :

A. Globular proteins—These are spherical in shape and occur mainly in plants, esp., in seeds and in leaf cells. These are bundles formed by folding and crumpling of protein chains.

e.g., pepsin, edestin, insulin, ribonuclease etc.

B. Fibrous proteins—These are thread-like or ellipsoidal in shape and occur generally in animal muscles. Most of the studies regarding protein structure have been conducted using these proteins.

e.g., fibrinogen, myosin etc.

Each protein molecule is characterized for its specific size (Fig. 11–1). For example :

(*a*) Hemoglobin has a diameter of 55 Å.

(*b*) Edestin has a diameter of 80 Å.

(*c*) Catalase has dimensions of 80 × 64 × 54 Å (of the axes).

(*d*) Human fibrinogen has a diameter of 38 Å and a length of 700 Å.

(*e*) Collagen is one of the longest proteins, with a length of 3,000 Å.

In general, the protein molecules are always very large, as can be seen in the following examples :

(*a*) Gliadin (from wheat)—$C_{685}H_{1068}N_{196}O_{211}S_5$

(*b*) Zein (from corn)—$C_{736}H_{1161}N_{184}O_{208}S_3$

(*c*) Casein (from milk)—$C_{708}H_{1130}N_{180}O_{224}S_4P_4$

(*d*) Beta-lactoglobulin (from milk)—$C_{1642}H_{2652}O_{492}N_{420}S_{18}$

3. Molecular Weight. The extraordinary size, poor stability, specific solubility conditions and high reactivity have rendered the determination of molecular weight of proteins as a difficult task (**Edsall,** 1953). However, the proteins generally have large molecular weights ranging between 5×10^3 and 1×10^6 (see Table 11–1). It might be noted that the values of molecular weights of many proteins lie close to or multiples of 35,000 and 70,000. Previously, this was interpreted as a regularity under the name *Svedberg's rule*. Also, it was then assumed that proteins are composed of units of molecular weight 17,500. This corresponds to about 145–150 amino acid residues, since the average molecular weight of an amino acid residue amounts to about 115–120. The discovery in recent times of too many exceptions to this rule, however, finally forced its abandonment.

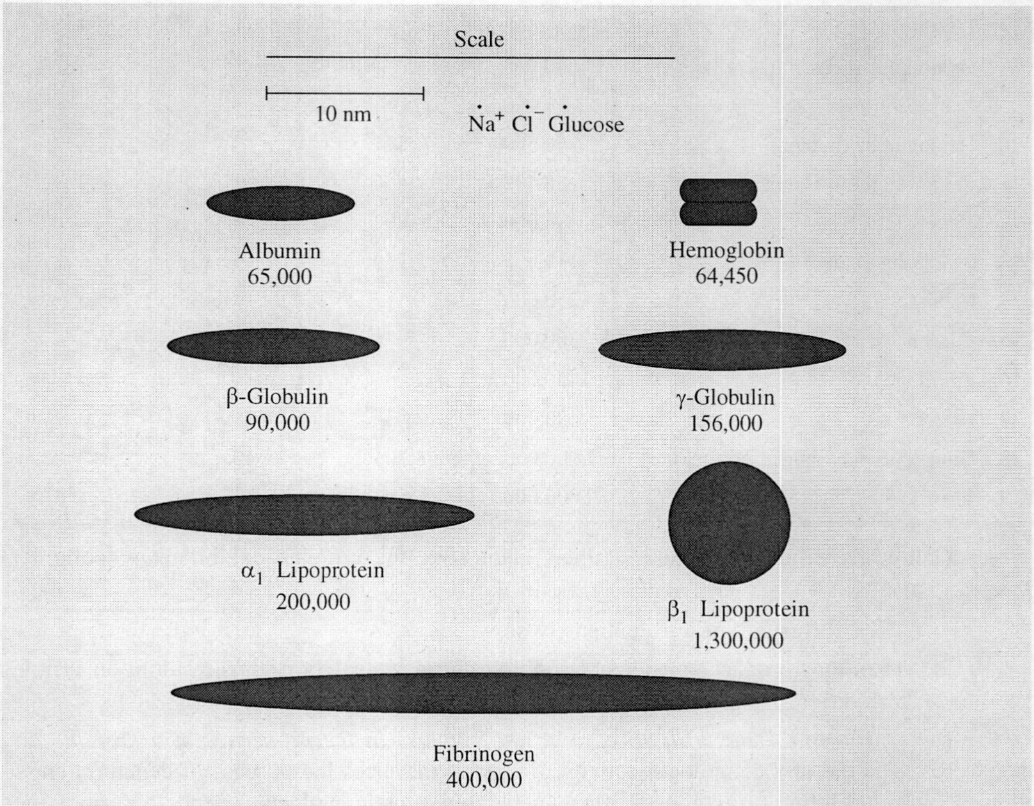

Fig. 11–1. Relative dimensions and molecular weights of some of the protein molecules in the blood

(*After Oncley JL, 1949*)

The approximate number of amino acid residues in a simple protein having no prosthetic group can be calculated by dividing its molecular weight by 110. The average molecular weight of the 20 amino acids is about 138. But as the smaller amino acids predominate in most proteins, hence the average molecular weight of an amino acid is nearer 128. Since a molecule of water (MW = 18) is eliminated to produce each peptide bond, the average molecular weight of the amino acid residue is about 128 – 18 = 110. Table 11–1 also gives the number of amino acid residues present in different proteins.

Table 11–1. Exact / Approximate molecular weights and the isoelectric points of some important proteins

Protein	*Molecular weight (MW)*	*Number of residues*	*Number of chains*	*Isoelectric point (pI)*
1. Insulin (bovine)	5,733	51	2	5.4
2. Cytochrome C	12,500	104	—	9.8
3. Ribonuclease (bovine)	14,000	124	1	7.8/9.5
4. Lysozyme (eggwhite)	14,600	129	1	11.0
5. Myoglobin (horse)	16,700	153	1	7.0
6. Chymotrypsin (bovine)	22,600	241	3	—
7. Pepsin	35,500	—	—	2.7
8. Ovalbumin (hen)	40,000	—	—	4.6
9. Zein	40,000	—	—	—
10. Hemoglobin (human)	64,500	574	4	—
11. Serum albumin (human)	68,500	~ 550	1	4.9
12. Hexokinase (yeast)	96,000	~ 800	4	—
13. γ-globulin (horse)	149,900	~ 1,250	4	6.6
14. Catalase	250,000	—	—	5.6
15. Edestin	300,000	—	—	6.9
16. Fibrinogen	450,000	—	—	5.5
17. Urease	480,000	—	—	5.0
18. Glutamate dehydrogenase (bovine)	1,000,000	~ 8,300	~ 40	—
19. Virus protein of TMV	60,000,000	—	—	—

4. Colloidal Nature. Because of their giant size, the proteins exhibit many colloidal properties, such as :

I. Their diffusion rates are extremely slow.

II. They may produce considerable light-scattering in solution, thus resulting in visible turbidity (Tyndall effect).

5. Denaturation. *Denaturation refers to the changes in the properties of a protein.* In other words, it is the loss of biologic activity. In many instances the process of denaturation is followed by **coagulation**— a process where denatured protein molecules tend to form large aggregates and to precipitate from solution.

Denaturation may be brought about by a variety of agents, both physical and chemical. The **physical agents** include mechanical action (like shaking), heat treatment cooling and freezing operations, rubbing, high hydrostatic pressures, (5,000 to 10,000 atm.), ultraviolet rays, etc. The **chemical agents**, that cause denaturation, are many ionizing radiations (like X-rays, radioactive and ultrasonic radiations), organic solvents (acetone, alcohol), aromatic anions (salicylates), some anionic detergents (like sodium dodecyl sulfate), etc. A common example of protein easily denatured by shaking or heat is the albumin of eggwhite.

$$H_3C—(CH_2)_{10}—CH_2OSO^-Na^+$$

Sodium dodecyl sulfate, SDS

It was suggested by **Wu** (1931) that denaturation leads mainly to the unfolding of the peptide chain, thus causing disorganization of the internal structure of protein (Fig. 11–2). This is evidenced by the fact that the denatured proteins are more easily hydrolyzed (**Mirsky**, 1935).

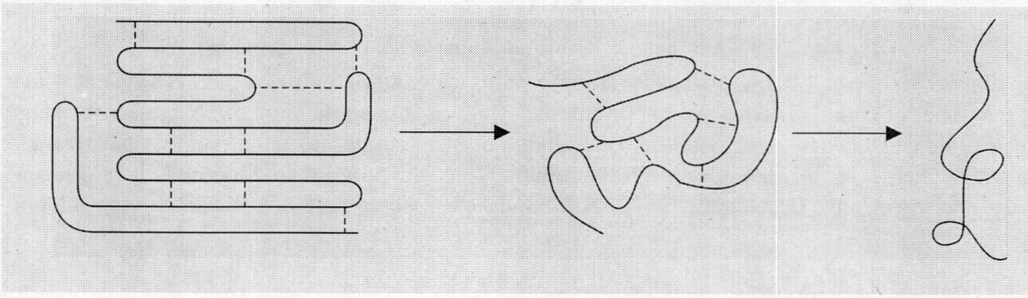

Fig. 11–2. Representation of denaturation of a protein
The process of denaturation breaks bonds.

When the peptide chains or the protein molecules are unrolled, certain bonds split and new sites of bundles are exposed to the action of certain proteolytic enzymes causing hydrolysis. Thus, the H-bonds linking the 2 peptide chains are partly freed and the disulfide (—S—S—) bonds also linking the two peptide chains split open to yield the free sulfhydryl (—SH) groups.

According to Putnam (1953), the proteins, on denaturation, undergo following changes :

1. Decrease in their solubility.
2. Cessation of their biochemical activity as enzymes or hormones.
3. Decrease in size and shape of the molecule.
4. Increased activity of some radicals present in the molecule such as —SH group of cysteine, —S—S— bond of cystine and phenolic group of tyrosine.

Further, on denaturation, new ionizable groups become available for acid-base titration (**Steinhardt** and **Zaiser,** 1955) and also there occurs a change in optical rotation in the direction of increased levorotation (**Simpson** and **Kauzmann**, 1953). Denaturation also leads to alteration in surface tension and loss of antigenecity.

Some proteins, when denatured, cannot be brought back to their original state. In that case denaturation is described as of '*irreversible*' type. On the other hand, denaturation in other proteins is of '*reversible*' type. For example, if trypsin is exposed to a temperature of 80–90°C, it denatures and when this solution is cooled at 37°C, the solubility and the activity of this protein-enzyme is regained. The process of regaining normal protein properties by a denatured protein is called **renaturation** or **refolding**. During renaturation, certain antibodies may cause a re-rolling of the protein bundles so that most of the original bonds are recovered (**Pauling,** 1940). The recovery of the renatured protein is, however, never complete.

6. Amphoteric Nature. Like amino acids, the proteins are *amphoteric, i.e.*, they act as acids and alkalies both (Fig. 11–3). These migrate in an electric field and the direction of migration depends upon the net charge possessed by the molecule. The net charge is influenced by the pH value. Each protein has a fixed value of isoelectric point (*pI*) at which it will move in

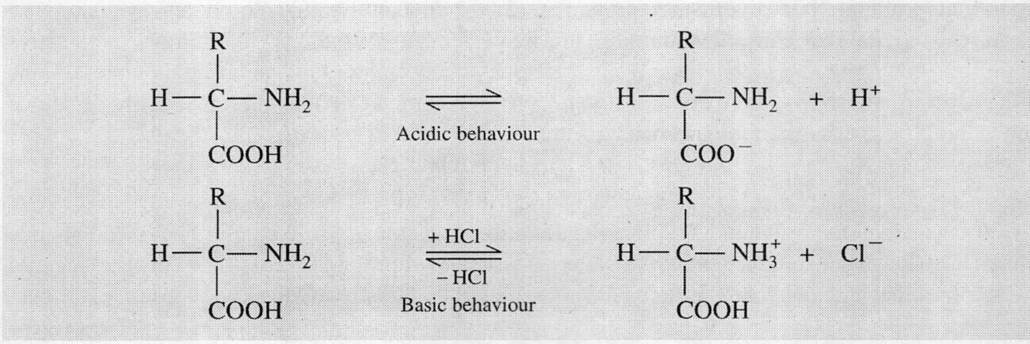

Fig. 11–3. Amphoteric nature of amino acids (and of proteins)

an electric field (refer Table 11–1). **Isoelectric point** (or **isoionic point**) is the pH value at which the number of cations is equal to that of anions. Thus, at isoelectric point, the net electric charge of a protein is always zero. But the total charge on the protein molecule (sum of positive and negative charges) at this point is always maximum. Thus, the proteins are **dipolar ions** or **internal salts** or **zwitterions** (German for 'ion of both kinds' ; amphoteric ions) at *pl* and exist in solution as :

$$(H_3N^+)_m—R—(COO^-)_n$$

This general structure of amino acids, represented as an inner salt, was originally proposed by **N. Bjerrum** in 1923. The formula depicts the carboxyl group as being dissociated while the amino group is protonated.

At pH values lower than *pl*, the protein will have a net positive charge and, as a cation, will migrate towards negative pole (cathode). Similarly, at pH values higher than *pl*, the protein will have a net negative charge and, as an anion, will move towards positive pole (anode).

It may be stated in general that those proteins having an excess of carbonyl groups will tend to have a lower *pl* while those having an excess of amino groups will tend to have a higher *pl*.

The osmotic pressure and viscosity of the protein solution are a minimum at the isoelectric point. Also at the isoelectric point, proteins are found to be least soluble and can be precipitated most easily.

7. Ion Binding Capacity. Being amphoteric in nature, the proteins can form salts with both cations and anions based on their net charge. In fact, a mixture of different proteins at a given pH (except at *pl*) will include cations and anions both and the salts of protein-protein combinations will be formed. This occurs in tissues since both acidic and basic proteins are present.

Many ions form insoluble salts with proteins and serve as excellent precipitating agents for proteins. For example, anions of some acids like phosphotungstic, trichloroacetic, picric etc., form insoluble salts with proteins when the latter behave **as cations** (acid side of their *pl*).

Heavy metals are used for precipitating proteins on the alkaline side of their *pl*, the proteins behaving **as anions**. Ions of Hg, Cu, Ag, Zn etc., are frequently used for this purpose. Many acid dyes find practical use for colouring the insoluble proteins like silk and wool.

8. Solubility. The solubility of proteins is markedly influenced by pH. *Solubility is lowest at isoelectric point and increases with increasing acidity or alkalinity.* This is because when the protein molecules exist as either cations or anions, repulsive forces between ions are high, since all the molecules possess excess charges of the same sign. Thus, they will be more soluble than in the isoelectric state.

A. '*Salting-in*' *effect.* Globulins are sparingly soluble in water but their solubility is greatly increased by the addition of neutral salts like NaCl. This phenomenon is commonly described as 'salting-in' effect.

B. '*Salting-out*' *effect*. Proteins are precipitated from aqueous solution by high concentrations of neutral salts. This is the 'salting-out' process. Divalent and trivalent ions are more effective than univalent ions. The salts commonly used for this purpose are $Na_2.SO_4$, $(NH_4)_2.SO_4$, magnesium salts and phosphates.

C. *Isoelectric precipitation*. Some proteins like casein of milk, however, are readily precipitated at or near their isoelectric point. This process is, therefore, described as isoelectric precipitation.

9. Optical Activity. All protein solutions rotate the plane of polarized light to the left, *i.e.*, these are *levoratotory*. For example, the specific rotation $[\alpha]_D$ for ovalbumin is near —30° over the *p*H range between 3.5 and 11. However, at lower or higher pH values the rotation becomes more negative, *e.g.*, at pH 13, the $[\alpha]_D$ is about —60°. The rotation is further increased by subjecting proteins to high temperatures.

CHEMICAL PROPERTIES

A. HYDROLYSIS

Proteins are hydrolyzed by a variety of hydrolytic agents.

1. By acidic agents. Proteins, upon hydrolysis with conc. HCl (6–12N) at 100–110°C for 6 to 20 hrs, yield amino acids in the form of their hydrochlorides. Undesirable side-effects of acid hydrolysis include the following:

(*a*) Tryptophan, serine and threonine are destroyed during acid hydrolysis and as such this reaction is not used for their isolation.

(*b*) Asparagine and glutamine are deamidated to aspartate and glutamate respectively.

(*c*) Glutamic acid undergoes *intramolecular dehydration* to pyrollidone 5-carboxylic acid.

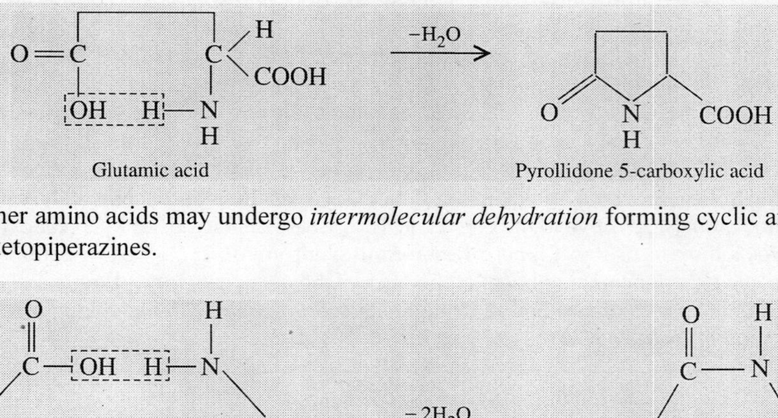

(*d*) Other amino acids may undergo *intermolecular dehydration* forming cyclic anhydrides or diketopiperazines.

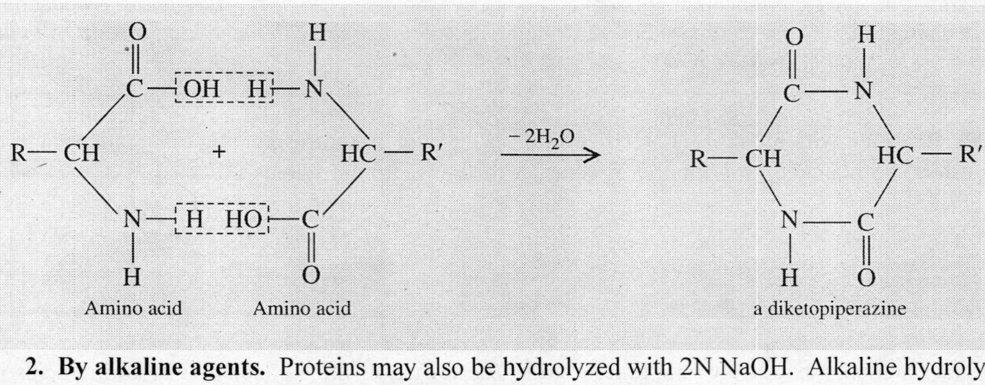

2. By alkaline agents. Proteins may also be hydrolyzed with 2N NaOH. Alkaline hydrolysis is, however, less used as it is highly disadvantageous:

(*a*) It leads to the destruction of certain amino acids like arginine, cysteine, cystine, serine, threonine etc.

(*b*) It also causes loss of optical activity (or racemization) of the amino acids.

3. By proteolytic enzymes. Under relatively mild conditions of temperature and acidity, certain proteolytic enzymes like pepsin and trypsin hydrolyze the proteins. Enzyme hydrolysis is used for the isolation of certain amino acids like tryptophan. Two important drawbacks with this type of hydrolysis are:

(*a*) It requires prolonged incubation.

(*b*) Hydrolysis is incomplete.

B. REACTIONS INVOLVING COOH GROUP

1. Reaction with alkalies (Salt formation). The carboxylic group of amino acids can release a H^+ ion with the formation of carboxylate (COO^-) ions. These may be neutralised by cations like Na^+ and Ca^{2+} to form salts. Thus, amino acids react with alkalies to form *salts*.

$$\underset{\displaystyle NH_2}{\underset{|}{R-CH}}-COO\boxed{H\ +\ HO}Na \longrightarrow \underset{\displaystyle NH_2}{\underset{|}{R-CH}}-COO\cdot Na + H_2O$$

Sodium salt of glutamic acid (monosodium glutamate) is used commercially as a flavouring agent. It imparts a meat-like flavour to soups, for example.

> **Monosodium glutamate (MSG)** or 'ajinomoto' in common parlance, or the 'taste powder' as some people call it, has a long history of use in China. Like sugar or salt, it is regarded as a fairly standard part of cuisine. It makes a great difference to the way the food tastes. MSG actually represents a new taste - the meaty, broth-like taste, common to foods as diverse as steak, lobster and tomato. *Virtually, all proteins contain glutamates, including mother's milk*. Most glutamates are `bound' and have no effect on the flavour but 'free' glutamates increase food's palatibility. Every 100 g of Parmesan cheese, for instance, contains 9,847 mg of bound and 1,200 mg of free glutamates. Tomatoes, peas, meat stock and canned foods and snacks also have it. Human body carries about 2 kg of glutamate in muscles, brain, organs and tissues. Oriental cooks have long added a large seaweed, *Laminaria japonica,* to their soup stocks to enhance that brothy flavour. The seaweed contains glutamate. MSG is used as a flavour enhancer not only by food processors but also by consumers as well as culinary experts in restaurants and hotels. In fact, MSG is one of the most ubiquitous of all food additives and is a popular ingredient of Chinese cuisine. Consuming food containing too much MSG has precipitated attacks of sweating, headaches and gastrointestinal disorders in individuals, sensitive to the chemical. The disease characterized by the above symptoms is called **Kwok's disease** or the '**Chinese Restaurant Syndrome (CRS)**. Kwok, a patron of eateries, described people experiencing numbness of the back of the neck, general weakness and palpitations after eating in Chinese restaurants (hence, its nomenclature). *People with high blood pressure should abstain foods added with MSG.*

2. Reaction with alcohols (Esterification). With alcohols, corresponding *esters* are produced. The esters, so obtained, are volatile in contrast to the free amino acids.

$$\underset{\displaystyle NH_2}{\underset{|}{R-CH}}-COO\boxed{H\ +\ HO}-C_2H_5 \xrightarrow{\text{Acid catalyst}} \underbrace{\underset{\displaystyle NH_2}{\underset{|}{R-CH}}-COO-C_2H_5}_{\text{Ethyl ester of amino acid}} + H_2O$$

The reaction was, for the first time, used by Emil Fischer for the isolation of amino acids in pure form from protein hydrolysates by the fractional distillation in vacuum of their ethyl esters.

3. Reaction with amines. Amino acids react with amines to form *amides*.

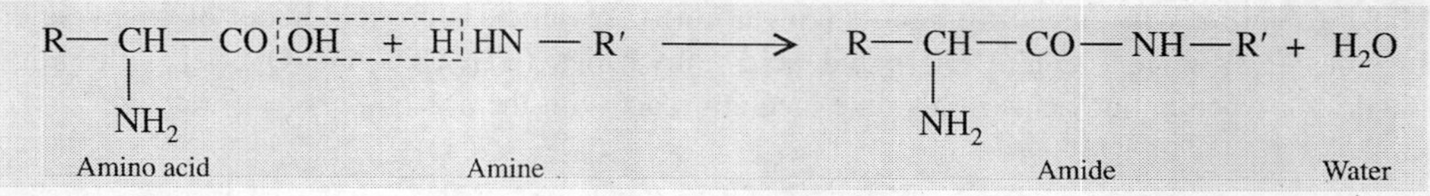

C. REACTIONS INVOLVING NH_2 GROUP

1. Reaction with mineral acids (Salt formation). When either free amino acids or proteins are treated with mineral acids like HCl, the *acid salts* are formed.

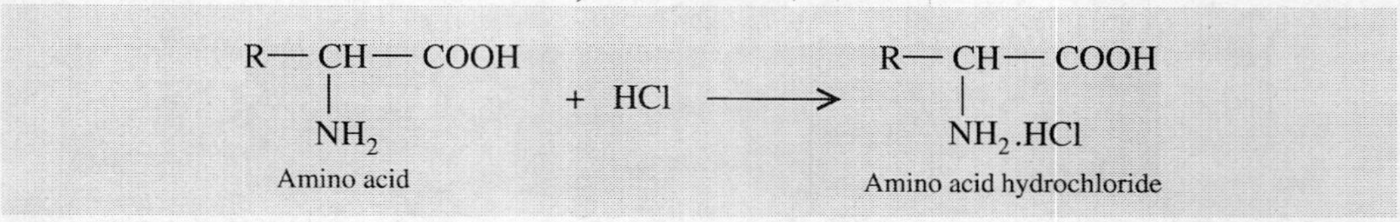

The basic amino acids, arginine and lysine react with CO_2 in the presence of air to form carbonate salts. Because of this property, these are usually stored and also sold in the form of their monochlorides.

2. Reaction with formaldehyde. With formaldehyde, the *hydroxy-methyl* derivatives are formed. These derivatives are insoluble in water and resistant to attack by microorganisms. Because of this action, formaldehyde is the principal reagent in embalming fluids and is used to harden and preserve certain fibres (Aralac, Vicara) obtained from globular proteins.

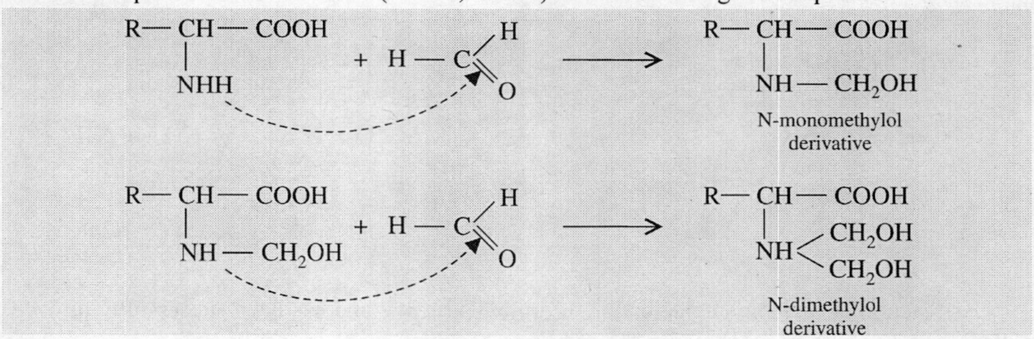

This reaction is the basis of the Sorensen titration method for determining the purity of the individual amino acids.

3. Reaction with benzaldehyde. *Schiff's* bases are formed.

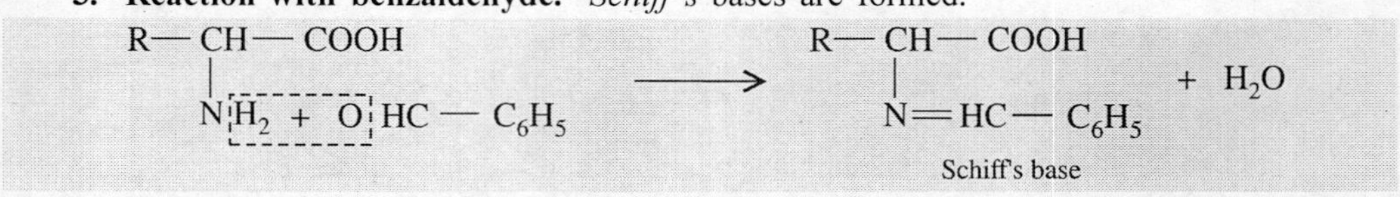

4. Reaction with nitrous acid (Van Slyke reaction). The amino acids react with HNO_2 to liberate N_2 gas and to produce the corresponding *α-hydroxy acids*.

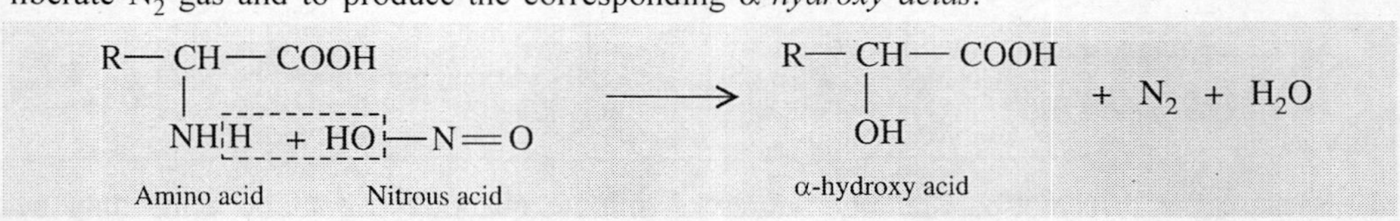

This reaction is characteristic of aliphatic primary amines and has been utilized by Van Slyke (1912) as the basis for his '*nitrous acid' method* for the estimation of amino acids by measuring the volume of N_2 gas liberated. The imino acids proline and hydroxyproline, however, do not respond to this reaction.

5. Reaction with acylating agents (Acylation). Acylation is brought about by many acid chlorides ($CH_3.COCl$, $C_6H_5.COCl$) and acid anhydrides ($CH_3.CO—O—OC.CH_3$, phthalic anhydride), when amino acids in alkaline medium react with them.

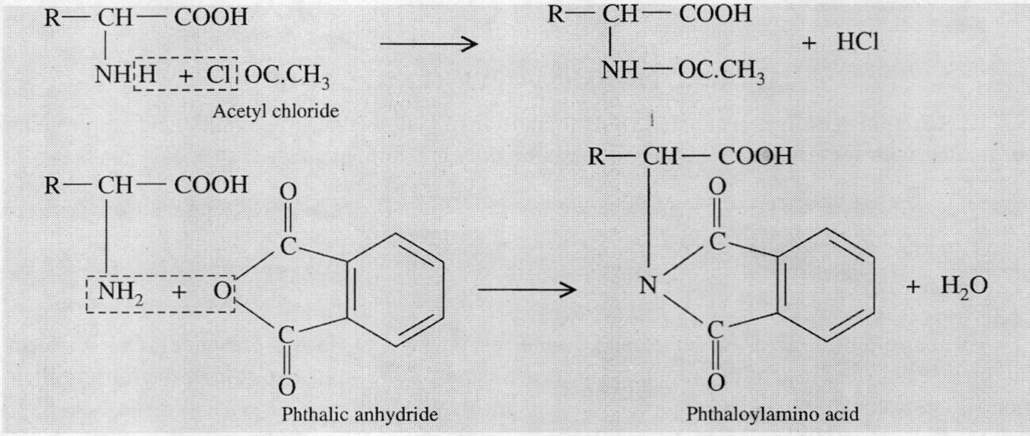

6. Reaction with FDNB or **Sanger's reagent.** In mildly alkaline solution, FDNB (1-fluoro- 2, 4-dinitrobenzene) reacts with α-amino acids to produce yellow coloured derivative, *DNB-amino acid.*

FDNB is also called as **DNFB** (2,4-dinitrofluorobenzene).

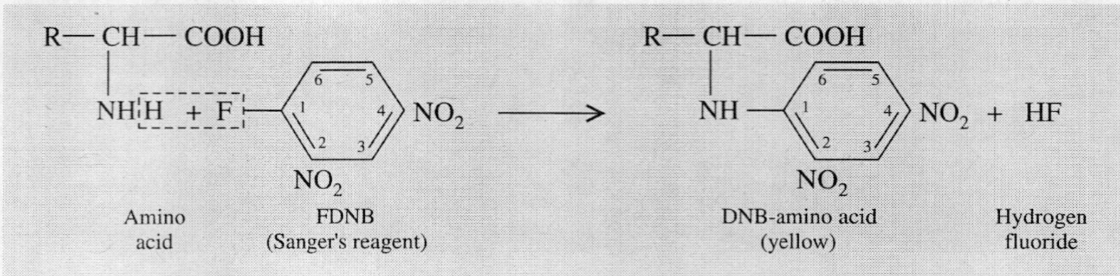

This reaction is valuable in elucidation of protein structure and has been successfully uitlized by Sanger in England in determining the sequence of amino acids in insulin.

7. Reaction with dansyl chloride. The N-terminal amino acid of the protein also combines with 1-dimethylaminonaphthalene-5-sulfonyl chloride (or dansyl chloride) to form a *dansyl amino acid derivative*.

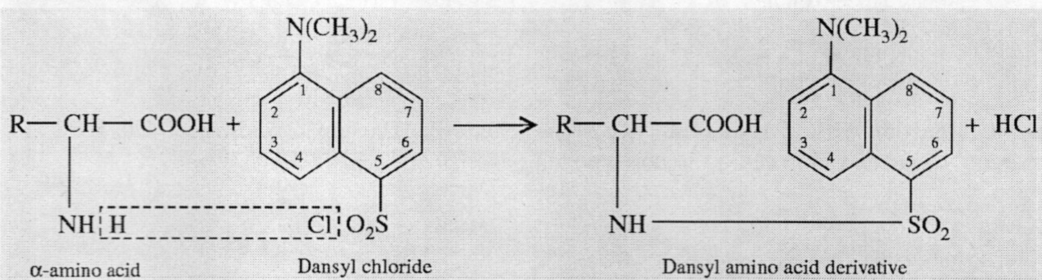

Because the dansyl group readily fluoresces, minute quantities of amino acids may be determined by this procedure.

D. REACTIONS INVOLVING BOTH COOH AND NH_2 GROUPS

1. Reaction with triketohydrindene hydrate (Ninhydrin reaction). Ninhydrin (= indane-1, 2, 3, -trione hydrate) is a powerful oxidizing agent and causes oxidative decarboxylation of α-amino acids producing CO_2, NH_3 and an aldehyde with one less carbon atom than the parent amino acid. The reduced ninhydrin (or hydrindantin) then reacts with the liberated NH_3 and a mole of ninhydrin, forming blue-coloured Ruheman's complex.

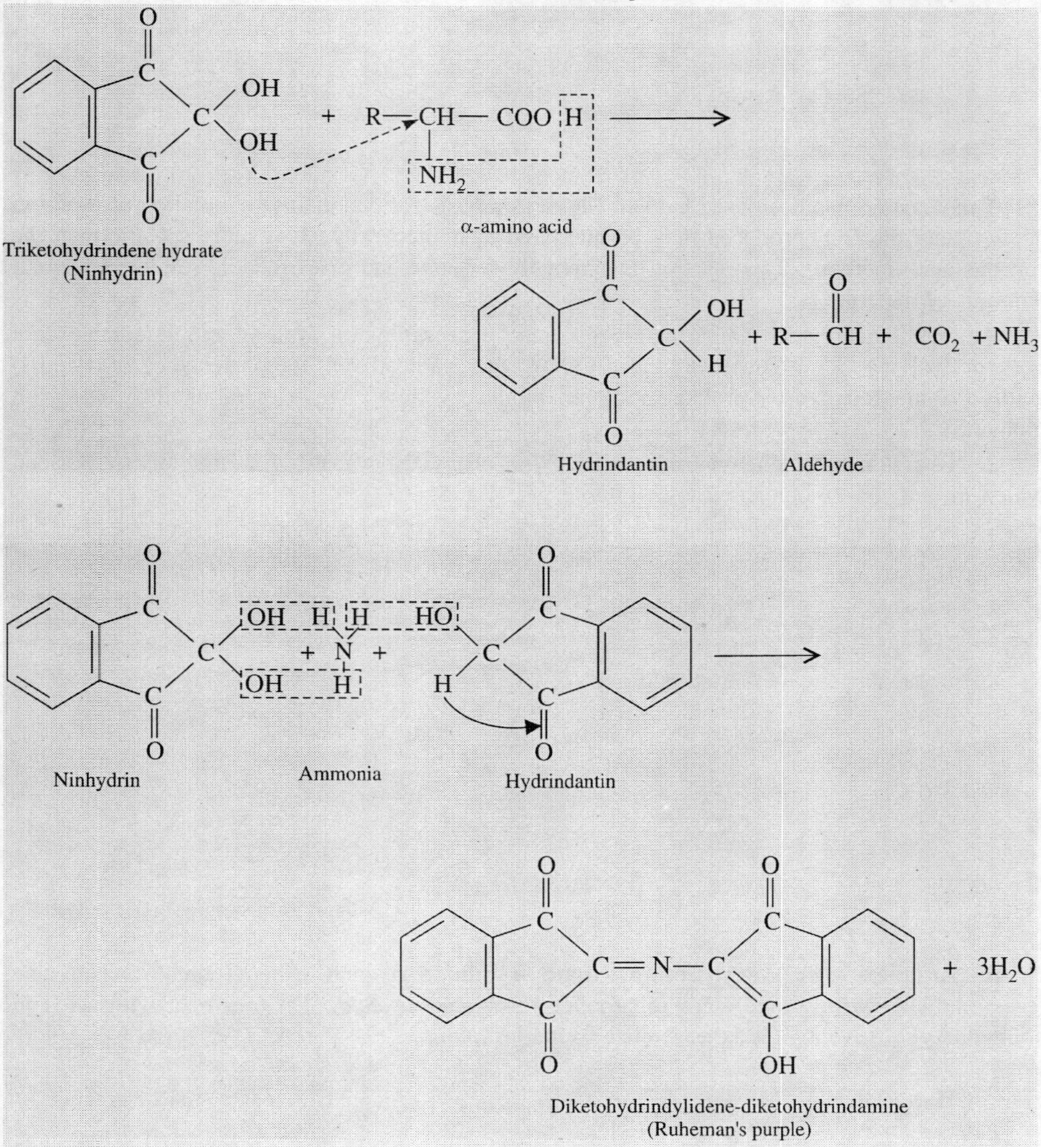

The net equation may, thus, be written as follows:

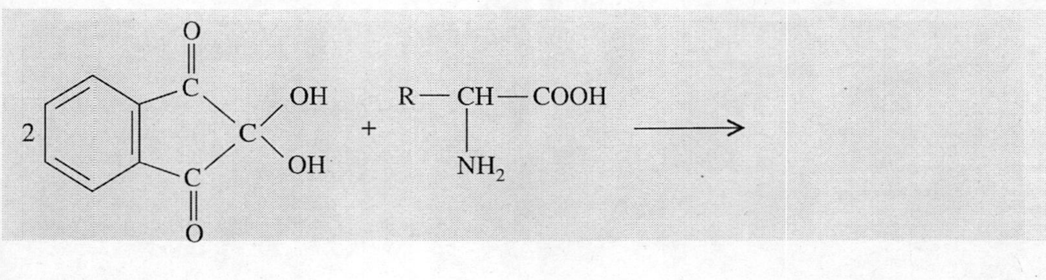

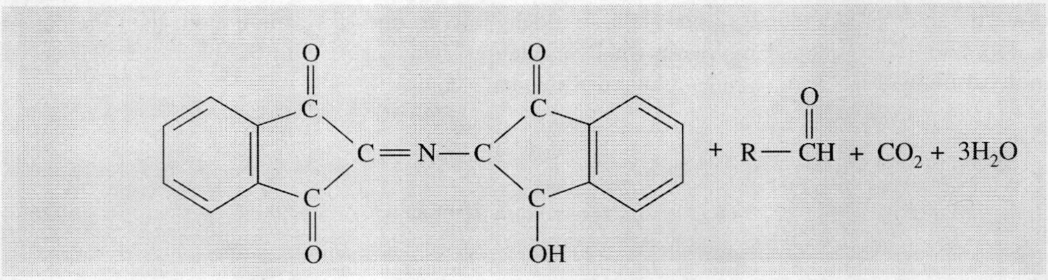

This reaction has been used by **Van Slyke** as a basis for quantitative estimation of α-amino acids. Here the CO_2 produced may be measured manometrically. It is more specific than the 'nitrous acid' method. This reaction is extremely sensitive and gives reliable results with small amounts of material.

Amines, other than α-amino acids, also react with ninhydrin forming a blue complex but without evolving CO_2. The evolution of CO_2 is, thus, indicative of the presence of an α-amino acid. Proline and hydroxyproline, however, produce yellow complexes rather than blue with ninhydrin.

2. Reaction with phenyl isocyanate. With phenyl isocyanate, *hydantoic acid* is formed which in turn can be converted to *hydantoin*.

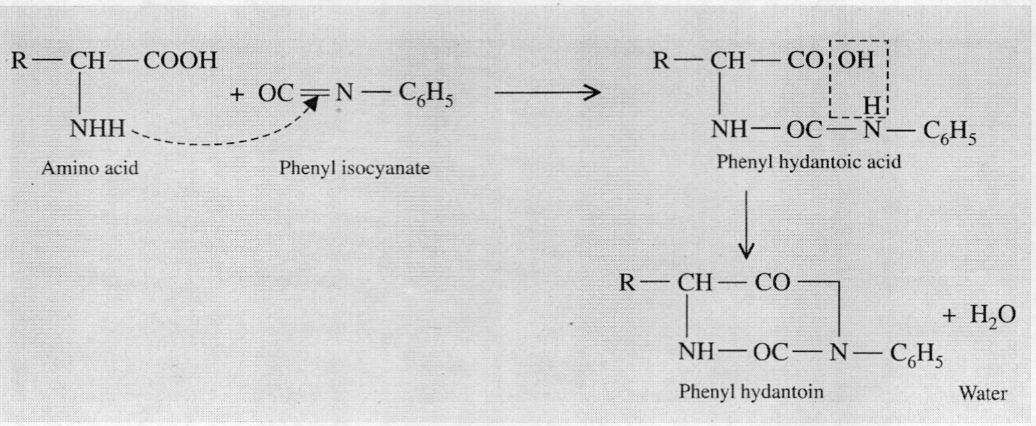

3. Reaction with phenyl isothiocyanate or Edman reagent. Phenyl isothiocyanate also reacts similarly with amino acids to produce *thiohydantoic acid.* On treatment with acids in nonhydroxylic solvents, the latter cyclize to *thiohydantoin.*

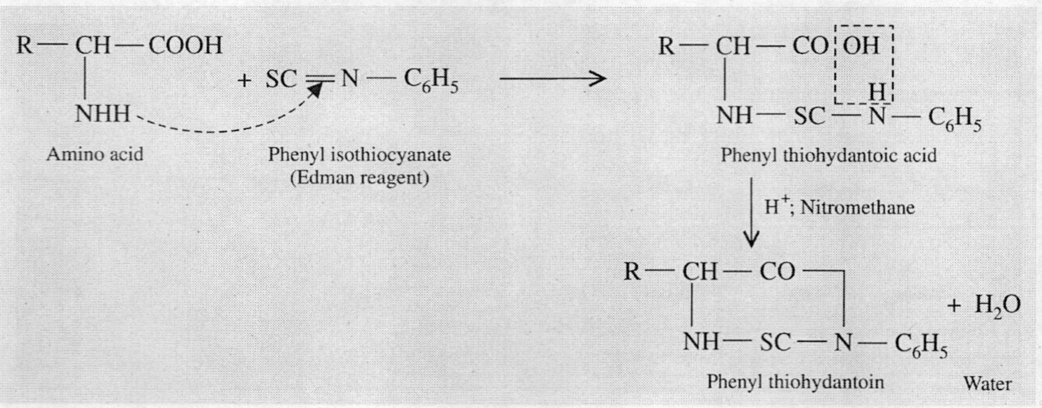

This reaction has proved useful in the studies of protein structure.

4. Reaction with phosgene. With phosgene, *N-carboxyanhydride is formed.*

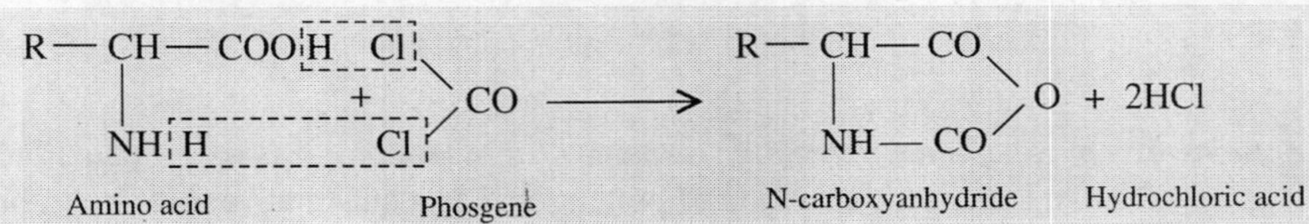

Phosgene (synonyms: carbonyl chloride, carbon oxychloride, carbonic dichloride), an acidic chloride, is a colourless gas and has a musty odour, resembling fresh mown hay or green corn. It liquefies at 8°C. Since it is heavier (3.43 times) than air, it was used extensively in gas warfare in World War I, and caused 80% of the deaths by gas in that conflict. Luckily, gas warfare was sparingly used in World War II, thus reducing the number of casualties. Phosgene does not occur in nature. It was first synthesized by Sir Humphrey Davy in 1812 by passing CO and chlorine through charcoal. *It is deadlier than both CO and* Cl_2 as it kills rapidly in as low a concentration as 550 ppm. Phosgene was also implicated in the *Bhopal tragedy* that occurred in India in December, 1984. However, phosgene is also an important industrial chemical being used in the synthesis or manufacture of isocyanates, polyurethane, polycarbonate resins, aniline dyes, pharmaceuticals, plastics and insecticides.

5. Reaction with carbon disulfide. With carbon disulfide, *2-thio-5-thiozolidone* is produced.

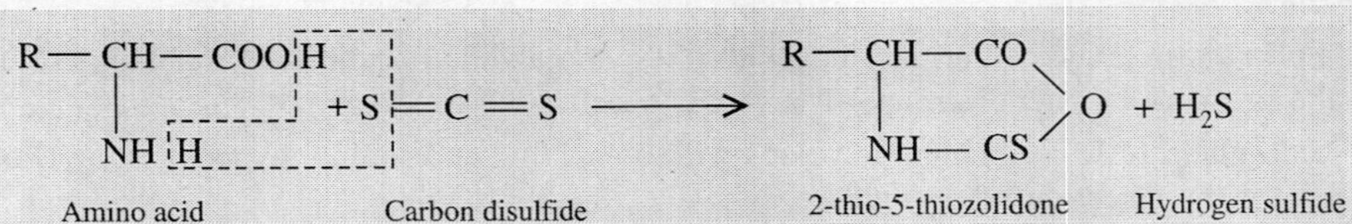

E. REACTIONS INVOLVING R GROUP OR SIDE CHAIN

1. Biuret test. Compounds containing peptide bonds produce a characteristic purple colour when treated with an alkaline 0.2% copper sulfate solution (or biuret reagent). This reaction is termed as 'biuret reaction' since it is also given by the substance biuret.

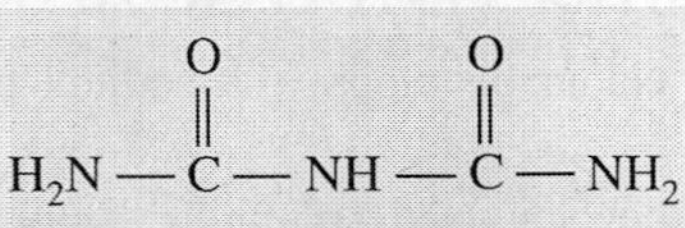

Biuret

The colour deepens as the number of peptide bonds is increased and the proteins produce a deep blue-violet colour due to the probable formation of a coordination complex whose structure is given below:

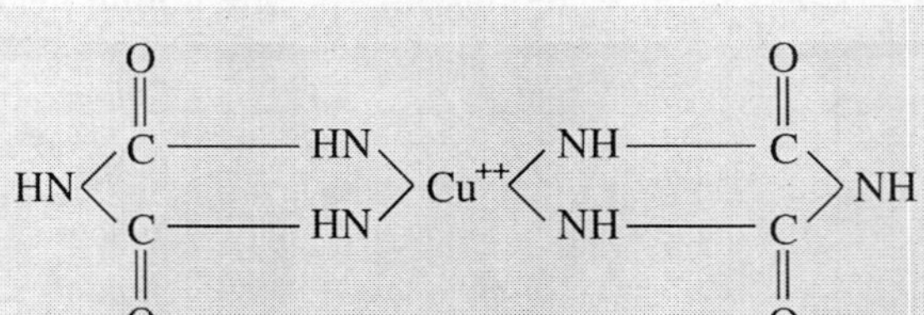

The test is, in fact, given by biuret as well as any similar structure having 2 amide or peptide bonds linked directly or through an intermediate carbon atom. The required unit is shown below between the two broken lines:

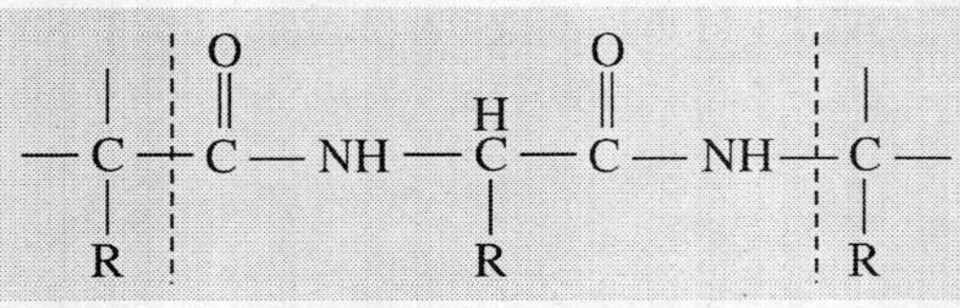

All proteins except dipeptides, therefore, respond to this reaction. This reaction is widely used both as a qualitative test for the detection of proteins and as a quantitative measure of protein concentration.

2. Xanthoproteic test. Yellow colour develops on boiling proteins with conc. HNO_3 due to the presence of benzene ring. This reaction is due to the nitration of the phenyl rings (of *tyrosine, tryptophan* and *phenylalanine*) to yield yellow substitution products, which turn orange upon addition of alkali due to the formation of salts of nitro derivatives.

$$\text{Protein} + HNO_3 \xrightarrow{\text{Xanthoproteic Reaction}} \underset{(=\text{ Picric acid})}{\text{Trinitrophenol}} + NO\uparrow$$

3. Millon's test. Red colour develops when proteins are heated with $Hg.NO_3$ in HNO_2. The reaction is specific for *tyrosine* and takes place between mercuric and mercurous nitrates and tyrosine residues of the protein. *Tryptophan* also responds to this reaction.

4. Hopkins–Cole test or Glyoxylic acid test. Violet ring develops on addition of conc. H_2SO_4 (36 N) at the junction of protein and glyoxylic acid solutions. The test is specific for *tryptophan*.

5. Folin's test. Blue colour develops with phosphomolybdotungstic acid in alkaline solution due to the presence of phenol group. The test is specific for *tyrosine*.

6. Sakaguchi test. Red colour develops with a-naphthol and sodium hypochlorite. The test is applied for the detection of *arginine*.

7. Pauly test. Red colour develops with diazotized sulfanilic acid in alkaline solution. The reaction is specific for *tyrosine* and *histidine*.

8. Ehrlich test. With *p*-dimethylaminobenzaldehyde in 12 N HCl, *tryptophan* develops a blue colour.

F. REACTIONS INVOLVING SH GROUP

1. Nitroprusside test. Red colour develops with sodium nitroprusside in dilute $NH_4.OH$. The test is specific for *cysteine*.

2. Sullivan test. *Cysteine* develops red colour in the presence of sodium 1, 2-naphthoquinone-4-sulfonate and sodium hydrosulfite.

The various colour reactions for different amino acids are summarized in Table 11–2.

Table 11–2. Colour reactions for specific amino acids

S.N.	*Test*	*Reagent*	*Colour*	*Amino acid(s)*
1.	Biuret test	Alkaline $Cu.SO_4$ soln.	Violet	Peptide bonds
2.	Xanthoproteic test	Conc. HNO_3	Yellow	Tyrosine Tryptophan Phenylalanine
3.	Millon's test	$Hg.NO_3$ in HNO_2; heat	Red	Tyrosine Tryptophan
4.	Hopkins–Cole test	Glyoxylic acid + conc.	Violet	Tryptophan

		H_2SO_4		
5.	Folin's test	Alkaline phosphomolybdotungstic acid	Blue	Tyrosine
6.	Sakaguchi test	α-naphthol + sodium hypochlorite	Red	Arginine
7.	Pauly test	Diazotized sulfanilic acid in alkaline soln.	Red	Tyrosine Histidine
8.	Ehrlich test	*p*-dimethylaminobenzaldehyde in 12 N HCl	Blue	Tryptophan
9.	Nitroprusside test	Sodium nitroprusside + dil. NH_4OH	Red	Cysteine
10.	Sullivan test	Sodium 1,2-naphthoquinone-4-sulfonate + sodium hydrosulfite	Red	Cysteine

PROBLEMS

1. After proper purification, the Edman reaction was used to sequence a dodecapeptide. The following data were obtained. The C-terminal amino acid is isoleucine; N-terminal amino acid is methionine; peptide fragments are Ala-Ala-Ile, Leu-Arg-Lys-Lys-Glu-Lys-Glu-Ala, Met-Gly-Leu, and Met-Phe-Pro-Met. What is the sequence of this peptide ?

2. Why is pepsin useful as a digestive aid in precooked foods ?

3. Why eggs become harder on boiling ?

4. What is Edman reagent ? Describe its use in the determination of amino acid sequence in protein.

ANSWERS TO PROBLEMS

1. You must have the correct N- and C-terminal amino acids and the correct total number (12) of amino acids in the peptide. Keeping these criteria in mind and overlapping the fragments, you should get the sequence: Met-Phe-Pro-Met-Gly-Leu-Arg-Lys-Glu-Ala-Ala-Ile. Note that the Lys-Glu fragment doesn't give any additional information for the sequencing.

2. Pepsin partially digests (hydrolyzes) the protein in foods.

3. The eggs become harder on boiling due to coagulation of natural proteins, present in egg.

4. Phenyl isothiocyanate (C_6H_5–N=C=S) is Edman reagent. It is used in the determination of N-terminal amino acid in a protein. Phenyl isothiocyanate reacts with the $-NH_2$ group of amino acid to produce phyenyl thiohydantoic acid which, on treatment with mild acids, results in the formation of phenyl thiohydantoin (see reaction on page 190). Phenyl thiohydantion can be separated and compared with the corresponding known derivative of a known amino acid. This method is quite useful because it results in the removal and identification of the NH_2^- terminoid amino acid together with the production of a polypeptide containing one amino acid less.

CHAPTER

12 Lipids–I
General Structure

INSIDE

- Importance
- Definition
- Alcohols
  - Saturated alcohols
  - Unsaturated alcohols
- Fatty Acids
  - Saturated Fatty Acids
  - Unsaturated Fatty Acids
  - Hydroxy Fatty Acids
  - Cyclic Fatty Acids
- Biological Roles of Lipids

Fur seals
Fur seals have an especially thick layer of fat, called blubber, to protect them from extreme cold.

IMPORTANCE

The lipids are important constituents of the diet because of their high energy value and also because of the fat-soluble vitamins and the essential fatty acids found with the fat of the natural foodstuffs. In the body, the fats serve as efficient source of energy which is stored in the adipose tissues. They also serve as an insulating material in the subcutaneous tissues and around certain organs. Fats combined with proteins (lipoproteins) are important constituents of the cell membranes and mitochondria of the cell.

Proteins, polysaccharides, DNA and RNA are macromolecules. *Lipids are not generally classed as macromolecules,* even though they share some of their features: for example, most are synthesized as linear polymers of a smaller molecule (the acetyl group on acetyl-CoA), and they self-assemble into larger structures (membranes). It is noteworthy that water and protein comprise most of the mass of both mammalian and bacterial cells (Table 12–1).

Table 12–1. Approximate chemical composition of a typical bacterium and a typical mammalian cell

Component	Percent of total cell weight	
	E. coli bacterium	*Mammalian cell*
H_2O	70	70
Inorganic ions (Na^+, K^+, Mg^{2+}, Ca^{2+}, Cl^- etc)	1	1
Miscellaneous small metabolites	3	3
Proteins	15	18
RNA	6	1.1
DNA	1	0.25
Phospholipids	2	3
Other lipids	—	2
Polysaccharides	2	2
Total cell volume :	2×10^{-12} cm^3	4×10^{-9} cm^3
Relative cell volume :	1	2,000

DEFINITION

The lipids (*lipos*[G] = fat) are a heterogeneous group of compounds related to fatty acids and include fats, oils, waxes and other related substances. These are oily or greasy organic substances, relatively insoluble in water and considerably soluble in organic solvents like ether, chloroform and benzene. They are, thus, **hydrophobic** in nature. These are variously called as *lipins* or *lipoids*. The latter term is, however, sometimes used to refer "fat-like" substances which may not actually be related to the fatty acids. The term 'lipid' was first used by the German biochemist **Bloor** in 1943 for a major class of tissue components and foodstuffs. The structure of some common fats and fatty acids is presented in Fig.12–1.

Chemical Abstracts and some biochemical journals use the spelling **'lipide'.**

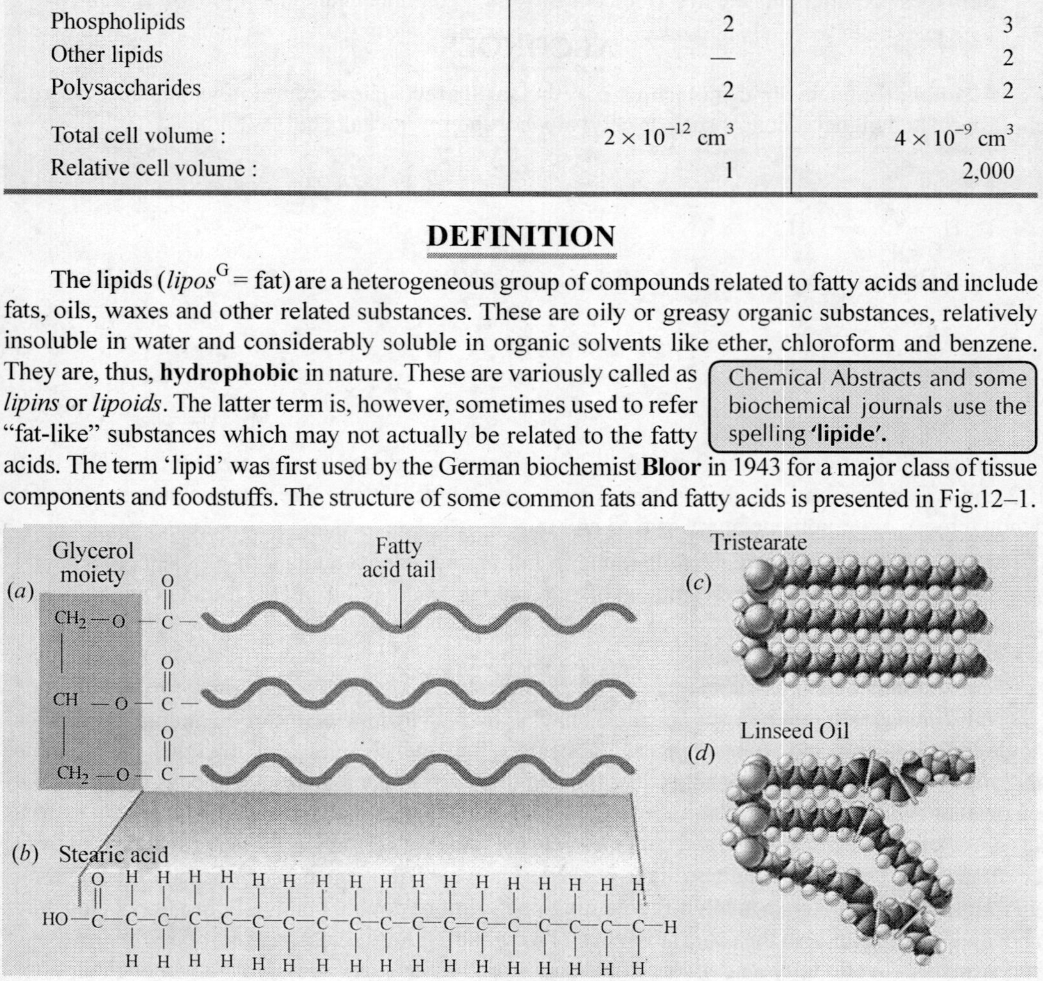

Fig. 12-1. Fats and fatty acids

(*a*) **Basic structure of a triacylglycerol (also called a triglyceride or a neutral fat)**. The glycerol moiety, indicated in *orange*, is linked by three ester bonds to the carboxyl groups of three fatty acids whose tails are indicated in *green*. (*b*) **Stearic acid.** It is an 18-carbon saturated fatty acid that is common in animals fats. (*c*) **Space-filling model of tristearate.** It is a triacylglycerol containing three identical stearic acid chains. (*d*) **Space-filing model of linseed oil.** It is a triaclglycerol containing two different unsaturated fatty acids, derived from flax seeds. The sites of unsaturation, which produce kinks in the molecule, are indicated by the *yellow-orange* bars.

Chemically, the fats are defined as the esters of glycerol and fatty acids or as the triglycerides of fatty acids.

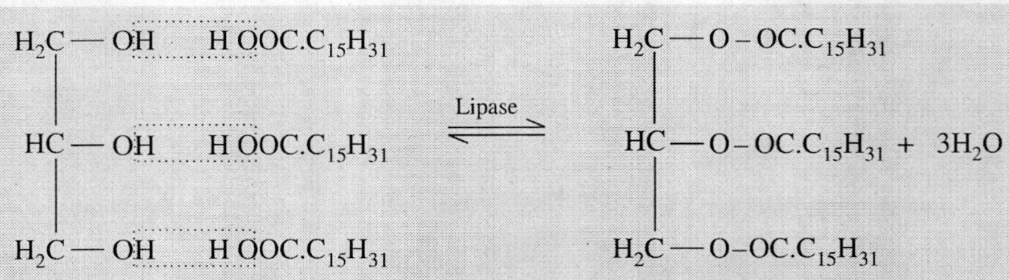

Glycerol (1 mole) | Palmitic acid (3 moles) | Tripalmitin (1 mole) | Water (3 moles)

Both these components are described below for a better understanding of the lipids.

ALCOHOLS

Alcohols found in lipid molecules may be **saturated.** These commonly include glycerol, cholesterol and higher alcohols such as cetyl alcohol and myricyl alcohol.

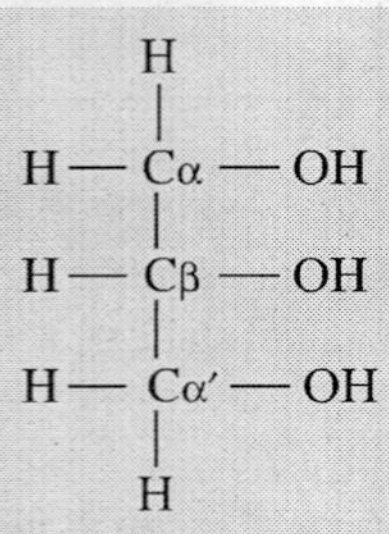

Glycerol

$CH_3(CH_2)_{14}CH_2OH$

Cetyl alcohol

$CH_3(CH_2)_{28}CH_2OH$

Myricyl alcohol

In the structural formula of glycerol, the C atoms are designated by *stereospecific numbers* (*sn*), 1, 2 and 3 from any end. As the C atoms I and 3 are identical, these three carbon atoms are better denoted by an older nomenclature, as α, β and α'.

Among the **unsaturated** alcohols found in fats are included a number of pigments like phytol and lycophyll. The former is a constituent of chlorophyll, whereas the latter is found in tomatoes as a purple pigment.

FATTY ACIDS

Fatty acids are long-chain organic acids having usually from 4 to 30 carbon atoms; they have a single carboxyl group and a long, nonpolar hydrocarbon 'tail', which gives most lipids their hydrophobic and oily or greasy nature. Fatty acids do not occur in free or uncombined state in cells or tissues but are present in covalently bound form in different classes of lipids. Fatty acids which occur in natural fats are usually monocarboxylic and contain an even number of C atoms as these are synthesized from 2 carbon units. These are usually straight-chain derivatives. The chain may be **saturated** (containing only single bonds) or **unsaturated** (containing one or more double bonds). Some fatty acids may have hydroxyl group(s) in the chain (**hydroxy** or **oxygenated** fatty acids) and still others may possess ring structure (**cyclic** fatty acids). Fatty acids are stored as an energy reserve (fat) through an ester linkage to glycerol to form *triglycerides*. If free, the *carboxyl group* of a fatty acid will be ionized.

But more usually, it is linked to other groups to form either *esters*

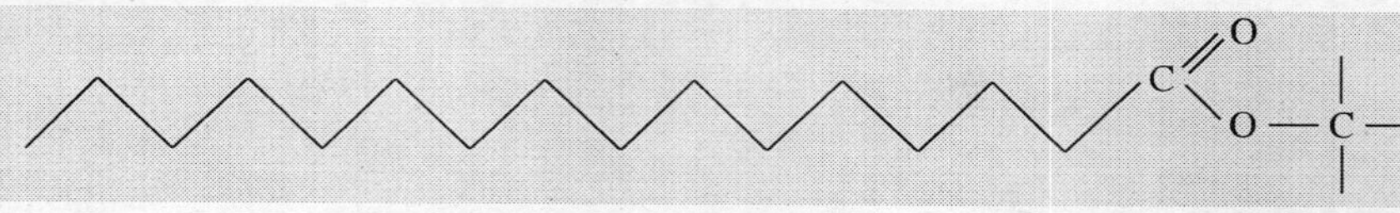

or *amides*.

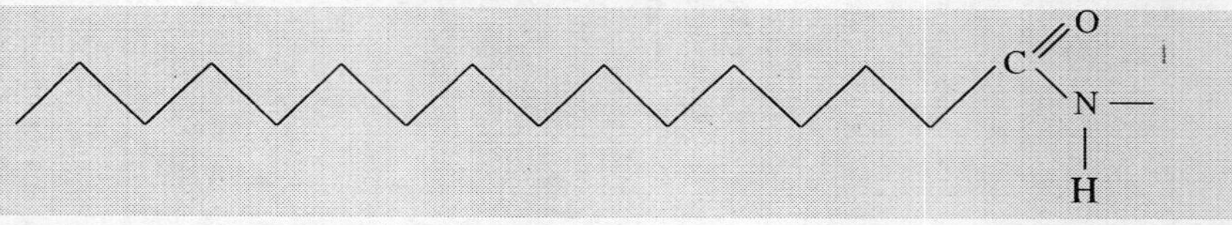

Nomenclature. The systematic nomenclature of the fatty acids is based on the **Genevan system.** According to this system, the fatty acid is named after the hydrocarbon with the same number of carbon atoms, the suffix -*oic* is written in place of the final letter *e* in the name of the hydrocarbon. The names of saturated fatty acids end with the suffix -*anoic* and those of unsaturated acids with the suffix -*enoic*.

> Generally, the numbers are used with the International Union of Pure and Applied Chemistry (IUPAC) **system of nomenclature** and the Greek letters when the common names are employed. For example :
>
> 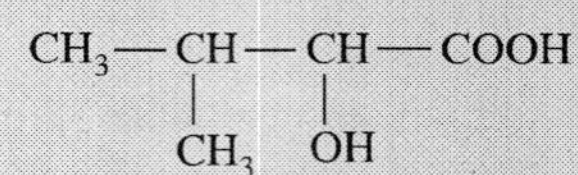
>
>
> 2-hydroxy-3-methylbutanoic acid
> or
> α-hydroxy-*iso*-valeric acid

The position of carbon atoms in the fatty acid chain is indicated either by **numbering** (in which case the carboxyl carbon is numbered as C1, the carbon adjacent to C1 as C2 and so on) or by the **use of Greek letters** (in which case C2 is denoted as α-carbon, C3 as β-carbon and so on). Thus :

$$R-\underset{\varepsilon}{\overset{6}{CH_2}}-\underset{\delta}{\overset{5}{CH_2}}-\underset{\gamma}{\overset{4}{CH_2}}-\underset{\beta}{\overset{3}{CH_2}}-\underset{\alpha}{\overset{2}{CH_2}}-\overset{1}{COOH}$$

A widely-used convention to indicate the number and position of the double bond(s) in the case of unsaturated fatty acids is to write the number of carbon atoms, the number of double bond(s) and the position of the double bonds(s) below the name of the acid. For example, oleic acid having 18 carbon atoms and a double bond between carbon atoms 9 and 10 is written as 18:1; 9. Similarly, linoleic acid (18 carbon atoms and 2 double bonds at C 9 and C 12) is written as 18:2; 9, 12.

> The **position of a double bond** is indicated by the lower number of the two carbon atoms involved in double bonding.

An alternative method to write the name of an unsaturated fatty acid is to write first the position of double bond(s) in numerals and then the total number of carbon atoms in Roman followed by the suffix -*enoic* acid. Thus, oleic acid may be written as 9-octadecenoic acid and linoleic acid as 9, 12-octadecadienoic acid.

A description of the various categories of fatty acids involved in lipid formation follows.

1. Saturated fatty acids

The general formula for these acids is $C_nH_{2n+1}COOH$. Table 12.2. lists some even-numbered straight chain saturated fatty acids, found distributed in both plant and animal worlds. In addition, lipids from all sources contain small amounts of saturated fatty acids with an odd number of carbon atoms (C 5 through C 17). Generally, these odd-carbon acids account for less than 1% of the total fatty acids.

In animal fats, palmitic and stearic acids (C_{16} and C_{18}) are the most abundantly found saturated fatty acids, next in order are shorter chain fatty acids (C_{14} and C_{12}) and longer chain fatty acids (C_{20}, C_{22} and C_{24}). Fatty acids of 10 carbon atoms or less are present in limited amounts in animal lipids except the milk fat which contains appreciable amounts of lower molecular weight fatty acids. The preponderance of these acids may, thus, be shown in the descending order as below :

Table 12–2. Straight chain saturated fatty acids, commonly found in natural fats

Trivial name	*Systematic name**	*Carbon skeleton*	*Structure†*	*Common source*
Butyric	*n*-Butanoic	4 : 0	$CH_3(CH_2)_2COOH$	Butter
Caproic	*n*-Hexanoic	6 : 0	$CH_3(CH_2)_4COOH$	Coconut and palm oils
Caprylic	*n*-Octanoic	8 : 0	$CH_3(CH_2)_6COOH$	Coconut and palm oils
Capric	*n*-Decanoic	10 : 0	$CH_3(CH_2)_8COOH$	Coconut and palm oils
Lauric (*laurus*[L] = laurel plant)	n-Dodecanoic	12 : 0	$CH_3(CH_2)_{10}COOH$	Laurel oil and spermaceti
Myristic (*Myristica*[L] = nutmeg genus)	*n*-Tetradecanoic	14 : 0	$CH_3(CH_2)_{12}COOH$	Butter and wool fats
Palmitic (*palma*[G] = palm tree)	*n*-Hexadecanoic	16 : 0	$CH_3(CH_2)_{14}COOH$	Animal and plant fats
Stearic (*stear*[G] = hard fat)	*n*-Octadecanoic	18 : 0	$CH_3(CH_2)_{16}COOH$	Animal and plant fats
Arachidic (*Arachis*[L] = legume genus)	*n*-Eicosanoic	20 : 0	$CH_3(CH_2)_{18}COOH$	Groundnut oil
Behenic	*n*-Docosanoic	22 : 0	$CH_3(CH_2)_{20}COOH$	Groundnut oil
Lignoceric (*lignum*[L] = wood; *cera*[L] = wax)	*n*-Tetracosanoic	24 : 0	$CH_3(CH_2)_{22}COOH$	Groundnut and Rapeseed oils
Cerotic	*n*-Hexacosanoic	26 : 0	$CH_3(CH_2)_{24}COOH$	Wool fat
Montanic	*n*-Octacosanoic	28 : 0	$CH_3(CH_2)_{26}COOH$	—

* The **prefix *n*-** indicates the "normal" unbranched structure. For instance, "octadecanoic" simply indicates 18 carbon atoms, which could be arranged in a variety of branched forms. Thus, *n*-octadecanoic specifies the linear, unbranched form.

† All **fatty acids** are shown in their unionized form. At pH 7, all free fatty acids have an ionized carboxylate. Note that the numbering of carbon atoms begins at the carboxyl-group carbon.

$C_{16}, C_{18} > C_{14}, C_{12}, C_{20}, C_{22}, C_{24} > C_{10}$ and less.

In addition to the straight chain fatty acids, a number of branched chain fatty acids having either an even or an odd number of carbon atoms have been identified as minor components of natural fats and oils. Table 12-3 lists a representative sample of these fatty acids.

Table 12–3. Some branched chain saturated fatty acids, identified in natural fats

Trivial	*Systematic atoms*	*No. of C atoms*	*Structure*	*Common source*
Isopalmitic	Isohexa-decanoic	16	$(CH_3)_2CH(CH_2)_{12}COOH$	Wool fat
Anteiso-palmitic	14-methyl-hexadecanoic	17	$(CH_3CH_2)(CH_3)CH(CH_2)_{12}COOH$	Wool fat
Tuberculo-stearic	D-(–) 10-methylocta-decanoic	19	$CH_3(CH_2)_7CH(CH_3)(CH_2)_8COOH$	Bacteria

2. Unsaturated fatty acids

These may be classified, based on the degree of unsaturation.

A. Monoethenoid acids — These contain one double bond and conform to the general formula, $C_nH_{2n-1}COOH$. The common example is oleic acid.

B. Diethenoid acids — Two double bonds;
$C_nH_{2n-3}COOH$; Linoleic acid.

C. Triethenoid acids — Three double bonds;
$C_nH_{2n-5}COOH$; Linolenic acid.

D. Tetraethenoid acids — Four double bonds;
$C_nH_{2n-7}COOH$; Arachidonic acid

Monoethenoid acids are commonly called as *monounsaturated fatty acids (MUFAs)* and the remaining ones as *polyunsaturated fatty acids (PUFAs)*. A few important unsaturated acids are listed in Table 12–4. A perusal of the table indicates that in most of the unsaturated fatty acids, there is a double bond (designated Δ^9) between carbon atoms 9 and 10. This is particularly true of the unsaturated fatty acids, commonly found in the plant world. If there are additional bonds, they usually occur between Δ^9 and the methyl-terminal end of the chain. It may, however, be generalized that in mammals, polyunsaturated fatty acids can have up to 22 carbon atoms and 6 double bonds, but in plants these acids do not exceed 18 carbon atoms and 4 double bonds.

> Linoleic acid was once known as **vitamin F**. Its function in this capacity is now discredited. It is a yellow oily liquid with a b.p. 229°C.

Human body can convert stearic acid to oleic acid by inserting a double bond but is incapable of inserting further double bonds so that the oleic acid cannot be converted to either linoleic, linolenic or arachidonic acid. For normal cell functioning esp., of skin tissues any one of these acids is needed. Since they cannot be synthesized by the cells, they must be obtained from diet. On account of the important physiological role, these 3 acids are collectively called as *essential fatty acids (EFAs),* a term introduced by **Burr** and **Burr** in 1930.

Vegetable oils contain two types of polyunsaturated fatty acids (PUFAs) – linoleic acid (*lin* with

Table 12–4. Unsaturated fatty acids*, commonly found in natural fats

Trivial name	*Systematic name*	*Carbon skeleton*	*Structure*	*Common source*
Crotonic	2-butenoic	4 : 1 ; 2	$CH_3CH{=}CHCOOH$	Croton oil
Myristoleic	9-tetradecenoic	14 : 1 ; 9	$CH_3(CH_2)_3CH{=}CH(CH_2)_7COOH$	*Pycnanthyus*
Palmitoleic	9-hexadecenoic	16 : 1 ; 9	$CH_3(CH_2)_5CH{=}CH(CH_2)_7COOH$	Animal and plant fats
Oleic (*oleum*^{G} = oil)	9-octadecenoic	18 : 1 ; 9	$CH_3(CH_2)_7CH{=}CH(CH_2)_7COOH$	Animal and plant fats
Vaccenic	11-octadecenoic	18 : 1 : 11	$CH_3(CH_2)_5CH{=}CH(CH_2)_9COOH$	Bacterial fat
Linoleic (*linon*^{G} = flax)	9,12-octadecadienoic	18 : 2 ; 9, 12	$CH_3(CH_2)_4CH{=}CHCH_2CH{=}CH(CH_2)_7COOH$	Linseed and cottonseed oils
Eleostearic	9,11,13-octadecatrienoic	18 : 3 ; 9, 11, 13	$CH_3(CH_2)_3CH{=}CH{-}CH{=}CH{-}CH{=}CH(CH_2)_7COOH$	Tung oil
Linolenic	9,12,15-octadecatrienoic	18 : 3 ; 9, 12, 15	$CH_3CH_2CH{=}CHCH_2CH{=}CHCH_2CH{=}CH(CH_2)_7COOH$	Linseed oil
Arachidonic	5,8,11,14-eicosatetraenoic	20 : 4 ; 5, 8, 11, 14	$CH_3(CH_2)_4CH{=}CHCH_2CH{=}CHCH_2CH{=}CHCH_2CH{=}CH(CH_2)_3COOH$	Animal fat
Nervonic	15-tetracosenoic	24 : 1 ; 15	$CH_3(CH_2)_7CH{=}CH(CH_2)_{13}COOH$	–

* All of the unsaturated fatty acids in this table are of the *cis* form with the exception of **eleostearic acid,** which is of the *cis*-9-*trans*-11-*trans*-13 form.

2 double bonds) and α-linolenic acid (*len* with 3 double bonds). In chemical jargon, *lin* and fatty acids derived from it are referred to as **n-6 fatty acids**, since the last double bond in their molecule occurs 3 carbon atoms away from the methyl end of the molecule. *Len* and fatty acids derived from it are referred to as **n-3 fatty acids** since the last double bond in their molecule is 3 carbon atoms away from the methyl end. *Lin* is more abundant in nature and is present in all vegetable oils, whereas *len* is present only in some vegetable oils such as mustard oil, rapeseed and soybean. Fish oils are, however, good sources of both *lin* and *len* fatty acids and hence are considered nutritionally quite rich. *Hydrogenated fat* or *'vanaspati'* is manufactured from vegetable oils by the process of hydrogenation. It is a saturated fat and, though derived from vegetable oils, behaves like other saturated fats of animal origin.

All foods contain small quantities of bound fat or invisible fat. Such bound fat is found in green leafy vegetables and pulses in plenty and is a rich source of *n*-3 fatty acids. About half of the daily human requirement of fat (40 g) can be derived through such bound fat. Thus, the minimum daily requirement of visible fat (oil etc) is only 20 g. For long, the 2 types of PUFAs, *n*-6 and *n*-3, were thought to have similar effects. But recent study shows that *n*-6 and *n*-3 PUFAs have different types of effects on blood lipids and blood clotting and that they should be present in the diet in a certain proportion. Too much of *n*-6 with very little *n*-3 is not desirable. A healthy cooking practice, therefore, would be to blend vegetable oils containing both *lin* and *len*, instead of patronizing any one type of oil, including even the much-publicized unsaturated oils like safflower or rice bran which are very effective in reducing blood cholesterol.

A most unusual unsaturated fatty acid, *nemotinic acid*, is excreted in the growth medium by a citrivorium mould. This fatty acid is unique in that it contains the single, double and triple C—C linkages. *Nemotinic acid is one of the few naturally-occurring compounds containing the allene group.*

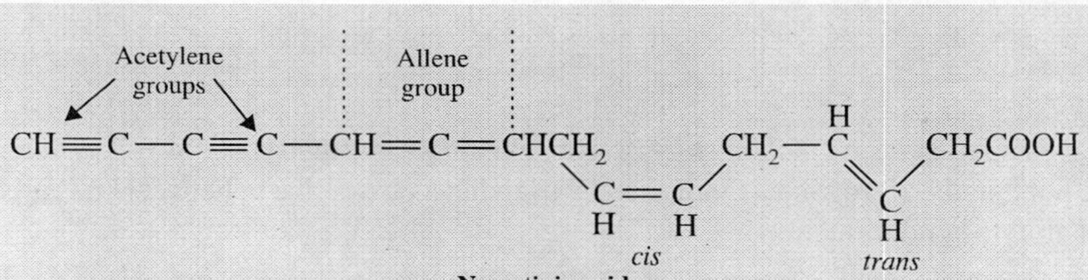

Nemotinic acid

The acetylene group has been detected in a number of unsaturated fatty acids, found in higher plants and microorganisms. For instance, *santalbic acid*, a major component of the seed oil of sandlewood contains 1 acetylene group as against 2 of nemotinic acid.

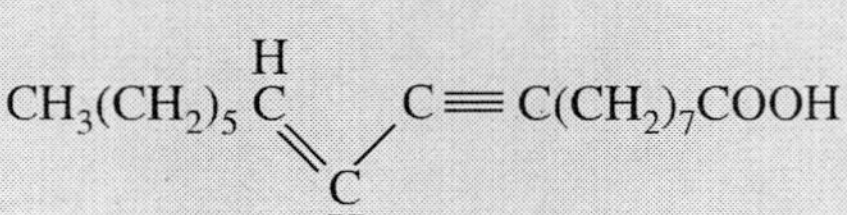

Santalbic acid

Geometric isomerism. On account of the presence of double bond(s), the unsaturated fatty acids exhibit geometric (or *cis-trans*) isomerism. *Most unsaturated fatty acids are found as the unstable cis isomer rather than as the more stable trans isomer.*

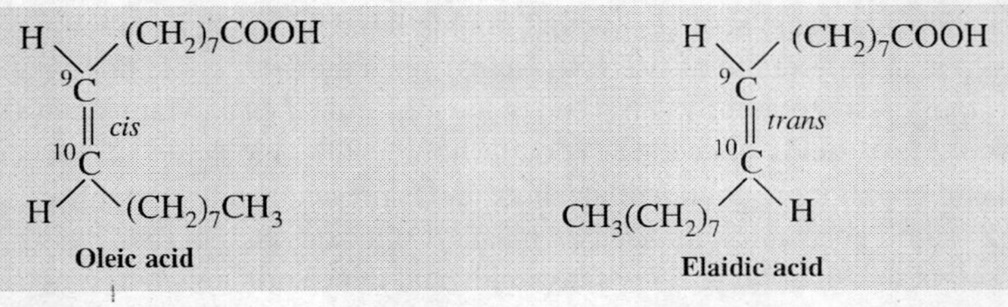

The hydrocarbon chain of saturated fatty acids (stearic acid, for example) has a zigzag configuration with the C—C bond forming a bond angle of 109° (Fig. 12–2).

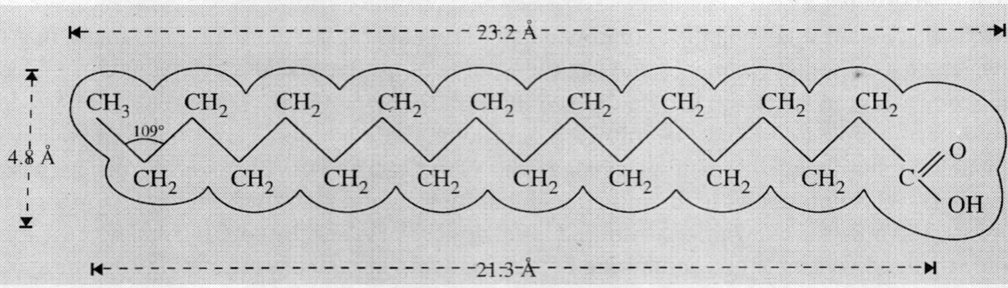

Fig. 12–2. Stearic acid

For the sake of simplicity, the long chains of CH_2 groups are represented by zigzag lines where each corner represents a C atom and the hydrogen atoms are left out. *The zigzag line represents the most stable configuration of such carbon chains.* The free end of this line, if not indicated otherwise, obviously represents the presence of a methyl group. The simplified formula for stearic acid would, thus, be as represented in Fig. 12–3.

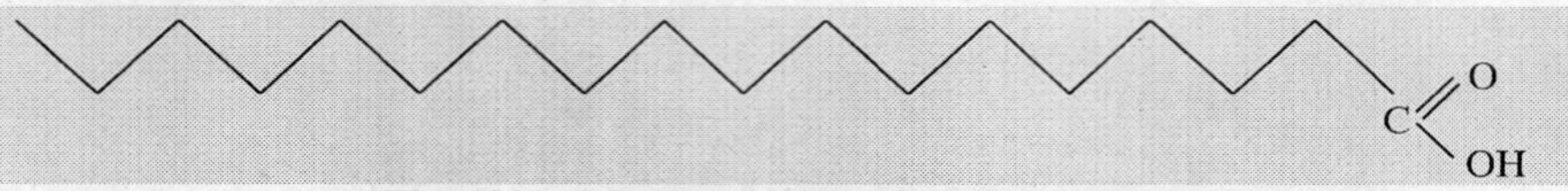

Fig. 12–3. Stearic acid (simplified formula)

The double bond in the zigzag line is indicated by drawing an extra line in between the carbon atoms involved in double bond formation.

When a *cis* double bond is inserted (as in oleic acid), the molecule bends assuming the shape, shown in Fig. 12–4. This double bond is *rigid* and creates a *kink* in the chain. The rest of the chain is free to rotate about the other C—C bonds.

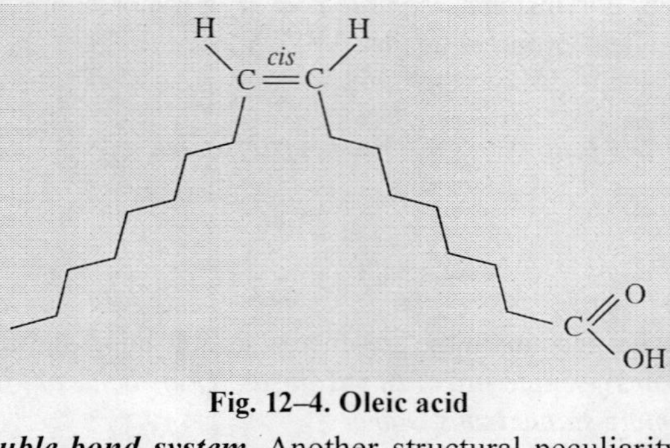

Fig. 12–4. Oleic acid

Nonconjugated double-bond system. Another structural peculiarity of naturally occurring polyunsaturated fatty acids is the presence of a **nonconjugated double-bond system.** It has a methylene group (—CH_2—) flanked by double bonds on both the sides as in linoleic, linolenic and arachidonic acids. *The conjugated double-bond system is, however, rarely present.* In it, the methylene group is not found in between the double bonds which, henceforth, occur one after the other as in eleostearic acid. This acid has valuable properties as a drying oil since it polymerizes readily. These two double-bond systems (Fig. 12–5) have different chemical reactivities.

$$-CH{=}CH-CH_2-CH{=}CH-CH_2-CH{=}CH- \quad \text{Nonconjugated system}$$

$$-CH{=}CH-CH{=}CH-CH{=}CH-CH{=}CH- \quad \text{Conjugated system}$$

Fig. 12–5. Two double-bond systems

3. Hydroxy or oxygenated fatty acids

One of these found in castor oil is *ricinoleic acid* (87%). It is a C 18 acid with a double bond at C_9 and an OH group on C_{12}.

$$CH_3(CH_2)_5\underset{}{\overset{OH}{\overset{|}{C}}}H\ CH_2CH=CH(CH_2)_7COOH$$

Ricinoleic acid

(12-hydroxyoctadec-9-enoic acid)

Cerebronic acid, a C 24 acid obtained from animal lipid, is another important hydroxy acid with an OH group on C_2.

$$CH_3(CH_2)_{21}\overset{OH}{\overset{|}{C}}H-COOH$$

Cerebronic acid

(2-hydroxytetracosanoic acid)

4. Cyclic fatty acids

These are of rare occurrence. Chaulmoogra oil, obtained from the plant *Hydnocarpus kurzil* and used in the treatment of leprosy, contains 2 such acids *hydnocarpic* and *chaulmoogric*. Chaulmoogric acid has a cyclopentenyl ring in its 18-carbon structure.

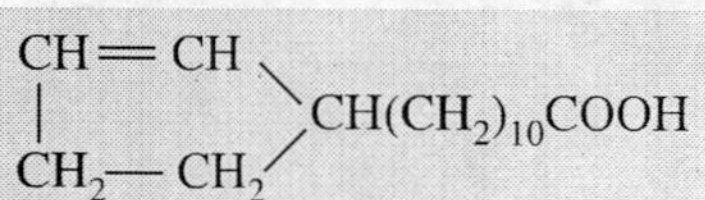

Hydnocarpic acid

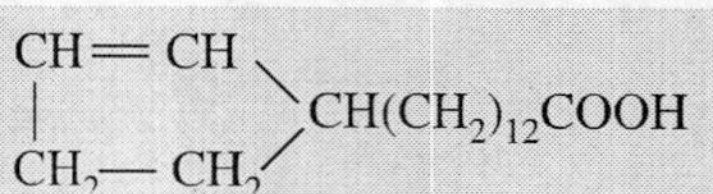

Chaulmoogric acid

CLINICAL IMPLICATIONS

Leprosy, the oldest disease known to mankind, is loathsome, endemic disease. The myth that leprosy was hereditary and a 'Curse of God' was exploded in 1873, when Dr. G. A. Hansen in Norway discovered rod-shaped bacteria in cells from leprosy nodules. *The leprosy bacillus* (*Mycobacterium leprae*), *also called Hansen's bacillus, is the slowest multiplying prokaryote known and takes about 2 weeks to double*. Once infected, it takes a long time for the disease to develop : from less than a year to as long as 30 years, though normally 3-7 years. About 20% of the infected people develop the disease, on an average. In patients with high immune resistance, the disease is relatively mild (*tuberculoid form*); those with low resistance develop highly infectious and more severe conditions (*lepromatous leprosy*), affecting not only nerves and skin but also lymph nodes, eyes, nose, mouth, larynx, spleen etc and the bacilli may appear in tears, nasal mucus, sputum, ulcerating nodules etc. *M. leprae is the only bacterium that can enter and destroy the nerves*. Because of sensory impairment due to nerve damage, the person does not feel pain. However, the primary deformities are pronounced on the face. The infected, untreated individual is the only source of infection. There is no known natural reservoir other than man. Leprosy is only weakly contagious : usually prolonged and close contact is considered necessary. The nose is the main portal of exit of the bacilli. Although leprosy affects individuals of both sexes, *males are more susceptible than females*, often in the ratio of 2 : 1. Although leprosy had assumed epidemic proportions in Europe in 13th and 14th century A.D., at present, Western Europe is virtually free from it. And these days, the leprosy 'hot spots' are India, Indonesia, Myanmar, Brazil, Columbia and few pockets of Central and Eastern Europe. Developed countries are, however, free from leprosy due to their improved socio-economic conditions. Of the estimated disease burden, Asia accounts for about 60%, followed by Africa with 35% and Latin America with 4%; the rest of the world accounting for about 1% of the disease load. India alone accounts for about 1/5th of the total lepers in the world.

Under chemotherapy, the drugs most commonly used against leprosy were *dapsone, rifampicin* and *clofazimine,* the last one being the best alternative as it had the much-desired anti-inflammatory effect and produced least resistance. However, in 1982, the WHO recommended the multi-drug therapy (MDT), combining all the 3 above drugs, as it has a very high potential for control of leprosy. The duration of treatment in most cases is about 3 years or more. A single dose of MDT kills about 95% of bacilli in a patient and there are no significant side effects. However, the overall long duration of the treatment regimen in chemotherapy, has led to the use of immunotherapy, using leprosy vaccines. Under the aegis of WHO, a vaccine was developed combining *M. leprae* (inactivated) derived from armadillo, and the BCG vaccine. Trials of this vaccine have shown that its application can restore cell-mediated immunity in patients and thus it could remarkably suppress the disease in about 18 months only.

Leprosy eradication programme has made spectacular progress worldover. From a long list of 122 endemic countries in 1982, the disease is now endemic only in 24 of them. Ninety per cent of leprosy is now confined to only 12 countries, including India, Brazil, Indonesia, Myanmar, Nepal, Mosambique, Guinea, and Angola.

BIOLOGICAL ROLES OF LIPIDS

The lipids perform a wide variety of functions (Fig. 12-6). These are briefly listed below:

1. Food material. Lipids provide food, highly rich in calorific value. One gram lipid produces 9.3 kilocalories of heat.

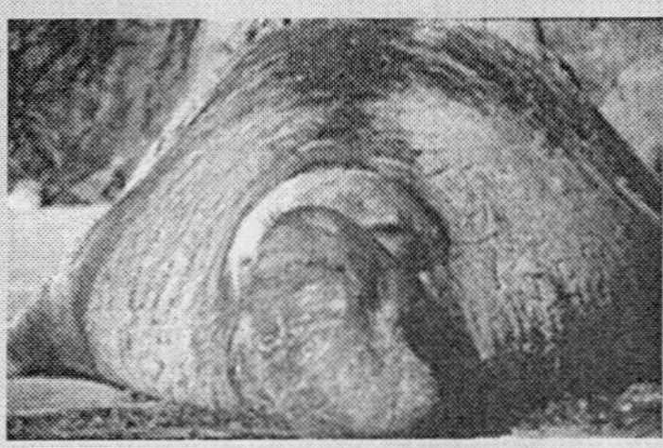

(*a*) The fat under the skin of this elephant seal protects it from the cold

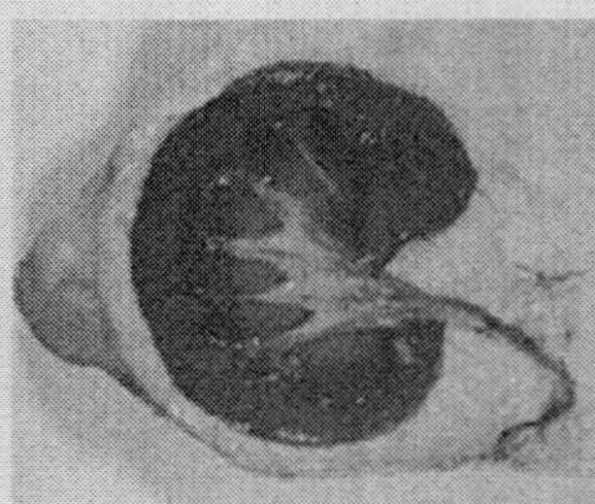

(*c*) Kidneys are embedded in fatty tissue

(*e*) Like water off a duck's back, fatty acids on a duck's feathers repel water

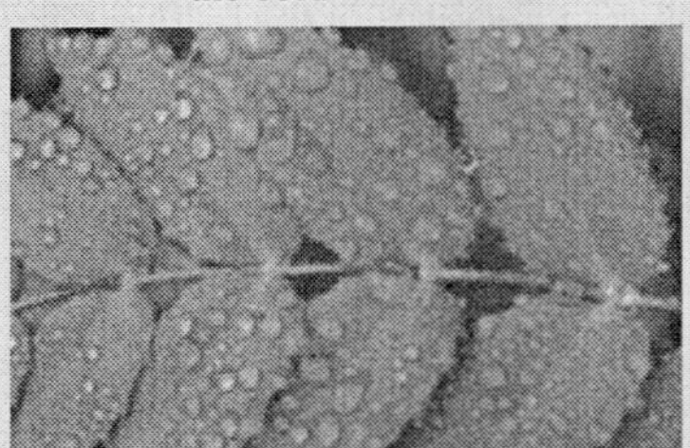

(*b*) The wax surface of leaves repels water, but also reduces water loss from the plant

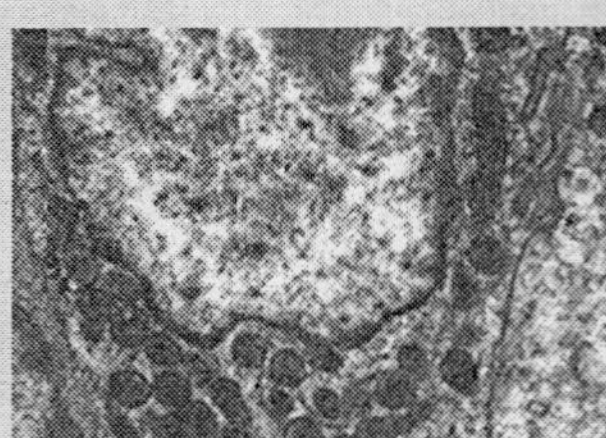

(*d*) Every cell membrane contains lipids

Fig. 12–6. Various functions performed by lipids

2. Food reserve. Lipids provide are insoluble in aqueous solutions and hence can be stored readily in the body as a food reserve.

3. Structural component. Lipids are an important constituent of the cell membrance.

4. Heat insulation. The fats are characterized for their high insulating capacity. Great quantities of fat are deposited in the subcutaneous layers in aquatic mammals such as whale and in animals living in cold climates.

5. Fatty acid absorption. Phospholipids play an important role in the absorption and transportation of fatty acids.

6. Hormone synthesis. The sex hormones, adrenocorticoids, cholic acids and also vitamin D are all synthesized from cholesterol, a steroidal lipid.

7. Vitamin carriers. Lipids act as carriers of natural fat-soluble vitamins such as vitamin A, D and E.

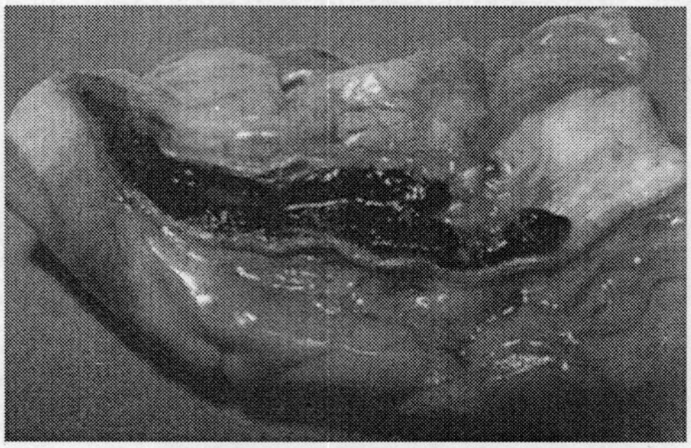

Fig. 12–7. A cholesterol deposit, (known as atheroma) filling almost all the space inside the artery.

8. Blood cholesterol lowering. Chocolates and beef, especially the latter one, were believed to cause many heart diseases as they are rich in saturated fatty acids, which boost cholesterol levels in blood and clog the arterial passage (Fig. 12–7). But researches conducted at the University of Texas by **Scott Grundy** and **Andrea Bonanome** (1988) suggest that at least one saturated fatty acid stearic acid, a major component of cocoa butter and beef fat, does not raise blood cholesterol level at all. The researchers placed 11 men on three cholesterol-poor liquid diets for three weeks each in random order. One formula was rich in palmitic acid, a known cholesterol booster; the second in oleic acid; and the third in stearic acid. When compared with the diet rich in palmitic acid, blood cholesterol levels were 14% lower in subjects put on the stearic acid diet and 10% lower in those on the oleic acid diet.

9. Antibiotic agent. *Squalamine*, a steroid from the blood of sharks, has been shown to be an antibiotic and antifungal agent of intense activity. This seems to explain why sharks rarely contract infections and almost never get cancer.

PROBLEMS

1. Why are the most unsaturated fatty acids found in phospholipids in the *cis* rather than the *trans* conformation ? Draw the structure of a 16-carbon fatty acid as saturated, *trans* monounsaturated, and *cis* monounsaturated.
2. Given these molecular components– glycerol, fatty acid, phosphate, long-chain alcohol, and carbohydrate– answer the following :
   (*a*) Which two are present in both waxes and sphingomyelin ?
   (*b*) Which two are present in both fats and phosphatidylcholine ?
   (*c*) Which are present in a ganglioside but not in a fat ?
3. The melting points of a series of 18-carbon fatty acids are stearic acid, 69.6°C, oleic acid, 13.4°C; linoleic acid –5°C; and linolenic acid, –11°C. What structural aspect of these 18-carbon fatty acids can be correlated with the melting point ? Provide a molecular explanation for the trend in melting points.
4. How is the definition of "lipid" different from the definitions of other types of biomolecules that we have considered, such as amino acids, nucleic acids, and proteins ?

5. Johann Thudichum, who practiced medicine in London about 100 years ago, also dabbled in lipid chemistry in his spare time. He isolated a variety of lipids from neutral tissue, and characterized and named many of them. His carefully sealed and labeled vials of isolated lipids were rediscovered many years later. How would you confirm, using techniques available to you but not to him, that the vials he labeled "sphingomyelin" and "cerebroside" actually contain these compounds ?
6. How would you distinguish sphingomyelin from phosphatidylcholine by chemical, physical, or enzymatic tests ?
7. Do all kinds of fibre help lower one's cholesterol ?
8. How much fat does one need in his diet ?
9. Is a high-protein diet good for one's health ?
10. Why is tea oil better for frying than olive oil ?
11. Why soaps cannot be manufactured from mineral oil or kerosene oil ?
12. How would you differentiate between hydrophilic and hydrophobic groups ?
13. Why fatty acids are called amphipathic molecules ?
14. Differentiate between prokaryotic and eukaryotic cells in terms of their lipids.

ANSWERS TO PROBLEMS

1. The presence of a *cis* double bond introduces a kink that prevents packing of the fatty acid chains. *Cis* double bonds maintain fluidity. *Trans* fatty acids have no structural effect, relative to saturated fatty acids, and so they are rare.

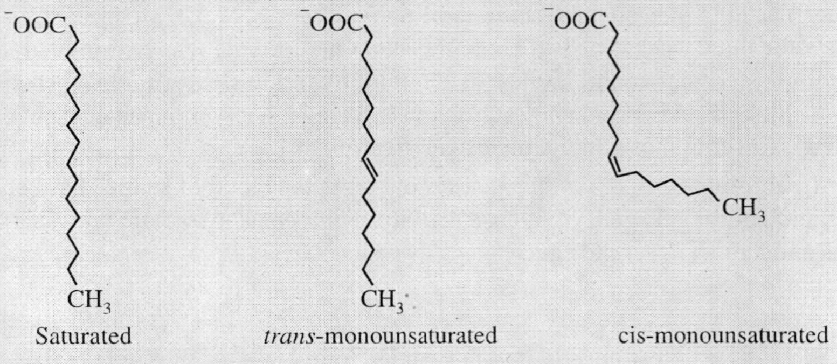

2. (*a*) Fatty acid, long-chain alcohol.
   (*b*) Glycerol, fatty acid.
   (*c*) Carbohydrate, long-chain alcohol.
3. The number of *cis* double bonds. Each *cis* double bond causes a bend in the hydrocarbon chain; it is more difficult to pack these bent chains in the crystal lattice.
4. The definition of substances such as lipids is based, not on a common structure, but on their solubilities in water and in nonpolar solvents.
5. Both compounds should yield sphingosine and a fatty acid upon strong alkaline hydrolysis. Phosphocholine is produced only from sphingomyelin; cerebroside produces one or more sugars, but no phosphate.
6. Strong alkaline hydrolysis of phosphatidylcholine released 2 fatty acids per phosphate in PC; the same treatment of sphingomyelin produces 1 fatty acid per phosphate, as well as sphingosine.

7. No, only soluble fibre (in oat bran, fruits and vegetables) does. Insoluble fibre (in wheat bran and beans) does not.
8. Not much. About one tablespoon of oil a day will supply all the fat one needs.
9. No. This diet is high in fat and low in complex carbohydrates. Just the opposite of good nutrition.
10. Tea oil, which is cold-pressed from tea seeds, is unrefined and low in saturated fat like olive oil. But tea oil is a better choice for frying because it resists breaking down into *trans* fats when heated. It also has a naturally neutral flavour.
11. The mineral and kerosene oils do not have glycerides and, hence, cannot be converted into soaps.
12. Colloids are of 2 types, when taking water as the dispersion medium :

    (a) Hydrophilic (water-loving) or lyophilic–These are capable of binding water and as such readily yield colloidal solutions when brought in contact with water. They are, therefore, also called *reversible colloids*. Agar-agar, gelatin, starch, cellulose, gums, and proteins are examples of hydrophilic colloids.

    (b) Hydrophobic (water-hating) or lyophobic–They are incapable of binding water and as such do not readily yield colloidal solutions when brought in contact with water. They are, henceforth, also called *irreversible colloids*. Hydrophobic colloids include metals, lipids, etc.
13. A fatty acid molecule has a polar end ($-COO^-$), and a nonpolar end of long carbon chain. The polar end is water-soluble, and is hydrophilic. The nonpolar end is water-insoluble and is, thus, hydrophobic. Molecules like these are called amphipathic, *i.e.*, they have both polar and nonpolar ends.
14. Prokaryotic cells (e.g., bacterial cells) have over 95% of its total components associated with its cell membrane and remaining 5% distributed between cytoplasm and cell wall. They are unable to synthesize steroid ring structure and are devoid of sterols. Barring mycobacteria, triacylglycerols are also missing in bacteria. Polyunsaturated fatty acids (PUFAs) are not synthesized by bacteria (with the exception of bacilli) and mostly saturated and branched-chain fatty acids are synthesized.

    Eukaryotic cells, such as plant cells, always contain triacylglycerol, PUFAs and other fatty acids like digalactosyl diacylglycerol, phosphatidyl glycerol etc. Animal cells, on the other hand, cannot synthesize PUFAs. However, phospholipids, sphingolipids, glyceryl ethers and plasmalogens play a vital role in maintaining the nervous system.

CHAPTER

13 Lipids–II Classification

INSIDE

- Simple Lipids
  - Fats and Oils
  - Waxes
- Compound Lipids
  - Phospholipids
    - Phosphoglycerides
    - Phosphoinositides
    - Phosphosphingosides
  - Glycolipids
    - Gangliosides
    - Sulfolipids
    - Sulfatids
- Derived Lipids
  - Steroids
  - Terpenes
  - Carotenoids

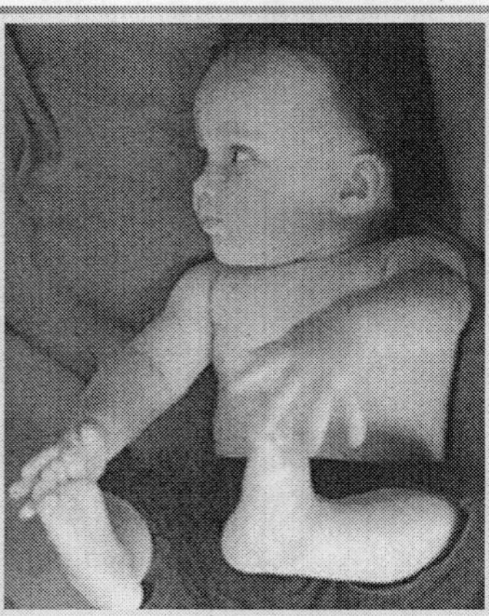

Adipose tissue in human babies

Brown fat owes its colour to the many large mitochondria it contains. Babies have a small surface area to volume ratio and so lose heat much more rapidly than adults. They cannot shiver (create heat from their muscles) but they can keep warm using triglyceride stores in their brown fat – a process called non-shivering thermogenesis.

Bloor (1943) has proposed the following classification of lipids based on their chemical composition.

A. Simple lipids or **Homolipids.** These are esters of fatty acid with various alcohols.

1. *Fats and oils (triglycerides, triacylglycerols).* These are esters of fatty acids with a trihydroxy alcohol, glycerol. A fat is solid at ordinary room temperature wheras an oil is liquid.

2. *Waxes.* These are esters of fatty acids with high molecular weight monohydroxy alcohols.

B. Compound lipids or **Heterolipids.** These are esters of fatty acids with alcohol and possess additional group(s) also.

1. *Phospholipids (phosphatids)*, These are compounds containing, in addition to fatty acids and glycerol, a phosphoric acid, nitrogen bases and other substituents.

2. *Glycolipids (cerebrosides).* These are the compounds of fatty acids with carbohydrates and contain nitrogen but no phosphoric acid. The glycolipids also include certain structurally-related compounds comprising the groups, gangliosides, sulfolipids and sulfatids.

C. Derived lipids. These are the substances derived from simple and compound lipids by hydrolysis. These include fatty acids, alcohols, mono- and diglycerides, steroids, terpenes and carotenoids.

Glycerides and cholesterol esters, because of their uncharged nature, are also called *neutral lipids.*
However, **Conn** and **Stumpf** (1976) have traditionally classified lipids into following 6 classes :

1. Acyl glycerols
2. Waxes
3. Phospholipids
4. Sphingolipids
5. Glycolipids
6. Terpenoid lipids including carotenoids and steroids

SIMPLE LIPIDS

FATS AND OILS
(= Triglycerides or Triacylglycerols)

The triglycerides are the most abundant of all lipids. They constitute about 98% of total dietary lipids ; the remaining 2% consists of phospholipids and cholesterol and its esters. They are the major components of storage or depot fats in plant and animal cells but are not normally found in membranes. They are nonpolar, hydrophobic molecules since they contain no electrically charged or highly polar functional groups. In animals, the *fat cells* or *adipocytes* contain very large quantities of triglycerides in the form of fat droplets, which fill almost the entire cell volume (Fig. 13-1). Adipocytes are abundantly found under the skin, in the abdominal cavity and in the mammary glands. Triglycerides can be stored in quantities, sufficient to supply the energy needs of the body for many months, as in the case of obese persons. On the contrary, the body can store the carbohydrate glycogen in meagre amounts, sufficient to supply energy need of a day only. *Triglycerides are much better adapted than glycogen to serve as storage form of energy.* They are not only stored in large amounts but also yield over twice as much energy as carbohydrates. Since fats tend to remain in the stomach longer than carbohydrates and are digested more slowly, they also have greater satiety value than carbohydrates. The arctic and antarctic animals (Fig. 13-2) such as whales, seals, walruses and penguins are amply padded with triglycerides to serve both as energy storage depots and as an insulation against very low temperatures. Most fats and oils, upon hydrolysis, yield several fatty acids as well as glycerol. However, the milk of spiny anteater is an exception in that it comprises almost pure triolein. Human body contains enough fat to make 7 bars of soap !

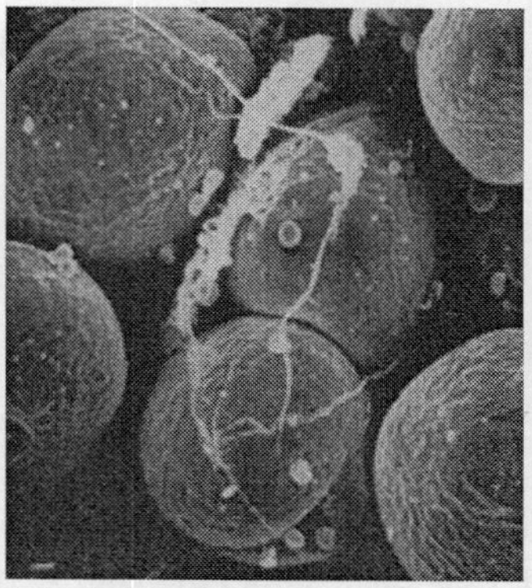

Fig 13-1. A scanning electron micrograph of adipocytes

Each adipocyte contains a fat globule that occupies nearly the entire cell.

(*Courtesy* : *Fred E. Hossler*)

Fig 13-2. Insulating blubber of whales and the fat molecules

Fat such as the triacylglycerol molecule (lower) are widely used to store excess energy for later use and to fulfill other purposes, illustrated by the insulating blubber of whales. The natural tendency of fats to exist in nearly water-free forms makes these molecules well-suited for these roles.

[*Courtesy* : (*Upper*) *Francois Gohier*]

In a normal man, weighing 70 kg, at least 10-20% of the body weight is lipid, the bulk of which is triacylglycerol (TAG). TAG is found in all organs of the human body, particularly in adipose tissue, in which droplets of triacylglycerols may represent more than 90% of the cytoplasm of the cells. Body lipid is a reservoir of potential chemical energy. About 100 times more energy is stored as mobilizable lipid than as mobilizable carbohyldrate in the normal human being. TAG is stored in a relatively water-free state in the tissue, in comparison to carbohydrate, which is heavily hydrated.

Chemically, triglycerides are esters of glycerol with 3 fatty acid molecules. Their generic formula is shown in Fig. 13–3.

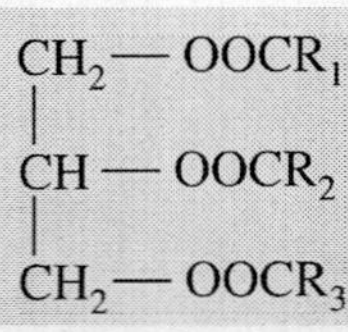

Fig. 13–3. Generic formula of triglycerides

Obviously, when the groups attached to carbon 1 and 3 differ, a centre of asymmetry is created at C 2. The 2 optical isomers may, thus, be represented as shown in Fig. 13–4. *The naturally-occurring fats are of L-type.*

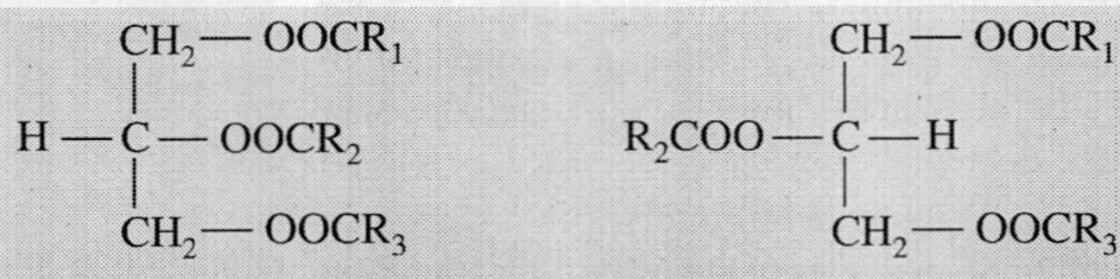

D-isomer L-isomer

Fig. 13–4. Two optical isomers of triglycerides

A fat molecule contains 3 moles of fatty acids which may be similar or dissimilar. Those containing a single kind of fatty acid in all 3 positions (α, β, α′) are called *simple* (or *symmetrical*) *triglycerides* ; they are named after the fatty acids they contain (Fig. 13–5). Examples are tripalmitin, tristearin

Tristearin (tristearoylglycerol) **Triolein** (trioleylglycerol)

Fig. 13–5. Two simple triglycerides

and triolein. They occur very infrequently in natural fats. Such simple triglycerides have, however, been synthesized in the laboratory *e.g.*, tristearin and triolein.

Most of the triglycerides of nature are mixed (or asymmetrical) triglycerides, i.e., they contain 2 or 3 different fatty acid units in the molecule. Representatives of such mixed triglycerides are oleodipalmitin and oleopalmitostearin (Fig. 13–6).

Oleodipalmitin (α-oleo-β, α′-dipalmitin) **Oleopalmitostearin** (α-oleo-β, palmito- α′-stearin)

Fig. 13–6. Two mixed triglycerides

Most **animal fats** such as those from meat, milk and eggs are relatively rich in saturated fatty acids but contain a rather low content of polyunsaturated fatty acids (Table 13–1) ; two exceptions are chicken fat and fish fat. The large proportion of saturated fatty acids (*esp*., palmitic and stearic) with high melting point confers solid state to the animal fats. The **plant fats**, on the other hand, contain a large proportion of unsaturated fatty acids (*esp*., polyunsaturated). The unsaturated fatty acids have low melting point and confer liquid state to the plant fats.

Table 13–1. Fatty acid composition of important animal and plant fats

Fat/Oil	*Percentage of total fatty acids*		
	Saturated	*Monounsaturated*	*Polyunsaturated*
Animal fats			
Butter fat	60	36	4
Pork fat	59	39	2
Beef fat	53	44	2
Chicken fat	39	44	21
Plant fats			
Olive oil*	20	26	54
Corn oil	15	31	53
Soybean oil	14	24	53
Soft margarine	23	22	52

* data not confirmed.

The differences between the animal and plant fats are presented in Table 13–2.

Table 13–2. Differences between animal and plant fats

Animal fat	*Plant fat*
1. Relatively rich in saturated fatty acids, *esp*., C_{16} and C_{18} acids.	1. Relatively rich in unsaturated fatty acids, *esp*., polyunsaturated acids.
2. Solid at ordinary room temperature.	2. Liquid at ordinary room temperaure.
3. These have usually low iodine number.	3. These have usually high iodine number.
4. These have usually high Reichert-Meissl . number	4. These have usually low Reichert-Miessl number.
5. These are stored mainly in liver and bone marrow	5. These are stored mainly in seeds and fruits.
6. Oxidative rancidity is observed more frequently.	6. Oxidative rancidity is observed less frequently.
Examples– Butterfat, Beef fat, Tallow	*Examples*– Olive oil, Castor oil, Soybean oil, Corn oil

WAXES

Far less spread but equally important are the waxes. The term `wax' originates in the Old English word *weax*, meaning "the material of the honeycomb", reminding of beeswax, the honeycomb is made of. These are esters of long-chain saturated and unsaturated fatty acids with long-chain monohydroxy alcohols. The fatty acids range in between C_{14} and C_{36} and the alcohols range from C_{16} to C_{36}. In vertebrates, waxes are secreted by cutaneous glands as a protective coating to keep the skin pliable, lubricated and water-proof. Hair, wool and fur are also coated with wax. Birds, particularly waterfowl, secrete waxes in their preen glands to make their feathers water-repellent. The leaves of many plants such as *Rhododendron, Calotropis* etc., are shiny because of the deposition of protective waxy coating. Waxes also serve as the chief storage form of fuel in planktons. Since marine organisms (whale, herring, salmon) consume planktons in large quantities, waxes act as major food and storage lipids in them.

Most of the waxes are mixtures of esters. Thus, *sperm whale wax* (spermaceti) and *beeswax* are composed mainly of palmitic acid esterified with either hexacosanol, $CH_3(CH_2)_{24}.CH_2OH$ or triacontanol, $CH_3(CH_2)_{28}.CH_2OH$.

$$\underset{\text{Palmitic acid}}{CH_3(CH_2)_{14}COO\vdots H} + \underset{\text{Hexacosanol}}{HO\vdots CH_2(CH_2)_{24}CH_3} \xrightarrow{-H_2O} \underset{\text{Hexacosanyl palmitate}}{CH_3(CH_2)_{14}COO—CH_2(CH_2)_{24}CH_3}$$

Moreover, sperm whale wax is also rich in cetyl palmitate and the beeswax in myricyl palmitate.

$$\underset{\textbf{Cetyl palmitate}}{CH_3(CH_2)_{14}COO—CH_2(CH_2)_{14}CH_3} \qquad \underset{\textbf{Myricyl palmitate}}{CH_3(CH_2)_{14}COO—CH_2(CH_2)_{28}CH_3}$$

Carnauba wax, the hardest known wax, consists mainly of fatty acids esterified with tetracosanol, $CH_3(CH_2)_{22}.CH_2OH$ and tetratriacontanol, $CH_3(CH_2)_{32}.CH_2OH$.

Waxes are unusually inert due to their saturated nature of the hydrocarbon chain. However, they can be split slowly with hot alcoholic KOH. They are insoluble in water and highly resistant to atmospheric oxidation. Hence, these are used in polishing furnitures and automobiles and also in wax-coated paper used to wrap perishable food products such as biscuits, cakes, etc. Lanolin (from lamb's wool), beeswax (from honeycomb), carnauba wax (from a Brazilian palm tree) and spermaceti oil (from sperm whales) are widely used in the manufacture of lotions, ointments and polishes.

COMPOUND LIPIDS

PHOSPHOLIPIDS
(= Phosphatids)

Phospholipids are the most abundant membrane lipids. They serve primarily as structural components of membranes and are never stored in large quantities. As their name implies, phospholipids contain phosphorus in the form of phosphoric acid groups. They differ from triglycerides in possessing usually one hydrophilic polar "head" group and usually two hydrophobic nonpolar "tails". For this reason, they are often called *polar lipids*. Thus, *phospholipids are amphipathic, whereas the storage lipids (triglycerides and waxes) are not*. In phospholipids, two of the OH groups in glycerol are linked to fatty acids while the third OH group is linked to phosphoric acid. The phosphate is further linked to one of a variety of small polar head groups (alcohols). Folch and Sperry (1955) have classified phospholipids into phosphoglycerides, phosphoinositides and phosphosphingosides.

A. Phosphoglycerides

These are the major phospholipids found in membranes and contain two fatty acid molecules or "tails" esterified to the first and second hydroxyl groups of glycerol. The third hydroxyl group of glycerol forms an ester linkage with phosphoric acid. In addition, phosphoglycerides contain a second alcohol, which is also esterified to the phosporic acid. This is referred to as 'head alcohol group' as it is present at one end ('head') of the long phosphoglyceride molecule.

The various phosphoglycerides differ in their head alcohol groups. However, all of them contain two nonpolar tails, each consisting of a long chain (usually C_{16} or C_{18}) fatty acid. Usually one of the fatty acids is saturated and the other unsaturated ; the latter is always esterified to the middle or β-hydroxy group of glycerol. A noteworthy feature of the phosphoglycerides is that they contain an asymmetric carbon atom at position 2 in the glycerol part of their molecule. It has the L-configuration since it is related to L-glyceraldehyde. All phosphoglycerides have a negative charge on phosphoric group at pH 7. In addition, the head alcohol group may also have one or more electric charges at pH 7.

1. *Lecithins* (= *phosphatidyl cholines*) – Lecithins (*likithos*[G] = yolk) are widely distributed in nature. Various oil seeds like soybean and the yeasts are important sources from plant world. In animals, the glandular and nervous tissues are rich in these lipids. The lecithins are required for the normal transport and utilization of as other lipids *esp.*, in the liver of animals. In their absence, accumulation of lipids occurs in the liver to as much as 30% against a normal value of 3-4%, giving rise to a condition called "*fatty liver*". This fatty infiltration may lead to fibrotic changes, characteristic of the liver disease *cirrhosis*.

In addition to glycerol and 2 moles of fatty acids, the lecithins (Fig. 13–7) also contain phosphoric acid and a nitrogen base choline at either the end or middle carbon atom of glycerol unit. Accordingly, two forms of lecithins, α and β are recognized.

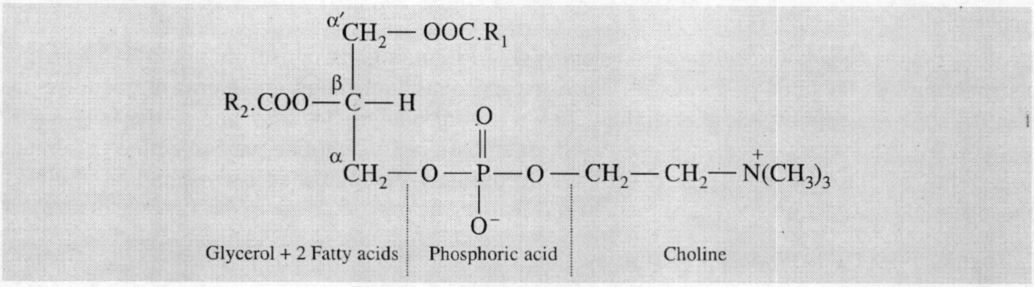

Fig. 13–7. Structure of α-lecithin

On complete hydrolysis, lecithin yields choline, phosphoric acid, glycerol and 2 moles of fatty acids. But partial hydrolysis of lecithins by *lecithinases* (active principles found in snake venoms) causes removal of only one fatty acid to yield substances called **lysolecithins** (Fig. 13–8). These, therefore, contain only one acyl radical. When subjected into the blood stream by sting as a result of snake bite or by needle, the lysolecithins cause rapid rupture (hemolysis) of the red blood corpuscles.

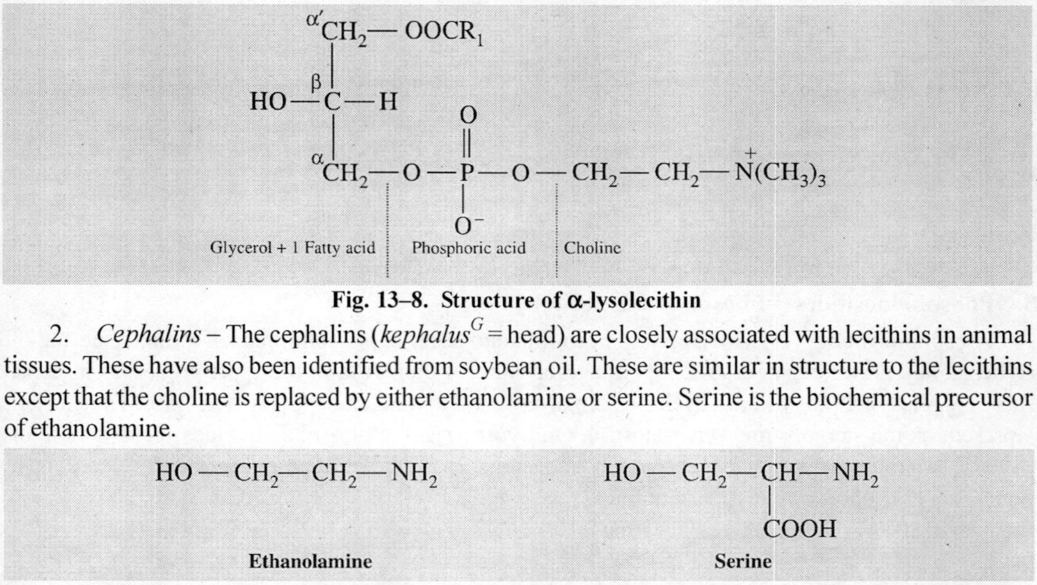

Fig. 13–8. Structure of α-lysolecithin

2. *Cephalins* – The cephalins (*kephalus*[G] = head) are closely associated with lecithins in animal tissues. These have also been identified from soybean oil. These are similar in structure to the lecithins except that the choline is replaced by either ethanolamine or serine. Serine is the biochemical precursor of ethanolamine.

Accordingly, two types of cephalins are recognized, phosphatidyl ethanolamine and phosphatidyl serine. Like lecithins, the cephalins (Fig. 13–9) also exist in 2 forms, α and β, depending upon the relative positions of the two substituent fatty acids.

Fig. 13–9. The two cephalins

Since the primary amino group of ethanolamine is a weaker base than the quaternary ammonium group of choline, *the cephalins are more acidic than lecithins.* Moreover, the cephalins are comparatively less souble in alcohol than lecithins.

Venoms containing lecithinases also split off fatty acids from cephalins, leaving hemolytic **lysocephalins.**

3. *Plasmalogens* (=*Phosphoglyceracetals*) – Plasmalogens constitute about 10% of the phospholipids of the brain and muscle. These are apparently not found in significant quantities in plant tissues. Structurally, these resemble lecithins and cephalins but have one of the fatty acids replaced by an unsaturated ether. Since the nitrogen base can be choline, ethanolamine or serine, three types of plasmalogens (Fig. 13–10) are accordingly distinguished : phosphatidal choline, phosphatidal ethanolamine and phosphatidal serine.

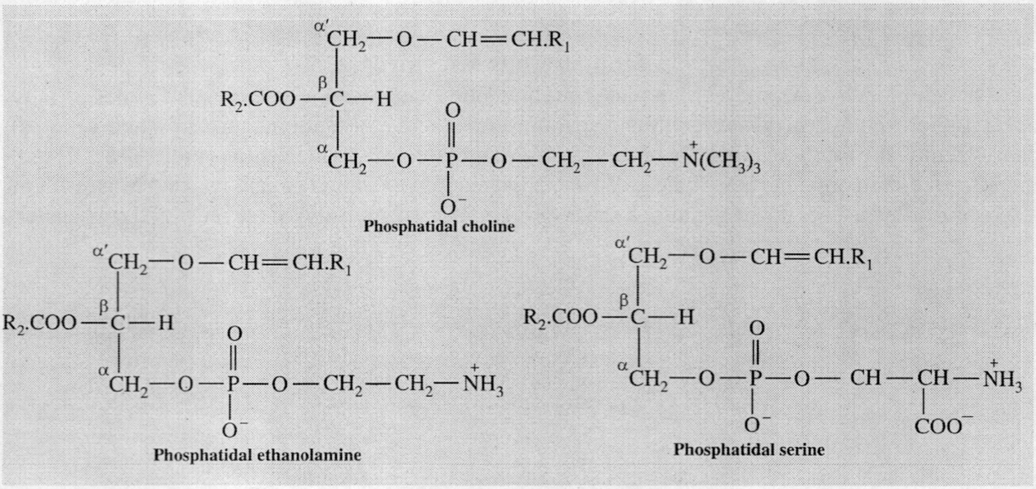

Fig. 13–10. The three plasmalogens

B. Phosphoinositides (=Phosphatidyl inositols)

Phosphoinositides (Fig. 13–11) have been found to occur in phospholipids of brain tissue and of soybeans and are of considerable importance because of their role in transport processes in cells. These are phospholipids where a cyclic hexahydroxy alcohol called inositol replaces base. The inositol is present as the stereoisomer, *myo*-inositol. On hydrolysis, the phosphoinositides yield 1 mole of glycerol, two moles of fatty acid, 1 mole of inositol and 1, 2, or 3 moles of phosphoric acid. Accordingly, mono-, di- or triphosphoinositides are found.

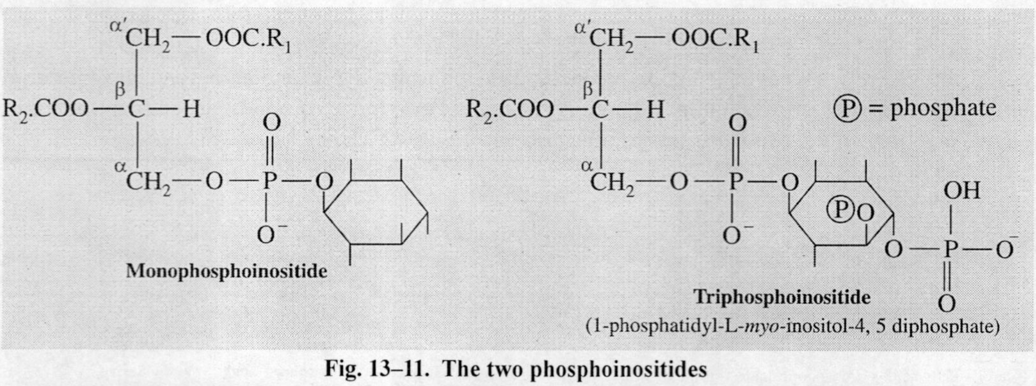

Fig. 13–11. The two phosphoinositides

Phosphoinositides are also classified as glycolipids, in as much as they contain carbohydrate residue.

C. Phosphosphingosides (=Sphingomyelins)

These compounds are commonly found in nerve tissue *esp.*, in the myelin sheath of the nerve (hence their name, sphingomyelins) and apparently lack in plants and the microorganisms. In a syndrome called *Niemann–Pick disease*, the sphingomyelins are stored in the brain in large quantities. These differ from other phospholipids in their lack of glycerol and the presence of another nitrogenous base sphingosine or a closely related dihydrosphingosine, besides choline, in place of glycerol. Sphingomyelins are electrically charged molecules and contain phosphocholine as their polar head groups.

> A ***syndrome*** (*syn*[G] = together; *dramein*[G] = to run) is a group of symptoms associated with a particular disease or abnormality, although not every symptom is necessarily shown by every patient with the disease.

CLINICAL IMPLICATIONS

Niemann–Pick disease (= sphingomyelin lipidosis) is a rare genetic disorder, inherited as an autosomal recessive condition. The disease is caused by a deficiency of the enzyme *sphingomyelinase*, which cleaves ceramide-phosphocholine bond of sphingomyelin. As a result, sphingomyelin accumulates in large amounts in the reticuloendothelial system since their synthesis is normal in rate but their degradation is interrupted. Niemann–Pick cells (prototype of 'foam cells') are found in the bone marrow, spleen, lymphoid tissues, liver, lung and tissues, of virtually any organ. Clinically, the disease manifests itself in two forms :

(a) *Early* or *infantile form*. This is characterized by hepatosplenomegaly, macular degeneration, cherry-red retinal spot (= macula) in nearly half of the cases, mental retardation and blindness. Death ensues by two years of age.

(b) *Late* or *adult form*. This perhaps is the most common of the two forms and reveals intellectual impairment during late infancy, slow evolution of the disease, no changes in the ocular fundi and survival beyond the age of five. However, death of the patient occurs by the second decade.

$$CH_3(CH_2)_{12}CH{=}CH{-}\underset{\displaystyle OH}{\underset{|}{CH}}{-}\underset{\displaystyle NH_2}{\underset{|}{CH}}{-}CH_2OH$$

Sphingosine or **4-sphingenine**
(1, 3-dihydroxy-2-amino-*trans*-octadec-4-ene)

$$CH_3(CH_2)_{14}CH{-}\underset{\displaystyle OH}{\underset{|}{CH}}{-}\underset{\displaystyle NH_2}{\underset{|}{CH}}{-}CH_2OH$$

Dihydrosphingosine or **sphinganine**

On hydrolysis, the phosphosphingosides yield equimolar amounts of fatty acid, phosphoric acid, choline and sphingosine or dihydrosphingosine but no glycerol. Thus, in these compounds the atomic ratio N/P is 2, in contrast to phosphoglycerides where this ratio equals unity. Phosphoinositides, however, do not contain a nitrogen base.

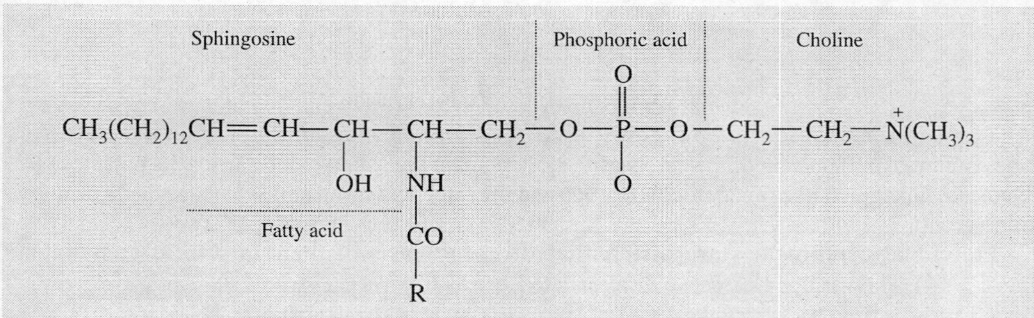

Fig. 13–12. Structure of sphingomyelin

It may be observed from the formula of sphingomyelin (Fig. 13–12) that sphingosine carries the phosphoric acid on its primary alcohol group and the fatty acid by amide linkage on its primary amino group.

GLYCOLIPIDS
(= Cerebrosides or Glycosphingosides)

The cerebrosides, as the name suggests, are important constituent of brain where they amount to about 8% of the solid matter. These may also occur in tissues other than brain. Since the head group characteristically consists of one or more sugar units, cerebrosides are often called glycosphingosides. Like phospholipids, glycolipids are composed of a hydrophobic region, containing 2 long hydrocarbon tails, and a polar region, which now contains one or more sugar residues and no phosphate. Both phospholipids and glycolipids form self-sealing lipid bilayers that are the basis for all cellular membranes. In Gaucher disease, the cerebrosides appear in relatively large amount in the liver and the spleen. They are also present in large amount in the brain in Niemann–Pick disease. These are present in much higher concentration in medullated than in nonmedullated nerve fibres. There is evidence that they also occur in some plant organs.

The structure of cerebrosides (Fig. 13–13) is somewhat similar to that of phosphosphingosides. They contain a high molecular weight fatty acid, sphingosine and either galactose or glucose instead of choline but no phosphoric acid. They have no electric charge since their polar head groups are neutral. In general properties, they resemble sphingomyelins.

CLINICAL IMPLICATIONS

Gaucher disease or **glucocerebrosidosis** is a not-so-rare, autosomal recessive, hereditary syndrome, often familial and of Jewish origin. It is 30 times more prevalent among the Ashkenazic Jews, with 1 case in 2,500 births. The disease is due to a deficiency of the enzyme *glucosyl ceramide hydrolase* (=*glucocerebrosidase*), which cleaves ceramide from ceramide trihexoside, causing deposition of the later. Gaucher's cells (pale lipid-containing cells) are found in abundance in all the organs specifically in the lungs and bone marrow, resulting in marked hepatosplenomegaly. Thus, the disease affects the lungs and causes destruction of bones. Clinically, two patterns of the disease have been reported :

(a) *Infantile or acute form.* In this, the symptoms appear early with hepatosplenomegaly, mental retardation, hypertonicity and signs of respiratory involvement. Most patients die during infancy.

(b) *Adult or chronic form.* This is more common a form of this disease. There is slow evolution of the disease with hepatosplenomegaly, hypersplenism, pathological fractures and often no neurological deficit.

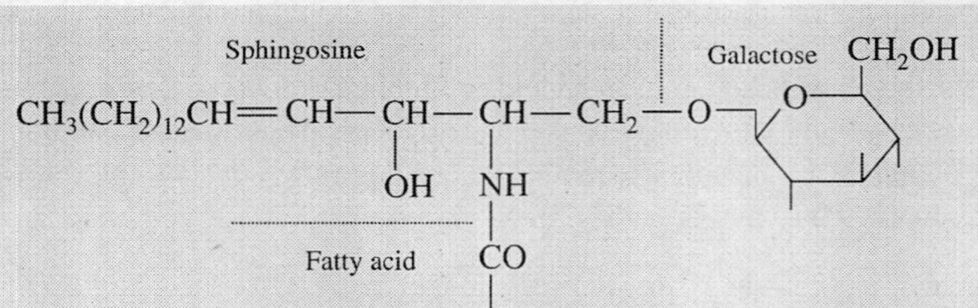

Fig. 13–13. Structure of a cerebroside

The name **ceramide** is commonly used to designate the sphingosine–fatty acid (or N-acylsphingosine) portion of the cerebrosides. Here again the sphingosine carries the galactose by glycosidic linkage on its primary alcohol group and the fatty acid by an amide linkage on its primary amino group.

Individual cerebrosides are differentiated on the basis of their fatty acid component (Fig. 13–14). The various classes, so differentiated, are as follows :

$CH_3(CH_2)_{22}COOH$	$CH_3(CH_2)_{21}\ CH(OH)COOH$
Lignoceric acid	**Cerebronic acid**
$CH_3(CH_2)_7\ CH{=}CH(CH_2)_{13}\ COOH$	$CH_3(CH_2)_7\ CH{=}CH(CH_2)_{12}\ CH(OH)COOH$
Nervonic acid	**Oxynervonic acid**

Fig. 13–14. Characteristic fatty acids of cerebrosides

(a) Kerasin – contains saturated C 24 lignoceric acid.

(b) Phrenosin (cerebron)– contains a 2 hydroxy derivative of lignoceric acid called cerebronic acid.

(*c*) Nervon– contains an unsaturated homologue of lignoceric acid called nervonic acid,

(*d*) Oxynervon– contains a 2-hydroxy derivative of nervonic acid called oxynervonic acid.

Gangliosides

In 1955, **Klenk** isolated a new type of glycolipid from brain tissue and named it as **ganglioside.** These are found in significant concentrations in ganglion cells of nervous tissue (hence so named) and also in most parenchymatous tissues like spleen and erythrocytes. They make up about 6% of the membrane lipids in the gray matter of the brain. They are also found in lesser amounts in the membranes of most nonneural tissues. Gangliosides are thought to act as receptors for toxic agents like the pathogens, *Vibrio cholerae* influenza virus and tetanus toxin. They are also implicated to play a role in cell-cell interaction.

The structure of gangliosides (Fig. 13–15) is complex and related to that of cerebrosides in that they contain a ceramide (N-acylsphingosine) linked to a carbohydrate (galactose or glucose).

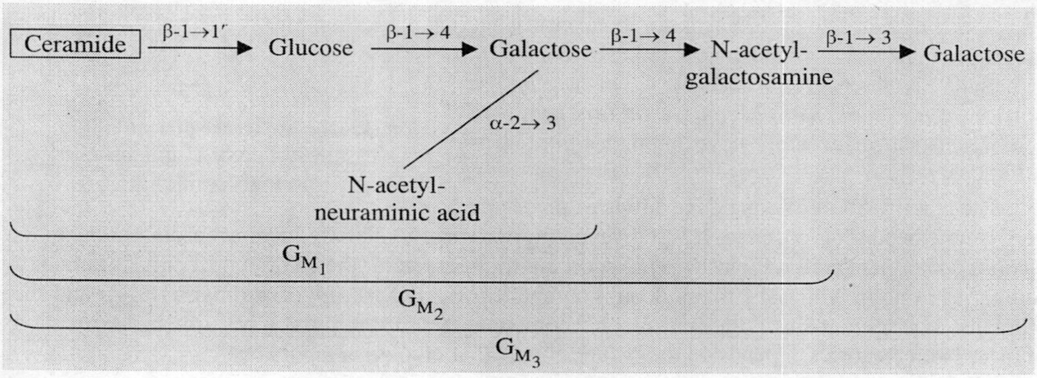

Fig. 13–15. The three gangliosides

In addition to these, the gangliosides also generally contain 2 additional moles of carbohydrates, 1 mole of N-acetylgalactosamine and from 1 to 5 moles of N-acetylneuraminic acid (NANA). In Tay–Sachs disease, the gangliosides are stored in relatively large amounts in the brain and spleen.

CLINICAL IMPLICATIONS

Tay-Sachs disease (named after an English physician **Warren Tay** who first described the disease and an American neurologist **Bernard Sachs** who worked out its details), also called **G_{M2} gangliosidosis,** was first described in 1881. This is the oldest medically described lipid storage disease or lipidosis. It is a common hereditary disorder. Although the disease is rare in the population at large (1 in 3,00,000 births), it has a very high incidence (1 in 3,600 births) in Ashkenazic Jews, who make up about 90% of the Jewish population in America. One in 30 Ashkenazic Jews carries the defective gene in recessive form (*i.e.*, the carrier rate is 1/30), whereas the carrier rate is 1/300 in non-Jewish Americans. Consequently, the incidence of the disease is about 100 times higher in Jewish Americans. When both parents are carrier, their children are likely to develop Tay-Sachs disease even though the parents show no symptoms of the disease. Because the disease is irreversible and needs intensive care, genetic counseling of parents has become important in averting its consequences. The disease is caused due to a deficiency of the enzyme N-*acetylgalactosaminidase* (= *hexosaminidase A*), which cleaves a specific bond (β-8→4) between an N-acetyl-D-galactosamine and a D-galactose residue in the polar head of the ganglioside, G_{M2}. In effect, G_{M2} is not degraded to G_{M1}. With the result, G_{M2}, which is also called as Tay–Sachs ganglioside, is accumulated in large amounts in the lysosomes, particularly in the brain cells to the tune of 100-300 times the normal G_{M2} content, causing degeneration of the nervous system. In Tay–Sachs disease, the symptoms are evident before an affected infant is a year old. Weakness, retardation in development and difficulty in eating are typical early symptoms. Mental retardation and blindness usually follow a few months later; death, between two and five years, is inevitable. Over 90% of the patients have a characteristic cherry-red spot in the retina. Tay–Sachs disease can be diagnosed during fetal development. Amniotic fluid is obtained by amniocentesis and assayed for b-N-acetylgalactosaminidase activity. As mentioned, Tay–sachs disease shares the tragic element of Niemann–Pick and Gaucher diseases in that the affected child seldom lives beyond age three.

. More than 15 different gangliosides have been characterized and their structures determined. These are commonly abbreviated by the letter G, with a subscript $_{M}$, $_{D}$ or $_{T}$ to designate that they contain one, two or three sialic acid and (N-acetylneuraminate or N-glycolyl neuraminate) residues respectively, and a number or letter to distinguish different members of a group from one another. In essence, they consist of an oligosaccharide chain attached to ceramide (N-acylsphingosine) by a sugar residue which is usually glucose. They lack phosphoric acid.

Sulfolipids

A glycolipid that contains sulfur is widely distributed in plants. It is localized in the chloroplasts but is also found in the chromatophores of photosynthetic bacteria. As the sulfur in this compound is present as a sulfonic group in a hexose, this may be included under a class of compounds called sulfolipids. Its structure is given in Fig. 13–16.

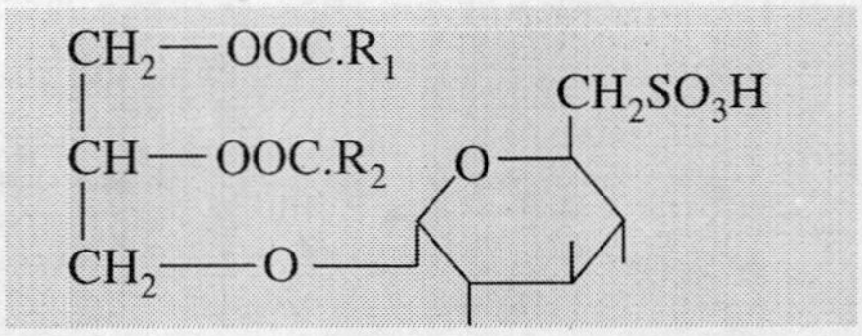

Fig. 13–16. Structure of a sulfolipid (6-sulfo-6-deoxy-α-glucosyl monoglyceride)

Sulfatides

A sulfate ester analogue of phrenosin, abundant in white matter of brain, is another sulfur-containing glycolipid. In it, the sulfate is present in ester linkage at C 3 of the galactose portion of the molecule. Members of this group of cerebroside sulfuric esters have been designated as sulfatides.

The classification of compound lipids as presented here is now becoming obsolete. It is, however, more logical to classify them according to the main alcohol component (refer Table 13–3). According to this system, the compound lipids fall under 2 categories : one in which the main alcohol component is glycerol **(glycerophosphatides)** and the other in which sphingosine is the main alcohol component **(sphingolipids).** The further classification of these two categories depends on the nature of the other alcohol components.

Table 13–3. Classification of compound lipids

Name	*Main alcohol component(s)*	*Other alcohol component(s)**	*N:P ratio*
Glycerophosphatides			
Phosphatidic acids	Diglyceride (= glycerol diester)	—	0 : 1
Lecithins	Diglyceride (= glycerol diester)	Choline	1 : 1
Cephalins	Diglyceride (= glycerol diester)	Ethanolamine, serine	1 : 1
Inositides	Diglyceride (= glycerol diester)	Inositol	0 : 1
Plasmalogens	Monoglyceride (= glycerol ester) + Enol ether	Choline, ethanolamine, serine	1 : 1
Sphingolipids			
Sphingomyelins	N-acylsphingosine	Choline	2 : 1
Cerebrosides	N-acylsphingosine	Galactose, glucose	1 : 0
Sulfatides	N-acylsphingosine	Galactose	($1H_2SO_4$)
Gangliosides	N-acylsphingosine	Galactose, glucose + N-acetylgalactosamine + N-acetylneuraminic acid	No P

* The other alcohol components in all the sphingolipids except sphingomyelins are not present as phosphoric esters, but rather in glycosidic linkages. For this reason, these sphingolipids are grouped under glycolipids.

DERIVED LIPIDS

The group derived lipids is a **"catch all" group** in Bloor's classification. It includes the hydrolysis products of simple and compound lipids and also various other compounds such as steroids, terpenes, fatty acids, alcohols, fatty aldehydes, ketones etc.

STEROIDS

The steroids (*stereos*[G] = solid) are one of the most studied classes of biological compounds and are often found in association with fat. Since they contain no fatty acids, they are *nonsaponifiable*, *i.e.*, cannot be hydrolyzed by heating with alkali to yield soaps of their fatty acid components. Fats, on the other hand, are saponifiable and form soaps when hydrolyzed with alkali. The steroids may be separated from the fat after the latter is saponified since they occur in 'nonsaponifiable residue'. All steroids may be considered as derivatives of a fused and fully saturated ring system called **cyclopentanoperhydrophenanthrene** or **sterane.** This system consists of 3 cyclohexane rings (A, B and C) fused in nonlinear or phenanthrene manner and a terminal cyclopentane ring (D). The sterane nucleus along with the conventional numbering of the carbon atoms is shown in Fig. 13–17.

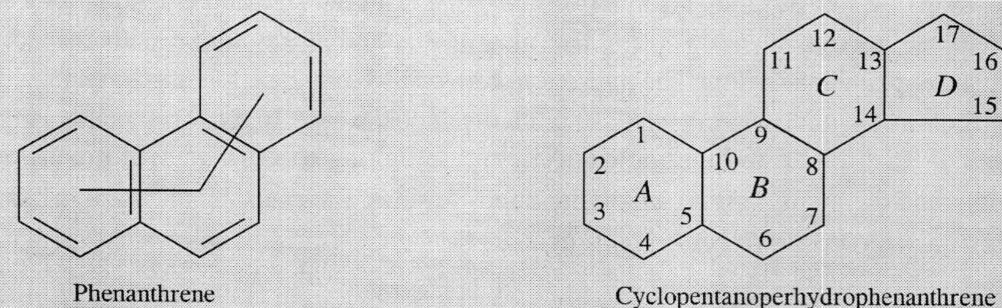

Fig. 13–17. Phenanthrene and sterane nucleus

It may, however, be emphasized that in steroids the hexagonal rings are not the benzene rings as in them the valences of C atoms are fully satisfied by hydrogen bonds, unless shown otherwise. The double bonds, if present, are shown as such. The 'angular' methyl groups occur typically at positions 10 and 13 and constitute carbon atoms 19 and 18 respectively. The steroids may have one or more hydroxyl groups, one OH group being present usually on C 3. A side chain at C 17 is usual. This side chain serves as a convenient basis for classification of steroids. For example, the side chain contains 8, 9 or 10 C atoms in sterols, 5 C atoms in bile acids, 2 C atoms in adrenal cortical steroids and in progesterone and none in naturally occurring estrogens and androgens.

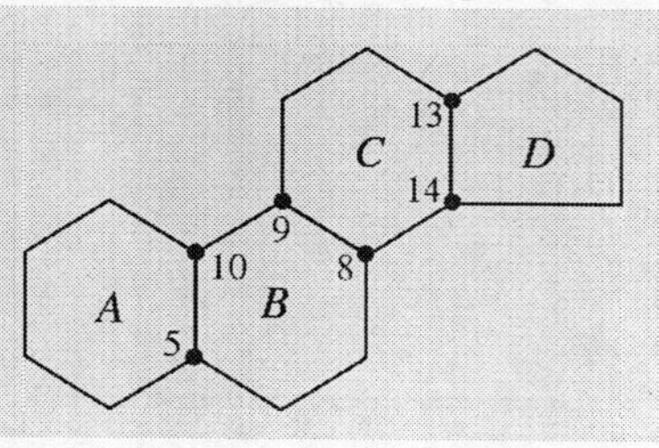

Fig. 13–18. Asymmetric centres in sterane (= gonane) nucleus

(shown by solid black circles)

Stereochemical considerations of steroids. Without even considering the substituent groups, there are 6 possible asymmetric centres present in the steroid nucleus (Fig. 13–18). Consequently, the steroids have many potential stereoisomers. Each of 3 cyclohexane rings, on the basis of the tetrahedral theory, is capable of existing in 3-dimensional conformation either of a **'chair'** or of a **'boat'** (Fig. 13–19). These 2 forms are aplanar and strainless and were first proposed by **Sachse** in 1890. The 'chair' form is *rigid* (in the

> Steroids containing hydroxyl groups (hydroxy steroids) are often referred to as **sterols**. Broadly speaking, the various sterols fall under 2 groups based on their source : the animal sterols or *zoosterols* (cholesterol, cholestanol, coprostanol) and the plant sterols or *phytosterols* (ergosterol, stigmasterol, spinasterol). A third group of sterols, which are obtained from fungi including yeasts, is sometimes separately referred to as *mycosterols*. This classification, however, is not rigid as some sterols are obtained from more than one of these sources.

sense that it resists distortion) and when it is changed to the 'boat' form, some angular deformity invariably takes place. The 'chair' conformation of cyclohexane itself is more *stable* than the 'boat' conformation. The 'chair' and 'boat' forms are both free from angle strain but because of differences in steric strain and bond opposition strain, the two forms differ in energy content (Finar, 1975).

Fig. 13–19. Two types of conformation of a cyclohexane ring

In naturally occurring steroids, all the rings are in 'chair' form. With respect to each other, the rings can be either *-cis* or *-trans* (Fig. 13–20). The junction between the rings A and B (A/B junction) in naturally-occurring steroids may be either *-cis* or *-trans*. But the B/C and C/D junctions are, with certain exceptions, always in *-trans* form. The idea of *-cis* and *-trans* forms with respect to A/B junction may be explained as below. The hydrogen atom on C 5 may be either on the same side of the plane of molecule as the methyl group on C 10 or on the opposite side. In the former case, rings A and B will be *-cis* to each other and the molecule is said to belong to ***normal*** **configuration.** If the hydrogen atom and the methyl group are on the opposite sides, the rings A and B are *-trans* to each other and the molecule is of ***allo*** **configuration.**

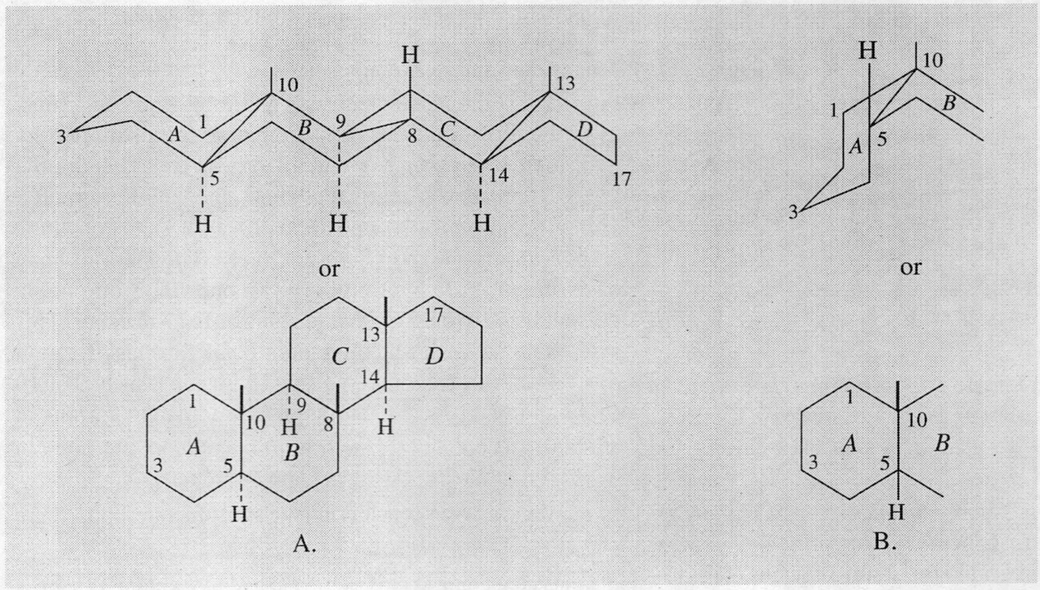

Fig. 13–20. Stereochemical aspects of steroid nucleus, A. all - *trans* configuration between adjacent rings, B. a -*cis* configuration between rings A and B

In the structural representation of the steroids, substituents that are above the plane of the ring (or **β-oriented**) are shown with solid lines, while those which are below the plane of the ring (or **α-oriented**) are represented with broken lines. The methyl groups at C 10 and C 13, the hydroxyl group on C 3 and the side chain at C 17 are all β-oriented.

Certain common steroids are described below.

C 29, C 28 and C 27 STEROIDS

1. Cholesterol (*chole*[G] = bile). Cholesterol is undoubtedly the most publicized lipid in nature, because of the strong correlation between high levels of cholesterol in the blood and the incidence of diseases of the cardiovascular system in humans. It is not only an important component of some cell membranes and of plasma lipoproteins but also the precursor of many other biologically important steroids, such as bile acids and various steroid hormones. It is the principal sterol of higher animals and is especially abundant in nerve tissues and in gallstones. It occurs either free or as fatty esters in all animal cells. It was first isolated in 1784, from human gallstones which consist almost entirely of cholesterol and hence so named (cholesterol literally means 'solid alcohol from bile'). Human body contains 1/4 lb of cholesterol ! Its main sources are fish liver oils and the brain and spinal cord of cattle. White matter contains as much as 14%, gray matter 5%, spinal cord 12% and liver about 1% cholesterol. *Cholesterol is , however, not found in plant fats.* Its parent hydrocarbon is cholestane, $C_{27}H_{48}$. The structure of cholesterol was determined by the German chemist, Adolph Windaus (LT, 1879 – 1959), who received 1928 Nobel Prize in Chemistry. Cholesterol (Fig. 13–21)

> **Cholesterol** is generally believed to be notorious as a major cause of heart disease. There are 2 types of cholesterol, the low-density lipoprotein cholesterol. (LDL-C) and the high-density lipoprotein cholesterol (HDL-C). The LDL-C grows with intake of eggs and dairy products and triggers heart trouble. However, in a study conducted by **Dr. Ian Craig** of the Goteborg University, Sweden in the early 70s showed that patients with heart trouble such as myocardial infarction had low levels of HDL-C, as compared to other people. He pointed out that most of the preventive heart attack studies have so far been aimed at lowering the LDL-C levels through drugs and diet, but it seemed that normal variations in diet made little difference in HDL-C levels. Also, HDL-C levels were generally found higher in women than men. Perhaps this is the reason why fewer women than men have heart attacks. There are, thus, hopeful signs that incidence of heart disease may be lowered by increasing the HDL-C level in heart patients.

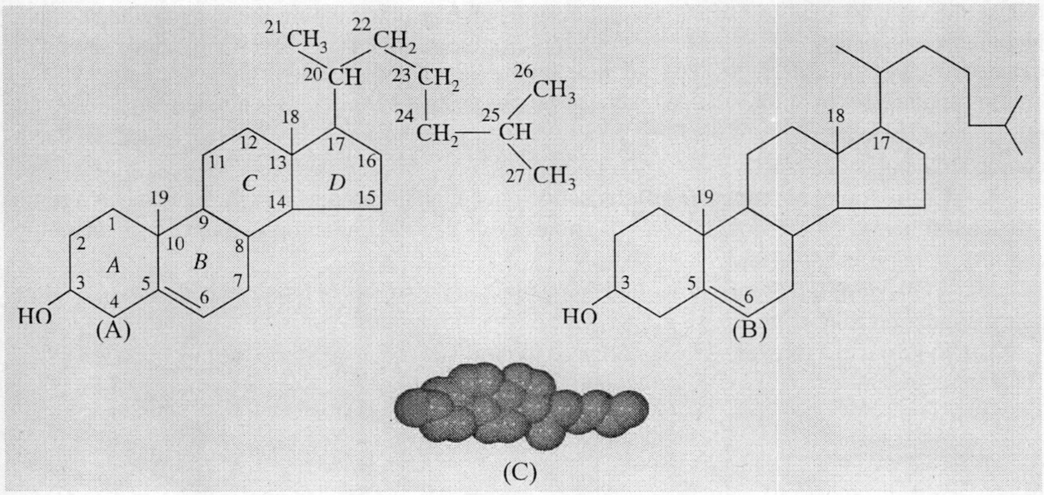

Fig. 13–21. Structure of cholesterol
(3-hydroxy-5, 6-cholestane)

(A) : **Molecular representation.** The individual rings in the fused-ring system are designated A through D.

(B) : **Simplified version.** It is customary to write steroid structure in the manner shown here, without, however, the methyl substituents.

(C) : **Space-filling model.** The C is shown with *grey*, vinyl C with *yellow*, O with *red*, N with *blue* and P with *green*. H is not shown.

[*Courtesy : (C) Richard Pastor and Richard Venable*]

has a molecular formula, $C_{27}H_{45}OH$. In addition to an OH group at C3, there is a double bond at C5. The hydroxyl group constitutes its polar head, the rest of the molecule is hydrophobic. It is a white crystalline solid and is optically active, $[\alpha]_D$ 39°. The crystals are rhombic plates with one of the angles broken. It has a melting point of 149°C. **Konrad S. Bloch**, a Germany-born American biochemist, elucidated the biosynthetic pathway of cholesterol synthesis, one of the most complex known, for which he received the coveted Nobel Prize in Chemistry along with **Feodor Lynen**, a German and **John Cornforth**, a Briton, in 1964. **Bloch** combined work and pleasure by traveling to Bimini for collection of shark livers.

On receiving the Nobel Prize in 1985 for their work on cholesterol, **Michael Stuart Brown** and **Joseph Goldstein** recounted in their lecture the extraordianry history of cholesterol :

> "Cholesterol is the most highly decorated small molecule in biology. **Thirteen Nobel Prizes** have been awarded to scientists who devoted major parts of their careers to cholesterol. Ever since it was isolated from gallstones in 1784, cholesterol has exerted an almost hypnotic fascination for scientists from the most diverse areas of science and medicine.... Cholesterol is a Janus-faced molecule. The very property that makes it useful in cell membranes, namely its absolute insolubility in water, also makes it lethal."

(Courtesy : The Nobel Foundation, 1985)

No vegetable oil contains any cholesterol; hence, it is misleading for oil manufacturers to advertise any oil as 'low cholesterol' oil. Saturated fats of animals origin (such as butter, ghee and lard) do contain cholesterol but *vanaspati,* which is derived from vegetable oils, does not contain cholesterol. Only a little portion of the body cholesterol is derived from diet. The bulk of it is synthesized in the body. Unsaturated fatty acids lower blood cholesterol levels by controlling cholesterol's synthesis as well as its elimination from the body. Thus, vegetable oils when consumed within the recommended amounts help to maintain proper levels of lipids in the blood than do the saturated fats like butter, ghee, lard and vanaspati. These saturated fats elevate blood lipids (cholesterol and triglycerides) and reduce the ratio of good to bad cholesterol even when consumed within the recommended limits. High intake of dietary cholesterol can probably raise blood cholesterol levels and hence should be avoided, esp., by the elderly people and those having high blood cholesterol.

In blood about two-thirds of the cholesterol is esterified mainly to unsaturated fatty acids, the remaining portion occurring as the free alcohol. Reduction of the double bond gives rise to 2 products, coprostanol and cholestanol.

2. Coprostanol (= coprosterol). It occurs in feces and is produced in the intestine as a result of bacterial action on the double bond of cholesterol. The A/B junction is *-cis* in coprostanol in contrast to *-trans* in cholesterol.

3. Cholestanol. It occurs as a minor constituent of the sterols of blood and other tissues. Here the A/B junction is in *-trans* form.

4. Ergosterol. It is present in ergot (hence its nomenclature), yeast and the mould *Neurospora*.

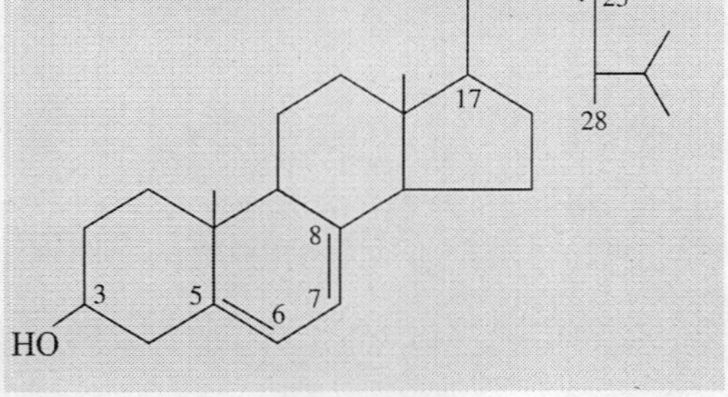

Fig. 13–22. Structure of ergosterol

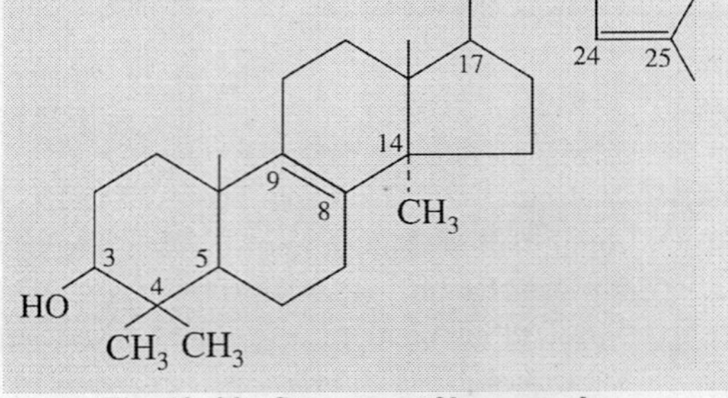

Fig. 13–23. Structure of lanosterol

Its parent hydrocarbon is ergostane, $C_{28}H_{50}$. Ergosterol (Fig. 13–22) has a molecular formula, $C_{28}H_{43}OH$ with one OH group at C_3 and 3 double bonds at C_5, C_7 and C_{22}. It is also optically active, $[\alpha]_D$ -135°.

Rupture of the ring B by UV radiation produces vitamin D_2 which is chemically known as ergocalciferol. A similar compound cholecalciferol (or vitamin D_3) is, however, obtained from 7-dehydrocholesterol on irradiation with UV light (refer Chapter 21).

5. **Lanosterol (= cryptosterol).** It is a major constituent of wool fat and is also present in minor quantities in liver and yeast. Lanosterol (Fig. 13–23) is a C 30 compound with twin methyl groups at C 4 and a third angular methyl group on C 14. There are 2 double bonds at C_8 and C_{24}. It is an intermediate in the biosynthesis of cholesterol.

A number of sterols have also been obtained from various plant and animal sources, for example **stigmasterol** (from soybean and wheat germ oils), **spinasterol** (spinach and cabbage), **sitosterol** (many higher plants), **ostreasterol** (oysters) and **chondrillasterol** (marine sponges).

C 24 STEROIDS or BILE ACIDS

The bile acids (Table 13–4) are the important end products of cholesterol metabolism in higher plants. About 20 natural bile acids have been characterized. All these are derived from a C 24 parent steroid, cholanic acid and resemble coprostanol in having rings A and B in cis form. The most abundant bile acids in human bile are cholic acid (25-60% of the total bile acids), chenodeoxycholic acid (30-50%) and deoxycholic acid (5-25%). Various bile acids differ from each other in the number and position of OH groups which are all in a configuration. In the bile acids, the number of OH group(s) may be 1, 2, or 3 and the position of OH group(s) may be any of the following : 3, 6, 7, 11, 12 and 23. The side chain is usually made up of 5 carbon atoms and bears the carboxyl group.

Table 13–4. Some important natural bile acids

Name	*Formula*	*Melting point, 0°C*	*Position of OH group(s)*	*Source*
Monohydroxycholanic acid	$C_{23}H_{38}(OH).COOH$			
Lithocholic acid		186	3	Man, Ox
Dihydroxycholanic acids	$C_{23}H_{37}(OH)_2.COOH$			
Deoxycholic acid		172	3, 12	Man, Ox
Chenodeoxycholic acid		140	3, 7	Man, Ox, Hen
α-hydrodeoxycholic acid		197	3, 6	Pig
Trihydroxycholanic acid	$C_{23}H_{36}(OH)_3.COOH$			
Cholanic acid		195	3, 7, 12	Man, Ox

The structure of two principal bile acids is given in Fig. 13–24.

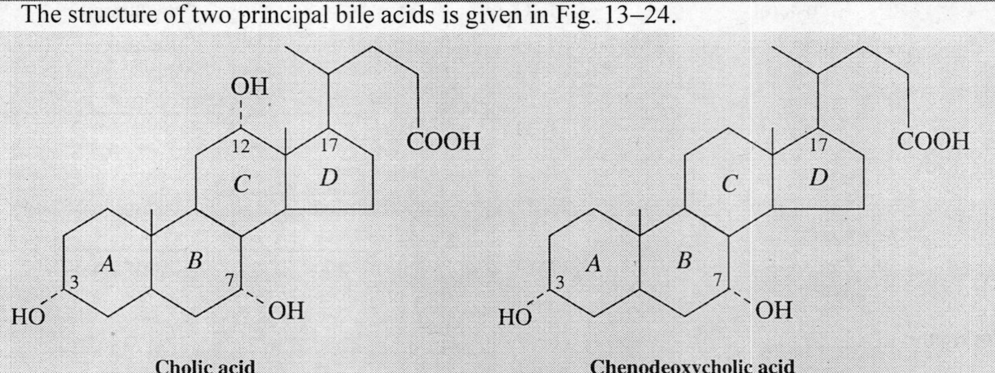

Cholic acid
(3α, 7α, 12α-trihydroxycholanic acid)

Chenodeoxycholic acid
(3α, 7α,-dihydroxycholanic acid)

Fig. 13–24. Two principal bile acids

C 21, C 19 and C 18 STEROIDS

On account of their hormonal role, these have been described in Chapter 19.

Several important poisons are based on the steroid structure. **Digitoxin** or **digitalin** (Fig. 13-25 *a*, *b*) is present in the foxglove plant (*Digitalis purpurea*) and has a powerful heart stimulating action. Another cardiac glycoside, **quabain**, (Fig. 13-25 *c*) is isolated from the East African quabio tree and has the sugar, rhamnose, attached to a modified sterol nucleus. It is of interest to note that it is a powerful inhibitor of the 'sodium pump', a device which normally ensures that the cell content of potassium is higher and that of sodium is lower than in the circumambient fluid.

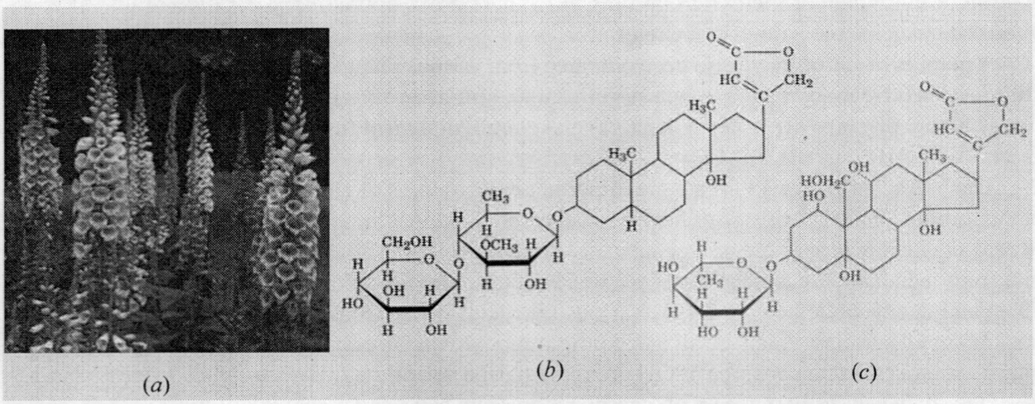

Fig. 13-25. The foxglove plant and two most-commonly prescribed cardiac drugs

(*a*) **The foxglove plant, *Digitalis purpurea*.** The leaves of purple foxglove plant are the source of the popular heart muscle stimulant, digitalis.

(*b*) **Digitoxin or digitalin .** It is extracted from the dried leaves of foxglove plant and is the major component of digitalis, one of the most-widely used cardiac muscle stimulant.

(*c*) **Quabain**. It is a cardiac glycoside, isolated from the East African Quabio tree.

Table 13–5 lists the occurrence and biological role of some important steroids.

Table 13–5. Some steroids and their biological role

Group	*Representative*	*Formula*	*Occurrence*	*Biological role*
Sterols				
C_{30}	Lanosterol	$C_{30}H_{51}O$	Wool fat	Biosynthesis of cholesterol
C_{28}	Ergosterol	$C_{28}H_{44}O$	Yeast	Provitamin D
C_{27}	Cholesterol	$C_{27}H_{46}O$	Animal fats	Structural component and precursor
	7-dehydrocholesterol	$C_{27}H_{44}O$	Skin	Provitamin D
Bile acids				
C_{24}	Cholic acid	$C_{24}H_{40}O_5$	Intestine and gallbladder	Absorption of fats
	Chenodeoxycholic acid	$C_{24}H_{40}O_4$	Intestine and gallbladder	Absorption of fats
Hormones				
C_{21}	Progesterone	$C_{21}H_{30}O_2$	Corpus luteum	Hormone

	Deoxycorticosterone	$C_{21}H_{30}O_3$	Adrenal cortex	Hormone
	Cortisol	$C_{21}H_{30}O_5$	Adrenal cortex	Hormone
C_{19}	Testosterone	$C_{19}H_{28}O_2$	Testes	Hormone
C_{18}	Estradiol	$C_{18}H_{24}O_2$	Ovaries	Hormone

TERPENES

Among the nonsaponifiable lipids found in plants are many hydrocarbons known as terpenes (from *turpentine*). In general, these hydrocarbons and their oxygenated derivatives have lesser than 40 carbon atoms. The simplest terpenes are called *monoterpenes* and conform to the formula $C_{10}H_{16}$ (equivalent to 2 isoprene units, Fig. 13–26), those with the formula $C_{15}H_{24}$ are called as *sesquiterpenes,* with $C_{20}H_{32}$ as *diterpenes* and with $C_{30}H_{48}$ as *triterpenes*. Terpenes with 40 carbon atoms (or *tetraterpenes*) include compounds called carotenoids (refer Chapter 21).

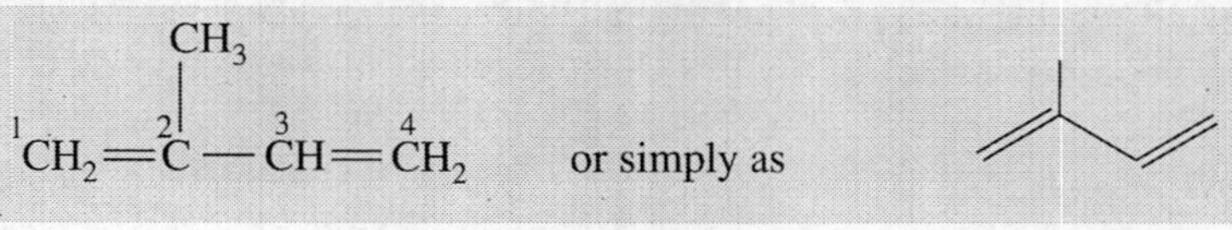

Fig. 13–26. Structure of isoprene
(2-methyl-1,3-butadiene)

Monoterpenes and Sesquiterpenes

The fragrances of many plants arise from volatile C_{10} and C_{15} terpenes. These and their oxygenated derivatives occur as components of the essential oils, some of which are used as perfumes. Important monoterpenes (Fig. 13–27) are *myrcene* (from oil of bay), *geraniol* (from rose oil), *limonene* (from lemon oil) and *menthol* (from peppermint oil).

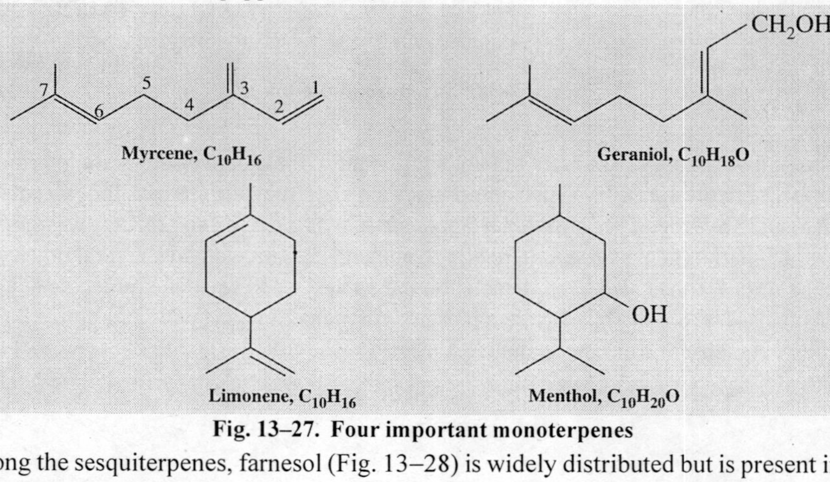

Fig. 13–27. Four important monoterpenes

Among the sesquiterpenes, farnesol (Fig. 13–28) is widely distributed but is present in essential oils in small amounts.

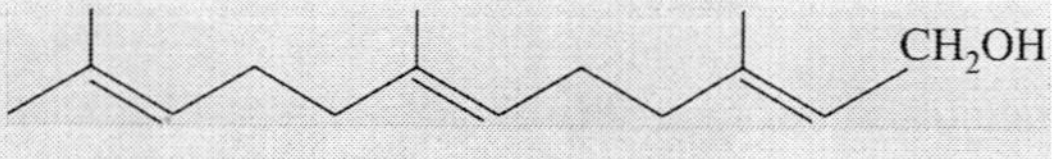

Fig. 13–28. Structure of farnesol

Diterpenes

These are usually found as substituents of the resins and balsams. The two resin acids, **abietic** and **sapietic** (Fig. 13–29) are the best known tricyclic diterpenes.

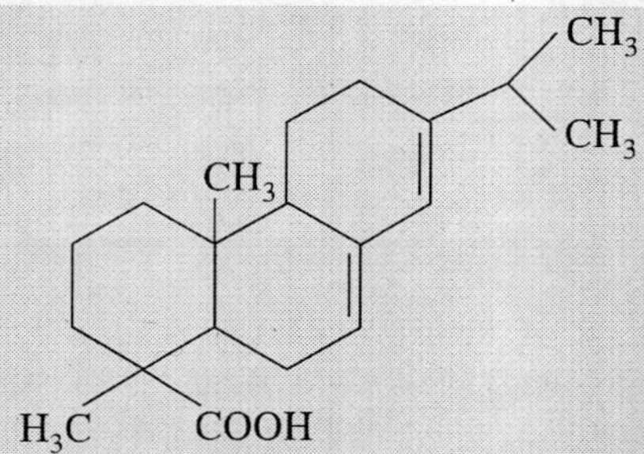

Abietic acid

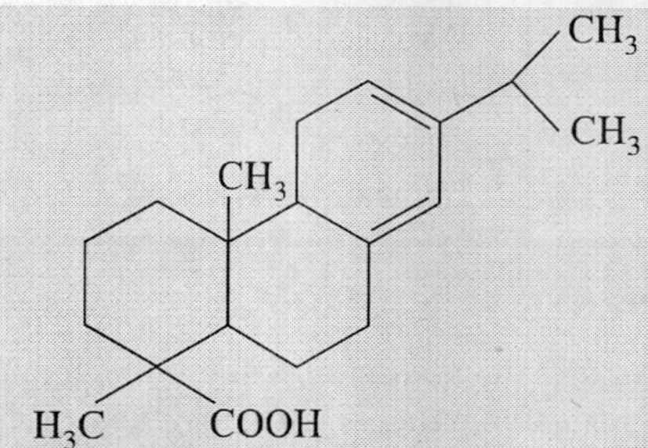

Sapietic acid

Fig. 13–29. Two best known diterpenes

Vitamins A_1 and A_2 and their aldehydes called retinenes are important monocyclic derivatives of diterpenes.

Phytol ($C_{20}H_{40}O$; b.p. 145°C) is an acyclic diterpene and is obtained from the hydrolysis of chlorophyll. It was isolated from nettles by **Paul Karrer** *et al* in 1943. The phytol molecule (Fig. 13–30) has 2 chiral (= asymmetric) centres, 7 and 11. Natural phytol is very weakly dextrorotatory.

> **Paul Karrer,** a Swiss, received his doctorate in 1911 from the University of Zurich. He was awarded **1937 Nobel Prize in Chemistry** for his researches into the constitution of carotenoids, flavins and vitamins A and B, along with Walter Norman Haworth of England. Karrer's contributions include: synthesis of squalene, structure and synthesis of α and β carotenes, and synthesis of riboflavin, α-tocopherol and perhydrovitamin A.

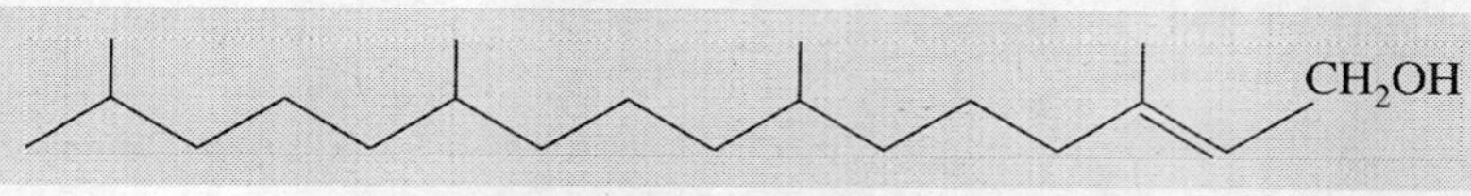

Fig. 13–30. Structure of phytol

Triterpenes

These are not widespread in nature but are significant in that some of them are intermediates in the biosynthesis of cholesterol. Two such compounds are : a tetracyclic alcohol, **lanosterol** (see page 223) and an acyclic hydrocarbon, **squalene**. Squlaene (C_{30} H_{50} ; b.p. 240–242°C) was first isolated from the liver of sharks (genus *Squalus*), hence so named. The olive oil and several other vegetable oils are other sources. It has also been detected in the leaves. Each squalene molecule (Fig. 13–31) has 6 double bonds. The double-bond system present is of nonconjugated type. The conjugated double bonds are, however, absent from the molecule.

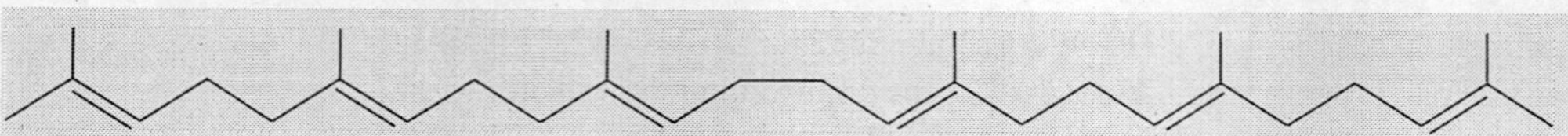

Fig. 13–31. Structure of squalene

Polyterpene

Mention, however, should also be made of **rubber** (MW = ca 3,00,000), a polyterpene present in the latex of many tropical plants. A molecule of rubber (Fig. 13–32) is composed of about 500 to 5,000 isoprene units joined in a long straight chain. When the bark of the rubber tree is cut, latex slowly exudes from the cut. Addition of acetic acid coagulates the rubber, which is then separated from the liquor and either processed into blocks or rolled into sheets, and finally dried in a current of warm air, or smoked.

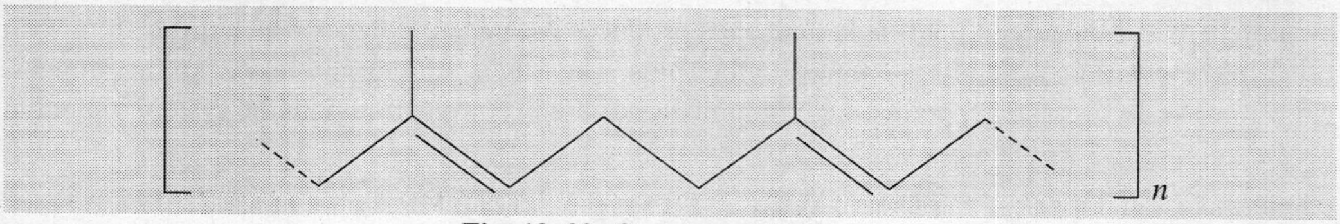

Fig. 13–32. Structure of rubber

(*cis* configuration about double bonds)

CAROTENOIDS

The carotenoids are tetraterpenes. These are widely distributed in both the plant and animal kingdoms but are exclusively of plant origin. These occur in unsaponifiable residue of plant and animal lipids. They are isoprene derivatives with a high degree of unsaturation. Because of the presence of many conjugated double bonds, they are coloured red or yellow. As an example, the pigment of tomato (*lycopene*) and that of carrot (α- and β-*carotene*) are red while many oxygen-containing carotenoids are yellow (*xanthophylls*). Since the olefinic (= double) bonds permit *cis-trans* isomerism, numerous forms are possible. *Most carotenoids, however, exist in all-trans form.* The presence of long hydrocarbon chain in carotenoids makes them lipid-soluble; they are hence also called *lipochromes* or *chromolipids*.

Lycopene

Lycopene ($C_{40}H_{56}$) is the main pigment of tomato, paprika and many other fruits. It is a highly unsaturated, unbranched, long chain hydrocarbon (a polyene) and is composed of two identical units ($C_{20}H_{28}$), joined by a double bond between C_{15} and $C_{15'}$. Each of these units may, in turn, be considered to have been derived from 4 isoprene units (C_5H_8). A molecule of lycopene (Fig. 13–33) in all contains as many as 13 double bonds, of which 11 are conjugated.

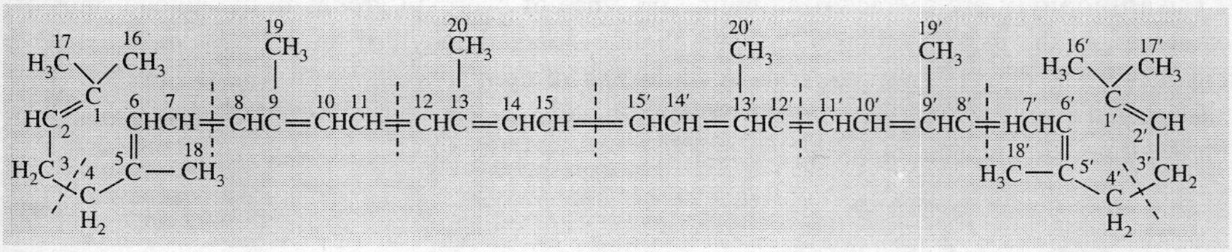

Fig. 13–33. Structure of lycopene

(Dotted lines mark the positions where various isoprene-like sements are joined with each other.)

Carotene

Another group of naturally occurring carotenoids, with the same molecular formula as that of lycopene, is **carotene**. Carotene was first isolated by **Wackenroder** (1831) from carrots (this was the origin of the name *carotin*, which was later on modified to *carotene*). The 3 types of carotenes are :

1. α-carotene– violet crystals ; m.p. 187°C ; optically active (dextrorotatory).
2. β-carotone– red crystals ; m.p. 183°C ; optically inactive.
3. γ-carotene– dark red crystals ; m.p. 152–154°C; optically inactive.

(The prefixes α, β and γ in the carotenes were assigned not on the basis of structural properties but rather as a result of observations of their elution patterns on a $CaCO_3$ chromatographic column, eluted with petroleum ether).

The carotenes are obtained commercially by chromatography. The two best sources are carrots and alfalfa. Their relative proportion, however, varies with the source; for example carrots contain 15% α-, 85% β- and 0.1% γ-form. These are unstable to air, heat, acids and alkalies.

It is to be noted that the ends of the lycopene chain can easily close up to form rings, with the disappearance of the double bonds. Ring closure on only one end results in the production of γ-carotene while closure of the ring at both ends produces β- and α-carotenes which differ from each other in the position of the double bonds in the rings. The ring is structurally related to the ionones. (Fig. 13–34). In β-carotene, both annular double bonds are in conjugation with the system of double

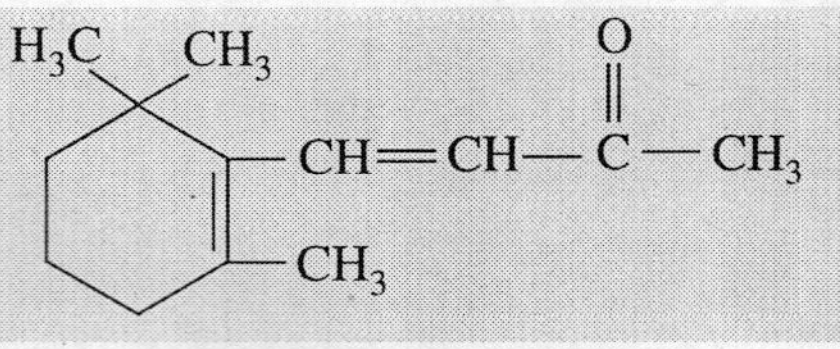

Fig. 13–34. Structure of β-ionone

bonds of the long chain (β-ionone structure). In α-carotene, one of the annular double bonds is removed from the system of conjugation by one position (α-ionone structure). Of particular interest is β-carotene (Fig. 13–35), which is precursor of vitamin A.

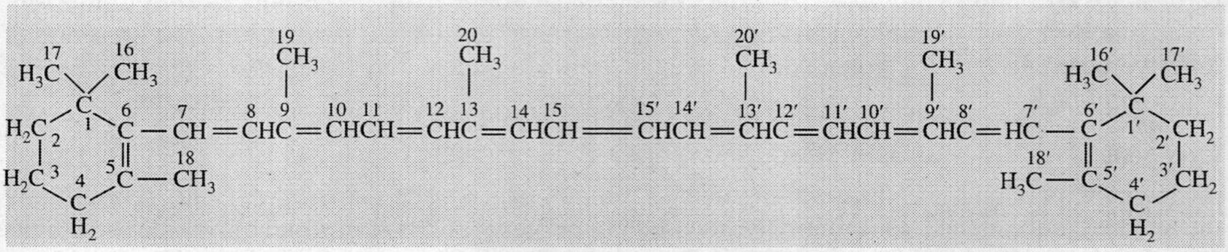

Fig. 13–35. Structure of β-carotene

Xanthophylls

Xanthophylls are characterized by the presence of hydroxyl groups in the ionone rings of carotenes, in the para position to the long chain. The xanthophyll of the leaf (*lutein*, $C_{40}H_{56}O_2$, Fig. 13–36) is derived from α-carotene while that of corn (*zeaxanthine*) from β-carotene. The xanthophyll of crustaceans (*astaxanthine*) is, however, more rich in oxygen. Astaxanthine is responsible for the appetizing redness of boiled lobsters.

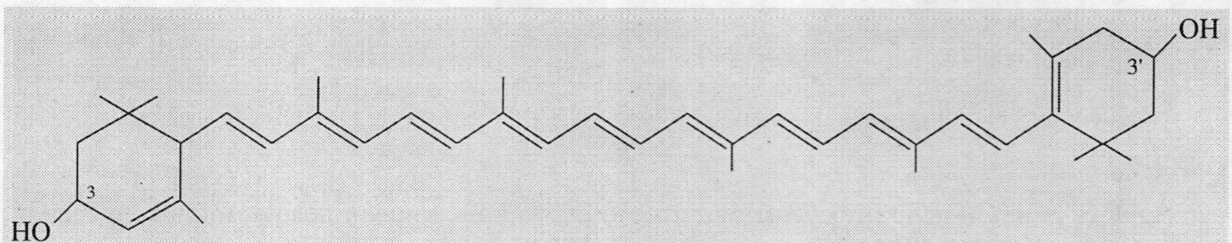

Fig. 13–36. Structure of lutein

(3, 3′-dihydroxy-α-carotene)

The classification of lipids described above may be summarized as depicted on page 229.

LIPIDS

- SIMPLE LIPIDS or HOMOLIPIDS
  - **Fats and oils** (Triglycerides)
    - Simple triglycerides
    - Mixed triglycerides
  - **Waxes**
    - Sperm whale wax
    - Beeswax
    - Carnauba wax
- COMPOUND LIPIDS or HETEROLIPIDS
  - **Phospholipids** (Phosphatids)
    - Phosphoglycerides
      - Lecithins
      - Cephalins
      - Plasmalogens
    - Phosphoinositides (Phosphatidyl inositols)
    - Phosphosphingosides (Sphingomyelins)
  - **Glycolipids** (Cerebrosides)
    - Kerasin
    - Phrenosin
    - Nervon
    - Oxynervon
- DERIVED LIPIDS
  - **Steroids**
    - C 29, C 28, C 27 steroids
    - C 24 steroids
    - C 21, C 19, C 18 steroids
  - **Terpenes**
    - Monoterpenes
    - Sesquiterpenes
    - Diterpenes
    - Triterpenes
    - Tetraterpenes
    - Polyterpenes
  - **Carotenoids**
    - Lycopene
    - Carotenes
    - Xanthophylls

PROBLEMS

1. A common structural feature of membrane lipid molecules is their amphipathic nature. For example, in phosphatidylcholine, the two fatty acid chains are hydrophobic and the phosphocholine head group is hydrophilic. For each of the following membrane lipids, name the components that serve as the hydrophobic and hydrophilic units :
   (*a*) phosphatidylethanolamine
   (*b*) sphingomyelin
   (*c*) galactosylcerebroside
   (*d*) cholesterol
2. Which of the following could not be classified as a lipid ?
   (*a*) steroids
   (*b*) waxes
   (*c*) photosynthetic pigments
   (*d*) anthocyanin pigments
3. What food has the most lycopene ?
4. At pH 7.0, what is the electrical charge of :
   (a) α-lecithin
   (b) α-lysolecithin
   (c) phosphatidyl ethanolamine
   (d) phosphatidyl serine
   (e) phosphatidal choline
   (f) phosphatidal ethanolamine
   (g) phosphatidal serine
   (h) sphingomyelin
5. Do tomato products lose their lycopene while being stored in the pantry ?
6. How sphingomyelin can be differentiated from a diacylglyceride ?
7. Is a glass of canned tomato juice equal to a serving of vegetables ?

ANSWERS TO PROBLEMS

1. *Hydrophobic units*: **(*a*)** 2 fatty acids; **(*b*)**, **(*c*)**, and **(*d*)** 1 fatty acid and the hydrocarbon chain of sphingosine
   Hydrophilic units: **(*a*)** phosphoethanolamine; **(*b*)** phosphocholine; **(*c*)** D-galactose; **(*d*)** several sugar molecules
2. **(*d*)** anthocyanin pigments
3. Contrary to popular belief, watermelon has 40% more lycopene than raw tomatoes. One-and-a-half cup of watermelon contains 9-13 mg of lycopene. If the tomatoes are cooked, their lycopene content is similar to that of watermelon.
4. (*a*) 0 (*b*) 0 (*c*) 0 (*d*) –1 (*e*) 0 (*f*) 0 (*g*) –1 (*h*) 0
5. Even after a year, tomato juice, spaghetti sauce and other tomato products maintain the same level of lycopene content.
6. The differences between a sphingomyelin and a diacylglyceride are given in tabulated form, as shown below :

Sphingomyelin	*Diacylglyceride*
1. This is a compound lipid and belongs to the category of phospholipids.	1. This is a simple lipid and is neutral in nature
2. This is also phosphodiester-like diacylglyceride but the first substituent is not a diglyceride but a sphingosine.	2. Here the glycerol forms ester with fatty acids.
3. Sphingomyelin can exist as a zwitterion over a wide range of pH	3. Zwitterionic form is not possible.

7. Yes. The lycopene in tomato juice helps fight prostate and breast concers. One glass of tomato juice raises blood lycopene by about 45%. The more veggies you eat, the lower your risk of strokes, heart attack and cancer.

CHAPTER

14

Lipids–III

Properties of Fats and Oils

INSIDE

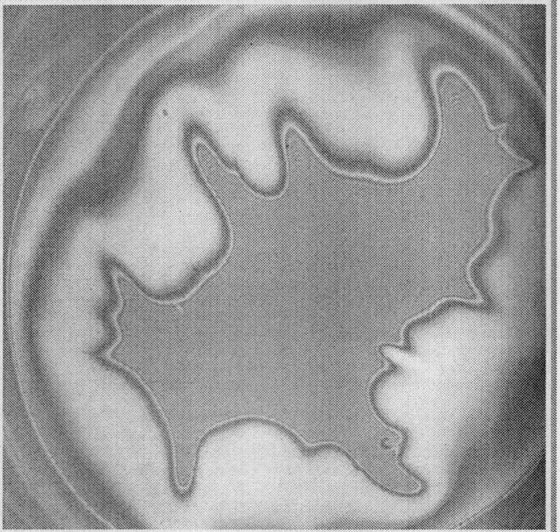

Oil and water do not mix.

The fats and oils owe the manifestation of their properties to the fatty acids and alcohols, the two component units.

PHYSICAL PROPERTIES

1. State. Fats containing saturated fatty acids are solid at ordinary room temperature. The *animal fats* belong to this category. Most *plant fats*, on the contrary, possess unsaturated fatty acids and are, henceforth, liquid at room temperature.

2. Colour, Odour and Taste. When pure, the fats are *colourless,* virtually *odourless* and *possess* an extremely *bland taste*. They are capable of absorbing a variety of odours and hence flavour during storage. For the same reason, a housewife knows that the flavour of an onion quickly permeates butter that is stored with it in a refrigerator. In some cases, however, this absorbing property of fats is of advantage. For example, the perfumes of some flowers can be isolated by placing their petals in contact with the fat for a certain period, then extracting the fat with alcohol and concentrating the essence.

3. Solubility. The fats are, however, only sparingly soluble in water. These are, therefore, described as *hydrophobic* in contrast to the water-soluble or hydrophilic substances like many carbohydrates and proteins.

However, these are freely soluble in organic solvents like chloroform, ether, acetone and benzene. These solvents, as they dissolve fats in them, are also known as '*fat solvents*'. The solubility of the fatty acids in organic solvents, in fact, decreases with the increase of chain length. *The introduction of hydroxyl groups, however, increases solubility.*

4. Melting point. The melting point of fats depends on the chain length of the constituent fatty acid and the degree of unsaturation. Fats containing saturated fatty acids from C 4 to C 8 are liquid at room temperature but those containing C 10 or higher saturated fatty acids are solid and their melting points increase with increasing chain length. With the introduction of double bond in the fat molecule, the melting point lowers considerably. It may be stated, in general, that *greater the degree of unsaturation* (*or higher the number of double bonds*) *of the constituent fatty acid, the lower is the melting point of the fat.* This may be easily visualized in terms of constituent fatty acids from the Table 14–1. In fact, short chain length and unsaturation enhance the fluidity of fatty acids and of their derivatives.

Table 14–1. Melting points of the common fatty acids*

Fatty acid	*No. of carbon atoms*	*No. of double bonds*	*M.P. (in °C)*
Saturated			
Caprylic	8	0	16.0
Capric	10	0	31.0
Lauric	12	0	44.2
Myristic	14	0	53.9
Palmitic	16	0	63.1
Stearic	18	0	69.6
Arachidic	20	0	76.5
Behenic	22	0	79.9
Lignoceric	24	0	86.0
Unsaturated			
Palmitoleic	16	1	11.0
Oleic	18	1	13.4
Linoleic	18	2	– 5.0
Linolenic †	18	3	–11.0
Arachidonic	20	4	– 49.5

* Here mention has been made only of even numbered fatty acids as they alone are one of the two constituents of the saturated fats. If, however, the entire homologous series of the saturated fatty acids is taken into consideration, the fact emerges that this series shows an alternation or oscillation of melting points, *i.e.*, the melting point of an 'even' acid is higher than that of the 'odd' acid immediately below and above it. Thus, the fatty acid series follows what is known as the **saw-tooth rule.** However, in many other homologous series the melting points of the n-members rise continuously, tending towards a maximum value.

† Once known as vitamin F, now the name is discredited.

5. Specific gravity. The specific gravity of the fats is less than 1 (about 0.86) and, therefore, they float on water surface. Solid fats are lighter than the liquid fats. Oils spread on water to form thin monomolecular layers. In general, either unsaturation of the fatty acid chains increase or increase in chain length of the fatty acid residues tend to increase the specific gravity.

6. Geometric isomerism. As stated earlier, the presence of double bond (*s*) in the unsaturated fatty acid part of the fat molecule produces geometric (or *cis-trans*) isomerism.

7. Insulation. The fats possess *high insulating power*, *i.e.*, they are bad conductor of heat. A layer of fat below the skin provides a sort of blanket for warm-blooded animals (or *homoiotherms*). This is especially important for whales and seals which have to maintain a high temperature in cold waters. The fishes are cold-blooded animals (*or poikilotherms*) and, therfore, do not require maintenance of high temperature and so have very little subcutaneous fat.

8. Emulsification. It is the process by which a lipid mass is converted into a number of small lipid droplets. The fats may be emulsified by shaking either with water or with emulsifying agents like soaps, gums, proteins etc. An emulsifying agent helps in the production of a finely divided suspension of a fat in an aqueous medium. The hydrocarbon portions of the two (the emulsifier and the fat) tend to aggregate. This leaves the water-soluble group of the emulsifier projecting into the aqueous phase. A fat droplet will associate with a number of molecules of the emulsifier, thus producing a new water-soluble surface. Water molecules, henceforth, tend to be held in a layer or '*cloud*' around each droplet, thus disallowing the aggregation of the fat droplets (Fig. 14–1).

The process of emulsification is of great metabolic significance. In fact, *the fats have to be emulsified before they can be absorbed by the intestinal wall.* The process is accomplished by the bile juice secreted from liver.

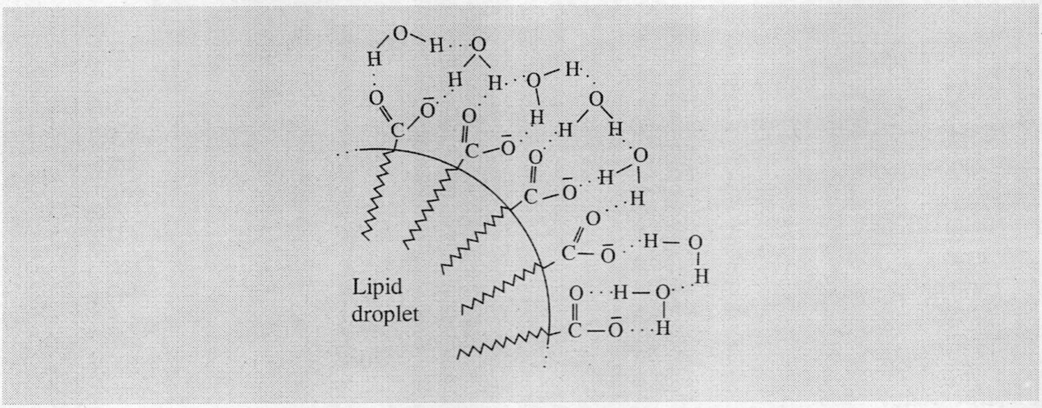

Fig. 14–1. A portion of the surface of an emulsified fat globule

[Here, the emulsifier selected is a simple anion of a fatty acid *i.e.*, a soap. The projecting COOH groups are linked with water molecules by H_2 bonds. Additional H_2 bonding between the water molecules leads to the further stability of the resulting water '*cloud*' around the fat droplet.]

(*Adapted from Fairley JL and Kilgour GL, 1966*)

9. Surface tension. The force with which the surface molecules are held together is called the surface tension. When liquid fat is poured on water, it spreads uniformly over the surface of water in the form of a unimolecular layer and thus reduces the surface tension of water.

CHEMICAL PROPERTIES

The chemical reactions of the fats reflect the reactivities of the ester linkage and the degree of unsaturation of the hydrocarbon chain.

REACTIONS INVOLVING COOH GROUP

1. **Hydrolysis.** The fats are hydrolyzed by the enzymes *lipases* to yield fatty acids and glycerol. The lipases catalyze this reaction at a slightly alkaline *p*H (7.5 – 8.5) in a stepwise manner. The fats first split to produce *diglycerides*, part of these are then split to *monoglycerides*. Finally, part of the monoglycerides split to yield *fatty acid* and *glycerol*.

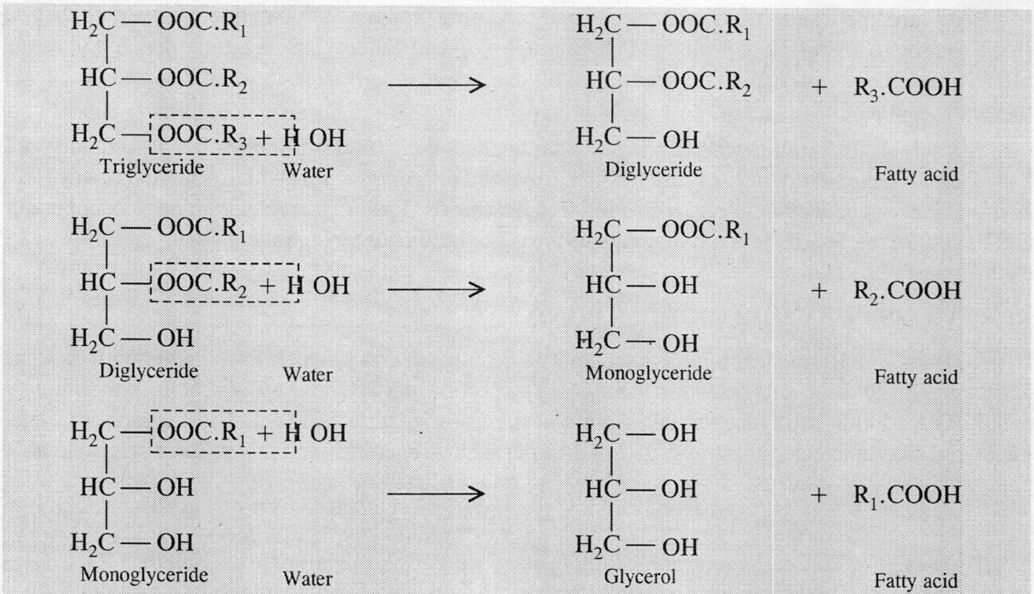

[Radicals R_1, R_2 and R_3 may be similar or dissimilar.]

In the intestine, the absorption of mono-, di-7and triglycerides is so rapid that very little free glycerol is formed.

2. Saponification. *The hydrolysis of fats by alkali is called saponification.* This reaction results in the formation of glycerol and salts of fatty acids which are called *soaps* (Fig. 14-2). The soaps are of two types : hard and soft. *Hard soaps* such as the common bar soaps are the sodium salts of the higher fatty acids. *Soft soaps* are the potassium salts of higher fatty acids and are marketed as semisolids or pastes. The fatty acid salts of calcium, magnesium, zinc and lead are, however, insoluble in water. Calcium soaps are used industrially as lubricating greases. Zinc soaps are employed in the manufacture of talcum powder and other cosmetics. Lead and magnesium soaps are used in paints industry to hasten the process of drying.

> **Acid hydrolysis of a fat,** however, yields the free fatty acid and glycerol.

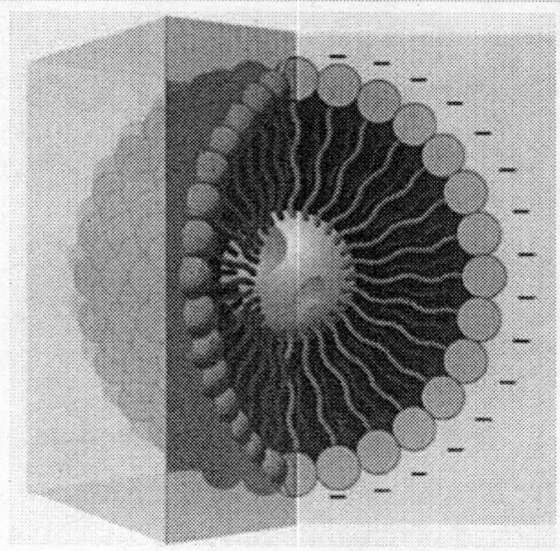

Fig. 14-2. Soaps consist of fatty acids.

In this schematic drawing of a soap micelle, the nonpolar tails of the fatty acids are directed inward, where they interact with the greasy matter to be dissolved. The negatively-charged heads are located at the surface of the micelle, where they interact with the surrounding water. Membrane proteins, which also tend to be insoluble in water, can also be solubilized in this way by extraction of membranes with detergents.

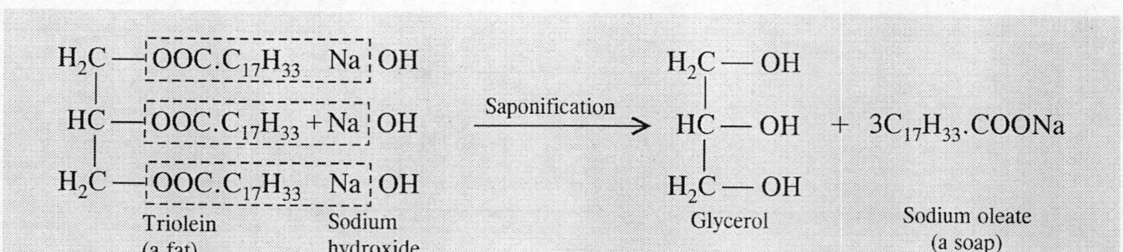

Soaps are important cleansing agents. Their cleansing property is due to their *emulsifying action* (*i.e.*, capacity to render more prolonged the mixing of oil and water). This is accomplished by means of negative charge the soap anion confers on oil droplets. The electrostatic repulsion then prevents the coalescence of soap and oil droplets into an oil phase.

3. Hydrolytic rancidity. When butter or other fats are stored, they often become rancid and hence unpalatable. Rancidity is caused by the growth of microorganisms which secrete enzymes like *lipases*. These split the fats into glycerol and free fatty acids. The fatty acids impart unpleasant odour and flavour to the fat. However, butter may be prevented from becoming rancid by refrigeration or by exclusion of water.

REACTIONS INVOLVING DOUBLE BOND

1. Hydrogenation. Unsaturated fatty acids, either free or combined in lipids, react with gaseous hydrogen to yield the *saturated fatty acids.* The reaction is catalyzed by platinum, palladium or nickel. The addition of hydrogen takes place at the C—C double bond (*s*). Thus, 1 mole of oleic, linoleic or linolenic acid reacts with 1, 2 or 3 moles of hydrogen respectively to form stearic acid.

$$\underset{\text{Oleic acid}}{CH_3(CH_2)_7\overset{10}{C}H{=}\overset{9}{C}H(CH_2)_7COOH} + H_2 \xrightarrow[\text{as catalyst}]{\text{Pt, Pd or Ni}} \underset{\text{Stearic acid}}{CH_3(CH_2)_7\overset{10}{C}H_2{-}\overset{9}{C}H_2(CH_2)_7COOH}$$

This reaction is of great commercial importance since it permits transformation of inexpensive and unsaturated liquid vegetable fats into solid fats. The latter are used in the manufacture of candles, vegetable shortenings and of oleomargarine.

2. Halogenation. Unsaturated fatty acids and their esters can take up halogens like Br_2 and I_2 at their double bond (*s*) at room temperature in acetic acid or methanol solution.

$$\underset{\text{Linoleic acid}}{CH_3(CH_2)_4\overset{13}{C}H{=}\overset{12}{C}HCH_2\overset{10}{C}H{=}\overset{9}{C}H(CH_2)_7COOH} + 2I_2$$

$$\xrightarrow[CH_3.COOH\ /\ CH_3.OH]{\text{Room temperature}} \underset{\text{9, 10, 12, 13-tetraiodostearic acid}}{CH_3(CH_2)_4\underset{I}{\overset{13}{C}H}{-}\underset{I}{\overset{12}{C}H}CH_2\underset{I}{\overset{10}{C}H}{-}\underset{I}{\overset{9}{C}H}(CH_2)_7COOH}$$

This reaction is the basis of the 'iodine number determination'.

3. Oxidation. Unsaturated fatty acids are susceptible to oxidation at their double bonds. Oxidation may be carried with ozone or $KMnO_4$.

(*a*) With ozone – An unstable *ozonide* is formed which later cleaves by water to give rise to 2 *aldehydic groups*.

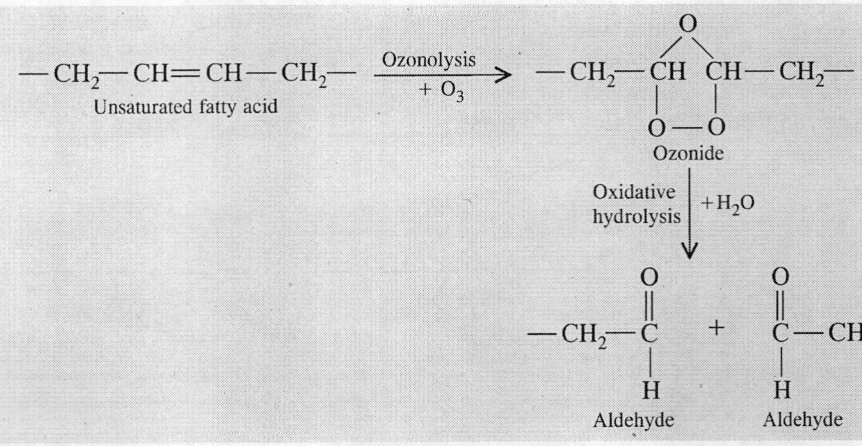

(*b*) With $KMnO_4$ – Under mild conditions, the *glycols* are formed at the sites of double bonds.

$$\underset{\text{Oleic acid}}{CH_3(CH_2)_7\overset{10}{C}H{=}\overset{9}{C}H(CH_2)_7COOH} \xrightarrow{KMnO_4} \underset{\text{9, 10-dihydroxystearic acid}}{CH_3(CH_2)_7\underset{OH}{\overset{10}{C}H}{-}\underset{OH}{\overset{9}{C}H}(CH_2)_7COOH}$$

Under vigorous conditions, the same reagent cleaves the molecule at the double bond and oxidizes the terminal portions to the *carboxyl group*.

$$\underset{\text{Oleic acid}}{CH_3(CH_2)_7\overset{10}{C}H{=}\overset{9}{C}H(CH_2)_7COOH} \xrightarrow{KMnO_4} \underset{\text{Pelargonic acid}}{CH_3(CH_2)_7COOH} + \underset{\text{Azelaic acid}}{HOOC(CH_2)_7COOH}$$

The oxidation reactions have been extensively used in establishing the position of double bond(s) in the fatty acid chain. This gives important clues regarding lipid structure.

4. Oxidative rancidity. Oils containing highly unsaturated fatty acids are spontaneously oxidized by atmospheric oxygen at ordinary temperatures. The oxidation takes place slowly and results in the formation of *short chain fatty acids* (C_4 *to* C_{10}) and *aldehydes* which give a rancid taste and odour to the fats. This type of rancidity or rancidification is called 'oxidative rancidity' and is due to a reaction called '*autoxidation*'. Autoxidation proceeds by a free radical mechanism in which the a- methylene group is primarily attacked. A hydrogen atom is removed from an α-methylene group. This initiates a chain of reactions leading to oxidation (**Holman**, 1954).

Oxidative rancidity is observed more frequently in animal fats than in vegetable fats. This is due to the presence, in the vegetable oils, of natural **'antioxidants'** such as tocopherols (= vitamin E), phenols, naphthols etc., which check autoxidation. Vitamin E is, therefore, some-times added to foods to prevent rancidity. Animal shortenings such as lard are nowadays protected against oxidative rancidity by the addition of synthetic antioxidants such as nordihydroguiaretic acid (NDGA), tertiary butyl hydroxy anisole (BHA) etc. Linseed oil, a plant oil used as a base for paints, is highly rich in unsaturated fatty acids. It undergoes autoxidation when exposed to air, followed by polymerization to a hard, resinous coating as it 'dries' or oxidizes.

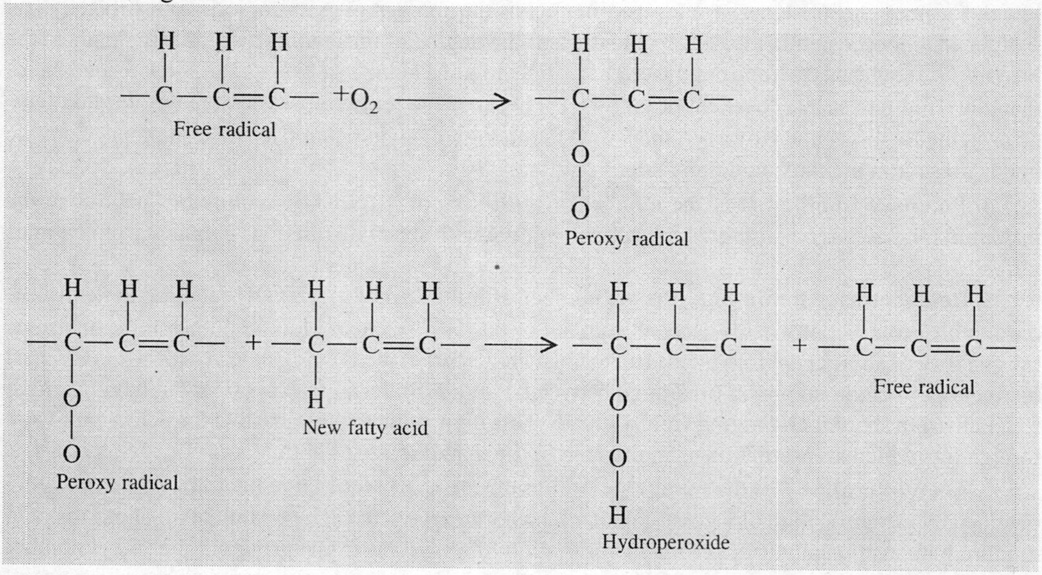

The action of antioxidants is opposed by a group of compounds present in the fats and oils. These accelerate the oxidation of the parent compound and are called *pro-oxidants.* Majority of these substances are formed during the processing and refining of fats. Among the noteworthy pro-oxidants are the copper, iron and nickel salts of organic acids like lactic, etc.

REACTION INVOLVING OH GROUPS

Dehydration (Acrolein test). Fats, when heated in the presence of a dehydrating agent, $NaHSO_4$ or $KHSO_4$ produce an unsaturated aldehyde called acrolein from the glycerol moiety. Acrolein is easily recognized by its pungent odour and thus forms the basis of the test for the presence of glycerol in fat molecule.

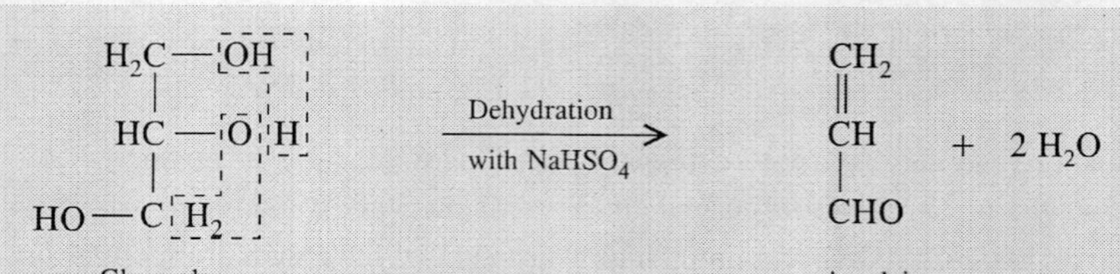

QUANTITATIVE TESTS

The reactions described above give valuable information about the chemical nature of fatty acids and the number of hydroxyl groups present in the fat molecule. Such chemical determinations involve various analytical tests. These are called *chemical constants* and include the following:

1. Acid value. It is the number of milligrams of KOH required to neutralize the free fatty acids present in 1 gm of fat. The acid number, thus, tells us of the quantity of free fatty acid present in a fat. Obviously, a fat which has been both processed and stored properly has a very low acid number.

2. Saponification number. It is the number of milligrams of KOH required to saponify 1 gm of fat. The saponification number, thus, provides information of the average chain length of the fatty acids in the fat. It varies inversely with the chain length of the fatty acids. *The shorter the average chain length of the fatty acids, the higher is the saponification number.*

3. Iodine value (or **Koettstorfer number).** It is the number of grams of iodine absorbed by 100 g of fat. The iodine number is, thus, a measure of the degree of unsaturation of the fatty acids in the fat. Oils like soybean, corn and cottonseed have higher iodine numbers (133, 127 and 109, respectively) than the solid fats such as beef fat or tallow (42) because the former possess more unsaturated fatty acids in the fat molecule. However, the iodine number gives no indication as to the number of double bonds present in the fatty acid molecule.

4. Polenske number. It is the number of millilitres of 0.1N KOH required to neutralize the insoluble fatty acids (*i.e.*, those which are not volatile with steam distillation) obtained from 5 gm of fat.

5. Reichert-Meissl number. It is the number of millilitres of 0.1N KOH required to neutralize the soluble, volatile fatty acids derived from 5 g of fat. The Reichert-Meissl number, thus, measures the quantity of short chain fatty acids (up to C 10 inclusive) in the fat molecule. The Reichert-Meissl numbers of coconut and palm oils range between 5 and 8. Butterfat is exceptional in having a high Reichert-Meissl number, ranging from 17 to 35. This high value makes possible the detection of any foreign fats which are, sometimes, adulterated in the manufacture of butter.

6. Acetyl number. It is the number of milligrams of KOH required to neutralize the acetic acid obtained by saponification of 1 gm of fat after it has been acetylated (The treatment of fat or fatty acid mixture with acetic anhydride results in acetylation of all alcoholic OH groups). The acetyl number is, thus, a measure of the number of OH groups in the fat. For example, the castor oil has a high acetyl number (146) because of high content of a hydroxy acid, ricinoleic acid, in it.

Some examples of the value of these numbers in defining fats are given in Table 14–2.

Table 14–2. Analytical values for some fats and oils

Fat/Oil	*Saponification number (mg KOH/g)*	*Iodine value (g I_2/100 g)*	*Reichert-Meissl number*	*Acetyl number*	*Acid value (mg KOH/g)*
Animal fats					
Halibut liver oil	170-180	120-135	–	–	1.0
Cod liver oil	180-190	140-180	–	–	0.5–5
Human fat	194-198	65-69	0.4	–	–
Lard	195-203	47-66	0.5–0.8	2.6	–
Beef fat	196-200	35-42	–	2.78.6	–
Butter	210-230	25-40	17-35	1.9-8.6	0.5-30
Cow's milk	216-235	26-45	–	–	–
Plant fats					
Castor oil	175-187	80-90	1.4	146-150	0.2-4
Corn oil	187-195	104-128	–	–	1.0-2
Linseed oil	188-195	170-195	1.0	4.0	1.0-4
Olive oil	190-195	80-90	0.6-1.5	10-11	0.2-3
Coconut oil	254-262	7-10	0.6-7.5	2.0	2.5-6

FATS – FACTS AND FANTASIES

Irrespective of the source, fat is the most concentrated form of dietary energy. While sugars (carbohydrates) and proteins provide 4 Calories (kilocalories) of energy per gram, fat of any type provides 9 Calories. *There is no such thing as low-calorie fat*. Thus, by replacing fats of animal origin like `ghee' and butter with vegetable oils like groundnut, sunflower, safflower, mustard, palm oil or any other, weight cannot be reduced. For weight reduction, the only rule is *to eat less and exercise more*. Despite the same Calorie content, the nutritional quality of different fats does differ. As already stated, the fats are composed of an alcohol component (usually glycerol) and a fatty acid component which may be saturated or unsaturated. Nutritionally, polyunsaturated fatty acids (PUFAs) are considered superior. While the human body can synthesize saturated fatty acids (SFAs) and monounsaturated fatty acids (MUFAs), it cannot synthesize the polyunsaturated fatty acids (PUFAs) and hence PUFAs are referred to as essential fatty acids (EFAs). Fats of animal origin (lard, cream, butter and 'ghee') are saturated whereas vegetable oils, except coconut, have varying degree of unsaturation. Fish oils too are unsaturated and have long-chain-length fatty acids.

It is long known that saturated fats raise blood cholesterol level whereas unsaturated fats lower blood cholesterol and also ensure a more favourable ratio of 'good' to 'bad' cholesterol. It is for this reason that the vegetable oils such as groundnut, mustard, rice bran, sunflower, safflower etc (except coconut oil, which is saturated) are regarded to be healthy, rather health-friendly fats. Therefore, these are prescribed for those suffering with high levels of blood cholesterol or those suffering from coronary artery disease.

PROBLEMS

1. Some fats used in cooking, such as olive oil, spoil rapidly upon exposure to air at room temperature, whereas others, such as solid shortening, remain unchanged. Why ?
2. Name the products of mild hydrolysis of the following lipids with dilute NaOH :
   (*a*) 1-stearoyl-2,3-dipalmitoylglycerol
   (*b*) 1-palmitoyl-2-oleoylphosphatidylcholine
3. A common procedure for cleaning the grease trap in a sink is to add a product that contains sodium hydroxide. Explain why this works.
4. Draw all the possible triacylglycerols that you could construct from glycerol, palmitic acid, and oleic acid. Rank them in order of increasing melting point.
5. Rank, in order of increasing solubility in water, a triacylglycerol, a diacylglycerol, and a monoacylglycerol, all containing only palmitic acid.
6. Why do eggs solidify after boiling while 'ghee' melts after heating ?
7. How can exercise make one thinner even though one does not lose any weight ?
8. Which type of diet is the more likely to lead to overconsumption and obesity, one rich in sugar or one rich in fat ?
9. Is it true that you will get rid of more fat if you exercise less strenuously ?
10. Which oils have greatest amount of the desirable monounsaturated fatty acid ?
11. Is margarine better for you than butter ?
12. Why can we blow a bubble with soap water but not with plain water ?
13. Why is it not possible to write on an oily paper ?
14. Why does butter turn rancid if left unrefrigerated for some time ?
15. Why do soap bubbles last longer than water bubbles ?
16. Name two lipids which exist in oil as well as fat form.
17. Why fats are found in the form of globules in cytoplasm of cells ?
18. Why fats float on water surface ?
19. Why fats are synthesized in a cell in which they occur ?
20. Why coconut oil and not mustard oil solidifies during winter season ?
21. Which oil is best for cooking ?
22. Why do different soaps irrespective of their colour produce only white foam ?
23. By which process the vegetable oils can be converted into vegetable 'ghee' (vanaspati 'ghee') ?
24. Milk contains water, sugars, fats and proteins. However, the colour of the milk is white. Which of these components is responsible for the colour ?
25. When a piece of paper is wet with water, it tears off easily. But the same paper if soaked in oil does not lose stength. Why ?

ANSWERS TO PROBLEMS

1. Unsaturated fatty fats (*e.g.*, olive oil) are susceptible to oxidation by molecular oxygen.
2. (*a*) The sodium salts of palmitic and stearic acids, plus glycerol
   (*b*) The sodium salts of palmitic and oleic acids, plus glycerol-3-phosphorylcholine.
3. The triacylglycerols of animal fats (grease) are hydrolyzed by NaOH, in a process known as saponification, to form soaps, which are much more soluble in water than are the triacylglycerols.

4. Eight different triacylglycerols can be constructed. All saturated (palmitic) fatty acids, all unsaturated (oleic) fatty acids, or any combination of oleic and palmitic acids can be used. Furthermore, positional isomers are possible, as the three carbons of glycerol are not equivalent. In order of increasing melting point: OOO, OOP and OPO; PPO and POP; PPP, where O = oleic, P = palmitic.
5. In order of increasing solubility: triacylglycerol; diacylglycerol; monoacylglycerol.
6. Eggs contain rich proteins composed of many amino acids. On heating eggs, their proteins coagulate into a solid mass. 'Ghee' is a mixture of more than one fatty acids, which may be saturated or unsaturated. These fatty acids soften at high temperatures and subsequently melt. In the case of eggs, the reaction is irreversible, but it is reversible in the case of 'ghee'.
7. Exercise gets rid of fat but increases muscle which weighs more. Muscles are denser and take less space than fat.
8. Obesity is the result of caloric intake in excess of body needs. On a per-gram-of-diet basis, a diet rich in fat is more likely to lead to obesity than a diet rich in sugar because the former contains a higher caloric content per gram.
9. No. The amount of fat that you lose depends on the number of calories you burn, not the intensity of the workout.
10. Olive, canola, peanut and sesame - seed oils. Olive oil has the most.
11. A little better, provided you use the tub or liquid type and use it sparingly.
12. Surface tension of water is so strong that it does not allow water to form stable bubbles when blown through a pipe. Soaps and detergents contain a basic cleaning agent called surfactant or surface-active agent. *Surfactants lower the surface tension of water sufficiently* to enable a film of soap solution to be blown as bubbles. So soapy water is used by children to blow bubbles.
13. As paper is made of cellulose, when we write with a fountain pen, the colouring substance in the ink enters the fine capillaries through capillary action. When the solvent evaporates, dry ink remains in the capillaries, thus making the visible mark on the paper. If the surface of the paper is oily, the oil repels the ink which as a result does not stick to the paper to produce a mark. So it is not possible to write on oily paper with a fountain pen.

    In the case of a ballpoint pen, when drawn over the oily surface, the ball on the tip of a ballpoint pen slips and does not rotate as there is no friction. As a result, the thick ink in it does not flow out and does not write.
14. Butter is nothing but solidified milk fat obtained from cream or whole milk by churning. Butter turns rancid if left unrefrigerated for some time because of the formation of butyric acid by the action of oxygen of air or due to bacterial action. Butyric acid is present in butter in the form of glycerol ester. This ester breaks down into butyric acid and butyl alcohol by the action of oxygen of air or due to bacterial action when kept at room temperature for a long time. It is the butyric acid that imparts the strong unpleasant smell to rancid butter. Refrigeration reduces the bacterial action and good packaging, the action of oxygen.
15. Bubbles are spherical in shape because of the surface tension of liquids. Soap bubbles are formed by blowing air into a film of soap water. *Soap is a surfactant, a chemical which when added to water reduces its surface tension. In other words, it makes the film to stretch more without breaking.* Therefore, the bubble formed from soap solution lasts longer. On the other hand, a water bubble bursts as soon as it is formed because of the higher surface tension of water.
16. Coconut oil and vegetable 'ghee'.
17. It is because the fats are insoluble in water.

18. It is because the specific gravity of fats is less than one (*i.e.*, around 0.86).

19. Being insoluble in water, the fats cannot diffuse from cell to cell. Hence, they remain in the cell in which they are synthesized.

20. The melting point of lipids depend on the degree of unsaturation. The more is the degree of unsaturation, less is its melting point. Coconut oil has greater degree of unsaturation compared to mustard oil and hence the melting point of the former falls rapidly with fall in temperature. Consequently, it solidifies in winter season.

21. Olive oil and canola oil are best for cooking. Although flaxseed oil is the best source of essential fatty acids (EFAs), it should not be used for cooking. Heat changes flaxseed oil into a toxic derivative called lipid peroxide.

22. Foam or lather is nothing but a large collection of small soap bubbles. A soap bubble is, in turn, a round ball of many thin film of soap solution enclosing some air. Because of the low surface tension of the soap solution, the film can stretch and spread. Thus, the total surface area in the lather is larger than that could be formed by water of equal volume. Because of this, whatever slight tint present in the thin film of the coloured soap solution gets subdued. Although the soap film is more or less transparent, the foam or lather looks white because the light striking this large collection of bubbles bounces in several different directions or becomes scattered. This is the reason why all kinds of foam look white.

23. Hydrogenation

24. Fats

25. Paper is made of thin sheets of cellulosic fibres and other ingredients mixed together. It is sized to make it suitable for writing. The cellulosic fibres are held together by weak forces of attraction called the van der Waal's forces. When paper is dipped in water, the sizing material used for refining the paper being water-soluble becomes soft. Water being a polar solvent weakens the van der Waal's forces that hold the fibres together. As result, the fibres come off easily when pulled and so a wet paper tears off easily.

On the contrary, oil being a nonpolar solvent does not affect the water-soluble sizing material. Nor does it weaken the forces holding the cellulosic fibres together. So paper soaked in oil does not tear off as easily as paper soaked in water does.

CHAPTER

15 Nucleic Acids

INSIDE

- Introduction
- Historical Resume
- Definition
- Components
- Nucleosides
- Nucleotides
- Deoxyribonucleic Acid
  - Variants of double helical DNA
  - DNAs with unusual structures
- Ribonucleic Acid
  - Ribosomal RNA
  - Transfer RNA
  - Messenger RNA
  - Heterogeneous Nuclear RNA
  - Informosome

Francis H.C. Crick and James D. Watson

The two, who propounded the DNA model, sharing a cup of tea in the early 1950's. The characterization of this "master molecule" transformed biology and medicine during the latter half of the twentieth century, and many aspects of everyday life are now affected by our knowledge and manipulation of its properties.

INTRODUCTION

The final class of biomolecules to be considered, the nucleotides plus molecules derived from them (*i.e.*, nucleic acids), represent a clear case in which last is not least. "Nucleotides themselves participate in a plethora of crucial supporting roles in cell metabolism, and the nucleic acids provide the script for everything that occurs in a cell" (**Lehninger, Nelson** and **Cox**, 1993).

Nucleotides are energy-rich compounds that drive metabolic processes (*esp*., biosynthetic) in all cells. They also serve as chemical signals, key links in cellular systems that respond to hormones and other extracellular stimuli, and are structural components of a number of enzyme cofactors and metabolic intermediates.

The **nucleic acids** (DNA and RNA) are the molecular repositories for genetic information and are jointly referred to as the 'molecules of heredity'. The structure of every protein, and ultimately of every cell constituent, is a product of information programmed into the nucleotide sequence of a cell's nucleic acids.

HISTORICAL RESUME

The nucleic acids have been the subject of biochemical investigations since 1869 when **Friedrich**

Miescher, a 25-year old Swiss chemist, isolated nuclei from pus cells (white blood corpuscles) and found that they contained a hitherto unknown phosphate-rich substance, which he named *nuclein*. This substance was quite different from the carbohydrates, proteins and fats. He continued his studies on salmon sperm, a prime source of nuclein and isolated it as a nucleoprotein complex in 1871, when he prophetically wrote :

> "It seems to me that a whole family of such phosphorus-containing substances, differing somewhat from each other, will emerge, as a group of nuclein substances, which perhaps will deserve equal consideration with the proteins."

FRIEDRICH MIESCHER (LT, 1844-1895), a Swiss and a student of the eminent German chemist, Felix Hoppe-Seyler, became interested in the chemistry of the nucleus in 1860s. He chose to work with white blood corpuscles (WBCs) because *they contained a large and easily observable nucleus.* A hospital at Tubingen supplied him with surgical bandages that had been peeled off purulent wounds. From the pus on the bandages, he obtained the WBCs, which he then treated with gastric juice (we now know that gastric juice contains an enzyme, pepsin, that digests protein). He observed microscopically that only the shrunken nuclei were left after the pepsin treatment; the remainder of the cell had dissolved. These nuclei were then analyzed and found to have different composition from any cellular component then known. It was called **nuclein.** Miescher continued his studies when he returned to his native city, Basel, in Switzerland. He soon found that a more convenient source of this material was salmon sperm. With this favourable material, he further purified nuclein. When all the protein was completely removed, it became clear that the new material was an acid. It was then referred to as **nucleic acid.**

DEFINITION

Although the name nucleic acid suggests their location in the nuclei of cells, certain of them are, however, also present in the cytoplasm. The nucleic acids are the hereditary determinants of living orgainsms.They are the macromolecules present in most living cells either in the free state or bound to proteins as nucleoproteins. Like the proteins, *the nucleic acids are biopolymers of high molecular weight with mononucleotide as their repeating units*, just as amino acids are the repeating units of proteins. As regards their elemental composition, the nucleic acids contain carbon, hydrogen, oxygen, nitrogen and, strangely enough, phosphorus ; the percentage of the last two elements being about 15 and 10, respectively.

COMPONENTS

There are two kinds of nucleic acids, deoxyribonucleic acid (DNA) and ribonucleic acid (RNA). Both types of nucleic acids are present in all plants and animals. Viruses also contain nucleic acids; however, unlike a plant or animal, a virus has either RNA or DNA, but not both. The previously held view, that the DNA occurred only in animals and the RNA only in plants, is now known to be incorrect. DNA is found mainly in the chromatin of the cell nucleus whereas most of the RNA (90%) is present in the cell cytoplasm and a little (10%) in the nucleous. It may be added that extranuclear DNA also exists; it occurs, for example, in mitochondria and chloroplasts. Upon hydrolysis, under different set of conditions (Fig. 15–1), the two nucleic acids yield 3

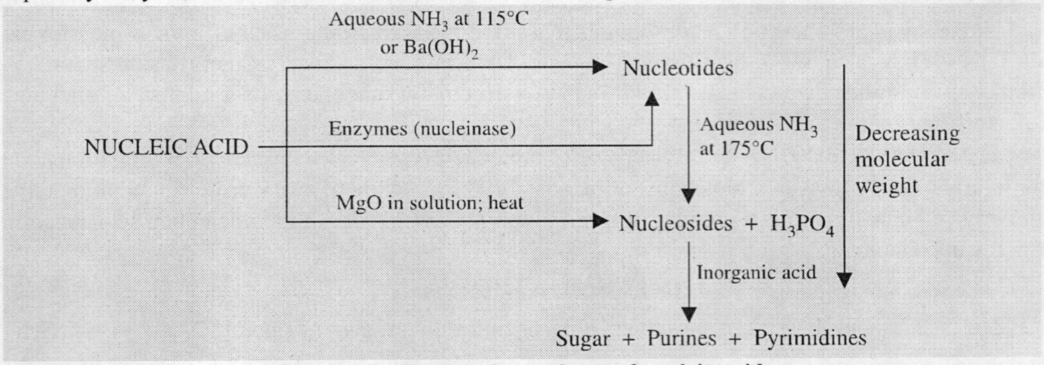

Fig. 15–1. Hydrolytic products of nucleic acid

components : phosphoric acid, a pentose sugar and nitrogenous bases. Table 15–1. summarizes the structural components of the two types of nucleic aicds, DNA and RNA.

Table 15–1. Structural components of RNA and DNA

Components	*Ribonucleic acid*	*Deoxyribonucleic acid*
Acid	Phosphoric acid	Phosphoric acid
Pentose sugar	D-ribose	D-2-deoxyribose
Nitrogenous bases		
Purines	Adenine Guanine	Adenine Guanine
Pyrimidines	Cytosine Uracil	Cytosine Thymine

Each of the 3 components of nucleic acids is discussed sepatately.

Phosphoric Acid

The molecular formula of phosphoric acid is H_3PO_4. It contains 3 monovalent hydroxyl groups and a divalent oxygen atom, all linked to the pentavalent phosphorus atom

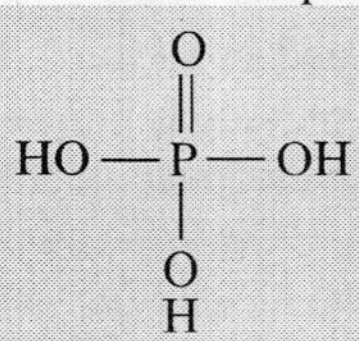

Phosphoric acid

Pentose Sugar

The two types of nucleic acids are distinguished primarily on the basis of the 5-carbon keto sugar or pentose which they possess. One possesses D-2-deoxyribose, hence the name deoxyribose nucleic acid or deoxyribonucleic acid, while the other contains D-ribose, hence the name ribose nucleic acid or ribouncleic acid. *Both these sugars in nucleic acids are present in the furanose form and are of* β *configuration.*

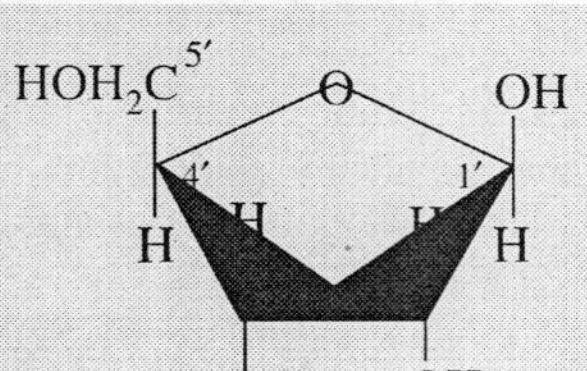

D-ribose
(β-D-ribofuranose)

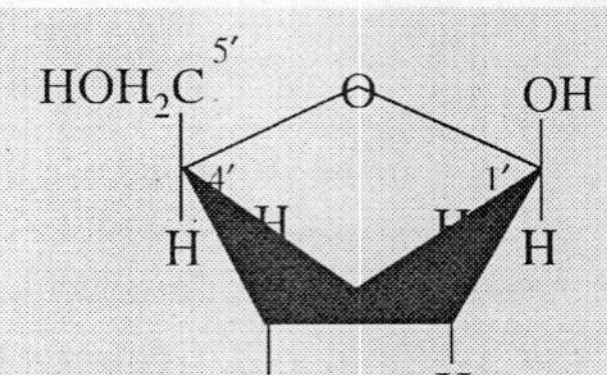

D-2-deoxyribose
(β-D-2-deoxyribofuranose)

[Note that in chemical nomenclature, the carbon atoms of sugars are designated by primed numbers, *i.e.*, C-1′, C-2′, C-3′ etc., while the various atoms in the bases lack the prime (′) sign and are designated by the cardinal numbers, *i.e.*, 1, 2, 3 etc.]

A perusal of the structure of the two types of sugars reveals that D-ribose is the parent sugar while D-2-deoxyribose is a derivative in which OH group on C2 has been replaced by an H atom.

The two sugars may be differentiated by means of specific colour reactions. Ribose reacts with orcinol in hydrochloric acid solution containing ferric chloride. Deoxyribose reacts with diphenylamine in acid solution.

An important property of the pentoses is their capacity to form esters with phosphoric acid. In this reaction the OH groups of the pentose, especially those at C3 and C5, are involved forming a 3′, 5′- phosphodiester bond between adjacent pentose residues. This bond, in fact, is an integral part of the structure of nucleic acids.

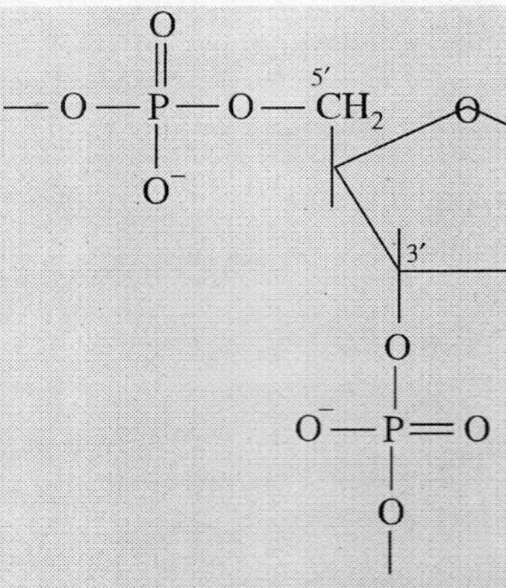

3′, 5′-phosphodiester bond

Nitrogenous Bases

Two types of nitrogenous bases are found in all nucleic acids. The base is linked to the sugar moiety by the same carbon (C1) used in sugar-sugar bonds. The nitrogenous bases are derivatives of pyrimidine and purine. Owing to their π electron clouds, both the pyrimidine and purine bases are planar molecules.

Pyrimidine Derivatives

These are all derived from their parent heterocyclic compound *pyrimidine*, which contains a six-membered ring with two nitrogen atoms and three double bonds. It has a melting point of 22°C and a boiling point of 123.5°C.

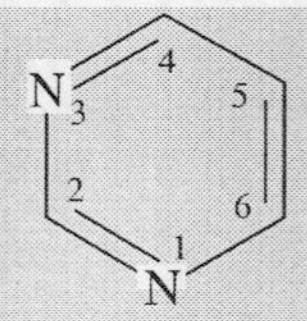

Pyrimidine

[For pyrimidine, the Chemical Abstracts use the numbering system, as depicted in the formula. This system is proposed by IUPAC. However, an older convention used a system, analogous to that indicated for the pyrimidine portion of purine, whose structural formula is given on the next page.]

The common pyrimidine derivatives found in nucleic acids are *uracil*, *thymine* and *cytosine* (Fig. 15–2).

Uracil ($C_4H_4O_2N_2$), found in RNA molecules only, is a white, crystalline pyrimidine base with MW = 112.10 daltons and a m.p. 338°C. Only rarely does uracil occur in DNA.

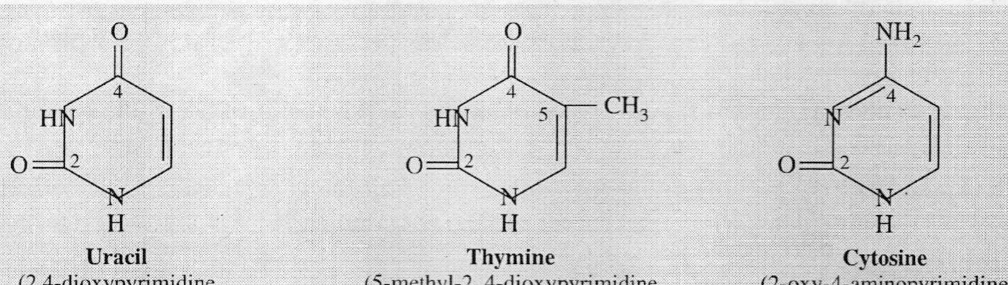

Uracil (2,4-dioxypyrimidine or pyrimidinedione) **Thymine** (5-methyl-2, 4-dioxypyrimidine or 5-methyluracil) **Cytosine** (2-oxy-4-aminopyrimidine or aminopyrimidone)

Fig. 15–2. Major pyrimidine derivatives

Thymine ($C_5H_6O_2N_2$), found in DNA molecules only, has MW = 126.13 daltons. It was first isolated from thymus, hence so named. Only rarely does thymine occcur in RNA.

Cytosine ($C_4H_5ON_3$), found in both RNA and DNA, is a white crystalline substance, with MW = 111.12 daltons and a m.p. 320-325°C.

Purine Derivatives

These are all derived from their parent compound *purine*, which contains a six-membered pyrimidine ring fused to the five-membered imidazole ring and is related to uric acid. It has a melting point of 216°C.

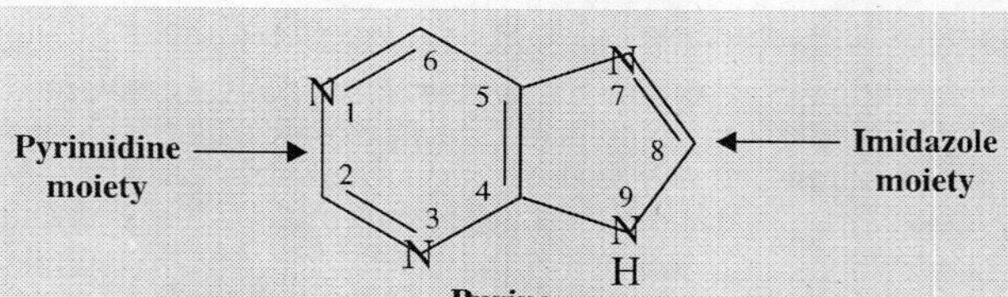

The prevalent purine derivatives found in nucleic acids are *adenine* and guanine (Fig. 15–3).

Fig. 15–3. Major purine derivatives

Adenine ($C_5H_5N_5$), found in both RNA and DNA, is a white crystalline purine base, with MW = 135.15 daltons and a m.p. 360-365°C.

Guanine ($C_5H_5ON_5$), also found in both RNA and DNA, is a colourless, insoluble crystalline substance, with MW = 151.15 daltons. It was first isolated from guano (bird manure), hence so named.

Other purine derivatives are also found in plants. For example, *caffeine* (1,3,7-trimethylxanthine) is present in coffee and tea and *theobromine* (3, 7-dimethylxanthine) is found in tea and cocoa. Both caffeine and theobromine have important pharmacological properties.

NUCLEOSIDES

> The 3 authors **J.L. Jain**, **Sunjay Jain** and **Nitin Jain** propose the abbreviation '**ns**' for the nucleoside(s). This is in conformity with the abbreviation 'nt' meant for the nucleotide(s).

The nucleosides are compounds in which nitrogenous bases (purines and pyrimidines) are conjugated to the pentose sugars (*ribose or deoxyribose*) *by a β-glycosidic linkage.* This is, in fact, the configuration in the polymeric nucleic acids. The β-glycosidic linkage involves the C-1′ of sugar and the hydrogen atom of N-9 (in the case of purines) or N-1 (in the case of pyrimidines), thus eliminating a molecule of water. Therefore, the purine nucleosides are N-9 glycosides and the pyrimidine nucleosides are N-1 glycosides. Like the O-glycosides, the nucleosides are stable in alkali. Purine nucleosides are readily hydrolyzed by acid whereas pyrimidine nucleosides are hydrolyzed only after prolonged treatment with concentrated acid.

The nucleosides are generally named for the particular purine or pyrimidine present. Nucleosides containing ribose are called *ribonucleosides*, while those possessing deoxyribose as *deoxyribonucleosides*. Table 15–2 lists the trival names of the purine and pyrimidine nucleosides which are related to the bases that occur in RNA and DNA.

Table 15–3. The nucleosides

Base	*Sugar*	*Nucleoside*	*Trivial name**	*Abbreviation*
Ribonucleosides				
Adenine	Ribose	Adenine ribonucleoside	Adenosine	AR
Guanine	Ribose	Guanine ribonucleoside	Guanosine	GR
Cytosine	Ribose	Cytosine ribonucleoside	Cytidine	CR
Thymine	Ribose	Thymine ribonucleoside	Thymidine	TR
Uracil	Ribose	Uracil ribonucleoside	Uridine	UR
Deoxyribonucleosides				
Adenine	Deoxyribose	Adenine deoxyribonucleoside	Deoxyadenosine	AdR
Guanine	Deoxyribose	Guanine deoxyribonuceloside	Deoxyguanosine	GdR
Cytosine	Deoxyribose	Cytosine deoxyribonucleoside	Deoxycytidine	CdR
Thymine	Deoxyribose	Thymine deoxyribonucleoside	Deoxythymidine**	TdR
Uracil	Deoxyribose	Uracil deoxyribonucleoside	Deoxyuridine	UdR

* Note that the trivial names for the pyrimidine nucleosides end with the suffix *-dine*, whereas those of purine nucleosides end with the suffix *-sine*.

** The deoxyribonucleoside of thymine is usually termed as thymidine instead of deoxythymidine since this pyrimidine is primarily found in DNA. However, as thymine also occurs in one type of RNA termed transfer RNA (tRNA), a growing tendency amongst the biochemists is to call this deoxyribonucleoside as deoxythymidine in conformity to the nomenclature and to restrict the use of the term thymidine to the ribonucleoside of thymine only.

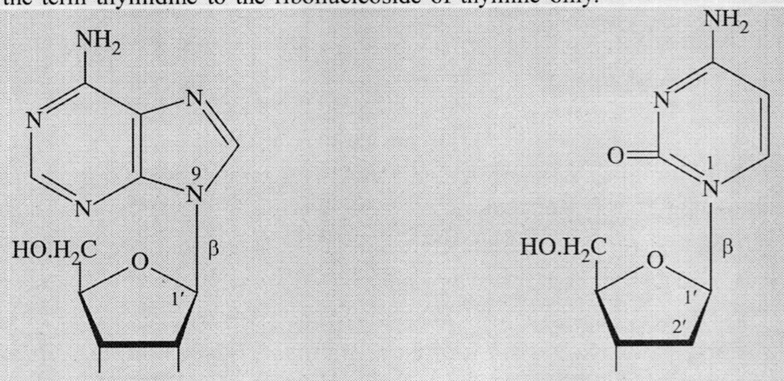

Adenosine (9-β-D-ribofuranosyladenine)

Deoxycytidine (1-β-2′-deoxy-D-ribofuranosylcytosine)

Fig. 15–4. A ribo- and a deoxyribonucleoside

The structure of two nucleosides, one ribonucleoside (also called riboside) and another deoxyribonucleoside (or deoxyriboside) is given in Fig. 15–4.

Pseudouridine (ψU), an unusual nucleoside, is present as a constituent of the transfer RNAs. In it, the β-glycoside linkage occurs between the C1 of ribose and C5 (instead of N3) of uracil (Fig. 15–5). The uracil moiety at physiological pH has one oxy group in the *keto* form and the other in the *enol* form.

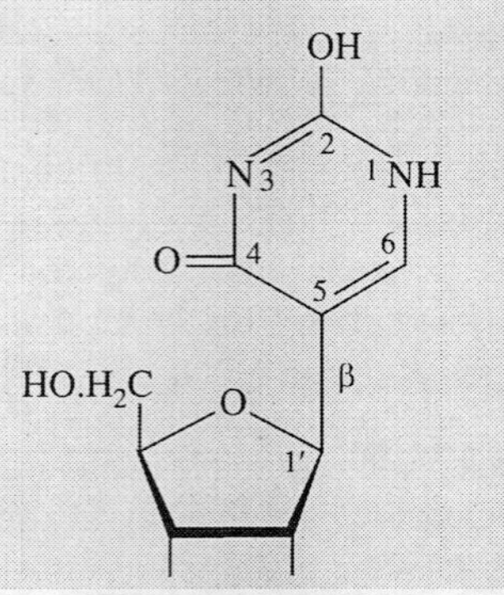

Fig. 15–5. Pesudouridine, ψU (5-ribosyluracil)
Note the absence of a true nucleosidic bond.

NUCLEOTIDES

Nucleotides are the phosphoric acid esters of nucleosides. These occur either in the free form or as subunits in nucleic acids. As mentioned earlier, the phosphate is always esterified to the sugar moiety. The component units of nucleotides are shown in Fig 15-6.

In the ribose moiety of a **ribonucleoside**, phosphorylation is possible only at *three* positions

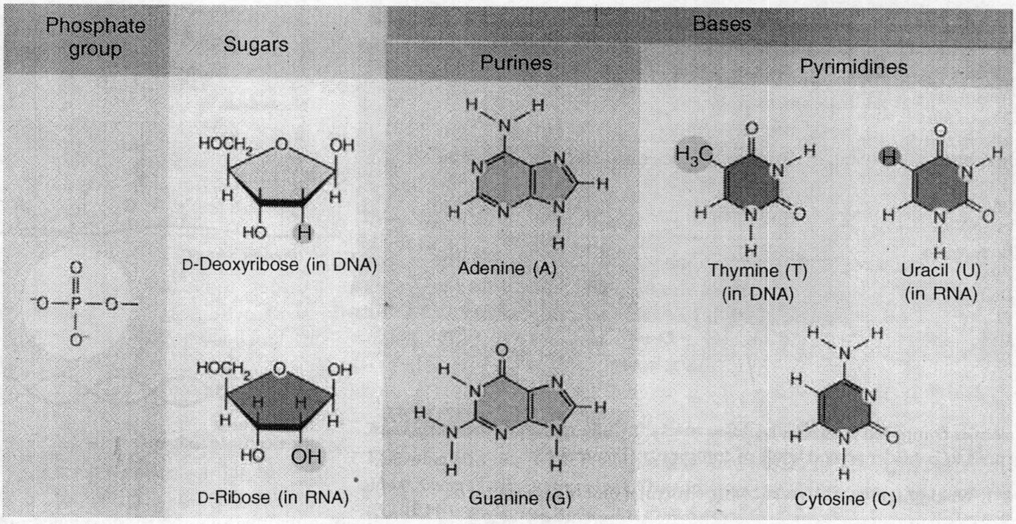

Fig. 15-6. The components of nucleotides

The three building blocks of a nucleotide are one or more phosphate groups, a sugar, and a base. The bases are categorized as purines (adenine and guanine) and pyrimidines (thymine, uracil, and cytosine).

(C2′, C3′, C5′) since C1′ and C4′ are involved in the furanose ring formation. In other words, the phosphate group could be esterified only at these three places. On the contrary, in the deoxyribose component of a **2′-deoxyribonucleoside**, only *two* positions (C3′, C5′) are available for phosphorylation, since in this sugar C1′ and C4′ are involved in the furanose ring and C2′ does not bear a hydroxyl gorup. Accordingly, hydrolysis of the two nucleic acids, RNA and DNA, by various methods and under different set of conditions gives rise to isomeric nucleotides of 3 types and 2 types respectively. Table 15–3 lists some important ribonucleotides (also called ribotides) and deoxyribonucleotides (or deoxyribotides). While names of nucleosides and nucleotides are generally derived from the corresponding bases, there is one exception to this rule : the base corresponding to the nucleoside called inosine (and the derived nucleotides) is called hypoxanthine. Fig. 15-7 depicts the structure of nucleotides found in DNA and RNA both

DEOXYRIBONUCLEIC ACID

A nucleic acid may be visualized as a polymer of a nucleotide monomer. In other words, it may be considered as a polynucleotide.

DNA as .... $(\text{base-deoxyribose-phosphate})_n$

RNA as .... $(\text{base-ribose-phosphate})_n$

The constituent units are coupled, as already stated, by means of 3′, 5′-phosphodiester bonds. The nature, properties and function of the two nucleic acids (DNA and RNA) depend on the exact order of the purine and pyrimidine bases in the molecule. This sequence of specific bases is termed the primary structure. *The purine and pyrimidine bases of DNA carry genetic information whereas the sugar and phosphate groups perform a structural role*. Interestingly enough, the human body contains about 0.5 gm of DNA!

Table 15–3. The common nucleotides

*Nucleotide**	*Trivial name*	*Abbreviations†*	
Ribouncleotides			
Adenosine-5′-monophosphate	Adenylic acid	AMP	Ado-5′-P
Guanosine-5′-monophosphate	Guanylic acid	GMP	Guo-5′-P
Cytidine-5′-monophosphate	Cytidylic acid	CMP	Cyd-5′-P
Uridine-5′-monophosphate	Uridylic acid	UMP	Urd-5′-P
2′-deoxyriobonucleotides			
Deoxyadenosine-5′-monophosphate	Deoxyadenylic acid	dAMP	dAdo-5′-P
Deoxyguanosine-5′-monophosphate	Deoxyguanylic acid	dGMP	dGuo-5′-P
Deoxyeytidine-5′-monophosphate	Deoxycytidylic acid	dCMP	dCyd-5′-P
Deoxythymidine-5′-monophosphate	Deoxythymidylic acid	dTMP	dThd-5′-P

* The table lists only the monophosphate derivatives at C5′ position of both ribonucleotides and 2′-deoxyribonucleotides. Among the other monophosphates of **ribonucleotides,** those at C2′ position are called as adenosine-2′-monophosphate (abbreviated as 2′-AMP or Ado-2′-P), cytidine-2′-monophosphate (2′-CMP or Cyd-2′-P) etc., and those at C3′ position as adenosine-3′-monophosphate (3′-AMP or Ado-3′-P), cytidine-3′-monophosphate (3′-CMP or Cyd-3′-P) etc. Similarly, in the case of **2′-deoxyribonulceotides,** the 3′-monophosphate derivatives are called as adenosine- 3′-monophosphate (3′-dAMP or dAdo-3′P), cytosine-3′-monophosphate (3′-dCMP or dCyd-3′-P) etc.

† In each case, the first abbreviation is more generally used. Note that in the case of 5′-monophosphate derivatives, the primed number 5′ is not written in the first system of abbreviation as these derivatives (as well as the 5′-diphosphates and 5′-triphosphates) are more important physiologically and are very frequently used in the literature. While adopting the second sytem of abbreviation, the diphosphates are written as Ado-5′-PP, Cyd-5′-PP etc., and the triphosphates as Ado-5′-PPP, Cyd-5′-PPP etc.

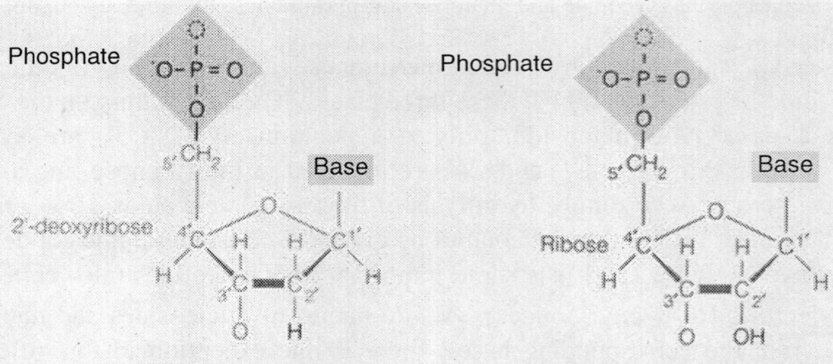

(*a*) Repeating unit of deoxyribonucleic and (DNA)

(*b*) Repeating unit of ribonucleic acid (RNA)

Fig. 15-7. The structure of nucleotides found in (*a*) DNA and (*b*) RNA

DNA contains deoxyribose as its sugar, and the bases A, T, G, and C. RNA contains ribose as its sugar, and the bases A, U, G, and C.

In a DNA or RNA strand, the oxygen on the 3′ carbon is linked to the phosphorus atom of phosphate in the adjacent nucleotide. This bond formation removes the hydrogen attached to the 3′-oxygen atom, and it removes an oxygen (shown with a dashed line) attached to phosphorus.

The terms, primary, secondary and tertiary structures for nucleic acids have much the same meaning as they do for proteins (Fig. 15-8). The ***primary structure*** is the order of the nucleotides in the chain; the term ***secondary structure*** relates to regions of regular conformation of the chain, stabilized by regular, repeating interactions (*e.g.*, the double helix of DNA) ; and the ***tertiary structure*** is the overall conformation of the chain.

Both DNA and protein are composed of a linear sequence of subunits. *The two sequences are colinear, i.e.*, the nucleotides in DNA are arranged in an order of the amino acids in the protein they specify. It is, thus, apparent that the DNA sequence contains a coded specification of the protein structure.

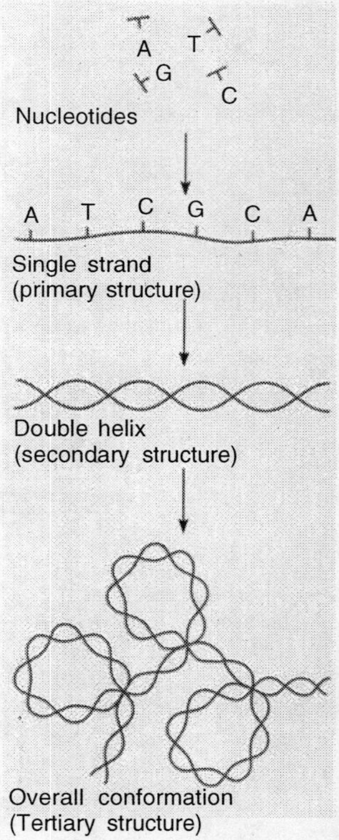

Fig. 15-8. Levels of nucleic acid structure

Internucleotide Linkages

The *backbone* of DNA (Fig. 15–9), which is constant throughout the molecule, consists of deoxyriboses linked by phosphodiester bridges. The 3′-hydroxyl of the adjacent sugar moiety of one deoxyribonucleotide is joined to the 5′-hydroxyl of the adjacent sugar by an internucleotide linkage called a phosphodiester bond. *The variable part of DNA is its sequence of bases.*

A short nucleic acid is referred to as an **oligonucleotide**. The definition of "short" is somewhat arbitrary, but the term oligonucleotide is often used for polymers containing 50 or fewer nucleotides. A longer nucleic acid is called a **polynucleotide**.

Base Composition of DNA

The base composition of DNA from a number of sources has been worked out by a number of investigators. Table 15–4 lists the DNA composition of various species. These analyses show a characteristic regularity of pattern as first noted by Erwin Chargaff and his coworkers in 1950. The important conclusions drawn by him are :

1. The sum of purines (Pu) is equal to the sum of pyrimidines (Py), *i.e.*, Pu/Py = 1. In other words, A + G = T + C (+ MC) where MC stands for methylcytosine, where it occurs.
2. The ratio of adenine to thymine is also one, *i.e.*, A/T = 1.
3. The ratio of guanine to cytosine (plus methylcytosine, where it occurs) is also one, *i.e.*, G/C(+ MC) = 1.
4. Bases with 6-amino groups are equal to bases with 6-keto (hydroxyl) groups, *i.e.*, A + C (+ MC) = G + T.

Erwin Chargaff

Chargaff, an Austrian refugee biochemist, was born in 1905 in Czernowitz, Austria (now Chernovtsy, the Ukraine). He received his Ph.D. from the University of Vienna in 1928, and then after periods at Yale, Berlin and Paris, joined Columbia University, New York city, in 1935. He has worked in many fields of biochemistry and his research papers include studies on subjects as diverse as lipid metabolism and blood coagulation. However, his most influential contribution was the demonstration during 1945-50 that the base ratios in DNA are constant. Using the simple but sensitive technique of paper chromatography, he along with his collaborators at Columbia University analyzed the base composition of DNA from various sources. Although this was one of the seminal discoveries that led to the structure of DNA, Chargaff has never claimed any allegiance to molecular biology, and indeed has become one of the sternest critics. He is the source of many pungent comments about contemporary science and scientists. He described Watson and Crick as two pitchmen in the search of a helix and explains his objection to molecular biology as 'by its claim to be able to explain everything, it acutally impedes the flow of scientific explanation'. His autobiography is ***Heraclitean Fire: Sketches From a Life Before Nature***, Rockefeller University Press, New York, 1978. He believed in the dictum, 'we can only seek what we have already found'. Chargaff is also famous for his following quote :

"What counts..... in science is to be not so much the first as the last."

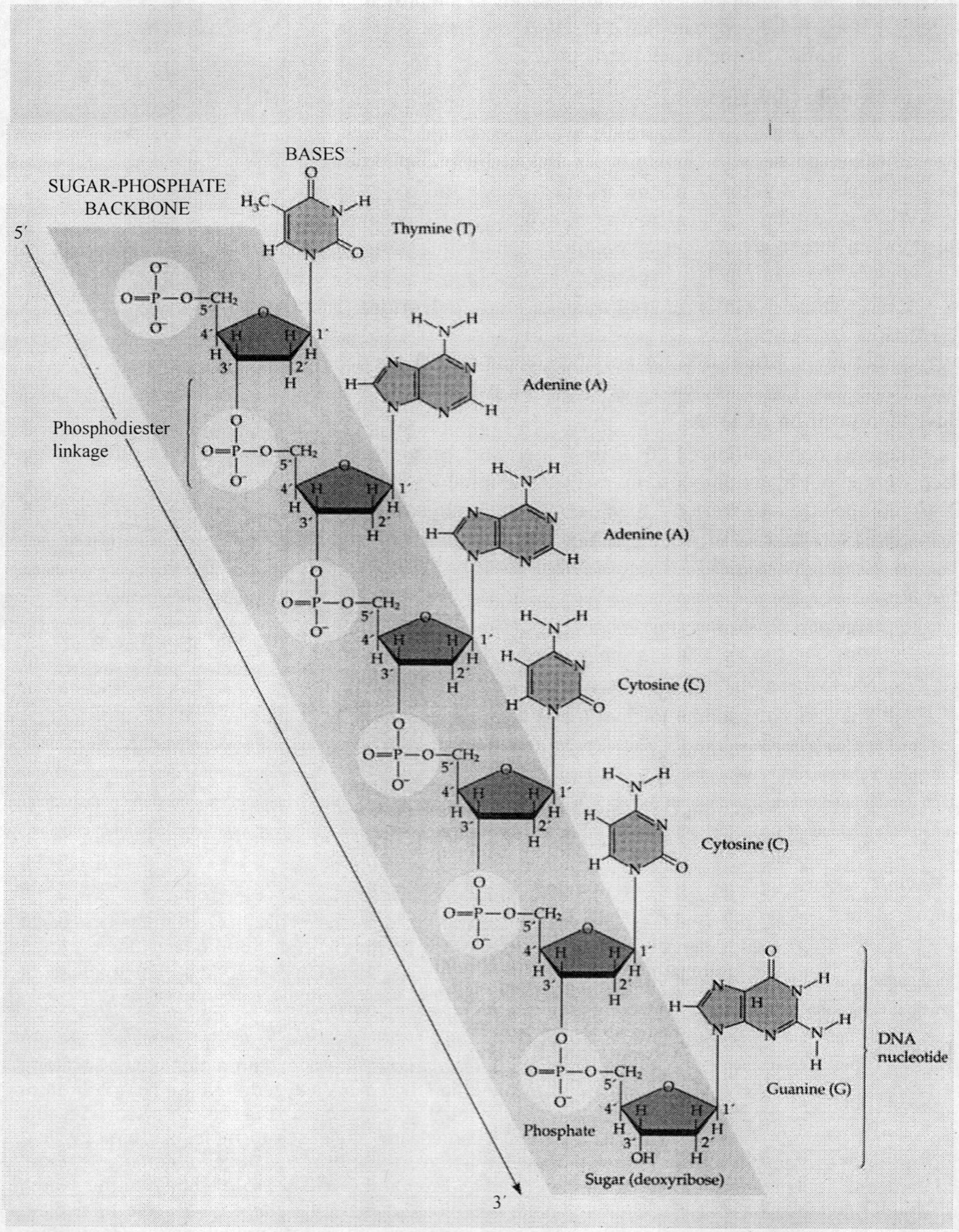

Fig. 15–9. A short strand of DNA containing six nucleotides

Nucleotides are covalently linked together to form a strand of DNA. Note the position of the internucleotide linkage between C3′ and C5′.

Table 15–4. DNA composition of various species

Species	Base proportions, moles%*					Pu/Py ratio		–	Dissymmetry ratio
	Adenine *A*	Guanine *G*	Thymine *T*	Cytosine *C*	Methyl-cytosine *MC*	$\frac{A+G}{T+C(+MC)}$	$\frac{A}{T}$	$\frac{G}{C(+MC)}$	$\frac{A+T}{G+C(+MC)}$
1. *Sarcina lutea*	13.4	37.1	12.4	37.1	—	1.02	1.08	1.00	0.35
2. *Acaligenes faecalis*	16.5	33.9	16.8	32.8	—	1.02	0.98	1.03	0.50
3. *Brucella abortus*	21.0	20.0	21.1	28.9	—	1.00	1.00	1.00	0.73
4. *Escherichia coli K 12*	26.0	24.9	23.9	25.2	—	1.08	1.09	0.99	1.00
5. *Salmonella paratyphi A*	24.8	24.9	25.3	25.0	—	0.99	0.98	1.00	1.00
6. Wheat germ	28.1	21.8	27.4	16.8	5.9	0.99	1.02	0.96	1.25
7. Rat bone marrow	28.6	21.4	28.4	20.4	1.1	1.00	1.00	0.99	1.33
8. Ox	29.0	21.2	28.7	19.9	1.3	1.00	1.01	1.00	1.36
9. *Staphyllococcus aureus*	30.8	21.0	29.2	19.0	—	1.07	1.05	1.11	1.50
10. Human thymus	30.9	19.9	29.4	19.8	—	1.03	1.05	1.01	1.52
11. Human liver	30.3	19.5	30.8	19.9	—	0.99	1.00	0.98	1.54
12. *Saccharomyces cerevisiae*	31.7	18.3	32.6	17.4	—	1.00	0.97	1.05	1.80
13. *Paracentrus lividus*	32.8	17.1	32.1	16.2	1.1	1.01	1.02	0.98	1.85
14. *Pasteurella tularensis*	32.4	17.6	32.9	17.1	—	1.00	0.98	1.03	1.88
15. *Clostridium perfringens*	36.9	14.0	36.3	12.8	—	1.04	1.02	1.09	2.70

* The mole per cent is the number of molecules of one base divided by the total number of molecules of all 4 bases, multiplied by 100. The data are accurate to ± 1 per cent.

Maurice Wilkins (b, 1916 at Pongaroa, New Zealand) and **Rosalind Franklin** (LT, 1920-1958)

Maurice Wilkins

Rosalind Franklin

Maurice Wilkins began as a physicist (BA Cambridge 1938, PhD Birmingham 1940) but his experiences on the atomic bomb project during the War turned him away from physical science and towards biology. He joined King's College, London, in 1947 and began studying the structure of chromosomes and genes using physical methods. In May 1950, he was given by **Rudolf Signer** of Bern a sample of what was probably the purest DNA available at the time, and from it obtained extremely detailed x-ray diffraction patterns. In January 1951, Rosalind Franklin joined King's from Paris and took over analysis of the Signer DNA, producing pictures that showed clearly that DNA is a helix. Unfortunately, there was a personality clash between Franklin and Wilkins and the intellectual discussion needed to solve the structure of DNA (which Watson and Crick enjoyed) never took place at King's. Nevertheless, Franklin was very close to a double helix structure when Watson and Crick announced their results.

Franklin moved to Birkbeck College to work on virus structure in 1953 and remained there until she died of breast cancer - at the premature age of 38. Most scientists agree that she had played a crucial but under-appreciated role in elucidation of the structure of DNA. Her portrayal in *The Double Helix* as 'Rosie' is now considered a fiction and it is clear from accounts of her contemporaries that, had she lived, she would probably have become one of the most eminent British scientists of her day (see *Rosalind Fanklin and DNA* by A. Sayre in 'Reading'). Wilkins continued with DNA analysis for a number of years and shared the **1962 Nobel Prize** with Watson and Crick. In 1969, he became the founding President of the British Society for Social Responsibility in Science.

5. The ratio of A + T/G + C(+ MC), known as *dissymmetry ratio*, varies greatly from one species of DNA to the other and is characteristic of that species. When the dissymmetry ratio exceeds one, such a DNA is called *AT type* ; when the value is less than one, such a DNA is designated as *GC type*. In bacteria, both AT and GC types of DNA are found. However, in the higher organisms, the range of dissymmetry ratio is more limited : in most animals the value ranges from 1.3 to 2.2 and in higher plants from 1.1 to 1.7. The value of dissymmetry ratio in human beings is 1.4 and that in *Mycobacterium tuberculosis* is 0.60.

In fact, **Watson** and **Crick** shared the Nobel Prize with **Wilkins**, although elucidation of the double helical structure of DNA was a concerted effort on the part of 5 scientists as **Watson** had, truly, once said, "actually it was a matter of 5 persons: Maurice Wilkins, Rosalind Franklin, Linus Pauling, F.H.C. Crick and me."

Only the 2 purines (adenine and guanine) and the 2 pyrimidines (thymine and cytosine) have been found in DNA molecule of most microorganisms such as bacteria, actinomycetes, algae, fungi and protozoa. However, the presence of methylcytosine, a pyrimidine, is a characteristic feature of DNA of higher plants and animals ; *the DNA of plants being richer in methylcytosine than the DNA of animals*. The richest source of methylcytosine, found as yet, is wheat germ which contains 6 moles of methylcytosine per 100 moles of bases (refer Table 15–4).

Double Helical Structure of DNA (Watson-Crick Model)

James D. Watson of the United States and **Francis H.C. Crick** of England proposed the now familiar three-dimensional model of DNA. For this epoch-making discovery, they were awarded the prestigious Nobel Price in 1962. The *salient features* of the **Watson-Crick model** (Fig. 15-10) for the commonly-found DNA (known as the B-DNA) are as follows :

Sometimes, one chain or strand of DNA is called the **Watson strand** (W strand) and its complement the **Crick strand** (C strand) as a gesture to the propounders of this model.

1. DNA molecule consists of two helical polynucleotide chains

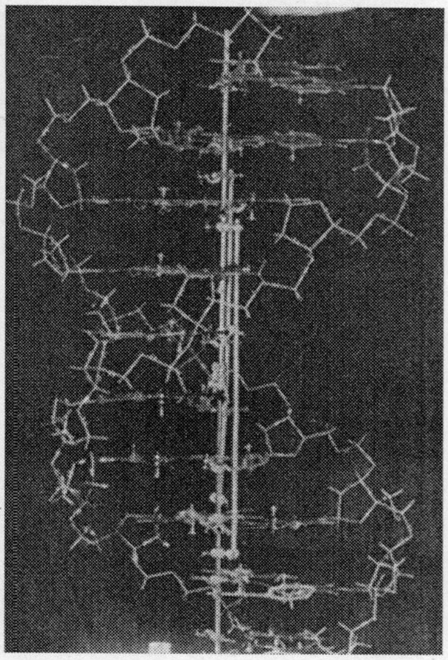

Fig. 15–10. The original molecular model of DNA double helix, proposed by Watson and Crick
The model shows A hydrogen-bonded to T and G hydrogen bonded with C.

which are coiled around (or wrapped about) a common axis in the form of a right-handed double helix (refer Fig. 15–11). The two helices are wound in such a way so as to produce 2 interchain spacings or grooves, a major or wide groove (width 12 Å, depth 8.5 Å) and a minor or narrow groove (width 6 Å, depth 7.5 Å). The two grooves arise because the glycosidic bonds of a base pair are not diametrically opposite each other. The minor groove contains the pyrimidine O-2 and the purine N-3 of the base pair, and the major groove is on the opposite side of the pair. Each groove is lined by potential hydrogen bond donor and acceptor atoms. It is noteworthy that the major groove displays more distinctive features than does the minor groove. Also, the larger groove is more accessible to proteins interacting with specific sequences of DNA.

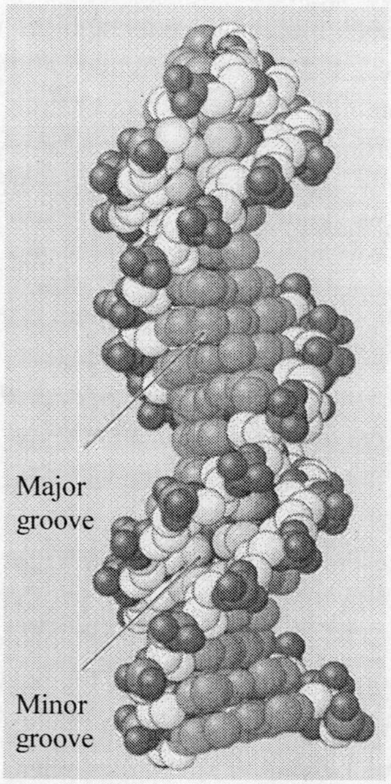

Fig. 15-11. Space-filing model of B-form DNA showing major and minor grooves
The major groove is depicted in *orange*, and the minor groove is depicted in *yellow*. The carbon atoms of the backbone are shown in *white*. Note that the bases fill the space between the two backbones. Atoms are shown at a size that includes their Van der Waals radii.

The two helices wind along the molecule parallel to the phosphodiester backbones. These may be visualized as to have formed when two wires are wound about a pencil or a common axis (Fig. 15-12). In these grooves, specific proteins interact with DNA molecules. Such double helices cannot be pulled apart and can be separated only by an unwinding process. They are called as *plectonemic coils*, *i.e.*, coils which are interlocked about the same axis. The two chains run in opposite direction, *i.e.*, they are *antiparallel* which means that in one strand, the ring oxygen of sugar moieties face upwards while in the other strand, the ring oxygen faces downwards. Also they are not identical but, because of base pairing, are *complementary* to each other.

2. The phosphate and deoxyribose units are found on the periphery of the helix, whereas the purine and pyrimidine bases occur in the centre (refer Fig. 15–13). The planes of the bases are perpendicular to the helix axis. The planes of the sugars are almost at right angles to those of the bases.

3. The diameter of the helix is 20 Å. The bases are 3.4 Å apart along the helix axis and are related by a rotation of 36 degrees. Therefore, the helical structure repeats after 10 residues on each chain, *i.e.*, at intervals of 34 Å. In other words, each turn of the helix contains 10 nucleotide residues.

4. The two chains are held together by hydrogen bonds between pairs of bases. Adenine always pairs with thymine by 2 hydrogen bonds and guanine with cytosine with 3 hydrogen bonds. This specific positioning of the bases is called *base complementarity*. The individual hydrogen bonds are weak in nature but, as in the case of proteins, a large number of them involved in the

DNA molecule confer stability to it. It is now thought that the stability of the DNA molecule is primarily a consequence of van der Waals forces between the planes of stacked bases.

5. The sequence of bases along a polynucleotide chain is not restricted in any way. *The precise sequence of bases carries the genetic information.*

As a corollary, the entire structure of a DNA molecule resembles a *winding staircase*, with the sugar and phosphate molecules of the nucleotides forming the railings and the linked nitrogen base pairs (A-T and G-C) forming the steps.

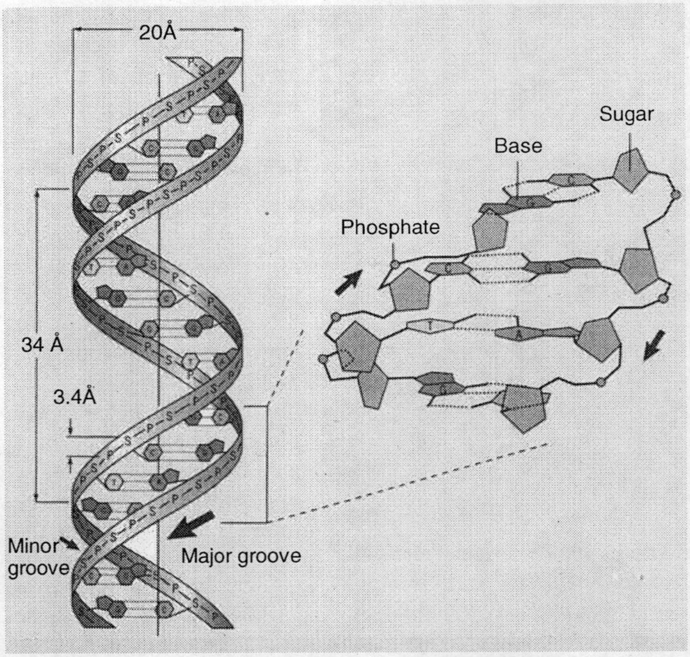

Fig. 15–12. Schematic representation of the Watson-Crick ribbon model of the B-DNA double helix

[P = phosphate; S = deoxyribose sugar ; A = adenine; G = guanine ; T = thymine ; C = cytosine]

The two ribbons represent the phosphate-sugar chains and the horizontal rods represent the bonding between the pair of bases. Two kinds of grooves, a major and a minor, are evident. The horizontal parallel lines symbolize hydrogen bonding between complementary bases. The vertical line represents the fibre axis. Note that the structure repeats at intervals of 34 Å, which corresponds to 10 residues on each chain.

Antiparallelity of the polynucleotide chain. The procedure of *nearest neighbour analysis* has shown that the two complementary chains do not run in the same direction with respect to their internucleotide linkages, but rather are *antiparallel, i.e.*, one strand runs in the 5′ → 3′ direction and the other in the 3′ → 5′ direction (refer Fig. 15–14). Thus, of the two chains, one ascends and the other descends. They have reverse polarity, that is to say the sequence of the atoms in the two chains is opposite to each other. This is analogous to 2 parellel streets, each running one way but carrying traffic in opposite directions.

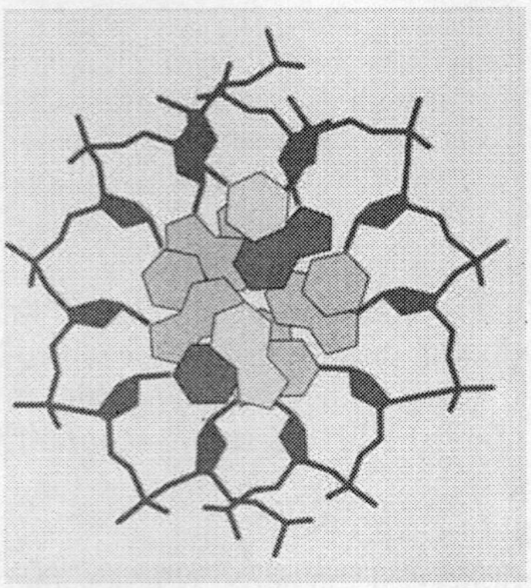

Fig. 15–13. Axial view of one of the strands of a DNA double helix

(as viewed down the helix axis)

Base pairs are stacked nearly one on top of another in the double helix. Note the position of the sugar-phosphate backbone and bases which is outward and inward respectively. The 10-fold symmetry is evident.

(*Adapted from Lubert Stryer, 2000*)

Base complementarity of the polynucleotide chain. An important feature of the double helix is the specificity of the pairing of bases. *Pairing always occurs between adenine and thymine and between guanine and cytosine* (Fig. 15-15). This means that a 6-aminopurine will always bond with a 6-ketopyrimidine, and a 6-ketopurine with a 6-aminopyrimidine. Consequently, the two helices of a DNA molecule are complementary to each other and not identical (Fig 15-15). Base-

JAMES DEWEY WATSON (born in 1928 at Chicago, USA) &
FRANCIS HARRY COMPTON CRICK (born in 1916 at Northampton, USA)

James D. Watson

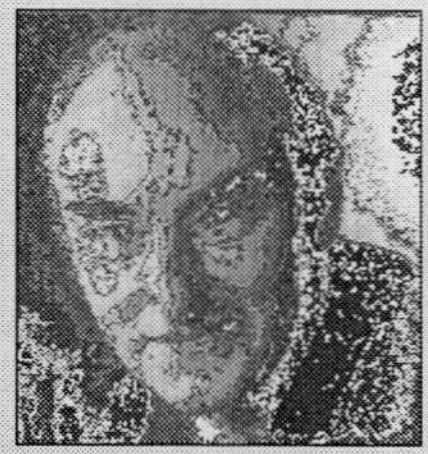

Francis H.C. Crick

James Watson was a child prodigy who entered the University of Chicago at the age of 15. He graduated in 1947 and obtained his PhD from the University of Indiana in 1950. He was then awarded a fellowship to study in Copenhagen but after a year there, he moved to the Cavendish Laboratory, Cambridge, with the specific objective of studying the gene. There, he met Francis Crick, who was working on protein structure for his PhD after starting as a physicist but becoming diverted to biology after the War. The famous fusion of minds that led Watson and Crick to deduce that DNA is a double helix has been described many times, notably by Watson in *The Double Helix.* The two shared the **1962 Nobel Prize** with Maurice Wilkins. The molecular details of DNA that continue to unfold are reminiscent of a statement once made by the great Albert Einstein (Nobel laureate in Physics, 1921) that, in science, for every milestone reached there is a signpost pointing to yet another.

After the double helix, Watson and Crick went separate ways. Watson returned to the USA in 1953 and eventually took up a professorship at Harvard before moving in 1976 to the Cold Spring Harbor Laboratory on Long Island, where he became Director. He has worked on RNA synthesis, protein synthesis and the role of viruses in cancer. Recently, he has been one of the prime motivators behind the Human Genome Project, the ambitious research programme that has as its goal the complete nucleotide sequence of the human genome. Crick stayed in Cambridge until moving to the Salk Institute in Southern California in 1977. During the 1950s and 1960s, he was the major influence in molecular biology and many of the great advances in understanding genes and gene expression are due directly to him. His autobiography is *What Mad Pursuit,* Weidenfield and Nicolson, London, 1989.

Crick named his house at Cambridge the '*Golden Helix*'. He had hung on the wall behind his desk in the Cavendish Lab the following motto : "Reading Rots the Mind."

He often says :

"It is very difficult to explain something scientifically when you cannot explain what it is."

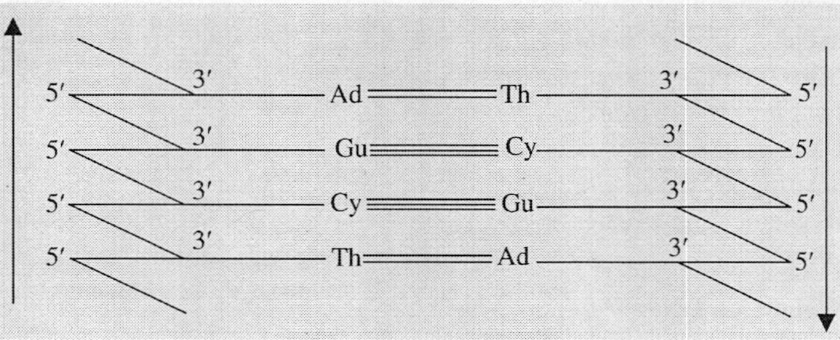

Fig. 15–14. Antiparallelity of the two polynucleotide chains of DNA molecule

pairing is due to steric and hydrogen-bonding factors.

(A) ***Steric factor.*** The steric restriction is imposed by the regular helical nature of the sugar-phosphate backbone of each polynucleotide chain. The geometry of the base pair has some special consequences in that the distances between the glycosidic bonds are the same for both the base pairs and also because the bond angle between the glycosidic bond direction and the line joining the C 1 atoms is the same in each pair. With the result, the glycosidic bonds of all nucleotides are arranged in an identical manner in relation to the axis of the helix, despite the differences in the bases.

The glycosidic bonds that are attached to a base pair are always 10.85 Å apart. A purine-pyrimidine base pair fits perfectly in this space. In contrary to this, there is insufficient space for two purines, whereas there is more than enough space for two pyrimidines so that they would be far apart to form hydrogen bonds. Hence, one member of a base pair in a DNA helix must invariably be a purine and the other a complementary pyrimidine because of steric reasons. Obviously, if the base order in one strand is known, the sequence of bases in the other strand can be predicted. For instance, if one strand has a portion which has arrangement, *adenine*, *guanine*, *thymine*, *cytosine*, the corresponding region in the complementary strand will have *thymine, cytosine, adenine, guanine*.

(B) ***Hydrogen-bonding factor.*** The base pairing is further restricted by hydrogen-bonding requirements. The hydrogen atoms in purine and pyrimidine bases have well-defined positions. Adenine cannot pair with cytosine because there would be two hydrogen atoms near one of the bonding positions and none at the other. Similarly, guanine cannot pair with thymine. In contrast, *adenine forms 2 hydrogen bonds with thymine whereas guanine forms 3 with cytosine* (refer Fig. 15–16). Thus, *the G-C bond is stronger by 50% than the A-T bond.* The higher the G-C content of a DNA moelcule, the greater is its buoyant density.

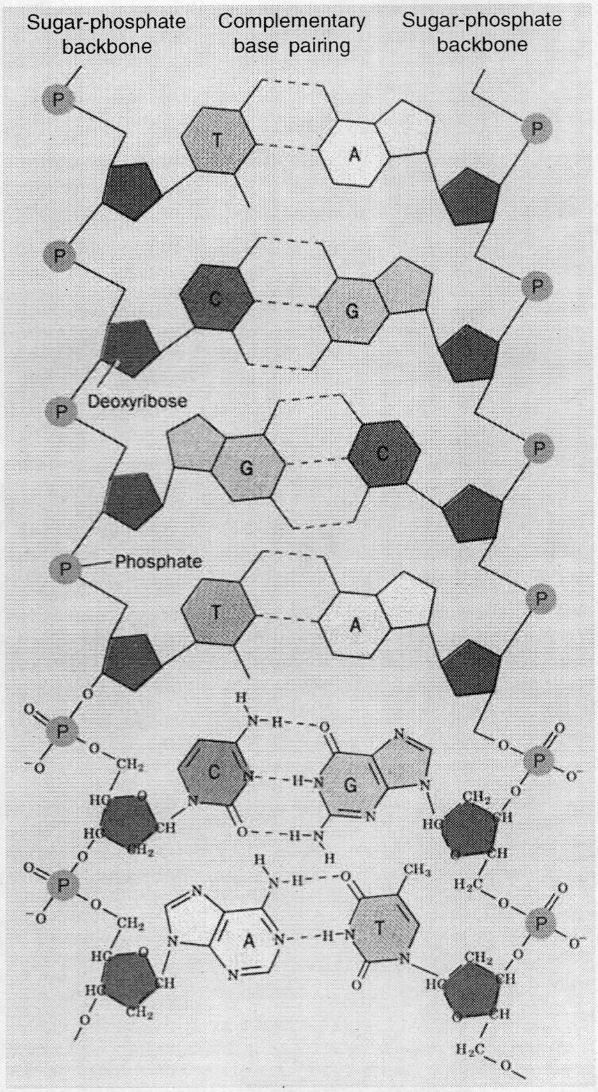

Fig. 15-15. Double-stranded DNA

The two polynucleotide chains associate through complementary hydrogen bonding.

Molecular Weight of DNA

DNA molecules are among the largest known. It is difficult to isolate DNA without fragmentation. The entire DNA of bacteria such as *Escherichia coli* has a molecular weight of 2.6×10^9. Viral DNAs range in size from about 1 to 350×10^6. The residue weight of a single nucleotide is 300 to 350. Thus, there are about 3,000 nucleotides per million molecular weight of DNA.

Length of DNA Molecule

The amount of DNA is usually measured by the microunit of the weight known as the **picogram, pg** (1 pg = 10^{-12} g). The amount of DNA has been found to be constant from cell to cell and species. *A piece of double helical DNA, 31 cm in length, weighs one picogram.* So by taking the weight of DNA amount in the nucleus, the length of the DNA molecule can be calculated

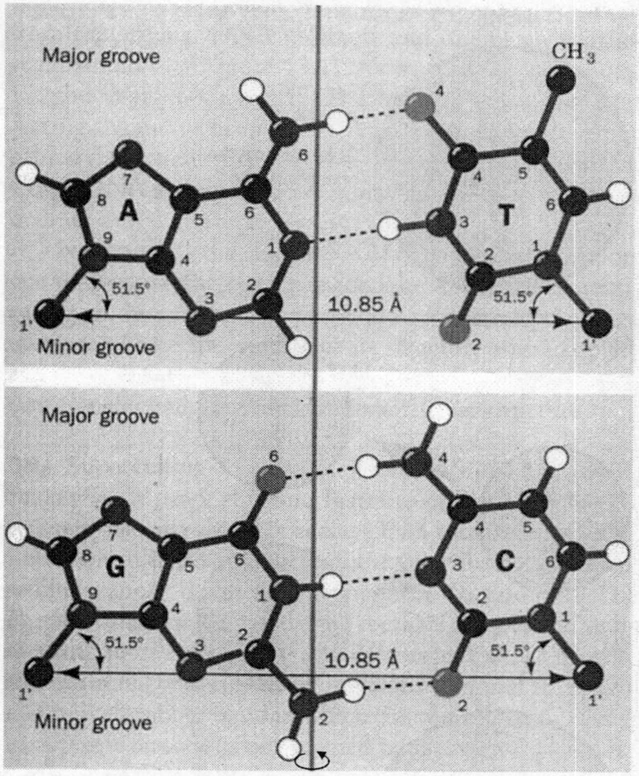

Fig. 15-16. Structure of the two Watson-Crick base pairs

The line joining the C1′ atoms is the same length in both base pairs and makes equal angles with the glycosidic bonds to the bases. This gives DNA a series of pseudo-twofold symmetry axes (often referred to as **dyad axes)** that pass through the centre of each base pair (*red line*) and are perpendicular to the helix axis. Note that A. T base pairs associate *via* two hydrogen bonds, whereas C. G base pairs are joined by three hydrogen bonds. (*After Arnott S, Dover, SD and Wonacott AJ, 1969*)

easily. By employing this method, **Dupraw** and **Bahr** (1968) have calculated that the human diploid cells contain 5.6 pg DNA amount which is equivalent to (5.6×31) or 173.6 cm-long DNA molecule. Similarly, they also calculated that a diploid cell of *Trillium* has dsDNA of 37.2 m (or 3,720 cm) length, which is equivalent to (3,720/31) or 120 pg DNA by weight. Also, the polytenic chromosomes of *Drosophila*, with 293 pg DNA contents, have (293×31) or 90.83 m-long DNA molecule.

Shape and Size of DNA Molecules

The DNA molecules are characteristically highly **elongate** structures. The *Escherichia coli* chromosome is a single molecule of double helical DNA, containing about 4 million base pairs. This DNA is highly asymmetric. Its contour length is 14×10^6 Å and its diameter is 20 Å. Whereas its contour length (1.4 mm) corresponds to a macroscopic dimension, its width (20 Å) is on the atomic scale. However, the largest chromosome of *Drosophila melanogaster* contains a single DNA molecule with 6.2×10^7 base pairs. These have a contour length of 2.1 cm. Such highly asymmetric DNA molecules are very susceptible to fragmentation by shearing forces.

Table 15–5 lists the dimensions of the various viral, bacterial and eukaryotic DNA molecules. A perusal of the table indicates that *even the smallest DNA molecules are highly elongate*. For instance, the DNA from polyoma virus contains 5,100 base pairs and has a contour length of 1.7 μm (or 17,000 Å). Furthermore, tropocollagen, the basic structural unit of collagen, has the shape of a rod which is 3,000 Å in length and 15 Å in diameter. *Tropocollagen is the longest known protein.*

Table 15–5. Dimensions of certain DNA molecules

Organism	*Number of base pairs (in thousands or kb)*	*Length* (in μm)*	*Molecular weight*
Viruses			
Polyoma or SV40	5.1	1.7	3.1×10^6
λ phage	48.6	17	31×10^6
T 2 phage	166	56	122×10^6
Vaccinia	190	65	157×10^6
Bacteria			
Mycoplasma	760	260	504×10^6
Escherichia coli	4,000	1,360	$2{,}320 \times 10^6$
Eukaryotes			
Yeast	13,500	4,600	—
Drosophila	1,65,000	56,000	—
Human	2,900,000	9,90,000	—

* 1 μm of double helix = 2.94×10^3 base pairs = 1.94×10^6 daltons.

(*Modified from A. Kornberg and T.A. Baker, 1992*)

The existence of a **circular** form of DNA was first demonstrated by **John Cairns** (1963) in *Escherichia coli* which was grown in the presence of tritium-labelled thymidine. The chromosomal material was then visualized by radioautography. This revealed a tangled circle that, when fully extended, had a circumference of 1,100 to 1,400 μ (refer Fig. 15–17). In fact, the finding that

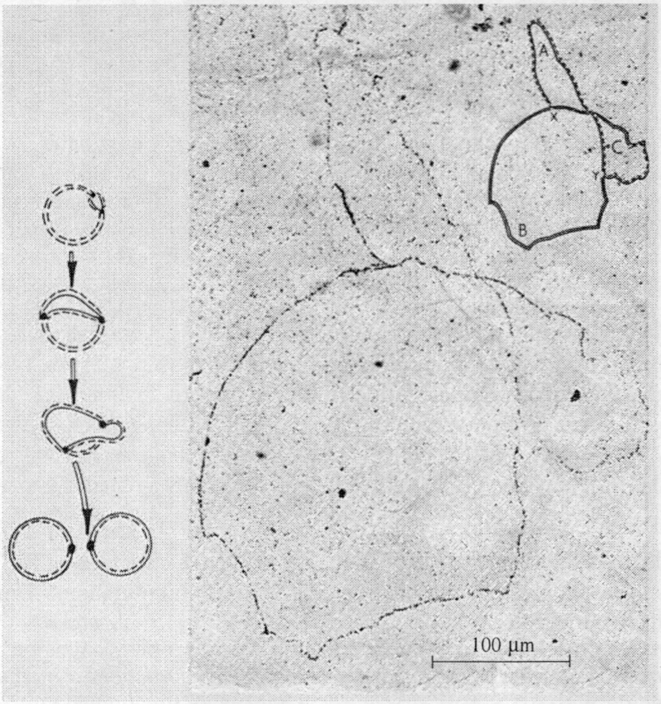

Fig. 15–17. Autoradiograph of the chromosome of *Escherihia coli*

***Right : E. coli* DNA in the process of replication**. The inset is a schematic representation of the autoradiogram. The dashed line in regions A and C represents one parent strand, one-fourth tritiated (solid portion) from the previous replication. The other parent strand and the new daughter strands (solid lines) are fully tritiated. X and Y are the "replication forks."

***Left :* The replication scheme suggested by autoradiography**. Here the addition of tritium is synchronized with the beginning of replication.

[*Courtesy of John Cairns, from Cold Spring Harbor Symp. Quant. Biol.*, **28** : *43, 1963.*]

E. coli has a circular chromosome was anticipated by genetic studies that revealed that the gene-linkage map of this bacterium is circular. *The term 'circular' refers to the continuity of the DNA chain, and not to its geometric form.*

DNA molecule from the T 7 bacteriophage is, however, **linear** in shape. Interestingly, the DNA molecules of some viruses such as λ bacteriophage interconvert between linear and circular forms. The linear form is present inside the virus particle whereas the circular form is present in the host cell.

VARIANTS OF DOUBLE HELICAL DNA

Under physiologic conditions, most of the DNA in a bacterial or eukaryotic genome is in the classic Watson-Crick form which is called the **B-form DNA** or simply **B-DNA**. The B-form is the most stable structure for a random-sequence DNA molecule, and is therefore the standard point of reference in any study of the properties of DNA. X-ray analyses of DNA crystals at atomic resolution have revealed that DNA exhibits much more structural diversity than formerly envisaged. Two such variations are :

(*a*) A DNA chain can be rotated about 6 bonds in each monomer—the glycosidic bond between the base and sugar, the $C_{4'}$ — $C_{5'}$, bond of the sugar and four bonds in the phosphodiester bridge joining $C_{3'}$ of one sugar and $C_{5'}$ of the next one — compared with just two for polypeptide chains.

(*b*) The DNA helix can be smoothly bent into an arc or it can be supercoiled with rather little change in local structure. This ease of deformation enables circular DNA to be formed and allows DNA to be wrapped around proteins. Deformability of DNA also allows compacting of DNA into a much smaller volume. *DNA can also be kinked, i.e.*, bent at discrete sites. Kinking can be induced by specific base sequences, such as a segment of at least 4 adenine residues or by the binding of a protein.

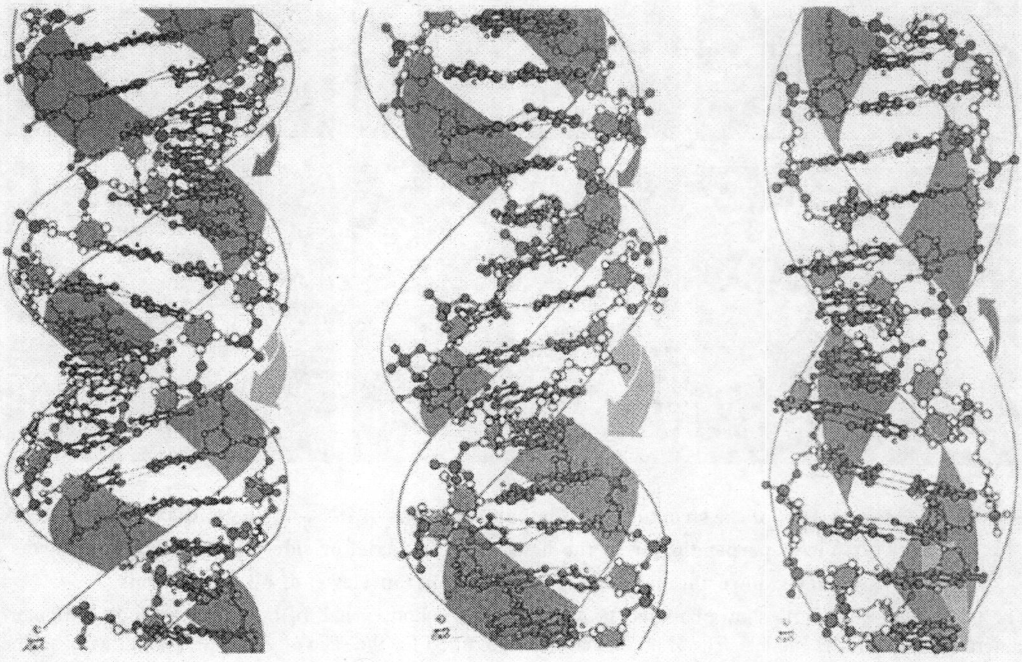

A-DNA B-DNA Z-DNA

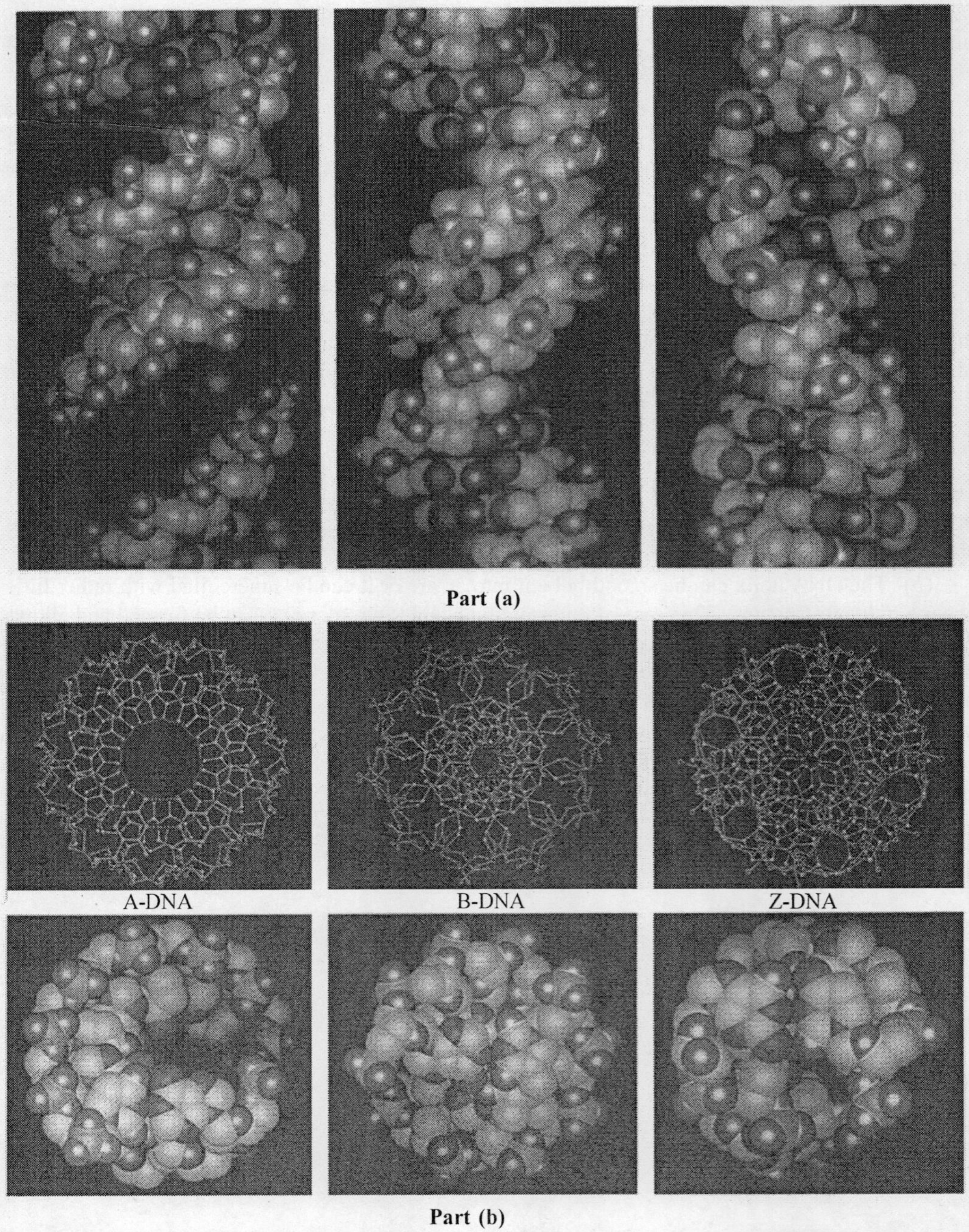

Fig. 15-18. A comparison of the structures of 3 important forms of DNA– A-DNA, B-DNA and Z-DNA

Part (a). Views perpendicular to the helix axis (*i.e.*, axial or side views) of all the 3 forms

Part (b). Views down the helix axis (*i.e.*, radial or top views) of all the 3 forms

The top half of both parts shows ball-and-stick drawings; the bottom half of both parts include computer-generated space-filling models. The H atoms have been omitted for the sake of clarity in both drawings. The highly detailed structures shown here were deduced by x-ray crystallography performed on short segments of DNA, rather than the less detailed structures obtained from DNA wet fibres. The diffraction pattern obtained from the crystallization of short segments of DNA provides much greater detail concerning the exact placement of atoms within a double helical structure. Alexander Rich at MIT, Richard Dickerson then at the California Institute of Technology, and their colleagues were the first reseachers to crystallize a short piece of DNA.

In the **A-DNA**, the based pairs have a marked propeller twist with respect to the helix axis. Note that the base pairs are inclined to the helix axis and that the helix has a hollow core. In the **B-DNA**, the base pairs lie in a plane that is close to perpendicular to the helix axis. Note that the helix axis passes through the base pairs so that the helix has a solid core. In the **Z-DNA ,** the helix is left-handed and in this respect differs from A-DNA and B-DNA, both of which are right-handed. Note that in Z-DNA, the sugar, phosphate chains follow a zigzag course (alternate ribose residues lie at different radii in Part *b*) indicating that Z-DNA's repeating motif is a dinucleotide. *Under physiologic conditions, all DNA molecules are in B form.*

It may, thus, be inferred that "DNA is a remarkably flexible molecule. Considerable rotation is possible around a number of bonds in the sugar-phosphate backbone and thermal fluctuations can produce bending, stretching and unpairing (melting) in the structure." (**Lehninger, Nelson** and **Cox**, 1993).

However, these structural variations do not affect the key properties of the DNA molecule as defined by **Watson** and **Crick** : strand complementarity, antiparallel nature of strands and base pairing specificity of A combining with T (A = T) and G combining with C (G ≡ C).

The DNA can, in fact, adopt several 3-'D' conformations and *can change reversibly from one form to the other*. These alternative forms of DNA differ in features like nature of the helix, the number of residues per turn ('n'), the spacing of the residues along the helix axis ('h') etc. Some significant alternative forms of DNA are described below (Fig. 15–18) :

A-DNA

A-DNA appears when the DNA fibre (B-DNA) is dehydrated, *i.e.*, relative humidity is reduced from 92 to 75% and Na^+, K^+ and Cs^+ ions are present in the medium. In other words, *in solution, DNA assumes the B form and under conditions of dehydration, the A form*. This is because the phosphate groups in the A-DNA bind fewer water molecules than do phosphates in B-DNA.

Like B-DNA, A-DNA is a right-handed double helix (Fig. 15–28) made of antiparallel strands held together by Watson-Crick base pairing. But the vertical rise per base pair (*bp*) is 2.3 Å and the number of base pairs per helical turn is 11, relative to the 3.4 Å rise and 10.4 base pairs per turn found in B-DNA. As such, the helix in A-DNA moves a total distance of (2.3 × 11) or 25.30 Å per turn which value is called the *helix pitch* (In B-DNA, the value of helix pitch is (3.4 × 10.4) or 35.36 Å. Obviously, the rotation per base pair in A-DNA is 360/11 or 32.72° while in B-DNA the value of rotation is 360/10.4 or 34.61° (Recall that during one turn, the helix rotates by an angle of 360°). The diameter of A-DNA helix is 25.5 Å whereas that of B-DNA helix is 23.7 Å. The helix of A-DNA is, therefore, wider and shorter than that of B-DNA.

The base pairs are tilted rather than normal to the helix axis. As far as puckering (wrinkles or small folds) of the ribose units is concerned, in A-DNA, $C_{3'}$ is out of the plane (called **C3′-endo**) formed by the other 4 atoms of the furanose ring of sugar. By contrast, in B-DNA, $C_{2'}$ is out of plane (called **C2′-endo**). In A-DNA, the base pairs are inclined at 19° with the axis of the helix but in B-DNA, the base pairs lie almost perpendicular (1° tilt) to the helix axis. Moreover, the minor groove is practically nonexistent in A-DNA, whereas it is quite deep in B-DNA.

C-DNA

C-DNA is formed at 66% relative humidity in the presence of Li^+ ions. This form of DNA is also right-handed, with an axial rise of 3.32 Å per base pair. There are 9.33 base pair per turn of the helix ; the value of helix pitch is, therefore, 3.32 ´ 9.33 Å or 30.97 Å. The rotation per base pair in C-DNA is 360/9.33 or 38.58°. The C-helix has a diameter of 19 Å, smaller than that of both B- and A-helix. The tilt of the base pairs is 7.8°.

D-DNA

D-DNA is an extremely rare variant with only 8 base pairs per helical turn. This form of DNA is found in some DNA molecules devoid of guanine. By contrast, *A-, B- and C-forms of DNA are found in all DNA molecules, irrespective of their base sequence.* There is an axial rise of 3.03 Å per base pair, with a tilting of 16.7° from the axis of the helix.

Z-DNA

Z-DNA is the more radical departure from B-DNA and is characterized by a *left-handed helical rotation.* It was discovered by **Rich, Nordheim** and **Wang** in 1984. They found that a hexanucleotide, CGCGCG, forms a duplex of antiparallel strands held together by Watson-Crick base pairing, as expected. Surprisingly, they found that this double helix was left-handed and the phosphates in the DNA backbone were in a *zigzag* manner ; hence, they termed this new form as Z-DNA. By contrast, other forms of DNA are right-handed, with a regular phosphate backbone. Another remarkable characteristic of Z-DNA is that in it the adjacent sugar residues have *alternating* orientation and it is because of this reason that in Z-DNA, the repeating unit is a *dinucleotide* as against the B-DNA, where the adjacent sugar residues have same orientation so that the repeating unit in B-DNA is a mononucleotide (Fig. 15-19). Unlike A- and B-DNAs, Z-DNA contains only one deep helical groove. There are 12 base pairs (or six repeating dinucleotide units) per helical turn, with an axial rise of 3.8 Å per base pair ; the bases are inclined at 9° with the axis of the helix. Because 12 base pairs are accomodated in one helix in Z-DNA, as against 10.4 in B-DNA, the angle of twist per repeating unit (dinucleotide) is (360/12 ′ 2) or 60° as against 34.61° per nucleotide in B-DNA. One complete helix is 45.60 Å in length in contrast to 35.36 Å in B-DNA. Since the bases get more length to spread out in Z-DNA and since the angle of tilt is 60°, they are more closer to the axis ; hence, the diameter of Z-DNA molecule is 18.4 Å hereas it is 23.7 Å in B-DNA.

The Z-DNA form occurs in short oligonucleotides that have sequences of alternating pyrimidine and purine bases. Methylation of C-5 of cytosyl residues in alternating CG sequences (as for example in the sequence CGCGCG) facilitates the transition of B-DNA to Z-DNA, because the added hydrophobic methyl groups stabilize the Z-DNA structure. The alternance of purines and pyrimidines permits at the glycosidic bonds an alternance of conformations, *syn* (base and sugar are near one another on the same side) and *anti* (base and sugar are distant, because they are on opposite sides). While in A- and B-DNA, all glycosidic bonds are *anti*, in Z-DNA, the purines are *syn* and the pyrimidines are *anti*.

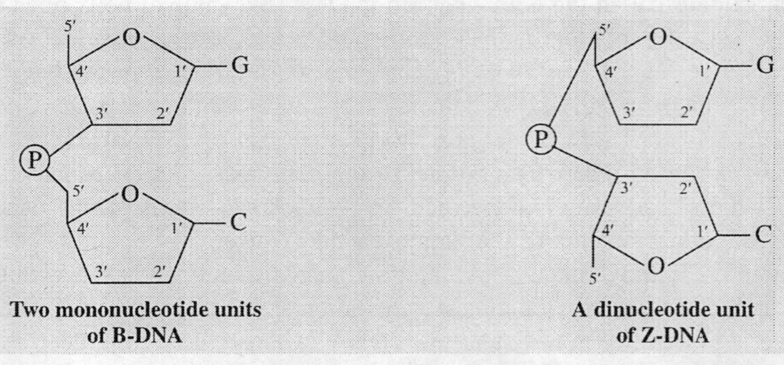

Two mononucleotide units of B-DNA **A dinucleotide unit of Z-DNA**

Fig. 15–19. Different orientation patterns of adjacent sugar residues in B-DNA and Z-DNA

Note that in B-DNA, the orientaion of adjacent sugar residues is same, while it is opposite in Z-DNA. The result is that in B-DNA, the repeating unit is a mononucleotide, while in Z-DNA, the repeating unit is a dinucleotide.

General characteristics of various forms of DNA are summarized in Table 15–6.

Table 15–6. Comparison of different forms of DNA

Characteristics	*A-DNA*	*B-DNA*	*C-DNA*	*Z-DNA*
Conditions	75% relative humidity ; Na^+, K^+, Cs^+ ions	92% relative humidity ; Low ion strength	60% relative humidity ; Li^+ ions	Very high salt concentration
Shape	Broadest	Intermediate	Narrow	Narrowest
Helix sense	Right-handed	Right-handed	Right-handed	Left-handed
Helix diameter	25.5 Å	23.7 Å	19.0 Å	18.4 Å
Rise per base pair ('h')	2.3 Å	3.4 Ã	3.32 Å	3.8 Å
Base pairs per turn of helix ('n')	11	10.4	9.33	12 (= 6 dimers)
Helix pitch ($h \times n$)	25.30 Å	35.36 Å	30.97 Å	45.60 Å
Rotation per base pair	+ 32.72°	+ 34.61°	+ 38.58°	–60° (per dimer)
Base pair tilt	19°	1°	7.8°	9°
Glycosidic bond	*anti*	*anti*	—	*anti* for C, T *syn* for A, G
Major groove	Narrow and very deep	Wide and quite deep	—	Flat
Minor groove	Very broad and shallow	Narrow and quite deep	—	Very narrow and deep

DNAs WITH UNUSUAL STRUCTURES

A number of other sequence-dependent structural variations in DNA have been detected that may serve locally important functions in DNA metabolism. These are :

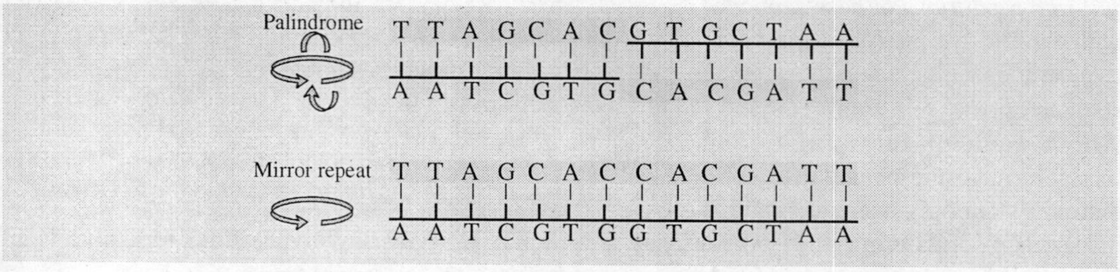

Fig. 15–20. Palindromic DNAs and mirror repeats

A. Palindromic DNA : It has sequences with twofold symmetry. In order to superimpose one repeat (shaded sequence) on the other (unshaded sequence), it must be rotated 180° around the horizontal axis and then again about the vertical axis, as shown by arrows.

B. Mirror repeat : It has a symmetric sequence on each strand. Superimposing one repeat on the other requires only a single 180° rotation about the vertical axis.

1. Bent DNA. Some sequences cause bends in the DNA helix. Bends are produced whenever 4 or more adenine residues appear sequentially in one of the two strands. Six adenines in a row produce a bend of about 18°. Bending may be important in the binding of some proteins to DNA.

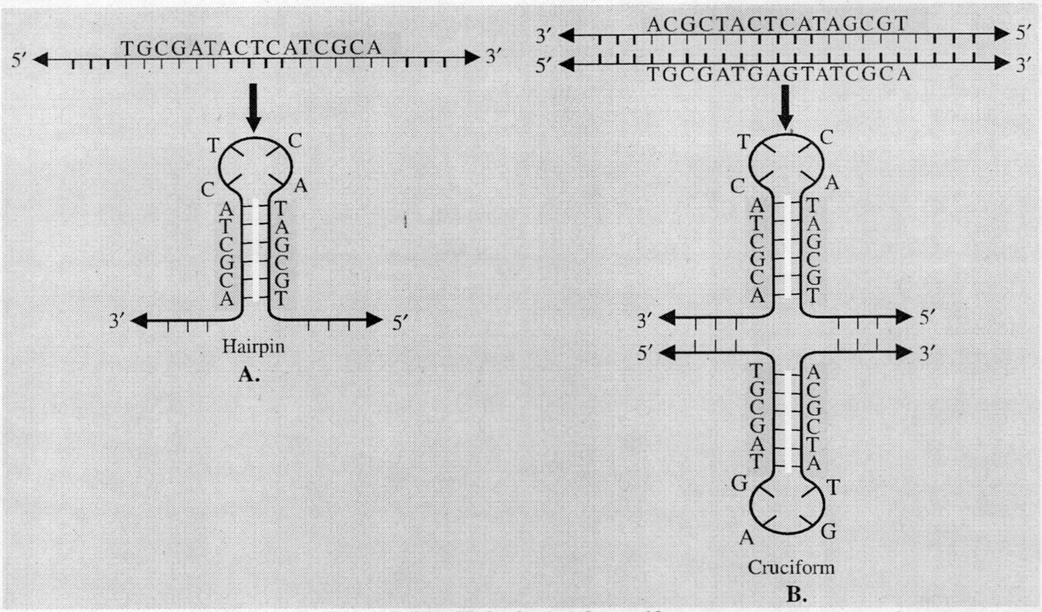

Fig. 15–21. Hairpins and cruciforms

Bases (enclosed in rectangle) are the asymmetric sequences that can pair alternatively with a complementary sequence in the same or opposite strand.

A. Hairpin : When only a single strand of palindromic DNA (or RNA) is involved, a hairpin is formed.

B. Cruciform : When both the strands of a double helical DNA are involved, a cruciform is formed.

2. Palindromic DNA. *The term palindromic DNA is applied to regions of DNA in which there are inverted repetitions* of base sequence with twofold symmetry occuring over two strands (Fig. 15–20A). Such sequences are self-complementary within each of the strands and therefore have the potential to form hairpin (when only a single strand is involved) or cruciform (when both strands of a duplex DNA are involved) structures, involving intrastrand base pairing (Fig. 15–21).

> A **palindrome** (*palindromos*[G] = running back again) is a name, word, phrase, sentence, or verse that reads the same in either direction; for examples names such as NITIN, words such as LEVEL and ROTATOR, and sentences such as SLAP NO GAG ON PALS and ABLE WAS I ERE I SAW ELBA. A spectacular, but less pure, example that ignores spaces and punctuation has been devised by Alastair Reid : "T. Eliot, top bard, notes putrid tang emanating, is sad. I'd assign it a name: gnat dirt upset on drab pot toilet." [Quoted by Brendan Gill in *Here at the New Yorker*, Random House (1975).]

When the inverted sequence occurs within each individual strand of the DNA, the sequence is called a **mirror repeat** (Fig. 15–20 B). Mirror repeats do not have complementary sequences within the same strand and cannot form hairpin or cruciform structures. Sequences of these types are found in virtually every large DNA molecule and can involve a few or up to thousands of base pairs.

3. H-DNA. H-DNA is usually found in polypyrimidine or polypurine segments that contain within themselves a mirror repeat. One simple example is a long stretch of alternating T and C residues (Fig. 15–22). A striking feature of H-DNA is the pairing and interwinding of 3 strands of DNA to form a triple helix. Triple-helical DNA is produced spontaneously only within long sequences containing only pyrimidines (or purines) in one strand. Two of the three strands in the H-DNA triple helix contain pyrimidines and the third contains purines.

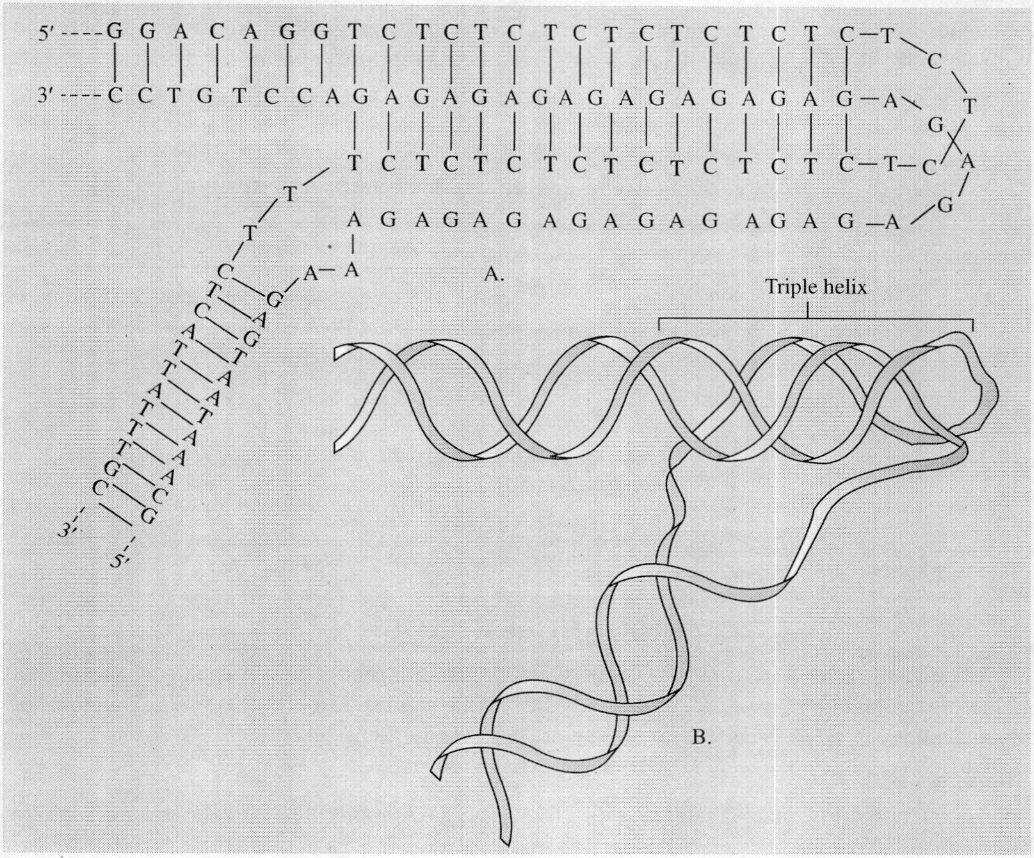

Fig. 15–22. H-DNA

A. A sequence of alternating T and C residues can be considered a mirror repeat centred about one of the central T or C residues.

B. The sequences of alternating T and C residues form an unusual structure in which the strands in one half of the mirror repeat are separated and the pyrimidine-containing strand folds back on the other half of the repeat to form a triple helix.

Single-stranded DNA

The DNA molecules are not always a double-helical structure. **Robert Sinsheimer**, in 1959, discovered that the DNA in ϕ X174, a small virus that infects *Escherichia coli*, is single-stranded. His finding was based on the following evidences :

1. The base ratios (A/T and G/C) of ϕ X174 DNA do not approach unity.

2. The ϕ X174 DNA behaves as a randomly coiled polymer whereas the DNA double helix behaves as a rigid rod.

3. The amino groups of the bases of ϕ X174 DNA react readily with formaldehyde. In contrast, the bases in the double-helical DNA do not react with this reagent.

The discovery of this single-stranded DNA raised doubts as to whether the semiconservative scheme of replication as proposed by **Watson** and **Crick**, is universally true. However, it was soon found that ϕ X174 DNA is single-stranded for only a part of its life span and the infected *E. coli* cells contain a double helical form of ϕ X174 DNA. This double helical DNA is called *replicative* *form*, since it acts as a template for the synthesis of DNA of the progeny virus. Viruses that contain single-stranded RNA also replicate *via* a double-stranded replicative form.

The main differences between double-stranded DNA and single-stranded DNA are given in Table 15-7.

Table 15–7. Differences between double-stranded and single-stranded DNAs

dsDNA	*ssDNA*
1. It is linear or filamentous types.	1. It is somewhat star-shaped (stellate).
2. DNA double helix behaves as a rigid rod.	2. ϕ X 174 DNA behaves as a randomly coiled polymer.
3. A/T ratio approaches unity.	3. A/T ratio is 0.77.
4. G/C ratio also approaches unity.	4. G/C ratio is 1.3.
5. It is resistance to the action of formaldehyde.	5. The amino groups of the bases of ϕ X 174 DNA react readily with formaldehyde.
6. Ultraviolet absorption increases from 0 to 80°C	6. Ultraviolet absorption increases from 20 to 90°C.

RIBONUCLEIC ACID

Ribonucleic acid (RNA), like DNA, is a long, unbranched macromolecule consisting of nucleotides joined by $3' \rightarrow 5'$ phosphodiester bonds (refer Fig. 15–23). The number of ribonucleotides in RNA ranges from as few as 75 to many thousands.

Differences with DNA

Although sharing many features with DNA, the RNA molecules possess several sepcific differences (Table 15-8) :

1. As apparent from its name, the sugar moiety in RNA, to which the phosphate and the nitrogen bases are attached, is *ribose* rather than 2′- deoxyribose of DNA. Ribose contains a 2′-hydroxyl group not present in deoxyribose.

2. RNA contains the pyrimidine *uracil* (U) in place of thymine which is characteristic of DNA molecule. Uracil, like thymine, can form a base pair with adenine by 2 hydrogen bonds. However, it lacks the methyl group present in thymine. It may be noted that in one case, however, RNA possesses thymine.

3. The native RNA is *single-stranded* rather than a double-stranded helical structure characteristic of DNA. However, given the complementary base sequence with opposite polarity, the single strand of RNA may fold back on itself like a hairpin and thus acquire the double-stranded pattern (refer Fig. 15–24). In the region of hairpin loops, A pairs with U and G pairs with C. Guanine can also form a base pair with uracil but it is less strong than the G-C base pair. The base pairing in RNA hairpins is frequently imperfect. Some of the apposing bases may not be complementary and one or more bases along a single strand may be looped out to facilitate the pairing of the others. The proportion of helical regions in various types of RNA varies over a wide range, although a value of 50% is typical.

4. Since the RNA molecule is single-stranded and complementary to only one of the two strands of a gene, *it need not have complementary base ratios*. In other words, its adenine content

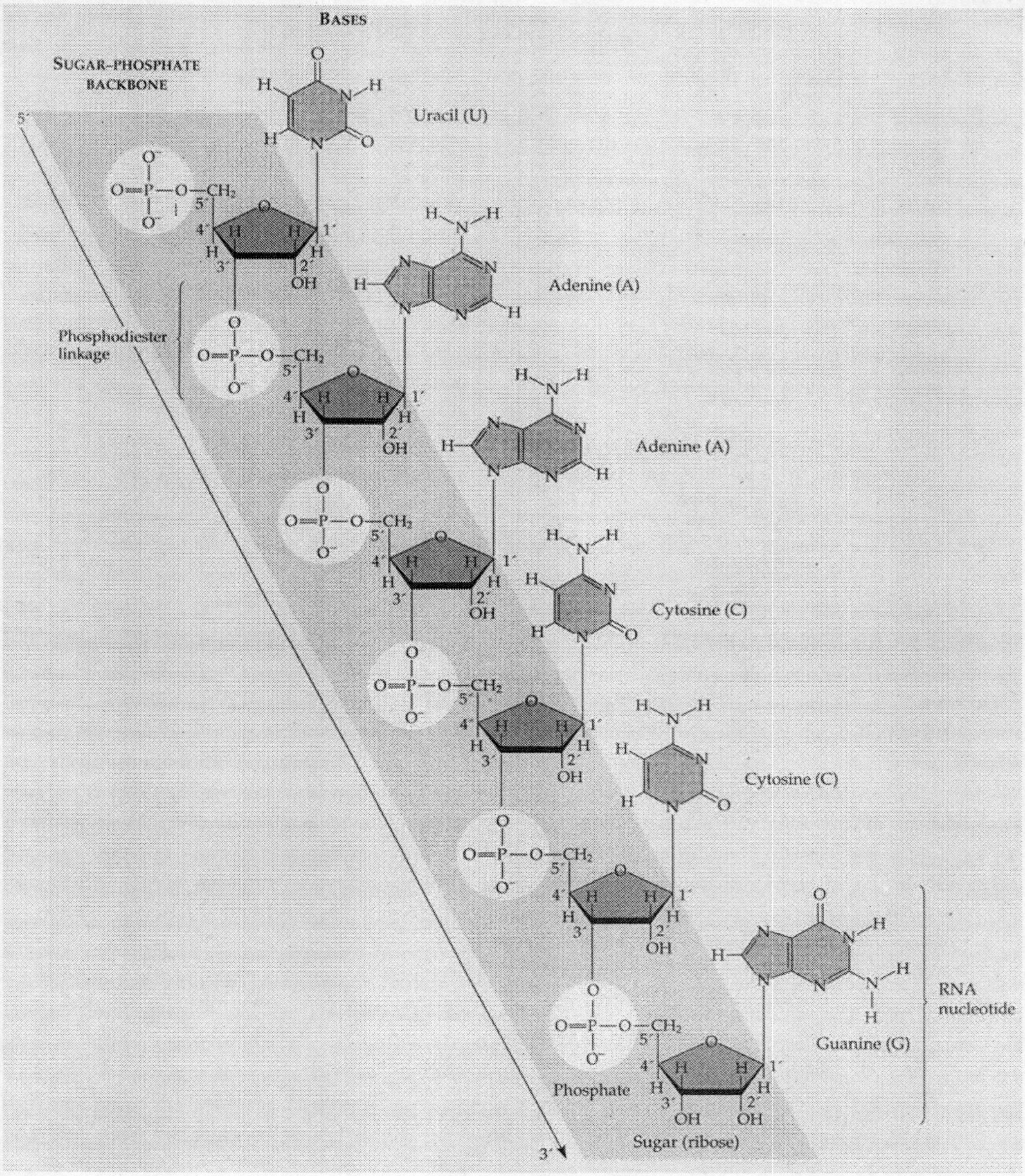

Fig. 15–23. A strand of RNA

This structure is very similar to a DNA strand (see Figure 15.9), except that the sugar is ribose instead of deoxyribose and uracil is substituted for thymine.

does not necessarily equal its uracil content, nor does its guanine content necessarily equal its cytosine content.

5. RNA can be hydrolyzed by weak alkali (pH 9 at 100′C) to 2′, 3′- cyclic diesters of the mononucleotides (shown on page 271) *via* an intermediate compound called 2′, 3′, 5′-triester. This intermediate, however, cannot be formed in alkali-treated DNA because of the absence of a 2′-hydroxyl group in its molecule. Thus, *RNA is alkali-labile whereas DNA is alkali-stable.*

Table 15–8. Comparison between DNA and RNA

DNA	*RNA*
1. Found mainly in the chromatin of the cell nucleus	1. Most of RNA (90%) is present in the cell cytoplasm and a little (10%) in the nucleolus.
2. Never present in free state in cytoplasm	2. May be present in free state.
3. Normally double-stranded and rarely single-stranded.	3. Normally single-stranded and rarely double-stranded.
4. DNA has both 'sense' and 'antisense' strands.	4. The sequence of an RNA molecule is the same as that of the 'antisense' strand.
5. Sugar moiety in DNA is 2′-deoxyribose (hence the nomenclature) which contains an H atom at C-2.	5. Sugar moiety in RNA is ribose (hence the nomenclature) which contains a 2′-hydroxyl group.
6. Sugars in DNA are in the $C_{2'}$ - endo form.	6. Sugars in RNA are in the $C_{3'}$ - endo form.
7. The common nitrogenous bases are adenine, guanine, cytosine and thymine (but not uracil).	7. The common nitrogenous bases are adenine, guanine, cytosine and uracil (but not thymine)
8. Base pairing is inevitable during which adenine pairs with thymine and guanine with cytosine.	8. In case pairing takes place, adenine pairs with uracil and guanine with cytosine.
9. The base ratios (A/T and G/C) are necessarily around one.	9. It need not have complementary base ratios.
10. Base pairing involves the entire length of DNA molecule.	10. Base pairing takes place in only the helical regions of RNA molecule, which amount to roughly half (50%) of the entire RNA molecule.
11. DNA contains only few unusual bases.	11. RNA contains comparatively more unusual bases.
12. DNA is of 3 types : filamentous (double helical or duplex), circular or single-stranded.	12. RNA is of 5 types : viral RNA, rRNA, tRNA, mRNA and double-stranded RNA.
13. It consists of a large number of nucleotides (up to 3-4 million) and has, therefore, high molecular weight.	13. It consists of fewer nucleotides (up to 12,000) and has, therefore, low molecular weight.
14. DNA is alkali-stable.	14. RNA is alkali-labile.
15. DNA stains blue with azureph thalate.	15. RNA stains red with azureph thalate.
16. DNA acts as a template for its synthesis.	16. RNA does not act as a template for its synthesis.
17. DNA on replication forms DNA and on transcription forms RNA.	17. Usually RNA does not replicate or transcribe.
18. During replicatin, exonuclease is needed.	18. During biosynthesis, exonuclease is not needed.
19. DNA is partially reversible only under certain conditions of slow cooling (= annealing).	19. RNA exhibits complete and practically instantaneous reversibility of the process of melting.
20. DNA undergoes mutation.	20. RNA does not undergo mutation.
21. DNA is the usual genetic material.	21. RNA is the genetic material of some viruses only.
22. DNA is stained green with a dye, pyronin.	22. RNA is stained red with pyronin.

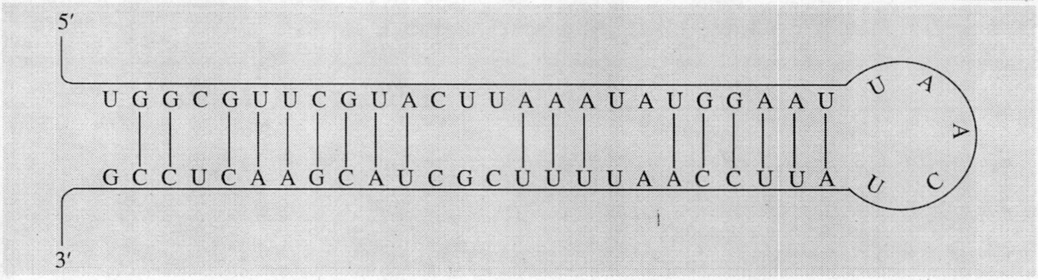

Fig. 15–24. Secondary structure of an RNA molecule

[Note the formation of a 'hairpin' which is dependent upon the intramolecular base pairing.]

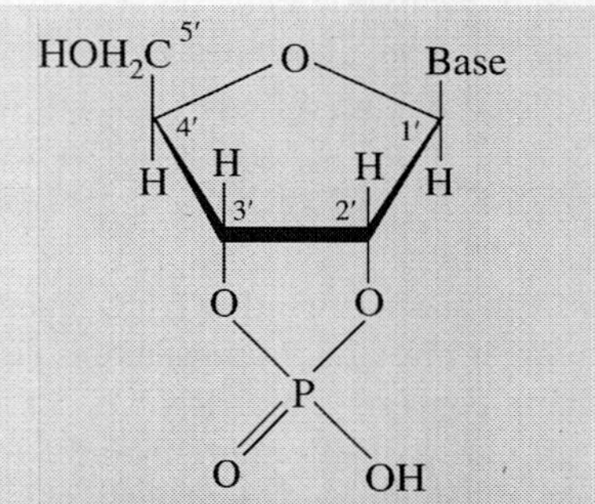

A 2′, 3′-cyclic monophosphate nucleotide

TYPES OF RNA

In all procaryotic and eucaryotic organisms, 3 general types of RNAs are found : *ribosomal*, *transfer* and *messenger RNAs*. Each of these polymeric forms serves as extremely important informational links between DNA, the master carrier of information and proteins. The 3 types of RNA molecules differ from each other by size, function and general stability. Their properties are summarized in Table 15–9.

Table 15–9. RNA molecules in *Escherichia coli*

Type	*Relative amount*	*Sedimentation coefficient*	*Molecular weight*	*Number of nucleotides*
Ribosomal RNA (rRNA)	80%	23 s	1.2×10^6	3,700
		16 s	0.55×10^6	1,700
		5 s	3.6×10^4	120
Transfer RNA (tRNA)	15%	4 s	2.5×10^4	75
Messenger RNA (mRNA)	5%	—	Heterogeneous	—

Ribosomal RNA (rRNA) or Insoluble RNA

It is the most stable form of RNA and is found in ribosomes. It has the highest molecular weight and is sedimented when a cell homogenate containing 10^{-2} M of Mg^{2+} is centrifuged at high speed (100,000 gravity for 120 minutes). In the bacterium, *Escherichia coli*, there are 3 kinds of RNA called 23 s, 16 s, and 5 s RNA because of sedimentation behaviour. These have molecular weights of 1,200,000, 550,000 and 36,000 respectively. One molecule of each of these 3 types of rRNA is present in each ribosome. *Ribosomal RNA is most abundant of all types of RNAs* and makes up about 80% of the total RNA of a cell. Ribosomal RNA represents about 40-60% of the total weight of ribosomes.

Table 15–10. Ribosomes and their RNAs

Ribosomes	*rRNA*
Procaryotic ribosomes	
30 s	16 s
50 s	5 s, 23 s
Eucaryotic ribosomes	
40 s	18 s
60 s	5 s, 28 s

The ribosomes of procaryotic and eucaryotic cells possess different species of RNA which have been summarized in Table 15–10. The ribosomes of procaryotes and those found in the plastids and mitochondria of eucaryotes contain 3 different types of RNA : 23 s RNA and 5 s RNA in the larger (50 s) subunit and 16 s RNA in the smaller (30 s) subunit. The ribosomes of the eucaryotic cells also contain 3 kinds of RNA : 28 s RNA and 5 s RNA in the larger (60 s) subunit and 18 s RNA in the smaller (40 s) subunit.

The mammalian ribosomes deserve special mention (refer Table 15–11). They also contain 2 major nucleoprotein subunits : a larger one of 2.7 megadaltons (60 s) and a smaller one of 1.3 megadaltons (40 s). The 60 s subunit contains a 5 s rRNA, a 5.8 s rRNA (formerly called 7 s rRNA) and a 28 s rRNA. There are also probably more than 50 specific polypeptides. The 40 s subunit contains a single 18 s rRNA and about 30 polypeptides chains. All types of rRNAs, except the 5 s rRNA, are processed from a single 45 s precursor RNA molecule in the nucleolus. The 5 s rRNA apparently has its own precursor which is independently transcribed.

Table 15–11. RNA components of mammalian ribosomes

Subunit		*RNA*	
Size	Mol. wt.	Size	Mol. wt.
60 s	2.7×10^6	5 s	3.5×10^4
(> 50 polypeptides)		5.8 s	4.5×10^4
		28 s	1.5×10^6
40 s	1.3×10^6	18 s	7.5×10^5
(> 30 polypeptides)			

It is remarkable to note that *rRNA from all sources has G-C contents more than 50%.* The rRNA molecule appears as a single unbranched polynucleotide strand (= *primary structure*). At low ionic strength, the molecule shows a compact rod with random coiling. But at high ionic strength, the molecule reveals the presence of compact helical regions with complementary base pairing and looped outer region (= *secondary structure*). The helical structure (Fig. 15–25) results from a folding back of a single-stranded polymer at areas where hydrogen bonding is possible because of short lengths of complementary structures.

The function of rRNA molecules in the ribosomes is not yet fully understood, but they are necessary for ribosomal assembly and seem to provide a specific sequence to which the messenger RNA moelcule can bind in order to be translated.

Transfer RNA (tRNA) or Soluble RNA (sRNA)

Transfer RNA is the smallest polymeric form of RNA. These molecules seem to be generated by the nuclear processing of a precursor molecule. In abundance, the tRNA comes next to rRNA and amounts to about 15% of the total RNA of the cell. The tRNA remains dissolved in solution after centrifuging a broken cell suspension at 100,000 X gravity for several hours (It is for this reason that previously it used to be called *soluble RNA*). The tRNA molecules serve a number of functions, the most important of which is to act as specific carriers of activated amino acids to specific sites on the protein- synthesizing templates. As the function of tRNA is to bind the specific amino acids, one might think that there are 20 types of tRNAs (*i.e.*, as many as the constituent amino acids of proteins). Since the code is **degenerate** (*i.e.*, there is more than one codon for an amino acid), there may also be more than one tRNA for a specific amino acid. In fact, their total number far exceeds than 20. In a bacterial cell, there are more than 70 tRNAs and in eukaryotic cells, this number is even greater, because there are tRNAs specific of mitochondria and chloroplasts (which usually differ from the corresponding cytoplasmic tRNAs). Therefore, there are generally several tRNAs specific of the same amino acid (sometimes up to 4 or 5) ; they are called **isoacceptor tRNAs.** These various tRNAs, capable of binding the same amino acid, differ in their nucleotide sequence; they can either have the same anticodon and therefore recognize the same codon or have different anticodons and thus permit the incorporation of the amino acid in response to multiple codons specifying the same amino acid.

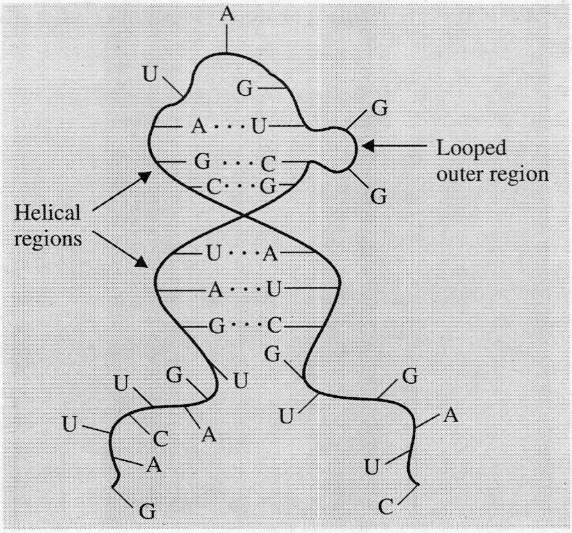

Fig. 15–25. Molecular strucutre of rRNA

Note the helical regions with complementary base pairs and a looped outer region.

(*After Davidson JN, 1972*)

> **Hydroxyproline (Hyp)** is a constituent amino acid of some proteins, but it is formed by a modification of proline after its incorporation into the polypeptide chain, so that there is no tRNA specific of hydroxyproline. There is no tRNA specific of **cystine** either, because the latter is formed by the union of 2 cysteine residues present in the same chain at places sometimes rather distant from one another or even in different polypeptide chains. But there are tRNAs specific of **asparagine (Asn)** and **glutamine (Gln)**, different from the tRNAs specific of the corresponding dicarboxylic amino acids, *i.e.*, aspartic acid (Asp) and glutamic acid (Glu).

As mentioned, each tRNA is specific of an amino acid, *i.e.*, it can bind (or "accept") only that particular amino acid. Thus, tRNA^{Ala} denotes a tRNA specific of alanine, capable of binding then transferring alanine ; tRNA^{Ser} denotes a tRNA specific of serine, etc. In fact, tRNA^{Ala} denotes a tRNA specific of alanine, but devoid of the amino acid (the terms "discharged" or "deacylated" are also used). To designate the ester bond formed by the binding of an amino acid to a specific tRNA, one must write alanyl-tRNA^{Ala} or simply alanyl-tRNA.

Because of its small size, tRNA has lent itself more readily to isolation procedures and to studies of nucleotide sequence. *Although tRNAs are quite stable in procaryotes, they are somewhat less stable in eucaryotic organisms.*

Primary structure (or base sequence) of a tRNA molecule. The base sequence of a tRNA molecule specific for the amino acid alanine (refer Fig. 15–26) was first determined by **Robert W. Holley** and coworkers in 1965, the culminaion of seven years of effort. In fact, this was the first determination of the complete sequence of a nucleic acid. **Holley** first suggested the '*cloverleaf*' model (*i.e.*, consisting of 3 folds) which is based on a secondary structure which would contain a maximum of intramolecular hydrogen bonding.

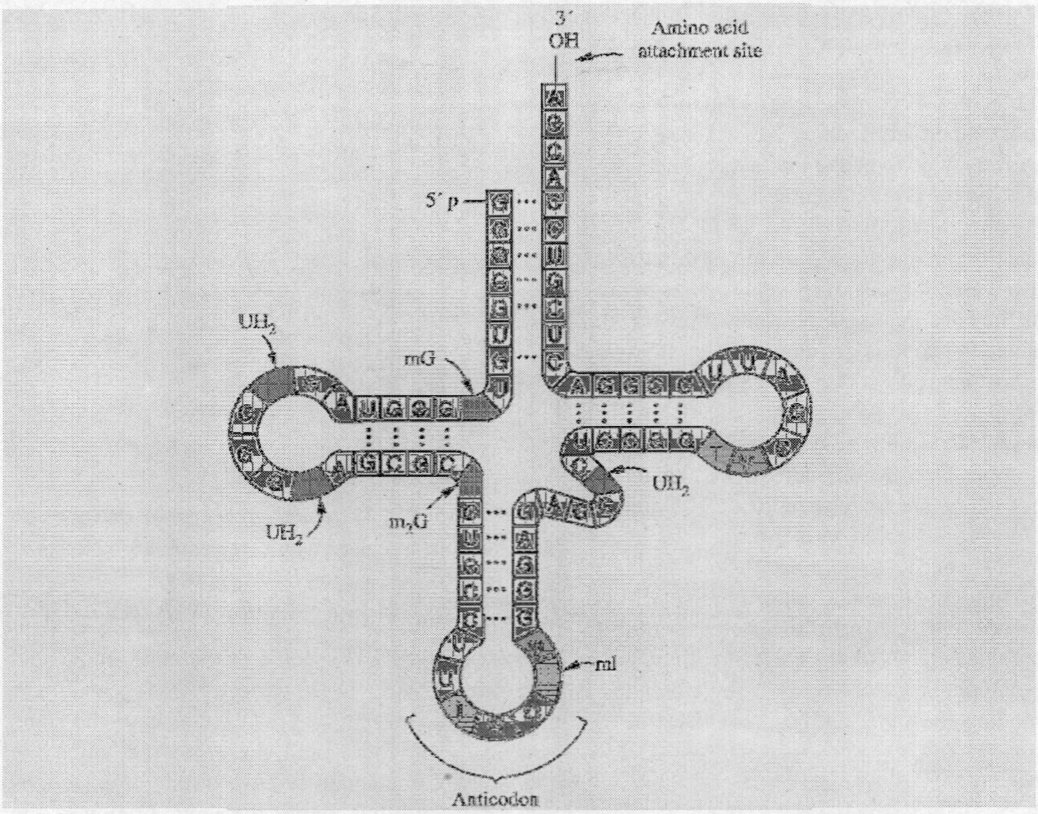

Fig. 15–26. Secondary structure, showing the base sequence of yeast alanine-specific tRNA, drawn in the cloverleaf form

The abbreviations used for modified nucleosides stand for I = inosine, mI = methylinosine, T = ribothymidine, UH_2 = dihydrouridine, ψ = pseudouridine, mG = methylguanosine and m_2G = dimethylguanosine. Note that a molecule of $tRNA^{Ala}$ contains a total of 77 amino acid residues, with the break-up : 8A + 11U + 25G + 23C + 10 modified bases.

(*Adapted from Lubert Stryer, 1995*)

The alanyl-tRNA of yeast consists of an unbranched chain of 77 ribonucleotides including 8A, 11U, 25G, 23C and 10 unusual nucleosides which are inosine, ribothymidine, dihydrouridine, pseudouridine and methylated derivatives of inosine and guanosine. The 77 residues represent a molecualr weight of 26,600 as the sodium salt. The 5′ terminus is phosphorylated whereas the 3′ terminus has a free hydroxyl group. While writing down the nucleotide sequence, the usual conventions are followed in that the *5′-phosphate is shown at the left hand and the 3′-hydroxyl on the right side.* The attachment site for the amino acid alanine is 3′-hydroxyl group of the adenosine residue at 3′ terminus of the molecule. The sequence *inosine-guanine-cytosine* in the middle of the molecule is the anticodon. It is complementary to *guanine-cytosine-cytosine*, one of the codons for alanine.

ROBERT W. HOLLEY (LT, 1922-1993)

After obtaining his Bachelor's degree from the University of Illinois, Robert Holley went to Cornell University at Ithaca, New York, gaining his PhD in 1947. From 1944 to 1946, he worked at Cornell University Medical College in New York City on penicillin synthesis. He started working with nucleic acids at the California Institute of Technology (CALTECH) in 1955 and continued in this area when he returned to Cornell a year later. He chose to study alanine tRNA *because it was relatively easy to purify*; nevertheless, it took 3 years to work out new procedures and isolate 1 g of tRNA from 140 kg of yeast. He developed new methods for the isolation and sequencing of large oligonucleotides in order to work out the nucleotide sequence of the tRNA, finally completing the work in March 1965. He states that the cloverleaf structure was not his own idea but was discovered by two of his associates, **Betty Keller** and **John Penswick**. In fact the cloverleaf was only one of several possible two-dimensional structures and was not confirmed until other tRNAs were found to fit the same pattern. In 1968, Holley received the **Nobel Prize** for determining the complete sequence of yeast alanyl-tRNA. He also became Professor of Molecular Biology at the Salk Institute in San Diego, the same year. He died in 1993 at Los Gatos, California.

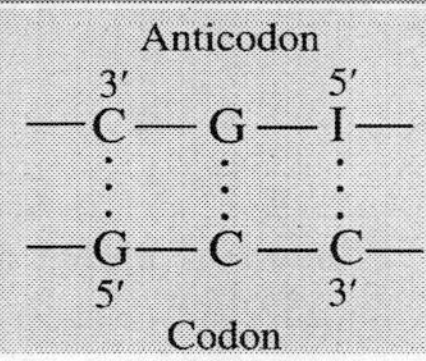

Common structural features of tRNAs. Soon after the determination of amino acid sequence of alanyl-tRNA by **Holley** *et al* in 1965, the sequences of amino acids of many other tRNA

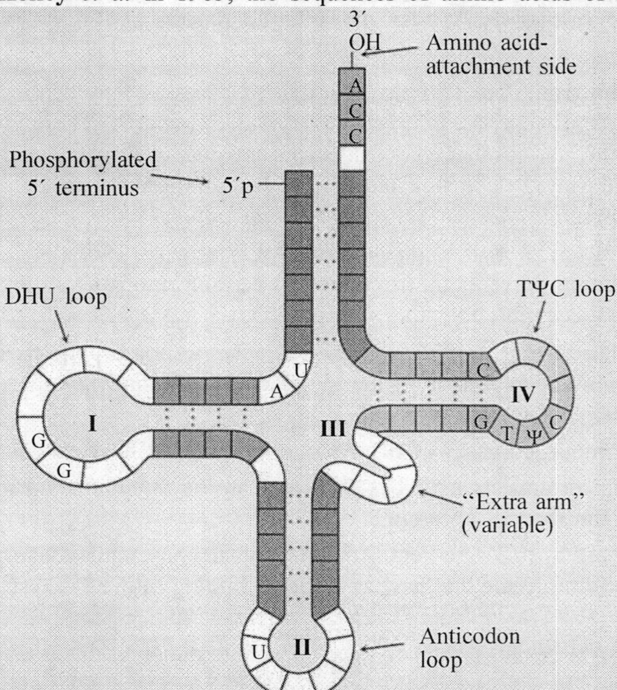

Fig. 15–27. Common structural features of tRNA molecules

Comparison of the base sequences of many tRNAs reveals a number of conserved features. Note the presence of an amino acid attachment site, the 4 'loops' and the 4 'arms' or stems.

molecules were determined by many workers. As a result, about 50 sequences are now known. These have been found to possess certain *common structural features* (refer Fig. 15–27). These are enumerated below :

1. All tRNA molecules have a common design and consist of 3 folds giving it a shape of the cloverleaf with four arms (refer Fig. 15–28) ; the longer tRNAs have a short fifth or extra arm. The actual 3-dimensional structure of a tRNA looks more like a twisted L than a cloverleaf (Fig. 15–43).

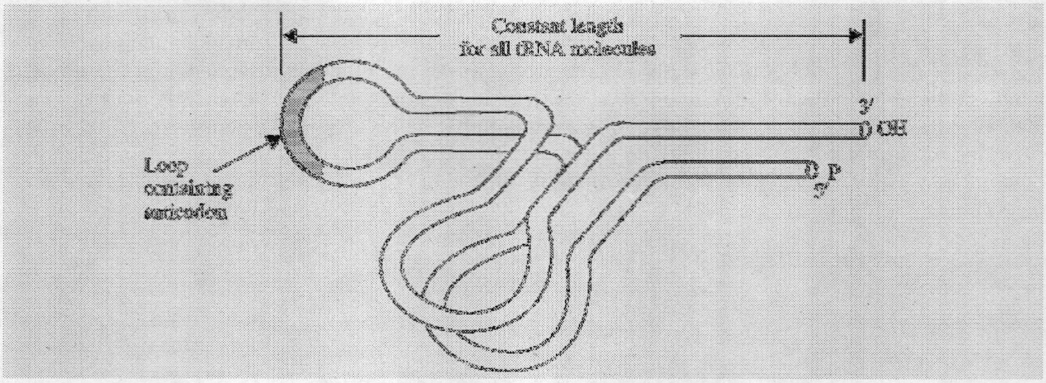

Fig. 15–28. Cloverleaf model of tRNA

2. All tRNA molecules are unbranched chains containing from 73 to 93 ribonucleotide residues, corresponding to molecular weights between 24,000 and 31,000 (Mitochondria contain distinct RNAs that are somewhat smaller). For example, $tRNA^{Ala}$ contains 77 nucleotides (**Holley** *et al*, 1965), $tRNA^{Tyr}$ contains 78 nucleotides (**Medison** *et al*, 1967) $tRNA^{Ser}$ contains 85 nucleotides (**Zachan** *et, al,* 1967).
3. They contain from 7 to 15 unusual modified bases. Many of these unusual bases are methylated or dimethylated derivatives of A, U, G and C. These include nucleotides of pseudouridine, various methylated adenines and guanines, methylated pyrimidines such as thymine and 5-methylcytosine and others. Not all these are present in any one source of tRNA but *pseudouridine* (*ϕU*) *is the most abundant and universally distributed.* Although, the role of these bases is uncertain, yet two roles seem certain :
   a. Methylation prevents the formation of certain base pairs so that some of the bases become accessible for other interactions.
   b. Methylation imparts hydrophobic character to some portions of tRNA molecules which may be important for their interaction with the *synthetases* and with ribosomal proteins.

   In fact, the modified bases of tRNAs prompted **Crick** to remark :

   > "It almost appears as if tRNAs were nature's attempt to make an RNA molecule play the role of a protein."

   What he meant by this statement is that the unusual bases are probably needed to stabilize a nucleic acid molecule that has the intricate folding of the 3-'D' structure of a globular protein.
4. The 5′ end of tRNAs is phosphorylated. The 5′ terminal residue is usually guanylate (pG).
5. The base sequence at the 3′ end of all tRNAs is CCA. All amino acids bind to this terminal adenosine *via* the 3′-OH group of its ribose.

   (5′ end) G........................tRNA........................CCA (3′ end)
6. About 50% of the nucleotides in tRNAs are base-paired to form double helices.

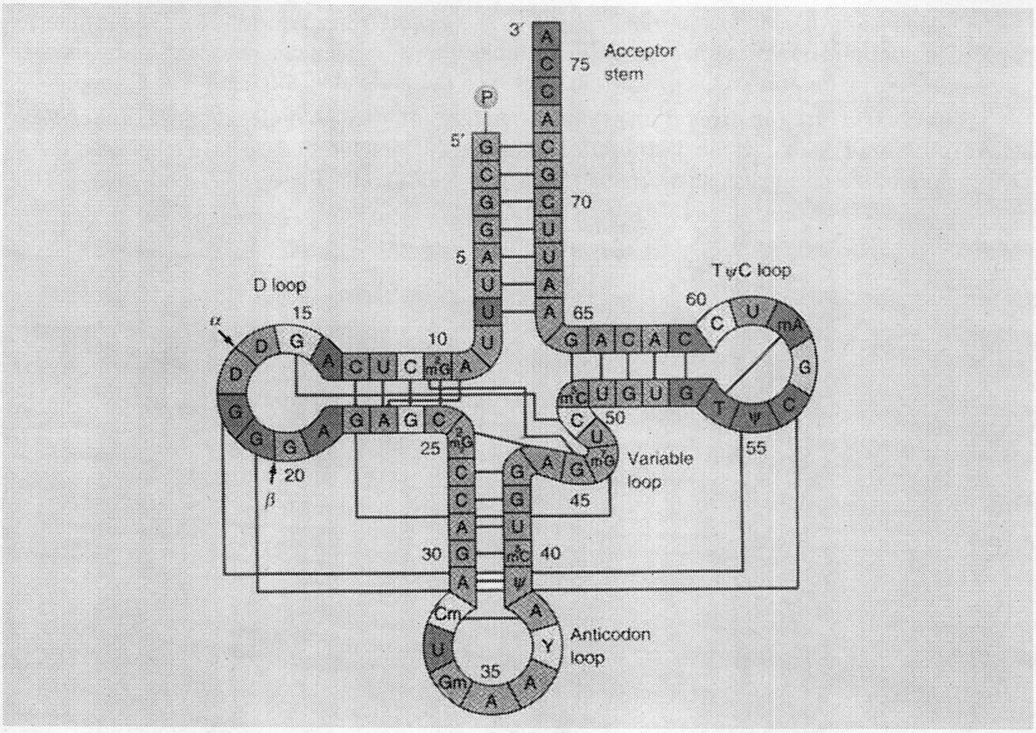

Fig. 15-29. Secondary or two-dimensional structure of yeast phenylalanine tRNA in the form of a cloverleaf

The conventional numbering of nucleotides begins at the 5′ end and proceeds toward the 3′ end. In all tRNAs, the nucleotides at the 3′ end contains the sequence CCA. Certain locations can have additional nucleotides that are not found in all tRNA molecules. The entire nucleotide sequence of yeast $tRNA^{Phe}$ is shown. *Red* lines connecting nucleotides indicate hydrogen-bonded bases. *Rose* squares indicate constant nucleotides; *tan* squares indicate a constant purine or pyrimidine. Insertion of nucleotides in the D loop occurs at positions α and β for different tRNAs.

(*Courtesy: Quigley GJ and Rich A, 1976*)

7. There are, however, 5 groups of bases which are not base-paired. These 5 groups, of which 4 form 'loops', are :
   a. the 3′ CCA terminal region,
   b. the ribothymine-pseudouracil-cytosine (= T ϕ C) loop,
   c. the 'extra arm' or little loop, which contains a variable number of residues,
   d. the dihydrouracil (= DHU) loop, which contains several dihydrouracil residues, and
   e. the anticodon loop, which consists of 7 bases with the sequence,
   5′ —pyrimidine—pyrimidine—X—Y—Z—modified purine—variable base — 3′
   This loop contains a triplet of bases which allows the tRNA to hydrogen-bond to a complementary sequence on mRNA attached to a ribosome.
8. The 4 loops are *recognition sites*. Each tRNA must have at least two such recognition sites : one for the activated amino acid-enzyme complex with which it must react to form the aminoacyl-tRNA and another for the site on a messenger RNA molecule which contains the code (codon) for that particular amino acid. It is interesting to note that the former involves recognition by bases of amino acid residues (either of the activated amino acid or of a site on the enzyme molecule) whereas the latter involves recognition by bases of bases (hydrogen bonding).
9. A unique similarity among all tRNA molecules is that *the overall distance from CCA at*

one end to the anticodon at the other end is constant (refer Fig. 15–28). The difference in nucleotide numbers in various tRNA molecules is, in fact, compensated for by the size of the "extra arm", which is located between the anticodon loop and T ϕ C loop.

Tertiary structure (or three-dimensional structure) of a tRNA molecule. Alexander Rich and **Aaron Klug** (1960s), on the basis of their x-ray crystallographic studies, have elucidated the 3-'D' structure of the phenylalanine-accepting tRNA from yeast (refer Figs. 15–29 and 15-30). This

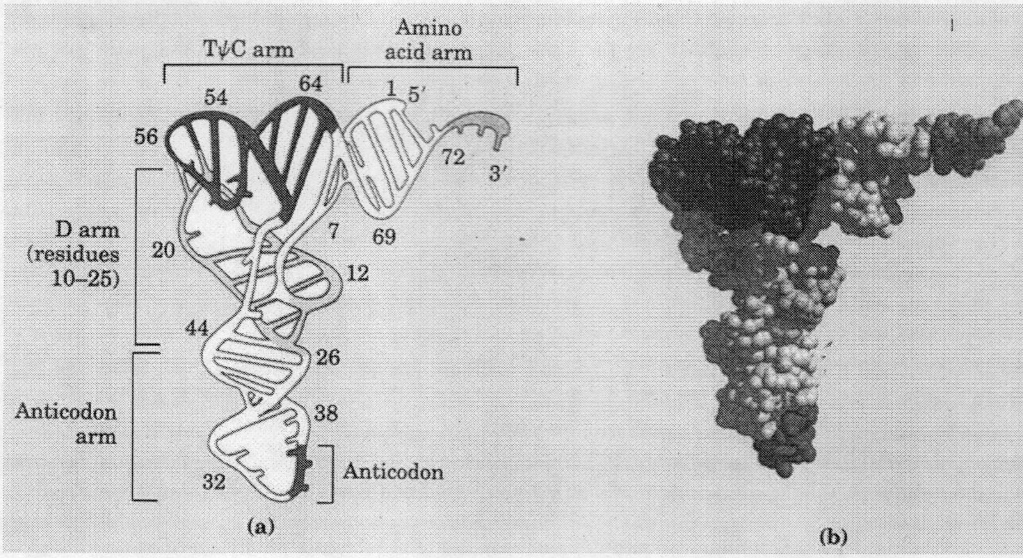

Fig. 15–30. The 3-dimensional tertiary structure of yeast phenylalanine tRNA, as deducted from diffraction analysis

(a) A schematic diagram. Various arms, identified in Fig. 15-29, are shaded in different colours.

(b) A space-filling model. Colour coding is the same in both [(a) and (b)] representations. The 3 bases of the anticodon are shown in *red* and the CCA sequence at the 3′ end (the attachment for the amino acid) in *orange.* The TψC and D arms are *blue* and *yellow*, respectively.

The specified amino acid is attached at the CCA 3′ terminus. The amino acid (or amino acceptor) arm and TΨC) arm form a continuous double helix, and the anticodon (AC) arm and the DHU (D) arm form the other partially continuous double helix. These 2 helical columns meet to form a twisted L-shaped molecule.

is probably closer to reality than the postulated cloverleaf structure. The important features observed by them are :

1. The molecule is L-shaped.
2. There are 2 segments of the double helix. Each of these helices contains about 10 base pairs which correspond to one turn of the helix. These helical segments are perpendicular to each other, thus forming an L.
3. The CCA terminus containing the attachment site for the amino acid is at one end of the L. The other end of the L is occupied by the anticodon loop. The DHU and TϕC loops form the corner of the L.
4. The CCA terminus and the adjacent helical region do not interact strongly with the rest of the molecule. This part of the molecule may change conformation during amino acid activation and also during protein synthesis on the ribosome.

Messenger RNA (mRNA) or Template RNA

The abundance of RNA in the cytoplasm and its role in protein synthesis suggested that the genetic information of nuclear DNA is transmitted to an RNA which functions at the sites of protein synthesis. In 1961, the two Nobel laureates, **Francois Jacob** and **Jacques Monod** postulated that control of protein formation, at least in certain microorganisms, is determined by the rate of synthesis of templates. This requires that the templates do not accumulate in contrast of the constant presence of DNA, rRNA and tRNA. They, therefore, suggested the transient existence of an RNA, which they called "the messenger" RNA (mRNA). Later, they proposed that "the messenger" should have the following properties :

1. The messenger should be a polynucleotide.
2. The base composition of the messenger should reflect the base composition of the DNA that specifies it.
3. The mRNA should be very heterogeneous in size because genes or groups of genes vary in length. They also correctly assumed that 3 nucleotides code for one amino acid and that the molecular weight of an mRNA should be at least a half million.
4. The messenger should be, for a short period, associated with ribosomes.
5. The messenger should be synthesized and degraded very rapidly.

All these properties are nowadays ascribed to the messenger RNA because the other 2 types of RNAs (rRNA and tRNA)are homogeneous and also their base composition is similar in species that have very different DNA base ratios.

Messenger RNA is most heterogeneous in size and stability among all the types of RNAs. It has large molecular weight approaching 2×10^6 and amounts to about 5% of the total RNA of a cell. It is synthesized on the surface of DNA template. Thus, it has base sequence complementary to DNA and carries genetic information or 'message' (hence its nomenclature) for the assembly of amino acids from DNA to ribosomes, the site of protein synthesis. In procaryotic cells, mRNA is metabolically unstable with a high turnover rate whereas it is rather stable in eucaryotes (*cf* tRNA). It is synthesized by *DNA-dependent RNA polymerase*.

On account of its heterogeneity, mRNA varies greatly in chain length. Since few proteins contain less than 100 amino acids, the mRNA coding for these proteins must have at least 100×3 or 300 nucleotide residues. In *E. coli*, the average size of mRNA is 900 to 1,500 nucleotide units.

If mRNA carries the codes for the synthesis of simple protein molecule, it is called *monocistronic type* and if it codes for more than one kind of protein, it is known as *polycistronic type* as in *Escherichia coli*.

Half-life is the period of time after which half is degraded.

The mRNAs are unstable in the bacterial systems with a half-life from a few seconds to about 2 minutes. In mammalian systems, however, mRNA molecules are more stable with a half-life ranging from a few hours to one day.

As stated above, the eukaryotic mRNA molecules, especially those of mammals, have some peculiar characteristics. The 5′ end of mRNA is 'capped' by a 7-methylguanosine triphosphate which is linked to an adjacent 2′-O-methylribonucleoside at its 5′-hydroxyl through the 3 phosphates (Fig. 15–31). Although the function of this capping of mRNA is not well understood, the cap is probably involved in the recognition of mRNA by the translating machinery. The translation of mRNA into proteins begins at the capped 5′ end. The other end of most mRNA molecules, the 3′ hydroxyl end, has attached a polymer of adenylate residues, 20–250 nucleotides in length. The specificity of the poly A "tail" at the 3′ hydroxyl end of mRNA is not understood. It probably serves to maintain the intracellular stability of the specific mRNA.

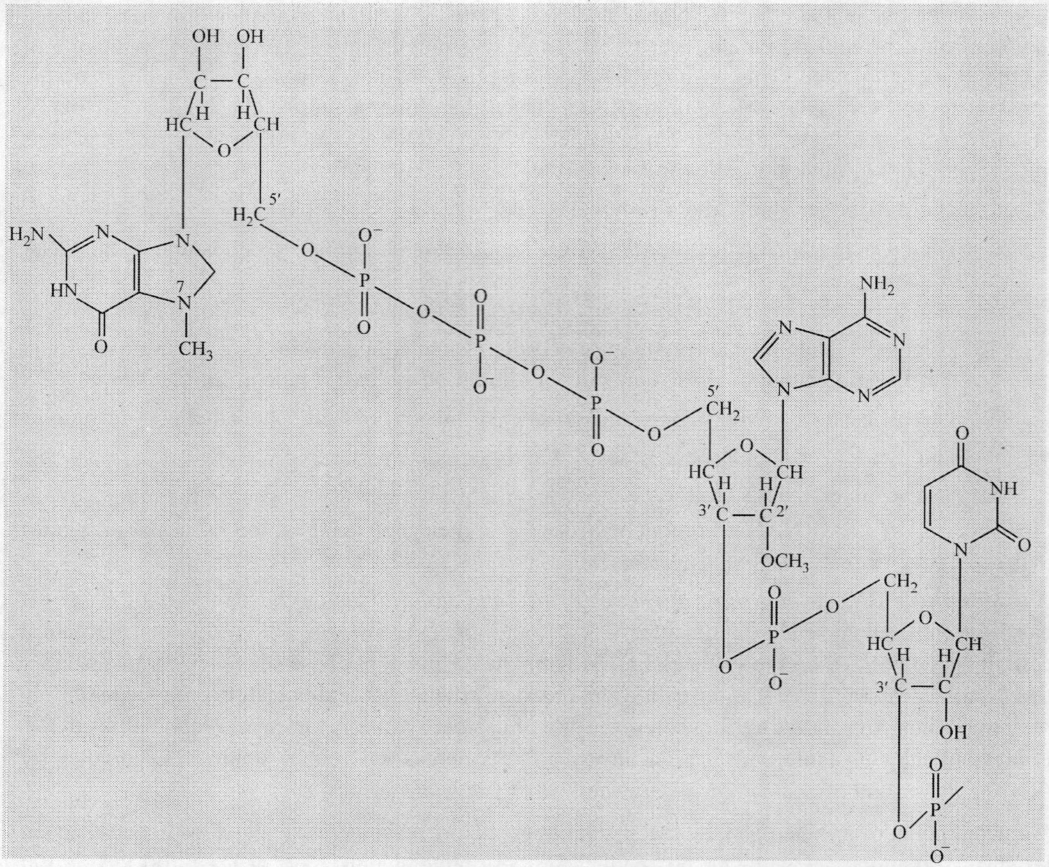

Fig. 15–31. The 'cap' structure attached to the 5′ end of most mRNA molecules

Note that a 7-methylguanosine triphosphate is attached at the 5′ terminus of the mRNA, which usually contains a 2′-O-methylpurine nucleotide.

Heterogeneous Nuclear RNA(hnRNA)

In mammalian cells including those of human beings, a precursor RNA is first synthesized in the nucleoplasm by *DNA-dependent RNA polymerase*. This precursor is then degraded by a *nuclear nuclease* to mRNA that is then translocated to the cytoplasm where it becomes associated to the ribosomal system. This precursor RNA constitutes the fourth class of RNA molecules and is designated as heterogeneous nuclear RNA (*hn*RNA). The *hn*RNA molecules may have molecular weights exceeding 10^7 daltons whereas the mRNA molecules are generally smaller than 2×10^6 daltons. Most mammalian mRNA molecules are 400–4,000 nucleotides in length whereas a *hn*RNA molecule possesses 5,000–50,000 nucleotides. Some uncertainty still exists concerning the precursor-product relationship between *hn*RNA and mRNA, the former being 10—100 times longer than the latter. Thus, the hnRNA molecules appear to be processed to generate the mRNA molecules which then enter the cytoplasm to serve as templates for protein synthesis.

The painstaking researches conducted by the two 1989 Nobel Laureates in chemistry, **Thomas Cech** from the University of Colorado and **Sidney Altman** from the Yale University, have recently shown that RNA not only carries the blueprint message from the imperious DNA to the factories that churn out proteins but it could catalyze the reactions in the same way that protein enzymes do. The particular type of RNA acting as a biological catalyst has, hence, been called by them as *riboenzyme* or simply *ribozyme* (Fig. 15-32). This has, in fact, demolished the firm impression to date that proteins alone can act as enzymes, the biological catalysts. They have, independently of each other, shown that pieces of RNA can slip themselves out of larger chains and then splice together the severed ends. Thus, their work has strengthened the claim of RNA as being the progenitor molecule of life in this world. And the 'plain-looking' RNA has been transformed by them into a fairy princess. They, in fact, showed that RNA is the Cinderella among the nucleic acids. In the words of **Thomas Cech**, the 'ribozymes have the ability to act as a sort of molecular scissors.'

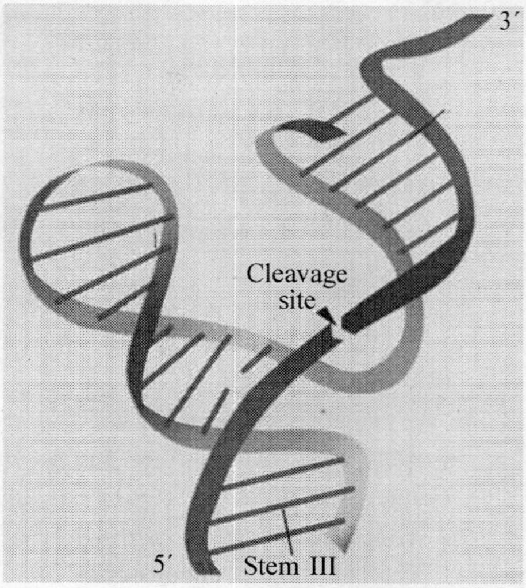

Fig. 15-32. Custom-designed ribozymes
The ribozyme shown here in *green*, called a hammerhead ribozyme, binds to RNAs having a complementary nucleotide sequence and catalyzes a reaction that cleaves the backbone of the substrate RNA (shown in *red*).

In fact, the discovery of ribozymes has led to the concept of an **"RNA world"**, in which the primordial biomolecule was RNA. Later researches have revealed that RNA cannot only catalyze excision of introns and splicing of exons, it can even break bonds between amino acids and nucleic acids. Cech has suggested that if RNA can break the bonds, it may also make them. And evidences indicate that RNA-driven protein synthesis may occur even now in certain bacterial systems. Nevertheless, an intricate problem with the notion of an RNA world is the origin of RNA in the first place. RNA is a complex molecule, notoriously difficult to synthesize under ideal conditions in the laboratory. How could it have arisen spontaneously under conditions prevailing in the prebiotic soup ?

THOMAS ROBERT CECH

Thomas Cech was born in 1947 at Chicago, USA. He received his PhD from the University of California, Berkeley, and then moved to the Massachusetts Institute of Technology (MIT), in 1975. In 1978, he joined the faculty of the University of Colorado at Boulder and began the research that led to the discovery of self-splicing RNA. The first suggestion that the *Tetrahymena* intron might mediate its own splicing was published in 1981, and was followed the next year by the demonstration that the intron RNA excises itself from the adjoining exons in the complete absence of any proteins. In 1986, Cech and his group engineered the intron RNA into an RNA enzyme, capable of cleaving other RNA molecules without being altered in the process. As well as the importance of these discoveries in understanding introns and how they are spliced, Cech's work has implications regarding the origins of life, as ribozymes may have served as both templates and catalysts to produce a primordial, self-replicating biochemical system. Cech shared the **1989 Nobel Prize for Chemistry** with Sydney Altman of Yale University.

Informosome

Spirin, Beltisina and **Lerman** (1965) have investigated that in certain eucaryotic cells, the mRNA does not enter the cytoplasm as a naked RNA strand but often remains encased by certain proteins. Spirin has proposed the term informosome for this mRNA-protein complex. The proteins bound to informosomes provide stability to mRNA molecule and also protect it from degradation by the enzyme *ribonuclease*.

PROBLEMS

1. Write the complementary sequence (in the standard 5′ → 3′ notation) for
   (*a*) GATCAA, (*b*) TCGAAC, (*c*) ACGCGT, and (*d*) TACCAT.
2. The composition (in mole-fraction units) of one of the strands of a double-helical DNA molecule is [A] = 0.30 and [G] = 0.24. (*a*) What can you say about [T] and [C] for the same strand ? (*b*) What can you say about [A], [G], [T], and [C] of the complementary strand ?
3. Write the sequence of the mRNA molecule synthesized from a DNA template strand having the sequence :
   5′-ATCGTACCGTTA-3′
4. What amino acid sequence is encoded by the following base sequence of an mRNA molecule ? Assume that the reading frame starts at 5′ end.
   5′-UUGCCUAGUGAUUGGAUG-3′
5. What is the sequence of the polypeptide formed on addition of poly(UUAC) to a cell-free protein-synthesizing system ?
6. RNA is readily hydrolyzed by alkali, whereas DNA is not. Why ?
7. The sequence of part of an mRNA is
   5′-AUGGGGAACAGCAAGAGUGGGGCCCUGUCCAAGGAG-3′
   What is the sequence of the DNA coding strand ? Of the DNA template strand ?
8. Why is RNA synthesis not as carefully monitored for errors as is DNA synthesis ?
9. Why is it advantageous for DNA synthesis to be more rapid than RNA synthesis ?
10. Given the following sequence for one strand of a double-strand oligonucleotide :
    $^{5'}$ACCGTAAGGCTTTAG$^{3'}$
    (*a*) Write the sequence for the complementary DNA strand.
    (*b*) Suppose you knew that the strand shown above had phosphate on both ends. Using an accepted nomenclature, write the sequence so as to show this.
    (*c*) Write the sequence of the RNA complementary to the strand shown above.
11. What positions in a purine ring have the potential to form hydrogen bonds, but are not involved in the hydrogen bonds of Watson–Crick base pairs ?
12. Write the base sequence of the complementary strand of double-helical DNA in which one strand has the sequence (5′)ATGCCCGTATGCATTC(3′).
13. In the cells of many eukaryotic organisms, there are highly specialized systems that specifically repair G–T mismatches in DNA. The mismatch is repaired to form a G≡C base pair (not A=T). This G–T mismatch repair system occurs in addition to a more general system that repairs virtually all mismatches. Can you think of a reason why cells require a specialized system to repair G–T mismatches ?
14. How would you measure the turnover of DNA ?
15. A sample of DNA containing 30% adenine and 20% guanine will probably contain what percentage of thymine and cytosine ?
16. In DNA, which carbon of the deoxyribose bears a phosphate, a hydroxyl group and the base ?
17. The molecular weight of a bacterial DNA molecule is 2.64×10^9. The average molecular weight of a nucleotide pair is 660 and each nucleotide pair contributes 0.34 nm to the length of DNA.

(*a*) Calculate the length of bacterial DNA molecule.

(*b*) Assume that average protein consists of a chain of 400 amino acids. What is the maximum number of proteins that can be coded by the bacterial DNA molecule ?

18. What happens to DNA in acidic solution ?

19. What is satellite DNA ?

20. How does the base composition of a particular DNA influence the melting temperature (T_m) ?

21. In a molecule of nuclear DNA from an eukaryote, there are 29,000 single nucleotides. The number of guanine nucleotides is 6,000. Assuming that normal purines and pyrimidines are present, spell out the number of each of the 4 types of nucleotides.

ANSWERS TO PROBLEMS

1. (*a*) TTGATC; (*b*) GTTCGA; (*c*) ACGCGT; (*d*) ATGGTA.

2. (*a*) [T] + [C] = 0.46, (*b*) [T] = 0.30, [C] = 0.24, and [A] + [G] = 0.46.

3. 5′ -UAACGGUACGAU-3′

4. Leu-Pro-Ser-Asp-Trp-Met

5. Poly(Leu-Leu-Thr-Tyr)

6. The 2′ -OH group in RNA acts as an intramolecular catalyst. In the alkaline hydrolysis of RNA, it forms a 2′-3′ cyclic intermediate. A 2′-OH group is, however, absent in DNA.

7. The sequence of the coding (+, sense) strand is :

5′–ATGGGGAACAGCAAGAGTGGGGCCCTGTCCAAGGAG–3′

and the sequence of template (–, antisense) strand is :

3′ –TACCCCTTGTCGTTCTCACCCCGGGACAGGTTCCTC–5′

8. An error will only affect one molecule of mRNA of many synthesized from a gene. In addition, the errors do not become a permanent part of the genomic information.

9. At any given instant, only a fraction of the genome (total DNA) is being transcribed. Consequently, speed is not necessary.

10. (*a*) $^{3'}$TGGCATTCCGAAATC$^{5'}$.

(*b*) $^{5'}$pApCpCpGpTpApApGpGpCpTpTpTpApGp$^{3'}$.

(*c*) $^{3'}$UGGCAUUCCGAAAUC$^{5'}$.

11. N-3, N-7, and N-9.

12. (3′) TACGGGCATACGTAAG (5′).

13. In eukaryotic DNA, about 5% of cytosine residues are methylated. 5-methylcytosine can spontaneously deaminate to form thymine, and the resulting G–T pair is one of the most common mismatches in eukaryotic cells.

14. Thymine is derived almost exclusively from DNA. The degradation of thymine leads to the unique product β-aminoisobutyrate, which is excreted. This can be measured in urine.

15. 30% thymine and 20% cytosine. Such a DNA is of AT type because the value of dissymmetry ratio is more than one, *i.e.*, 1.5.

16. Phosphate → 5′ carbon atom Hydroxyl group → 3′ carbon atom Base → 1′ carbon atom.

16. Base stacking in nucleic acids tends to reduce the absorption of UV light. Denaturation of DNA involves the loss of base stacking, and UV absorption increases.

17. (*a*) Mol. wt. of bacterial DNA molecule= 2.64×10^9

Average mol. wt. of nucleotide pair $= 660$

No. of nucleotide pair present $= \dfrac{2.64 \times 10^9}{660}$

Each nucleotide pair $= 0.34$ nm

$\therefore$ For above number of nucleotides, the total length of bacterial DNA molecule would be $\Big\} = \dfrac{2.64 \times 10^9}{660} \times 0.34$

$= 13,60,000$ nm

(*b*) Again, mol. wt. of nucleotide pair $= 660$

Mol. wt. of one nucleotide $= \dfrac{660}{2} = 330$

No. of nucleotides present $= \dfrac{2.64 \times 10^9}{330} = 8 \times 10^6$

No. of triplets present $= 266666.66$

Again, 1 amino acid is produced $\rightarrow$ 1 triplet

400 amino acids are produced $\rightarrow$ 400 triplets

400 triplets coded $\rightarrow$ 1 protein (or 400)

1 triplet coded $\rightarrow \dfrac{1}{400} \times 266666.66$

$= 6667$ proteins

6667 is the maximum number of proteins to be coded by the bacterial DNA.

18. Gentle acid hydrolysis at pH 3.0 cleaves the *N*-β-glucosidic bonds between the deoxyribose sugar unit and the purine bases, *i.e*., adenine and guanine. The resulting DNA derivative is devoid of DNA bases and is called as *apuminic acid.*

19. Eukaryote DNAs have 3 classes of sequences : highly repetitive, moderately repetitive and non-repetitive. The highly repetitive sequence of chromosomal DNA can be separated by isopycnic centrifugation in $CsCl_2$ after shearing the DNA into segments. The DNA distributes into a main band and a set of smaller bands termed satellite bands of DNA. In humans, 4 satellite DNAs constitute 6% of the chromosomal DNA. They contain very simple sequences that are repeated manyfold. They are generally present in centromeric region and may act in pairing and segregation of chromosomes.

20. The T_m of a DNA depends on the G-C base pair content. As G-C base pairs contribute more to the stability of DNA than do the A-T pairs, the T_m of different DNAs increase linearly as a function of the percentage of G-C base pairs. In fact, G-C base pair is 50% more stronger than A-T base pair because of the presence of 3 hydrogen bonds in it, as opposed to 2 in A-T base pair.

21. Total number of nucleotides = 29,000

Number of guanines = 6,000

Applying Erwin Chargaff's equivalence rule G = C,

$\therefore$ Number of cytosines = 6,000

As the total number of nucleotides = 29,000

and G + C = 6,000 + 6,000 = 12,000

$\therefore$ A + T = Total nucleotides – (G + C)

= 29,000 – (6,000 + 6,000)

= 29,000 –12,000 = 17,000

As A = T, so A = 17,000/2 = 8,5000

Also T = 8,500

CHAPTER

16 Enzymes-I Nomenclature and Classification

INSIDE

- Definition
- Historical Resume
- Nomenclature and Classification
- Isoenzymes or Isozymes
- Multienzyme Systems
- Biological Roles of Enzymes

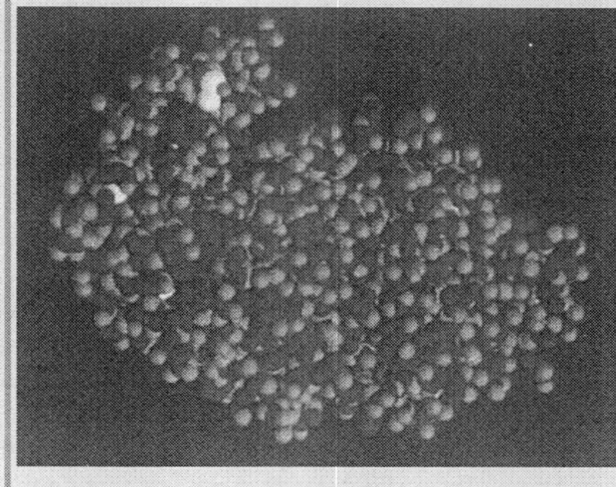

A computer-generated model of pepsin
This globular-protein enzyme digests the food which is ingested.

DEFINITION

Life is an intricate meshwork involving a perfect coordination of a vast majority of chemical reactions. Some of these reactions result in synthesizing large molecules, others in cleaving large molecules and all of them either utilize energy or liberate energy. All these reactions occur very slowly at the low temperatures and the atmospheric pressures—the conditions under which living cells carry on their life processes. Yet in the living cells these reactions proceed at extremely high rates. This is due to the presence of some catalysts produced and synthesized inside the body of the organisms. The term 'enzyme' was coined in 1878 by **Friedrich Wilhelm Kuhne** to designate these **'biological catalysts'** that had previously been called 'ferments'. As they quicken most of the chemical reactions occurring in the body, the enzymes have been designated as the "*manifestations of nature's impatience*".

> In fact, Kuhne intended for the name **enzyme** to apply to both yeast ferment and the extracellular catalysts of more complex animals, e.g., *pepsin and trypsin*. Prophetically, his definition specifically implied that lower and higher forms of life are not so fundamentally different.

The name **'enzyme'** (en^G = in ; $zyme^G$ = yeast) literally means 'in yeast'. This was referred to denote one of the most noteworthy reactions wherein the production of ethyl alcohol and carbon dioxide through the agency of an enzyme, the *zymase*, present in yeast takes place. This reaction is most popularly known as *alcoholic fermentation.* Sumner and Myrbäck (1950) have beautifully defined the enzymes as 'simple or combined proteins acting as specific catalysts'. They are soluble, colloidal molecules which are produced by living cells. All enzymes are globular proteins with a complex 3-'D' structure, capable of binding substrate molecules to a part of their surface. They affect the life of an organism to such an extent that life has aptly been called as "an orderly function of enzymes."

As an analogy, if one considers a living cell as a factory, the individual enzymes might be compared to the machines that work together to cause the transformation of a raw material (like steel) into parts of a finished product (like an automobile).

Difference from catalysts. Like catalysts, the enzymes do not alter the chemical equilibrium point of a reversible reaction but only the speed of the reaction is changed. They, however, differ from catalysts in being the biological products, *i.e.*, produced from the living cells. Moreover, the enzymes are all protein and, unlike catalysts, cannot last indefinitely in a reaction system since they, being colloidal in nature, often become damaged or inactivated by the reactions they catalyze. Henceforth, they must be replaced constantly by further synthesis in the body. Furthermore, unlike catalysts, most individual enzymes are very specific in that they act either on a single or at the most on some structurally-related substrates.

Difference from vitamins. All enzymes are proteinaceous in nature and differ from vitamins in the fact that the latter are not synthesized by the animal cells, in contrast to the former.

Endoenyzmes and exoenyzmes. Most of the enzymes usually act within the cells in which they are produced and hence are called **intracellular enzymes** or **endoenzymes**, *e.g.* most of the plant enzymes. As these enzymes catalyze the metabolic reactions of the cell, they are also referred to as *metabolic enzymes*. On the other hand, certain enzymes which are liberated by living cells, catalyze useful reactions outside the cell in its environment and hence are known as **extracellular enzymes** or **exoenzymes**, *e.g.*, enzymes found in bacteria, fungi and some insectivores like *Drosera* and *Nepenthes*. They act chiefly as digestive enzymes, catalyzing the breakdown of complex substances to simpler ones which can readily be absorbed by the cell.

LOUIS PASTEUR (LT, 1822–1895)

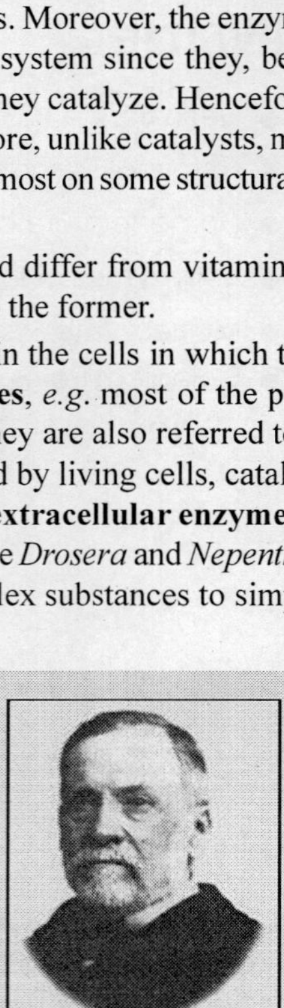

Louis Pasteur, a French chemist and biologist, was born on December 27,1822 in Dole, France into a family of poor tanners. His father Joseph Pasteur was a sergent-major in Nepoleon's army and his valour had earned him the Legion of Honour. Louis did not seem to be particularly outstanding at school but was an excellent artist and made numerous portraits of his family. He studied at the Paris school, Ecole Normale Superieure, a school founded by Napoleon to train professors, and graduated as a Ph.D. in Sciences. In 1848, he achieved distinction in organic chemistry for his discovery that tartaric acid, a C-4 compound, forms two different types of crystals and successfully separated the two forms while looking through a microscope. He formulated a fundamental law : asymmetry differentiates the organic world from the mineral world. His work became the basis of a new science– **stereochemistry**. In January 1849, he became an Assistant Professor of Chemistry at the University of Strasborough and soon after on May 29, 1849 married Marie Laurent, daughter of the University principal. She often discussed his work, spurred his thinking, and was one of the best scientific collaborators. The couple had 5 children, but only 2 survived to adulthood. And in September 1854, at the age of 32, he was appointed full-fledged Professor of Chemistry and Dean at the University of Lille in Northern France. Three years later (*i.e.*, in 1857), he was called back and appointed as Research Director at the Ecole Normal on the Rue d'Ulm in Paris. As a researcher, he showed remarkable observational skill which allowed him to discover what other people looked at but did not see.

The same year, 1857, he unfolded the mystery of why local wines were turning sour. The prevailing theory held that wine fermentation results from the chemical breakdown of grape juice to alcohol. Pasteur, however, saw yeast cells under microscope and believed that yeasts played a major role in fermentation. In a classic series of experiments, he clarified the role of **yeasts** in fermentation and showed that the sticks and rod-like structures (now known as bacteria) were responsible for making the wine sour. Pasteur's work also indicated that the bacteria could be a cause of disease, for if they could sour the wine, perhaps they could also make the body ill or diseased. He, henceforth, held the view that microorganisms are responsible for infectious diseases. He, thus, set down the foundation for the **germ theory of disease** and also founded the science of **Bacteriology.**

Pasteur also recommended a practical solution to the sour wine problem. He suggested that grape juice be heated to 55°C for several minutes to destroy all the evidence of life, after which yeasts could be added to begin the fermentation. Acceptance of this technique, known as **pasteurization** (which is also applied to milk), gradually ended the problem. Pasteur's elation was tempered with sadness as his daughter Jeanne died of typhoid fever in 1859.

Finally, in an elegant series of experiments using swan-neck flasks, Pasteur successfully performed his meticulous experiments in public for the Academy of Science and silenced all but the ardent supporters of spontaneous generation hypothesis. By now, Pasteur became a national celebrity. And the theory of spontaneous generation was given a burial after holding sway for 20 centuries! In 1862, he was elected to the Academy. Once again, tragedy struck him as his two-year-old daughter Camille developed a tumour and died of blood poisoning in September 1865. Pasteur, thus, realized that he was no closure to solving the riddle of disease.

The same year (*i.e.*, 1865) **cholera** engulfed Paris, killing about 200 people daily! Pasteur tried to capture the responsible bacterium by filtering the hospital air and trapping the bacterium in cotton. Unfortunately, he was unable to cultivate one bacterium apart from the others as he was using broth (Later, Robert Koch used solid culture media, instead of broth media).

In order to help the ailing French industry, Pasteur turned to pebrine, the disease of silkworms. Late in 1865, he identified a protozoan infesting the sick silkworms and the mulberry leaves fed to them. Then, he separated the healthy silkworms from the diseased ones and their food, and he managed to check the spread of disease. His success further endorsed the germ theory of disease. It was again a terrible blow to Pasteur when twelve-year-old Cecille, another of his daughters died of typhoid fever in May 1866. This diverted his attention again to the study of human diseases, and he worked still harder. At around the same time, he himself suffered a brain hemorrhage which left him paralyzed on the left side of his body. But all this could not keep him down for long. In 1873, he was elected to the Academy of Medicine. Later, he developed **vaccines** against chicken cholera in poultry (1880), anthrax in sheep (1881) and most notably rabies in dogs (1885). He also discovered 3 bacteria responsible for human illnesses : *Staphyllococcus, Streptococcus* and *Pneumococcus.*

In fact, Pasteur's genius reached a peak with his development of a vaccine for rabies ('rage' or 'madness' in Latin). To prepare the vaccine, he inoculated rabbits with the brain tissue of rabid animals. He then removed the spinal cords and dried them. Next, he inoculated experimental animals with 15-day-old cord tissue and followed this up the next day with 14-day-old cord tissue and so on for two weeks. Animals so treated did not develop rabies.

Pasteur was a sturdy and unitiring worker who drove himself and his subordinates mercilessly. At heart, however, he was quite sentimental as the following event displayed.

On July 6, 1885, a 9-year-old boy named Joseph Meister was brought to Pasteur. Only two days ago, the boy had suffered as many as 18 bites from a rabid dog, and physicians assured Pasteur of the imminent horrible death, to the boy. The boy's only hope was the vaccine. But Pasteur's vaccine had never been tested on humans, and the scientist, now 63, could not bring himself to make an immediate decision. In his writing, he recalls the incident :

> "The child's death appeared inevitable. I decided not without acute and harrowing anxiety, as may be imagined, to apply to Joseph Meister the method I had found consistently successful with dogs."

The next day, Pasteur began the treatment. He turned the first sample of vaccine over to two physicians from the Academy of Medicine and watched as they gave 12 successive injections in 10 days of the vaccine to the terrified, crying child. Each day his concern lessened as the injections proceeded smoothly. The vaccine appeared to be working. Meister survived his ordeal and the scientist heaved a sigh of relief. When Meister left Paris for his village, Pasteur gave him stamped envelopes so that he could write often. Pasteur's viewpoint in dealing with a disease was well known. In his own words :

> "When meditating over a disease, I never think of finding a remedy for it, but, instead, a means of preventing it."

The defeat of rabies was Pasteur's crowing glory. After Pasteur saved the boy Joseph Meister from the dreaded rabies in 1885, French government on November 4,1888, established the **Pasteur Institute** in Paris to treat cases of rabies, one of the world's foremost scientific establishments. The institute was inaugurated by the then President of France Sadi Carnot. Many monetary rewards were floated towards the institute, including a generous gift from the Russian government after Pasteur immunized 20 Russians against rabies. Louis continued working till 1887, when he suffered another paralytic stroke which prevented him from personally doing experimental work but his dialogue with his pupils and his collaborators never ceased. Pasteur presided over the Institute until his death on September 28, 1895 in Villeneuve-lEtang. He was given a State Funeral at the Cathedral of Notre Dame and his body placed in a permanent crypt at the Pasteur Institute. He would teach his disciples :

> "Do not put forward anything that you cannot prove by experimentation."

For nearly half a century, he had dominated the scientific world; for a quarter of a century, he had surged ahead despite a half-paralyzed body. Such was his genius! In later years, a grateful Meister returned to Paris as caretaker of the Pasteur Institute and in 1940 preferred commiting suicide rather than obey the demands of some occupying Hitler's Nazi soldiers to open Pasteur's crypt.

Pasteur was perhaps best known to the French nation as the **'saviour of the wine industry'** because his pasteurization salvaged an ailing industry beset with problems of microbial contamination. He is truly called as the **'father of microbiology'**.

The life of Pasteur stands as a supreme testimony to the fact that the best and most far-reaching applications of science come from passionate studies of seemingly esoteric subjects. He evolved into a prophet whose vision accelerated the progress of science. He was a true **benefactor of mankind**.

Pasteur's famous quote reads as follows :

"Dans les champs de l'observation, le hasard ne favorise que les esprits prepares."

The English transliteration is :

"In the field of observation, chance favours only prepared minds."

HISTORICAL RESUME

The history of enzymes may be regarded as commencing with the work of **Dubrunfaut** (1830) who prepared malt extract from germinating barley seeds. This extract possessed the power of converting starch into sugar. Later in 1833, **Payen** and **Persoz** prepared an enzyme, the *diastase* (now known as *amylase*), from malt extract by precipitation from alcohol. The same year, **Horace de Saussure** prepared a substance from germinating wheat which acted like diastase, *i.e*., converted starch into sugar. Within the next few years, **Theodor Schwann** succeeded in extracting *pepsin*, which digests meat (protein), from gastric juice and he later identified *trypsin*, a peptidase in digestive fluids. Hence, the notion of diastases (the early name for enzymes) was soon extended to animals. By 1837, the famous chemist **Jönes Jacob Berzelius** recognized with remarkable foresight the catalytic nature of these biological diastases.

> Anselme Payen, a chemist-industrialist was the owner of the sugar factory in Paris. When he and Jean-Francois Persoz called the first known biological catalyst "diastase", they introduced the now historical use of the suffix **-ase** in the naming of most enzymes.

Later, **Pasteur** (1857) demonstrated that alcoholic fermentation was brought about by the action of living yeast cells. It, thus, became apparent that such catalytic actions could be induced by the action of living cells (as in alcoholic fermentation) or by nonliving substances (*as diastase* and *emulsin*). Obviously, such catalysts or ferments were regarded as forming two classess : **organized ferments** (like yeasts and certain bacteria) which were living cells and **unorganized ferments** (like diastase and emulsin) which acted independently of living cells.

In fact, the term 'enzyme' was later proposed for the unorganized ferments by **Kuhne**. On the contrary, **Eduard Buchner** and his brother **Hans Buchner** (1897) showed that, besides the living yeast cells, even their extract (which **Eduard** named as *zymase*) could bring about alcoholic fermentation. In fact, the Buchners discovered cell-free fermentation when they attempted to preserve

EDUARD BUCHNER
(LT, 1860–1917)

Buchner, a German biochemist, studied under Prof. Adolf von Baeyer and later became his assistant. He received the **Nobel Prize in Chemistry in 1907** "for his biochemical researches and his discovery of cell-less fermentation." In fact, Buchner, in 1897, reported that he had prepared from brewer's yeast, a cell-free press juice that caused CO_2 and ethyl alcohol to form in solutions of various sugars (sucrose, glucose, fructose, maltose). And he concluded :

"the initiation of the fermentation process does not require so complicated an apparatus as is represented by the yeast cell. The agent responsible for the fermenting action of the press juice is rather to be regarded as a dissolved substance, doubtless a protein; this will be denoted **zymase.**"

In fact, his discovery of zymase was the first proof that fermentation was caused by enzymes and did not require the presence of living cells. Edward, a major in the German army, died in action on the Romanian front (1917) during World War I.

JAMES BATCHELLER SUMNER
(LT, 1887-1955)

After Sumner lost his left arm as a result of hunting at age 17, he was discouraged by his teachers from pursuing a career in Chemistry. They felt that Sumner was too handicapped for the profession. However, his illustrious career belied all this. He is truly called as the **'father of modern enzymology'.** Sumner received the **1946 Nobel Prize in Chemistry** for crystallizing the first enzymes, along with his compatriots **John H. Northrop** and **W.M. Stancey** for their work on preparation of enzymes and virus proteins in pure form.

their yeast extracts with sugar, the preservative of jellies and jams. The words—ferment and enzyme—thus became synonym and the latter is now frequently used in the literature. The use of the term ferment (*fermentare*[L] = to agitate) is, however, misleading as it also denotes the fermenting microorganisms.

James B. Sumner (1926), at Cornell University, for the first time isolated and purified an enzyme, the *urease*, in crystalline form from jack bean (*Canavalia ensiformis*), thus confirming the proteinaceous nature of the enzymes. Since then, some 250 enzymes have been obtained in pure crystalline form such as :

(*a*) Trypsin from beef pancreas by **John H. Northrop** and **Kunitz** (1936)
(*b*) Catalase from beef liver by **Sumner** and **Dounce** (1937)
(*c*) RNase from beef pancreas by **Kunitz** (1940)
(*d*) Pepsin from swine stomach by **John Howard Northrop** (1946)
(*e*) DNase from beef pancreas by **Kunitz** (1950)

NOMENCLATURE AND CLASSIFICATION

With the continuous increase in our knowledge of enzymology, various systems have evolved to name and classify the enzymes, using one or the other criterion as the basis. However, many of the enzymes were known before these systems of naming enzymes were adopted. The names of such enzymes were not changed under the new systems. In this category belong : bromelin, chymotrypsin, diastase, emulsin, papain, pepsin, ptyalin, rennin, trypsin etc.

A. Substrate acted upon by the enzyme. The substance upon which an enzyme acts is called the substrate. **Duclaux** (1883) named the enzymes by adding the suffix *-ase* in the name of the substrate catalyzed. For example, enzymes acting upon carbohydrates were named as *carbohydrases*, upon proteins as *proteinases*, upon lipids as *lipases*, upon nucleic acids as *nucleases* and so on. A few of the names were even more specific like *maltase* (acting upon maltose), *sucrase* (upon sucrose), *urease* (upon urea), *lecithinase* (upon lecithin), *tyrosinase* (upon tyrosine) etc.

B. Type of reaction catalyzed. The enzymes are highly specific as to the reaction they catalyze. Hence, this has necessitated their naming by adding the suffix-*ase* in the name of the reaction; for example *hydrolases* (catalyzing hydrolysis), *isomerases* (isomerization), *oxidases* (oxidation), *dehydrogenases* (dehydrogenation), *transaminases* (transamination), *transaldolases* (transaldolation), *transketolases* (transketolation), *phosphorylases* (phosphorylation) etc.

> In fact, the consensus at that time in Europe was that enzymes were not proteins, because of the findings of the influential German chemist **Richard Willstätter** (**Nobel Laureate, 1915**), who, in 1920s, reported that he could not detect protein in purified enzyme preparations from yeast. With hindsight, it is realized that the protein assays used in that era were not sensitive enough to detect the small amounts present in Willstätter's purified preparations. The nonprotein nature of enzyme was so entrenched in scientific thinking that decades passed before their polypeptidyl composition was unequivocally accepted.

Although these two systems are quite simple and easy to follow, there are certain discrepancies present in them. The former system does not take into account the type of the reaction catalyzed, whereas in the latter system no idea can be derived regarding the nature of the substrate acted upon by the enzyme.

C. Substrate acted upon and type of reaction catalyzed. The names of some enzymes give clue of both the substrate utilized and the type of reaction catalyzed. For example, the enzyme *succinic dehydrogenase* catalyzes the dehydrogenation of the substrate succinic acid. Similarly, *L-glutamic dehydrogenase* indicates an enzyme catalyzing a dehydrogenation reaction involving *L-glutamic* acid.

D. Substance that is synthesized. A few enzymes have been named by adding the suffix *-ase* to the name of the substance synthesized, *viz.*, *rhodonase* that forms rhodonate irreversibly from hydrocyanic acid and sodium thiosulphate, and also *fumarase* that forms fumarate irreversibly from L-malate.

E. Chemical composition of the enzyme. Based on their chemical composition, the enzymes have been classified into following three categories :

1. Enzyme molecule consisting of protein only— *e.g.*, pepsin, trypsin, urease, papain, amylase etc.
2. Enzyme molecule containing a protein and a cation— *e.g.*, carbonic anhydrase (containing Zn^{2+} as cation), arginase (Mn^{2+}), tyrosinase (Cu^{2+}) etc.
3. Enzyme molecule containing a protein and a nonprotein organic compound known as prosthetic group—Tauber (1950) has further subdivided them, on the basis of the nature of prosthetic group involved :
   (*a*) Iron prophyrin enzymes— catalase, cytochrome *c* peroxidase I and II.
   (*b*) Flavoprotein enzymes— glycine oxidase, pyruvate oxidase, histamine.
   (*c*) Diphosphothiamin enzymes — β-carboxylase, pyruvate mutase.
   (*d*) Enzymes requiring other coenzymes— phosphorylase, amino acid decarboxylase.

F. Substance hydrolyzed and the group involved.

1. Carbohydrate-hydrolyzing enzymes
   (*a*) Glycosidases—cellulase, amylase, sucrase, lactase, maltase
   (*b*) β-glucorinidase
2. Protein-hydrolyzing enzymes
   (*a*) Peptide bonds
      I. Endopeptidases
         Animals— pepsin, trypsin, rennin
         Plants—papain ficin, bromolin
      II. Exopeptidases—dipeptidase, tripeptidase
   (*b*) Nonpeptide C—N linkages (amidases)
   urease, arginase, glutaminase
3. Lipid-hyrolyzing enzymes
   lipases, esterases, lecithinases
4. Other ester-hydrolyzing enzymes
   (*a*) Phosphatases
   (*b*) Cholinesterases
   (*c*) Chlorophyllases
   (*d*) Sulfatases
   (*e*) Pectinesterases
   (*f*) Methylases
5. Oxidation-reduction enzymes
   hydrases, mutases, oxidases, dehydrogenases, peroxidases
6. Miscellaneous enzymes
   catalase, carboxylase, carbonic anhydrase, thiaminase, transpeptidase

The **I.U.B. system** is based on the report of the Commission on Enzymes, International Union of Biochemistry Symposium, Vol. 20, Pergamon Press, New York, 1961. In fact, the international symbolic language of chemistry was originally developed by **Jönes Jacob Berzelius**, who proposed that an element be identified by the initial letter or first two letters of its Latin name.

G. Over-all chemical reaction taken into consideration. The chemical reaction catalyzed is the specific property which distinguishes one enzyme from another. In 1961, *International Union of Biochemistry* (*I.U.B.*) used this criterion as a basis for the classification and naming of enzymes. Although complicated, the I.U.B. system is precise, descriptive and informative.

The major features of this system of classification of enzymes are as follows :

(*a*) The reactions and the enzymes catalyzing them are divided into 6 major classes, each with 4 to 13 subclasses.

(*b*) Each enzyme name has 2 parts—the first part is the name of the substrate(s) and the second part which ends in the suffix *-ase*, indicates the type of reaction catalyzed.

(*c*) Additional information regarding the nature of the reaction, if needed, is given in parenthesis. For example, the enzyme *malate dehydrogenase* catalyzes the following reaction :

$$\text{L-malate} + NAD^+ \longrightarrow \text{Pyruvate} + CO_2 + NADH + H^+$$

This enzyme has now been designated as L-malate : NAD oxidoreductase (decarboxylating).

(*d*) Each enzyme has been alloted a systemic code number called Enzyme Commission (E.C.) number. The E.C. number for each enzyme consists of a series of numbers at 4 places : the first place numbers representing the major class to which the enzyme belongs, the two median numbers denoting the subclass and the sub-subclass of the enzyme within the major class. The last place number or the fourth digit represents the serial number of the enzyme within the sub-subclass. Thus E.C. 2.7.1.1 represents class 2 (a transferase), subclass 7 (transfer of phosphate), sub-subclass 1 (an alcohol group as phosphate acceptor). The final digit denotes the enzyme, *hexokinase* or *ATP*: *D-hexose-6-phosphotransferase*. This enzyme catalyzes the transfer of phosphate from ATP to the hydroxyl group on carbon 6 of glucose.

$$\text{ATP} + \text{D-hexose} \xrightarrow{\text{Hexokinase}} \text{ADP} + \text{Hexose-6-phosphate}$$

(*e*) Where no specific category has been created for an enzyme, it is listed with a final figure of 99 in order to leave space for new subdivisions. For example, 4.2.99 refers to "other carbon-oxygen lyases."

The 6 major classes of enzymes with some important examples from some subclasses are described below :

1. Oxidoreductases. This class comprises the enzymes which were earlier called dehydrogenases, oxidases, peroxidases, hydroxylases, oxygenases etc. The group, in fact, includes those enzymes which bring about oxidation-reduction reactions between two substrates, S and S′.

$$S_{reduced} + S'_{oxidized} \longrightarrow S_{oxidized} + S'_{reduced}$$

More precisely, they catalyze electron transfer reactions. In this class are included the enzymes catalyzing oxidoreductions of CH—OH, C=O, CH—CH, $CH—NH_2$ and CH=NH groups. Some important subclasses are :

1.1 Enzymes acting on CH—OH group of electron donor. For example :

1.1.1.1 Alcohol : NAD oxidoreductase

[common or trivial name, Alcohol dehydrogenase]

This enzyme catalyzes the following reaction :

$$\text{Alcohol} + \text{NAD} \longrightarrow \text{Aldehyde or Ketone} + NADH + H^+$$

1.3 Enzymes acting on CH—CH group of electron donor. For example :

1.3.2.2 Acyl-CoA : cytochrome c oxidoreductase

[Acyl-CoA dehydrogenase]

Acyl-CoA + oxidized cytochrome c ⟶ 2,3-dehydroacyl-CoA + reduced cytochrome c

1.9 Enzymes acting on the heme groups of electron donors. For example :

1.9.3.1 Cytochrome c : O_2 oxidoreductase

[Cytochrome oxidase]

4. reduced cytochrome + O_2 + $4H^+$ ⟶ 4 oxidized cytochrome c + $2H_2O$

1.11 Enzymes acting on H_2O_2 as electron acceptor. For example :

1.11.1.6 H_2O_2: H_2O_2 oxidoreductase

[Catalase]

$$H_2O_2 + H_2O_2 \rightleftharpoons 2H_2O + O_2$$

> Although not an oxidizing enzyme, **catalase** is usually classified with oxidases because its action is closely connected with physiological oxidation.

2. Transferases. Enzymes which catalyze the transfer of a group, G (other than hydrogen) between a pair of substrates, S and S′ are called transferases.

$$S—G + S' \longrightarrow S + S'—G$$

In these are included the enzymes catalyzing the transfer of one-carbon groups, aldehydic or ketonic residues and acyl, glycosyl, alkyl, phosphorus or sulfur-containing groups. Some important subclasses are :

2.3 Acyltransferases. For example :

2.3.1.6 Acetyl-CoA : choline O-acetyltransferase

[Choline acetyltransferase]

Acetyl-CoA + choline ⟶ CoA + O-acetylcholine

2.4 Glycosyltransferases. For example :

2.4.1.1 α-1, 4-Glucan : orthophosphate glycosyl transferase

[Phosphorylase]

$(\alpha\text{-1, 4-Glucosyl})_n$ + orthophosphate ⟶ $(\alpha\text{-1, 4-Glycosyl})_{n-1}$ + α-D-glucose-1-phosphate

2.7 Enzymes catalyzing the transfer of phosphorus-containing groups. For example :

2.7.1.1 ATP : D-hexose-6-phosphotransferase

[Hexokinase]

ATP + D-hexose ⟶ ADP + D-hexose-6-phosphate

3. Hydrolases. These catalyze the hydrolysis of their substrates by adding constituents of water across the bond they split. The substrates include ester, glycosyl, ether, peptide, acid-anhydride, C—C, halide and P—N bonds. Representative subclasses are :

3.1 Enzymes acting on ester bonds. For example :

3.1.1.3 Glycerol ester hydrolase

[Lipase]

A triglyceride + H_2O ⟶ A diglyceride + a fatty acid

3.2 Enzymes acting on glycosyl compounds. For example :

3.2.1.23 β-D-galactoside galactohydrolase

[β-galactosidase]

A β-D-galactoside + H_2O ⟶ An alcohol + D-galactose

3.4. Enzymes acting on peptide bonds

Here the classical trivial names (pepsin, trypsin, thrombin, plasmin etc.) have been largely retained due to their consistent long usage and also due to dubious specificities which make systematic nomenclature almost impractical at this time.

3.5 Enzymes acting on C—N bonds, other than peptide bonds. For example :

3.5.3.1 L-arginine ureohydrolase

[Ariginase]

L-arginine + H_2O ⟶ L-ornithine + urea

4. Lyases (= Desmolases). These are those enzymes which catalyze the removal of groups from substrates by mechanisms other than hydrolysis, leaving double bonds.

$$\begin{array}{cc} X & Y \\ | & | \\ C - & C \end{array} \longrightarrow C{=}C + X{-}Y$$

In these are included the enzymes acting on C—C, C—O, C—N, C—S and C—halide bonds. Important subclasses include :

4.1 Carbon-carbon lyases. For example :

4.1.2.7 Ketose-1-phosphate aldehyde-lyase

[Aldolase]

A ketose-1-phosphate $\longrightarrow$ Dihydroxyacetone phosphate + an aldehyde

4.2 Carbon-oxygen lyases. For example :

4.2.1.2 L-malate hydro-lyase

[Fumarase]

L-malate $\longrightarrow$ Fumarate + H_2O

4.3 Carbon-nitrogen lyases. For example :

4.3.1.3 L-histidine ammonia-lyase

[Histidase]

L-histidine $\longrightarrow$ Urocanate + NH_3

5. Isomerases. These catalyze interconversions of optical, geometric or positional isomers by intramolecular rearrangement of atoms or groups. Important subclasses are :

5.1 Racemases and epimerases. For example :

5.1.1.1 Alanine racemase

L-alanine $\longrightarrow$ D-alanine

5.2 *Cis-trans* isomerases. For example :

5.2.1.3 *All trans*-retinene 11-*cis-trans* isomerase

[Retinene isomerase]

All *trans*-retinene $\longrightarrow$ 11-*cis*-retinene

5.3 Intramolecular oxidoreductases. For example :

5.3.1.9 D-glucose-6-phosphate keto-isomerase

[Glucosephosphate isomerase]

D-glucose-6-phosphate $\longrightarrow$ D-fructose-6-phosphate

6. Ligases (*ligare*[L] = to bind) **or Synthetases.** These are the enzymes catalyzing the linking together of two compounds utilizing the energy made available due to simultaneous breaking of a pyrophosphate bond in ATP or a similar compound. This category includes enzymes catalyzing reactions forming C—O, C—S, C—N and C—C bonds. Important subclasses are :

6.2 Enzymes catalyzing formation of C—S bonds. For example :

6.2.1.1 Acetate : CoA ligase (AMP)

[Acetyl-CoA synthetase]

ATP + acetate + CoA $\longrightarrow$ AMP + pyrophosphate + acetyl-CoA

6.3 Enzymes catalyzing formation of C—N bonds. For example :

6.3.1.2 L-glutamate : ammonia ligase (ADP)

[Glutamine synthetase]

ATP + L-glutamate + NH_3 $\longrightarrow$ ADP + orthophosphate + L-glutamine

6.4 Enzymes catalyzing formation of C—C bonds. For example :

6.4.1.2 Acetyl-CoA : CO_2 ligase (ADP)

[Acetyl-CoA carboxylase]

$$ATP + acetyl\text{-}CoA + CO_2 + H_2O \longrightarrow ADP + orthophosphate + malonyl\text{-}CoA$$

To date, over 2,000 different enzymes are known, of which the oxidoreductases, transferases and hydrolases predominate. Because official names are often lengthy, the trivial names of enzymes are generally used after initial identification.

ISOENZYMES OR ISOZYMES

Many enzymes occur in more than one molecular form in the same species, in the same tissue or even in the same cell. In such cases, the different forms of the enzyme catalyze the same reaction but since they possess different kinetic properties and different amino acid composition, they can be separated by appropriate techniques such as electrophoresis. Such multiple forms of the enzymes are called **isoenzymes** or **isozymes**. Isozymes are of widespread nature. Over a hundred enzymes are now known to be of isozymic nature and consequently occur in two or more molecular forms. *Lactic dehydrogenase*, LDH (E.C. No 1.1.1.27), for example, is an enzyme wihch exists in 5 possible forms in various organs of most vertebrates. LDH catalyzes the reversible oxidation-reduction reaction :

> **Isoenzymes** is in fact, the preferable spelling according to the International Union of Biochemistry.

$$Lactate + NAD^+ \rightleftharpoons Pyruvate + NADH + H^+$$

Two basically-different types of LDH are found :

(*a*) heart *LDH*. This predominates in the heart and is active at low levels of pyruvate. This has 4 identical subunits called *H* subunits (H for heart).

(*b*) *muscle LDH*. This is characteristic of many skeletal muscles and maintains its activity in much higher concentrations of pyruvate. This also has 4 identical subunits called M *subunits* (*M* for muscle) which are enzymatically inactive.

The two types of subunits, H and M, have the same molecular weight (35,000) but differ in amino acid composition and in immunological properties. The two subunits are produced by two separate genes. LDH can be formed from two types of polypeptide chains designated as H and M subunits to yield a pure H tetramer and a pure M tetramer. Combinations of H and M subunits will, however, produce 3 additional types of hybrid enzymes, thus making the total number of possible forms as five (refer Fig. 16-1). This is confirmed by the fact that when the 2 subunits are mixed in equal proportions, a sequence of 5 bands is obtained by electrophoresis. The 5 different LDH enzyme forms are designated as H_4, H_3M, H_2M_2, HM_3 and M_4. The various isozyme forms of LDH differ significantly in the maximum activities V_{max}, in the Michaelis constant Km for their substrates, especially for pyruvate and in the degree of their allosteric inhibition by pyruvate.

The relative concentrations of the LDH enzymes differ from one type of tissue to another, and the relative composition is characteristic of the tissue. For example, LDH-1 plus LDH-2 makes up about 60% of the LDH of cardiac muscle, while LDH-4 plus LDH-5 constitutes about 80% of the LDH of liver. Thus, a diagnosis based on elevated serum LDH can be made more specific with respect to the organ involved by an electrophoretic analysis of the LDH isozymes. For example, the elevated serum level of LDH in a myocardial infarct is primarily due to an increase in LDH-1 and LDH-2; while in liver disease the elevated serum LDH is primarily due to increases in LDH-4 and LDH-5.

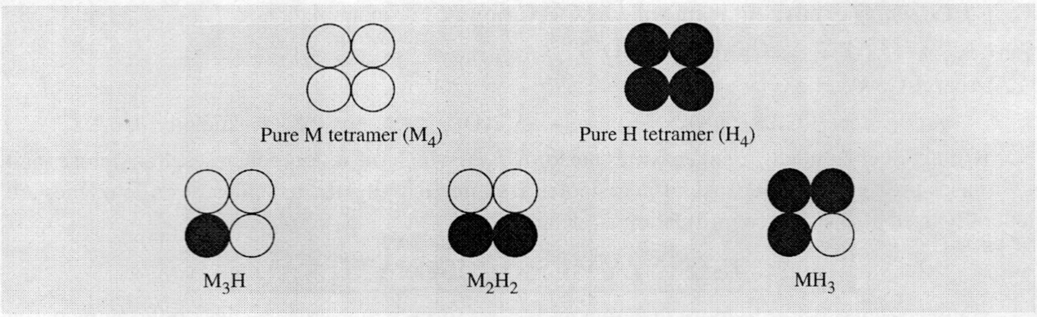

Fig. 16-1. Possible combinations of H and M subunits of lactic dehydrogenase

Damages to tissues and cell death both result in a release of intracellular enzymes from damaged cells into the blood. Therefore, the concentrations of these enzymes increase in the blood serum and can serve as a valuable diagnostic aid in a number of diseases, including myocardial infarction, pancreatitis, liver disease and prostatic cancer.

The serum concentrations of 3 enzymes are often used in diagnosis of myocardial infarction (MI). These are creatinine phosphokinase (CPK), glutamate oxaloacetate transaminase (GOT), and lactate dehydrogenase (LDH). In more than 95% of MI patients, serum levels of GOT rise rapidly and then return to normal in 4—5 days. The serum levels of CPK characteristically rise and fall even more rapidly than GOT levels. However, the serum levels of LDH typically rise and fall more slowly, returning to normal levels in about 10 days.

There seems to be an interesting relation between LDH and the physiological role of flight in birds. In those birds grouped as *short flyers*, over 90% muscle LDH is present in breast muscle. In *long flyers*, breast muscle contains over 95% heart LDH which suggests a possible relation to sustained heart contraction. Those birds classed as *intermediate flyers* contain mixtures of both heart LDH and muscle LDH.

Other noteworthy examples of isozymes are *malic dehydrogenase* (*MDH*), *hexokinase, esterase and glycol dehydrogenase*.

MULTIENZYME SYSTEMS

A few examples of complex enzyme systems are known to exist. These are not independent molecules but occur as aggregates in a mosaic pattern involving several different enzymes. *Pyruvic acid dehydrogenase* of *E. coli* is one such example. This complex molecule has a molecular weight 4,800,000 and consists of 3 enzymes : 24 moles of pyruvate decarboxylase (90,000), 24 moles of pyruvate decarboxylase (90,000), 24 moles of dihydrolipoic dehydrogenase (55,000) and 8 subunits of lipoyl reductase transacetylase (120,000). Each component of this complex enzyme is so arranged as to provide an efficient coupling of the individual reactions catalyzed by these enzymes. In other words, the product of the first enzyme becomes the substrate of the second and so on.

BIOLOGICAL ROLES OF ENZYMES

The enzymes find many applications in our daily life. Enzymatic processes such as baking, brewing and tanning have been known from antiquity. The manifold applications of the enzymes are described below :

1. **Wine manufacturing.** Much of the early interest in enzymology was developed by scientists like **Pasteur, Payen** and **Persoz**, who were associated with food, wine, and beer industries. Pasteur was perhaps best known to the French nation as the "*saviour of the wine industry*" because his pasteurization process salvaged an ailing industry beset with problems of microbial contamination. *Papain* is used in brewing industry as a stabilizer for chill-proof beer, because it removes small amounts of protein that cause turbidity in chilled beer.

2. **Cheese making.** Since long the animal rennin (or *rennet*) is employed in making cheese. The enzyme rennet is obtained on a commercial scale from the fourth or true stomach of the unweaned calves which are specifically slaughtered for this purpose. One calf produces only 5 to 10 gm of rennet. The enzyme helps in coagulating the casein of milk. Certain preservatives (boric acid, benzoic acid or sodium chloride) are, sometimes, added to prevent decomposition of the enzyme preparations by bacteria. An enzyme *lipase* is added to cheese for imparting flavour to it. Many vegetarians are unaware that the cheese made in india contains animal rennet. However, an international charitable trust concerned with the welfare of animals, the Beauty Without Cruelty (BWC) has, with the help of Aurey Dairy, Mumbai, undertaken successful experimental trials in cheese making using nonanimal rennet.

> **Rennin** (=chymase) is a milk-curdling enzyme found in the stomach of ruminant mammals and is probably widely distributed among other mammals. By clotting and precipitating milk proteins, it apparently slows the movement of milk through the stomach. Human infants, who lack rennin, digest milk proteins with acidic pepsin, just as adults do. Rennin should not be confused with another phonetically similar enzyme called **renin** which is produced by the kidney. Renin catalyzes the synthesis of angiotensins which cause vasoconstriction in the kidneys, thereby causing electrolyte and water retention in the body.

3. **Candy making.** An enzyme, *invertase* helps preventing granulation of sugars in soft-centred candies. Another enzyme, *lactase* prevents formation of lactose crystals in ice cream which would otherwise not allow the product seem sandy in texture.
4. **Bread whitening.** *Lipoxygenase* is used for whitening the bread.
5. **Clarifying fruit-juices.** The enzymes are being used in processing of fruit juices such as apple juice and grape juice. The juices are clarified by adding a mixture of *pectic enzymes* which hydrolyze the pectic substances causing turbidity.
6. **Tenderizing meat.** Because hydroxyprolyl residues create bends in collagen helices, which contribute to the tough and rubbery texture often associated with cooked meat, treating the meat with a protease (*bromelain* or *papain*) prior to its cooking hydrolyzes peptide bonds, and thus tenderizes it.
7. **Desizing fabrics.** The woven fabrics are sized by applying starch to the warp (lengthwise) threads to strengthen the yarn before weaving. But when these fabrics are printed or dyed, the sizing should be removed. Desizing may be done by acids, alkalies or enzymes. Enzymatic desizing is, however, preferred as it does not weaken the fabrics. Enzymes for this purpose are obtained from a variety of sources including bacteria, fungi and malt.
8. **Destaining fabrics.** In drycleaning, the stains due to glue, gelatin or starch are removed by employing certain enzymes, such as *alcalase*.
9. **Dehairing hide.** In the manufacture of leather, the hide is made free from hair. This is done by *employing pancreatic enzymes* which hydrolyze the proteins of the hair follicles, thus freeing the hair so that it may be easily scraped off from the hide.
10. **Recovering silver.** *Pepsin* is used to digest gelatin in the process of recovering silver from photographic films.
11. **Correcting digestion.** When the enzymes are present insufficiently in the body, certain digestive disorders come up. These may be cured by supplying the lacking enzymes. *Pepsin, papain* and *amylases* aid digestion in the stomach while *pancreatic enzymes* act in the duodenum.
12. **Wound healing.** *Proteolytic enzymes* from pig pancreas are used to alleviate skin diseases, bed sores and sloughing wounds. These enzymes act by destroying proteolytic enzymes of man, that prevent the healing of such wounds. The enzymes commonly used for wound debridement are the proteases such as *streptodornase*, *ficin* and *trypsin*.

13. **Analyzing biochemicals.** Certain enzymes are used in clinical analysis. For example, *uricase* and *urease* are employed in the determination of uric acid and urea respectively in blood. Besides, sucrose and raffinose contents in sugar mixtures are determined by polarimetry before and after treating the solutions with the enzymes, *sucrase* and *melibiase*.
14. **Dissolving blood clot.** The enzyme *urokinase*, which is manufactured from urine, is being used effectively in Japan in the treatment of blood clot in brain, artery and other circulatory diseases. A team of Soviet scientists led by **Yevgeni Chazov**, Director of the National Cardiological Research Centre, Moscow, have, in 1982, developed an effective enzyme *streptodekase*, which can dissolve blood clots in vessels. The new enzyme is particularly useful in preventing heart attacks as clots are responsible for 9 out of 10 fatal cases of cardiac arrests.
15. **Changing the blood type.** In 1981, **Prof. Ken Furukawa** and his associates of the Gunma University's Medical School, Japan have successfully employed several types of specific enzymes in an epoch-making experiment to freely change human blood types. They found that the composition of polysaccharide on the surface of blood corpuscles determines each person's type of blood. Different kinds of sugar characteristics of each blood type form on the surface of RBCs due to the function of a synthetic enzyme. If the sugar is separated from the surface of RBCs by using a specific decomposition enzyme, type A blood and type B blood can be reverted to type O, the prototype of the two blood types. If this breakthrough can be put to practical use, it will fulfill a long-cherished dream of doctors to administer blood transfusions irrespective of the type of blood a patient has by merely changing the patient's blood type to match the blood available.
16. **Diagnosing hypertension.** A new method called radio immunoassay procedure for diagnosing cases of hypertension has been developed by Bhabha Atomic Research Centre (BARC). In it, the activity of *renin*, a proteolytic enzyme secreted by the kidneys, is calculated indirectly by measuring angiotensin-I which is formed by the action of renin. Renin acts as part of a complex feedback mechanism for regulating blood volume and pressure.
17. **Augmenting surgery.** A technique using the enzyme *trypsin* as an adjunct to cataract surgery has been developed in 1980 by **Dr. Joseph Spina Jr.** of Philadelphia. With older techniques it required an incision about 2.5 cm long in the white of the eye to remove the clouded lens. Modern microsurgery has, however, reduced this cut to only 0.3 cm. But Dr. Spina's method involves a still smaller cut wide enough for a needle 0.025 cm wide. The hollow needle is used to inject a microscopic amount of trypsin, a digestive enzyme secreted by the pancreas. Trypsin digests and liquefies the semisolid interior of the lens without harming other parts of the eye. Once the enzyme liquefies the lens—which takes from a few hours to overnight—the lens is removed by suction through the same hollow needle. This eliminates the necessity of intervention in the eye, the constant passing in and out of the instruments and suturing. The lesser the tissue is wounded, the quicker it recovers. This enzyme surgery for cataracts could be done as an outpatient operation. The patients would come in one day to have the enzyme injected and return the next day to have the cataract removed.

CLINICAL IMPLICATIONS

A **cataract** is a clouding of eye lens which can cause blindness. It is the most common eye disease and comes with age. The lens of the eye is made of protein and as time goes by, there occurs denaturation process and the protein gets cloudy and one cannot see. India, at present, has 18 million blind people, out of which 12 million are cataract patients.

Table 16-1 outlines some of industrial applications of enzymes. This is a rapidly changing field, and new applications of enzyme technology appear all the time.

Table 16.1 Some Applications of Enzymes

Enzyme	*Reaction*	*Source of enzyme*	*Application*
Industrial applications			
α-amylase	breaks down starch	bacteria	converts starch to glucose in the food industry
Glucose isomerase	converts glucose to fructose	fungi	production of high fructose syrups
Proteases	digest protein	bacteria	washing powder
Rennin	clots milk protein	animal stomach linings; bacteria	cheese making
Catalase	splits hydrogen peroxide into $H_2O + O_2$	bacteria; animal livers	turns latex into foam rubber by producing gas
β-galactosidase	hydrolyzes lactose	fungi	in dairy industry, hydrolyzes lactose in milk or whey
Medical applications			
L-asparginase	removes L-asparagine from tissues - this nutrient is needed for tumour growth	bacteria (*E. coli*)	cancer chemotherapy- particularly leukaemia
Urokinase	breaks down blood clots	human urine	removes blood clots, *e.g.*, in heart disease patients
Analytical applications			
Glucose oxidase	oxidizes glucose	fungi	used to test for blood glucose, *e.g.*, in Clinistix™ diabetics
Luciferase	produces light	marine bacteria; fireflies	binds to particular chemicals indicating their presence, *e.g.*, used to detect bacterial contamination of food
Manipulative applications			
Lysozyme	breaks 1-4 glycosidic bonds	hen egg white	disrupts bacterial cell walls
Endonucleases	break DNA into fragments	bacteria	used in genetic manipulation techniques, *e.g.*, gene transfer, DNA fingerprinting

(Source : Biochemical Society Guidance Notes 3, Enzymes and their role in biotechnology)

PROBLEMS

1. Which of the reactions listed below are catalyzed by an isomerase, lyase, hydrolase or transferase ?

   (*a*) Protein $\longrightarrow$ Amino acids

   (*b*) Histidine $\longrightarrow$ Histamine + CO_2

   (*c*) Glucose + ATP $\longrightarrow$ Glucose-6-phosphate + ADP

   (*d*) Glucose-6-phosphate $\longrightarrow$ Glucose + H_3PO_4

   (*e*) $CH_3COCOOH + 2H^+ \longrightarrow CH_3CHOHCOOH$

(*f*) $CH_3COCOOH \longrightarrow CH_3CHO + CO_2$
(*g*) 3-phosphoglycerate ⟶ 2-phosphoglycerate
(*h*) Tryptophan ⟶ Tryptamine + CO_2
(*i*) L-lysine ⟶ D-lysine
(*j*) Acetycholine ⟶ Acetic acid + choline
(*k*) 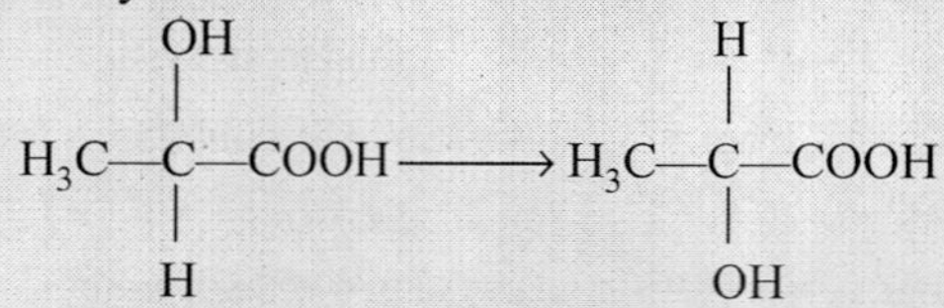
(*l*) Glutaric acid + oxaloacetate ⟶ α-ketoglutarate + aspartic acid

2. How are enzymes different from catalysts ?
3. Why does our mouth water when we see or smell something spicy ?

ANSWERS TO PROBLEMS

1. (*a*) hydrolase
   (*b*) lyase
   (*c*) transferase
   (*d*) hydrolase
   (*e*) reductoisomerase
   (*f*) lyase
   (*g*) transferase
   (*h*) lyase
   (*i*) isomerase
   (*j*) hydrolase
   (*k*) isomerase
   (*l*) transferase
2. Some important differences between enzymes and catalysts are as follows :
   (*a*) Enzymes are organic compounds and are, without exception, proteins while catalysts are inorganic in nature.
   (*b*) Some of the enzyme molecules are inactivated or destroyed during the reaction, while catalysts are not
3. Our mouth starts watering on seeing or smelling a spicy food because of what is known as *reflex action.* The salivary glands in our mouth become active when we eat. The enzymes present in the saliva help in digestion of the food. But sometimes the glands start secreting copious amounts of saliva even when there is no food in the mouth. This happens in anticipation of eating a tasty or spicy food when we see or smell it.

   On seeing or smelling a spicy food, our nerve cells in the brain are stimulated and signals are sent which activate the salivary glands. The latter start secreting enzymes and our mouth starts watering.

CHAPTER 17

Enzymes–II Characteristics and 3 'D' Structure

INSIDE

- Chemical Nature
- Characteristics
  - Colloidal Nature
  - Catalytic Nature
  - Specificity of Enzyme Action
  - Thermolability
  - Reversibility
  - pH Sensitivity
- Three-Dimensional Structure of the Enzymes
  - Ribonuclease
  - Lysozyme

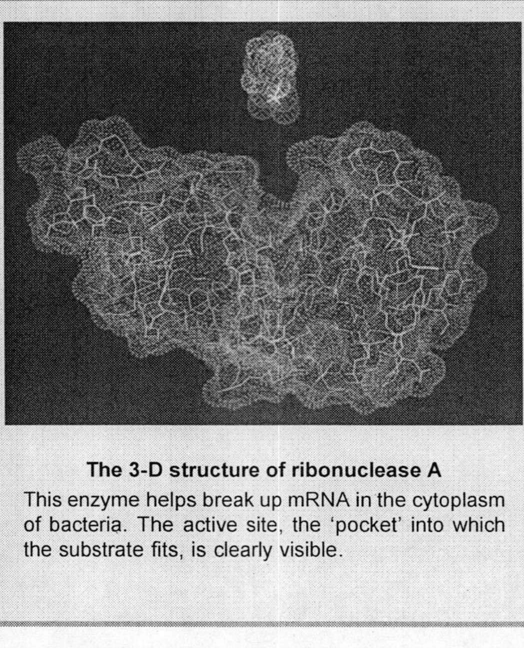

The 3-D structure of ribonuclease A
This enzyme helps break up mRNA in the cytoplasm of bacteria. The active site, the 'pocket' into which the substrate fits, is clearly visible.

CHEMICAL NATURE

All the enzymes are essentially proteins and possess properties characteristic to these. **Dixon** and **Webb** (1964) have stressed the protein nature of an enzyme by defining it as "*a protein with catalytic properties due to its power of specific activation*".

Evidences Proving the Protein Nature of the Enzymes

(*a*) **Elementary composition.** In their elementary composition, the enzymes show the usual proportion of C, H, N and S, as found in the proteins. Some crystalline enzymes, however,also contain minute quantities of P or metal ions such as Cu^{2+}, Mg^{2+}, Zn^{2+} etc. On hydrolysis, the crystalline enzymes yield the amino acids.

(*b*) **Identical action of some enzymes over other enzymes and the proteins.** Enzymes are subjected to the action of those enzymes which are specifically meant for the breakdown of peptide bonds of proteins.

(*c*) **Amphoteric nature.** Like other proteins, the enzymes behave as ampholytes in an electric field. The isoelectric point (*pI*) for various enzymes has also been determined.

(*d*) **Denaturation.** Enzymes, like other proteins, also undergo denaturation. If the crystalline *proteinase chymotrypsin* is subjected to an unfavourable *p*H, some part of protein becomes denatured. This percentage of denatured protein is usually found to be equal to the per cent loss in enzymic activity, thus proving a sort of correlation between the enzymes and the proteins..

(*e*) **Formation of antibodies.** Many purified enzymes, on injection into animal body, produce the specific antibodies. Since many nonprotein materials have been shown to serve as antigens, this cannot be treated as an evidence in support of the protein nature of enzymes but simply a further support to it.

Chemically, the enzymes may be divided into 2 categories :

1. **Simple-protein enzymes.** These contain simple proteins only *e.g., urease, amylase, papain* etc.
2. **Complex-protein enzymes.** These contain conjugated proteins *i.e.,* they have a protein part called *apoenzyme* (*apo*[G] = away from) and a nonprotein part called *prosthetic group* associated with the protein unit. The two parts constitute what is called a *holoenzyme*, *e.g., catalase, cytochrome c* etc.

The activity of an enzyme depends on the fact that the non-proteinaceous prosthetic group is intimately associated with the proteinaceous apoenzyme. But sometimes the prosthetic group is loosely bound to the protein unit and can be separated by dialysis and yet indispensable for the enzyme activity. In that case, this dialyzable prosthetic group is called as a coenzyme or cofactor. Thus :

$$\text{Conjugated-protein enzyme} \rightleftharpoons \text{Protein part + Prosthetic group}$$

or

$$\text{Holoenzyme} \rightleftharpoons \text{Apoenzyme + Coenzyme}$$

Coenzymes are thermostable, dialyzable organic compounds. They may be either attached to the protein molecules or may be present in the cytoplasm. *The coenzyme accounts for about 1% of the entire enzyme molecule.* Sometimes, a distinction is made between coenzyme and cofactor : the former includes the organic prosthetic groups and the latter the metal ions (**Fairley** and **Kilgour**, 1966).

CHARACTERISTICS

The enzymes possess many outstanding characteristics. These are enumerated below :

1. Colloidal Nature. Enzyme molecules are of giant size. Their molecular weights range from 12,000 to over 1 million. They are, therefore, very large compared with the substrates or functional group they act upon (Fig. 17–1).

It has been observed that the molecular weights of many enzymes prove to be approximately an *n*-fold multiple (where *n* is an integer) of 17,500 which is found to be an unit in most proteins (Table. 17–1).

On account of their large size, the enzyme molecules possess extremely low rates of diffusion and form colloidal systems in water. Being colloidal in nature, the enzymes are nondialyzable although some contain dialyzable or dissociable component in the form of coenzyme.

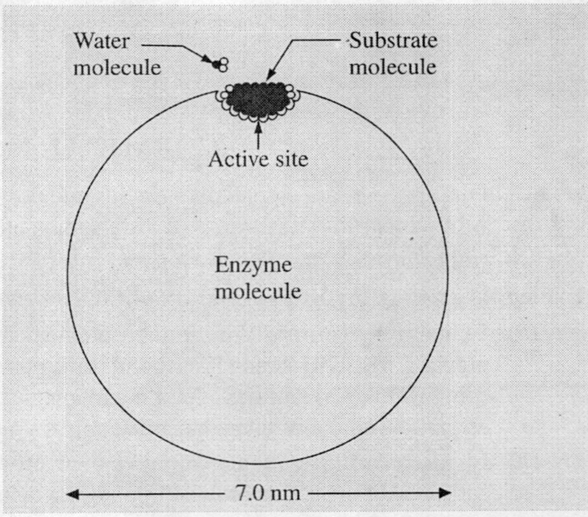

Fig. 17–1. Relative dimensions of a medium-sized enzyme molecule (MW 1,00,000 ; diameter 7 nm) and a typical substrate molecule (MW 250 ; diameter 0.8 nm)

[Note that the active site occupies only a small fraction of the surface area of the enzyme molecule. Also shown for comparison is a water molecule.]

2. Catalytic Nature or Effectiveness. *An universal feature of all* *enzymatic reactions is the virtual absence of any side products.* Therefore, just as hemoglobin is precisely tailored to transport oxygen, an enzyme is precisely adapted to catalyze a particular reaction. They act catalytically and accelerate the rate of chemical reactions occurring in plant and animal tissues. They do not normally participate in these reactions or if they do so, at the end of the reaction, they are recovered as such without undergoing any qualitative or quantitative change. This is the reason why they, in very small amounts, are capable of catalyzing the transformation of a large quantity of substrate. Thus, *the catalytic potency of enzymes is exceedingly great.*

Table 17–1. Molecular weight of some enzymes

Enzyme	*Molecular weight*	*n**
Pepsin	35,500	2
Catalase	250,000	14
Urease	480,000	27

*n = an integer, which is a multiple of 17,500

The catalytic power of an enzyme is measured by the "**turnover number**" (a term devised by **Wechselzahl**) or **molecular activity** (a term devised by **Norman Arthur Edwards** and **Kenneth Arnold Hassall,** 1980) which is defined as– *the number of substrate molecules converted into product per unit time, when the enzyme is fully saturated with substrate*. For example, a single molecule of *catalase* can convert 50,00,000 H_2O_2 molecules into H_2O and O_2 in a minute (**Sumner** and **Somers**, 1947). The value of turnover number varies with different enzymes and depends upon the conditions in which the reaction is taking place. However, *for most enzymes, the turnover numbers fall between 1 to 10^4 per second* (refer Table 17–2). The turnover number of 600,000 sec^{-1} for *carbonic anhydrase* is one of the largest known. Carbonic anhydrase (Fig. 17-2) catalyzes the hydration of carbon dioxide to produce 3,60,00,000 molecules of carbonic acid per minute. This catalyzed reaction is 6×10^7 times faster than the uncatalyzed one.

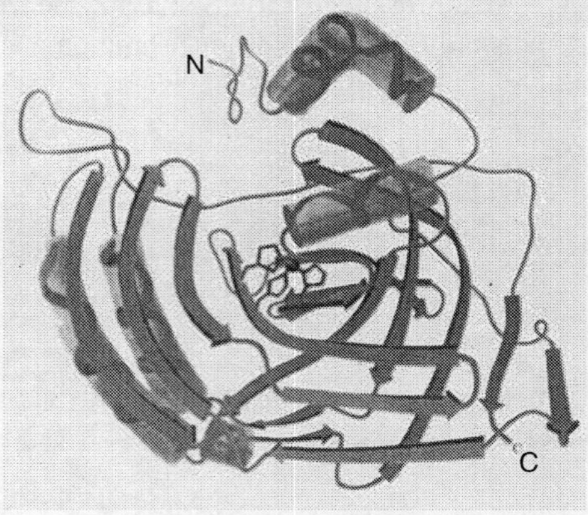

Fig. 17-2. Ribbon model of the tertiary structure of human carbonic anhydrase

The α helices are represented as cylinders and each strand of β sheet is drawn as an *arrow* pointing towards the polypeptide's C-terminus. The *grey* ball in the middle represents a Zn^{2+} ion that is coordinated by three His side chains (*blue*). Note that the C-terminus is tucked through the plane of a surrounding loop of polypeptide chain so that carbonic anhydrase is one of the rare native proteins in which a polypeptide chain forms a knot.

(Courtesy: Kannan KK et al, 1971)

$$CO_2 + H_2O \longrightarrow H_2CO_3$$

Table 17–2. Maximum turnover numbers of some enzymes

Enzyme	*Turnover number (per second)*
1. Lysozyme	0.5
2. Tryptophan synthetase	2
3. DNA polymerase I	15
4. Phosphoglucomutase	20.5

5. Chymotrypsin	100
6. β–galactosidase	208
7. Lactate dehydrogenase	1,000
8. Penicillinase	2,000
9. β–amylase	18,333
10. Acetylcholinesterase	25,000
11. Carbonic anhydrase	600,000

3. Specificity of Enzyme Action. With few exceptions, the enzymes are specific in their action. Their specificity lies in the fact that they may act (*a*) on one specific type of substrate molecule or (*b*) on a group of structurally-related compounds or (*c*) on only one of the two optical isomers of a compound or (*d*) on only one of the two geometrical isomers. Accordingly, four patterns of enzyme specificity have been recognized :

A. Absolute specificity. Some enzyme are capable of acting on only one substrate. For example, *urease* acts only on urea to produce ammonia and carbon dioxide.

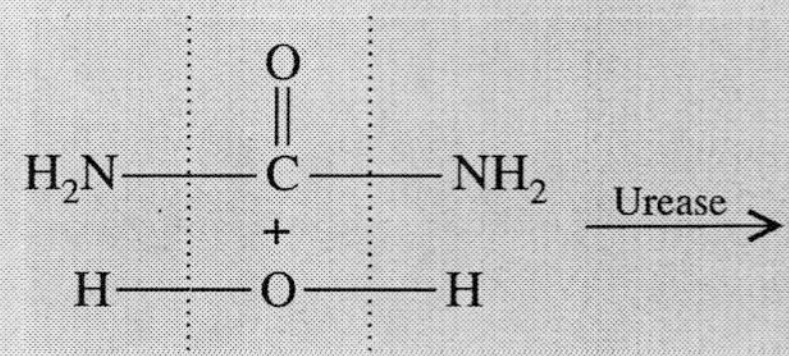

Similarly, *carbonic anhydrase* brings about the union of carbon dioxide with water to form carbonic acid.

$$H_2O + CO_2 \xrightarrow{\text{Carbonic anhydrase}} H_2CO_3$$

B. Group specificity. Some other enzymes are capable of catalyzing the reaction of a structurally-related group of compounds. For example, *lactic dehydrogenase* (LDH) catalyzes the interconversion of pyruvic and lactic acids and also of a number of other structurally-related compounds.

$$\underset{\text{Pyruvic acid}}{CH_3.CO.COOH} + NADH + H^+ \xrightleftharpoons[]{\text{Lactic dehydrogenase}} \underset{\text{Lactic acid}}{CH_3.CHOH.COOH} + NAD^+$$

C. Optical specificity. The most striking aspect of specificity of enzymes is that a particular enzyme will react with only one of the two optical isomers. For example, *arginase* acts only on L-arginine and not on its D-isomer. Similarly, D-*amino acid oxidase* oxidizes the D-amino acids only to the corresponding keto acids.

Although, the enzymes exhibit optical specificity, some enzymes, however, interconvert the two optical isomers of a compound. For example, *alanine racemase* catalyzes the interconversion between L- and D-alanine.

$$\text{L-alanine} \xrightleftharpoons[]{\text{Alanine racemase}} \text{D-alanine}$$

D. Geometrical specificity. Some enzymes exhibit specificity towards the *cis* and *trans* forms. As an example, *fumarase* catalyzes the interconversion of fumaric and malic acids :

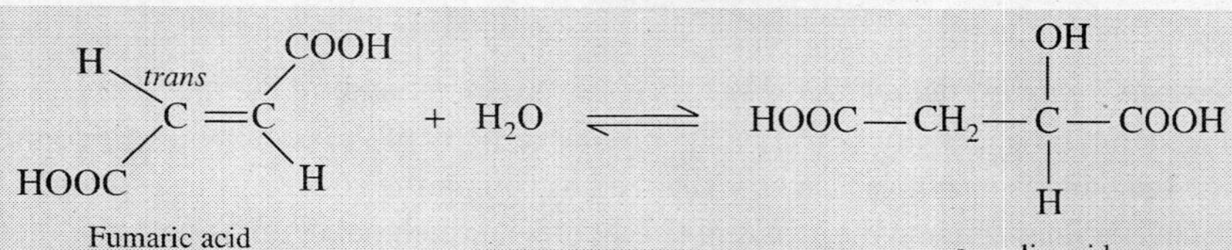

It does not react with maleic acid which is the *cis* isomer of fumaric acid or with D-malic acid.

The degree of specificity of the enzymes for substrate is usually high and sometimes virtually absolute. Proteolytic enzymes, for instance, catalyze the hydrolysis of a peptide bond :

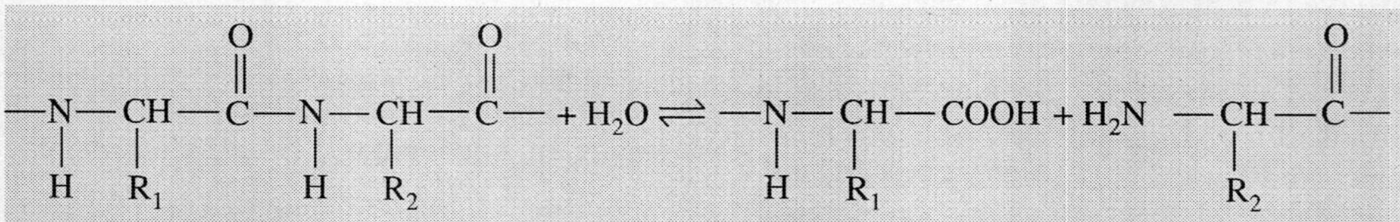

Many proteolytic enzymes (pepsin, trypsin, chymotrypsin) catalyze a different but related reaction, namely the hydrolysis of an ester bond.

$$\underset{\text{Ester}}{R_1-\overset{\displaystyle O}{\overset{\|}{C}}-O-R_2} + H_2O \rightleftharpoons \underset{\text{Acid}}{R_1-COOH} + \underset{\text{Alcohol}}{HO-R_2}$$

There are, however, instances where one enzyme acts on more than one substrate and conversely a substrate may also be catalyzed by more than one enzyme (Fig. 17–3). For example, *sucrase* acts on

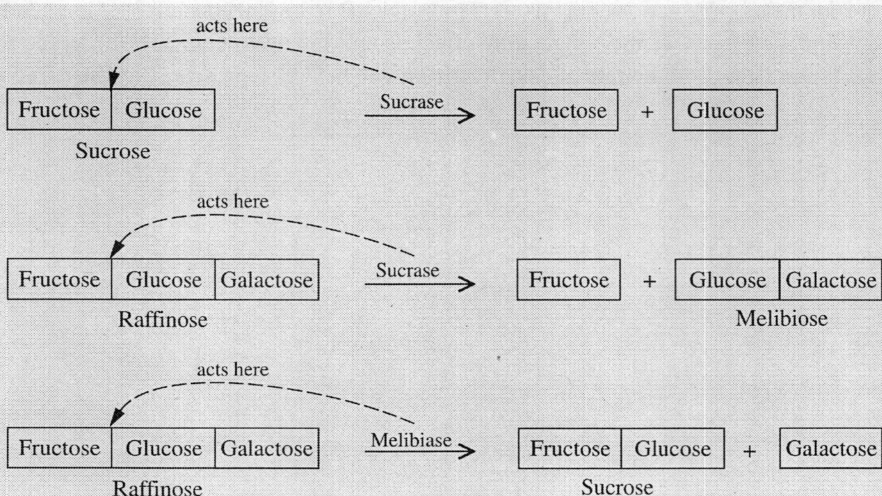

Fig. 17–3. Enzyme specificity of sucrase and melibiase

both sucrose (a disaccharide sugar, containing one mole of glucose and fructose each) and raffinose (a trisaccharide sugar, containing one mole each of glucose, fructose and galactose). But in both these cases, the enzyme sucrase attacks only the glucose-fructose linkage resulting in the production of fructose and glucose (in the case of sucrose) or fructose and melibiose (in the case of raffinose). However, raffinose is also acted upon by another enzyme, the *melibiase*. But this enzyme, unlike sucrase, breaks up glucose-galactose linkage so that at the end of the reaction sucrose and galactose are produced.

The coenzymes possess much less specificity. For example, among the *hydrolases*, NAD^+ and $NADP^+$ act as common coenzymes. The relative nonspecificity of the coenzymes, in contrast to the absolute specificity of the enzymes, can be visualized by comparing coenzyme to a common hammer, used equally by various apoenzyme workers (ironworker, watchmaker, shoemaker or electrician). Although some differences may occur in the nature of hammers (*e.g.*, between NAD^+ and $NADP^+$), the apoenzyme worker is strictly specific, corresponding to the nature of the substrate concerned.

4. Thermolability (= Heat sensitivity). Being proteinaceous in nature, the enzymes are very sensitive to heat. The rate of an enzyme action increases with rise in temperature, the rate being frequently increased 2 to 3 times for a rise in temperature of 10°C, *i.e.*, the value of temperature quotient or Q_{10} is 2 to 3. But at higher temperatures, the value of coefficient does not remain constant and decreases rapidly. Above 60°C, the enzymes coagulate and thus become inactivated, because there occurs an irreversible change in their chemical structure. The enzymes of dry tissues like seeds and spores, however, can endure still higher temperatures of about 100° to 120°C.

A temperature difference of 10 °C has become a standard that is used to measure the temperature sensitivity of a biological function. This value, called the **temperature quotient (Q_{10})**, is determined (for temperature intervals of exactly 10 °C) simply by dividing the value of a rate function (such as metabolic rate or the rate of an enzymatic reaction) at the higher temperature by the value of the rate function at the lower temperature. In general, metabolic reactions have Q_{10} values about 2 to 3. Purely physical processes, such as diffusion, have much lower Q_{10} values, usually close to 1.

The observed effect of temperature on enzyme action is the net result of the effect of temperature on the rate of enzyme action and their destruction as well. There will, thus, be obtained an optimum temperature for the enzyme action (Fig. 17–4). The cruve AB represents the effect of temperature on action alone and the curve CD, the temperature effect on enzyme destruction. The observed relation between the temperature and the rate of enzyme action will then be represented by the curve AE.

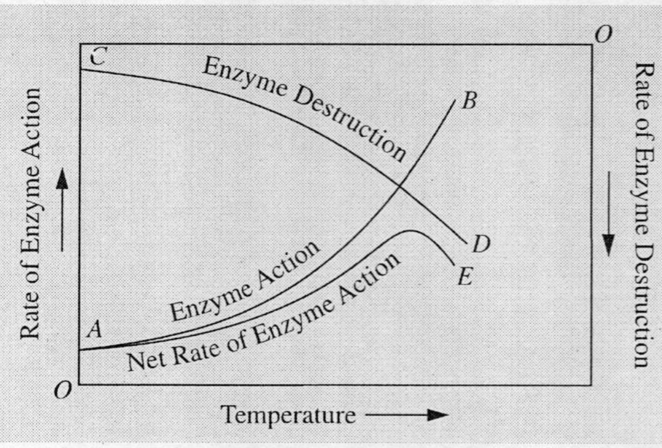

Fig. 17–4. Graph representing the relation between temperature and the rate of enzyme activity
(After Duclaux, 1883)

In general, the optimum temperature range for most enzymes varies between 30 and 45°C. For instance, it is 30°C for *catalase*. Because enzymes are globular proteins, most are thermolabile and begin to denature (indicated by loss of enzyme activity) at temperatures between 45° and 50°C (Fig. 17–5).

At low temperatures, the catalytic activity of the enzyme predominates, although some thermal denaturation takes place during this period. Decreasing temperatures to near or below 0°C although inactivate the enzyme but this is a reversible type of change and the enzyme regains its catalyzing power upon increasing the temperature to optimum. At higher temperaure, although the catalytic activity of the enzyme increases, yet its denaturation predominates. Henceforth, all the enzyme is denatured in a very short time. The enhanced enzyme activity with the rise in temperature is due to the fact that the energy of molecule becomes greater which, in turn, enhances the inherent reactivity of the molecules and the frequency of their collisions. It is because of the high rate of enzyme destruction at increasing temperatures that an enzyme is stable for weeks at 0°C, for days at 10°C, for hours at 30°C but for fraction of seconds at 70°C.

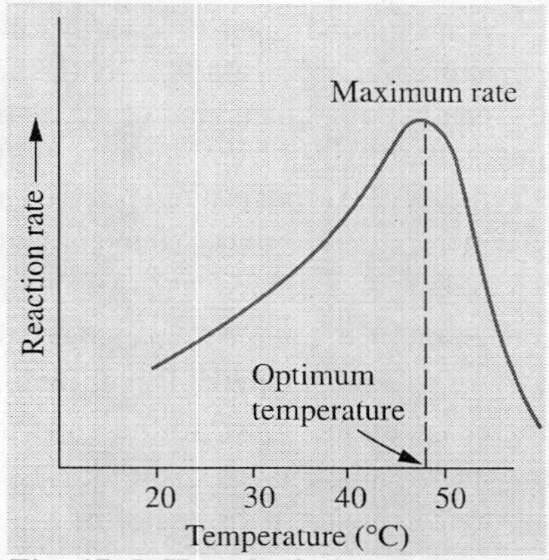

Fig. 17–5. Hypothetical temperature activity profile of an enzyme

The effect of heat also mainfests itself in the preservation of enzyme activity during storage. *The best preservation of enzyme preparations is by refrigeration or quick freezing.* This has been shown by **Nord** (1932) in the case of *zymase*.

5. Reversibility of a Reaction. The enzymes are capable of bringing about reversion in a chemical reaction. The digestive enzymes catalyze the hydrolytic reactions which are reversible. For instance, *lipase*, which catalyzes the synthesis of fat from glycerol and fatty acid, can also hydrolyze them into their component units.

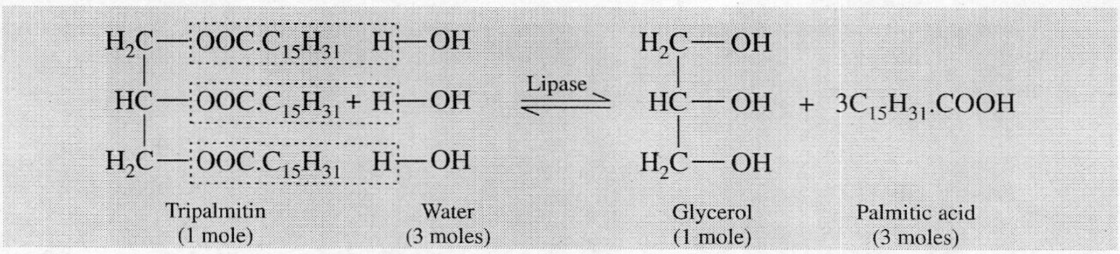

Fig. 17–6 shows the results of experiments on the action of *lipase* from castor on a fat, triolein. The final equilibrium mixture is the same whether one starts with the ester or with its individual components.

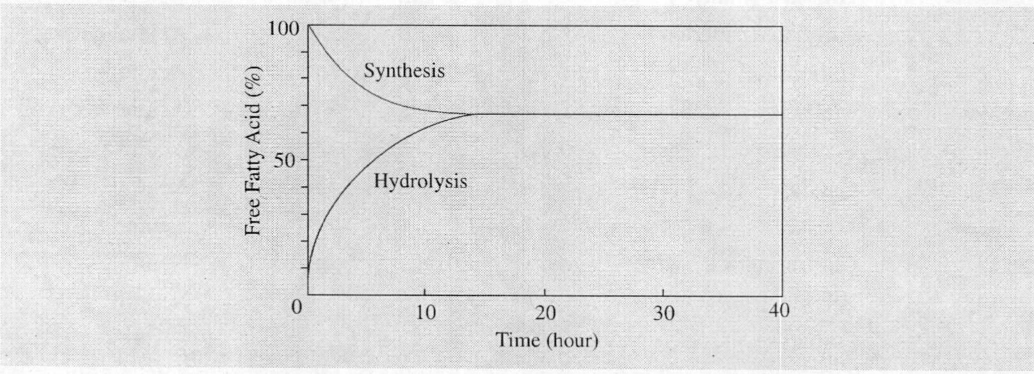

Fig. 17–6. Reversibility of action of lipase (from castor) on triolein

The direction in which the reaction proceeds depends upon many factors like –

(*a*) the *p*H of the cell sap,

(*b*) the presence of reacting substances, and

(*c*) the accumulation of end products.

It does not, however, necessarily follow that the same enzyme invariably catalyzes both the synthesis and degradation of a given kind of molecule. For instance, urea is synthesized from arginine by the action of the enzyme, *arginase* but is hydrolyzed by action of another enzyme, *urease* to produce ammonia and carbon dioxide.

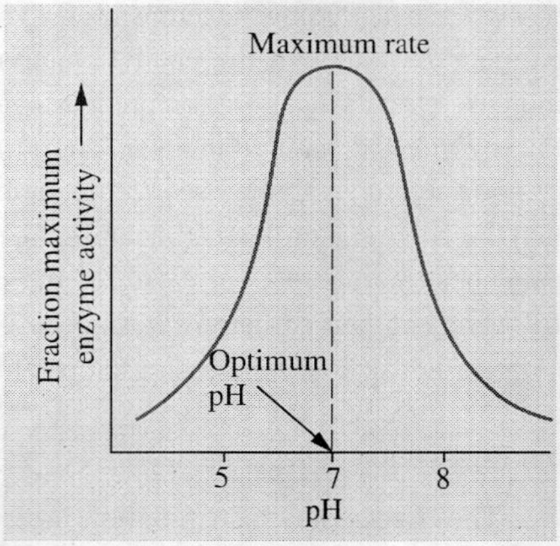

Fig. 17–7. Hypothetical pH activity profile of an enzyme

6. pH Sensitivity. The *p*H value or the H^+ ion concentration of the medium controls the activity of an enzyme to a great extent. This is mainly related to the degree of dissociation, to the electric charge of the enzyme and, through this, to the formation of the enzyme-substrate complex (a discussion of which will follow in the succeeding chapter). Each enzyme, thus, acts best in a certain pH value which is specific to it and its activity slows down with any appreciable change (increase or decrease) in the H^+ ion concentration. In fact, the *p*H will affect the efficiency of an enzyme and usually there will be a *p*H at which the activity is at a maximum. The activity will fall off on either side of this value. Fig. 17–7 depicts the effect of *p*H on an enzyme-catalyzed reaction.

Table 17–3. pH optima for various enzymes

S.No.	*Enzyme*	*Optimum pH of the medium*	*Nature of the medium*
1.	Pepsin	1.5–1.6	Highly acidic
2.	Invertase	4.5	Acidic
3.	Lipase (stomach)	4.0–5.0	Acidic
4.	Lipase (castor oil)	4.7	Acidic
5.	Lipase (pancreas)	8.0	Alkaline
6.	Amylase (malt)	4.6–5.2	Acidic
7.	Amylase (pancreas)	6.7–7.0	Acidic–neutral
8.	Cellobiase	5.0	Acidic
9.	Maltase	6.1–6.8	Acidic
10.	Sucrase	6.2	Acidic
11.	Catalase	7.0	Neutral
12.	Urease	7.0	Neutral
13.	Cholinesterase	7.0	Neutral
14.	Ribonuclease	7.0–7.5	Neutral
15.	Fumarase	7.8	Alkaline
16.	Trypsin	7.8–8.7	Alkaline
17.	Adenosine triphosphatase	9.0	Alkaline
18.	Arginase	10.0	Highly alkaline

Some optimum *p*H values for various enzymes are given in Table. 17–3. A perusal of the table indicates that *the approximate optimum pH value for most enzymes lies near neutrality.* This value depends on many factors such as :

(*a*) the nature of buffer system,
(*b*) the presence of other colloids, activators or inhibitors,
(*c*) the age of the cell tissue, and
(*d*) the nature of the substrate.

Usually maximum enzyme activity is obtained at or near the isoelectric point (*p*I) of the enzymes. Thus *trypsin*, whose *p*I value is 10.1, shows maximum activity at *p*H range between 7 and 9.

The correction between the enzymic activity and the *p*H value for 3 enzymes has been graphically represented in Fig. 17–8.

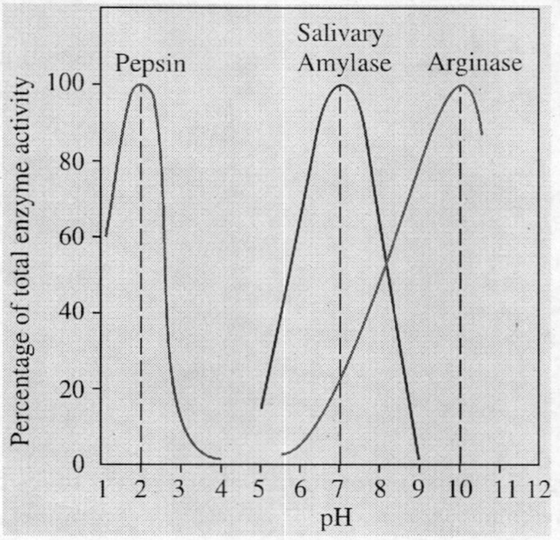

Fig. 17–8. Effect of pH on enzyme action
(Modified from Fruton and Simmonds, 1958)

THREE-DIMENSIONAL STRUCTURE OF THE ENZYMES

A single crystal of protein or the protein fibres will deflect x-rays and the resultant image formed on a photographic plate can give certain important clues regarding the structure of the crystal or the fibres. This techinque is called **x-ray crystallography** and has been widely used for the elucidation of protein structure at micro level. X-ray crystallography has so far revealed the structure of many enzymes, namely, *ribonuclease, lysozyme, chymotrypsin, trypsin* etc. The structure of *two* of them is described below :

1. Ribonuclease (RNase)

Ribonuclease (Fig. 17–9), a small globular protein, is an enzyme secreted by the pancreas into the small intestine, where it catalyzes the hydrolysis of certain bonds in ribonucleic acids present in ingested food.

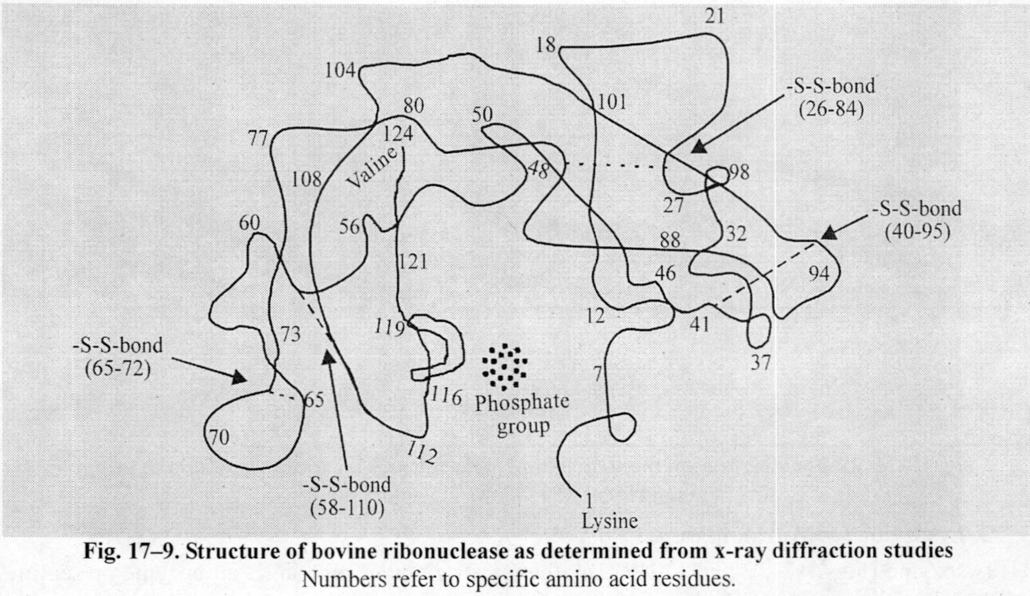

Fig. 17–9. Structure of bovine ribonuclease as determined from x-ray diffraction studies
Numbers refer to specific amino acid residues.
(From Harper HA and Rodwell, VW 1973)

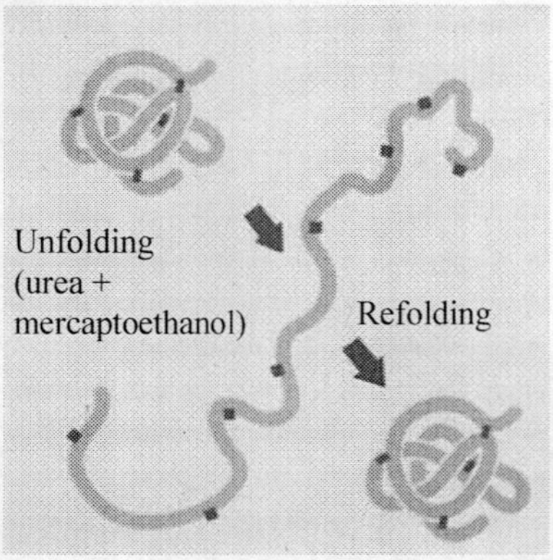

Fig. 17–10. Denaturation and refolding of ribonuclease

A native ribonuclease molecule (with intramolecular disulfide bonds indicated) is reduced and unfolded with β-mercaptoethanol and 8 M urea. After removal of these reagents, the protein undergoes spontaneous refolding.

(From. CJ Epstein, RF Goldberger and CB Anfinsen, 1963)

The molecule of ribonuclease is reniform (kidney-shaped) and has dimensions of about 3.2, 2.8 and 2.2 nm. Ribonuclease, like myoglobin, contains a tightly packed, highly nonpolar interior. This enzyme-protein (as already described on page 152) consists of a single polypeptide chain of 124 amino acid residues with lysine at the N-terminal and valine at the C-terminal (**Hirs, Moore and Stein**, 1960). It has a molecular weight of 13,700. There are 8 cysteine residues, thus apparently forming 4 disulfide linkages– 26-84, 40-95, 58-110 and 65-72. These serve to hold the tertiary structure firmly in place. There is very little (26%) α-helix structure ; many of its segments are present in β conformation which amounts to 35%. Only 4 turns of the helix, two each at residues 5-12 and 28-35, are present. The chain assumes a complex configuration with a deep depression in the middle of one side. The active site is believed to be on the edge of this depression and the residues forming the active site are 6-8, 11, 12, 41, 42, 46-48 and 117-119. A phosphate ion is associated directly with the active site of the enzyme. The amino acid residues 12 and 119 (both histidine) are nearest to the phosphate ion.

The bacterial enzyme, *subtilisin*, cleaves the chain into 2 inactive fragments : the shorter one (S-peptide) consisting of first 21 amino acids from N-terminal and the longer one (S-protein) with remaining residues. On reunion of the two segments, the enzyme molecule regains its full activity upon treatment with 8 M urea and mercaptoethanol, the native ribonuclease molecule become reduced and unfolded. After removal of these reagents, the enzyme undergoes spontaneous refolding (Fig. 17–10).

It is interesting to note that there is remarkable similarity in structure between the ribonucleases from cows and humans beings (Fig. 17–11). The strucutural similarity is often followed by functional similarity.

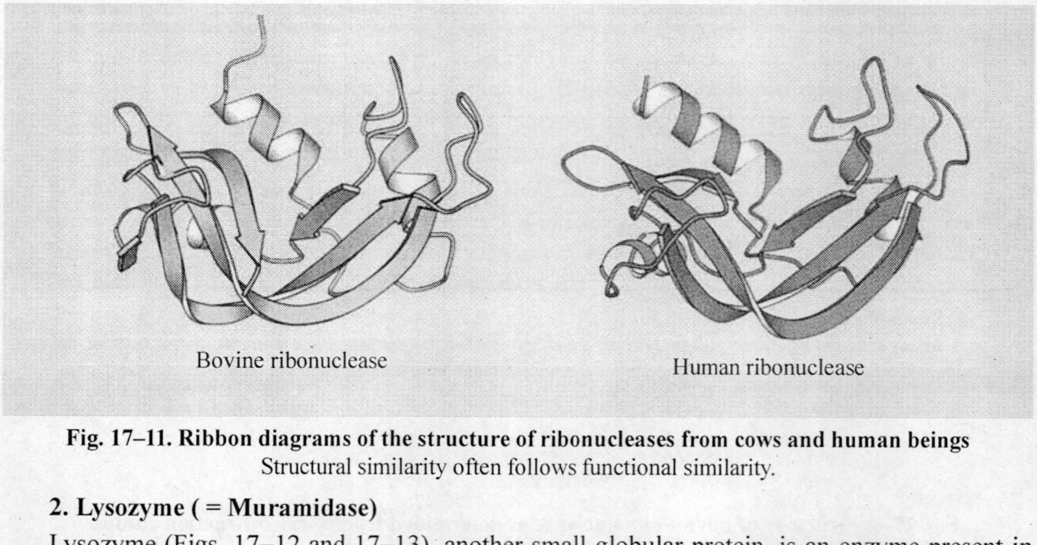

Fig. 17–11. Ribbon diagrams of the structure of ribonucleases from cows and human beings

Structural similarity often follows functional similarity.

2. Lysozyme (= Muramidase)

Lysozyme (Figs. 17–12 and 17–13), another small globular protein, is an enzyme present in tears, nasal mucus, gastric secretions, milk and egg white. Lysozyme is so named because it can

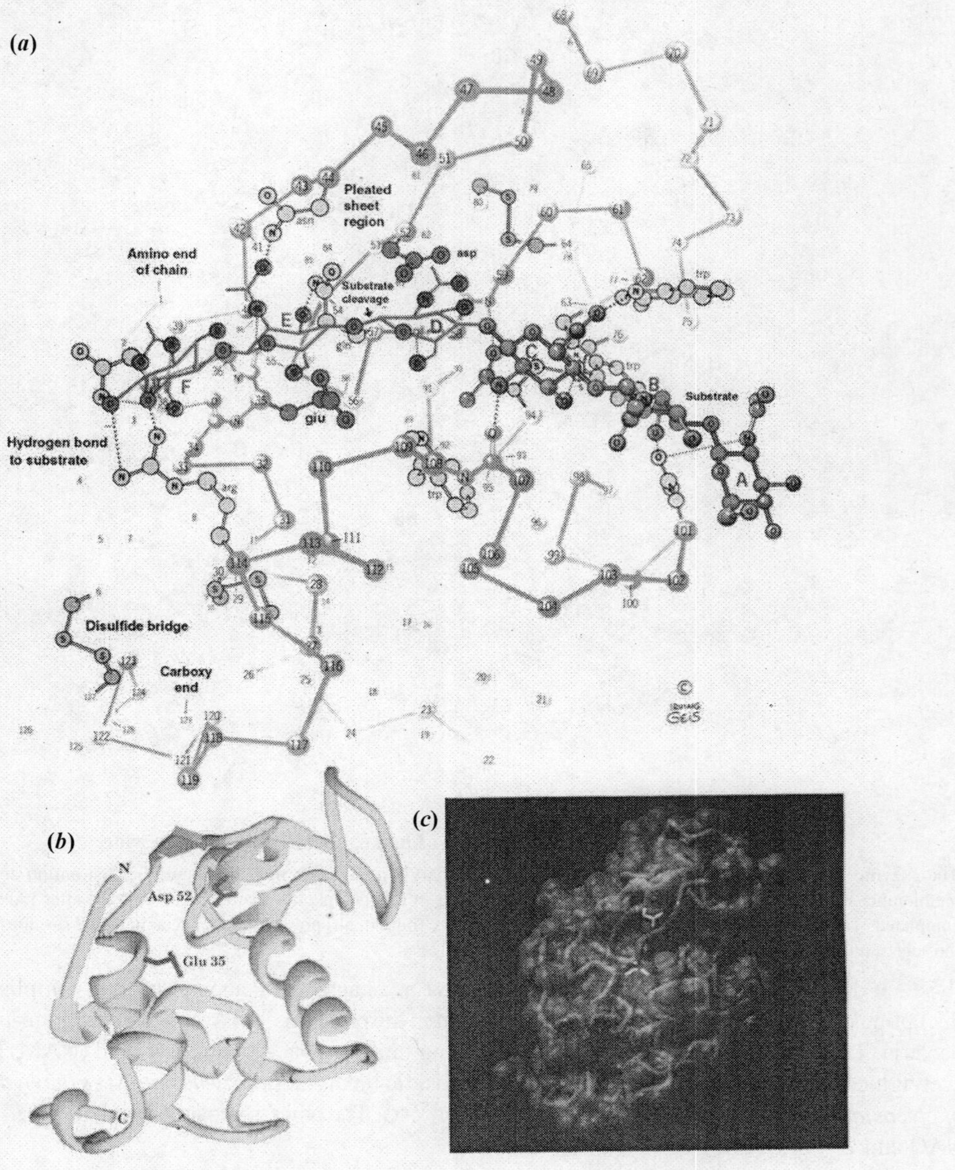

Fig. 17–12. The x-ray structure of hen egg white (HEW) lysozyme

(*a*) **The ball-and-stick model.** Each circle represents a single amino acid residue. Numbers refer to specific amino acid residues. The segment between residues 43 and 54 has a pleated sheet structure. The polypeptide chain is shown with a bound $(NAG)_6$ substrate (*green*). The positions of the backbone C_α atoms are indicated together with those of the side chains that line the substrate-binding site and form disulfide bonds. The substarate's sugar rings are designated A, at its nonreducing end (*right*), through F, at its reducing end (*left*). Lysozyme catalyzes the hydrolysis of the glycosidic bond between residues D and E. Rings A, B, and C are observed in the x-ray structure of the complex of $(NAG)_3$ with lysozyme; the positions of rings D, E, and F were inferred from model studies.

(*b*) **The ribbon model.** It highlights the protein's secondary structure and indicates the positions of its catalytically important side chains. The letters N and C represent the amino and carboxyl terminals of the protein molecule, respectively.

(*c*) **A computer-generated model.** It shows the protein's molecular envelope (*purple*) and C_α backbone (*blue*). The side chains of the catalytic residues, Asp 52 (*above*) and Glu 35 (*below*) are coloured *yellow*. Note the enzyme's prominent substrate-binding cleft. (Parts (*a*), (*b*) and (*c*) have approximately the same orientation.)

(Courtesy : (a) Irving Geis, (b) & (c) Arthur Olson)

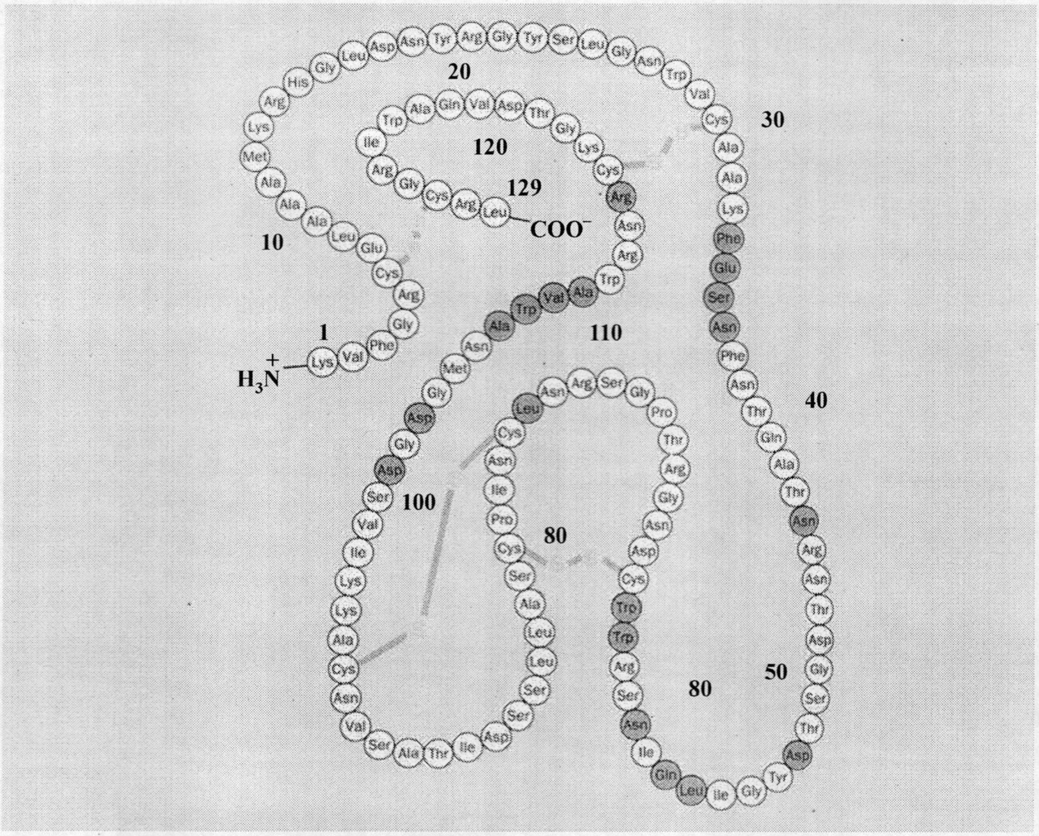

Fig. 17–13. The primary structure of hen egg white (HEW) lysozyme

The enzyme lysozyme contains 129 amino acids in its primary structure. As may be noted, the first amino acid is not methionine; instead, it is lysine. The first methionine residue in this polypeptide sequence is removed after translation is completed. The removal of the first methionine occurs in many (but not all) polypeptides. The amino acid residues that line the substrate-binding pocket are shown in *dark purple*.

'lyse', or dissolve, bacterial cell walls and thus serve as a bactericidal agent. This is accomplished by disrupting certain polysaccharide molecules present in the protective cell walls of many gram-positive bacteria. These polysaccharides consist of repeating units of N-acetylmuramic acid (NAM) and N-acetylglucosamine (NAG), joined by β-1→4 glycosidic linkages. *Lysozyme catalyzes the hydrolysis of glycosidic bond between C-1 of NAM and C-4 of NAG.* The other glycosidic bond, between C-1 of NAG and C-4 of NAM, is not cleaved.

In 1965, **David C. Phillips** and his colleagues determined the three-dimensional structure of lysozyme. It is a relatively small, compact molecule, roughly ellipsoidal in shape and with dimensions 45 ×30 × 30 Å. It has a molecular weight of 14,600. Its molecule consists of a single polypeptide chain of 129 amino acid residues with 4 intra-chain disulfide linkages : 6-127, 30-115, 64-80 and 76-94. It has lysine at the N-terminal and leucine at the C-terminal. *Lysozyme is devoid of coenzyme or metal ion cofactors* and thus lacks a built-in marker at its active site, in contrast with such proteins as myoglobin and hemoglobin. Like myoglobin and cytochrome c, lysozyme has a compactly-folded conformation and has most of its hydrophobic R groups inside the globular structure, shielded from water, and its hydrophilic R groups outside, facing the aqueous medium. The enzyme has only 12%β conformation and 40% α-helical segments which line a long deep cleft in the side of the molecule. This central cleft is the active site of the enzyme molecule. The interior of lysozyme, like that of myoglobin and hemoglobin, is almost entirely nonpolar. Hydrophobic interactions evidently play an important role in the folding of lysozyme, as they do for most proteins.

The active site has 6 subsites (A to F ; Fig. 17–14) which bind various substrates or inhibitors. The amino acid residues located at the active sites are 35, 52, 59, 62, 63 and 107. It is, thus, apparent that the active site may include amino acid residues which are distantly placed, as shown in Fig. 17–15. The residues which bring about bond cleavage lie between the subsites D and E, close to the COOH groups of glutamic acid (35) and aspartic acid (52). It is thought that glutamic acid protonates the acetal bond of the substrate while the aspartic acid stabilizes the resulting carbonium ion from the back side (**Harper** and **Rodwell**, 1973).

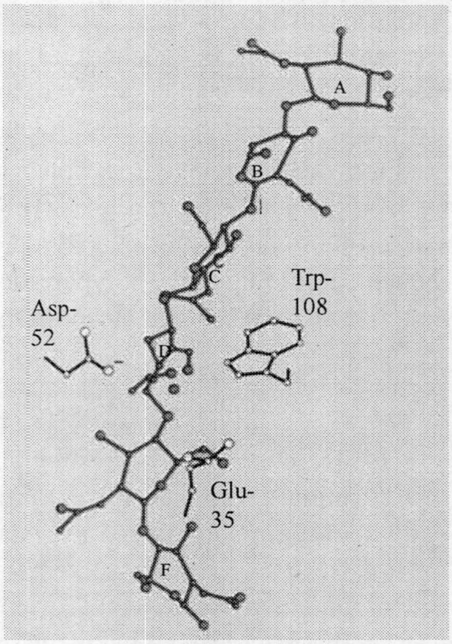

Fig. 17–14. Simplified model of a lysozyme molecule showing a hexasaccharide bound in the cleft of the enzyme

A to F represent the subsites of the enzyme. The locations of key amino acid residues of the enzyme are indicated.

Lysozyme binds 6 of the monomeric units of its polysaccharide substrate, and the strain induced by the binding facilitates the formation of the unstable carbonium ion, C^+–, intermediate. The proposed mechanism of lysozyme catalysis employs (**Charles J. Flickinger** *et al*, 1979) :

1. orientation and approximation through formation of ES complex,
2. strain,
3. general acid-base catalysis, and
4. electrostatic stabilization of a carbonium ion intermediate.

The importance of the ability of proteins to structure precisely a volume of space is evident.

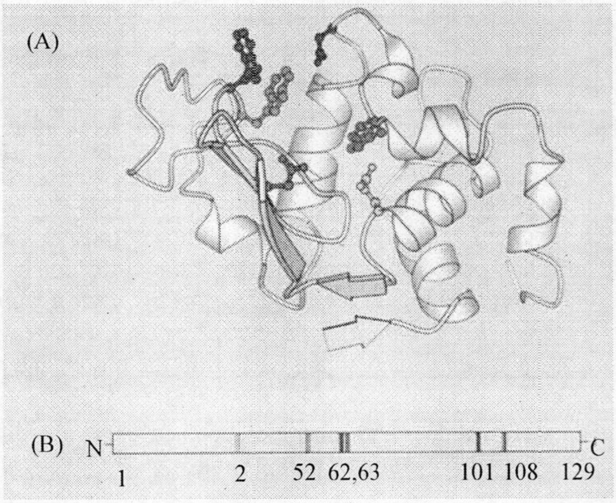

Fig. 17–15. The constituent amino acid residues of the lysozyme active site

(A) Ribbon diagram of the enzyme lysozyme with several components of the active site shown in color.

(B) A schematic of the primary structure of lysozyme showing that the active site is composed of residues that come from different parts of the polypeptide chain.

It is to be noted that in the enzymes described above, the active site is present in the groove or depression. The groove is, in fact, the ideal place for the active site as it provides a nonpolar microenvironment in which alone the van der Waal's forces and the hydrogen bond formation can operate between the polar groups of the active site and the substrate. In the case of an `exposed' active site, on the contrary, the water molecules will interfere with such activity.

From the foregoing discussion, it may be concluded that the chemical nature of an enzyme depends much on the active site of the enzyme molecule. Apart from the various kinds of constituent amino acid residues comprising an enzyme, its properties further depend on the pattern of '3-D' structure. The typical folding of the polypeptide chain, for example, ensures that the constituent amino acids of the active site are brought closer in the grooves.

PROBLEMS

1. The sweet taste of freshly picked corn is due to the high level of sugar in the kernels. Store-bought corn (several days after picking) is not as sweet because about 50% of the free sugar of corn is converted into starch within one day of picking. To preserve the sweetness of fresh corn, the husked ears are immersed in boiling water for a few minutes ("blanched") and then cooled in cold water. Corn processed in this way and stored in a freezer maintains its sweetness. What is the biochemical basis for this procedure ?
2. Which of the following statements is not universally applicable to enzymes :
   (*a*) they generally work very rapidly.
   (*b*) they can catalyze a reaction in both directions
   (*c*) they are not used up during a reaction.
   (*d*) they will bind one substrate only.
   (*e*) they are proteins.
3. Why does a wound on tongue heal faster than those on other parts of the body ?
4. Cite the factors that contribute to the catalytic power of enzymes.

ANSWERS TO PROBLEMS

1. The activity of the enzyme that converts sugar to starch is destroyed by heat denaturation.
2. (*d*) they will bind one substrate only.
3. Cuts on the tongue heal fast because *the tongue is highly vascularized, that is, has plenty of blood vessels*. The rich supply of blood means that white blood corpuscles can easily reach the site of the wound and carry out defense functions preventing infection by bacteria. The secondary infection can prolong healing. *The saliva too helps in this as it bathes the tongue in a chemical called lysozyme that has mildly antiseptic properties.* Besides, *the epithelial cells of the tongue divide very rapidly* and so the wounds in the mouth and on the tongue heal fast.
4. Precise positioning of catalytic residues and substrates, geometrical strain (distortion of the substrate), electronic strain and desolvation of the substrate.

CHAPTER

18

Enzymes-III
Mechanism of Enzyme Action

INSIDE

- Energy Mechanics of Enzymatic Reactions
- Michaelis-Menten Hypothesis
- Active Site
  - Fisher's Lock and Key Model
  - Koshland's Induced Fit Model
- Enzyme Reaction Rates
- Modifiers of Enzyme Activity
  - Inorganic Modifiers
  - Organic Modifiers
  - Reversible Enzyme Inhibition
  - Irreversible Enzyme Inhibition
- Allosteric Enzymes
  - Simple Sequential Model
  - Concerted or Symmetry Model

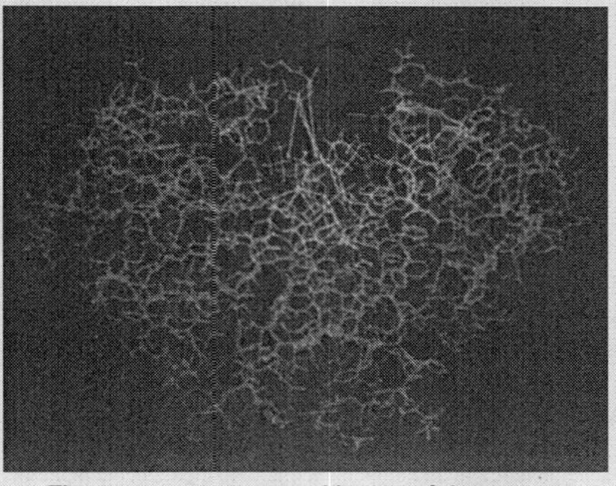

The computer-generated image of the enzyme penicillopepsin

The active site of this enzyme is clearly visible as the cleft at the top in this model, as is the fit of the substrate (*blue*)in the active site. Only a substrate whose shape and structure are complementary to those of the active site can fit into the enzyme.

ENERGY MECHANICS OF ENZYMATIC REACTIONS

A simple enzymatic reaction might be written :

$$E + S \rightleftharpoons ES \rightleftharpoons EP \rightleftharpoons E + P$$

where, E, S and P represent enzyme,substrate and product, respectively. ES and EP are complexes of the enzyme with substrate and with the product, respectively.

For a better understanding of the kinetics of a chemical reaction, the two terms, reaction equilibria and reaction rates must be differentiated. The function of a catalyst is to increase the *rate* of a reaction. Catalysts do not affect reaction *equilibria*. Any reaction, such as S $\rightleftharpoons$ P, can be described by a **reaction coordinate diagram** (Fig. 18–1). Energy in biological systems is described in terms of free energy, G. In the coordinate diagram, the free energy of the system is The free energy of the system is plotted against the progress of the reaction. Energy diagrams describe the energetic course of the reaction (vertical axis) and the progressive chemical changes (horizontal axis) as the substrate is converted to product.

S = free energy of the substrate
P = free energy of the product
$\neq$ = transition state

$\Delta G^{\neq}$ = activation energies for the two reactions, $S \rightarrow P$ and $P \rightarrow S$

$\Delta G^{\circ\prime}$ = overall standard free-energy change in moving from S to P

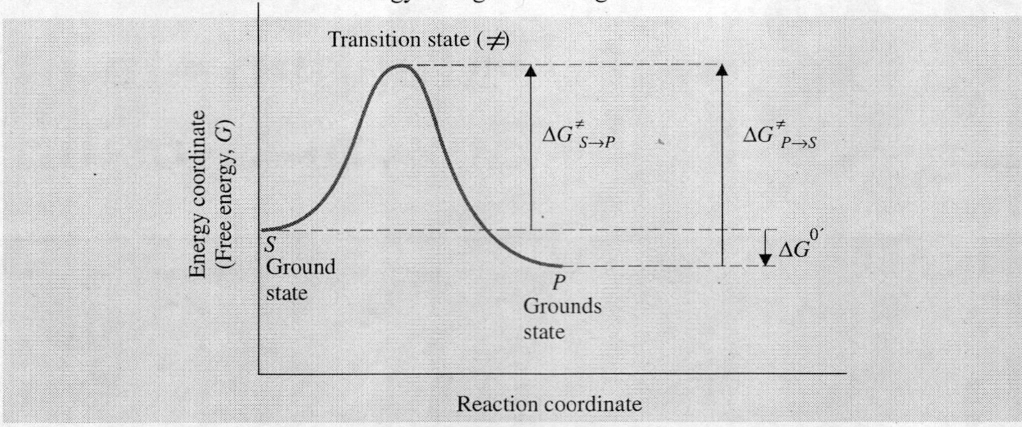

Fig. 18–1. Free energy diagram for a simple chemical reaction, S ⇌ P

described in terms of free enery, *G*. In the coordinate diagram, the free energy of the system is plotted against the progress of the reaction (or reaction coordinate). In its normal stable form or **ground state**, any molecule contains a specific amount of free energy. Chemists express the free energy change for this reacting system under standard set of conditions (temperature, 298 K; partial pressure of gases, each 1 atm or 101.3 kPa ; pH = 0; concentration of solutes, each 1M) and call this as **standard free-energy change, ΔG°**. Because biochemical systems commonly involve H+ concentrations far from 1M, biochemists define a constant $\Delta G^{\circ\prime}$, the standard free-energy change at pH 7.0.

The equilibrium between S and P reflects the difference in the free energy of their ground states. In the example shown in Fig., the free energy of the ground state of P is lower than that of S, hence $\Delta G^{\circ\prime}$ for the reaction is negative and the equilibrium favours P. This equilibrium is *not* affected by any catalyst. A favourable equilibrium, however, *does not* mean that $S \rightarrow P$ conversion is fast. The rate of a reaction, in fact, depends on an entirely different parameter. There exists an energetic barrier between S and P that represents the energy required for alignment of reacting groups, bond rearrangements and other changes needed for the reaction to occur in either direction. To undergo reaction, the molecules must overcome this barrier or "energetic hill" and therefore must be raised to a higher energy level. At the top of energy hill is a point at which decay to the S or P state is equally probable, which is downhill either way. This is called the **transition state** and should not be confused with a reaction intermediate. It is simply a moment of fleeting molecules in which certain events (bond breakage, bond formation, charge development etc) have proceeded to decide the future course of the reaction, *i.e.*, a collapse to either substrate or product. The difference between the energy levels of the ground state and the transition state is called the **Gibbs free energy of activation** or simply **activation energy** and is symbolized by $\Delta G^{\neq}$. The double dagger (≠) denotes a thermodynamic quantity of a transition state. The rate of reaction reflects this activation energy; *a higher activation energy corresponds to a slower reaction.* Reaction rates can be increased by raising the temperature, thereby increasing the number of molecules with higher energy to overcome this energy barrier. As an alternate, the activation energy can be lowered by adding a catalyst (Fig. 18–2). Thus, *catalysts enhance reaction rates by lowering activation energies.*

The ES and EP intermediates occupy minima in the energic process curve of the enzyme-catalyzed reaction. The terms $\Delta G_{NE}^{\neq}$ and $\Delta_{GE}^{\neq}$ correspond to the activation energies for the nonenzymatic and enzymatic reactions. The activation energy for the overall process is lower when the enzyme catalyzes the reaction. Note that for both reactions, the free energy for S and P, called the $\Delta G^{o\prime}$ of the reaction, is the same.

The catalysts affect the reaction rates, not the reaction equilibria. And so are the enzymes, the bidirectional arrows put in the equation on page 315 make the point clear: any enzyme that catalyzes the reaction, S → P also catalyzes the reverse reaction, P → S. Its only role is to accelerate the interconversion of S and P. The enzyme is not consumed in the process, and the equilibrium point remains unaffected. However, the reaction reaches equilibrium much faster when the appropriate enzyme is present because the rate of the reaction is increased.

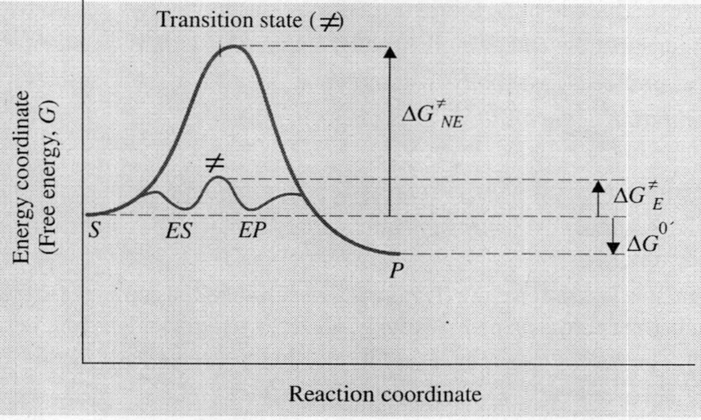

Fig. 18–2. Energy diagram, comparing the nonenzymatic and enzymatic reactions, S → P

Practically, any reaction may have several steps involving the formation and decay of transient (unstable) chemical compounds called **reaction intermediates.** When the S → P reaction is catalyzed by an enzyme, the ES and EP complexes are intermediates and they occupy valleys in the reaction coordinate diagram (Fig. 18–2). When several steps occur in a reaction, the overall rate is determined by the step(s) with the maximum activation energy. This is called the **rate limiting step.** In practice, the rate limiting step can vary with reaction conditions. It may, however, be inferred from the above discussion that reaction *equilibria* are intimately linked with $\Delta G^{\circ\prime}$ and reaction *rates* are linked with $\Delta G^{\neq}$.

The energy of activation is a measure of the energy needed for the conversion of molecules to the reactive state. An inorganic catalyst lowers the activation energy and the organic catalyst (i.e., an enzyme) further decreases this. For example, decomposition of H_2O_2 requires 18,000 cal/mol ; this is lowered to 11,700 when colloidal platinum acts as catalyst and is further lowered to 2,000 when an enzyme catalase *catalyzes* the reaction. It is , thus, evident that enzymic catalysts are far more efficient in lowering the energy of activation than the nonenzymic catalysts for the same reaction. A perusal of Table 18–1 supports the above statement.

Table 18–1. Activation energy for enzymic and nonenzymic catalyses

S.No.	*Reaction*	*Catalyst*	*Activation Energy (expressed as cal/mol)*
1.	Decomposition of hydrogen peroxide	None	18,000
		Colloidal platinum	11,700
		Catalase	2,000
2.	Hydrolysis of ethyl butyrate	Hydrogen ion	16,800
		Hydroxyl ion	10,200
		Lipase	4,500
3.	Hydrolysis of sucrose	Hydrogen ion	25,600
		Invertase	10,000
4.	Hydrolysis of casein	Hydrogen ion	20,600
		Trypsin	12,000

(Adapted from White, Handler and Smith, 1964)

MICHAELIS-MENTEN HYPOTHESIS

Building upon earlier work by **Victor Henri, Leonor Michaelis** and **Maud L. Menten** (1913), while studying the hydrolysis of sucrose catalyzed by the enzyme *invertase*, proposed this theory. Their theory is, however, based on the following assumptions :

> The discovery by **Michaelis** that **keratin**, the major ingredient of hair, is soluble in thioglycolic acid made possible the development of the home permanent industry.

1. Only a single substrate and a single product are involved.
2. The process proceeds essentially to completion.
3. The concentration of the substrate is much greater than that of the enzyme in the system.
4. An intermediate enzyme-substrate complex is formed.
5. The rate of decomposition of the substrate is proportional to the concentration of the enzyme-substrate complex.

The theory postulates that the enzyme (E) forms a weakly-bonded complex (ES) with the substrate (S). This enyzme-substrate complex, on hydrolysis, decomposes to yield the reaction product (P) and the free enzyme (E). These reactions may be symbolically represented as follows :

$$E + S \rightleftharpoons ES \longrightarrow E + P$$

Although one may not feel any difficulty (at least theoretical) in describing the kinetics of these reactions, yet the difficulty is encountered when one starts to determine the concentration of ES or even S practically. The same difficulty was experienced by **Michaelis** and **Menten,** who devised an equation where these immeasurable quantities were replaced by those which could be easily measured experimentally.

Following symbols may be used for deriving Michaelis-Menten equation :

$$(E_t) = \text{total concentration of enzyme}$$
$$(S) = \text{total concentration of substrate}$$
$$(ES) = \text{concentration of enzyme-substrate complex}$$
$$(E_t) - (ES) = \text{concentration of free enzyme}$$

The rate of appearance of products (*i.e.*, velocity, V) is proportional to the concentration of the enzyme-substrate complex.

$$V = k \neq (ES) \qquad ...(1)$$

The maximum reaction rate, V_m will occur at a point where the total enzyme E_t is bound to the substrate. Then the maximum concentration of ES will be equal to the total enzyme concentration, E*t*. Thus :

$$V_m = k \times (Et) \qquad ...(2)$$

Dividing equation (1) by (2,) we get :

$$\frac{V}{V_m} = \frac{(ES)}{(E_t)} \qquad ...(3)$$

With the help of this equation, one can easily measure the immeasurable quantities, (ES) and (E_t), in terms of the reaction rates experimentally.

Now coming back to the reversible reaction, $E + S \rightleftharpoons ES$, one can write the equilibrium constant for dissociation of ES as K_m which is equal to :

$$K_m = \frac{(E_t) - (ES) \times (S)}{(ES)} \qquad ...(4)$$

or $$(ES) \times K_m = (E_t) \times (S) - (ES) \times (S)$$

$$\text{or} \quad (ES) \times K_m + (ES) \times (S) = (E_t) \times (S)$$

$$\text{or} \quad (ES) \times [K_m + (S)] = (E_t) \times (S)$$

$$\text{or} \quad \frac{(ES)}{(E_t)} = \frac{(S)}{K_m + (S)} \qquad ...(5)$$

Substituting the value of $\frac{(ES)}{(E_t)}$ from equation (3) to equation (5), we get :

$$\frac{V}{V_m} = \frac{S}{K_m + (S)}$$

$$\text{or} \quad V = \frac{V_m \times (S)}{K_m + (S)} \qquad ...(6)$$

$$\text{or} \quad K_m = (S)\left[\frac{V_m}{V} - 1\right] \qquad ...(7)$$

Equation (7) is called as **Michaelis-Menten equation.** This can be used to calculate K_m after experimentally determining the reaction rates at various substrate concentrations. This equilibrium constant, K_m, is usually called **Michaelis constant.** *It is a measure of the affinity of an enzyme for its substrate.* Referring to the equation (4), the greater the concentration of ES complex, the lower is the concentration of free enzyme and consequently the lower is the value of K_m.

For experimental determination of K_m, the velocity of the reaction (relative activity of the enzyme) is measured as a function of substrate concentration. When V = ½V_m, it will be seen from equation (6) that K_m is numerically equal to the substrate concentration or in other words K_m is equal to the concentration of the substrate which gives half the numerical maximal velocity, V_m. Thus, it is possible to determine K_m. The K_m, shown in Fig. 18–3, is indicated to be 0.017 M. It is noteworthy that *for any enzyme-substrate system, K_m has a characteristic value which is independent of the enzyme concentration.*

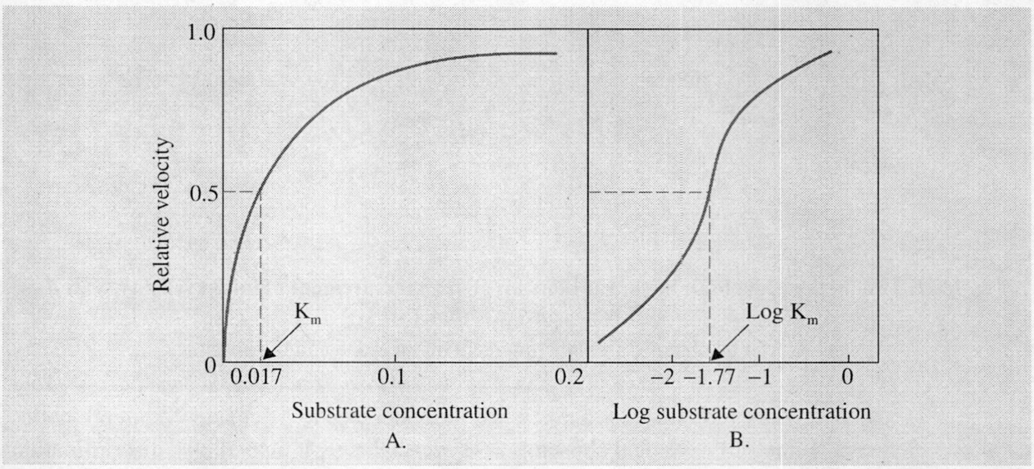

Fig. 18–3. Relative initial velocity as a function of substrate concentration (A) and as a function of the logarithm of the substrate concentration (B) for the action of yeast invertase on sucrose

(Adapted from White, Handler and Smith, 1964)

Lineweaver-Burk equation. The method described above for the detertmination of K_m is somewhat complex and, therefore, simpler methods have been devised. Two such methods (Fig. 18–4) are given below :

First method. A convenient means of evaluating K_m and V_m is to plot kinetic data as the reciprocals of V and [S]. Such a double reciprocal plot was proposed by Hans Lineweaver and Dean Burk in 1934. If one takes the reciprocal of Michaelis-Menten equation, the following equation is obtained :

$$\frac{1}{V} = \frac{K_m + (S)}{V_m \times (S)}$$

or

$$\frac{1}{V} = \frac{K_m}{V_m} \times \frac{1}{(S)} + \frac{1}{V_m} \qquad ...(8)$$

This is known as **Lineweaver-Burk equation.** This equation is of the form, $y = mx + b$, if one considers the variables to be $\frac{1}{V}$ and $\frac{1}{(S)}$. When one plots the graph against these two variables, a straight line is obtained. The slope of this line corresponds to $\frac{K_m}{V_m}$ and the $\frac{1}{V}$ intercept corresponds to $\frac{1}{V_m}$. Since V_m can be determined from the intercept, the K_m may also be calculated.

Second Method. Another graphical method for the measurement of K_m from experimental data on V as a measure of (S) makes use of the above Lineweaver-Burk equation. Multiplication of both sides of this equation by (S) gives :

$$\frac{(S)}{V} = \frac{K_m}{V_m} + \frac{(S)}{V_m} \qquad ...(9)$$

A plot of $\frac{(S)}{V}$ versus (S) gives a straight line. The intercept of the line on $\frac{(S)}{V}$ axis is $\frac{K_m}{V_m}$ and the slope is $\frac{1}{V_m}$.

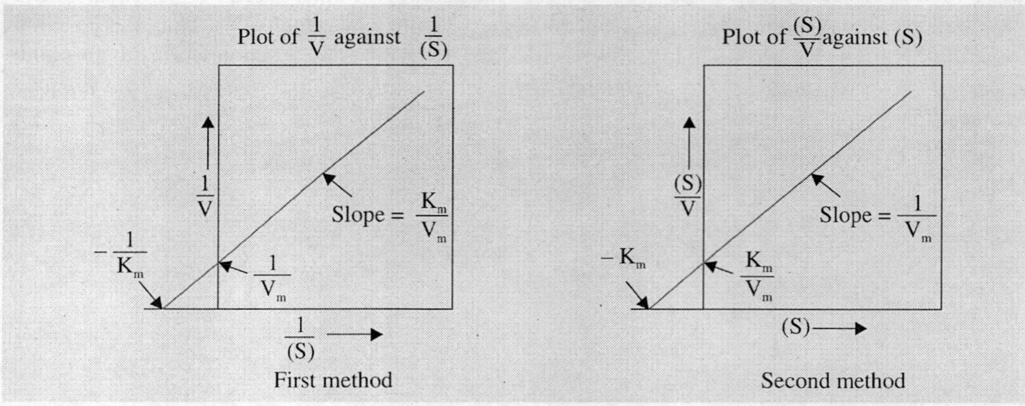

Fig. 18–4. Two methods of obtaining linear plots involving measurement of initial velocity, V and substrate concentration, (S)

V_m and K_m can be obtained from the intercept on the slope and the abscissa.

(Adapted from Fruton JS and Simmonds S, 1960)

A Lineweaver-Burk plot provides a quick test for adherence to Michaelis-Menten kinetics and allows easy evaluation of the critical constants. As we shall see, it also allows discrimination between different kinds of enzyme inhibition and regulation. A disadvantage of Lineweaver-Burk plot is that a long extrapolation is often required to determine K_m, with corresponding uncertainty in the result. Consequently, other ways of plotting the data are sometimes used. One such alternative is to rearrange equation (6) into the form,

$$V = V_m - K_m \frac{V}{[S]} \qquad ...(10)$$

and graph V versus V/[S]. This yields what is called an **Eadie-Hofstee plot** (Fig. 18–5).

Significance of K_m and V_m Values

The **Michaelis constant (K_m)** values of the enzymes differ greatly. However, for most of the enzymes, the general range is between 10^{-1} and 10^{-6} M (refer Table 18–2). The K_m value depends on the particular substrate and on the environmental conditions such as temperature and ionic concentration.

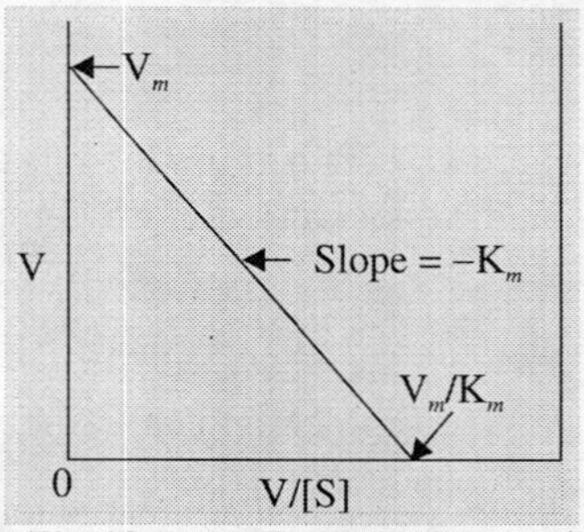

Fig. 18-5. An Eadie-Hofstee plot

Graphing V versus V/[S], one obtains V_m at (V/[S]) = 0 and K_m from the slope of the line.

The K_m value signifies two meanings :

(*a*) K_m is the concentration of substrate at which half the active sites are occupied. If K_m is known, the fraction of the sites filled, f_{ES}, at any substrate concentration, can be calculated from :

$$f_{ES} = \frac{V}{V_m} + \frac{(S)}{K_m + (S)} \quad ...(11)$$

(*b*) K_m is related to the rate constants of the individual steps in the catalytic scheme,

Table 18–2. Values of Michaelis constant (Km) of some enzymes

Enzyme	*Substrate*	K_m*
1. Lysozyme	Hexa-N-acetylglucosamine	6×10^{-6} M
2. Penicillinase	Benzylpenicillin	5×10^{-5} M
3. β-galactosidase	Lactose	4×10^{-3} M
4. Chymotrypsin	Acetyl-L-tryptophanamide	5×10^{-3} M
5. Carbonic anhydrase	Carbon dioxide	8×10^{-3} M
6. Pyruvate carboxylase	pyruvate	4×10^{-4} M
	ATP	6×10^{-5} M
7. Arginine-tRNA synthetase	tRNA	4×10^{-7} M
	Arginine	3×10^{-6} M
	ATP	3×10^{-4} M
8. Hexokinase	ATP	4×10^{-4} M
	D-glucose	5×10^{-5} M
	D-fructose	15×10^{-4} M
9. Sucrase	Sucrose	0.016 – 0.04 M
10. α-glucosidase	Methyl-α-D-glucoside	0.037 – 0.075 M
11. β-glucosidase	Methyl-β-D-glucoside	0.060 – 1.12 M
12. Pepsin	Ovalbumin	4.5%
13. Trypsin	Casein	2.0%
14. Urease	Urea	0.025 M
15. Catalase	Hydrogen peroxide	0.025 M
16. Lipase (Esterase)	Ethyl butyrate	> 0.03 M
17. Phosphatase	Glycerophosphate	< 0.003 M
18. Xanthine oxidase	Xanthine	4×10^{-7} M
19. Succinic dehydrogenase	Succinate	5×10^{-7} M
20. Amylase	Starch	0.8 – 0.25%
21. Dipeptidase	Glycylleucine	0.02 – 0.07 M

* K_m values are not absolute constants but depend on the temperature, substrate and source of enzyme.

$$E + S \underset{k_2}{\overset{k_1}{\rightleftharpoons}} ES \xrightarrow{k_3} E + P$$

where k_1, k_2 and k_3 are the rate constants for the 3 reactions. If k_2 is much greater than k_3 ($K_2 >> k_3$), *i.e.*, the dissociation of ES complex is more rapid than formation of P and regeneration of E, then

$$K_m = \frac{k_2}{k_1} \quad ...(12)$$

The dissociation constant of ES complex is given by,

$$K_{ES} = \frac{(E)(S)}{(ES)} = \frac{k_2}{k_1} \quad ...(13)$$

In other words, K_m is equal to the dissociation constant of the ES complex if k_3 is much smaller than k_2 ($k_3 << k_2$). When this condition is reached, K_m is a measure of the strength of ES complex. The high K_m value indicates weak binding whereas the low K_m value signifies strong binding. It may, however, be emphasized that K_m indicates the affinity of the ES complex only when $k_2 >> k_3$.

The **maximal rate (V_m)** represents the turnover number of an enzyme, if the concentration of the active sites (E*t*) is known, since,

$$V_m = k_3 (E_t) \quad ...(14)$$

A 10^{-6} M solution of *carbonic anhydrase*, for instance, catalyzes the formation of 0.6 M carbonic acid per second when it is fully saturated with substrate. Hence, k_3 is 6×10^5 sec^{-1}. *The kinetic constant, k_3 is known as the turnover number.*

HERMANN EMIL FISCHER

(LT, 1852-1919)

Emil Fischer, a German chemist, was born on October 9, 1852. He was the son of a wealthy merchant. He graduated from the Gymnasium of Bonn in 1869. After an abortive foray into the business world, he entered the University of Bonn in 1871 to study chemistry under **Frederick August Kekule** (LT, 1829-1867), master of structural chemistry. After receiving his doctoral degree from the University of Strasbourg in 1874, he taught at the Universities of Erlangen and Wurzburg, eventually becoming Professor of Chemistry at the University of Berlin in 1892 while succeeding **Hofmann.** He specialized in synthetic organic chemistry and published his discovery of phenylhydrazine in 1875. Later in 1899, he turned to a detailed study of proteins. It was in the study of structural chemistry to the study of proteins that Fischer saw the possible future collaboration of biology and chemistry. He developed the *theory of the 'peptide bond',* the chemical linkage by which all amino acids are joined together to form their respective proteins. He also chemically synthesized an *octadecapeptide,* composed of 3 leucine + 15 glycine residues. Frequently referred to as the **father of biochemistry**, Fischer received the second-ever **Nobel Prize in Chemistry (1902)** for his work on the synthesis of purines, peptides and sugars. Fischer's collection of 9,000 reference compounds is housed in the Department of Biochemistry, the University of California at Berkeley. The prized collection was a gift of **H.O.L. Fischer**, a biochemist and the only one of Fischer''s three sons to survive World War I. Following the deaths of his two sons in World War I, Emil Fischer committed suicide.

ACTIVE SITE

As the substrate molecules are comparatively much smaller than the enzyme molecules, there should be some specific regions or sites on the enzyme for binding with the substrate. Such sites of attachment are variously called as **'active sites'** or **'catalytic sites'** or **'substrate sites'**.

Although the enzymes differ widely in their properties, the active site present in their molecule possesses some common features (Fig. 18–6). These are litsed below :

1. The active site occupies a relatively small portion of the enzyme molecule.
2. The active site is neither a point nor a line or even a plane but is a 3-dimensional entity. It is made up of groups that come from different parts of the linear amino acid sequence. For example, as already stated, *lysozyme* (refer Fig. 17–14) has 6 subsites in the active site. The amino acid residues located at the active site are 35, 52, 59, 62, 63 and 107.
3. Usually the arrangement of atoms in the active site is well defined, resulting in a marked specificity of the enzymes. Although cases are known where the active site changes its configuration in order to bind a substance which is only slightly different in structure from its own substrate.

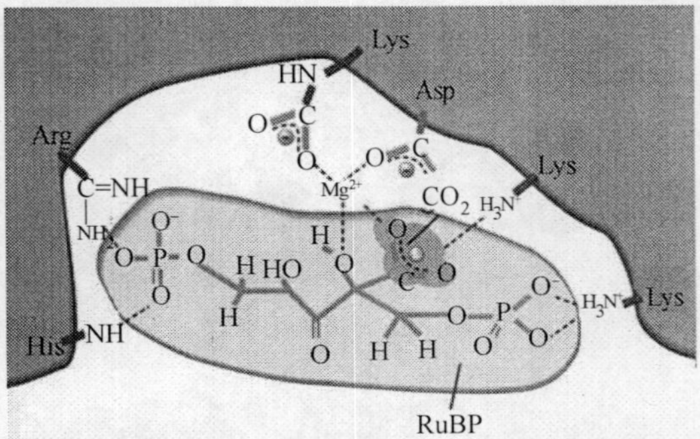

Fig. 18-6. Diagrammatic representation of the active site of the enzyme ribulose bisphosphate carboxylase (Rubisco) showing the various sites of interaction between the bound substrates (RuBP and CO_2) and certain amino acid side chains of the enzyme

In addition to determining the substrate-binding properties of the active site, these noncovalent interactions alter the properties of the substrate in ways that accelerate its conversion to products.

(*After Harris DA, 1995*)

4. The active site binds the substrate molecule by relatively weak forces.
5. The active sites in the enzyme molecules are grooves or crevices from which water is largely excluded. It contains amino acids such as aspartic acid, glutamic acid, lysine serine etc. The side chain groups like —COOH, —NH_2, —CH_2OH etc., serve as catalytic groups in the active site. Besides, the crevice creates a micro-environment in which certain polar residues acquire special properties which are essential for catalysis.

Fischer's Lock and Key Model

Previously, the interaction of substrate and enzyme was visualized in terms of a lock and key model (also known as *template model*), proposed by **Emil Fischer** in 1898. According to this model, the union between the substrate and the enzyme takes place at the active site more or less in a manner in which a key fits a lock and results in the formation of an enzyme-substrate complex (Fig. 18–7).

> To quote **Fischer** (1898) himself, "There is a relation between the unknown structure of an active enzyme and that of substrate, they are complementary and the one may be said to fit the other **as a key fits a lock.**"

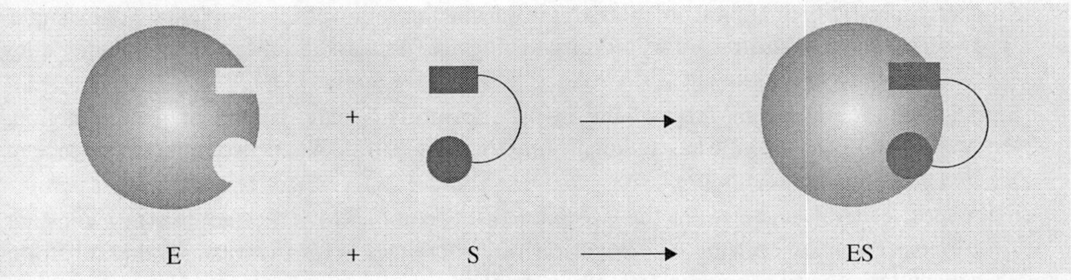

Fig. 18–7. Formation of an enzyme-substrate complex according to Fischer's lock and key model

In fact, the enzyme-substrate union depends on a *reciprocal fit* between the molecular structure of the enzyme and the substrate. And as the two molecules (that of the substrate and the enzyme) are involved, this hypothesis is also known as the **concept of intermolecular fit.** The enzyme-substrate complex (Fig. 18–8) is highly unstable and almost immediately this complex decomposes to produce

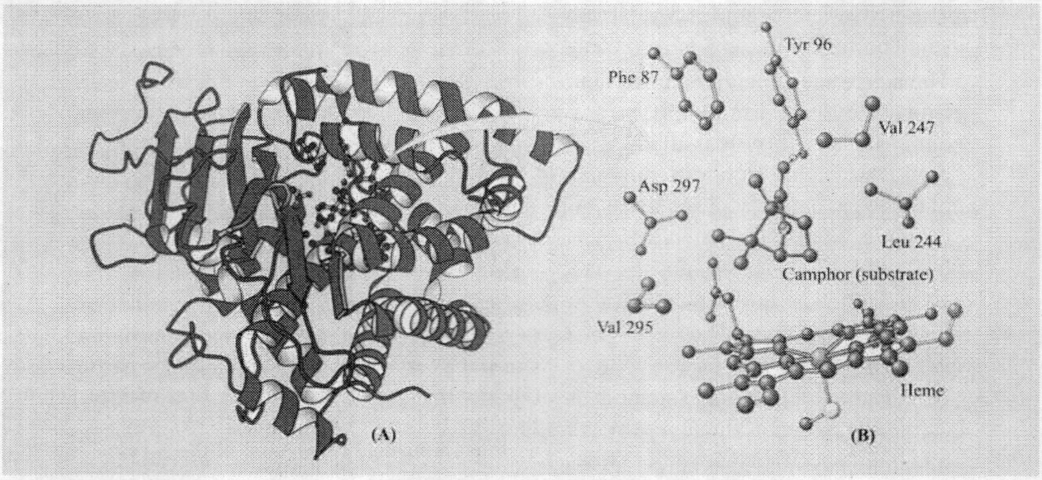

Fig. 18–8. Structure of an enzyme–substrate complex

(A) The enzyme cytochrome P-450 is illustrated bound to its substrate camphor.

(B) In the active site, the substrate is surrounded by residues from the enzyme. Note also the presence of a heme cofactor.

the end products of the reaction and to regenerate the free enzyme. The enzyme-substrate union results in the release of energy. It is this energy which, in fact, raises the energy level of the substrate molecule, thus inducing the *activated state*. In this activated state, certain bonds of the substrate molecule become more susceptible to cleavage.

Evidences Proving the Existence of an ES Complex:

The existence of an ES complex during enzymatically-catalyzed reaction has been shown in many ways :

1. The ES complexes have been directly observed by electron microscopy and x-ray crystallography.
2. The physical properties of enzymes (*esp*., solubility, heat sensitivity) change frequently upon formation of an ES complex.
3. The spectroscopic characteristics of many enzymes and substrates change upon formation of an ES complex. It is a case parallel to the one in which the absorption spectrum of deoxyhemoglobin changes markedly, when it binds oxygen or when it is oxidized to ferric state.
4. Stereospecificity of highest order is exhibited in the formation of ES complexes. For example, D-serine is not a substrate of tryptophan synthetase. As a matter of fact, the D-isomer does not even bind to the enzyme.
5. The ES complexes can be isolated in pure form. This may happen if in the reaction, $A + B \rightleftharpoons C$, the enzyme has a high affinity for the substrate A and also if the other reactant B is absent from the mixture.
6. A most general evidence for the existence of ES complexes is the fact that at a constant concentration of enzyme, the reaction rate increases with increase in the substrate concentration until a maximal velocity is reached.

Koshland's Induced Fit Model

An important but unfortunate feature of Fischer's model is the rigidity of the active site. The active site is presumed to be pre-shaped to fit the substrate. In order to explain the enzyme properties more efficiently, **Koshland**, in 1958, modified the Fischer's model. Koshland presumed that the enzyme molecule does not retain its original shape and structure. But the contact of the substrate *induces* some configurational or geometrical changes in the active site of the enzyme molecule. Consequently, the enzyme molecule is made to *fit* completely the configuration and active centres of the substrate. At the same time, other amino acid residues may become buried in the interior of the molecule. Koshland's hypothesis has recently been confirmed by **Lipscomb.**

To explain the theory, a hypothetical illustration may be given (Fig. 18–9). The hydrophobic and charged groups both are involved in **substrate binding.** A phosphoserine (-P) and the -SH group of cysteine residue are involved in **catalysis.** Other amino acid residues not involved in either substrate binding or catalysis are lysine (Lys) and methionine (Met). In the absence of substrate, the substrate binding and catalytic groups are far apart from each other. But the proximity of the substrate induces a conformational change in the enzyme molecule aligning the groups for both substrate binding and catalysis. Simultaneously, the spatial orientation of other regions is also changed so that the lysine and methionine are now much closer.

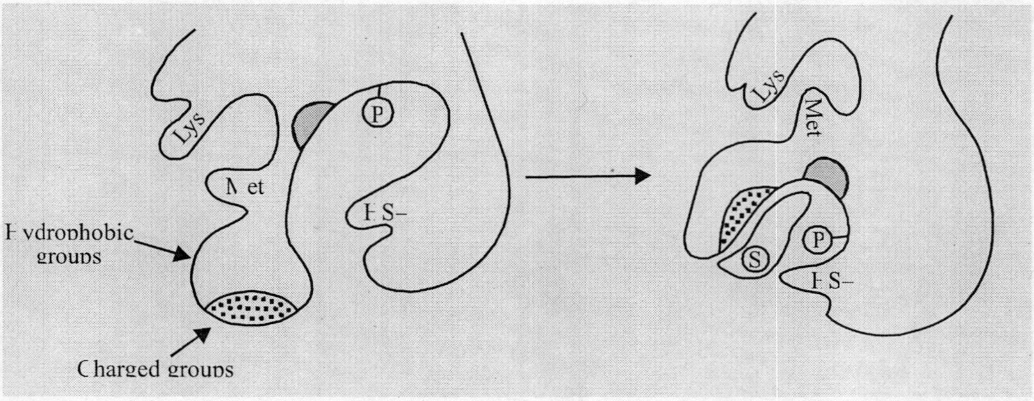

Fig. 18–9. Conformational changes brought about by induced fit in an enzyme molecule

(Modified from Daniel E. Koshland Jr., 1964)

An illustration of the competitive inhibitor or substrate analogue may also be given (Fig. 18–10). On contact with the true substrate, all groups are brought into correct spatial orientation. But attachment of a competitive inhibitor, which is either too "slim" or too "bulky", induces incorrect alignment.

As to the sequence of events during the conformational changes, 3 possibilities exist (refer Fig. 18–11). The enzyme may first undergo a conformational change (A), then bind substrate (B). An alternative pathway is that the substrate may first be bound (C) and then a conformational change (D) may occur. Thirdly, both the processes may occur simultaneously (E) with further isomerization (F) to the final conformation.

Originally little more than an attractive hypothesis, Koshland's model has now gained much experimental support. Conformational changes during substrate binding and catalysis have been demonstrated for various enzymes such as *phosphoglucomutase, creatine kinase, carboxypeptidase* etc.

ENZYME REACTION RATES

A plot of V (velocity, or reaction rate) versus substrate concentration in an enzymatically- catalyzed reaction produces a hyperbolic curve (Fig. 18–12) which is representative of the reaction kinetics of a noncooperative enzyme. The hyperbolic curve is analogous to the oxygen-dissociation curve of myoglobin. The plot shows that the velocity increases with substrate concentration until maximum V (V_{max}) is approached asymptotically, after which larger concentrations of substrate do not significantly enhance the reaction rate. In the lower region of the curve, the reaction approaches *first-order kinetics*, which means that v is a direct function of substrate concentration because

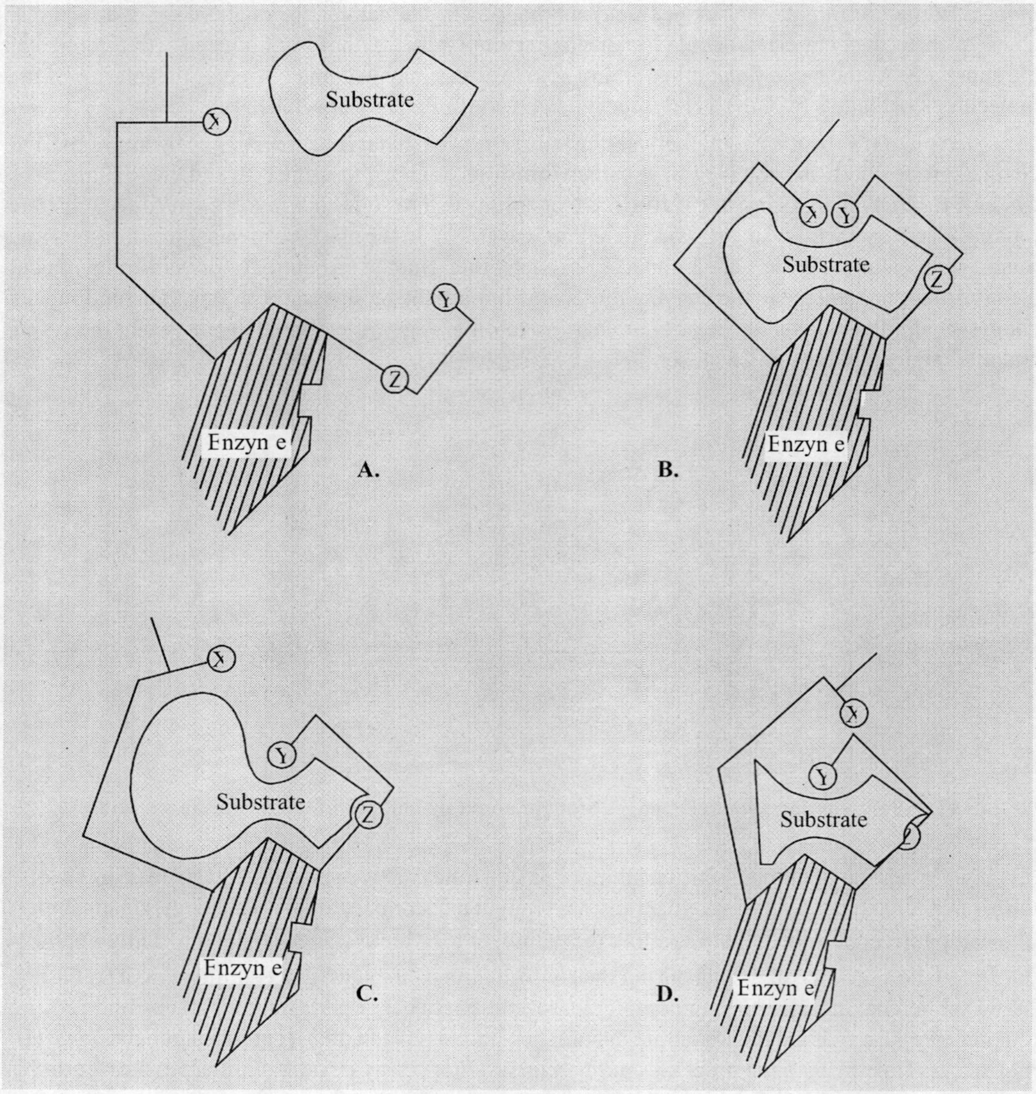

Fig. 18–10. Schematic model of flexible active site mechanism

Black lines indicate protein chains containing catalytic groups X and Y and binding groups Z.

A. Substrate and enzyme dissociated.

B. Substrate with induced change of protein chains to bring X and Y into proper alignment for reaction.

C. Bulky groups added to substrate prevent proper alignment of X and Y.

D. Deleted group eliminates buttressing action on chain containing X so that the complex has incorrect alignment of X and Y.

(From Daniel E. Koshland Jr., 1964)

the active sites of the enzyme molecules are not saturated. At the plateau at the upper portion of the plot, the reaction approaches *zero-order kinetics* because the active sites of all the enzyme molecules are saturated and the reaction rate is, therefore, independent of further increases in substrate concentration. For the intermediate portion of the curve, as the enzyme approaches substrate saturation,

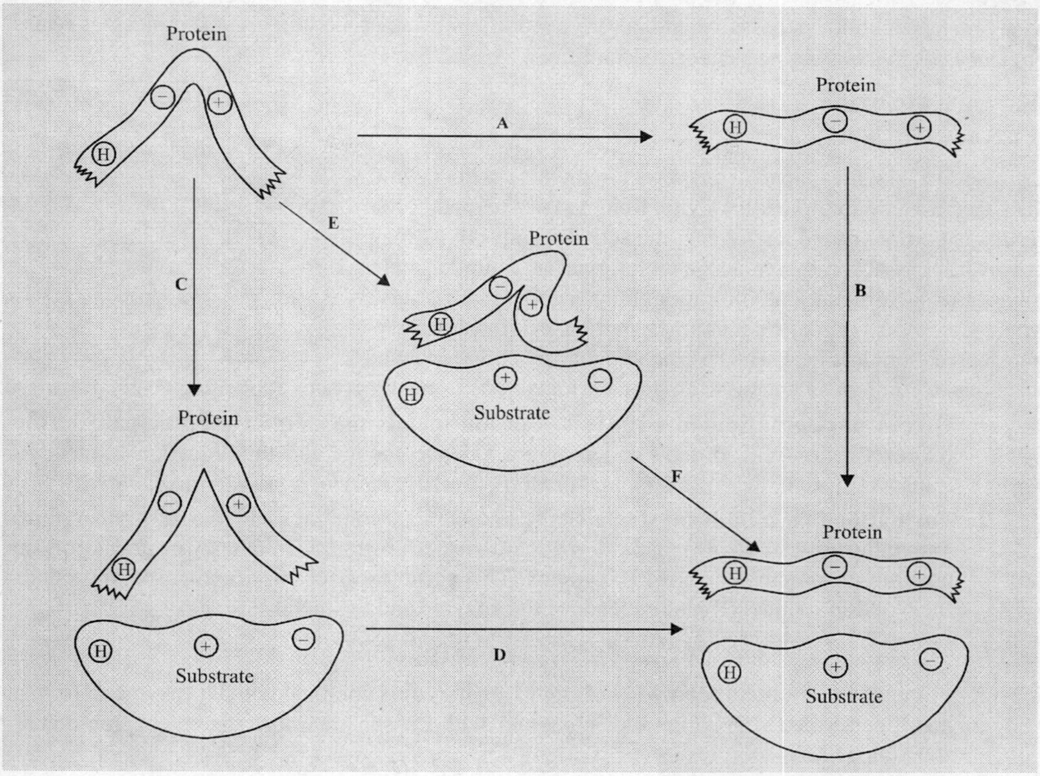

Fig. 18–11. Alternative pathways for a substrate-induced conformational change

(Adapted from Koshland and Neet, 1968)

kinetics are mixed zero and first order in substrate concentration. Routine enzyme assays are designed to follow zero-order kinetics to avoid the influence of substrate concentration on reaction velocity.

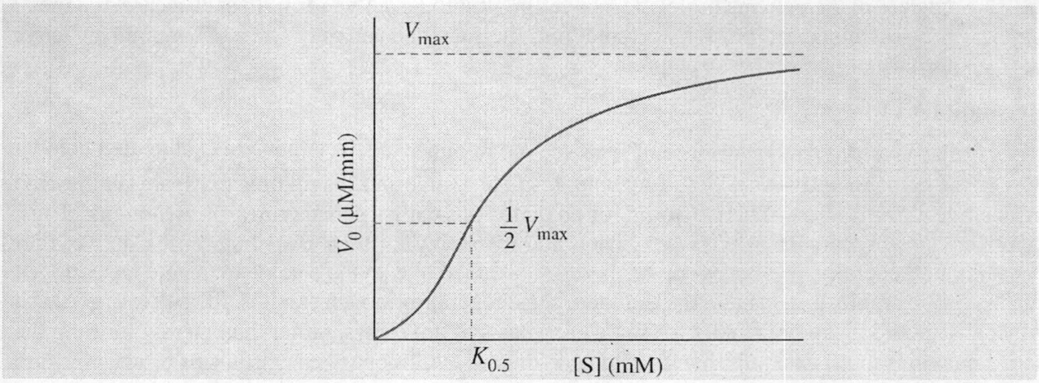

Fig. 18–12. Effect of substrate concentration on the velocity of an enzyme-catalyzed reaction

Under such conditions, measured rates are directly proportional to the concentration of the enzyme itself. In contrast, *uncatalyzed reactions do not show this saturation effect.*

MODIFIERS OF ENZYME ACTIVITY

The catalytic activity of certain enzymes is reversibly altered by certain inorganic and organic molecules called modifiers. Those molecules which increase the enzyme activity are called *positive modifiers* or *activators* and those which decrease the enzyme activity as *negative modifiers* or *inhibitors*.

Many metals act both as postitive and negative modifiers, whereas certain organic molecules retard enzyme activity, thus acting as negative modifiers.

Inorganic Modifiers (Enzyme Activators)

Certain enzymes, apart from a requirement of a coenzyme, also need a metal ion for full activity. Removal of the metal often results in partial or total loss of enzyme activity. The activity may, however, be restored by replacing the original or a similar metal ion. Some such metal ions (or cations) are K^+, Cu^+, Fe^{2+}, Mg^{2+}, Mn^{2+}, Ca^{2+}, Cu^{2+}, Zn^{2+}, Fe^{3+} etc. Mg participates in phosphate-transfer reactions and Fe, Cu and Mo are required in oxido-reduction reactions. Certain mechanisms as to how the metal ions bring about activation are given beolw :

> **Metals** are the common inorganic modifiers. Besides accelerating the rate of enzymatically-catalyzed reactions, metals also inhibit the rate of such reactions.

1. **Direct participation in catalysis.** Certain metals may directly participate in the oxidoreduction reactions by undergoing a valence change and thus function in electron transport system. Fe, for exmaple, functions similarly in *cytochromes* or in *catalase*.
2. **Formation of a metallosubstrate.** Sometimes a metal combines with the substrate to form a metallosubstrate (MS) which, in fact, is the true substrate for the enzyme and forms an enzyme-metal-substrate (EMS) complex. This complex, later on, decomposes to produce the reaction product (P) and regenerate the enzyme and the metal.

$$S + M \rightleftharpoons MS \xrightarrow{+E} EMS \longrightarrow E + M + P$$

3. **Formation of a metalloenzyme.** A metal ion may first combine with an enzyme to form a metalloenzyme (ME) which then combines with the substrate forming an enzyme-metal-substrate (EMS) complex.

$$E + M \rightleftharpoons EM \xrightarrow{+S} EMS \longrightarrow E + M + P$$

4. **Alteration of equilibrium constant.** Metals may also change the nature of the reactants so that the apparent equilibrium constant of the reaction is also altered.
5. **Conformational change in the enzyme.** Metal ions may also bring about conformational change in the enzyme molecule, converting it into an active form. In such a case, the metal may be linked at a point far remote from the substrate and may serve to maintain an active tertiary or quaternary structure.

Regulation of Enzyme Activity

The activity of the enzymes is controlled by certain mechanisms which are enumerated below :

1. **Zymogen activation.** Some enzymes (esp., digestive and coagulating ones) are synthesized in inactive forms. These inactive forms of enzymes are called **proenzymes** or **zymogens**. Thus, zymogens are precursors of enzymes. They are not active on their own but can be active after conversion. Generally, prefix *pro-* or suffix *-ogen* is added to enzyme's name to denote its zymogen such as prothrombin, proelastase, trypsinogen, pepsinogen, fibrinogen etc. It is difficult to argue as to why cells should make precursors for some enzymes and not others, rather than merely forming the active enzymes in all cases. Perhaps, the precursors are evolutionary anachronisms or else they are required at times or places when the active enzymes would injure cells.

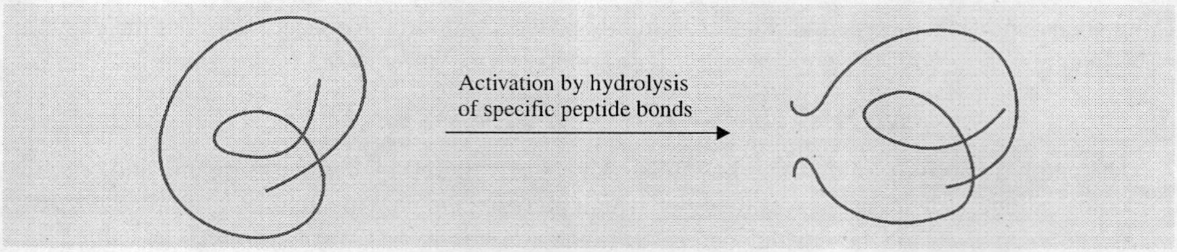

Fig. 18–13. Zymogen activation by hydrolysis of specific peptide bonds

The *precursors* should be converted to the active forms before they can catalyze a chemical reaction. *This type of activation usually involves cleavage of the peptide bond* (Fig. 18–13). Often, a part of the zymogen is also removed. Examples of some common precursor systems are given below :

(A) Chymotrypsin is synthesized by the exocrine cells of the pancreas in its precursor form, chymotrypsinogen. Hydrolysis by trypsin converts chymotrypsinogen into the active form, chymotrypsin.

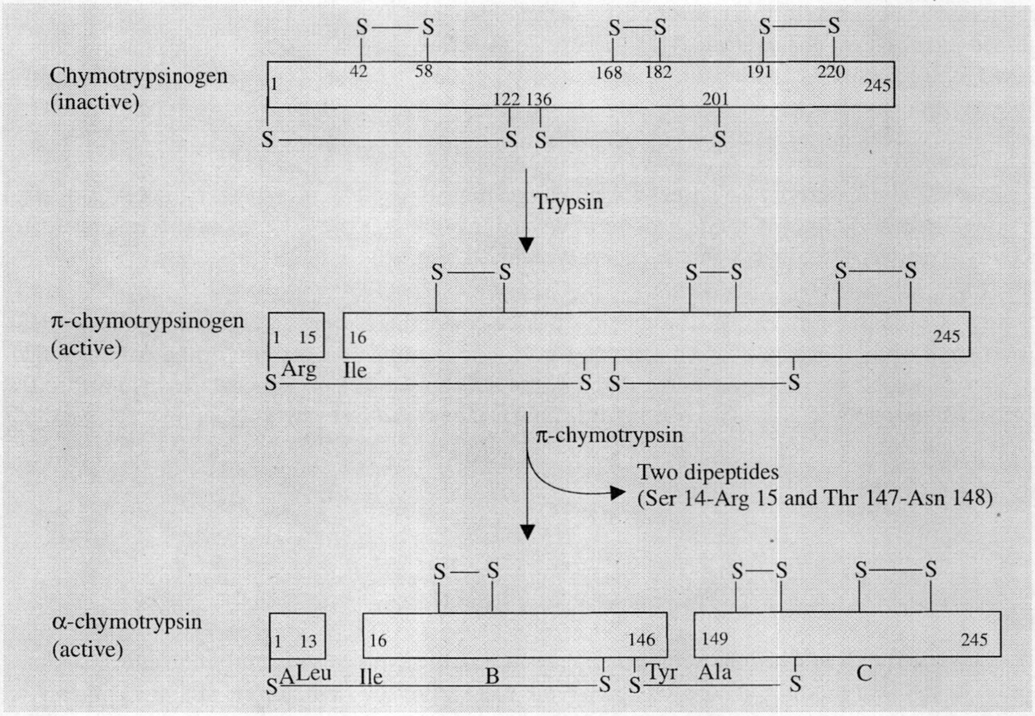

Fig. 18–14. Proteolytic activation of chymotrypsinogen

The 3 chains of α-chymotrypsin are linked by two interchain disulfide bonds (A to B and B to C).

$$\text{Chymotrypsinogen} \xrightarrow{\text{Trypsin}} \text{Chymotrypsin}$$

Chymotrypsinogen is composed of a single polypeptide chain with 245 amino acid residues (representing all the 20 standard amino acids) and 5 intrachain disulfide bridges (Fig. 18–14). It is virtually devoid of enzymatic activity and is converted into a fully active enzyme when the peptide bond linking Arg^{15} and Ile^{16} is cleaved by trypsin. The resulting enzyme, called π- **chymotrypsin**, then acts on other π-chymotrypsin molecules. Two dipeptides (between residues 14 and 15 and residues 147 and 148) are removed to yield α-**chymotrypsin**, the stable form of the active enzyme. The 3 resulting chains (designated A, B and C) in α-chymotrypsin remain covalently linked to each other by two interchain disulfide bonds, although 3 more (intrachain) disulfide bonds, one in chain B and two in chain C, are also present in the molecule. The additional cleavages made in converting π-chymotrypsin into α -chymotrypsin are superfluous, because π- chymotrypsin is already fully active. The striking feature of this activation process is that *cleavage of a single specific peptide bond transforms the protein from a catalytically inactive form into one that is fully active.*

(B) An intestinal enzyme termed enterokinase converts *trypsinogen*, a proenzyme also secreted by pancreas, to trypsin. Trypsin acts upon peptide linkages involving the carboxyl group of arginine and lysine.

$$\text{Trypsinogen} \xrightarrow{H^+ \text{ or Pepsin}} \text{Trypsin}$$

(C) Another instance is the proteinase of the gastric juice called as pepsin. This is derived from its proenzyme *pepsinogen,* secreted by gastric mucosa. Pepsinogen is converted to the active form, pepsin, both by the acidity of the gastric juice and by pepsin itself. This process is, therefore, called as **autocatalysis.** During this process, a polypeptide is liberated from the proenzyme.

$$\text{Pepsinogen} \xrightarrow{\text{Enterokinase}} \text{Pepsin} + \text{a polypeptide}$$

Autocatalytic activity of an ezyme precursor may be graphically represented. If a graph of time versus enzyme activity is plotted, a sigmoid (S-shaped) curve is obtained (Fig. 18–15).

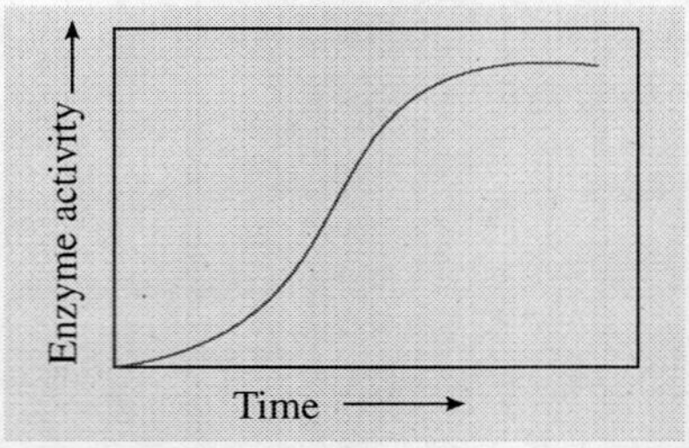

Fig. 18–15. Autocatalytic activation of an enzyme precursor

2. **Covalent modification.** The activity of serine-containing enzymes is regulated by the covalent insertion of a small group on an enzyme molecule. For example, the activity of the enzymes, that synthesize and cleave glycogen, is regulated by the attachment of a phosphoryl group to a specific serine residue on these enzymes. This modification can be reversed by hydrolysis.

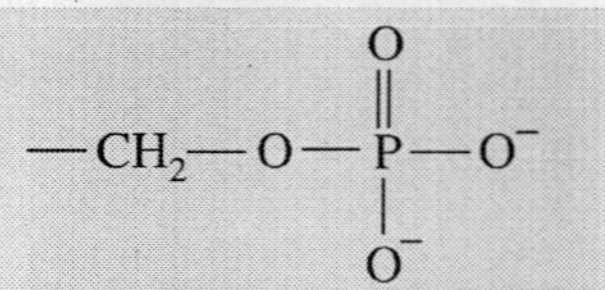

Phosphorylated derivative of a serine residue

3. **Feedback inhibition.** Some enzymes catalyze the synthesis of small molecules (such as amino acids) in a number of steps. The enzyme catalyzing the first step in this biosynthesis is inhibited by the end product of the reaction. Such type of regulatory mechanism, which is called feedback inhibition (Fig. 18–16), is beautifully illustrated by the biosynthesis of isoleucine from threonine.

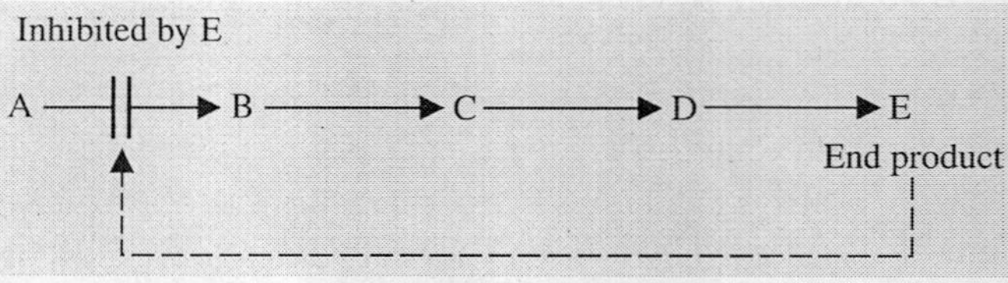

Fig. 18–16. Feedback inhibition of the first enzyme in a multistep reaction by reversible binding of the end product

The reaction completes in 5 steps. The first step reaction is catalyzed by the enzyme *threonine aminase.* The activity of this enzyme is inhibited upon accumulation of high quantities of isoleucine. Isoleucine binds to a different site from threonine. This is called *allosteric interaction.* However, when the level of isoleucine drops sufficiently, the enzyme reactivates and isoleucine is resynthesized.

Organic Modifiers (Enzyme Inhibitors)

Compounds which convert the enzymes into inactive substances and thus adversely affect the

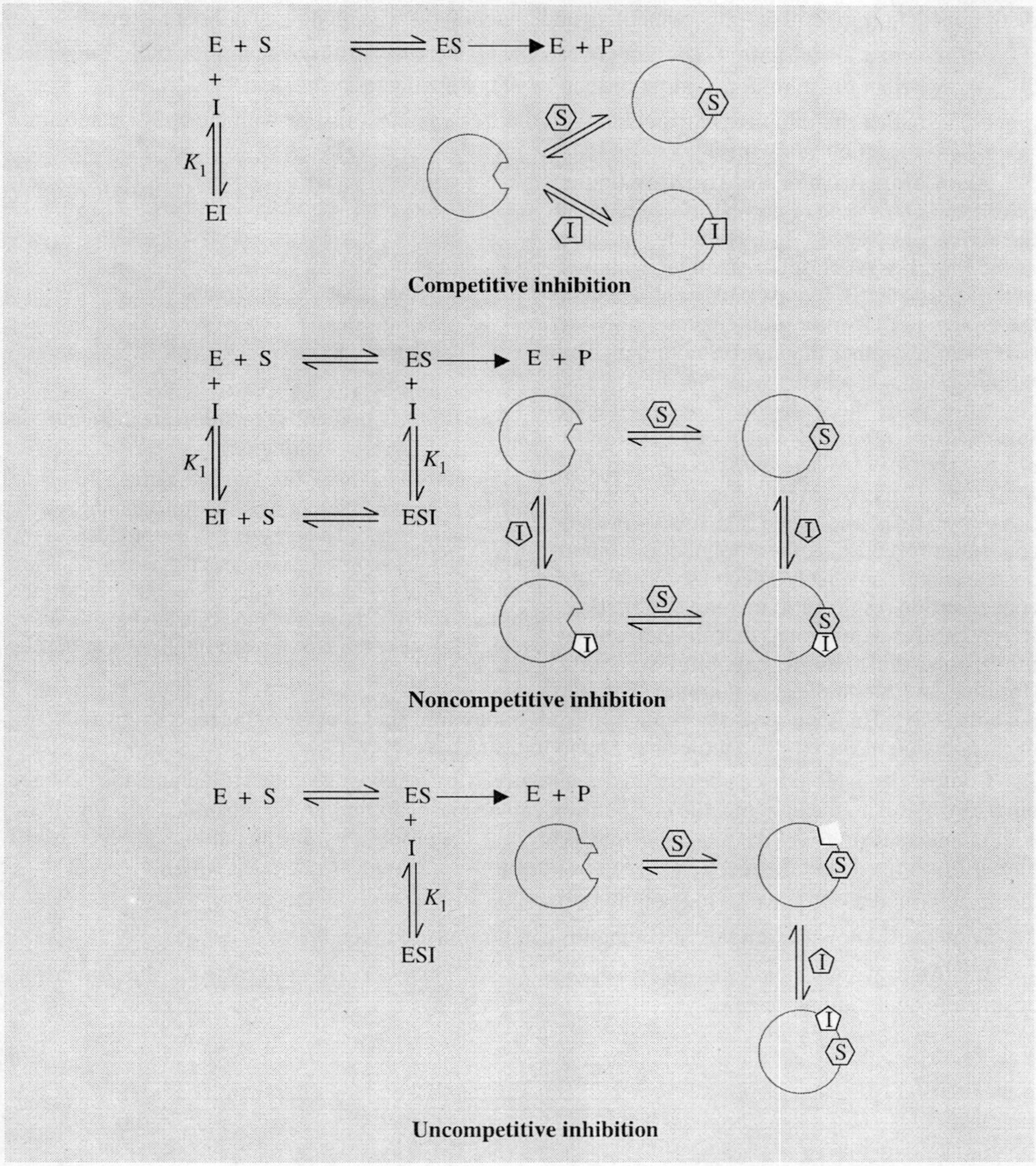

Fig. 18–17. Three types of reversible enzyme inhibition

Competitive inhibitors bind to the enzyme's active site. Noncompetitive inhibitors generally bind at a separate site termed allosteric site. Uncompetitive inhibitors also bind at an allosteric site but they bind only to the ES complex.

rate of enzymatically-catalyzed reaction are called as *enzyme inhibitors*. Such a process is known as *enzyme inhibition*. Two broad classes of enzyme inhibitors are generally recognized : reversible and nonreversible, depending on whether the enzyme-inhibitor (EI) complex dissociates rapidly or very slowly.

Reversible Enzyme Inhibition

A reversible inhibitor dissociates *very rapidly* from its target enzyme because it becomes very loosely bound with the enzyme. Three general types of reversible inhibition (Fig. 18–17) are distinguished : competitive, noncompetitive and uncompetitive, depending on three factors:

a. whether the inhibition is or is not overcome by increasing the concentration of the substrate,

b. whether the inhibitor binds at the active site or at allosteric site, and "

c. whether the inhibitor binds either with the free enzyme only, or with the enzyme substrate complex only or with either of the two.

A. Competitive or Substrate analogue inhibition. This type of competition occurs at the active site (Fig. 18-18). Here the structure of the inhibitor (I) closely resembles with that of the substrate (S). It may, thus, combine with the enzyme (E), forming an enzyme-inhibitor (EI) complex rather than an ES complex. The inhibitor, thus, competes with the substrate to combine with the enzyme. *The degree of inhibition depends upon the relative concentrations of the substrate and the inhibitor.* Thus, by increasing the substrate concentration and keeping the inhibitor concentration constant, the amount of inhibition decreases and conversely a decrease in substrate concentration results in an increased inhibition. It may, however, be noted that in competitive inhibition, the enzyme can bind substrate (forming an ES complex) or inhibitor (EI), but not both (ESI). Thus, we see that a competitive inhibitor diminishes the rate of the reaction by reducing the proportion of the enzyme molecules bound to a substrate.

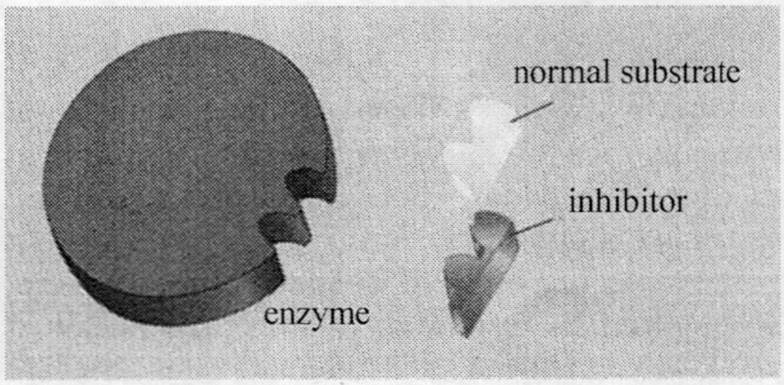

Fig. 18–18. Competitive or substrate analogue inhibition

A competitive inhibitor fits into the active site of the enzyme, preventing the real substrate from gaining access. The inhibitor cannot be converted to the products of the reaction and so the overall rate of reaction is slowed down. If an inhibitor is present in equal concentrations to the substrate, and if both types of molecule bind to the active site equally well, the enzyme can only work at half its normal rate.

Competitive inhibition can be analyzed quantitatively by steady-state kinetics. Because the inhibitor binds reversibly to the enzyme, the competition can be biased to favour the substrate only simply by adding more substrate. When more substrate is present, the probability that an inhibitor molecule will bind is minimized, and the reaction exhibits a normal V_{max}. However, the [s] at which $V_0 = 1/2\ V_{max}$, the K_m will increase in the presence of inhibitor.

Some well-known examples of competitive inhibition are given below :

1. An enzyme, *succinic acid dehydrogenase* (= *succinodehydrogenase*) catalyzes the conversion of succinic acid to fumaric acid.

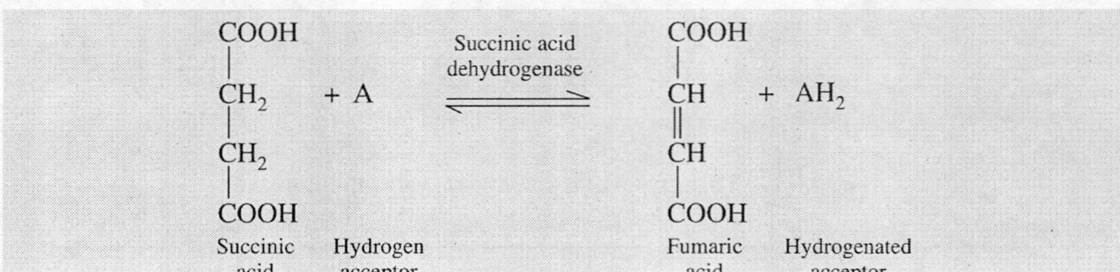

Many organic compounds, which are structurally related to succinic acid, combine with the enzyme, thus inhibiting the reaction. A few inhibitors of this reaction are :

Malonic acid is most efficient of all these inhibitors. When the inhibitor : enzyme ratio is 1 :50, the enzyme is inhibited 50%. Malonic acid differs from succinic acid in having one rather than two methylene groups.

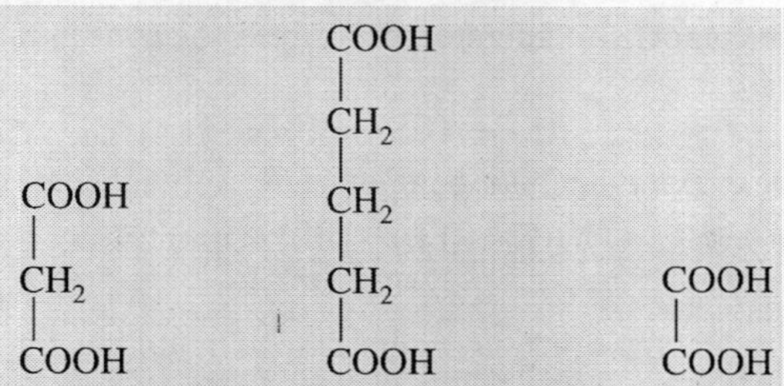

Malonic acid | Glutaric acid | Oxalic acid

2. Many microorganisms, like bacteria, synthesize the vitamin folic acid from para-aminobenzoic acid (PABA). Sulfanilamide and other sulfa drugs are structural analogues of PABA. Hence,

***p*-aminobenzoic acid, PABA** (Substrate) | ***p*-aminobenzenesulfonamide** or **Sulfanilamide** (Inhibitor)

sulfa drugs act as enzyme inhibitor and occupy the active site of some *bacterial enzymes* catalyzing this reaction. Failure of PABA to combine with the bacterial enzymes at the active site results in blocking off the folic acid synthesis. The resulting deficiency of this vitamin is fatal to these microorganisms. Since man lacks the enzymes necessary for folic acid synthesis from PABA, folic acid is needed as a vitamin in the diet. Thus, the sulfa drugs inhibit growth of the bacteria in man by competing with PABA for the active centres of the bacterial enzymes.

3. Competitive inhibition is used therapeutically to treat patients who have ingested methanol ($CH_3.OH$), a solvent found in gas-line antifreeze. Methanol is converted to formaldehyde by the action of the enzyme alcohol dehydrogenase (E.C. No. 1.1.1.1) . Formaldehyde damages many tissues esp., the optic ones, causing blindness. Ethanol competes effectively with methanol as a substrate for alcohol dehydrogenase. Ethanol, thus, acts as an inhibitor for the substrate methanol and competes with it to occupy the active site of the enzyme. Methanol-poisoning may, thus, be cured by an intravenous infusion of ethanol to the patients so that formaldehyde formation is considerably lowered. Most of the methanol can be extracted harmlessly in the urine.

4. A physiologically important example of competitive inhibition is found in the formation 2, 3- bisphosphoglycerate (BPG) from 1,3-bisphosphoglycerate. *Bis-phosphoglycerate mutase*, the enzyme catalyzing this isomerization reaction, is completely inhibited by even low levels of 2,3-bisphosphoglycerate. In fact, it is not uncommon for an enzyme to be completely inhibited by its own product because of the product's structural resemblance to the substrate. However, increasing the concentration of substrate checks the inhibitory effect.

Michaelis-Menten equation may also be applied to the competitive inhibition of enzymes. Here, besides the normal equation :

$$E + S \rightleftharpoons ES \longrightarrow E + P$$

one must also consider the equilibrium state between the enzyme and the inhibitor, I as follows :

$$E + I \rightleftharpoons EI$$

In the presence of the competitive inhibitor, henceforth, the concentration of the free enzyme would be expressed as :

$$[(E) — (ES) — (EI)]$$

The dissociation of the enzyme-inhibitor compound, K_i would then be defined as :

$$K_i = \frac{[(E) - (ES) - (EI)]\ (1)]}{(EI)}$$

Deriving the Michaelis-Menten equation for this case of inhibition, one obtains as follows :

$$V = \frac{V_m \times (S)\ K_i}{K_m K_i + K_m\ (I) + K_i (S)} \quad ...(15)$$

Reversal of this equation gives the modified form of Lineweaver-Burk equation :

$$\frac{1}{V} = \frac{K_m K_i + K_m (I) + K_i\ (S)}{V_m \times (S)\ K_i}$$

or

$$\frac{1}{V} = \frac{K_m}{V_m}\left(1 + \frac{(I)}{K_i}\right) \times \frac{1}{(S)} + \frac{1}{V_m} \quad ...(16)$$

When $\frac{1}{V}$ is plotted against $\frac{1}{(S)}$ (Fig. 18–19), the intercept $\frac{1}{V_m}$ remains the same as in the case of non-inhibited reaction but the slope, which is now $\frac{K_m}{V_m}\left(1 + \frac{(I)}{K_i}\right)$, is increased by the factor,

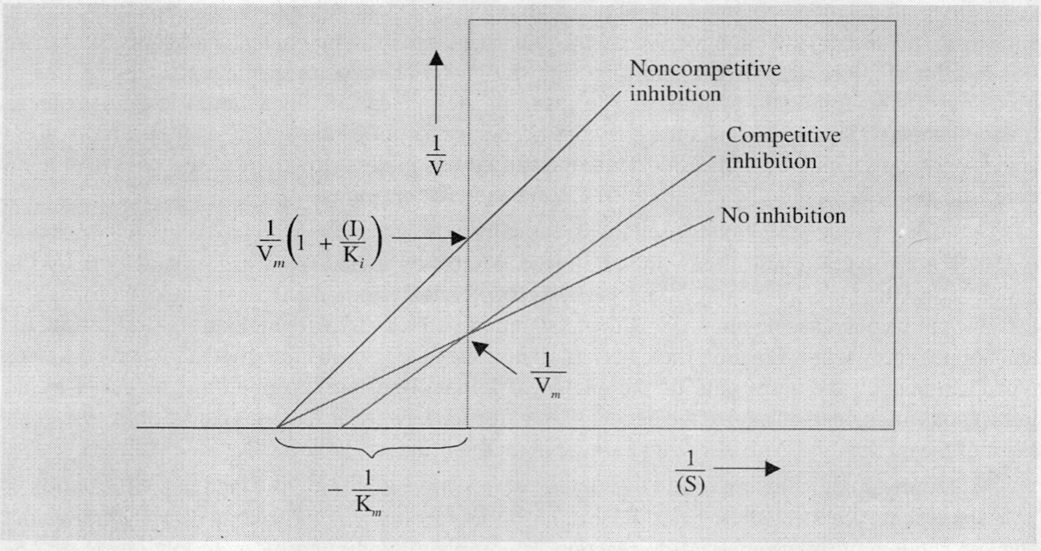

Fig. 18–19. Double reciprocal graph (Lineweaver-Burk plots)

A plot of $\frac{1}{V_m}$ versus $\frac{1}{(S)}$ for an enzyme reaction with (*i*) no inhibitor (*ii*) a competitive inhibitor and (*iii*) a noncompetitive inhibitor.

(Modified from Fairley JL and Kilgour GL, 1966)

$1 + \frac{(I)}{K_i}$. Thus, if the substrate concentration is large enough, the effect of the competitive inhibitor can be overcome and V_m may be reached.

The phenomenon of competitive inhibition is of practical value. It may be employed to prevent the growth of one organism in the presence of other, for example: (*a*) bacterial growth inhibited in the presence of an animal (*b*) growth of the insect inhibited on a fruit tree.

B. Noncompetitive inhibition. Here no competition occurs between the substrate, S and the inhibitor, I (Fig. 18-20). The inhibitor has little or no structural resemblance with the substrate and it binds with the enzyme at a place other than the active site (*i.e.*, at the allosteric site). Since I and S may combine at different sites, formation of both EI and ESI complexes takes place. Both ES and ESI may break down to produce the reaction product (P). It may, however, be noted that in noncompetitive inhibition, the inhibitor and substrate can bind simultaneously to an enzyme molecule since their binding sites are different and hence do not overlap. The enzyme is inactivated when inhibitor is bound, whether or not substrate is also present. Thus, it is apparent that a noncompetitive inhibitor acts by lowering the turnover number rather than by decreasing the proportion of enzyme molecules that are bound to the substrate. Noncompetitive inhibition, in contrast to competitive inhibition, cannot be overcome by increasing substrate concentration. The inhibitor effectively lowers the concentration of active enzyme and hence lowers V_{max}. There is almost no effect on K_m, however.

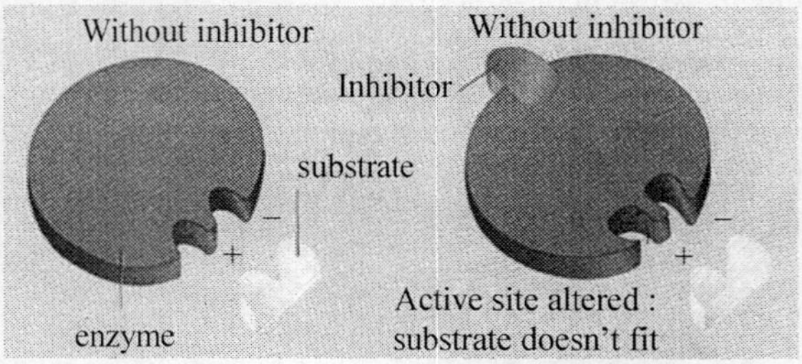

Fig. 18–20. Noncompetitive inhibition

Non-competitive inhibitors attach to enzyme molecules and alter the overall shape, so that the active site cannot function. Although the substrate may still bind, forming an *enzyme–inhibitor substrate* complex, the substrate cannot be turned into product. When the inhibitor molecule is removed, normal function is restored.

Certain noteworthy examples of noncompetitive inhibition are as follows :

1. Various heavy metals ions (Ag^+, Hg^{2+}, Pb^{2+}) inhibit the activity of a variety of enzymes. *Urease,* for example, is highly sensitive to any of these ions in traces. Heavy metals form mercaptides with sulfhydryl (—SH) groups of enzymes :

$$\text{Enz—SH} + Ag^+ \rightleftharpoons \text{Enz—S—Ag} + H^+$$

The established equilibrium inactivates enzymes that require a—SH group for activity. Because of the reversibility of mercaptide formation, the inhibition can be relieved by removal of the heavy metal ion. In the treatment of lead poisoning, advantage is taken of the metal's affinity for —SH groups. Therefore, the sulfhydryl compounds are administered to interact with the metal in the circulatory system and form mercaptides, which are then excreted.

2. Similarly, cyanide and hydrogen sulfide strongly inhibit the action of iron-containing enzymes like *catalase* and *peroxidase*.

The Lineweaver-Burk equation for this type of inhibition would be :

$$\frac{1}{V} = \left[1+\frac{(I)}{K_i}\right]\left[\frac{1}{V_m}+\left(\frac{K_m}{V_m}\right)\times\frac{1}{(S)}\right] \quad ...(17)$$

Thus, here both the slope and the intercept are altered, rather increased by the factor $1+\frac{(I)}{K_i}$ in contrast to the competitive inhibition where only the slope is changed. Moreover, the maximal velocity attained is less than that found in noninhibited case.

C. Uncompetitive inhibition. An uncompetitive inhibitor also binds at an allosteric site (like the noncompetitive inhibitors) but the binding takes place only with the enzyme-substrate (ES) complex, and not the free enzyme molecule.

Irreversible Enzyme Inhibition

Although irreversible inhibition was once categorized and tested as noncompetitive inhibition, it is now recognized as a distinct type of inhibition. Irreversible inhibitors are those that combine with or destroy a functional group on the enzyme that is essential for its activity. In fact, an irreversible inhibitor dissociates *very slowly* from its target enzyme because it becomes very tightly bound to its active site, thus inactivating the enzyme molecule. The bonding between the inhibitor and enzyme may be covalent or noncovalent.

Two common examples of irreversible inhibition are discussed below :

1. Alkylating reagents, such as **iodoacetamide,** irreversibly inhibit the catalytic activity of some enzymes by modifying cysteine and other side chains. Iodoacetamide is a widely-used agent for the detection of sulfhydryl group.

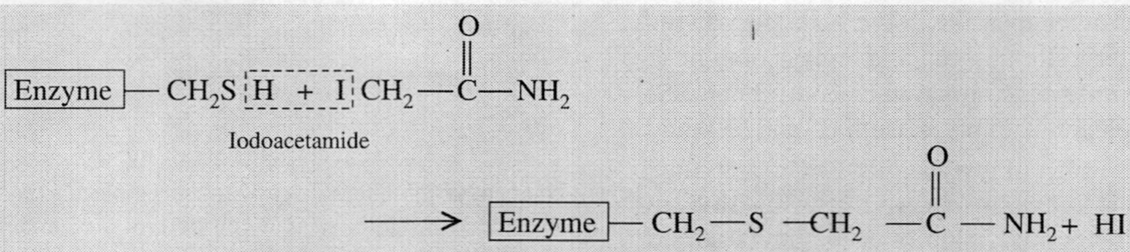

Inactivation of an enzyme (with a critical cysteine residue) by iodoacetamide

2. Organophosphorus compounds, such as diisopropyl-phosphofluoridate, DIPF, are potent irreversible inhibitors of enzymes that have active seryl residues at their catalytic sites. DIPF is closely

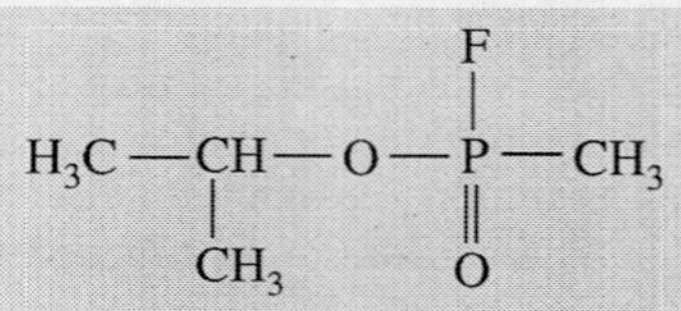

Sarin, a nerve gas

related chemically to nerve gas, whose lethality is due to the inactivation of *acetylcholinesterase*, an enzyme critical for the transmission of nerve impulses. Acetylcholinesterase cleaves the neurotransmitter acetylcholine, an essential step in normal functioning of nervous system.

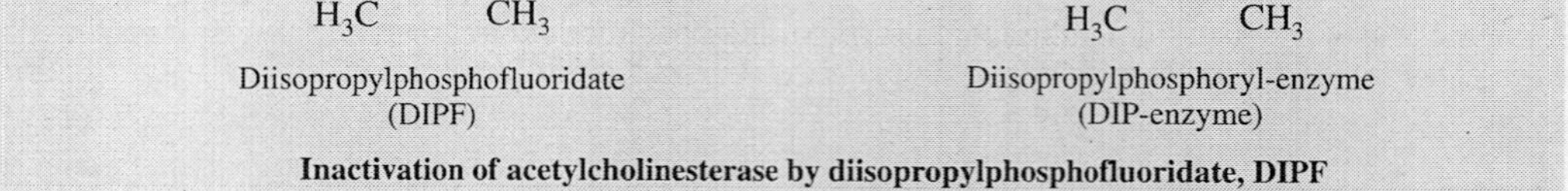

Diisopropylphosphofluoridate (DIPF) Diisopropylphosphoryl-enzyme (DIP-enzyme)

Inactivation of acetylcholinesterase by diisopropylphosphofluoridate, DIPF

A special class of irreversible inhibitors are the **suicide inhibitors.** These compounds are relatively unreactive until they bind to the active site of a specific enzyme. A suicide inhibitor is designed to carry out the first few chemical steps of the normal enzyme reaction. Instead of being transformed into the normal product, however, the inhibitor is converted to a very reactive compound that combines irreversibly with the enzyme. These are also called **mechanism-based inactivators**, because they utilize the normal enzyme reaction mechanism to inactivate the enzyme.

ALLOSTERIC ENZYMES

A **modulator** is a metabolite which, when bound to the allosteric site of an enzyme, alters its kinetic characteristics. The modulators for allosteric enzymes may be either stimulatory or inhibitory. A stimulator is often the substrate itself. The regulatory enzymes for which substrate and modulator

FRANCOIS JACOB
(Born, 1920)

JACQUES MONOD
(LT, 1910–1976)

Both Jacob and Monod served heroically during World War II, Jacob as member of the Free French Forces, injured seriously in Normandy during August 1944, and Monod as a leader of the Paris Resistance. Monod had already obtained his PhD, Jacob had to wait until 1947 to gain his MD from the University of Paris. Both spent the major parts of their careers at the Pasteur Institute, the famous Paris research centre set up around Louis Pasteur in the late nineteenth century and still one of the most influential European laboratories. Jacob contributed to some of the early work on bacteriophages, with **Andre Lwoff** and **Elie Wollman**, before beginning the collaboration with Monod that led to the **operon theory** and the **concept of messenger RNA** in 1961. The two shared the **1965 Nobel Prize** with Lwoff. Both have published important books : *Chance and Necessity* by Monod is a powerful vindication of natural selection, *The Logic of Life* by Jacob an illuminating account of changes in scientific thinking over the past three centuries. Jacob's autobiography. *The Statue Within,* was published in 1987.

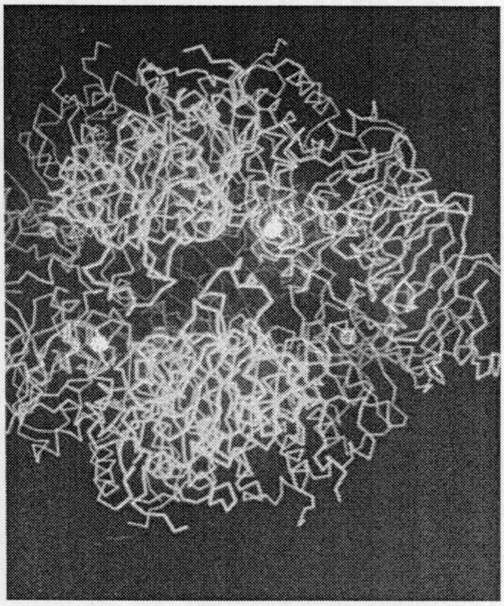

Fig. 18–21. The structure of the enzyme aspartate transcarbamoylase (ATCase) in the T from.

Catalytic subunits are shown in *yellow* and regulatory subunits in *green.*

The term **allosteric** (*allos*[G] = other ; *stereos*[G] = space or site) **site** has been introduced by the two Nobel Laureates, **Monod** and **Jacob** to denote an enzyme site, different from the active site, which noncompetitively binds molecules other than the substrate and may influence the enzyme activity.

are identical are called **homotropic.** When the modulator has a structure different than the substrate, the enzyme is called **heterotropic.** Some enzymes have more than one modulators.

The allosteric enzymes are structurally different from the simple nonregulatory enzymes. Besides the presence of an active or catalytic site, the allosteric enzymes also have one or more regulatory or allosteric sites for binding the modulator. Just as an enzyme's active site is specific for its substrate, the allosteric site is specific for its modulator. Enzymes with several modulators generally have different specific binding sites for each. By contrast, in homotropic enzymes, the active site and allosteric site are the same. Most of them have two or more polypeptide chains or subunits, for example, *aspartate transcarbamoylase, ATCase* (Fig. 18–21) has 12 subunits.

Other differences between nonregulatory enzymes and allosteric enzymes involve kinetic properties. The allosteric enzymes often display sigmoidal plots (Fig. 18–22) of the reaction velocity, V versus substrate concentration, [S], rather than the hyperbolic plots predicted by Michaelis-Menten equation. Recall that the oxygen-binding curve of myoglobin is hyperbolic, whereas that of hemoglobin is sigmoidal. The binding of enzymes to substrates is analogous. In allosteric enzymes, the binding of substrate to one active site can affect the properties of other active sites in the same enzyme molecule. A possible outcome of this interaction between subunits is that the binding of substrate becomes cooperative. In addition, the activity of an allosteric enzyme may be altered by regulatory molecules that are bound to the allosteric sites, just as oxygen binding to hemoglobin is affected by H^+, CO_2 and BPG.

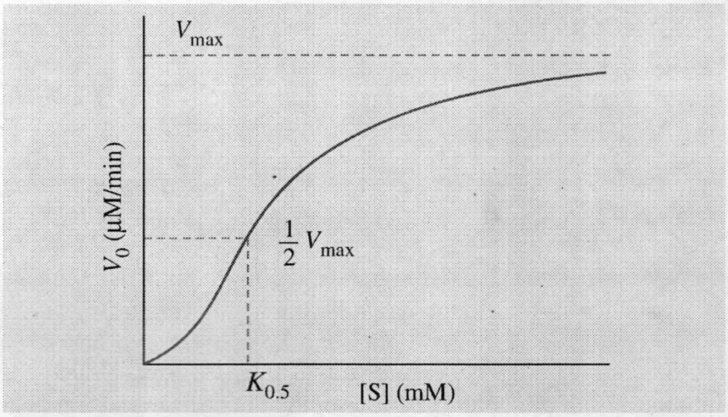

Fig. 18–22. Schematic of an allosteric enzyme's activity by its substrate

Note the sigmoid curve given by a homotropic enzyme, in which the substrate also serves as a positive (stimulatory) modulator. The analogous curve for the nonregulatory enzymes is hyperbolic, as also predicted by the Michaelis-Menten equation. The allosteric enzymes, however, do not obey the Michaelis-Menten kinetics. Also note that a relatively small increase in [S] in the steep part of the curve can cause a very large increase in V_o.

With heterotropic enzymes, however, it is difficult to generalize about the shape of the substrate-saturation curve (Fig. 18–23). An activator may cause the substrate-saturation curve to become more nearly hyperbolic with a decrease in $K_{0.5}$ but no change in V_{max}, thus resulting in an increased reaction velocity at a fixed substrate concentration. On the contrary, a negative modulator (= an inhibitor) may produce a more sigmoid substrate-saturation curve, with an increase in $K_{0.5}$. In other words, allosteric activators shift the substrate-saturation curve to the left (to higher saturation), whereas allosteric inhibitors shift it to the right (to lower saturation).

[Although one can find a value of [S] on the sigmoid-saturation curve at which V_0 is half-maximal, one cannot put the designation K_m to it because the enzyme does not follow the hyperbolic Michaelis-Menten relationship. Instead, the symbol $[S]_{0.5}$ or $K_{0.5}$ is often used to represent the substrate concentration giving half-maximal velocity of the reaction catalyzed by an allosteric enzyme.]

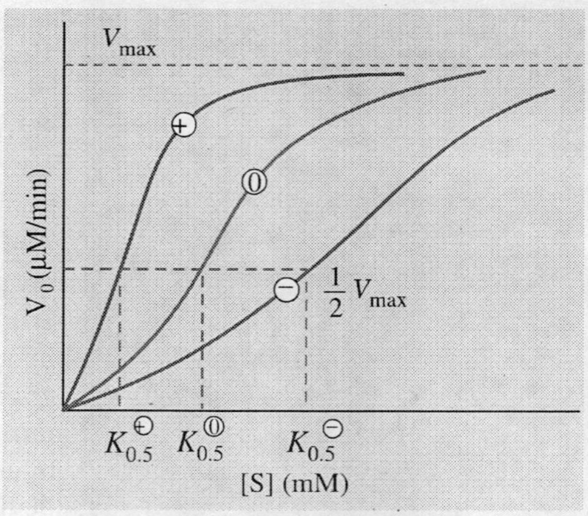

Fig. 18–23. Effects of an activator ⊕, and inhibitor ⊖ and no modulator ◎ on an allosteric enzyme in which $K_{0.5}$ is modulated without a change in V_{max}

Mechanism of Kinetic Behaviour of Allosteric Enzymes

Two general models for the interconversion of inactive and active forms of allosteric enzymes have been proposed.

A. Simple sequential model

Proposed by **Daniel E. Koshland Jr.** in 1966, the model envisages that the allosteric enzyme can exist in only two conformational change individually. Binding of substrate increases the probability of the conformational change. A conformational change in one subunit makes a similar change in an adjacent subunit, as well as the binding of a second substrate molecule, more likely. There are more potential intermediate states in this model than in the other (symmetry) model.

Consider an allosteric enzyme consisting of two identical subunits, each containing an active site

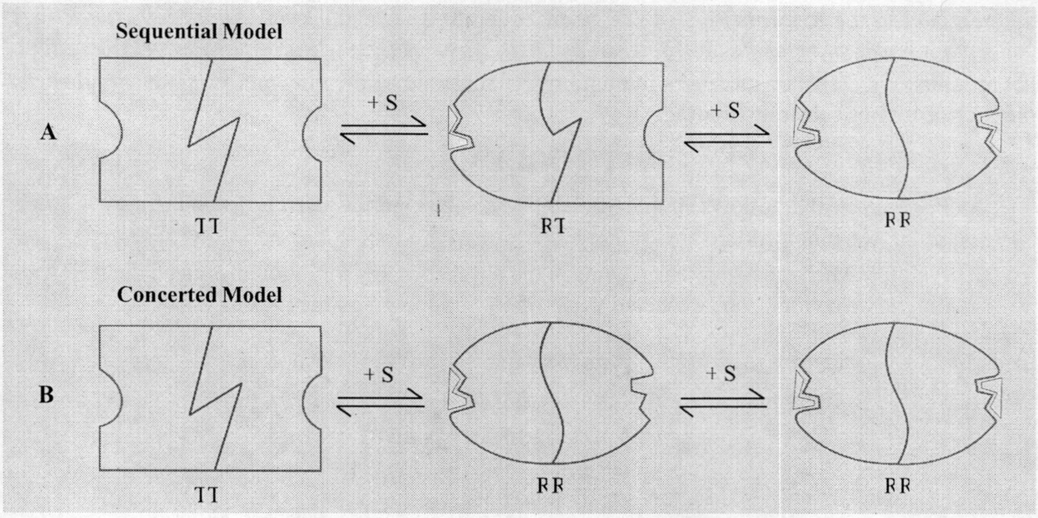

Fig. 18–24. Comparison of the two allosteric models
A. Simple sequential model
B. Concerted or symmetry model

(Adapted from Lubert Stryer, 1995)

(Fig. 18–24A). The T (tense) form has low affinity and the R (relaxed) form has high affinity for substrate. In this model, the binding of substrate to one of the subunits induces a T → R transition in that subunit but not in the other. The affinity of the other subunit for substrate is increased because the subunit interface has been altered by the binding of the first substrate molecule.

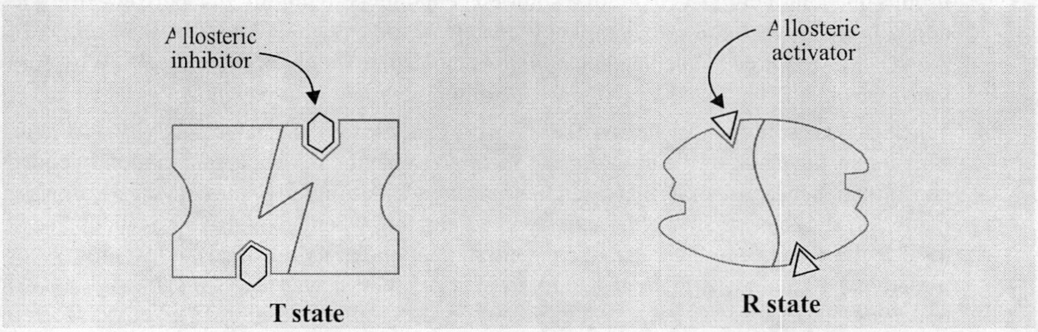

Fig. 18–25. Effect of allosteric activator and allosteric inhibitor on substrate binding

Note that in the concerted model, an allosteric inhibitor stabilizes the T state, whereas an allosteric activator stabilizes the R state.

B. Concerted or Symmetry model

Proposed by **Jacques Monod** and colleagues in 1965, the model envisages that an allosteric enzyme can exist in still two conformations, active and inactive. All subunits are in the active form or all are inactive. Every substrate molecule that binds increases the probability of transition from the inactive to the active state.

In the concerted model, the binding of substrate to one of the subunits increases the probability that both switch from the T to the R form (Fig. 18–24B). Thus, symmetry is conserved in the concerted model but not in the sequential model.

The effects of allosteric activators and inhibitors can be explained quite easily by the concerted model. An allosteric inhibitor binds preferably to the T form, whereas an allosteric activator binds preferentially to the R form (Fig. 18–25). Consequently, *an allosteric inhibitor shifts the R → T conformational equilibrium toward T, whereas an allosteric activator shifts it toward R.* The result is that an allosteric activator increases the binding of substrate to the enzyme, whereas an allosteric inhibitor decreases substrate binding.

PROBLEMS

1. For enzyme that follows simple Michaelis-Menten kinetics, what is the value of V_{max} if V_0 is equal to 1 µmol/minute at 1/10 K_M?
2. A simple Michaelis-Menten enzyme, in the absence of any inhibitor, displayed the following kinetic behaviour. The expected value of V_{max} is shown on the *y*-axis.

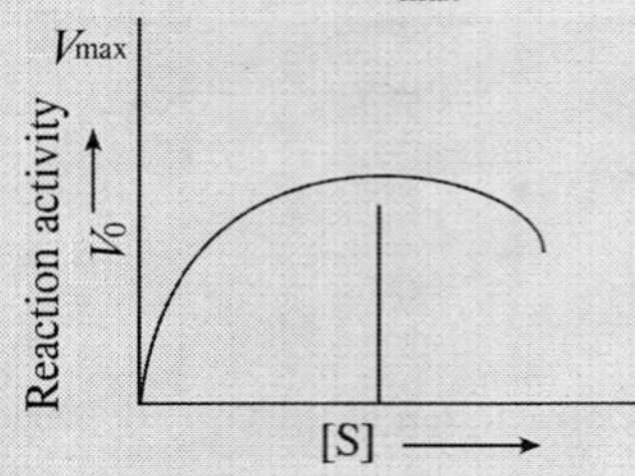

   (*a*) Draw a double-reciprocal plot that corresponds to the velocity-versus-substrate curve.
   (*b*) Provide an explanation for the kinetics results.
3. Many enzymes are inhibited irreversibly by heavy-metal ions such as Hg^{2+}, Cu^{2+}, or Ag^+, which can react with essential sulfhydryl groups to form mercaptides :

$$\text{Enz—SH} + Ag^+ \longrightarrow \text{Enz—S—Ag} + H^+$$

   The affinity of Ag^+ for sulfhydryl groups is so great that Ag^+ can be used to titrate —SH groups quantitatively. To 10 mL of a solution containing 1.0 mg/mL of a pure enzyme was added just enough $AgNO_3$ to completely inactivate the enzyme. A total of 0.342 µmol of $AgNO_3$ was required. Calculate the *minimum* molecular weight of the enzyme. Why does the value obtained in this way give only the minimum molecular weight ?
4. When enzyme solutions are heated, there is a progressive loss of catalytic activity with time. This loss is the result of the unfolding of the native enzyme molecule to a randomly coiled conformation, because of its increased thermal energy. A solution of the enzyme hexokinase incubated at 45°C lost 50% of its activity in 12 min, but when hexokinase was incubated at 45°C in the presence of a very large concentration of one of its substrates, it lost only 3% of its activity. Explain why thermal denaturation of hexokinase was retarded in the presence of one of its substrates.
5. An experiment measuring velocity versus substrate concentration was run, first in the absence of substance A, and then in the presence of substance A. The following data were obtained.

[S] (*µM*)	*Velocity in Absence of A* (*µmol min^{-1}*)	*Velocity in Presence of A* (*µmol min^{-1}*)
2.5	0.32	0.20
3.3	0.40	0.26
5.0	0.52	0.36
10.0	0.69	0.56

   Is substance A an activator or an inhibitor ? If it is an inhibitor, what kind of inhibitor is it ?
6. For the experiment in Question 5, calculate the K_m and V_{max} both in the absence and in the presence of substance A. Are these results consistent with your answer for Question 5 ?
7. Bacterial growth is inhibited by sulphanilamide and related sulfa drugs, and there is a

concomitant accumulation of 5-aminoimidazole-4-carboxamide ribonucleotide. This inhibition is reversed by the addition of *p*-aminobenzoate. Propose a mechanism for the inhibitory effect of sulfanilamide.

ANSWERS TO PROBLEMS

1. 11 μmol minute^{-1}.

2. (*a*)

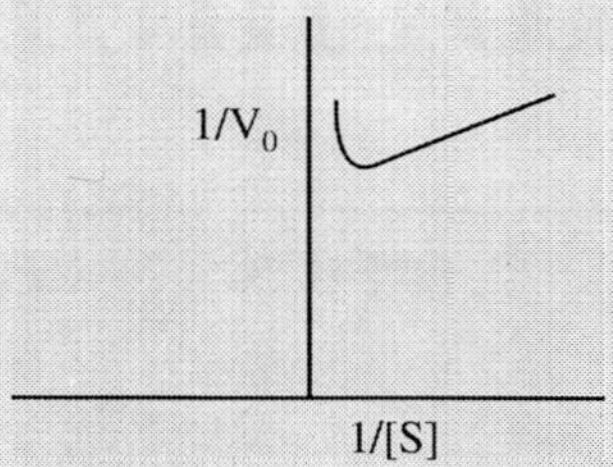

(*b*) This behaviour is substrate inhibition—at high concentrations, the substrate forms unproductive complexes at the active site. The drawing below shows what might occur, Substrate normally binds in a defined orientation, shown in the drawing as red to red and blue to blue. At high concentrations, the substrate may bind at the active site such that the proper orientation is met for each end of the molecule, but two different substrate molecules are binding.

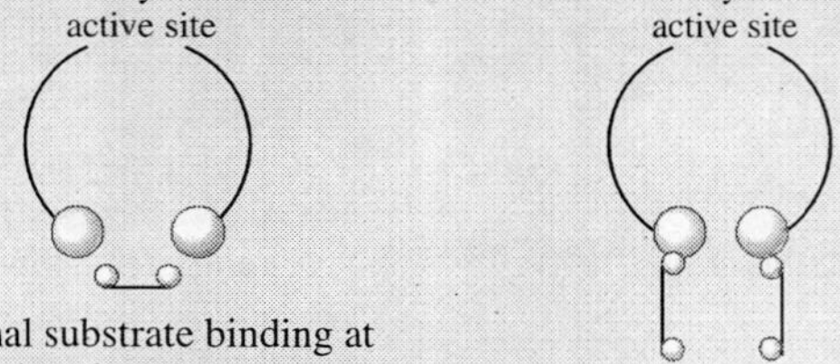

3. 29,000; it is assumed that each enzyme molecule contains only one titratable sulfhydryl group.

4. The enzyme–substrate complex is more stable than the enzyme alone.

5. The best way to handle the data is to take the reciprocals of both [S] and *v* and construct a Lineweaver–Burk plot. You should find that the two curves cross the *y*-axis at the same point but the curve in the presence of A crosses the *x*-axis closer to the origin. This pattern indicates that A is a competitive inhibitor.

6.

	$-1/K_m$	K_m	$1/V_{max}$	V_{max}
Absence of A	–0.14	7.1	0.8	1.25
Presence of A	–0.08	12.5	0.8	1.25

With a competitive inhibitor V_{max} remains constant (be sure you understand why) but the apparent K_m is larger. It takes more substrate to reach a given velocity because the substrate has to compete with the inhibitor

7. There is a deficiency of N^{10}-formyltetrahydrofolate. Sulphanilamide inhibits the synthesis of folate by acting as an analogue of *p*-aminobenzoate, one of the precursors of folate.

CHAPTER

19 Animal Hormones

INSIDE

- Definition
- General Functions
- Vertebrate Hormones
  - Steroid Hormones
    - C_{18} Steroids
    - C_{19} Steroids
    - C_{21} Steroids
  - Peptide Hormones
  - Amino Acid Hormones
- Parahormones or Tissue Hormones
- Vasoactive Peptides
- Hormone from Thymus
- Pheromones or 'Social' Hormones
- Mechanisms of Hormone Action

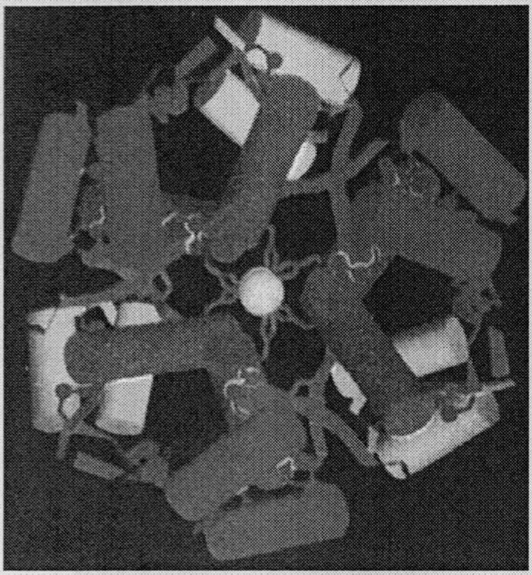

A computer-generated model of insulin

This enzyme-protein, whose structure was first of all ascertained, controls blood sugar levels. Its molecule is one of the smallest protein molecules.

DEFINITION

The term hormone (*hormao*[G] = to excite) was first used by **William M. Bayliss** and his brother-in-law **Ernest H. Starling**, both of London University College, in 1904, who showed that a chemical substance (*secretin*) from the intestine could stimulate the action of a pancreatic secretion. These substances were then called as 'chemical messengers'. **Went** and **Thimann** (1937) defined a hormone as *"a substance which, produced in any one part of an organism, is transferred to another part and there influences a specific physiological process."* The tissues or organs where they are produced are called as **effectors** and those where they exert their influence as **targets.** These have low molecular weight and diffuse readily. As they are readily oxidized, their effects do not remain permanent unless these are supplied continuously.

Based on their site on action, the hormones are of two types : local and general. The *local hormones,* obviously, have specific local effects, whence their nomenclature. These may be exemplified by acetylcholine, secretin, cholecystokinin etc. The *general hormones*, on the other hand, are secreted by specific endocrine glands and are transported in the blood to cause physiologic actions at points remote from their place of origin. A few of the general hormones affect almost all cells of the body, *e.g.*, growth hormones (GH) and thyroid hormones ; whereas other general hormones, however, affect specific tissues far more than other tissues, *e.g.*, adrenocorticotropin (a hormone secreted from adenohyprophysis and stimulating the adrenal cortex) and ovarian hormones (affecting the uterine endometrium).

Non-specificity and Cross Effects — Hormones are not sepcific for the organisms in which they are produced and may influence bodily processes in other individuals also. Adrenalin, for example, also influences protozoans and crustaceans besides man and other vertebrates. Their cross effects have also been found between plants and animals. For example, auxin, a growth-promoting plant hormone, can also stimulate a protozoan, *Euglena viridis*. Likewise, some animal hormones also stimulate growth in root tips of many plants after decapping.

SIR WILLIAM H. BAYLISS (LT, 1860-1924)

ERNEST H. STARLING (LT,1866-1927)

Bayllis and Starling, both of London University, are called ***'founders of endocrinology'***.
(Photo Courtesy: Fulton JF and Wilson LG, 1966)

Differences with Nutrient Materials and Enzymes — A distinction between hormones and **nutrient materials** may be made, as the former are utilized in extremely minute quantities in comparison to the latter. For example, a 5×10^{-10} molar solution of ocytocin (about 1 mg in 2,000 litres of solution) will cause the contraction of uterine muscles.

They also differ from **enzymes**, which are essential to initiate and continue reactions, in that the hormones cannot initiate reactions but can influence the rate at which they proceed. Furthermore, the hormones are usually consumed in the process of growth, whereas the enzymes are not.

GENERAL FUNCTIONS

The hormones conduct a wide variety of functions ranging from growth, vegetative and sexual development, cellular oxidation to thermal production and the metabolism of carbohydrates, proteins and fats. The various functions performed by hormones may, in general, be discussed under following heads :

1. Regulatory or homeostatic function. The hormones have regulatory effects on the composition of the body fluids, the rate of gaseous exchange and the activity of the vascular system and the central nervous system (CNS). There always exists a high degree of precision and constancy in the composition of the body fluids in a normal individual for the conduction of various activities. Such an environment within the cell has been termed the *millieu intérieur* (or internal enivirnoment) in 1857 by **Claude Bernard**, a French physician. Throughout his research, he had been impressed by the way in which organisms were able to regulate physiological parameters, such as body temperature and water content, and maintain them within fairly narrow ranges. This concept of self-regulation leading to physiological stability was summed up by **Bernard** in the now classic statement, '*La fixité du milieu intérieur est la condition de la vie libre.*' (The constancy of the internal environment is the condition of the free life). In other words, it means that for an organism to function optimally, its component cells must be surrounded by a medium of closely-regulated composition. Bernard went on to distinguish between the *external environment* in which organisms live and the *internal environment* in which the individual cells live (in mammals, this is tissue, or interstitial, fluid). He realized the importance of conditions in the latter being continuously stable. He concluded that an organism is the sum of its constituent cells and the optimum functioning of the whole depends upon the optimum functioning of its parts.

CLAUDE BERNARD
(LT, 1813-1878)

Claude Bernard, a French physiologist, had parents who were vineyard workers. From his early experiences, Bernard retained an enthusiasm for the Beaujolais region of his native country and for the countryside in general, throughout a working life that kept him in a Paris laboratory. He trained as an apothecary, and later as a doctor, but was no more than an average student. He never worked as a medical practitioner, but did make his mark in experimental medicine.

Bernard's work centred on the physiology of mammals. Working with dogs and rabbits he established that :

(*a*) an enzyme is present in the gastric juice.
(*b*) the digestion of dietary carbohydrates to sugar occurs prior to their absorption.
(*c*) the digestion of fat involves bile and pancreatic juices.
(*d*) glycogen and sugar are interconverted in the liver.
(*e*) respiration produces heat in all body tissues.

Bernard was a 'modern' researcher, combining experimental skills with an appreciation of the theory behind his work.

He believed in the constancy of the *milieu intérieur* ("internal environment"), which is the extracellular fluid bathing the cells. He pointed out that it is through the *milieu intérieur* that foods and wastes and gases are exchanged and through which chemical messengers are distributed. He wrote :

"The living organism does not really exist in the external environment (the outside air or water) but in the liquid *milieu intérieur*... that bathes the tissue elements."

He is best remembered for his idea that animal life is dependent upon a constant internal environment– that cells function best in a narrow range of conditions of solute potential and temperature, and when bathed in a constant concentration of chemical constituents. Truly , Bernard was one of the most influential of nineteenth century physiologists.

For performing various metabolic functions, an organism should maintain a normal, constant internal state or **homeostasis** (*homoios*^{G} = same or similar ; *stasis*^{G} = state or standing), a term coined by an American physiologist, **Walter B. Cannon** in 1932. Homeostasis can be defined as the tendency to maintain uniformity or stability in the internal environment of the organism and to maintain the normal composition of the body fluids. In other words, homoeostasis is the maintenance of a constant internal environment in the face of changes in the external environment. Hormones play an important and decisive role in homeostatic regulation of internal mileu.

2. Permissive function. Not only does each endocrine gland affect a number of processes, but these glands also affect the functioning of one another. Thus certain hormones require the presence (or 'permission') of another hormone for the expression of their activity. This helps in maintaining a perfect hormonal balance. Derangements of this balance, either clinical or experimental, lead to a variety of metabolic aberrations.

The scientist **James Lovelock** argues that the Earth with its atmosphere is 'alive', in the same that the world shares with other living things the capacity to self-regulate (homeostasis). He calls his idea the **gaia hypothesis.** In fact, *gaia is the homeostasis at planet level.*

3. Integrative function. The integrative function of the hormones is reflected in the fact that they support the role of nervous system. However, the integrative properties of the endocrine system are slow and steady whereas those of the nervous system are rapid. This close tie between the two systems has led to the emergence of a new discipline of science called *neuroendocrinology.*

WALTER BRADFORD CANNON
(LT, 1871–1945)

(Photo Courtesy : Fulton JF and Wilson LG, 1966)

Cannon, an American scientist, was George Higginson Professor of Physiology, Emeritus at Harvard University Medical School from 1906 till 1942. He introduced the first radio-opaque agent, barium (later, he used barium salts) into the gut of a living animal to create a 'shadow' of the alimentary tract so that the mechanics of digestion could be observed. He devised this technique first in the year 1896, the year after Roentgen discovered x-rays. The technique has been used in diagnostic radiography ever since. He developed the concept originated by French physiologist Claude Bernard. Developed out of his studies of the nervous system and reactions to stress, he described the ceaseless balancing and rebalancing of physiological processes that maintain stability and restore the normal state when it has been disturbed. And he gave it the name, **homeostasis.** He spend his life in the study of homeostatic mechanisms. The term soon flooded the medical literature of 1930s. Physicians spoke of getting their patients back into homeostasis. Even politicians and sociologists saw what they considered to be deep nonphysiological implications. Cannon enjoyed this broadened application of the concept and later suggested that democracy was the form of government that took the homeostatic middle course. Despite the enduring importance of the homeostasis concept, Cannon never received the Nobel Prize – *one of several acknowledged oversights of the Nobel Committee.* In his terminal year (*i.e.*, 1945), Cannon expressed his ideas about scientific research in his autobiography, *The way of an Investigator.* This absorbing book describes the resourceful career of a homespun man whose life embodied the traits that favour successful research.

4. **Morphogenetic function.** The hormones govern the ontogenetic development of an individual from the embryonic to the adult state.

An account of **vertebrate hormones** is given below.

VERTEBRATE HORMONES

Most glands of the body release their secretions by means of ducts. These are called as the **exocrine** (*exos*[G] = outside ; *krinein*[G] = to separate) **glands** or **duct glands**, *e.g.*, salivary, sebaceous, sweat glands etc. Other glands in the body, on the contrary, produce chemical substances which are released directly into the nearby blood and lymph vessels which carry them to various organs (target or effector organs) where they exert their characteristic effect. These glands have been termed as the **endocrine** (*endon*[G] = within) **glands** or **ductless glands** and their secretions are called as the hormones. The science of the study of structure and function of these endocrine glands and their secretions (*i.e.*, hormones) is designated as **endocrinology**, a term introduced by **Pende**.

A noteworthy point about the endocrines is their similarity, except in minor details, in all the vertebrates. They are highly vascular as the blood not only acts as a vehicle for transport of their secretions but also serves as a source of the chemical raw materials from which these hormones are synthesized. Ontogenetically, all these glands are derived from the epithelial tissues. The amount of hormones produced by the endocrines in 24 hours measures fractions of a milligram. The principal endocrine glands (refer Fig. 19–1) include :

(*a*) two in the head region— the *pineal* and the *pituitary,*
(*b*) three in the neck region— the *thymus,* the *thyroid* and the *parathyroid,* and

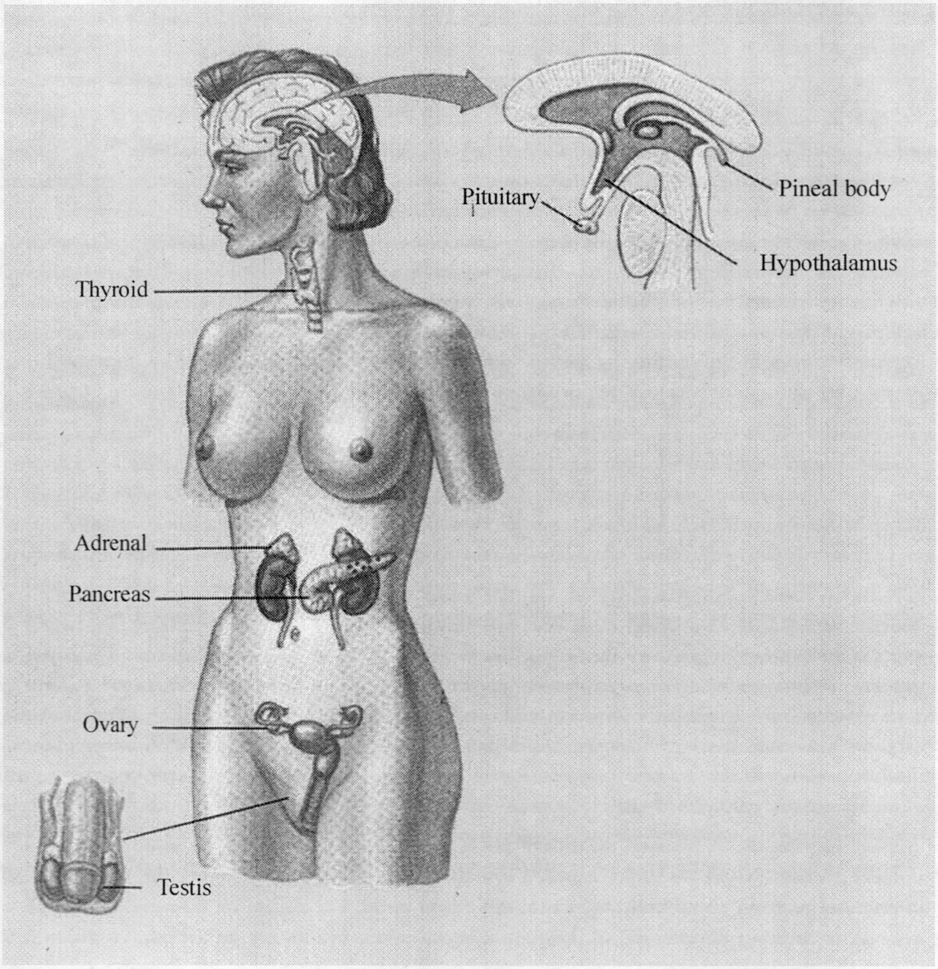

Fig. 19–1. Endocrine glands in human body

Location of major endocrine glands in humans. The hypothalamus regulates the anterior pituitary, which regulates the hormonal secretions of the thyroid, adrenals, and gonads (ovary in the female and testis in the male).

(*After Lehninger AL, 1984*)

(*c*) four in the abdominal region— the *pancreas,* the *gastrointestinal mucosa,* the *adrenals* and the *gonads.*

Barring the parathyroid, all the remaining 8 glands are found in all the vertebrates. There is, however, controversy regarding the endocrinal nature of the two glands, the pineal and the thymus. But it is customary to include them in the endocrine system.

A gland may be **temporary**, lasting only for a limited period and then dying away, *e.g.*, thymus. Some glands are, however, **seasonal** being periodic and recurrent, *e.g.*, placenta and corpus luteum Chemically, a hormone may be any kind of organic molecule. Most known hormones are either **steroids** or **peptides** with usually high molecular weights. A third group of hormones, which is less common, consists of **amino acid derivatives** (or phenolic derivatives) with relatively low molecular weights. Thus, three categories of hormones may be recognized: steroids, peptides and amino acid derivatives. A classification of vertebrate hormones, based on their chemical composition, is outlined on page 347.

VERTEBRATE HORMONES

STEROID HORMONES

C_{18} Steroids

1. **Ovarian Hormones**
   β-estradiol
   Estriol
   Estrone

C_{19} Steroids

2. **Testicular Hormones**
   From testes
   Testosterone
   Androsterone
   Dehydroepiandrosterone

   From adrenal gland
   Androst-4-ene-3,17,dione
   Androst-4-ene-3,11,17-trione

C_{21} Steroids

3. **Adrenal Cortical Hormones**
   Mineralocorticoids
   Aldosterone
   Deoxycorticosterone

   Glucocorticoids
   Cortisone
   Cortisol
   Corticosterone

4. **Corpus Luteal Hormone**
   Progesterone

PEPTIDE HORMONES

1. **Hormones of the Pancreas**
   Insulin
   Glucagon
2. **Hormones of the Hypophysis**
   Pars distalis
   Thyrotropin, TSH
   Corticotropin, ACTH
   Gonadotropins, GTH
      FSH
      LH
      LTH
   Somatotropin, SH
   Pars intermedia
   Intermedins, MSH
      α-MSH
      β-MSH

   Pars nervosa
   Ocytocin or pitocin
   Vasopressin or pitressin
3. **Hormone of the Parathyroid**
   Parathormone, PTH
4. **Hormones of the Gastrointestinal Tract**
   Gastrin
   Secretin
   Cholecystokinin
   Pancreozymin
   Enterogastrone
   Enterokrinin
   Hepatocrinin
   Duicrinin
   Villikinin
   Parotin
5. **Hormone of the Corpus Luteum**
   Relaxin

AMINO ACID HORMONES

1. **Thyroidal Hormones**
   Triiodothyronine, T_3
   Tetraiodothyronine T_4
2. **Adrenal Medullary Hormones**
   Adrenalin
   Noradrenalin

STEROID HORMONES

These include the sex hormones and the hormones from adrenal cortex. These are synthesized in mammals by the ovary (or testis), adrenal cortex, corpus luteum and the placenta. The activity of sex hormones appears to be controlled by the hormones secreted by the anterior lobe of the hypophysis

(= adenohypophysis). Because of this, the sex hormones are, sometimes, referred to as *secondary sex hormones* and the hormones of the adenohypophysis, which are of proteinaceous nature, are called as *primary sex hormones*. Three types of sex hormones are recognized :

(*a*) the estrogens (female or ovarian or follicular hormones)

(*b*) the androgens (male or testicular hormones)

(*c*) the gestogens (corpus luteal hormones).

The sex hormones are concerned with the sexual processes and the development of secondary characteristics which differentiate males from females. The adrenal cortical hormones perform a variety of important functions related to cell metabolism.

Based on the number of carbon atoms present in the molecule, the steroid hormones may be named as C_{18}, C_{19} or C_{21} steroids.

C_{18} STEROIDS

1. Ovarian Hormones

Structure. Mammalian ovary contains ovarian follicles and corpus lutea. Hormones produced mainly in the follicles are known as estrogens (*oistros*^{G} = a gadfly, hence sting or frenzy). "Estrogen is a generic term for a substance that induces estrus, which is a cyclic phenomenon of the female reproductive system. The stages and timings differ in various species but, in general, first a *proestrus* period occurs, during which the follicle repens and the organs of reproduction develop. This is followed by *estrus*, the period of heat, in which the female will receive the male. Ovulation takes place toward the end of estrus, either spontaneously or, as in rabbit, after mating. Then follows a period of retrogression of the accessory reproductive organs and a period of sexual inactivity" (**Orten** and **Neuhaus**, 1979).

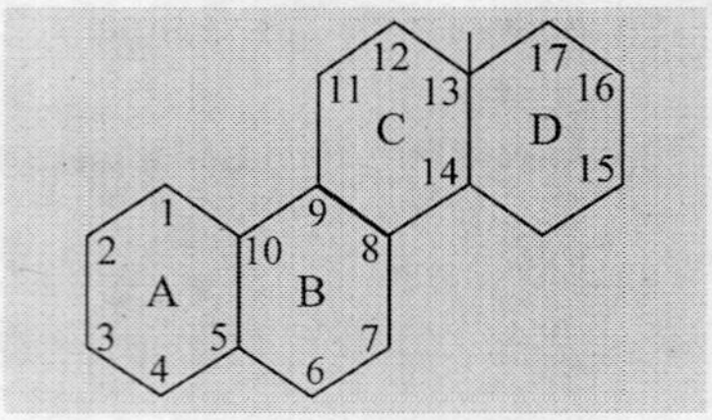

Fig. 19–2. Estrane

(Parent hydrocarbon of estrogens)

Chemically, the estrogens are derivatives of a C_{18} hydrocarbon, estrane (Fig. 19–2).

The three compounds of this group (Fig. 19–3) with hormonal activity are :

1. β-estradiol (= dihydrotheelin), $C_{18}H_{24}O_2$
2. Estriol (= theelol), $C_{18}H_{24}O_3$
3. Estrone (= theelin), $C_{18}H_{22}O_2$

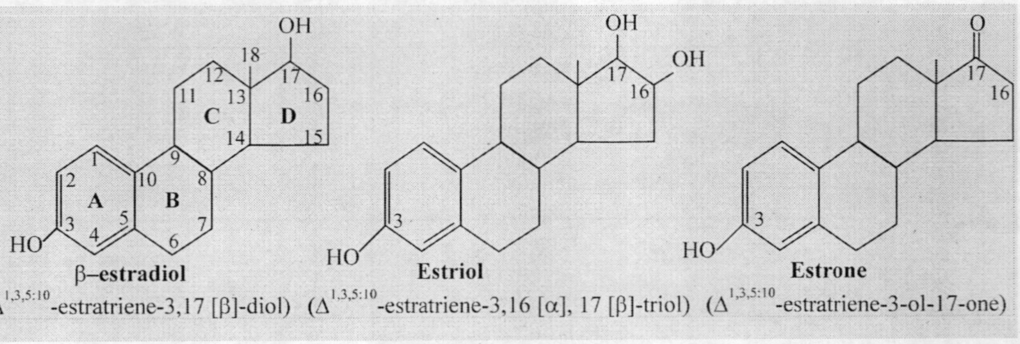

β–estradiol ($\Delta^{1,3,5:10}$-estratriene-3,17 [β]-diol) **Estriol** ($\Delta^{1,3,5:10}$-estratriene-3,16 [α], 17 [β]-triol) **Estrone** ($\Delta^{1,3,5:10}$-estratriene-3-ol-17-one)

Fig. 19–3. Principal ovarian hormones

All these are characterized by the absence of a CH_3 group at carbon 10 and by the aromatic nature of ring A, making the OH group phenolic in character. Of all these, β-estradiol is most potent physiologically, estrone less potent and estriol is least active. Their relative potencies are 50 : 5 : 1, respectively. Although ovary is the chief source of estrogens, they are in smaller amounts also produced by the testis and the adrenal cortex.

Estrogen production is highest when a woman is young and slows down with age, giving rise to menopausal symptoms. The symptoms vary with each women. By 40's, women enter perimenopause, when menstruation becomes less regular, skin becomes dryer, hair turn brittle and sparser; women may feel a loss of libido and may suffer fluctuations in mood. Menopause follows, on an average between 45 and 50 years of age. Heavy women have an advantage over the slim ones. Their fat cells manufacture a form of estrogen called estrone, even after estrogen from ovaries shuts off.

Cholesterol
↓
Testosterone
↓
19-hydroxytestosterone
↙ ↘
19-Oxotestosterone Estradiol-17
↓
10 -carboxy-17 -hydroxyestr-4-ene-3-one ⟶ 19-nortestosterone

Fig. 19–4. Biosynthesis of estrogen

Biosynthesis. In nonepregnant females, estrogen is mainly synthesized in the ovary. The estrogen (as well as the androgen) are, in part, transported by binding to a specific plasma protein called sex steriod binding protein SBT. The amount of this protein increases in pregnancy or estrogen therapy which results in reduced androgenic action.

Curiously enough, testosterone, a male hormone, is the precursor of estrogens. Eve was made from more than Adam's rib ! Fig. 19–4 depicts the probable pathway of estrogen synthesis.

Metabolism. Most of the metabolic reactions of the estrogens (*i.e*, the interconversion reactions of all the three forms) take place in the liver as follows :

Estradiol-17β Estrone ⟶ 16 α-hydroxyestrone
↓ ↓ ↓
2-hydroxyestradiol-17β ⟶ 2-hydroxyestrone Estriol
↓
2-methoxyestradiol-17β ⟶ 2-methoxyestrone

Estriol is the principal estrogen found in the placenta and urine of pregnant females. It is produced by hydroxylation of estrone at C_{16} and reduction of keto group at C_{17}.

Functions. "In women, the follicular hormones (estrogens) prepare the uterine mucosa for the later action of the progestational hormones (produced by the corpus luteum). The changes in the uterine include proliferative growth of the lining of the endometrium, deepening of uterine glands, and increased vascularity; changes in the epithelium of the fallopian tubes and of the vagina also occur. All of these changes begin immediately after menstrual bleeding has ceased." (**Harper** and **Grodsky**, 1973).

Estrogen preserves the elasticity of the skin and possibly improves the memory in women at postmenopausal stages. It also protects women from osteoporosis by slowing the rate at which calcium is leached from their bones. Estrogen supplement also preserves the flexibility of blood vessels, thus helping to prevent cardiac diseases. Nowadays, estrogen in combination with the hormone progestin is considered an important tool for helping women remain healthy. The combination is known as **hormone replacement therapy (HRT)**, which has indeed become the closest in medicine to a woman's elixir of youth. HRT, when used by menopausal women, relieves hot splashes, dry sweats and vaginal dryness. However, the long-term use of HRT or estrogen therapy hightens the risk of ovarian cancer.

The estrogens are also effective in the development of *secondary sex characters in females*. These are listed in Table 19–1.

Table 19–1. Body changes at puberty in girls (=Secondary sex characters in females)

Characters	*Changes*
1. External genitalia	Enlargement of uterus and vagina; Widening of pelvis.
2. Internal genitalia	Periodic vaginal bleeding that occurs with the shedding of the uterine mucosa (i.e., menstruation).
3. Voice	Larynx retains its prepubertal proportions, i.e., small in size; Voice stays high-pitched.
4. Hair growth	Less body hair and more scalp hair; Hair line on scalp resembles that of a child and does not recede anterolaterally (Fig. 19–5); Hair appear in axillae (axillary hair) and around vagina; Pubic hair have a characteristic female pattern, i.e., flat-topped; Hair on face absent i.e., no beard.
5. Mental	Less aggressive; Passive attitude; Interest in opposite sex less pronounced.
6. Body conformation	Narrow shoulders and broad hips, which are popularly called as 'hip pads'; Thighs that converge and arms that diverge, i.e., a wide carrying angle; Distribution of fat in the breasts and buttocks takes place, leading to their enlargementÅthe breasts also have high pigmentation in the areola which becomes even more intense during the first pregnancy; Muscles not pronounced.
7. Skin	Sebaceous gland secretions become more fluid and thus counter the effect of testosterone and inhibit formation of comedones ('black-heads') and acne (a hard, red inflamed pimple)
8. Weight gain	Tend to gain weight from the waist down.

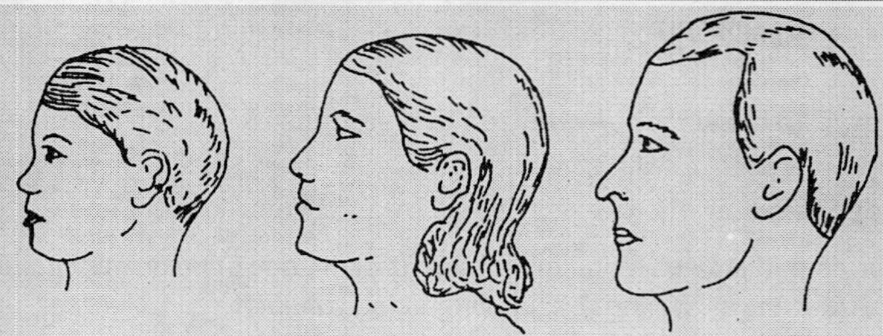

Fig. 19–5. Hair line pattern of child, woman and man

The hair line of the women is like that of the child, whereas that of the man is indented in the lateral frontal region.

(*Adapted from Grenlich et al, 1942*)

Estrogens also influence to a great deal (*a*) the inorganic metabolism of Ca and P and (*b*) the organic metabolism of proteins and lipids.

C_{19} STEROIDS

2. Testicular Hormones

Structure. These hormones are secreted mainly by the testes, the male reproductive organs and are called as androgens (*andros*[G] = male). Chemically, these are derivatives of a C_{19} hydrocarbon, *androstane* (Fig. 19–6).

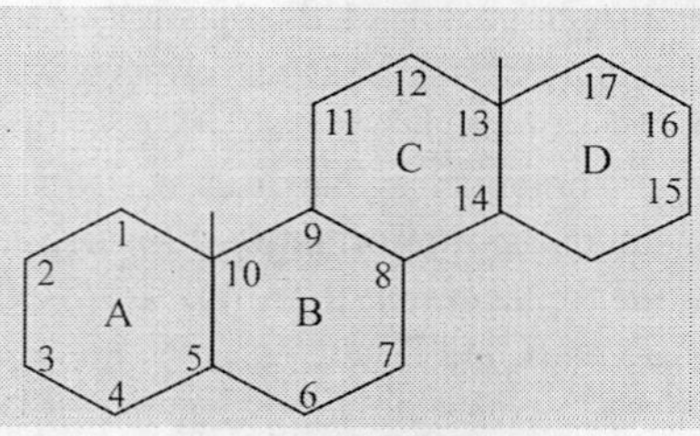

Fig. 19–6. Androstane
(parent hydrocarbon of androgens)

There are many hormones secreted from testes with androgenic activity. The three important ones (Fig. 19–7) are :

1. Testosterone, $C_{19}H_{28}O_2$
2. Androsterone, $C_{19}H_{30}O_2$
3. Dehydroepiandrosterone, $C_{19}H_{25}O_2$

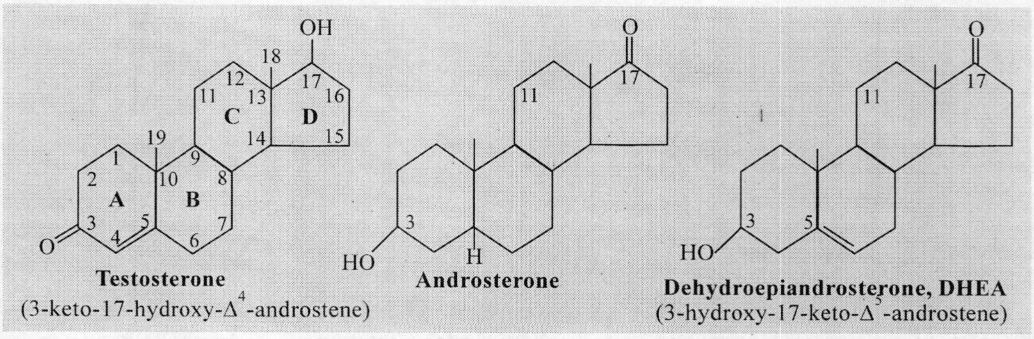

Fig. 19–7. Principal testicular hormones

Androsterone was first isolated by **Adolf Butenandt** *et al* in 1931 from male urine (about 15 mg from 15,000 litres of urine).

Testosterone is most potent of all these and dehydroepiandrosterone is least active. The relative potency ratio of these three forms is 20 : 7 : 1. Testosterone has a tendency to rise during late summer and early fall to peak in September. DHEA production peaks between ages 25 and 30 and wanes with age. Restoring DHEA levels to peak is said to boost the immune system.

A few testicular hormones are also produced by the adrenal gland. The structure of two such hormones is given in Fig. 19–8.

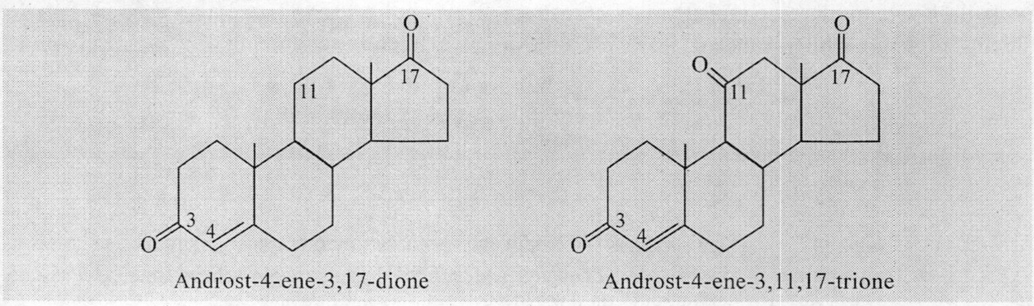

Fig. 19–8. Two testicular hormones produced from adrenal cortex

Biosynthesis. In Fig. 19–9, biosynthesis of testosterone from cholesterol has been depicted.

The principal male hormone, testosterone, is synthesized by the Leydig cells of the testes from cholesterol through pregnenolone, progesterone and hydroxyprogesterone. The latter is then converted to a C_{19} ketosteroid called androstenedione which is the immediate precursor of testosterone. It is presumed that the same sequence of events also takes place in the adrenal gland, ovary and placenta.

In addition to testosterone, adrostenedione and DHEA are also synthesized in the testes, although in amounts far less than that of testosterone.

In normal male, 4 to 12 mg of testosterone are secreted each day. The amount of DHEA secreted is, however, greater than that of testosterone (approximately 15 to 50 mg/day).

Metabolism. Most of the metabolic transformations of androgens takes place in the liver. In some mammals like rats, these reactions occur mainly in the bile and urine. Two major reactions occurring in the liver are :

(*a*) Conversion of testosterone to androst-4-ene-3,17-dione.

(*b*) Interconversion of 3-hydroxy and 3-keto derivatives.

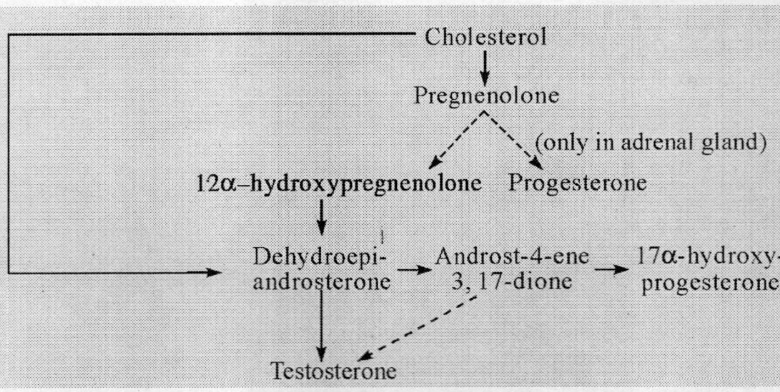

Fig. 19–9. Biosynthesis of testosterone

Functions. Testosterone has often been considered to be a 'youth horomone', because of its effects on the musculature, and it is occasionally used for treatment of persons who have poorly developed muscles. Because of the ability of testosterone to increase the size and strength of bones, it is often used in old age to treat osteoporosis. Like estrogens, the androgens are also responsible for the development of secondary sex characters in males. These are listed in Table 19-2.

Androgens regulate the activities of the male reproductive system; estrogens stimulate the growth, maturation and maintenance of the female reproductive system and accessory sex tissues. However, both androgens and estrogens also have significant effects on the non-reproductive tissues of the body. For instance, androgens stimulate the growth of skeletal muscles. Androgens and certain androgen derivatives (collectively called as **anabolic steroids**) are often used by weight lifters, wrestlers and football players to increase muscle mass and strength. Anabolic streroids are also used by female athletes, probably with advantage ; however, they produce other masculinizing effects as well.

Table 19–2. Body changes at puberty in boys (= Secondary sex characters in males)

Characters	*Changes*
1. External genitalia	Penis increases in length and width; Scrotum becomes pigmented and rugose.
2. Internal genitalia	Seminal vesicles enlarge and secrete– they also begin to form fructose ; Prostate and bulbourethral glands also enlarge and secrete.
3. Voice	Larynx enlarges and the vocal cords increase in length and thickness; Voice becomes deeper.
4. Hair growth	General body hair increases; Hair line on scalp recedes anterolaterally (Fig. 19–10); Hair appear in axillae (axillary hair) and around anus; Pubic hair have a characteristic male pattern, i.e., triangle with apex up; Hair on face grow as beard.
5. Mental	More aggressive; Active attitude; Interest in opposite sex more pronounced.
6. Body conformation	Shoulders broaden and hips remain unaltered, i.e., narrow; Thighs that diverge and arms that converge, i.e., a narrow carrying angle; No distribution of body fat in the chests and buttocks; Muscles enlarge, leading to a muscular body contour.
7. Skin	Sebaceous gland secretion thickens and increases (predisposing to acne).
8. Weight gain	Tend to gain weight in the abdominal region.

C_{21} STEROIDS

3. Adrenal Cortical Hormones

Secretory gland. The adrenals (*ad* = at; *renal* = pertaining to kidneys)or suprarenal glands in all tetrapods are a pair of glands, so named because of their position very close to or at the top of the kidneys. Each of the two adrenals among mammals is actually a '*double gland*' and is composed of 2 distinct parts : namely an outer barlike covering called the *cortex*, surrounding an inner corelike dark-coloured mass called the *medulla*. The cortex is derived from the mesodermal glandular tissue and the medulla originates from the cells of neural crest. Both these parts secrete hormones which differ from each other chemically as well as physiologically. Hence, these 2 components are discussed separately. In man, the adrenals are two small structures sitting like 'cocked hats' over the apical end of the kidneys (refer Fig. 19–10) and each gland weighs about 3 grams.

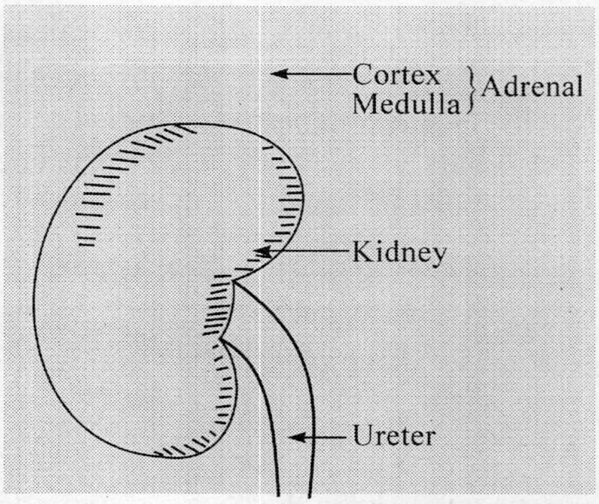

Fig. 19–10. The adrenal gland in man
(*After D. Marsland, 1964*)

Histologically, the adrenal cortex is made up of three layers (Fig. 19–11) :

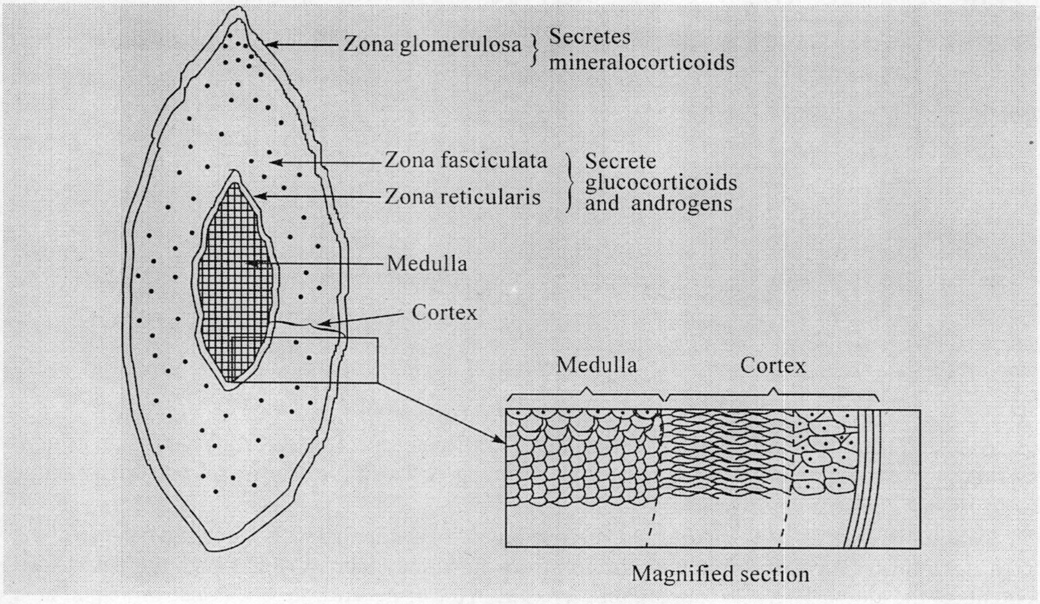

Fig. 19–11. Microscopic structure of the adrenal gland

(*a*) an outer narrow *zona glomerulosa*, believed to be the site of biosynthesis of the mineralocorticoid hormones.

(*b*) a middle, comparatively broader *zona fasciculata*, responsible for the production of glucocorticoid hormones and the adrenal androgens.

(*c*) an inner narrow *zona reticularis*, secreting glucocorticoids along with the middle zone.

When there is prolonged stimulation of the adrenal cortex by adrenocorticotropic hormone (ACTH), the middle and inner zones both hypertrophy; but a total lack of ACTH causes these two zones to atrophy almost entirely, leaving the outer zona glomerulosa partially intact. On the other hand, enhanced aldosterone production causes hypertrophy of the zona glomerulosa, while the other two zones remain almost unaffected.

Structure. Adrenal cortex secretes some 40-50 closely related C_{21} steroids, collectively called as corticosteroids (refer Fig. 19–12). From physiological viewpoint, the corticosteroids may be grouped under two categories :

A. *Mineralocorticoids* — concerned primarily with the transport of electrolytes and the distribution of water in tissues, *e.g.*, aldosterone and deoxycorticosterone.

B. *Glucocorticoids*—concerned primarily with the metabolism of carbohydrates, proteins and fats in the body, *e.g.*, cortisone (= compound E), cortisol (= hydrocortisone) and corticosterone.

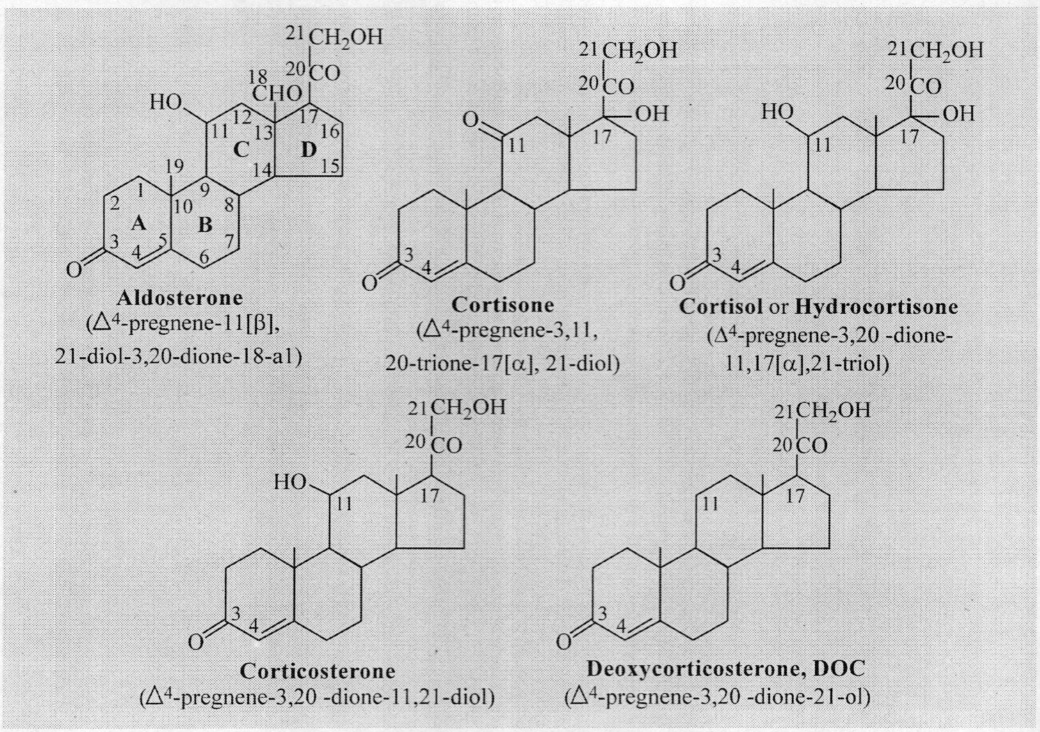

Fig. 19–12. Principal corticosteroids

Aldosterone is 30 times more active than deoxycorticosterone. Deoxycorticosterone, in its turn, is 4 times more potent than cortisone and cortisol in maintenance of life. Corticosterone is least active in this regard.

Functions. A *mineralocorticoid*, aldosterone is chiefly concerned with water-salt balance of the body. It stimulates the reabsorption of Na^+ ion from the kidney tubules and as such regulates NaCl contents of the blood. This also causes excretion of K in the urine. Aldosterone is also more potent in maintaining the life of adrenalectomized animals.

Glucocorticoids, on the contrary, govern many other processes. They perform the following physiological functions :

1. Influence the carbohydrate metabolism firstly by increasing release of glucose from the liver and secondly by promoting the transformation of amino acids to carbohydrates.
2. Inhibit protein synthesis in muscle tissues.
3. Control eosinophil cells of the blood.
4. Regulate lipogenesis.
5. Reduce the osteoid matrix of bone, thus favouring osteoporosis (weak bones) and heavy loss of calcium from the body.

6. Decrease immune responses associated with infection and anaphylaxis (immunosuppressive effects).
7. Cause increased secretion of hydrochloric acid and pespinogen by the stomach and that of trypsinogen by the pancreas (exocrine secretory effects).
8. Cause retention of sodium(and water) and loss of potassium to some extent. In this respect, it resembles aldosterone in action.

Hypoadrenocorticism. A decrease in the amount of corticosteroids in the body (hypoadrenocorticism) leads to the decreased metabolic rate, excessive pigmentation, loss of appetite (anorexia), muscular weakness, deficiency of blood (anemia), eosinophilia and decreased blood sugar (hypoglycemia) with fasting.

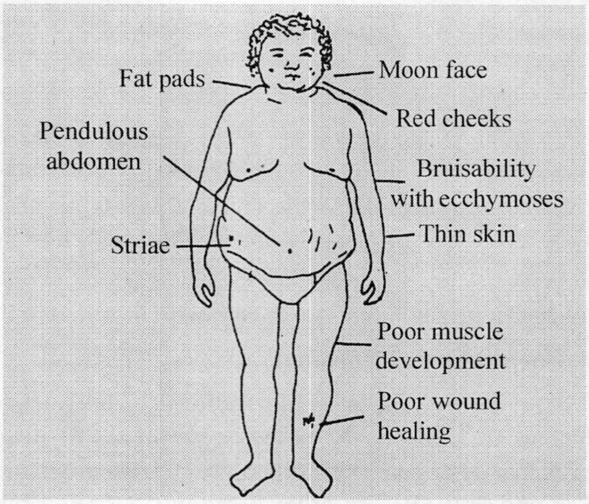

Fig. 19–13. Typical findings in Cushing's syndrome (*Adapted from Forsham and Di Raimondo in : Traumatic Medicine and sufery for the Attorney. Butterworth 1960*)

Hyperadrenocorticism. The excessive supply of adrenal cortical steroids (hyperadrenocorticism) results from cortical cell tumours which may arise in or outside the adrenal gland. Oversecretion of cortisol in man leads to a rare disease, **Cushing's syndrome** (Fig. 19–13), after its discoverer, **Harvey Cushing.** The most common cause of the symptoms of Cushing's syndrome is the prolonged administration of glucocorticoids for medical treatment and henceforth the disease is also known by its synonym **glucocorticoid excess.** The syndrome is characterized by profound disturbance of carbohydrate, protein, fat and calcium metabolism. There occurs mobilization of fat from the lower part of the body, with the concomitant extra deposition of fat in the thoracic region. The obesity becomes visible on the neck (buffalo hump) and on the face (moon face). Weakness and muscle wastings with marked osteosis become evident. Hypertension, pigmentation of the hair and excessive growth of hair are other symptoms. In men, there is impotence, and in women, amenorrhea and masculinization. Thus, Cushing's syndrome resembles somewhat adrenogenital syndrome.

Hypersecretion of aldosterone leads to a marked Na^+ and water retention, resulting in edema and hypertension causing heart failure.

Adrenal decortication. Removal of adrenal cortex (*adrenalectomy*) leads to a fatal human disease known as **Addison's disease** (Fig. 19–14), named after its discoverer **Thomas Addison**. Many of the symptoms of this syndrome resemble those of adrenalectomized animals. **Addison** (1855) described these in his own words as follows :

> "The leading and characteristic features of the morbid state to which I would direct attention are anemia, general langour and debility, remarkable feebleness of the heart's action, irritability of the stomach, and a peculiar change of colour in the skin, occurring in connection with a *diseased* condition of the 'supra-renal' capsules".

Thus, the chief symptoms of this syndrome are: anorexia, emesis (= vomiting), diarrhea, anemia, deep pigmentation of the buccal cavity and nipples, rapid loss of weight, excessive loss of NaCl in the urine and low blood pressure. As the extracellular fluid becomes depleted, the plasma volume falls, the concentration of R.B.C. rises markedly, the cardiac output lowers down and the patient dies in shock. The grip of Addisonian patients on their life is tenuous and any stress, infection, cold or even

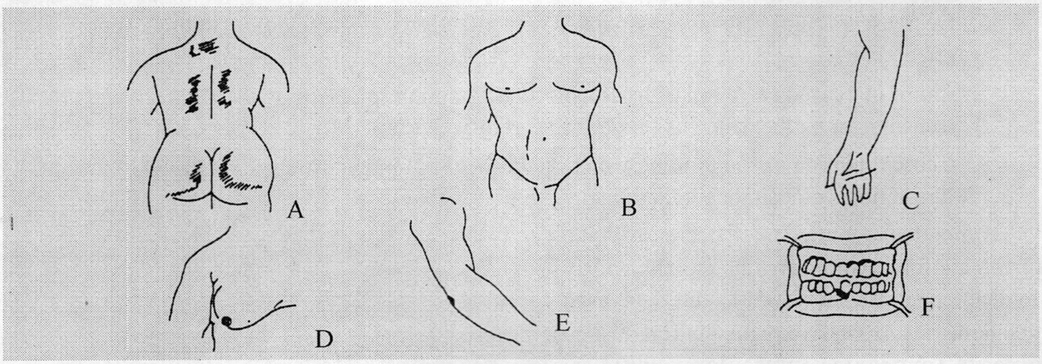

Fig. 19–14. Permentation in Addison's disease

(A) Tan and vitiligo (B) Pigmentation of scars (C) Pigmentation of skin creates (D) Darkening of areolas (E) Pigmentation of pressure points (F) Pigmentation of the gums.

(Adapted from Forsham and Di Raimondo, 1960)

noise can precipitate a crisis leading to death.

4. Corpus Luteal Hormones

Structure. The hormones secreted by the ovarian bed, corpus luteum are collectively called as **gestogens** or **progestins**. The principal gestogen is progesterone (the two pregnenolones, 20α-OH and 20β-OH are other hormones secreted by the corpus luteum). Progesterone (Fig. 19–15) is a C_{21} steroid and is secreted by the corpus luteum during the second half of the menstrual cycle. This was first isolated in pure form by **Adolf Butenandt** *et al* (1934) from corpus lutea of pregnant sows. It has also been isolated from adrenal cortical extracts. But its presence in the adrenal tissue is a consequence of its role as an intermediate in the biosynthesis of the typical adrenal cortical hormones. Chemically, progesterone is one of the pregnane derivatives and lacks the ketol group. Its molecular formula is $C_{21}H_{30}O_2$. It closely resembles deoxycorticosterone in structure. It is, therefore, not surprising to find progesterone with certain adrenocortical properties, *viz*., those influencing salt and water. Indeed, it serves as a procursor of the steroidal adrenocorticoids. It is soluble in most organic solvents and in vegetable oils but is insoluble in water.

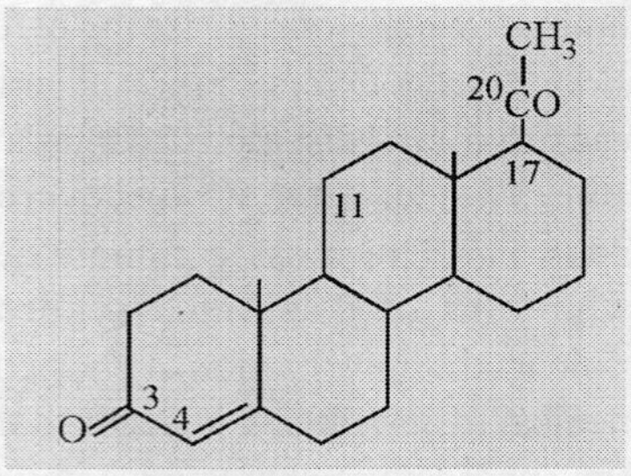

Fig. 19–15. Progesterone or "pregnancy hormone"

(Δ^4-pregnene-3, 20-dione or 3,20-diketo-$\Delta^{4,5}$-pregnene, *cis*)

The 3-dimensional structures of some steroid hormones determined by x-ray crystallography is presented in Fig. 19-16.

Biosynthesis. Progesterone is synthesized in the corpus luteum, placenta and the adrenal cortex from its immediate precursor, pregnenolone by a combined dehydrogenase and isomerase reaction. Unlike testosterone and estradiol, the progesterone is bound in plasma to the corticosteroid-binding globulin (CBG).

Metabolism. The metabolic fate of progesterone has been studied by injecting C^{14} - labelled hormone. About 75% of the injected progesterone is translocated to the intestine *via* bile and passed out in feces. A 20-hydroxy compound, pregnanediol, is the main urinary product of progesterone metabolism in human beings and rabbits. Such a conversion of progesterone into pregnanediol is not carried out in rat liver.

Functions. Progesterone has manifold functions :

1. The primary function of progesterone is to promote the proliferation of uterine mucosa so

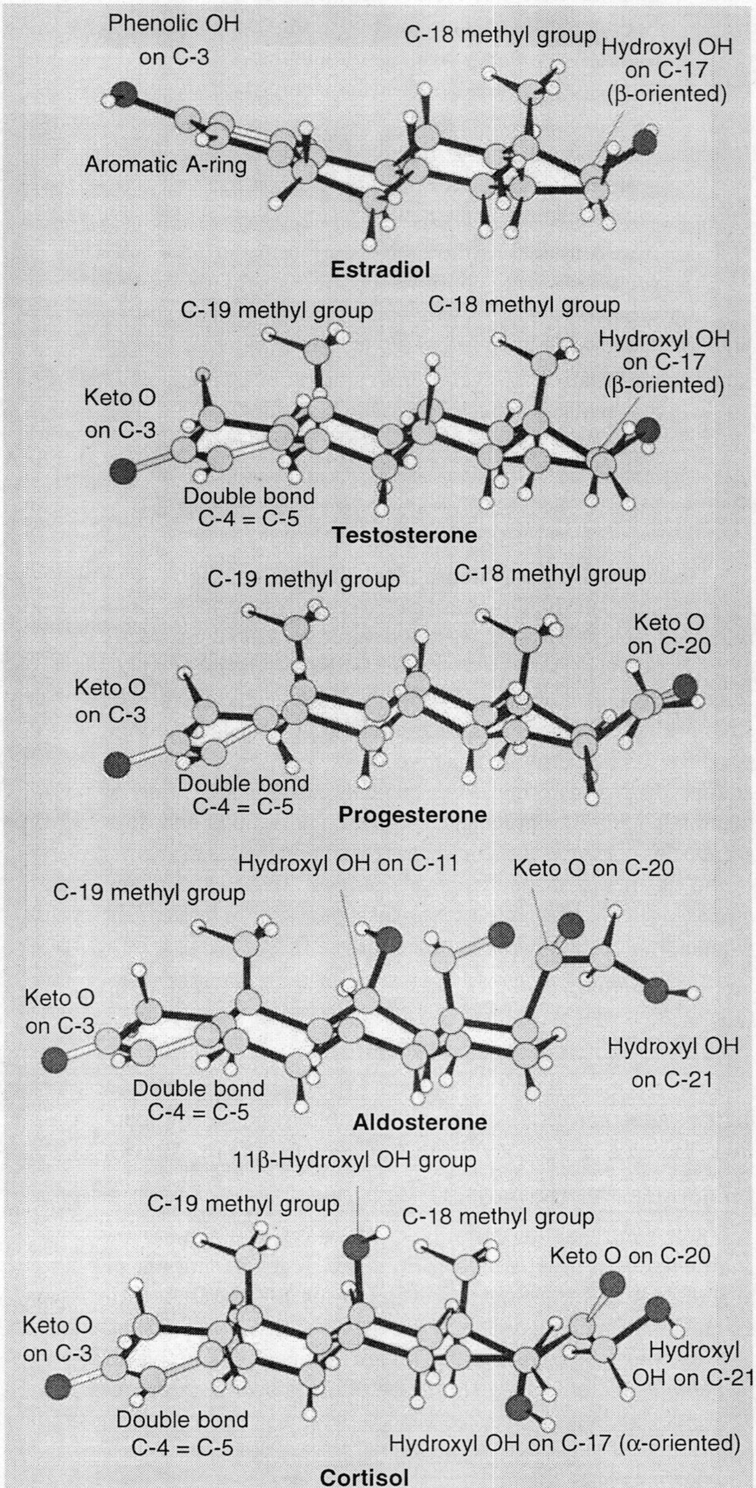

Fig. 19-16. Ball-and-stick representation of some steroid hormones

Details of each structure are labelled. In aldosterone, the acetal grouping is

$$R—CH\begin{matrix} OR_1 \\ OR_2 \end{matrix}$$

and the hemiketal grouping is

$$\begin{matrix} R_1 & & OR_3 \\ & C & \\ R_2 & & OH \end{matrix}$$

where R_1, R_2 and R_3 refer to different substituents.

(Courtesy: Glusker JP, 1979)

that the latter may receive the fertilized ovum. It, thus, serves for implantation of the fertilized ovum. If pregnancy ensues, continued secretion of progesterone is essential for completion of term.

2. It brings mammary glands to full maturity during gestation (pregnancy) for their onward use in breast-feeding by the newborn (Fig. 19–17)
3. It also maintains the uterus quiescent during pregnancy (*i.e.*, inhibits contraction of the uterus).
4. If given between 5th and 25th day of the normal menstrual cycle, progesterone exerts an antiovulatory effect. This is the basis for the use of certain progestins as oral contraceptive agents.
5. It serves as precursor of cortisol and corticosterone in the adrenal glands.

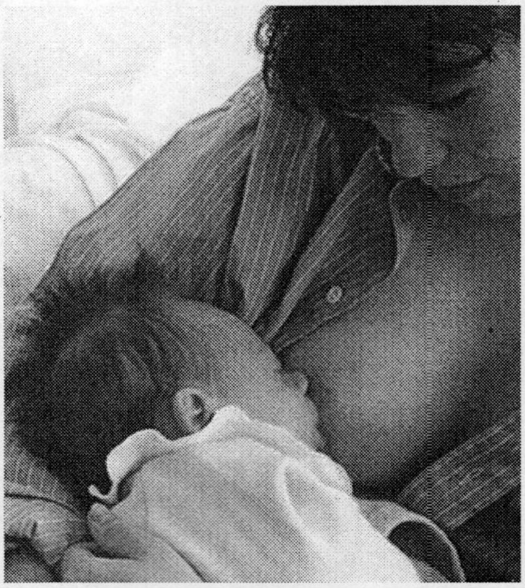

Fig. 31–17. A mother breast-feeding her baby

Progesterone brings mammary glands to full maturity and ocytocin helps in the ejection of milk which is suckled by the baby during infancy.

[**Relaxin.** Corpus luteum also secretes another hormone (or group of hormones) called relaxin or uterine-releasing factor (URF). It was discovered by **Hisaw** in 1926. Relaxin is a polypeptide of molecular weight about 9,000. It has not yet been obtained in pure form and, consequently, its structure has remained undetermined as yet. It is produced during pregnancy. The hormone is active only when injected into an animal which is in normal estrus (sexual heat). A wide variety of mammals including pregnant women, cows, rabbits and dogs secrete this hormone. Relaxin causes relaxation (hence the name) of the ligaments of the symphysis pubis in the estrus rat and guinea pig. This effect is, however, least pronounced in pregnant woman. It performs two other functions :

I. softening of the cervix of the pregnant woman at the time of delivery.

II. inhibition of the uterine motility.]

We, thus, see that many aspects of sexual reproduction in humans are controlled by hormones, including gamete production, puberty, pregnancy, birth and lactation (Fig. 19-18).

Fig. 19–18. The effect of hormones on sexual reproduction in humans

Hormones control gamete production, puberty, pregnancy, birth and lactation in humans.

Prohormones or Hormogens

Some polypeptide hormones, like certain enzymes, are synthesized in an inactive form, *i.e.*, as prohormones or hormogens. Prohormones are examples of proteins with extracellular functions, such as the enzymes chymotrypsin and trypsin, whose biological activities are dormant until activated by *peptidases*. Table 19–3 lists some of the known prohormones.

Table 19-3. Some prohormones

Prohormone	*Source*
Proinsulin	Pancreas– β cells
Proparathormone	Parathyroid
Angiotensinogen	Liver
Progastrin	Stomach

PEPTIDE HORMONES

1. Hormones of the Pancreas

Secretory gland. The pancreas (from pankreas : *pan*[G] = all, *kreas*[G] = flesh ; referring to the fleshy nature of the tissue) is both an exocrine and endocrine gland. It is situated transversely below and behind the stomach between the curve of duodenum and spleen. It is a compact and lobulated organ. It weighs about half a pound and resembles an elongated cone lying on its side. The broad end or '*head*' of the pancreas is located next to a curve of the duodenum, the part of the small intestine just beyond the stomach. The gland tapers off to the left in the direction of the spleen and left kidney and ends in a portion called '*tail*' (Fig. 19-19).

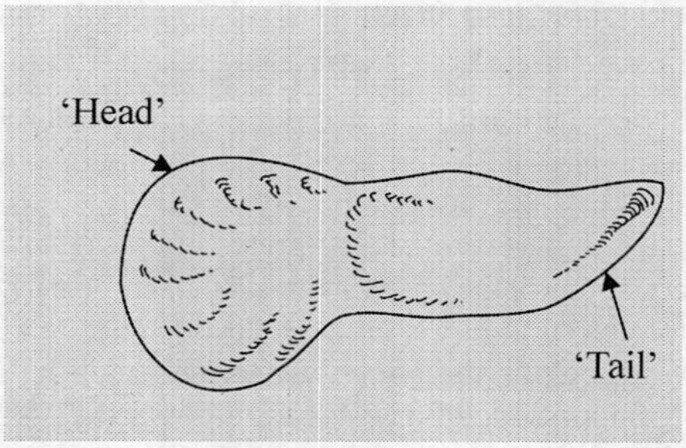

Fig. 19–19. Morphologic structure of pancreas

Barring cyclostomes, the pancreas in all the vertebrates is composed of two types of cells (Fig. 19–20).

(*a*) the glandular cells or *acinar* or *acini* (exocrine), which make up the bulk of the pancreatic tissues and secrete digestive juices into the duodenum by the pancreatic duct.

(*b*) the polygonal cells or *islets of Langerhans* or *islet tissue* (endocrine), which do not have any means for emptying their secretions externally but instead pour their secretions (*i.e.*, insulin and glucagon) directly into the blood. These were discovered by **Langerhans** in 1867, hence so named.

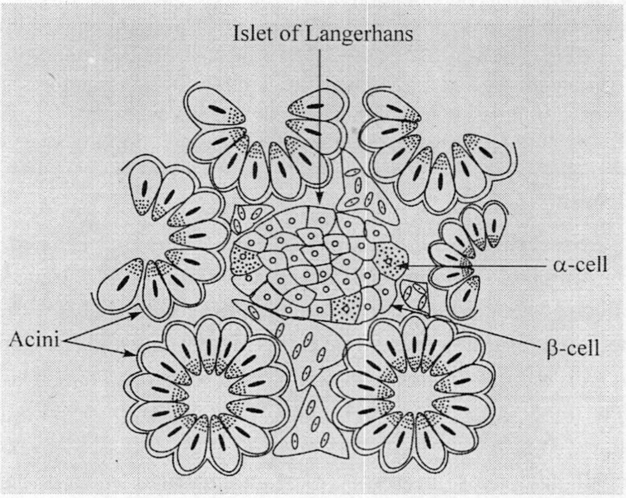

Fig. 19–20. Histologic structure of pancreas

Each pancreas has about 1,00,000 islets of Langerhans, which are clusters of various types of cells. These islets in mammals contain 3 major types of cells: α cells, β cells and δ cells. Each islet contains between 1,000 and 2,000 **β cells,** which were first described in 1869. The β cells contain granules which are insoluble in alcohol and manufacture a hormone insulin, store it and eventually release it directly into the bloodstream at the appropriate times. These amazing cells (*i.e.*, β cells) are also capable of measuring the blood glucose level within seconds to within a range of 2 mg %. Using this information, they can determine how much insulin is needed, and, within a minute or so, secrete the precise amount of required insulin. The **α-cells** contain granules which are soluble in alcohol and produce another hormone, glucagon. The α-cells tend to be arranged about the periphery of the islets. The existence of α-cells is dubious in cyclostomes and urodeles. Of the two cell types, the α-cells predominate in the reptiles and birds, whereas in amphibians and mammals, the β-cells are more abundant. Teleost fishes, however, often exhibit seasonal variations between the two cell types. However, a 3rd cell type **δ cells** secrete another hormone, *somatostatin*.

> The islets of Langerhans are also known to contain one more type of cells designated as **D** or **α_1-cells**. These cells probably secrete *gastrin* which is a recognized gastrointestinal hormone and is derived from pyloric mucosa.

FREDERICK SANGER

(Born, 1918 at Rendcombe, England)

Sanger, a British chemist, received his doctorate in 1943 on the metabolism of lysine from Cambridge University. During his service tenure, he preferred working in a laboratory with his own hands to teaching and administrative jobs. Sanger has been responsible for two of the most important technical advances of the past 45 years. First, in the early 1950s, he developed a method for determining the amino acid sequence of a polypeptide, and put it to use by working out the sequence of amino acids in insulin. This work, completed in 1955, provided the first real proof that the amino acids in a protein are present as a constant, genetically determined sequence. This gave molecular biologists the confidence to tackle other problems such as the genetic code. Then, in 1977, Sanger perfected a rapid method for sequencing DNA, opening the door to precise examination of gene structure and organization. Both advances have been invaluable and both were recognized by Nobel Committee. Sanger is *one of the two scientists to have ever won Nobel Prizes twice in the same subject :* **First in 1958** for determining the amino acid sequence of insulin on his own, and **later in 1980** for determining the base sequences in nucleic acids, along with **W. Gilbert,** thus both times in chemistry [the other being an American physicist John Bardeen-in 1956 and 1972, both time jointly]. Sanger has spent his entire career at the University Department of Biochemistry and the Medical Research Council Laboratory for Molecular Biology at Cambridge, U.K. The Principle behind his success, in his own words, is :

"If the planned experiment doesn't work, don't worry, start planning the next experiment".

Insulin

Structure. Insulin (*insula*^{L} = island) was first isolated in 1922 from the pancreas of dogs by **Banting** and **Best**, both of the University of Toronto, Canada. They also demonstrated the curative effect of pancreatic extract in dogs ailing with diabetes mellitus. Abel and his associates (1926) obtained insulin in crystalline form (Fig. 19–21) and also demonstrated its protein nature. *Insulin is, in fact, the first hormone to be recognized as a protein.* The chemical structure of insulin has been determined by **Sanger** and his coworkers at Cambridge, England in a painstaking 10-year (1944-54) study. It has a molecualr formula, $C_{254} H_{377} O_{75} N_{65} S_6$ (consult Chapter 9 for structure). Just to recall, bovine insulin consists of 51 amino acid residues dispersed in two chains. The two polypeptide chains are held together by cross linkages of two disulfide bonds. The acidic chain A contains 21 residues and the peptide chain B having 30 residues. It has a molecular weight of 5,733 and is isoelectric at pH 5.4. Human insulin, however, has a molecular weight of 5,808. Insulin is destroyed by alkali but is relatively stable in acid solutions. Reduction of the disulfide bond results in a loss of biologic activity. Zinc is always found with this hormone but is not a part of the insulin molecule.

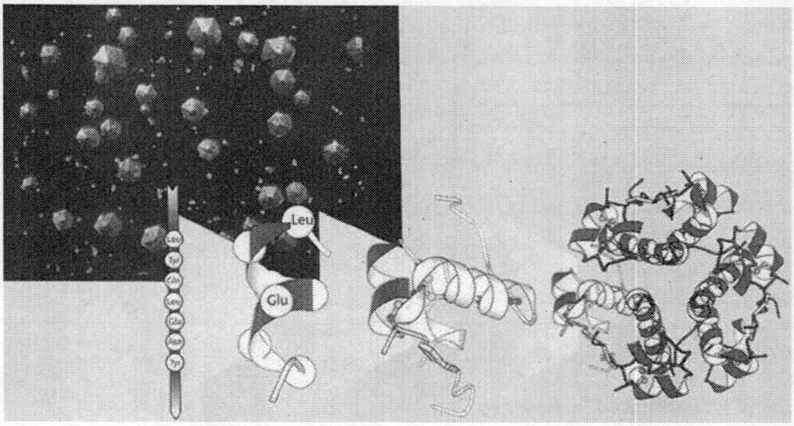

Fig. 19–21. Crystals of human insulin

Insulin is a protein homone, crucial for maintaining blood sugar at appropriate levels. (Above) Chains of amino acids in a specific sequence (the primary structure) define a protein like insulin. These chains fold into well-defined structures (the tertiary structure)– in this case a single insulin molecule. Such structures assemble with other chains to form arrays such as the complex of six insulin molecules shown at the far from right (the quartenary structure). These arrays can often be induced to form well-defined crystals (photo at left), which allows determination of these structures in detail,

[Courtesy: (Left) Alfred Pasieka/Peter Arnold]

It has been established that the insulins from various species differ only slightly, although their biologic activities are identical. **Sanger** and **Smith** (1957) have determined the amino acid sequence of insulins from 5 different species. They found the sequence to be identical except the identity and sequence of only three amino acids at positions 8, 9 and 10 in the chain A (refer Table 19–4).

SIR FREDERICK GRANT BANTING (LT, 1891–1941)
AND
CHARLES HERBERT BEST (LT, 1899–1978)

Banting and Best were both Canadian physicians who are credited with curing dogs ailing with diabetes mellitus by giving them pancreatic extract containing insulin. Later, insulin purified from pig or cow pancreas was successfully tried on humans but it produced immunogenic reactions in them. However, in 1982, using genetic engineering (recombinant DNA technology), insulin was manufactured and marketed for human use. Thus, *insulin became the first hormone product of genetic engineering.* The new recombinant insulin has the exact structure of human insulin and hence will not produce immunogenic reactions.

Sir Banting (and not Best) was awarded the **1923 Nobel Prize in Medicine or physiology** for demonstrating the curative effect of insulin, along with his physiology professor JJR Macleod, also of the University of Toronto.

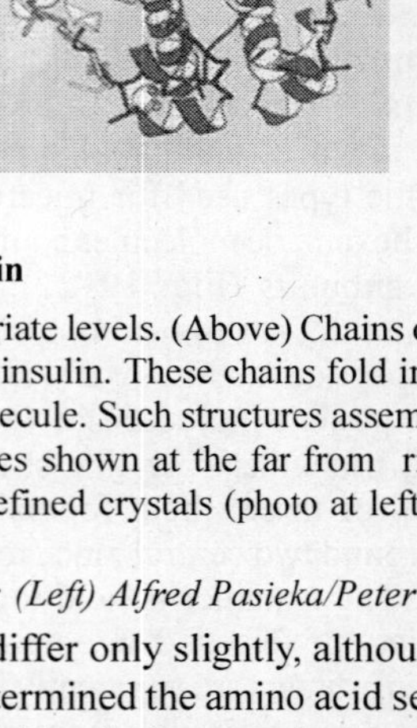

Best (left) and **Banting** with their dog, which they experimented upon, on November 19, 1921
The dog was depancreatized prior to administration of the pancreatic extract. Following injection, the blood sugar fell from 330 mg per cent to 170 mg per cent in one hour. This experiment demonstrated that the active principle could be extracted from the fetal pancreas with acetone and alcohol and that it was not destroyed by chloroform and ether.

(*Photo courtesy: Fruton JF and Wilson LC, 1966*)

Table 19-4. Differences in amino acid sequences in insulin of various species

Species	*Amino acid present in chain A at position*		
	8	9	10
1. Beef or Cattle	Ala	Ser	Val
2. Pig or Hog	Thr	Ser	Ile
3. Sheep	Ala	Gly	Val
4. Horse	Thr	Gly	Ile
5. Whale	Thr	Ser	Ile

Dorothy Crowfoot Hodgkin and coworkers of Oxford have shown that porcine and presumably also bovine insulin (the types used for injection in man) are hexameric molecules composed of three subunits (Fig. 19–22). Each subunit, in turn, is a dimer composed of two polypeptide chains, A and B. This multicomponent molecule of insulin, thus, is a triangular ring like structure consisting of three tilted, football like dimers around two central zinc atoms in the core. The hexamer is held together by a variety of forces. The ends of the overlapping dimers are apparently joined by interlocking phenylalanine groups. There is also a hydrogen bonding between dual glutamic acid groups midway in the dimer. The two chains of the monomers are linked by hydrogen and hydrophobic interactions and by two disulfide bridges between the chains. The role of two zinc atoms in the quaternary structure is yet to be ascertained.

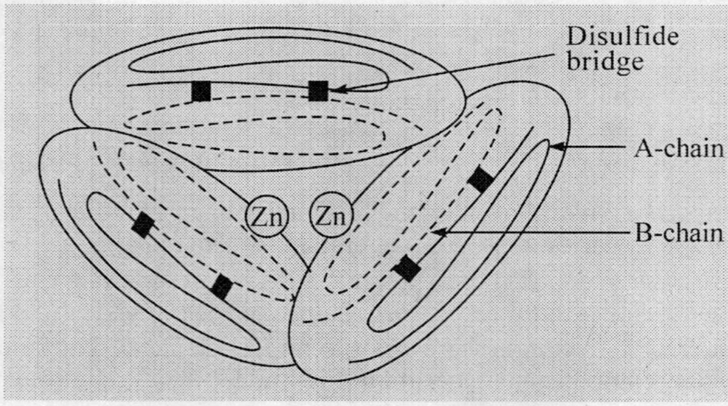

Fig. 19–22. Schematic representation of the three-dimensional structure of insulin

[The zinc atoms are apparently involved in the formation of the quaternary structure]

(*Adapted from Orten and Neuhaus, 1970*)

Biosynthesis. β-cells of the pancreas synthesize insulin by the ribosomes of the endoplasmic reticulum. Previously, it was suggested that the two chains of insulin are synthesized independently and later these combine by disulfide bonds. But now **Donald F. Steiner** *et al* (1967) have shown that it is formed from its precursor, *proinsulin*. Proinsulin has been isolated and purified from pancreatic extracts. It is a linear protein with 84 amino acid residues and has a molecular weight of about 9,100.

Functions. Insulin has a profound influence on carbohydrate metabolism. It facilitates entry of glucose and other sugars into the cells, by increasing penetration of cell membranes and augmenting phosphorylation of glucose. This results in lowering the sugar content of the blood – a fact leading to its common name, **hypoglycemic factor**, which may be abbreviated as **hG-factor.**

> One of the present authors (JL Jain) proposes the abbreviation **'hG-factor'** for the hypoglycemic factor insulin, where the small letter 'h' justifiably denotes 'small' or hypo function. It should, however, not be abbreviated as 'HG- factor', which is aptly kept reserved for the hyperglycemic factor glucagon, where the capital letter 'H' signifies 'big' or hyper function.

Insulin administration promotes protein synthesis (proteogenesis) by assisting incorporation of amino acids into proteins. This effect is not dependent on glucose utilization. At the same time, it also acts as an *antiproteolytic agent*, *i.e.*, discourages excessive breakdown of tissue protein. This action is similar to that of GH and testosterone.

It may be said, in general, that insulin promotes anabolic processes (synthesis of glycogen, fatty acids and proteins) and inhibits catabolic ones (breakdown of glycogen and fat).

Insulin deficiency. The deficiency of insulin caused either by inadequate insulin production or by accelerated insulin destruction, leads to *diabetes mellitus* in man. This disease is characterized by :

1. an increase in blood sugar or glucose (hyperglycemia) from a normal value of 80 mg/100 ml of plasma to abnormal value ranging between 150—200 mg/100 ml of plasma.
2. the appearance of sugar in the urine (glycosuria); with the result, the victim's urine tastes sweet.
3. an increase in concentration of ketone bodies in the blood (ketonemia) and in the urine (ketonuria).
4. the excretion of large quantities of urine (polyuria), frequently at night (nocturia), leading to dehydration.
5. the excessive drinking of water (polydipsia) on account of an unrelenting thirst.
6. the excessive eating (polyphagia) due to feeling constant hunger. This is because the tissues cannot utilize glucose normally, even though they need fuel.
7. the lack of energy (asthenia) which is apparently caused mainly by loss of body protein.

DOROTHY CROWFOOT HODGKIN

(LT, 1910-1994)

Dorothy Hodgkin was one of the outstanding British scientists of the 20th century, endowed with a brilliant mind and an iron will to succeed and in many ways proved herself a trendsetter. One of her students is Margaret Thatcher, the 'Iron Lady' and ex Prime Minister of Great Britain. Born on 12 May, 1910 in Cairo, Dorothy Mary Crowfoot (her name before she married) was educated at Sir John Leman School, Beccles, Suffolk and later at Sommerville College at Oxford where she also taught. She was elected a fellow of the Royal Society at age 37 (in 1947) and was also awarded the Gold Medal of the Royal Society in 1956. She took to x-ray crystallography and elucidated the x-ray diffraction patterns of pepsin and insulin. Insulin's structure was the one she laboured with perseverance of a hedgehog. She also researched on the structure of a number of antibiotics including penicillin and cephalosporins. In 1957, she established the unusually distinct structure of the complex molecule of vitamin B_{12} and was awarded the **Nobel Chemistry Prize in 1964** for the same. In 1960, Hodgkin became the Royal Society Wolfson Professor. She was made a member of the Order of Merit in 1965. She remained the President of the International Union of Crystallography from 1969 to 1975. Dorothy was the founding mother of the *"Pugwash" movement*, which helped scientists maintain communication with the countries behind the 'Iron Curtain'. She remained the President of the Pugwash Movement of scientists for peace between 1976-1988. (Pugwash Organization later got Nobel Peace Prize in 1995). A 14-year-old boy (Leonard Thompson) in Canada was treated with insulin, the first-ever diabetic patient to be so treated (along with Joseph Rotblat, a scientist turned antinuclear weapon compaigner, who nursed the Organization for about 4 decades by the time he got the award.) She held the Sir C.V. Raman Professorship of the Indian Academy of Sciences, of which she was an honorary fellow. During the last few years, she was bodily crippled by arthritis but this affliction had had no debilitating effect on her indomitable will. Only a year before her death on 29 July, 1994, although she was wheelchair-bound, she flew to China to attend an International Crystallography Conference. Such was Hodgkin ! Once she helped her three-year-old son search for a lost toy when she said :

"It must be somewhere; it can't be nowhere.'

CLINICAL IMPLICATIONS
DIABETES MELLITUS

The Greek physician Arteus (ca A.D. 200) apparently was the first to call insulin deficiency disease as **diabetes** (to follow through; urine), referring to a large volume of urine passed in it. Later, the Latin term **mellitus** (sweetened or honey-like) was added to it and dates from the time when urine was tested by tasting, and the urine in this condition is sweet to the taste. Used alone, diabetes is taken to mean diabetes mellitus, by far the most common (80–90%) of the two types of diabetes (*diabetes mellitus* and *diabetes insipidus*). Normally, some sugar (glucose) is in the blood at all times. These terms, glucose and sugar, are used interchangeably, although with regard to diabetes we almost always mean glucose. The level of glucose is usually between 60 and 120 milligrams per cent (mg %), which means that there are 60 to 120 milligrams of glucose in each 100 cubic centimeters (cc) of blood.

Milligrams per cent is sometimes expressed as milligrams per deciliter (mg/dL). *Diabetes mellitus is characterized by a level of glucose in the blood that is above normal, most of the time.*

Glucose is the major fuel of the body. We ingest a great deal of glucose each day, as carbohydrates (starches) are made of glucose. When we eat glucose, it gets into the bloodstream and ultimately into each of the individual cells of the body to provide them with energy. Glucose cannot just flow into the cells, however. Cells are enclosed in membranes that separate what is inside the cell from what is outside. Somehow, the cells must be told that the glucose waiting outside the cell should be allowed in. That is what insulin does. It is needed to signal to the cells that they should allow the glucose, that normally is in the blood, to penetrate their outer layer. If the glucose cannot enter the cells, it "backsup" in the blood. The level of glucose in the blood increases and a state of diabetes is produced. Such a condition occurs if there is a lack of insulin for whatever reason. Perhaps the pancreas cannot produce enough insulin.

Diabetes mellitus is of 2 types : type I (insulin-dependent) and type II (noninsulin-dependent). The use of Roman numerals probably dignifies the importance of this classification.

Type I (Insulin-dependent diabetes mellitus, IDDM) : IDDM occurs because the insulin producing β-cells are destroyed and there is not enough insulin produced. In the past, there was treatment that had some measure of success. This was *relative starvation.* People with IDDM had very short careers. The former synonyms for this disease are *brittle diabetes* or *juvenile-onset diabetes*.

Type II Noninsulin-dependent diabetes mellitus, NIDDM) : NIDDM is caused by a relative insulin insufficiency due to insulin resistance– the inability of the insulin to tell the cells to use glucose– plus insufficient insulin to overcome this resistance. Patients with NIDDM often lived for many years, if they heroically reduced their weight to live with the small amount of insulin that might be available. In a sense, this was *organised starvation*. NIDDM is also known by the former names, *stable diabetes* or *adult-onset diabetes*. In case of type II diabetes, it may be said that heredity may load the cannon, but stress or obesity pulls the trigger.

Table given below draws comparison between type I and type II diabetes.

Comparison of two types of diabetes mellitus

Item	*Type I (IDDM)*	*Type II (NIDDM)*
1. Percentage of all people with diabetes	20%	80%
2. Dependence on insulin	Dependent	Not dependent
3. Age of discovery	In children or in adults, usually under 40	In adults, usually over 40
4. Cause of diabetes	Reduced or none insulin production	Insulin resistance and relative or absolute deficiency of insulin
5. Pancreatic insulin content	0	> 5% of normal
6. Weight	Often underweight or normal weight	Usually overweight but 20% are normal weight
7. Primary insulin resistance	Minimal	Marked
8. Anti-islet antibodies	85%	> 5%
9. Concordance rate of identical twins for diabetes mellitus	25–50%	~ 100%
10. Condition when discovered	Usually moderately to severely ill	Often not ill at all or having mild symptoms
11. Acute complications	Ketoacidosis	Nonketotic, hyperosmolar, hyperglycemic comma, not usually prone to ketoacidosis.
12. Usual treatment	Insulin, eating plan, exercise	Diet, exercise, insulin, if needed oral agents

The control of diabetes, however, depends on the triad of food, activity and insulin or oral agents. All of these, however, must be balanced properly by the person with diabetes.

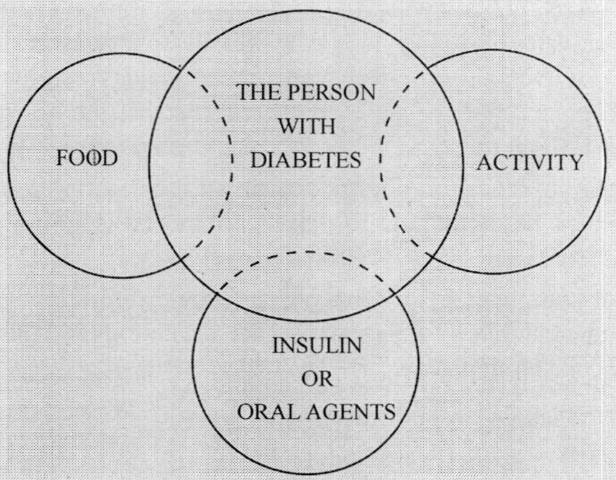

Diabetes and its controlling triad

The diabetic person gradually becomes weaker and loses weight due to the failure of glucose utilization by the body despite a voracious appetite. Diabetic comma ensues, the plasma volume decreases and the kidney fails. This eventually leads to death. In England, the disease for centuries was approximately called the "*pissing evil*".

Since insulin is a protein, it is not amazingly digested and thus inactivated by the proteolytic enzymes, pepsin and trypsin. Hence, insulin has no effect when taken orally and must be, therefore, administered parenterally in diabetic patients.

Parenteral is any route outside the gastrointestinal tract, including subcutaneous, intramuscular, intraperitoneal and intravenous injections or infusions. In practice, parenteral feeding is done intravenously.

Hypoglycemic agents. There are many hypoglycemic drugs, effective when taken orally, that control diabetes. Certain of these drugs, which belong to the group sulfonamide, are tolbutamide (orinase), chlorpropamide (diabinese) and tolazamide (tolinase).

Glucagon

Structure. Glucagon was first isolated in crystalline form by **Behrens** and others. This peptide hormone has a molecular weight of 3,485 and is isolectric at pH 8. It has 29 amino acid residues (of 15 different types) arranged in a linear row. Histidine is the N-terminal amino acid and threonine, the C-terminal amino acid. Unlike insulin, it contains no cystine, proline or isoleucine, but possesses methionine and tryptophan in appreciable amounts (Fig. 19–23). The small amount of sulfur present is, thus, in the form of methionine rather than cystine.

```
 1   2   3   4   5   6   7   8   9   10
His.Ser.Gln.Gly.Thr.Phe.Thr.Ser.Asp.Tyr.Ser 11
                                        Lys 12
                                        Tyr 13
                                        Leu 14
                                        Asp 15
                                        Ser 16
                                        Arg 17
                                        Arg 18
                                        Ala 19
                                        Gln 20
                                        Asp.Phe.Val.Gln.Trp.Leu.Met.Asn.Thr
                                        21  22  23  24  25  26  27  28  29
```

Fig. 19–23. Structure of glucagon

Functions. Like insulin, glucagon also influences carbohydrate metabolism but in an opposing way (refer Fig. 19–24). Glucagon (as also the other hormone, epinephrine) activates the enzyme adenyl cyclase which converts ATP to cyclic AMP. The latter compound activates phosphorylase *b* kinase which, in its turn, activates phosphorylase *b* to yield phosphorylase *a*. This releases glucose-l-phosphate from glycogen of liver. Glucose-l-phosphate then yields free glucose in blood, whereby increasing blood sugar contents. It is because of this reason that the hormone is also termed as **Hyperglycemic factor** or **HG-factor**.

In contrast to epinephrine, glucagon does not cause an increase in blood pressure. Therefore, glucagon and not epinephrine has found clinical applications and is administered in patients with acute hypoglycemia.

Acting in the liver, it stimulates glycogenolysis (glycogen breakdown) and gluconeogenesis (production of glucose from noncarbohydrate sources such as proteins and fats). The former function is similar to that of ACTH and epinephrine.

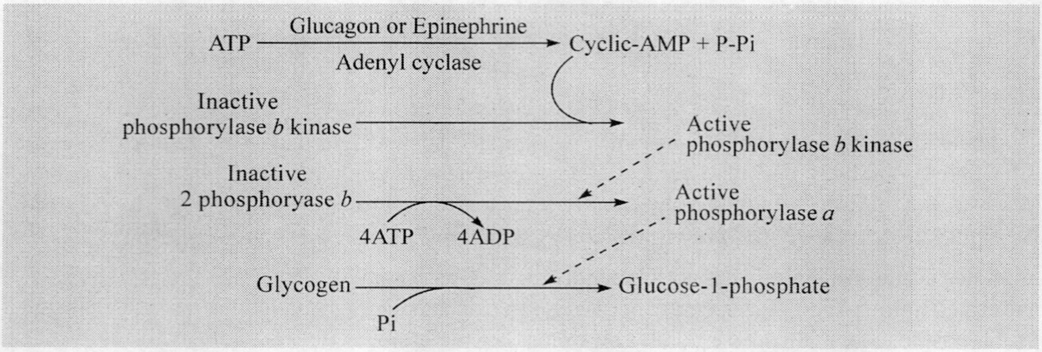

Fig. 19–24. Role of glucagon and epinephrine in glycogen synthesis

(*After Fairley JL and Kilgour GL, 1966*)

Hyperglycemic agent. Overdosages of insulin, given to diabetic patients, often result in acute hypoglycemia. This may be cured by giving crystalline glucagon (in the form of glucagon hydrochoride) either intramuscularly or intravenously. Response may be observed within 10–15 minutes after administration of 0.5 to 1.0 mg of glucagon.

2. Hormones of the Hypophysis or Pituitary Gland

Secretory gland. Hypophysis (meaning *undergrowth*) is so named because of its location below the brain as an undergrowth. Its synonym pituitary gland is, however, misleading as the gland is not concerned with the secretion of mucus or phlegm (*pituita*[L] = phlegm), as was thought previously. This is an unpaired small ovoid gland and is no larger than the end of the little finger. It is located at the base of the brain and lies below the diencephalon in a depression of basisphenoid bone of the skull called sella turcica. It is a complex structure formed of ectodermal outgrowth of the mouth cavity and downgrowth of the infundibulum.

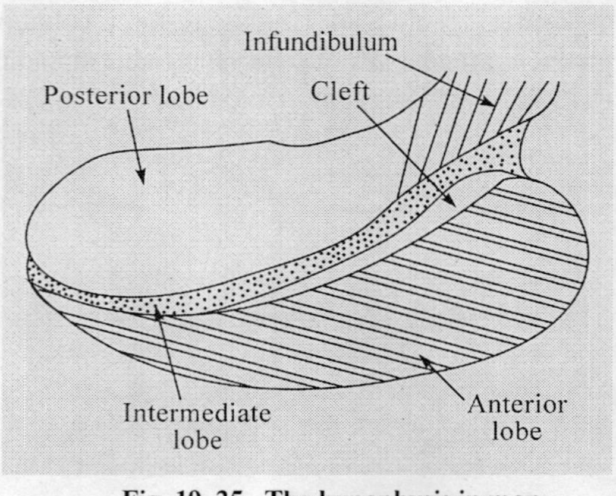

Fig. 19–25. The hypophysis in man

(*After D. Marsland, 1964*)

The human hypophysis is a reddish- grey oval structure and measures about 10 mm in diameter. Its average weight varies in the two sexes : 0.5—0.6 g in males and 0.6—0.7 g in females. It consists of 3 lobes (Fig. 19–25) :

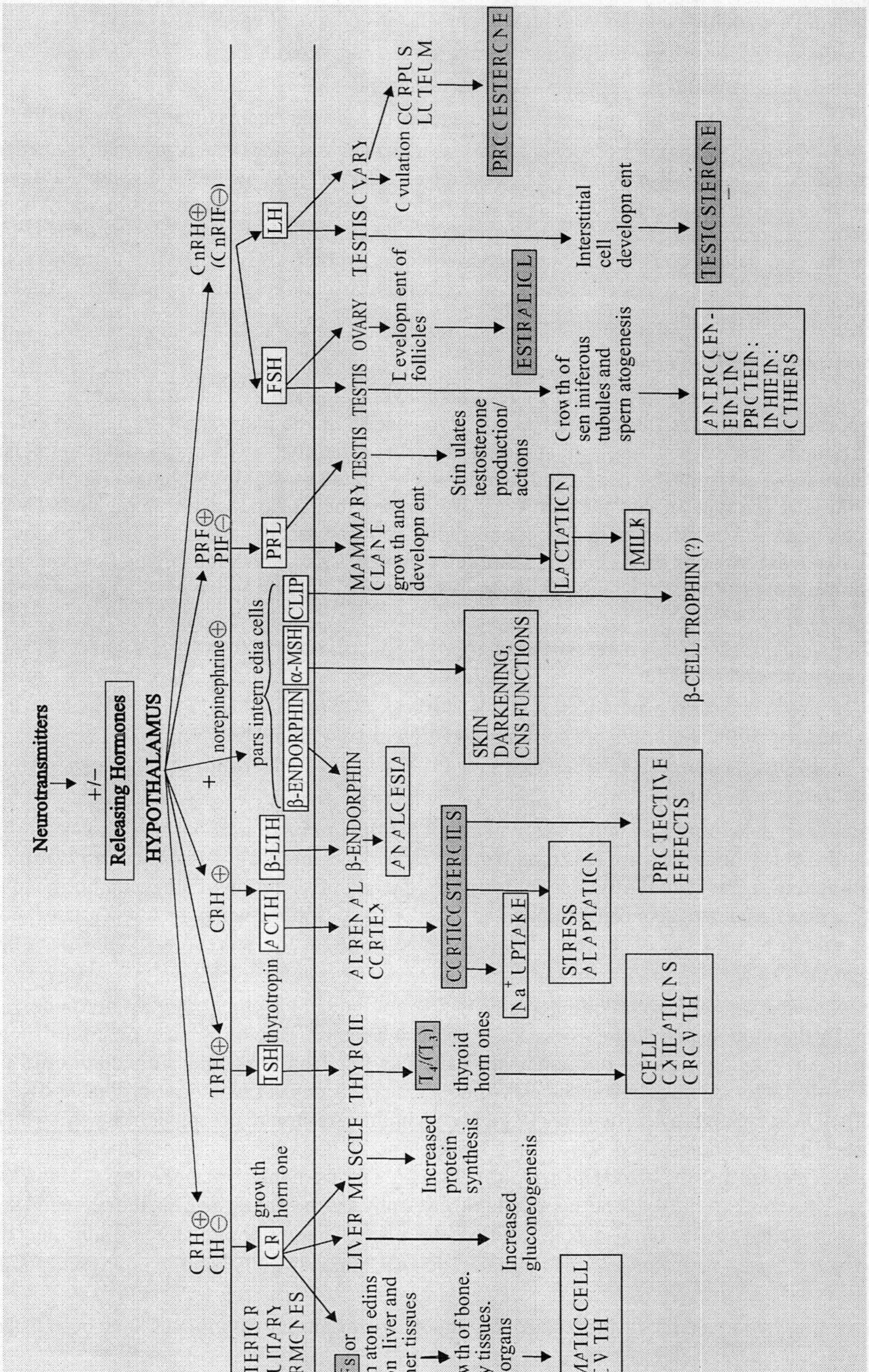

Fig. 19–26. Overview of anterior pituitary hormones with hypothalamic releasing hormones and their actions

(*a*) an anterior richly vascular largest lobe, pars distalis or adenohypophysis.

(*b*) an intermediate relatively avascular smallest lobe, pars intermedia.

(*c*) a posterior neural lobe, pars nervosa or neurohypophysis.

While all the three parts are important, it is the anterior lobe that seems to be essential to life.

This gland plays perhaps the most dominant role as it secretes hormones which govern the secretion of other endocrine glands (like thyroid, adrenal and gonads) and also its secretions have a direct effect on the metabolism of nonendocrine tissues (refer Fig. 19–26). It has, therefore, aptly been described as the '*master gland*' of the system or the 'master (= *conductor*) *of endocrine orchestra*'. However, the view held by some biochemists is that since the hypophysis is subservient to the nervous system and some of the other endocrine glands, it is erroneous to call this gland as the master gland of the body.

Pars distalis or Adenohypophysis

The anterior lobe is the largest and shares about 70% of the total weight of hypophysis. The adenohypophysis originates from an embryonic invagination of the pharyngeal epithelium called Rathke's pouch. This explains the epithelioid nature of its cells. It consists of glandular epithelial cells of varying shapes and sizes, arranged in columns separated by sinusoids containing blood. In general, there is one type of cell for each type of hormone that is produced in this gland. These various cell types can be differentiated from one another on the basis of special staining techniques (the only likely exception to this is that the same cell type may secrete both luteinizing hormone and follicle-stimulating hormone). Using acid-base histological stains, only 3 types of cells may be differentiated in this region :

(*a*) *acidophils* or *α cells*—these stain strongly with acidic dyes (acid fuchsin) and secrete growth hormone and prolactin.

(*b*) *basophils* or *β cells*—these stain strongly with basic dyes (aniline blue) and secrete luteinizing hormone, follicle-stimulating hormone and thyroid-stimulating hormone.

(*c*) *neutrophils* or *chromophobe cells*—these do not stain with either and are believed to secrete adrenocorticotropic hormone.

The anterior lobe produces hormones which govern the production of hormones secreted by other glands. These hormones are called as **tropins** (*tropos*^{G} = turning) or **trophic** (*trophikos*^{G} = nursing) **hormones**. Evidences available indicate that the rate of secretion of a trophic hormone is inversely proportional to the concentration, in the blood, of the hormone with which it is related. For example, a high blood level of thyroid hormone tends to inhibit the secretion of TSH from the adenohypophysis and a low level causes an increased production of it.

Neurohormones— The secretion of thyrotropin (as of almost all other pituitary hormones) is controlled by the hormones (or factors) released from hypothalamus, a region of the brain immediately proximal to the pituitary. These hormones are called as **hypothalamo-releasing hormones** or **hypothalamic factors.** These have been classified as *neurohormones*, *i.e.*, those produced by the nerve cells. These are unlike *neurohumors* (*e.g.*, acetylcholine, serotonin, norepinephrine etc) which are released at nerve endings and activate the adjacent nerve bodies. The neurohormones, on the contrary, are released into the blood and activate cells a little far from their point of release. These are introduced into the capillary of the hypothalamo-hypophyseal portal system at the floor of the hypothalamus called median eminence. In addition, the release of anterior pituitary hormone may be inhibited by *a release inhibiting factor* which passes down the same hypothalamo-hypophyseal portal veins.

The various hypothalamic factors controlling the release of pituitary hormones have been listed in Table 19–5.

Table 19–5. Pituitary hormones and the hypothalamic factors controlling their secretion

S.N.	Pituitary Hormones	Hypothalamic Releasing Factors*
1.	Thyrotropin, TSH	Thyrotropin-releasing factor, TRF
2.	Corticotropin, ACTH	Corticotropin-releasing factor, CRF
3.	Follicle-stimulating hormone, FSH	Follicle-stimulating hormone-releasing factor, FSH-RF
4.	Luteinizing hormone, LH	Luteinizing hormone-releasing factor, LH-RF
5	Prolactin, PL	Prolactin-releasing factor, PRF
6.	Growth hormone, GH	Growth hormone-releasing factor, GH-RF
7.	Melanocyte-stimulating hormone, MSH	Melanocyte-stimulating hormone-releasing factor, MRF

* Also, sometimes called hormones, so in some books, the abbreviations may have an H instead of an F at the end.

(*After Schally et al, 1973*)

Some 30 tropins are secreted from the anterior lobe. The four important tropins are described below.

1. Thyrotropin or thyroid-stimulating hormone, TSH. It is a glycoprotein with molecular weight about 30,000. Each molecule has 8-9 cystine residues and the disulfide groups are present as intrachain linkages rather than interchain linkages.

In general, it stimulates the activity of thyroid gland and enhances the rate of certain reactions such as :

(*a*) removal of iodide from blood by thyroid

(*b*) conversion of iodide to thyroid hormones

(*c*) release of hormonal iodine from thyroid.

The release of thyrotropin is controlled by another hormone from hypothalamus called thyrotropin-releasing factor, TRF.

2. Corticotropin or adrenocorticotrophic hormone, ACTH. Corticotropin (Fig. 19–27) is a straight chain polypeptide with a molecular weight of about 4,500 and consists of 39 amino acid residues in mammals like man, ox, sheep and pig. The most potent segment of activity is from residue 15 to 18. ATCH molecule in these species differs from each other only in the constituents present from residue 25 to 33. Thus far, no differences in residues 1-24 and 34-39 have been reported.

ATCH, in general, has a stimulatory effect on the hormone-producing capacity of the adrenal cortex. ATCH administration leads to accelerated gluconeogenesis (= neoglucogenesis) with accompanied retardation of protein synthesis in all tissues except liver. It also possesses an intrinsic melanocyte-stimulating activity, causing darkening of the skin in a manner similar to that of another hormone, MSH.

Oversecretion of ACTH results in Cushing's disease, already described earlier.

Certain peptides found in hypothalamus and also in neurohypophysis have ACTH-releasing activity. These have been termed as corticotropin-releasing factor, CRF.

3. Gonadotropins or gonadotrophic hormones, GTH. These hormones control the development and functioning of the gonads which remain dormant until the age of 12-14 years in the human beings. Damage to certain areas of the hypothalamus greatly decreases the secretion of gonadotrophic hormones by the anterior pituitary. If this occurs prior to puberty, it causes typical eunuchism. The damage often causes simultaneous overeating because of its effect on the feeding centre of the hypothalamus. Consequently, the person develops severe obesity along with the eunuchism. This condition is called **adiposogenital syndrome** or **Frohlich's syndrome** or **hypothalamic eunuchism**. Three gonadotropins are known :

Ser(1).Tyr(2).Ser(3).Met(4).Glu(5).His(6).Phe(7).Arg(8).Trp(9).Gly(10).Lys(11).Pro(12).Val(13).Gly(14).Lys(15).Lys(16).Arg(17).Arg(18).Pro(19).Val(20).Lys(21).Val(22).Tyr(23).Pro(24).Asp(25).Ala(26).Gly(27).Glu(28).Asp(29).Gln(30).Ser(31).Ala(32).Glu(33).Ala(34).Phe(35).Pro(36).Leu(37).Glu(38).Phe(39)

Most potent segment of activity: Lys 15 · Lys 16 · Arg 17 · Arg 18

Fig. 19–27. Structure of human adrenocorticotropin

A. Follitropin or follicle-stimulating hormone, FSH. It is a glycoprotein that contains galactose, mannose, galactosamine, glucosamine, sialic acid, fucose and uronic acid. It has a molecular weight of about 30,000 in man. In human females, it induces the growth of graafian follicles resulting in an increased weight of the ovary. In males, however, FSH promotes spermatogenesis by stimulating the development of seminiferous tubules, thus leading to the formation of a large number of spermatocytes.

The release of this hormone is controlled by another hormone from hypothalamus called follicle-stimulating hormone-releasing factor, FSH-RF.

B. Luteinizing hormone, LH or interstitial cell-stimulating hormone, ICSH. It is a peptide hormone with molecular weight of about 26,000 (in man) or 100,000 (in swine). It lacks tryptophan but has a high content of cystine and proline. Each molecule contains 10 glucosamine and 3 galactosamine residues. In females, LH is concerned with the ripening and rupturing of ovarian follicles, which later transform into corpus lutea. It also induces the development of interstitial cells of both the ovaries and the testes— a fact responsible for its nomenclature.

The secretion of LH is controlled by luteinizing hormone-releasing factor, LH-RF, a secretion from hypothalamus. The long-acting analogue of LH-RF has been found useful in the treatment of precocious puberty in females by **Florence Comite** *et al* (1983), the American researchers at the National Institute of Health. Puberty is initiated by pulsed nocturnal secretions of gonadotropins that result from the release of the hormone LH-RF by the hypothalamus gland in the brain. The administration of LH-RF analogue "initially stimulated but subsequently inhibited" the release of LH and FSH, the two sex hormones that initiate puberty.

C. Luteotropin or luteotrophic hormone, LTH. Because of its broad spectrum of effects on vertebrates in general, *luteotropin is the most versatile of all the adenohypophyseal hormones.* It is also the first anterior pituitary hormone to be obtained in pure form. This is also a peptide hormone with 198 amino acid residues and a molecular weight of about 23,500. It has 3 disulfide bonds between cysteine residues at 4-11, 58-173 and 190-198 (**Li** *et al*, 1971). *It differs from FSH and LH in that it contains no carbohydrate.* It is thermolabile and is destroyed by tryptic digestion. In pure form, it has no growth-promoting, thyrotropic, diabetogenic, adrenotropic or gonadotropic activities. However, in association with estrogen, luteotropin promotes the growth of the mammary glands (mammogenesis) and also induces secretion of milk (lactation) at the time of child birth (parturition). Henceforth, this hormone is variously called as *prolactin*, PL or *lactogenic hormone or mammotrophic hormone*, MH. It also stimulates glucose uptake and lipogenesis. Along with androgens, it causes the development of secondary male sex characters. In rat, at least, prolactin also has gonadotrophic activities in that it maintains functional corpora lutea in hypophysec tomized animals. It also acts as an anabolic agent mimicking the effects of growth hormone, although it is less active in this respect. In pigeons, however, it stimulates enlargement of crop glands and formation of 'crop milk'. In fact, prolactin is credited with performing some more than 80 functions and it is for the same reason that it has been jocularly termed as a "*jack-of-all-trades*." According to **Nicoll** and **Bern** (1971), these various functions of prolactin fall under 5 major categories : reproduction, osmoregulation, growth, integument and synergistic effect with the steroid hormones.

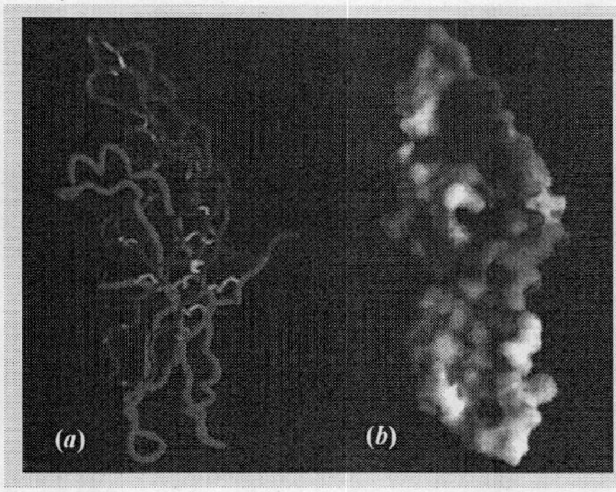

Fig. 19–28. The x-ray structure of human chorionic gonadotropin (hCG)

(*a*) A C_α diagram showing the α subunit in *red* and the β subunit in *blue* with disulfide linkages in *yellow*. The pseudo-twofold axis relating the two subunits is horizontal.

(*b*) The surface electrostatic potential of the hCG as viewed towards its receptor-binding surface. The surface is coloured according to its electrostatic potential, with the most positive regions *dark blue* and the most negative regions *dark red.* The view is related to that in Part *a* by a 60° rotation about the vertical axis.

[*Courtesy : Hao Wu and Wayne Hendrickson*]

[Human chorionic gonadotropin

A hormone chorionic gonadotropin, although of placental origin, resembles hypophyseal hormones in its biological effects. It was previously recognized as anterior pituitary like factor (APL factor). *Human chorionic gonadotropin, hCG* (Fig 19–28) is a glycoprotein with molecular weight of about 30,000. The carbohydrate moiety contains some 6 components, *viz*., D-galactose, D-mannose, *N*-acetylglucosamine, *N*-acetylgalactosamine, L-fucose and *N*-acetylneuraminic acid. Their sequence is not fully known. *N*-acetylneuraminic acid is essential to the biologic activity of this hormone. Apparently, all the common amino acids are present in the protein moiety, proline being especially high (about 20%).

As to its biological role, hCG factor supplements hypophysis in maintaining growth of the corpus luteum during pregnancy. This hormone, when administered, stimulates Leydig tissue and hence the male accessory organ.

The hormone appears sharply after pregnancy in the urine. This fact has become the basis for a pregnancy test known as *Aschheim-Zondek test*. This involves injecting urine or an alcoholic precipitate of urine from a woman into immature female rats. Urine from a pregnant woman containing this hormone will cause rupturing (hemorrhage) of ovarian follicles within 48 hours.

> The name **anterior pituitary-like factor** is, however, inaccurate because the hormone is not effective in hypophysectomized rats whereas the hypophyseal hormones are.

In another pregnancy test known as *Friedman test*, a virgin rabbit is used and the test urine is injected into an ear vein. After 24 hours, the ovaries are examined for ruptured or hemorrhagic follicles.]

4. Somatotropin or somatotrophic hormone, STH or growth hormone, GH. Somatotropin obtained from human hypophysis (Fig. 19–29) is a protein with molecular weight 27,000 (41,600 in pig) and an isoelectric point 4.9. It has 190 amino acids and consists of 2 disulfide bridges between adjacent cysteine residues at 53-164 and 181-188 (**Niall**, 1971 ; **Li**, 1972). The N-terminal and C-terminal residues are both phenylalanine.

Unlike other adenohypophyseal hormones, the various effects of somatotropin are not due to its influence on other endocrine glands. It acts rather directly upon various tissues to produce diverse effects. *It is, therefore, not a true tropic hormone*. The various metabolic activities particularly attributed to this hormone are listed below.

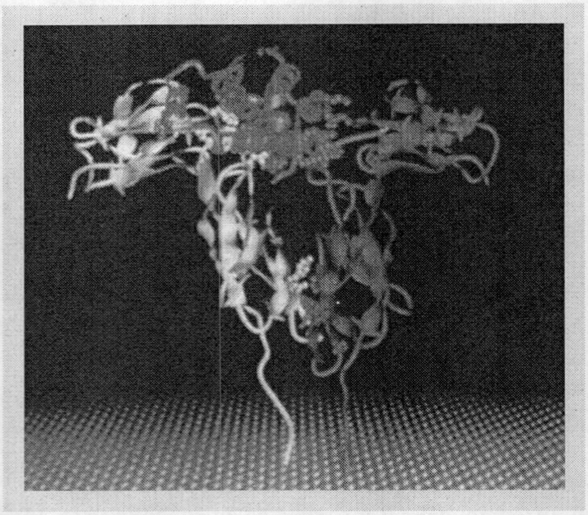

Fig. 19–29. The x-ray structure of human growth hormone (hGH) in complex with two molecules of its receptor's extracellular domain (hGHbp)

The proteins are shown in ribbon form, with the two hGHbp molecules, which together bind one molecule of hGH, *green* and *blue* and with the hGH *red*. The side chains involved in intersubunit interactions are shown in *space-filling form*. The *yellow pebbled surface* represents the cell membrane, through which the C-terminal ends of the hGHbp molecules are shown penetrating as they do in the intact hGH receptor.

[*Courtesy : Abraham de Vos and Anthony Kossiakoff, Genentech Inc., South San Francisco, California.*]

(*a*) It affects the rate of skeletal growth and gain in body weight. In adult animals with closed epiphyses, STH stimulates chondrogenesis followed by ossification.

(*b*) It causes abnormal increase in blood sugar by producing degenerative changes in islets of Langerhans (*diabetogenic effect*).

(*c*) It stimulates the growth of the islets of Langerhans (*pancreatotropic effect*).

(*d*) It controls the production of fat in the body and its deposition in the liver (*ketogenic effect*).

(*e*) It prevents the fall of muscle glycogen in fasting and hypophysectomized animals.

(*f*) It also stimulates milk secretion in cows and also the growth of mammary glands in hypophysectomized rats (*galactopoietic effect*).

(*g*) STH is also known to cause adrenal enlargement (*corticotropic effect*). The adrenal enlargement can be greatly augmented by simultaneous treatment with low doses of thyroxine, thus indicating the participation of thyroid gland along with the hypophysis in this action.

It is now recognized that as people live to their 70s and beyond, the effect of declining hormones may contribute significantly to chronic, debilitating and costly illnesses. Among the well-known **attributes of ageing** that hormone loss may bring about are loss of muscle mass and strength, an increase of body fat, particularly fat around the abdomen, a weakening of the bones, a decline in immune responses and a general loss of energy. In 1990s, attention has now been diverted to the study of growth hormone and other trophic hormones, substances that promote growth or maintenance of tissues. They may, at least, have a promise for halting or reversing degenerative changes in bones, muscles, nerves and cartilage.

Pars Intermedia

In human beings, the intermediate lobe is poorly developed but is much larger and much more functional in some lower animals. It is, however, absent in certain mammals such as whale, Indian elephant and armadillo.

The intermediate lobe secretes **melanocyte** (or melanophore)-**stimulating hormones**, **MSH** or **melanotrophins.** These are also known as **intermedins.** In chicken and whales, where intermediate lobe is absent, MSH fucntion is taken over by adenohypophyseal hormones. In mammals, MSH is a vestigeal hormone. The structure of 2 types of intermedins, α-MSH and β-MSH is given in Fig. 19–30.

Acetyl-Ser. Tyr. Ser. Met. Glu. His. Phe. Arg. Trp. Gly. Lys. Pro. Val amide
1 2 3 4 5 6 7 8 9 10 11 12 13

α–**MSH** (from monkey)

Asp. Glu. Gly. Pro. Tyr. Arg. Met. Glu His. Phe. Arg. Trp. Gly. Ser. Pro. Pro. Lys. Asp
1 2 3 4 5 6 7 8 9 10 11 12 13 14 15 16 17 18

β–**MSH** (from monkey)

Fig. 19–30. Two types of intermedins

[Note that the strucutre of α-MSH is identical with the first 13 residues of ACTH except that in α-MSH terminal serine residue is acetylated.]

α-MSH contains 13 amino acid residues with N-terminal residue acylated and the C-terminal valine is in the amide form. b-MSH has been found to possess 18 amino acid residues with aspartic acid residue at both N- and C-terminals. Thus, in a-type the terminal groups are blocked whereas the β-type has both the end groups free. *The α-MSH has greater biologic activity than β-MSH.*

MSH causes dispersal of black pigment (melanin) in melanophore cells of certain lower vertebrates. This leads to darkening of the skin. This hormone is particularly used in colour adaptation in lower vertebrates. The animal becomes pale on a light background and dark on a dark background. Its function in higher vertebrates including man is not known but it is, however, thought to participate in darkening the areas already pigmented.

Pars Nervosa or Neurohypophysis

The neurohypophysis develops from an outgrowth of the hypothalamus. This explains the presence of glial type cells in the gland. The posterior pituitary secretes 2 hormones : ocytocin and vasopressin. These were separated by **Kamm** and others.

1. Ocytocin (*ocy*^{G} = quick ; tokos^G = birth) or **pitocin.** It is a nonapeptide amide. A disulfide bond (—S—S—) is present to link the two cysteine residues present in the molecule (refer Fig. 9–15). Ocytocin preparations from man, cow and pig are identical. Besides mammals, it is also found in chondrichthyes, lung fishes, amphibians and aves.

Ocytocin stimulates the contraction of smooth muscles, *esp.*, those of uterus, thus facilitating childbirth. Commercial form of ocytocin is frequently used to induce 'labour'. In general, it also causes contraction of other smooth muscles like those of intestine, urinary bladder and the ducts of mammary glands resulting in milk ejection. It is, therefore, also called as *milk-let-down-ejection factor*. Ocytocin levels are increased by suckling which is necessary for the continued formation of milk by the breasts. Optimum milk secretion lasts for about 8 to 10 months after which it gradually falls and ultimately ceases.

> **Ocytocin** is often called oxytocin. However, the name ocytocin is preferred since its Greek origin (meaning *quick birth*) tallies with it and also it avoids confusion with the prefix ***oxy*** = oxygen-containing, as the hormone refers in no way to oxygen.

2. Vasopressin or **pitressin.** It is also a cyclic nonapeptide amide and resembles ocytocin except that isoleucine is replaced by phenylalanine and leucine by arginine (Fig. 19–31). The hormone was synthesized, in 1953, by **Vincent du Vigneaud** and colleagues from U.S.A. and **du Vigneaud** received the Nobel Prize in 1955 for the first synthesis of a polypeptide hormone.

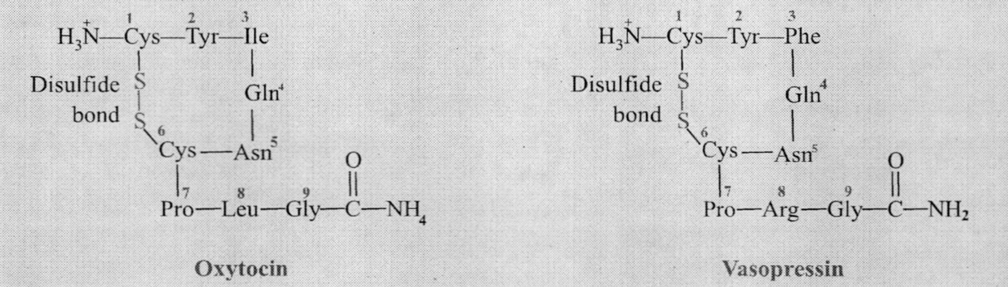

Fig. 19–31. Structure of the two neurohyphyseal hormones

Vasopressin (VP), obtained from pig, differs from that of human beings (shown above) in arginine being replaced by lysine at position 8. It is, therefore, better called as *lysine vasopressin* and that from human beings as *vasopressin.*

Vasopressin regulates many functions :

I. Circulatory or pressor action. Vasopressin causes a rise in blood pressure by contraction of peripheral blood vessels. It is, thus, a vasopressor substance and for this reason it is termed 'vasopressin'. It has been used in surgical shock as an adjuvant in elevating blood pressure. It may also be used at the time of delivery to overcome uterine inertia. The effect is similar to that caused by adrenalin. However, the rise in blood pressure caused by vasopressin lasts much longer than that caused by adrenalin which wears off in a few minutes.

II. Antidiuretic action. It brings about a reduction in the urine volume by causing renal tubules to withhold more water. Consequently, the urine passed is rich in sodium chloride, phosphate and total nitrogen. It was earlier thought that this antidiuretic effect was due to a different hormone called *antidiuretic hormone*, *ADH*. But it is now established that the two hormones (vasopressin and ADH) are one and the same. Vasopressin, therefore, finds use against persons suffering from **diabetes insipidus**, a disease characterized by excretion of large quantities of urine (polyuria) and a marked thirst (polydipsia).

> The term **'insipidus'** (= tasteless) dates from the time when the only method of testing urine was to taste it after diluting, and in this condition the urine is tasteless.

Table 19–6. Summarizes the biological effects of the two neurohypophyseal hormones.

Table 19–6. Comparison of the biological effects of ocytocin and vasopressin

S. N.	*Biological Effect*	*Ocytocin*	*Vasopressin*
1.	Water diuresis	no effect	inhibits
2.	Blood pressure	slightly lowers	raises
3.	Coronary arteries	slightly dilates	constricts
4.	Intestinal contractions	?	stimulates
5.	Uterine contractions	stimulates	stimulates
6.	Ejection of milk	stimulates	slightly stimulates

Hypopituitarism. Insufficient secretion of pituitary hormones (or hypopituitarism) may occur as a result of pituitary tumours or atrophy of the gonad. Hypoactivity of this gland may lead to the following disorders :

1. **Dwarfism.** It refers to the arrested growth of the individuals (Fig 19–32). It is of 2 types :

(*a*) Lorain type— short statured individuals with a head large in comparison to the rest of the body ; usually intelligent but unattractive.

(*b*) Fröhlich type— obesity and arrested sexual development ; mentally below normal and usually lethargic.

2. **Panhypopituitarism.** It is caused because of the destruction of the gland, thus leading to cessation of all hypophyseal functions.

3. **Pituitary myxedema.** It is caused due to the lack of TSH and produces symptoms similar to those described for primary hypothyroidism (see page 382).

Fig. 19–32. Two physiologic conditions arising from abnormal pituitary secretion

Here we see dwarfism and gigantism both.

Hyperpituitarism. It refers to the overproduction of hypophyseal hormones. Hypersecretion of this gland leads to **gigantism** during childhood or adolescence, *i.e.*, before closure of the epiphyses. The disease is characterized by overgrowth of the bones, especially at joints, leading to a tall individual with 2 to 2.5 M height. The limbs usually become disproportionately large.

> **Obesity** is "that state in which the accumulation of reserve fat becomes so extreme that the functions of the organism are interfered with". It can be quantitated by measuring skinfold thickness with calipers. Obesity may arise from either overeating or diminished utilization or a combination of both. Obesity results from a low basal metabolic rate (BMR) in obese and therefore, even with a normal intake of food, an excess of calorie is available. The BMR diminishes slightly as the individuals grow older but often the food consumption is not decreased proportionally. As a consequence, obesity may result. Obesity is usually due to an excess intake of food. Appetite may be influenced by a variety of factors such as psychologic disturbances; hypothalamic, pituitary, or other brain lesions; and hyperinsulinism. Obesity may also develop from an increase in number or in size of fat cells, adipocytes. Adipocytes increase in number when caloric intake is increased, especially in the gestational months and during the first year of life. The obese may become resistant to insulin, resulting in an increase in the levels of circulating insulin. Insulin decreases lipolysis and increases fat synthesis and uptake.
>
> Obesity may become evident at any stage, but it appears most frequently in the first year of life, at 5 to 6 years of age, and during adolescence. The adiposity in the mammary regions of **boys** is often indicative of breast development and hence an embarrassing feature. The abdomen tends to be pendulous, and white or purple striae are often present. The external genitalia of boys appear disproportionately small but actually are most often of average size; the penis is often embedded in the pubic fat. Puberty may occur early, with the result that the ultimate height of the obese may be less than that of their slower-maturing peers. The development of external genitalia is normal in the majority of **girls,** and menarche is usually not delayed. The obesity of the extremities is usually greater in the upper arm and thigh and is sometimes limited to them. The hands may be relatively small and the fingers tapering. Obesity remains an enigma today as it has for many centuries – the earliest recorded case of human obesity dating back to the Willendorf Stone Age Venus of about 22,000 B.C.

In human adults with closed epiphyses, excessive secretion of pituitary hormones causes **acromegaly** (*acron*[G] = extremity). The chief symptoms of this disease are (Fig. 19–33) :

1. consistent overgrowth of the bones of face, hands and feet (*acral* parts ; hence the term acromegaly) so that the patient often complains of having have to change globes and shoes frequently as they no longer fit.
2. protrusion of the lower jaw (prognathism). Overgrowth of the malar, frontal, and basal bones combines with prognathism to produce the coarse facial features called *acromegalic facies.*

3. bowing of the spine (kyphosis)
4. overgrowth of the body hair
5. thickening of the soft tissues of nose, lips and forehead
6. enlargement of the visceral organs such as lungs, heart, liver and spleen
7. increased sexual activity in the beginning which is ultimately followed by atrophy of the gonads. This leads to impotence in man and amenorrhea in woman.

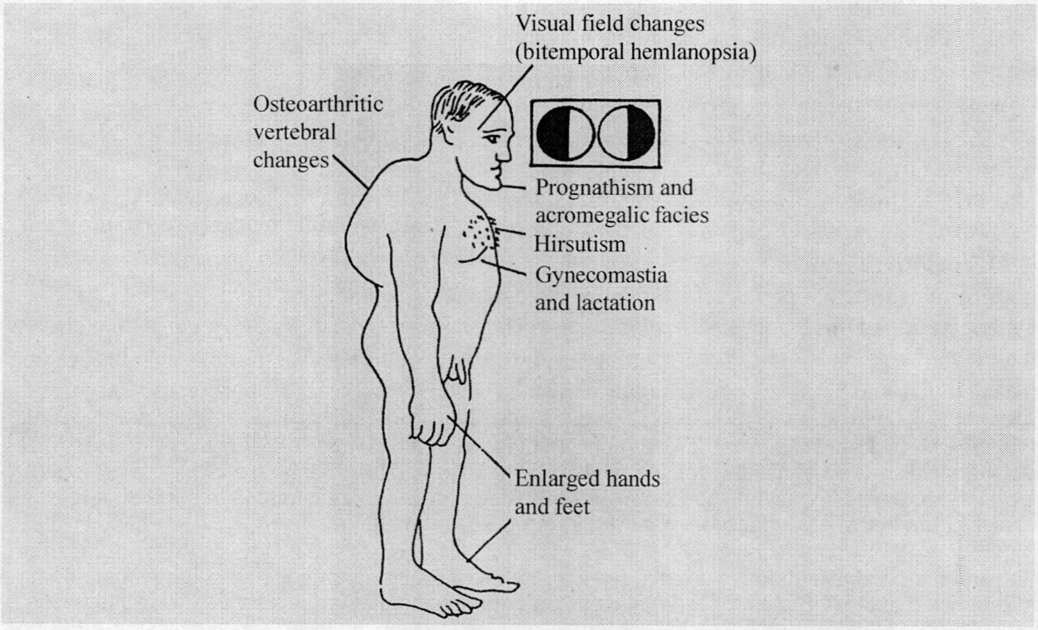

Fig. 19–33. Typical finding in a acromegaly

3. Hormone of the Parathyroid

Secretory gland. The parathyroids were first discovered by Sandström in 1880. In the human beings, there are usually four small parathyroid glands so closely associated with the thyroid gland that they remained undiscovered for some time. Each parathyroid gland is a reddish or brownish, oval body, measuring about 5—7 mm in length, 3—5 mm in width and 1—2 mm in thickness. The four glands together weigh only 0.1 to 0.2 gm. The glands have a macroscopic appearance of dark brown fat, therefore these are difficult to locate and hence often removed during thyroidectomy (removal of the thyroid). Of the four parathyroids, two lie embedded in the thyroid and are called as *internal parathyroids* ; the other two lie close and behind the thyroid and are known as *external parathyroids.* There may be fewer than four or as many as eight parathyroids. The extra ones are scattered along the trachea and are called as accessory parathyroids. Occasionally, the parathyroids are located in the anterior mediastinum or, rarely, in the posterior mediastinum. The parathyroids are, however, lacking in the fishes.

Histologically, the parathyroid of the adult human being consists mainly of chief cells (= principal cells) and oxyphil cells (refer Fig. 19–34). The oxyphil cells are usually absent in young human beings. The chief cells are concerned with the secretion of the parathyroid hormone. The oxyphil cells are rich in mitochondria but lack glycogen. Their function is uncertain. They are regarded as probably aged chief cells that still secrete some hormones.

Structure. Parathyroids secrete a hormone called parathyroid hormone (*parathormone, PTH*), which is also known as **Collip's hormone.** PTH has been isolated in pure form. It is a linear polypeptide consisting of 84 amino acid residues and has a molecular weight of about 7,000 to 8,500. **Potts** and others (1971) have, however, indicated that the physiologic activity of this hormone on both skeletal and renal tissues is contained within the first 34 amino acids from the N-terminal of the chain. As is also true for many other hormones (such as α-MSH and corticotropin), the PTH can be cleaved to form smaller but still active units. Oxidation inactivates the parathormone rapidly.

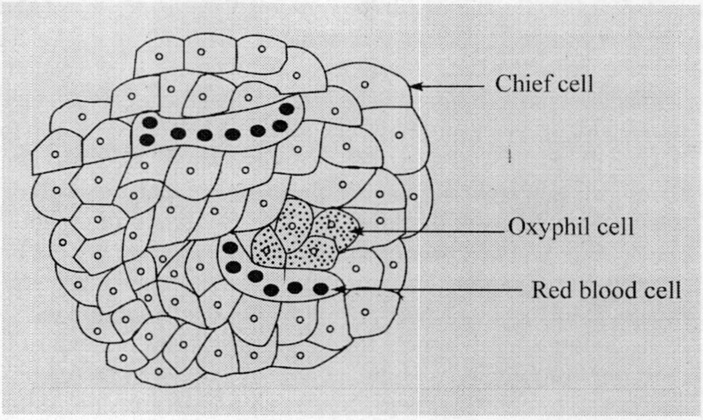

Fig. 19–34. Macroscopic structure of a parathyroid gland

Functions. The principal sites of parathyroid action are bones, kidney and gastrointestinal tract. Following physiological functions are attributed to this hormone :

1. *Bone resorption*— It exerts a direct influence on the metabolism of bone, leading to an increased release of bone Ca^{2+} into the blood. The exact mechanism behind this phenomenon is not truly known. It has, however, been suggested that the hormone stimulates the production of citric acid in the bone tissues and an increased concentration of citrate ions leads to the removal of phosphate from calcium phosphate, the bone material. The bone is, thus, made soluble.

2. *Renal reabsorption of calcium*— In the kidney, parathormone affects renal tubular reabsorption of calcium and reabsorption or secretion of phosphate. It increases the elimination of calcium and phosphorus in the urine.

It is interesting to note that the secretion of this hormone is controlled by Ca^{2+} ion concentration of the blood itself. As the Ca^{2+} ion concentration increases, PTH secretion decreases tending to preserve the original condition. This affords an excellent example of feedback mechanism of metabolic control.

3. *Increase in osteoelastic activity*— Parathormone increases osteoelastic activity with augmented growth of the connective tissue.

4. *Calcium homeostasis*— An optimum Ca^{2+} concentration is necessary for various functions of the body, *viz.*, normal transmission of impulses, the contraction of the muscles, the formation of the bones, the coagulation of the blood etc.

Hypoparathyroidism. Undersecretion of PTH causes a decrease in Ca contents of the blood from the normal 10 mg per 100 ml to 7 mg per 100 ml (hypocalcemia) which leads to excessive contraction of the muscles (convulsions). In fact, convulsions occur when the calcium is further decreased to 4 mg per 100 ml of plasma. As the calcium decreases in the blood, there is a decrease in the urine. However, during this period, the phosphorus in plasma increases from a normal 5 mg per 100 ml to 9 mg per 100 ml and even higher (hyperphosphotemia). These changes develop into a fatal disease called **muscular twitchings** or **tetany**. It is characterized by locking up of the jaw, rapid breathing, increased heart beat, rise in temperature and ultimately death due to asphyxia. The signs of tetany in man include

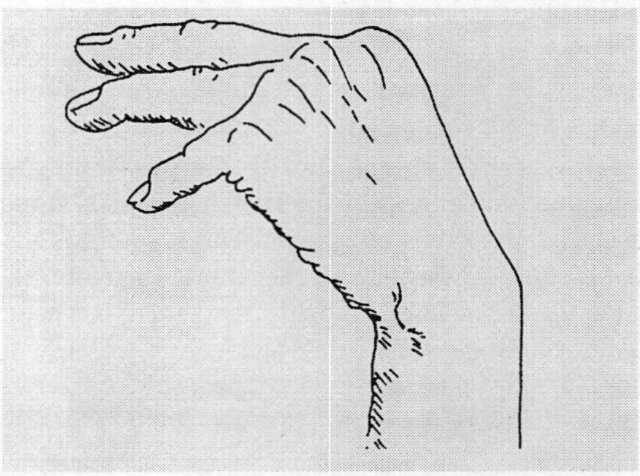
Fig. 19–35. Position of the hand in hypocalcemic (Trousseau's sign)

(*After Aub. Drawn from a photograph in Adams : Physical Diagnosis, 14th eddition, Williams and Wilkins, 1958*)

CLINICAL IMPLICATIONS

Neurofibromatosis is one of the most common autosomal dominant genetic disorders. It occurs roughly in 1 of 4,000 live births. It is seen equally in every racial and ethnic group throughout the world. There is virtually a complete dominant penetrance of the gene localized at the centromeric region of chromosome 17. The gene responsible for the disease, isolated in 1990, is huge and actually includes 3 smaller genes. This is only the second time that *nested genes* have been found in humans. This is a tumour-suppressor gene active in controlling cell division. When it mutates, a benign tumour develops. At birth or later, the affected individual may have 6 or more large tan spots on the skin. Such spots may grow in size and number and become darker. Small benign tumours (or lumps), called *neurofibromas,* may occur under the skin or elsewhere. Neurofibromas are made up of nerve cells and other cell types. This genetic disorder shows *variable expressivity*. In most cases, the symptoms are mild and patients live a normal life. In some cases, however, the effects are severe. Skeletal deformities, including a large head, become evident; eye and ear tumours can lead to blindness and hearing loss. Many children with neurofibromatosis have learning disabilities and are hyperactive. In the most extreme form of neurofibromatosis, the connective tissue of the bones and the lining of the nerves produce disfiguring tumorous growths. A devastating expression of the gene occurred in Joseph Merrick, a severely-deformed late 19th century Londoner, whose tragic life became the subject of a contemporary play and movie, *The Elephant Man*. However, researchers today believe that Merrick actually suffered from a much rare disorder called *Proteus syndrome*.

Chvostek's sign, a quick contraction of the ipsilateral facial muscles elicited by tapping over the facial nerve at the angle of the jaw; and **Trousseau's sign**, a spasm of the muscles of the upper extremity that causes flexion of the wrist and thumb with extension of the fingers (Fig. 19–35). Tetany can be relieved either by the administration of a soluble calcium salt or of PTH.

Hyperparathyroidism. An increase in PTH production is usually due to a tumour of the gland (parathyroid adenoma). Oversecretion of PTH in man results in a cystic bone disease variously called as **osteitis fibrosa cystica** or **von Recklinghausen's disease** or **neurofibromatosis.** It is an autosomal dominant condition. The disease is characterized by increased calcium contents of the blood (hypercalcemia) usually up to 20 mg calcium per 100 ml palsma, decreased phosphate concentration and increased renal excretion of calcium. Overproduction of parathormone causes calcium and phosphorus to move out of the bones and teeth, making them soft and fragile. Such patients, therefore, suffer fractures of the bones very frequently. Cysts in the bones are another characteristic of this disease.

4. Hormones of the Gastrointestinal Tract

The mucosal cells of the gastrointestinal tract, GIT secrete a number of hormones. A few of them, whose structure has been studied and are found to be proteinaceous in nature, are described below.

1. Gastrin. It is produced by the pyloric mucosa, which is apparently stimulated by the proteins present in food or possibly by HCl. Mechanical stimulation caused by distention of the stomach also results in the secretion of gastrin. This hormone is absorbed into the blood stream and is carried to the fundic cells, causing them to secrete HCl actively. The hormone stimulates the gastric glands to pour out more gastric juice. Gastrin (Fig. 19–36) is a heptadecapeptide with a molecular weight 2,100. It dialyzes through a membrane and has an isolectric point of 5.5.

```
1                                      10                               17
Glu.Gly.Pro.Trp.Leu.Glu.Glu.Glu.Glu.Glu.Ala.Tyr.Gly.Trp.Met.Asp.Phe–NH2
                                                |
                                               HSO3
```

Fig. 19–36. Gastrin

2. Secretin. *It is the first compound to be designated as hormone*. It is formed by the upper intestinal mucosa and is liberated by HCl present in the acid chyme (semifluid material from the stomach). It is carried by the blood stream to the pancreas which it stimulates. Secretin, thus, causes the flow of pancreatic juice rich in bicarbonate. This occurs even if the nerves supplying the pancreas are cut. Hence, it is a true hormonal action. Secretin also stimulates the flow of bile and intestinal juices, although to a lesser extent. It has been obtained in crystalline form. Secretin (Fig. 19–37) is a polypeptide of 27 amino acid residues (14 of which are identical to those found in glucagon) with molecular weight of 3,056. The molecule has no structural homology with gastrin or cholecystokinin.

His(1).Ser.Asp.Gly.Thr.Phe.Thr.Ser.Glu.Leu(10).Ser.Arg.Leu.Arg.Asp.Ser.Ala.Arg.Leu.Gln(20).
|
Arg.Leu.Leu.Gln.Gly.Leu.Val(27)—NH_2

Fig. 19–37. Secretin

3. Cholecystokinin, CCK. It is also secreted by the upper part of the small intestine. It sitmulates contraction of the gall bladder so as to release its contents into the duodenum. Cholecystokinin (Fig. 19–38) is a polypeptide with 33 amino acid residues and a molecular weight of 3,883. It is, however, noteworthy that the last 5 amino acids towards the C-terminal in the gastrin and cholecystokinin are exactly the same. It is in this terminal portion of these hormones that the principal activity resides.

Lys(1).(Ala, Gly, Pro, Ser).Arg.Val.(Ile, Met, Ser). Lys.Asn.(Asn, Gln, His, Leu_2, Pro, Ser_2).

Arg.Ile.(Asp, Ser).Arg.Asp.Tyr.Met.Gly.Trp(30).Met.Asp.Phe(33)–NH_2
|
HSO_3 (on Tyr)

Fig. 19–38. Cholecystokinin

4.. Pancreozymin, PZ. It is another polypeptide hormone elaborated by the intestinal mucosa. Along with secretin, it functions to stimulate the release of pancreatic juice, which is rich in enzymes or their zymogens as well as in bicarbonate. The enzymes or zymogens include trypsinogen, lipase, carboxypeptidase etc. The release of pancreozymin is believed to be brought about in the presence of any one of a variety of compounds such as peptone, casein, dextrin, maltose, lactose, saline water and also even distilled water. Pancreozymin has been obtained in pure form. It is thermostable and is not destroyed by acid but it is destroyed by alkali. Like cholecystokinin, it also contains 33 amino acid residues. The 5 C-terminal amino acid residues are the same as those of gastrin and cholecystokinin. Jorpes (1968) believes that the two gastrointestinal hormones, cholecystokinin and pancreozymin are identical. They are now thought to be a single factor and represented as CCK-PZ.

5.. Enterogastrone. It is present in duodenal mucosa and is secreted from it only when fat (derived from food) reaches the duodenum. The hormone reaches the stomach where it inhibits gastric secretion and motility of the stomach. There is secretion of lesser volume of juice with a lower concentration of HCl and a smaller amount of pepsin. The effect is to permit digestion to be more completely accomplished.

6. Enterokrinin. It is also isolated from intestinal mucosa and controls the intestinal secretion. It does not influence pancreatic secretion but may increase the volume as well as the enzyme concentrations of succus entericus. It is distinct from secretin in that it stimulates the secretion of both fluid and enzymes by the intestinal mucosa. It is also proteinaceous in nature.

Although evidences exist to indicate the presence of all the humoral agents mentioned above, yet gastrin, secretin and cholecystokinin are the only three gastrointestinal hormones which have been fully established with regard to their biochemical and physiological behaviour.

AMINO ACID HORMONES

1. Thyroidal Hormones

Secretory gland. *The thyroid is the largest endocrine gland in the body.* It was first described by **Whartonin** in 1659 who gave it the descriptive name, thyroid because of its resemblance to a shield (*thyreoides*[G] = shield-shaped). In man, the gland consists of two lobes on either side of and anterior to the trachea just below the larynx. The two lobes are connected across the ventral surface of the trachea with a narrow bridge called isthmus, making the entire gland more or less H-shaped in appearance (refer Fig. 19–39). The isthmus crosses infront of the 2nd, 3rd and 4th tracheal ring. In the adult, the gland weighs about 25 to 30 gm. The thyroid gland receives blood flow about 5 times its own weight per minute. It has a blood supply as rich as that of any other area of the body barring probably the adrenal cortex. The thyroid is presumed to be homologous with the endostyle of the early vertebrates.

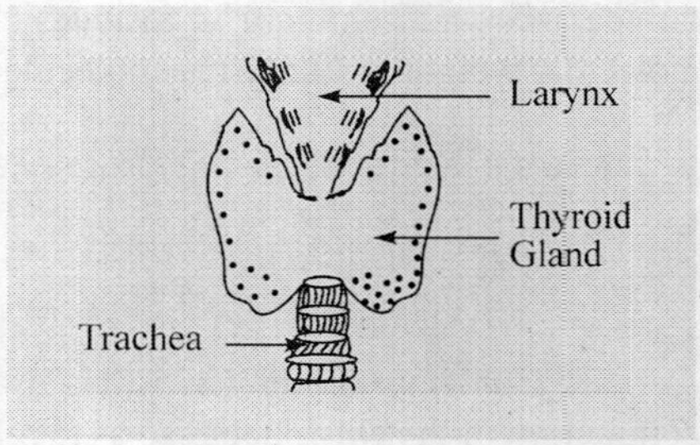

Fig. 19–39. The thyroid in man
(After D. Marsland, 1964)

Histologically, the thyroid gland (refer Fig. 19–40) is composed of a large number of tiny closed vesicles called *follicles*, 150 to 300 microns in diameter. The follicles are held together by areolar tissue and are surrounded by a rich network of capillaries. Each follicle is lined with a single peripheral layer of columnar or cuboidal *epithelioid cells* that secrete into the interior of the cells. Its lumen is filled with a secretory substance called colloid. The major constituent of colloid is a large protein called thyroglobulin, which contains the thyroid hormones. Once the secretion has entered the follicles, it must be absorbed back through the follicular epithelium into the blood before it can perform its function in the body. The thyroid gland is, thus, unique amongst the endocrine glands in that *it stores its hormone as a colloid in small vesicles in the gland.* The other endocrine glands, however, store their hormones in the cells themselves.

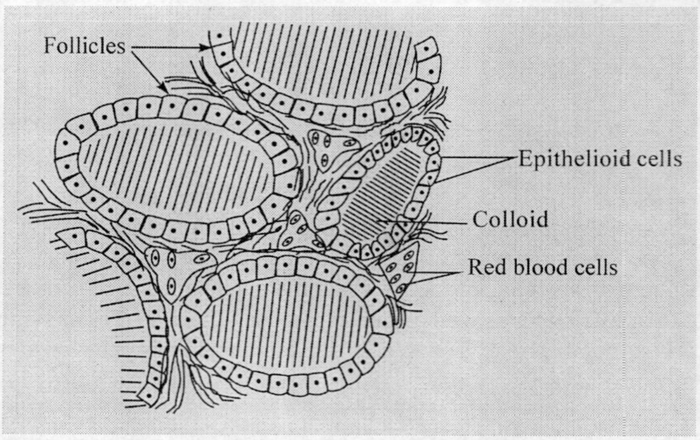

Fig. 19–40. Histologic structure of thyroid gland

As age advances, the thyroid activity tapers off. That's why elderly persons feel colder, since their body does not produce sufficient heat.

Structure. Thyroid contains large amounts of elemental iodine which is bound to a protein named iodothyroglobulin or simply thyroglobulin. It is a glycoprotein with a molecular weight of about 650,000 and iodine content from 0.5 to 1.0%. This protein represents the storage form of the hormone in the gland.

Evidences available at present indicate that thyroglobulin is hydrolyzed, in the presence of thyrotropin, to release thyroxine (= 3, 5, 3′,5′-tetraiodothyronine) in the blood (Fig. 19–41). The release of thyrotropin is, in turn, controlled by the level of thyroxine in the blood.

Thyroxine is one of the earliest recognized hormones. It was so named and isolated first by **Kendell** in 1915. **Harington** and **Barger** (1925) established its chemical formula. It is an iodine-containing aromatic amino acid and closely resembles tyrosine in structure. Diiodotyrosine is believed to be the precursor of thyroxine.

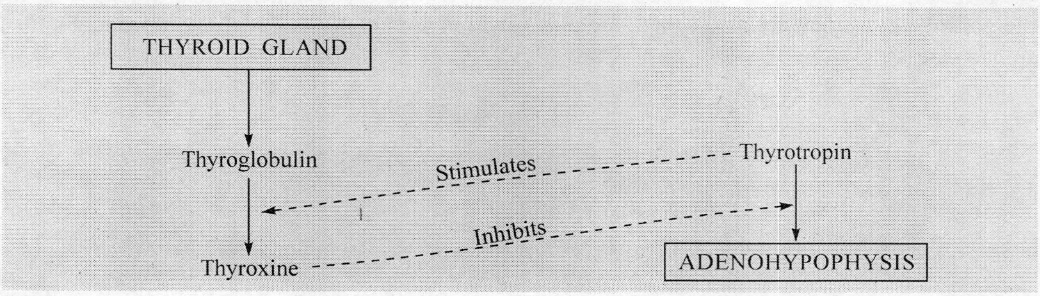

Fig. 19–41. Interdependence of throxine and thyrotropin

Besides thyroxine, 3,5,3′-triodothyronine is also produced from enzymatic hydrolysis of thyroglobulin. It is 5 to 10 times more potent in biologic activity than thyroxine. This may, possibly, be due to the fact that triodothyronine is bound loosely by serum proteins and hence diffuses much more rapidly into the tissues. It is present in the blood in much smaller quantities and persists for a much shorter time than does thyroxine. The structure of triiodothyronine and tetraiodothyronine is given in Fig. 19–42.

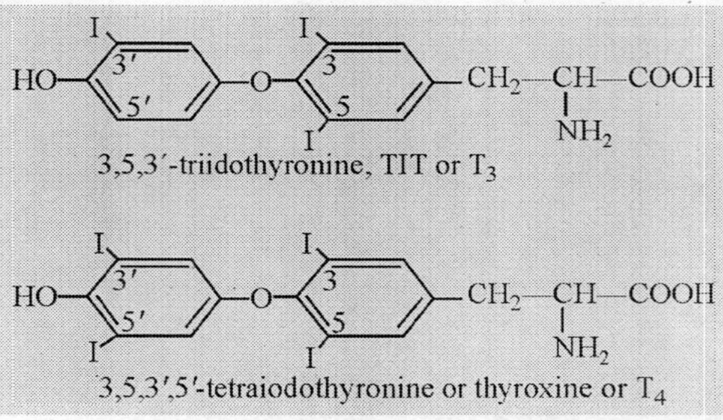

Fig. 19–42. Thyroidal hormone

Functions. Thyroid hormones have widespread effects on the ossification of cartilage, the growth of teeth, the contours of the face and the proportions of the body. They also carry out following functions :

1. They bring about deamination reactions in the liver.
2. They also carry on deiodination in the extrahepatic tissues.
3. These influence oxidative phosphorylation by altering the permeability of the mitochondrial membrane.
4. Their presence accelerates metamorphosis in amphibians. This is so a sensitive test that tadpole has been widely used for the assay of the potency of these hormones.
5. These may increase the level of cytochrome *c* in the tissues.

It is, thus, apparent that these hormones affect the general metabolism, regardless of the nature of its specific activity. It is for this reason that the thyroid gland has rightly been called as the '*pace setter*' of the endocrine system.

Thyroid inhibitors or **goitrogens.** Certain substances act as antithyroid agents by inhibiting the production of thyroxine. This is explained by the fact that these compounds prevent the oxidation of iodide, which means that the iodine cannot be used for the synthesis of thyroxine. Some important thyroid inhibitors (Fig. 19–43) are 6-propyl-2-thiouracil, 2-thiouracil, thiourea and the sulfonamides. Thiouracil itself is relatively toxic ; less so are propylthiouracil and thiourea. Of these, propylthiouracil has been used most extensively. It is 3 to 5 times more active than thiouracil. The antithyroid drugs or the goitrogens, as they are also called, are also found in common foods such as cabbage, turnips, spinach, peas etc.

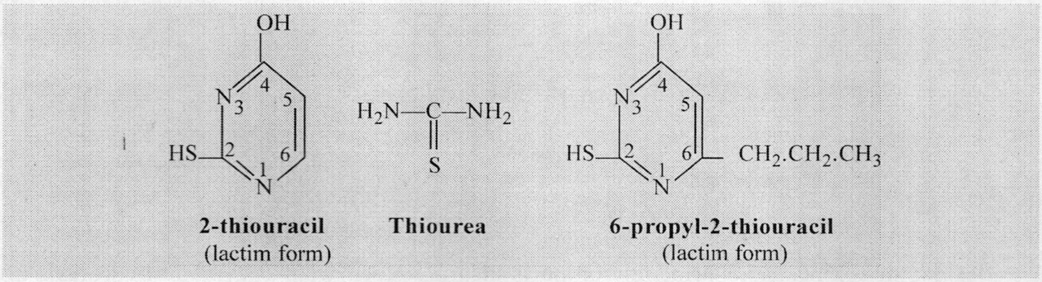

Fig. 19–43. Some thyroid inhibitors

Hypothyroidism. Underactivity of the thyroid may result from two causes : a degeneration of thyroid cells or a lack of sufficient iodide in the diet. The disease that results from thyroid cell degeneration is cretinism in children and myxedema in adults. **Cretinism** (=feeble-mindedness) is characterized by dwarfism and mental suppression. The cretinic children have infantile features (Fig. 19–44). They possess a large head and an apathetic face ; their teeth erupt late and the speech is retarded. Early treatment with thyroid-active compounds partially prevents this disease.

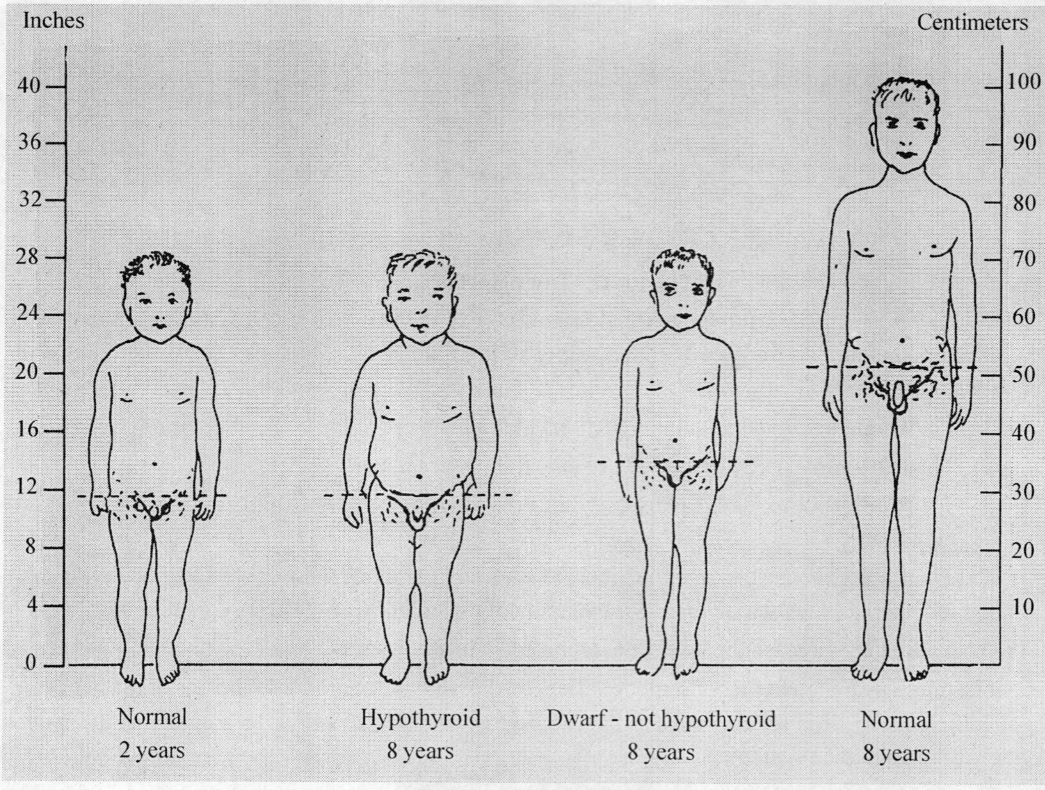

Fig. 19–44. Normal and abnormal growth

Hypothyroid dwarfs retain their infantile proportions, whereas dwarfs of the constitutional type and of the hypopituitary type have proportions characteristics of their chronologic age.

(*Adapted from Wilkins : The Diagnosis and Treatment of Endocrine Disorders in Childhood and Adolescence, 34d ed., Thomas. 1966*)

Myxedema is characterized by an abnormally low basal metabolic rate (BMR). In it, the adults become mentally lethargic and possess thick puffy skin (edema) and dry hair. The patient shows bagginess under the eyes and swelling of the face. The hair thin on the eyebrows and scalp. As there is deposition of semi-fluid material under the skin, the name myxedema (*myxa*^{G} = mucus; *oidema*^{G} = swelling) is given to this condition. Myxedema also responds well to administration of thyroid-active compounds. *Myxedema is less severe than cretinism.*

Lack of sufficient iodide in the diet results in thyroid gland enlargement, known as **simple goiter.** It is also associated with a low BMR. This type of goiter is also known as **endemic goiter**, since it is prevalent in areas where the soil and drinking water lack iodide. Simple goiter was once fairly common in some mountainous parts of Switzerland and the United States, where soil and water are deficient in iodine compounds.

Hyperthyroidism or **thyrotoxicosis.** Abnormally high activity of this gland may occur due to either oversecretion of the gland or an increase in size of the gland. Swelling of the gland results in an **exophthalmic goiter** (Fig. 19–45), characterized by protrusion of the eye balls. In it, the BMR increases considerably above the normal figures ; 80% above normal is not unusual. Consequently, appetite is increased in hyperthyroid individuals. In spite of this, they lose weight and often feel hot because of the increased heat production. Their pulse rate is also increased and excessive sweating occurs. Other symptoms of this disease are dilated pupils, mental excitement, irritability and cardiac dilatation. Hyperthyroid individuals are, in general, characterized by an above-normal rate of many physiological activities. The clinical syndrome is generally termed **Graves' disease**, after its discoverer **Robert James Graves. Basedow's disease** and **thyrotoxic exophthalmos** are other names of this disease. Not surprisingly, people with hyperactive thyroids (*i.e.*, with Graves' disease), thus, show many symptoms opposite those in hypothyroidism.

Hyperthyroidism can be cured by surgical removal of the thyroid (thyroidectomy), treatment with x-rays, injection of radioactive iodide (^{119}I) or by treating with antithyroid drugs or with agents like thiocyanate or perchlorate which compete with iodide for the uptake mechanism. Propylthiouracil is being particularly used against the Graves' disease.

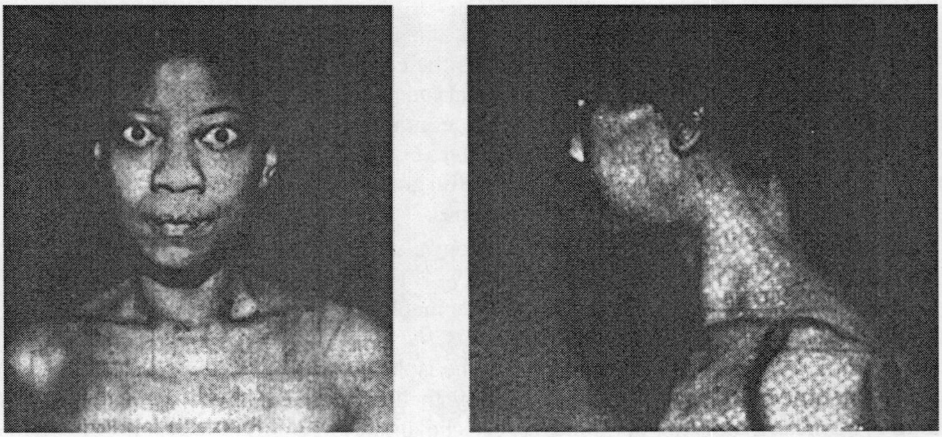

Fig. 19–45. Graves' disease

Note the goitre and exophthalmos.

(*Courtesy of PH Forsham*)

[Thyrocalcitonin or Calcitonin

A quite different type of hormone which has effects of decreasing calcium ion concentration in the blood has been discovered by Copp et al in 1961 as an impurity in the commercial parathyroid extracts. They termed this hypocalcemic factor as calcitonin (CT). *Later researches revealed that this factor originated in the thyroid, rather than in the parathyroids.* This substance was, henceforth, also named as thyrocalcitonin (TCT). It is secreted by the parafollicular cells or C-cells in the interstitial tissue between the follicles of the human thyroid gland. Later, it was discovered that CT is secreted by the ultimobranchial glands of fishes, amphibians, reptiles and birds. In mammals, however, these glands do not exist as such but have become incorporated into either the parathyroids or the thyroid. The parafollicular cells of human thyroid glands are remnants of the ultimobranchial glands of lower animals.

Thyrocalcitonin (Fig. 19–46) is a large, unbranched polypeptide containing 32 amino acids and has a molecular weight of about 3,600. It is unique in having no isoleucine and lysine residues. Its amino acid sequence has been determined by **Foster** in 1968.

Thyrocalcitonin acts by causing a transfer of calcium from blood into bone either by increasing calcification of the bones or by diminishing decalcification or by both processes. In other words, it

```
S———————————————————————S
|                       |
Cys—Gly—Asn—Leu—Ser—Thr—Cys—Met—Leu—Gly—Thr—Tyr—Thr—Gln—Asp—
1                       7           10
Phe—Asn—Lys—Phe—His—Thr—Phe—Pro—Glu—Thr—Ala—Ile—Gly—Val—Gly—
                20                                  30
Ala—Pro—NH2
    32
```

Fig. 19–46. Human calcitonin

Note the presence of an intrachain disulfide bond between two cysteinyl residues placed at positions No. 1 and No. 7.

rapidly inhibits calcium withdrawal from bones. This characteristic property, however, promises it to be a therapeutic agent for the treatment of certain types of bone diseases. This action of thyrocalcitonin is counterbalanced by the hypercalcemic hormone, the parathormone which is secreted by the parathyroids. *TCT, thus, aids in the maintenance of calcium homeostasis.*]

2. Adrenal Medullary Hormones

As already stated, the adrenal medulla forms the central core of adrenal gland and originates from the neural canal. It is composed of densely packed polyhedral cells containing chromaffin granules. It is highly vascular and receives 6—7 ml of blood per gram of tissue per minute. The chromaffin granules store large quantities of adrenal medullary hormones.

> **Adrenalin** and **noradrenalin** are also spelt as adrenaline and noradrenaline, respectively (**J.H. Green**, 1978).

Structure. Adrenal medulla, whose secretion is under nervous control, produces two hormones (Fig. 19–47) : (*a*) epinephrine or adrenalin ($C_9H_{13}O_3N$) and (*b*) norepinephrine or noradrenalin ($C_8H_{11}O_3$ N). *Epinephrine was the first hormone to be isolated in the crystalline form.* The isolation was, however, done by **Abel**. It has been produced synthetically by **Stoltz**. Chemically, these two hormones are catecholamines (= dihydroxyphenylamines) and are closely related to tyrosine and phenylalanine. Norepinephrine†, however, differs from epinephrine structurally in having a hydrogen atom in place of the methyl group. The commonly available epinephrine, therefore, is a mixture of these two hormones (usually 10—20% norepinephrine) and its effects are a resultant of the combined actions of these two hormones.

> The word '**nor**' in norepinephrine (and noradrenalin) was originally coined to indicate nitrogen (**N**) without (**O**-Ohne) a radical (**R**), in this case a methyl group.

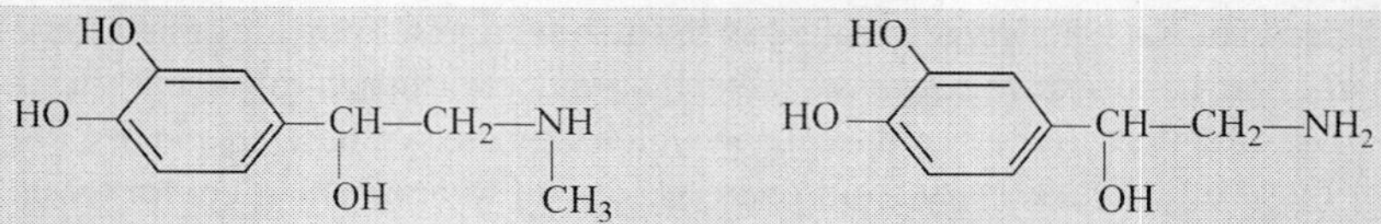

Epinephrine (Adrenalin) **Norepinephrine** (Noradrenalin or arterenol)

Fig. 19–47. Adrenal medullary hormones

Since these hormones possess an asymmetric carbon atom, two stereoisomers are possible for each one of them. The naturally-occurring epinephrine is the L-isomer and is levorotatory. It is 15 times more active than the D-form. Similarly, the natural D (–) form of norepinephrine is about 20 times more potent than the unnatural isomer. Their relative proportion in the adrenal medulla differs from species to species (refer Table 19–7).

Table 19–7. Percentage composition of the two adrenal medullary hormones in various organisms

S.N.	*Organism*	*Epinephrine (Adrenalin)*	*Norepinephrine (Noradrenalin)*
1.	Man	75-90	25-10
2.	Cat	50	50
3.	Whale	10	90

Functions. In general, the adrenal medullary hormones reinforce the functions performed by the sympathetic nervous system. Although both these hormones exert similar effects in regulating carbohydrate metabolism and blood pressure, yet epinephrine is more closely related to carbohydrate metabolism and norepinephrine to blood pressure.

Epinephrine conducts a wide variety of functions, which are as follows :

1. It promotes glycogenolysis in muscles and liver, resulting in an increase of blood glucose level and an increased lactic acid formation in muscles. These changes are then followed by an increase in oxygen consumption.
2. It causes an increase in blood pressure because of arteriolar vasoconstriction of the skin and splanchnic vessels.
3. It brings about an increase in the heart rate and in the cardiac output.
4. It causes dilation of vessels (= vasodilation) of skeletal muscles, corona and the viscera. This results in an increase of blood flow in these areas.
5. It relaxes the muscles of gastrointestinal tract and bronchials of the lungs but causes contraction of the pyloric and ileocecal sphincter muscles.
6. It also serves in cases of emergency. Under emotional stress, fear or anger, it is secreted in the blood stream and the blood is shifted from the viscera to the brain and the muscles so that the individual becomes ready for fight. It is for this reason that the adrenals are frequently referred to as the '*emergency glands*' or the '*glands of flight*, fright and fight' and the two adrenal medullary hormones as **'emergency hormones'**.

Norepinephrine, on the other hand, does not relax bronchiolar muscles and has little effect on cardiac output. It augments both systolic and diastolic blood pressure.

PARAHORMONES OR TISSUE HORMONES

In addition to the hormones discussed in the preceding pages, a number of others have been found to possess hormonal properties. These compounds are effective at low concentrations; unlike global hormones, they are not transported between tissues in the blood, but act on the tissue in which they are produced. They are, hence, also called as **tissue hormones**. They are at best local hormones because they are short-lived and are *paracrine* rather than endocrine in nature, but they fit the definition of hormone. They alter the activities of the cells in which they are synthesized and of adjoining cells. The nature of these effects may vary from one type of cell to another, in contrast with the more uniform actions of global hormones such as insulin or glucagon. Four such hormones (or groups of hormones), whose hormonal function has been established, are : *melatonin, renal hormones, eicosanoids and opiate peptides.*

1. Pineal Hormone or Melatonin

Secretory gland : The pineal gland or epiphysis (*epi*[G] = upon ; *phyein*[G] = to grow) is a tiny conical or pine-like organ (hence its nomenclature, pineal) on the dorsal side of brain in vertebrates, lying deep in the groove between the cerebellum and cerebral hemispheres. The gland was long associated with the folklore of a mysterious "third eye". As the gland works round the clock, **Bruce Fellman** (1987) calls this as a '*clockwork gland*'. Pineal has a back-dated history. **Herophilus** (4th century BC) called this as the *'sphincter of thought'* the mind's valve. **Rene Descartes**, a 17th century French physiologist called it as *'the seat of the rational soul'.* Scientists of the early 20th century, wearied by a vain search for the gland's function, cast it as a neurological vestige, *an appendix of the brain.* The pineal, however, was not easily relegated to the cerebral scrap heap. In the early 1960s, **Richard Wurtman** of the Massachusetts Institute of Technology (MIT), USA, aptly coined a phrase to describe the gland's function, **"a neurological transducer"** : a system that converts a nerve-type signal – one set off by dark and light – into an endocrine signal – that of a hormone whose levels rise and fall in the bloodstream.

Nearly all modern fish, amphibians reptiles, birds and mammals have a pineal. Even the few exceptions–notably alligators crocodiles and armadillos–have cells that act like the gland. And the organ itself has occupied most vertebrates' skulls since the first backboned creatures joined the earth's fossil record 500 million years ago. A structure as this ubiquitous and ancient is surely not just going along for the evolutionary ride. In man, this gland is very well developed in children and shrinks after the 7th year and is reduced to a little knot of tissue after puberty. It weighs less than a button. Many functions have been attributed to this gland but they, however, remain to be verified.

Structure. Pineal secretes a hormone called **melatonin** which inhibits the secretion of an adenohypophyseal hormone called luteinizing hormone. Melatonin is synthesized (Fig. 19–48) from its precursor, *serotonin* under the influence of two enzymes, N-acetyltransferase (NAT) and hydroxyindole-o-methyltransferase (HIOMT) which are present in rich quantity in the pineal gland. Serotonin was first isolated and crystallized by **Page** and others in 1948. In fact, like the clockwork, the pineal manufactures melatonin only in the night. However, during daytime, it does not sit idle and converts the amino acid tryptophan (Trp) into serotonin which accumulates and is stored in an inactive

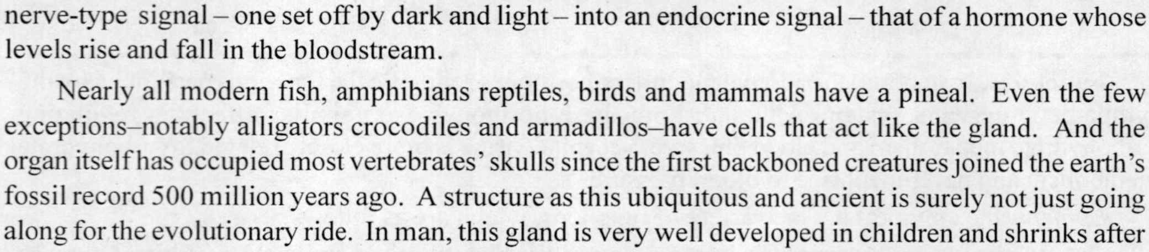

Fig. 19–48. Synthesis of melatonin

form until night. Neurotransmitters (such as norepinephrine) are released by the nerve cells connected to the pineal and cause increased concentration in the gland of the enzymes NAT and HIOMT. These enzymes convert serotonin into melatonin which then flows into the bloodstream. Thus, exposure to light slows while exposure to darkness stimulates the production of melatonin in the pineal gland.

David Klein argues that the pineal is a slave and it is under the control of a master clock. The master clock is a concentration of specialized nerve cells found in the hypothalamus, close to where the optic nerve crosses. Called the suprachiasmatic nuclei (SCN), this structure was also identified late, like the pineal gland but its function was a mystery. Klein has found that the SCN turns melatonin production on and off in the pineal by directing the nightly increase in NAT.

Functions. In humans, the gland's rhythm may affect the sleep cycle. Lieberman and Vollrath domonstrated that melatonin makes people drowsy. Hence, nowadays melatonin is being consumed as a sleep aide by insomniacs and people suffering from jet-lag, and also by those having trouble in adaptation to work at odd hours. The hormone is also being used for prevention of cancer and for strengthening the immune system. This *'miracle hormone'* keeps a person using it, away from growing old. As melatonin production diminishes naturally with age, it has been linked to a possible role in the onset of puberty. Studies have also indicated that melatonin may regulate menstrual cycles in women and sperm production in men.

> **Erythropoietin** is also spelt as erythropoetin (**W.S. Hoar,** 1975).

2. Renal Hormones

The kidney secretes two hormones, erythropoietin and renin.

Erythropoietin or erythrocyte stimulating factor, ESF. It is secreted by the kidney. Its secretion is stimulated by tissue anoxia and also by androgenic hormones and cobalt. In fact, erythropoietin is secreted as an inactive protein called renal erythropoietin factor (REF) which is enzyme like in behaviour. The REF converts a plasma globulin to the active erythropoietin.

$$\text{Plasma globulin} \xrightarrow{\text{REF (Enzyme) ?}} \text{Active erythropoietin}$$

Erythropoietin has been prepared from the plasma of anemic sheep and also from the urine of anemic human beings. It is a glycoprotein with 8-12% total hexose and has a molecular weight of about 60,000. Its molecule contains all the common amino acids except methionine. Its activity is much retarded by proteolytic enzymes and also by the antibiotic actinomycin-D.

Erythropoietin stimulates the differentiation of the stem cell (hemohistioblast) of the bone marrow into the erythroid series, increasing the numbers of proerythroblasts in the bone marrow. This is followed by increases in other nucleated erythrocytes and finally, by increases in reticulocytes (erythrocyte precursor cells) and mature erythrocytes, in the peripheral circulation. Recent studies indicate that the earliest effect of erythropoietin is stimulation of the synthesis of a very large RNA (150 S) by bone marrow cells.

Renin. It catalyzes the synthesis of angiotensins which cause vasoconstriction in the kidneys, thereby causing electrolyte and water retention in the body. This system has been referred to as the *renin-angiotensin system.* An increase in the renal pressure and the levels of plasma Na^+ and angiotensins cause a decrease in renin production.

3. Eicosanoid Hormones

A. Structure and Metabolic Roles

Eicosanoid hormones (or simply eicosanoids) are fatty acid derivatives with a variety of extremely potent hormonelike actions on various tissues of vertebrates. Eicosanoids, in general, are known to be involved in reproductive function ; in the inflammation, fever and pain ; in the formation of blood clots and the regulation of blood pressure ; in gastric acid secretion ; and in a variety of other human processes.

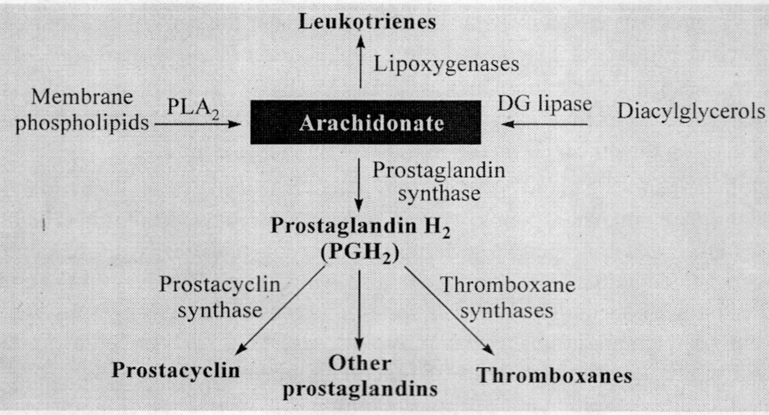

Fig. 19–49. Arachidonic acid and some of its eicosanoid derivatives

In response to certain hormonal signals, *phospholipase A_2* releases arachidonic acid (arachidonate at pH 7) from membrane phospholipids. Arachidonic acid then serves as a precursor of various eicosanoids. *Prostaglandin synthase* catalyzes the first step in a pathway leading to prostaglandins, prostacyclins and thromboxanes. *Lipoxygenase* catalyzes the initial step in a pathway leading to leukotrienes.

Eicosanoids are all derived form 20-carbon polyunsaturated fatty acid, **arachidonic acid** (20 : 4; 5, 8, 11, 14), from which they take their general name (*eikosi*[G] = twenty). There are 3 classes of eicosanoids (or the *signal molecules*, as they are also called) : prostaglandins, prostacyclins and thromboxanes, and leukotrienes (Fig. 19–49).

Prostaglandins

Kurzrok and **Lieb** (1930), for the first time, observed that the human semen is able to bring about strong muscular contraction or relaxation when placed in the uterus. This is due to the presence in the semen of a number of structurally-related compounds collectively termed prostaglandins. The name prostaglandin (abbreviated as PG) was first given by a Swedish chemist **U.S. von Euler** in 1935 to this lipid-soluble acidic substance, initially found by him in human seminal fluid and in extracts from ovine vesicular glands. Although prostaglandins were originally found in the seminal fluid of man and other species (hence the nomenclature) but now these have been found to occur in a wide variety of mammalian tissues including brain, spinal cord, thymus, lungs, pancreas, kidneys, menstrual fluid and placenta. However, *semen remains one of the richest sources of prostaglandins as yet.*

Chemically, the prostaglandins resemble prostanoic acid (Fig. 19–50). They are hydroxy derivatives of the polyunsaturated C-20 cyclic fatty acids. In its molecule, the carbon atoms 8 to 12 are involved in the formation of a 5-carbon ring called cyclopentane ring. Variations in the double bonds and in the hydroxyl and ketone groups give rise to

ULF SVANTE VON EULER (LT, 1905-1983)

Von Euler was born in Stockholm and received an M.D. from the Karolinska Institute, where he remained as a member of the faculty after receiving his degree. He first identified prostaglandins– from semen– in the early 1930s, and named them for their source, the prostate gland. By the time it was realized that all cells except red blood cells synthesize prostaglandins, their name had become entrenched.

He discovered noradrenaline and identified its function as a chemical intermediate in nerve transmission. For this work, he shared the **1970 Nobel Prize in Medicine or Physiology** with **Julius Axelrod** and **Sir Bernard Katz.**

prostaglandins that can be divided into 9 groups designated as A through I (and accordingly, the prostaglandins are designated as PGA through PGI). The PG, obviously, stands for prostaglandin and the third capital letter, in most cases, indicates the type of the substituents found on the hydrocarbon chain. In fact, two groups of prostaglandins were originally recognized : **PGE** which is ether-soluble (hence the nomenclature, E from ether) and has a keto group at C_9, and **PGF** which is phosphate-buffer-soluble (hence the nomenclature, F from *fosfat* in Swedish) and has a hydroxyl group at C_9. There are 3 compounds from each of these two groups, arising from eicosanoic (*i.e.*, C_{20}) fatty acids with 3, 4 or 5 double bonds. These are the 6 primary prostaglandins and are abbreviated as PGE_1, PGE_2, PGE_3, $PGF_{1\alpha}$, $PGF_{2\alpha}$ and $PGF_{3\alpha}$. In the abbreviations of the prostaglandins, the number 1, 2, 3 is added as subscript to indicate the number of carbon-carbon double bonds outside the ring. When there is more than one member, the group is subdivided into α, β, γ etc. These occur in most cells. Their structure appears in Fig. 19–51 along with their biosynthetic origin. In addition to these, there are several secondary prostaglandins that are derived from PGE types through enzymic conversions.

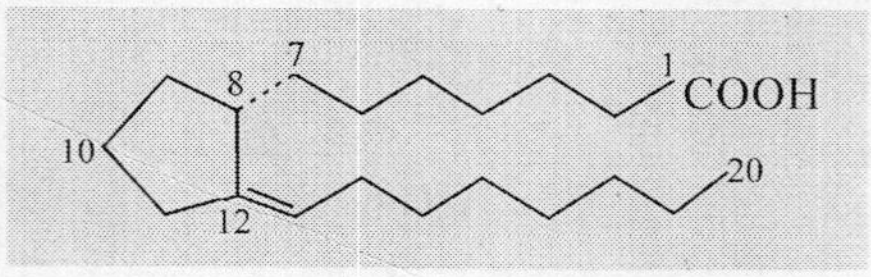

Fig. 19–50. Prostanoic acid

Prostaglandins have a half life of 5 minutes or less, hence they are destroyed very rapidly in the body. They have, therefore, a *high turnover rate.* The short half-life of these tissue hormones is thought to ensure their transient and limited response at the intermediate site of production. From a medical viewpoint, they are potentially the most revolutionary therapeutic substances yet discovered. Prostaglandins affect smooth muscles and blood pressure and often the activities of individual prostaglandins oppose one another. For example, prostaglandin E_2 (PGE_2) dilates blood vessels and bronchi, and prostaglandin $F_{2\alpha}$ ($PGF_{2\alpha}$) constricts these smooth muscle tissues.

The prostaglandins perform a wide variety of biologic activities :

1. As mentioned earlier, they bring about contraction or relaxation of the smooth muscles of the uterus, *esp.*, at the time of ovulation. This may be due to a chelation of calcium ions. As little as 1 ng/ml can cause contraction of the smooth muscles. They, thus, resemble ocytocin in this regard. They are modulators of hormone action.
2. They lower down blood pressure.
3. They inhibit lipolysis in adipose tissue, possibly by inhibiting the conversion of ATP to cyclic AMP and inhibition of platelet aggregation. The prostaglandins, thus, have the opposite effect of epinephrine, norepinephrine, glucagon and corticotropin on the release of fatty acids from adipose tissue.
4. They behave both as pressor and depressor agents under different conditions and thus affect the cardiovascular system.
5. They appear to control the secretion of gastric hydrochloric acid.
6. They also have some beneficial effect in the control of the acid-induced gastric ulcers.
7. Prostaglandins are best known for their effects on reproductive system. There seems to be a strong link between male fertility and seminal prostaglandin content. Human semen is rich in prostaglandins, which when deposited in vagina through coitus, facilitate conception. Thus, low prostaglandin content in the human semen is related to infertility.
8. They are effective labour inducers in pregnant women.
9. Recent work indicates that the prostaglandins are also involved in the inflammatory reaction and pain. Anti-inflammatory drugs such as aspirin, in part, act by inhibiting the synthesis of

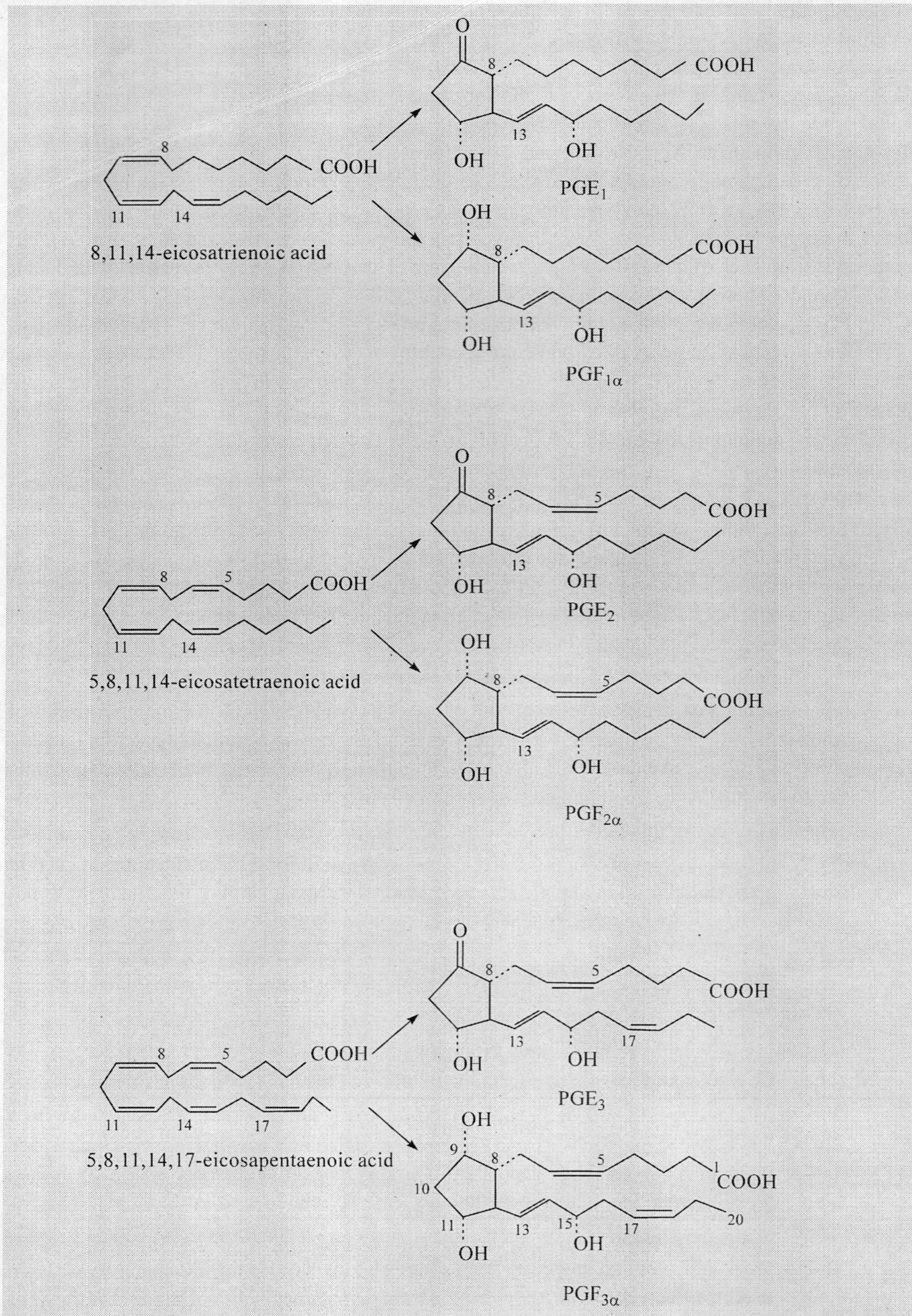

Fig. 19–51. The 6 primary prostaglandins and their biosynthetic origins

prostaglandins. However, paracetamol (an analgesic drug like aspirin) is not anti-inflammatory as it does not inhibit the synthesis of prostaglandins.

The widespread distribution of prostaglandins and their capacity to carry out varied metabolic effects have, however, led some to question the propriety of their being called hormones.

Thromboxanes and Prostacyclins

Thromboxanes (TXAs) and prostacyclins (PGIs) are structurally-related compounds that arise from a nascent prostaglandin. In compounds of both these categories, carbons 8 and 12 are joined and an oxygen atom is added to form the six-membered ring (*cf* prostaglandins, where a five- membered ring is formed). **Thromboxane A_2** (TXA_2) was first isolated, in 1975, by **Samuelsson** *et al* from blood platelets (also known as thrombocytes, hence the nomenclature). The following year, **John Vane** and colleagues at the Royal College of Surgeons, London identified yet another type of eicosanoid, **prostacyclin I_2** (PGI_2), which is produced primarily in vascular tissues, *i.e.*, blood vessels.

The striking similarity and diversity in the physiological roles of thromboxanes and prostacyclins displays a critical balance required for the normal functioning in the body. TXA_2 and PGI_2 are medically important examples of how such a balance operates *in vivo*. TXA_2 is a highly effective vasoconstrictor (blood vessel constrictor) and platelet aggregator ; conversely, PGI_2 is a potent vasodilator and inhibitor of platelet aggregation. Platelets are the blood cells that first appear and aggregate at the site of injury to produce a temporary plug that serves as a base on which the strong fibrin clot ultimately forms. However, for maintenance of normal blood flow, TXA_2-induced aggregation of platelets would quickly prove fatal. Thus, a vital opposing role of PGI_2 is to prevent platelets from aggregating on blood vessel walls, a site of PGI_2 production.

In **1982, Nobel Prize in Physiology or Medicine** was awarded to **Sune Bergström** (born,1916) and **Bengt Ingemar Samuelsson** (born, 1934), both from Sweden, along with **Sir John Robert Vane** (born,1927) from England, for their work on eicosanoids (prostaglandins). Bergström and Samuelsson are both at the Karolinska Institute, while Vane is at the Wellcome Foundation in Beckenham, England.

Unlike other eicosanoids, PGI_2 is not metabolized during passage through the lungs. Thus, TXA_2 and PGI_2 are continuously engaged in a 'tug of war' with respect to platelet aggregation (Fig. 19–52).

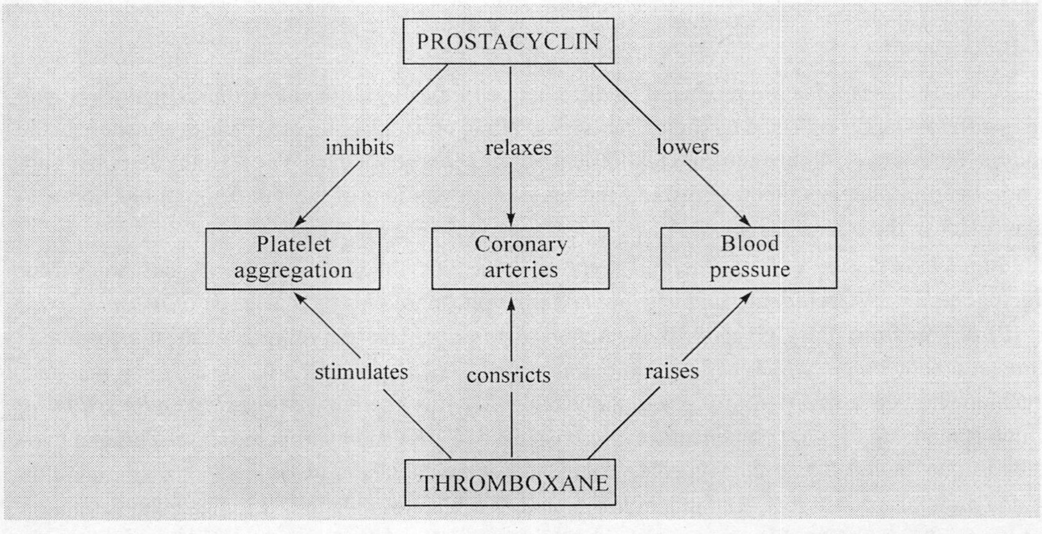

Fig. 19–52. Opposing effects of prostacyclin and thromoboxane on the cardiovascular system

Leukotrienes

Leukotrienes (LTs), first found in leukocytes, are cysteinyl-containing derivatives of arachidonic acid with a series of three conjugated double bonds in its molecule (Fig. 19–53), hence their nomenclature.

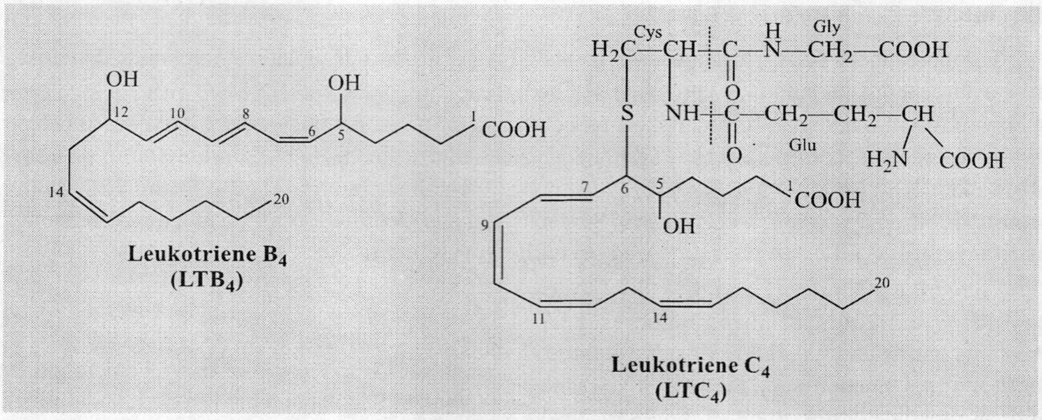

Fig. 19–53. Two common leukotrienes

Leukotriene B_4 (LTB_4) has two hydroxyl groups at C-5 and C-12 and three conjugated double bonds at C-6, C-8 and C-10. An additional double bond is present between C-14 and C-15. **Leukotriene C_4** (LTC_4) contains the tripeptide glutathione (γ-Glu-Gly-Cys) covalently bonded to a derivative of arachidonic acid. **Leukotriene D_4** (LTD_4) possesses the dipeptide, Gly-Cys (Glu residue is eliminated) and **leukotriene E_4** (LTE_4), the amino acid Cys (Gly residue eliminated).

Neutrophils make one class of leukotrienes to alter mobility and act as chemotactic agents. *Mast cells* make another class, formerly known as **slow-reacting substances**, which is responsible for bronchial constriction and other anaphylactic allergic reactions.

Leukocytes are powerful biological signals ; for example, they induce contraction of the muscle lining the airways to the lung. They also cause a slow and persistent contraction in the smooth muscle of blood vessels. Overproduction of leukotrienes causes asthmatic attacks and also stimulates mucus secretion.

B. Biosynthesis

Various eicosanoids are produced in different cell types by different synthetic pathways, and have different target cells and biological activities. While prostaglandins are made everywhere in the body, synthesis of thromboxanes and leukotrienes occurs in restricted locations. Both compounds are synthesized in platelets, neutrophils and the lung. Some amount of thromboxane production also takes place in the brain.

Fig. 19–54 depicts various reactions involved in the biosynthesis of 3 types of eicosanoids from arachidonic acid. Arachidonic acid is generated from phospholipids by the action of *phospholipase A_2* (PLA_2), or from diacylglycerol by the action of a *lipase*. The biosynthesis of most eicosanoids starts at arachidonate, which has 4 double bonds at C_5, C_8, C_{11} and C_{14}. In the first key reaction, arachidonate gets converted to prostaglandin H_2 (PGH_2) by the enzyme *prostaglandin synthase*, which is made up of two components, cyclooxygenase (COX) and hydroperoxidase. This is a two-step reaction. In *first step*, cyclooxygenase component of prostaglandin synthase catalyzes the addition of one mole of oxygen to C-9 of arachidonate and of a second mole to C-15. The bond formation between C-8 and C-12 accompanying this oxygenation produces the 5-membered endoperoxide ring

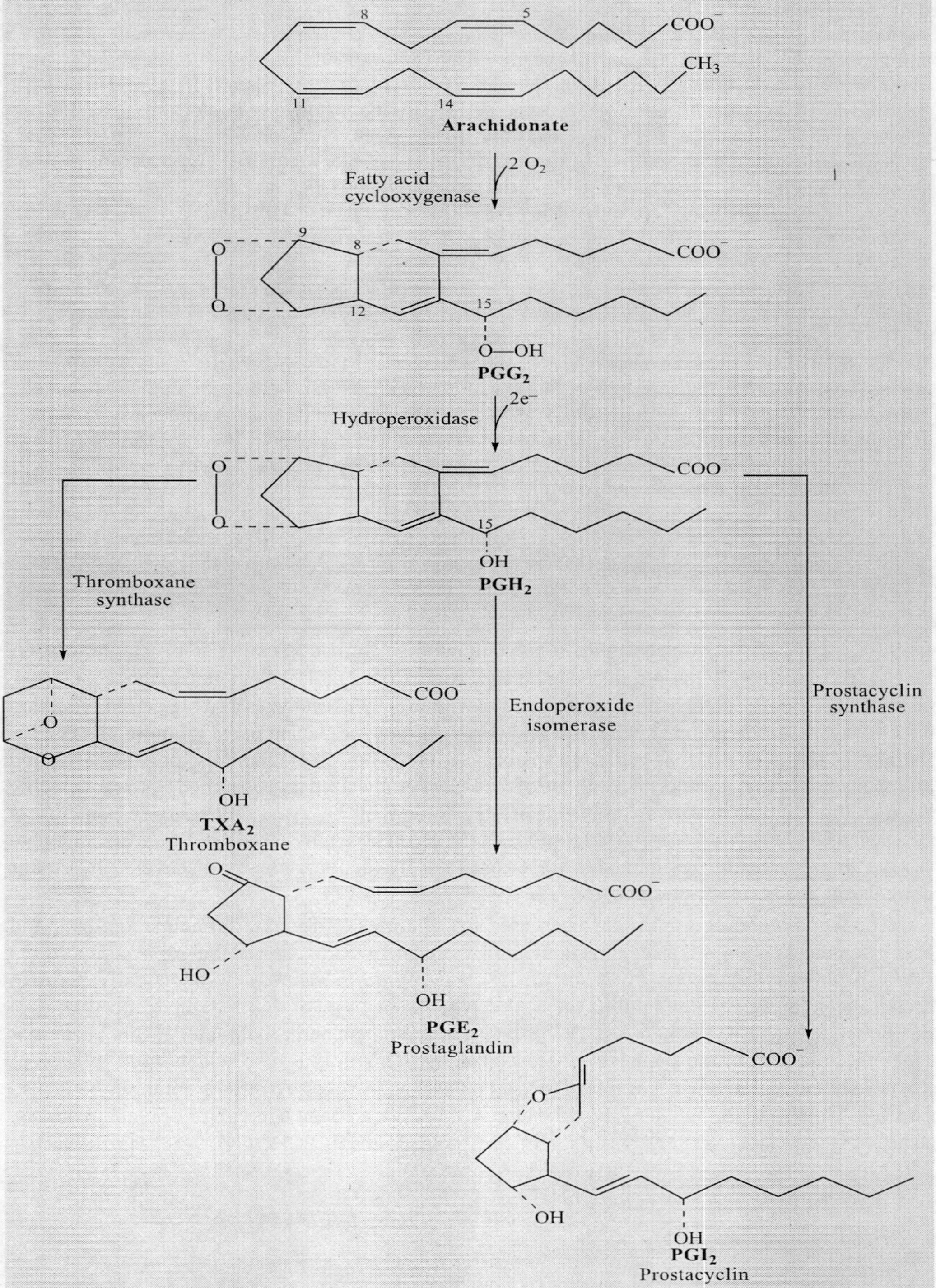

Fig. 19–54. Biosynthesis of 3 types of eicosanoids (PGE_2, PGI_2 and TXA_2) from arachidonic acid

The chemical name of **aspirin** is acetylsalicylic acid, $CH_3CO_2C_6H_4COOH$. The name 'aspirin' is derived from a plant genus *Spirea* (from a Latin word which means spiral, which refers to the fact that most members of this genus are used for making wreaths and garlands, which are coiled or somewhat spiral in shape), whose members contain salicylic acid, and thus are its natural sources. However, salicylic acid also occurs in many other plants such as jasmine, patridge berry, legumes (pea, bean, clover), grasses (wheat, rye, sugarcane) and many trees (beech, birch, olive, poplar, willow). The chemical term salicylic acid itself comes from *Salix*, the Latin name for willow tree. The active ingredient in willow bark is a bitter glycoside called **salicin.** Salicin was first identified by the Italian chemists, **Fontana** and **Brugnatelli** in 1826, and isolated in pure form by the French pharmacist, **Henri Leroux** in 1829. On hydrolysis, salicin yields glucose and salicylic alcohol. The latter gets converted in the body to **salicylic acid**, which actually brings the fever down. Salicylic was first successfully extracted, in 1838, by an Italian chemist **Rafelle Piria** (the second name of this chemist, as one can probably note, has an earie similarity to the plant genus *Spirea*). Later in 1874, the same scientist **Adolf Kolbe**, who had earlier synthesized acetic acid, came out with an economical procedure for the laboratory synthesis of salicylic acid as well. In pure form, however, salicylic acid is too corrosive to the tissues in that it peels off the upper horny layers of the skin and that is why it is an excellent agent for the removal of corns, callosities and warts. But taken orally, it produces severe vomiting and sometimes even comma. For internal administration, however, it had to be taken in the form of a salt. The most significant and potent salt, **acetylsalicylic acid** was, however, produced in 1899 by **Felix Hoffmann** (LT, 1868–1946) who also looked to its usefulness for a compelling reasons to try the same on his father suffering from rheumatoid arthritis and both remedies of pain (sodium, phenyl salicylate), available at that time, having had failed to give him relief. The drug fortunately worked on him. **Heinrich Dreser**, Director of the Friedrich Bayer & company, gave it the tradename '**aspirin**'. Aspirin is widely used to suppress platelet function in humans at risk of coronary heart disease, or deep vein thrombosis, and to treate patients that have suffered a heart attack or stroke.

structure, characteristic of eicosanoids. The compound so formed is called as **prostaglandin G$_2$** (PGG_2). The 4 oxygen atoms introduced into PGG_2 come from 2 moles of oxygen. In the second step, the hydroperoxidase component of prostaglandin synthase, then, catalyzes a two-electron reduction of the 15-hydroperoxy group of PGG_2 to a 15-hydroxyl group, producing **prostaglandin H$_2$** (PGH_2). The highly unstable PGH_2 is rapidly transformed into other prostaglandins, prostacyclins and thromboxanes. In fact, the biochemical fate of the PGH_2 synthesized is determined by tissue-specific enzymes. For example, in a tissue producing prostaglandin E_2, the enzyme endoperoxide isomerase is present and converts PGH_2 into **prostaglandin E$_2$** (PGE_2). The synthesis of PGE2 from arachidonate was the first-ever pathway elucidated for eicosanoid production and was discovered by **Bengt Samuelsson** and his associates in 1964.

The wondrous drug aspirin has been used for centuries to decrease inflammation, pain and fever. Itsmode of action was an enigma until **John Vane**, in 1975, discovered that *aspirin inhibits the synthesis of prostaglandins by inactivating prostaglandin synthase*. Specifically, aspirin (acetylsalicylate) irreversibly inhibits the cyclooxygenase activity of this enzyme by acetylating a specific serine hyydroxyl group (Fig. 19–55). Aspirin is a potent antiinflammatory agent because it blocks the first step in the synthesis of prostaglandins. This drug is also widely used to prevent excessive blood clotting, which can lead to heart attacks and strokes. Aspirin is antithrombotic also because it blocks the formation of thromboxane A_2 (TXA_2), a potent aggregator of blood platelets. Inhibition of the cyclooxygenase blocks the formation of prostaglandin H_2, PGH_2 (Fig. 19–55). PGH_2, which is

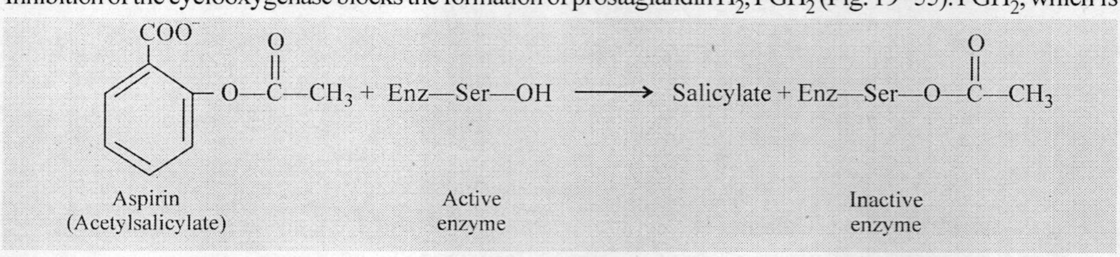

Fig. 19–55. Inactivation of prostaglandin synthase by aspirin

produced in the platelets, is also the precursor of thromboxane A_2 (TXA_2) ; the reaction being catalyzed by the enzyme *thromboxane synthase*. **Prostacyclin I_2** (PGI_2) is synthesized from PGH_2 and the reaction is mediated by *prostacyclin synthase*. Thus, we see that the tissues are differently endowed with enzymes that transform endoperoxides into sepcific types of eicosanoids.

Leukotrienes are made from arachidonate by another pathway, beginning with the addition of oxygen to C-5 of arachidonate ; the reaction being catalyzed by *lipoxygenase*. This reaction is not affected by antiinflammatory drugs.

4. Opiate Peptides

Opiate peptides are compounds that bind to specific receptors and exhibit hormone-like actions. The various opiate peptides fall under 3 families, enkephalins, dynorphins and endorphins, which are generated in the body by the action of protease.

(*a*) **Enkephalins.** These consist of 5 to 7 amino acids and are derived from a precursor, *proenkephalin*. These have been isolated from extracts of brain and pituitary and exhibit morphine-like properties.

(*b*) **Dynorphins.** These consist of 10 to 17 amino acids and are derived from a precursor, *prodynorphin*. These are endogenous compounds made in several locations, but the major locations are brain, pituitary, adrenal medulla and peptidergic neurons.

(*c*) **Endorphins.** These consist of 16 to 27 amino acids and are derived from a precursor, proopiomelanocortin (POMC). These have also been isolated from extracts of brain and pituitary and exhibit morphine-like properties.

It has been suggested that there are at least 3 types of receptors, δ, μ and *k*, through which these drugs mediate their physiological effects.

Though these opoids produce major effects on the central nervous system, they also function as neurotransmitters and neurohormones to modulate neurotransmission. Some of the effects that opiate peptides produce are analgesia, drowsiness, nausea, vomiting, respiratory depression, decreased gastrointestinal motility and modulations of endocrine and autonomic nervous systems.

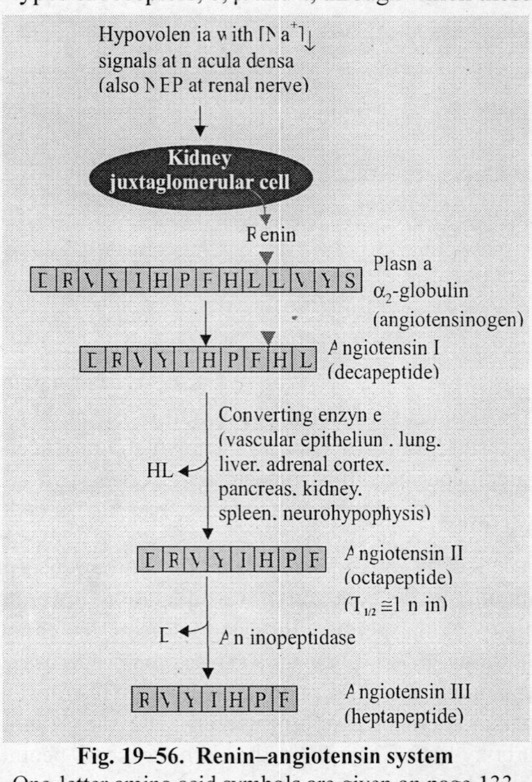

Fig. 19–56. Renin–angiotensin system
One-letter amino acid symbols are given on page 133.
NEP = norepinephrine

VASOACTIVE PEPTIDES

There are three groups of well defined peptides that possess vasoactive properties. These are : *neurohypophyseal hormones* (ocytocin and vasopressin), *angiotensins* and *kinins*. Ocytocin and vasopressin have already been described on pages 373 and 374. The remaining two groups of vasoactive peptides are described below.

1. Angiotensins

Goldblatt (1947) demonstrated that the partial occlusion of the renal artery in a dog results in permanent hypertension. Obviously, slowing the circulation causes the formation of a substance that produces vasoconstriction. The presence of such a substances has been demonstrated in the blood from such ischemic kidneys. The mechanism behind this operates as follows (Fig. 19-56). An enzyme renin is formed in the kidney and is released into the blood. It is a proteinase and acts on *angiotensinogen*, a plasma protein synthesized in the liver, to split off a decapeptide called *angiotensin I*. This decapeptide, in turn, is acted upon by a peptidase called converting enzyme, present in serum, to form an octapeptide called *angiotensin II*. Another peptidase, termed angiotensinase, produced by kidney, hydrolyzes angiotensin II to produce *inactive fragments*. Angiotensinase, thus, serves as a balancing antipressor agent. The sequence of biochemical steps leading to its release is as follows :

Angiotensinogen (an α_2-globulin of plasma) —Renin→ Angiotensin I (decapeptide) —Converting enzyme→ Angiotensin II (octapeptide)

Angiotensin II (octapeptide) —Angiotensinase→ Inactive fragments

Various synonyms exist in the literature for the above substances. Angiotensinogen has been variously called as hypertensinogen or renin activator. Likewise, angiotensin II is also known as hypertensin or angiotonin and angiotensinase as hypertensinase.

The two angiotensins, I and II are peptides and their amino acid sequence (refer Fig. 19–57) was determined by **Page** and coworkers in 1960.

Asp. Arg. Val. Tyr. Ile. His. Pro. Phe. His. Leu. Leu. Val. Tyr. Ser–Glycoprotein
Angiotensinogen (?)

↓ Renin

Asp. Arg. Val. Tyr. Ile. His. Pro. Phe. His. Leu + Leu. Val. Tyr. Ser–Glycoprotein
Angiotensin I

↓ Peptidase

Asp.Arg. Val. Tyr. Ile. His. Pro. Phe + His. Leu
Angiotensin II

Fig. 19–57. Formation of angiotensin II

Angiotensin I has only a slight effect on blood pressure, whereas angiotensin II is the most powerful vasoconstrictor now known and produces hypertension. It is 200 times more active than norepinephrine in this respect.

2. Kinins

"Kinins is the generic name for a group of peptides with potent biologic activities in causing smooth muscle contraction, vasodilatation, lowering of blood pressure, increasing blood flow and microvascular permeability, and inducing the emigration of granulocytic leukocytes" (**Orten** and **Neuhaus**, 1970). They, thus, have some properties similar to and other properties different from the remaining vasoactive peptides. Kinins are liberated from plasma protein called *kininogen* exposed to snake venom or to the proteolytic enzyme. Kininogen is an α_2-globulin containing about 18% carbohydrate. Chemically, the kinins are small peptides of 9 to 11 amino acid residues. Since they are derived from protein precursors, these peptides contain only protein amino acids. They are rapidly inactivated by the *kininases* of tissues. The biological significance of the kinins is not clear as yet. However, they appear to be the chemical mediators of inflammation. They are also powerful hypotensive agents. *Bradykinin* and *kallidin* are typical examples of kinins.

Bradykinin. It is formed from its precursor *bradykininogen*, an α_2-globulin in serum, in the presence of an enzyme kallikrein. Bradykininogen has been found in avariety of tissues. Heart contains the most, followed by liver, kidney and brain. The amount of bradykininogen in the blood is 6 to 8 μg per ml. Bradykinin is a nonapeptide with the amino acid sequence shown in Fig. 19–58.

Arg. Pro. Pro. Gly. Phe. Ser. Pro. Phe. Arg

Fig. 19–58. Bradykinin

Bradykinin is released by active sweat and salivary glands. It is a powerful vasodilator that enormously increases the blood flow locally and thus promotes secretion of sweat and saliva. But its role as a vasodilator is mainly conjectural.

Kallidin. Kallidin (Fig. 19–59) is a decapeptide and contains lysine, in addition, on the N-terminal of the peptide chain of bradykinin. Hence, it is also called *lysylbradykinin.*

Lys. Arg. Pro. Pro. Gly. Phe. Ser. Pro. Phe. Arg

Fig. 19–59. Kallidin

HORMONE FROM THYMUS

The thymus is present in all jawed vertebrates. It is a flat, pinkish, bilobed structure, located in the chest behind the sternum. The gland arises as a proliferation of the gill pouch epithelium which becomes infiltrated with lymphocytes. It grows rapidly to acquire a large size in the young animals and atrophies in the adults. In man, the gland reaches its greatest development at the age of 14 to 16, after which it atrophies because of the activity of the sex cells. Thymus is, however, absent in the hagfishes and in the lamprey, it is represented by a group of cells beneath the gill epithelium. This is called as a *prothymus*.

Structure. Purification of thymus extracts have yielded an active glycopeptide.

Functions. Certain biochemical functions are attributed to this gland. These are as follows :

1. It is the primary source of lymphocytes in the mammals. The lymphocytes are responsible for the immunological functions of the body by producing antibodies.
2. It helps the body to distinguish between its own tissue proteins and the foreign proteins such as those present in tissues transplanted from another animal. This has been inferred from the fact that when the thymus is removed at birth, skin grafts in animals persist for a long period. However, when the thymus is left intact, the graft or the transplanted tissue is rejected.
3. It is concerned with accelerating growth and as such metamorphosis is postponed.

> **Antibodies** are substances that help the animals to combat with the invading bacteria and viruses.

Thymectomy. Removal of the thymus gland has resulted in the following dysfunctions :

1. reduction in lymphocyte population (lymphopenia)
2. atrophy of the lymphoid tissue
3. loss of immunological competence.

The renal hormones, the prostaglandins, the angiotensins, the kinins and the thymus hormone are all sometimes collectively referred to as **parahormones** or **tissue hormones** as many of them act very locally. They are all produced endogenously by specialized groups of cells. These are different from hormones as they produce their effects without entering the blood stream. Moreover, their secretion is controlled by hormones, metabolic products, enzymes etc.

PHEROMONES OR 'SOCIAL' HORMONES

Definition. The pheromones or smell signals, as they are also called, are species-specific chemical substances or 'odours' which are released from animals (insects to mammals) into the environment and evoke behavioural, developmental or reproductive responses. The term '*pheromones*' was first used by **Karlson** and **Luscher** to designate active substances that mediate humoral correlations among individuals of a given species. *The pheromones work on other members of the same species (i.e., they are species-specific) whereas the hormones proper confine their activities to the animal which produces them.* Unlike the hormones, which are produced in glands to act on target tissues *internally*, the pheromones are produced in glands and discharged *externally* to influence other members of the same species. The pheromones, for the same reason, have also been christened as `social hormones'. These include substances responsible for olfactory attraction between the sexes, alarm substances which warn other members of the species of danger and 'markers' which establish territories and trials or mark rich sources of food. They are well known sex-attractants also. These peculiar 'odours' or 'scents' are elaborated by different epithelial glands in the epidermis, the oro-anal and the urinogenital regions.

Structure. Chemically, they belong to a diverse category of compounds ranging from amino acids to lipids, alcohols and organic acids. Female silkworm moth (*Bombax mori*) secretes a sex-attractant which excites male moths (Fig 19–60). The sex-attractant of the silk moth (Fig. 19–61) is a long-chain alcohol ($C_{16}H_{30}O$) with two double bonds at C_{10} and C_{12} (**Adolf Butenandt, Hecker** and **Stamm**).

In fact, **Adolf Butenandt** and his colleagues obtained only 12 mg of the compound from half a million glands of the silkworm moth. Only as low as 10^{-18} g per millilitre of this substance placed in the vicinity of the male silk moth is enough to induce flutter, dancing movements and other symptoms of sexual excitation. The signals could be detected even 1.5 kilometres away by the male silk moth making it one of the most active bioligical substances known. The *trans*-12 isomer requires 10^{12} times the quantity for equal effectiveness as an attractant.

Fig. 19–60. The male silkworm moth (*Bombax mori*)

The female moth releases sex attractants such as the pheromone bombykol into the air in minute quantities. The male is extremely sensitive to these pheromones, because his antennae are large and crammed with many specialist receptors. This male has about 17,000 receptors in each antenna that respond only to bombykol.

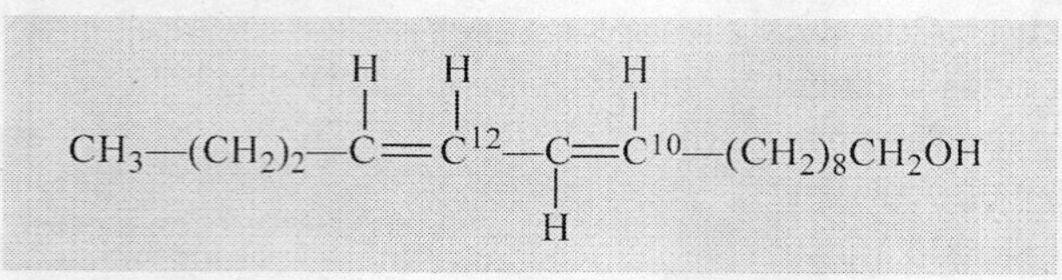

Fig. 19–61. Sex-attractant of the silk moth

(*trans*-10-*cis*-12-hexadecadien-1-ol)

Types. Based on their functional aspect, the pheromones have been grouped under two categories.

1. *Releaser pheromones*—They initiate specific patterns of behaviour. They serve as powerful sex-attractants, mark territories or trails, initiate alarm reactions or bring about aggregation of individuals.
2. *Primer pheromones*—They trigger physiological changes in endocrine activity, esp., related to sexual maturation, growth or metamorphosis.

Examples. Many of the insects live in a world dominated by odours which greatly affect their social life. Two such interesting examples are given below.

In the honeybee colony, the queen bee secretes from its mandibular glands a pheromone called '*queen substance*' which inhibits the growth of the ovaries in the worker bees. The latter obtain this substance by occasionally licking queen's body. If, however, an accident occurs to the queen or else she is removed from the hive, the ovaries of the worker bees start developing. The 'queen substance' has been isolated and synthesized by **Englemann** in 1970. Chemically, it is 9-oxodecenoic acid.

The female gypsy moths, *Porthetria dispar* cannot fly and have to depend, for luring the winged males for copulation, on sex-attractants which they secrete. Only very little amount of this substance (*i.e.*, 1×10^{-12} μg) is sufficient to attract males from some distance.

Many social insects produce alarm substances which excite other members of the species to attack intruders. The honeybee, for instance, injects a pheromone called 2-heptanone in the body of the victim it stings. This marks the invader so that its chances of getting more stings increase (**Free** and **Simpson**, 1968).

Pigs have been known to detect truffles burried as deep as 3 feet below the ground by scent alone. It has been discovered that truffles contain a pig sex pheromone called *5α-androst-16-en-3α-ol*. This might also explain why we like the fungus which is said to taste like a cross between musk, nuts and ozone. The above steroid is synthesized by human males in the testes and secreted by axillary sweat glands !

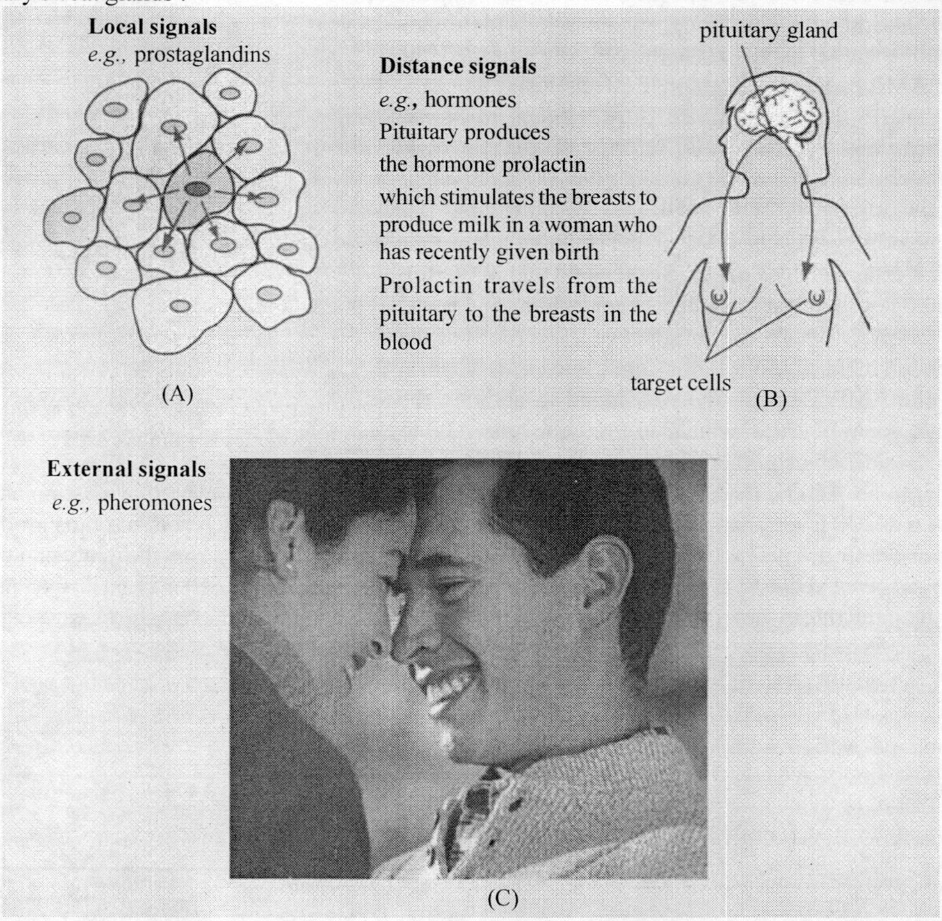

Fig. 19–62. A summary of the different types of chemical signalling that exist within and between organisms

(*A*) Local signals include neurotransmitters such as acetylcholine and local messengers such as prostaglandins, endorphins and histamines. (*B*) Distant signals such as hormones allow communication between different organs, *via* the blood. (*C*) Externally-produced messengers allow communication between organisms.

A human baby even when 2-days old has already learnt to recognize the odour of the breast of the mother, given off probably by the areola of the breast. Thus, if two wads of cloth–one infiltrated with the odour of the mother's breast and the other with the odour of the breast of another woman–are placed on either side of an infant's head, the child instinctively turns towards the wad with the odour of its mother's breast.

A general emerging trend is now to classify the various chemical signals, that exist within and between organisms, into 3 discrete categories : local signals, distant signals and external signals (Fig 19–62).

MECHANISMS OF HORMONE ACTION

The function of different hormones is to *control* the activity of levels of target tissues. To achieve this, the hormones may alter either the permeability of the cells or they may activate some other specific cellular mechanism. Although the exact site of action of any hormone is not established, five *general sites* have been proposed.

A. Hormonal Action at Cyclic Nucleotides Level. Many hormones exert their effect on cells by first causing the formation of a substance, cyclic 3′, 5′-adenosine monophosphate (Fig. 19–63) in the cell. Once formed, the cyclic AMP causes the hormonal effects inside the cell. Thus, *cyclic AMP acts as an intracellular hormonal mediator*. It is also frequently referred to as the *second messenger* for hormone mediation ; the *first messenger* being the original hormone itself.

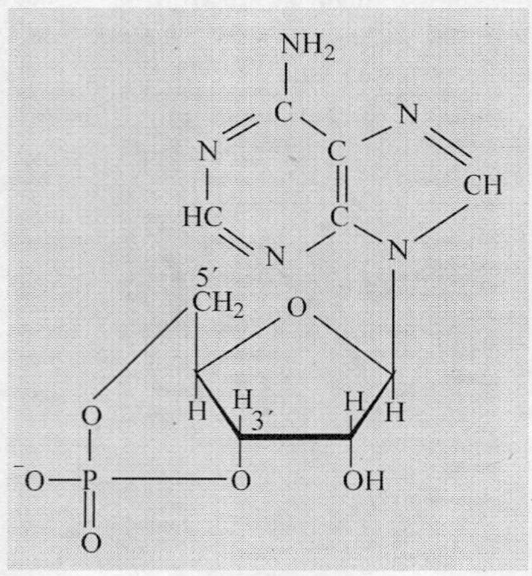

Fig. 19–63. Cyclic 3′, 5′-adenosine monophosphate, cAMP

The effects of cyclic AMP on the action of a hormone was first described by **Earl W. Sutherland** and **T.W. Rall** in 1960. They found that the effect of epinephrine on hepatic glycogenolysis (breakdown of glycogen) is a result of the conversion of inactive phosphorylase b into an active form by cyclic AMP. Epinephrine was found to activate the enzyme, adenyl cyclase which, in turn, converts ATP to cAMP. Besides epinephrine, other hormones like glucagon, parathormone, ACTH, TSH, ICSH, LH, α-MSH and vasopressin are now known to have a stimulatory effect on cAMP levels. Several hormones, on the contrary, decrease cAMP levels and thus produce an opposite effect. These include insulin, melatonin and the prostaglandins. From the many names of hormones given above, it appears that hormone action not mediated by cAMP may be an exception rather than the rule.

EARL WILBUR SUTHERLAND Jr.
(LT, 1915-1974)

Sutherland was an American pharmacologist and physiologist. He was awarded the coveted **Nobel Prize in Physiology or Medicine in 1971,** for his discoveries in cellular signal transduction. He showed that hormones such as adrenalin goad phosphorylase into action, indirectly, by increasing the production of another biochemical called cyclic AMP (cAMP). The latter is known as the `second messenger,' the first messenger being the hormone itself.

Fig. 19–64 depicts, in a schematic way, the effect of cAMP on hormone action. The cell contains receptor for hormones in the plasma membrane. The stimulating hormone acts at the plasma membrance

of the target cell and combines with a specific receptor for that particular type of hormone. The specificity of the receptor determines which hormone will affect the target cell. The combination of the hormone with its receptor leads to the activation of the enzyme, adenyl cyclase, which is also bound to the plasma membrane. The portion of the adenyl cyclase that is exposed to the cytoplasm causes immediate conversion of cytoplasmic ATP into cAMP. The reaction representing cAMP synthesis may, thus, be written as :

$$\text{ATP} \xrightarrow{Mg^{2+}} \text{Cyclic AMP} + PP_i + H^+$$

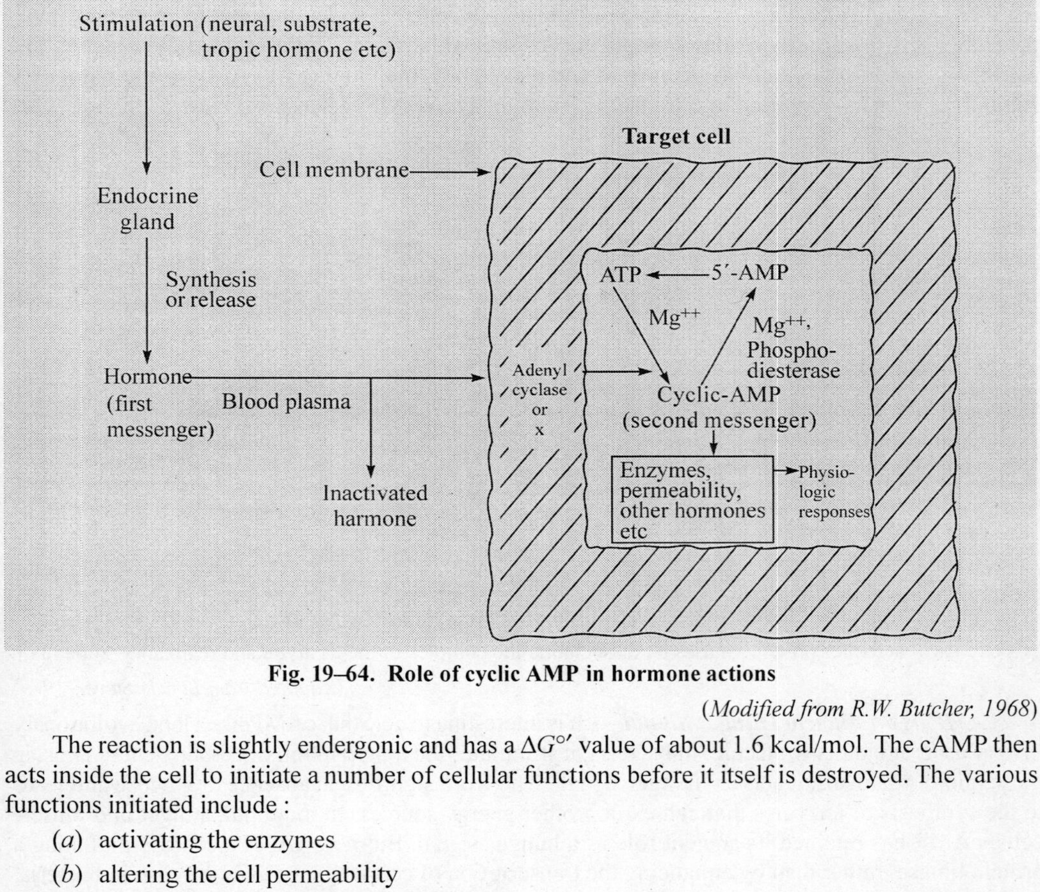

Fig. 19–64. Role of cyclic AMP in hormone actions

(*Modified from R.W. Butcher, 1968*)

The reaction is slightly endergonic and has a $\Delta G^{\circ\prime}$ value of about 1.6 kcal/mol. The cAMP then acts inside the cell to initiate a number of cellular functions before it itself is destroyed. The various functions initiated include :

(*a*) activating the enzymes
(*b*) altering the cell permeability
(*c*) synthesizing the intracellular proteins
(*d*) contracting or relaxing the muscles
(*e*) releasing other hormones (*third messengers*).

It should, however, be emphasized that *what cAMP does in a particular effector cell is determined by the cell itself, rather than by cAMP.*

Cyclic AMP is, however, destroyed (or inactivated) by a specific enzyme called *phosphodiesterase*, which hydrolyzes it to AMP. Like adenyl cyclase, the phosphodiesterase is present in practically all tissues.

$$\text{Cyclic AMP} + H_2O \xrightarrow{Mg^{2+}} \text{AMP} + H^+$$

This reaction is highly exergonic, having a $\Delta G^{\circ\prime}$ value of about –12 kcal/mol. Cyclic AMP is a very stable compound unless hydrolyzed by a specific phosphodiesterase.

An important feature of the second messenger model is that *the hormone need not enter the cell and its impact is made at the cell membrane*. The biological effects of the hormone are mediated inside the cell by cAMP rather than by the hormone itself.

cAMP and the Protein Kinases — Cyclic AMP elicits many of its effects by activating protein kinases. Protein kinases are ubiquitous in nature and are activated by cAMP at extremely low concentrations of 10^{-8} M. These kinases molecule the activities of different proteins in different cells by phosphorylating them. The enzyme *protein kinase* (Fig. 19–65) consists of two subunits : a catalytic subunit and a regulatory subunit which can bind cAMP. In the absence of cAMP, the catalytic and regulatory subunits form a complex that is enzymatically inactive. In the presence of cAMP, however, the complex disintegrates, freeing the catalytic subunit which now becomes catalytically active. The regulatory subunit binds cAMP to form a complex. Thus, the binding of cAMP to the regulatory subunit relieves its inhibition of the catalytic subunit. The cAMP acts as an allosteric effector.

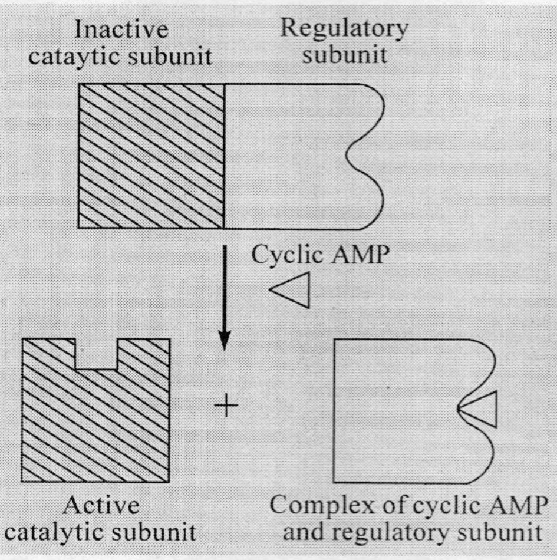

Fig. 19–65. Activation of protein kinase by cAMP

[Cyclic AMP activates protein kinases by dissociating the complex of the catalytic and regulatory subunits.]

(*Adapted from Lubert Stryer, 1975*)

cAMP as an Ancient Hunger Signal — It is interesting to note that cAMP has a long evolutionary history as a regulatory molecule. In bacteria, it stimulates the transcription of certain genes. Thus, in these microorganisms, it acts as 'hunger signal'. The word signifies an absence of glucose and leads to the synthesis of enzymes that can exploit other energy sources. In mammalian liver and muscle cells, cAMP has retained its ancient role as a hunger signal. But here cAMP acts by stimulating a protein kinase, rather than by enhancing the transcription of certain genes as is the case in bacteria. Moreover, cAMP has become a second messenger in higher organisms.

The querry emerging apparently from this discussion is that why was cyclic AMP chosen during evolution to be a second messenger ? This is because of the following three important reasons :

1. cAMP is derived from ATP which is an omnipresent molecule.
2. cAMP is very stable except for the presence of phosphodiesterase.
3. cAMP has a number of active groups that can bind it tightly to the receptor proteins such as the regulatory subunit of the protein kinase is muscles.

Other Intracellular Hormonal Mediators — It has been postulated that, besides cAMP, other types of intracellular hormonal mediators also exist.

1. One almost-certain mediator is **cyclic guanosine monophosphate** (= cyclic GMP). Cyclic GMP is a nucleoside similar to cAMP and is found in most tissues. It can probably catalyze some intracellular functions in a manner similar to that of cAMP.
2. Another type of intracellular hormonal mediator is a group of compounds referred to as

prostaglandins (refer page 388). These substances frequently cause intracellular inhibition, in contrast to the activation usually caused by cAMP.

B. Induction of Enzyme Synthesis at the Nuclear Level. A second major mechanism by which the hormones, *esp.*, the steroidal and thyroidal ones, act is to cause synthesis of proteins in the target cell. These proteins are presumably the enzymes which, in turn, activate other functions of the cells. The mechanism behind the **steroidal hormones** is depicted in Fig. 19–66. The sequence of events is as follows :

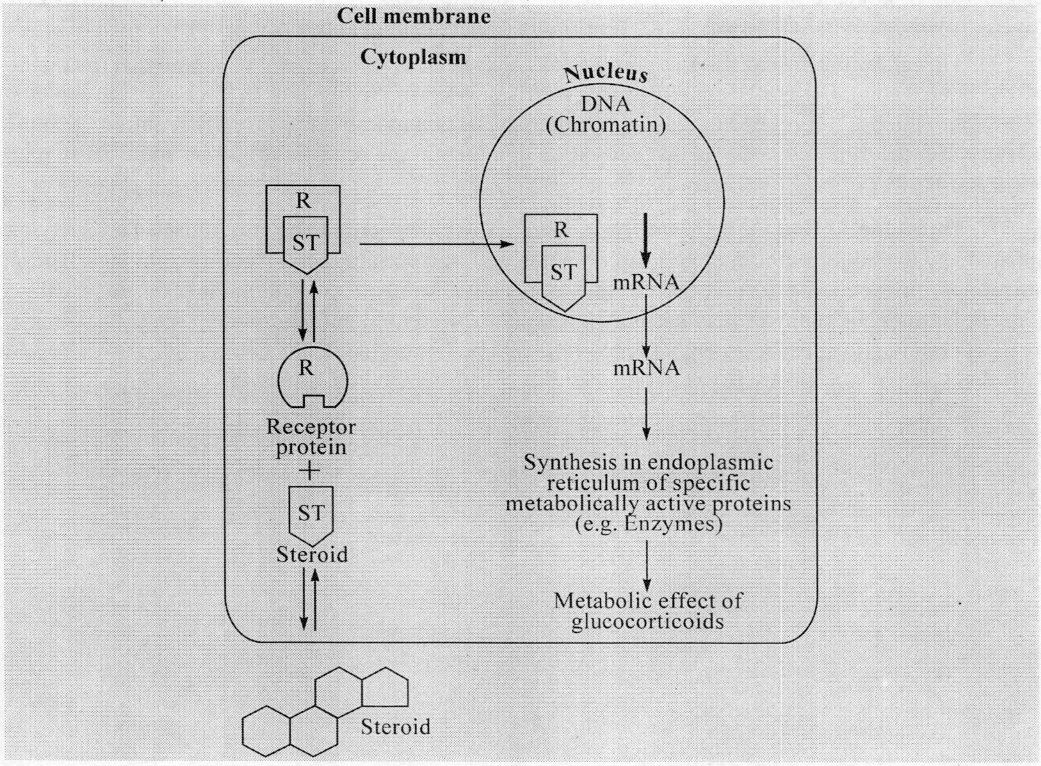

Fig. 19–66. Mechanism of action of steroidal hormones
ST = Steroid ; R = specific receptor protein
[The dissimilar shapes of R are intended to represent different conformations acquired by this protein.]

(*Adapted from Baxter and Forsham, 1972*)

1. The steroidal hormone enters the cytoplasm of the target cell where it binds with a specific, high- affinity receptor protein.
2. The receptor protein- hormone complex, so formed, then diffuses into (or is transported into) the nucleus, where it reacts with the nuclear chromatin.
3. Somewhere along this route, the receptor protein is structurally altered to form a smaller protein with low molecular weight. Or else the steroid hormone is transferred to a second smaller protein.
4. The combination of the small protein and hormone is now the active factor that stimulates the specific genes to form messenger RNA (mRNA) in the nucleus.
5. The mRNA diffuses into the cytoplasm where it accelerates the translation process at the ribosomes to synthesize new proteins.

It is, however, noteworthy that a direct chemical reaction of the hormone with DNA or RNA polynucleotide is not likely. Instead, the hormone must first combine with a specific receptor protein and it is this combination that acts on DNA chromatin. It is possible that the chromatin proteins may influence hormonal activity by modifying the ability of the receptor complex to bind with DNA.

The **thyroidal hormones** act similarly to enchance RNA and enzyme synthesis but may do so by directly binding with the sepcific receptor proteins present in the nuclear chromatin. The receptors present in the cytoplasm are less effective in this regard.

C. Stimulation of Enzyme Synthesis at Ribosomal Level. In the case of some hormones, the activity is at the level of translation of information carried by the mRNA on the ribosomes to the production of enzyme protein. For example, the ribosomes taken from animals, which have been given growth hormone, have a capacity for protein synthesis in the presence of normal mRNA.

D. Direct Activation at the Enzyme Level. It has been experimentally observed that treatment of the intact animal (or of isolated tissue) with some hormones results in a change in enzyme behaviour which is not related to *de novo* synthesis. The cell membrane is usually required for such activity. Henceforth, it is possible that activation of a membrane receptor might be an initial step in hormone action.

E. Hormone Action at the Membrane Level. Many hormones appear to transport a variety of substances, including carbohydrates, amino acids and nucleotides, across cell membranes. These hormones, in fact, bind to cell membranes and cause rapid metabolic changes in the tissues. Catecholamines and many protein hormones stimulate different membrane enzyme systems systems by direct binding to specific receptors on cell membrane rather than in the cytoplasm.

A schematic representation of the two principal mechanisms of action involving water soluble hormones and steroid hormones is presented in Fig. 19-67.

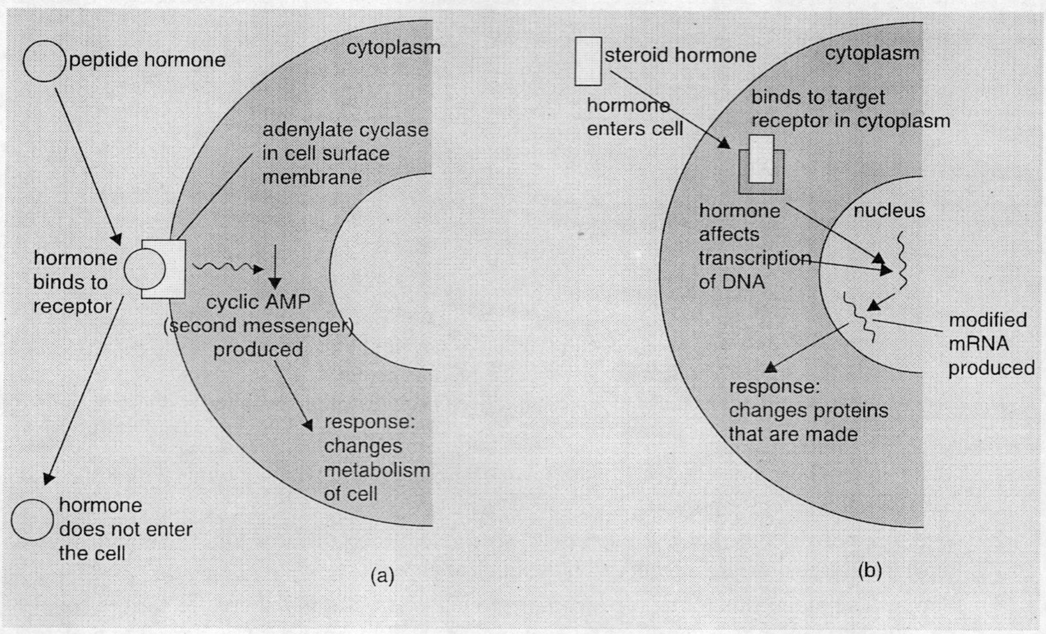

Fig. 19–67. Two principal mechanisms of hormone action

(*a*) Water-soluble hormones work *via* a second messenger which is formed inside the cell when they bind to receptors on the cell surface.

(*b*) When steroid hormones bind to a receptor inside the cytoplasm, they form a complex similar to an enzyme–substrate complex. This enters the nucleus of the cell and binds directly to the DNA, interfering with the cell's ability to read some of its genes. A steroid hormone can either switch on or switch off protein synthesis of particular genes. The overall effect of hormone action is different for each steroid hormone.

A summary chart of the various vertebrate hormones (Table 19-8) is given on the next page.

Table 19–18 Vertebrate Hormones

S.No.	Sectretory Organ	Hormone	Chemical Nature	Physiological Role	Hypofunction	Hyperfunction	Effect of Removal of Secretory Organ
1.	GONADS Ovary	Estrogens β-estradiol Estriol Estrone	C_{18} steriods (Derivative of estrane)	Proliferation of endometrium Deepening of uterine glands; Increase in vasculature; Development of secondary sex characters	—	—	Suspension of menstrual and reproductive cycles; Atrophy of uterus & vagina; Involuted mammary glands
	Testes	Androgens Testosterone Androsterone DHEA Androstenedione* Androstenetrione	C_{19} steriods (Derivative of androstane)	Development of secondary sex characters	—	—	Ossification of long-bone epiphyses; Increased adiposity; Extrusion of chest; Suppressed penis
2.	ADRENALS Cortex	Corticosteroids Aldosterone DOC Cortisone Cortisol Corticosterone	C_{21} steriods	Water-salt balance of the body; Regulate NaCI contents of blood; Control carbohydrate metabolism; Inhibit protein synthesis; Regulate lipogenesis	Decreased metabolism; Excessive pigmentation; Anorexia; Anemia; Hypoglycemia	Cushing's syndrome	Addison's disease
	Medulla	Epinephrine Norepinephrine	Amino acid derivatives	Regulate carbohydrate metabolism; Regulate blood pressure	—	—	Pheochromocytoma
3.	CORPUS LEUTEUM	Progesterone	C_{21} steriod (Derivative) of pregnane)	Proliferation of uterine mucosa; Growth of mammary glands; Inhibition of uterine contraction	—	—	—
		Relaxin	Polypeptide	Relaxation of pelvic ligaments during childbirth	—	—	—

Table 19–8. (Continued)

S.N.	Secretory Organ	Hormone	Chemical Nature	Physiological Role	Hypofunction	Hyperfunction	Effect of Removal of Secretory Organ
4.	ISLETS OF LANGERHANS						
	β-cells	Insulin	Polypeptide – 51 AARs†	Lowers sugar level in blood; Promotes protein synthesis; Promotes lipid synthesis	Diabetes mellitus	—	—
	α-cells	Glucagon	Polypeptide –29 AARs	Increases sugar level in blood Catabolic action on proteins; Accelerates ketogenesis	Hypoglycemia	—	—
5.	HYPOPHYSIS						
	Pars distalis	Thyrotropin	Polypeptide	Stimulates thyroid activity	Dwarfism	Gigantism (in children)	Atrophy of gonads, thyroid and adrenal cortex; Loss of body tissue
		Corticotropin	Polypeptide	Stimulates adrenal cortex	Panhypo-pituitarism		
		Gonadotropins					
		FSH	Polypeptide	Induces growth of graafian follicles	Pituitary myxedema	Acromegaly (in adults)	
		LH	Polypeptide	Induces ripening of ovarian follicles			
		LTH	Polypeptide	Lactation during parturition			
		Somatotropin	Polypeptide	Controls skeletal growth			
	Pars intermedia	Intermedins	Polypeptide	Darkening of the skin			
	Pars nervosa	Ocytocin	Polypeptide –9 AARs	Stimulates contraction of smooth muscles *esp.* of uterus			
		Vasopressin	Polypeptide –9 AARs	Causes a rise in blood pressure; Antidiuretic action			

Table 19–8. (Continued)

S.N.	Secretory Organ	Hormone	Chemical Nature	Physiological Role	Hypofunction	Hyperfunction	Effect of Removal of Secretory Organ
6.	GASTROINTES-TINAL TRACT	Secretin Pancreozymin Gastrin	Polypeptide Polypeptide Polypeptide	Stimulates flow of pancreatic juice Stimulates flow of pancreatic juice Stimulates flow of gastric	—	—	—
7.	PARATHYROID	Parathormone	Polypeptide –84 AARs	Bone resorption; Renal reabsorption of calcium	Tetany	Osteitis fibrosa	—
8.	THYROID	TIT or T_3 Thyroxine, T4	Amino acid derivative Amino acid derivative	Deamination reactions in the liver; Deiodination in extrahepatic tissues	Cretinism (in children) Myxedema (in adults)	Exophthalmia; Goiter	—

* Secreted from adrenal gland

† AARs = Amino acid residues

PROBLEMS

1. Ingesting large amounts of glucose before a marathon might seem to be a good way of increasing the fuel stores. However, experienced runners do not ingest glucose before a race. What is the biochemical reason for their avoidance of this potential fuel ? (**Hint** : Consider the effect of glucose ingestion on the level of insulin.)
2. Insulin-dependent diabetes is often accompanied by hypertriglyceridemia, which is an excess blood level of triacylglycerides in the form of very low density lipoproteins. Suggest a biochemical explanation.
3. Sildenafil (Viagra) is a drug widely used to treat male impotence. Sildenafil exerts its effect by inhibiting a cGMP phosphodiesterase isozyme (PDE5) that is especially prevalent in smooth muscle. Interestingly, certain airlines restrict pilots from flying for 24 hours after using sildenafil. Suggest a reason for this restriction.
4. Marathon runners preparing for a race engage in "carbo loading" to maximize their carbohydrate reserves. This involves eating large quantities of starchy foods. Why is starch preferable to candy of sugar-rich foods ?
5. During a "fight or flight" situation, the release of epinephrine promotes glycogen breakdown in the liver, heart, and skeletal muscle. The end product of glycogen breakdown in the liver is glucose. In contrast, the end product in skeletal muscle is pyruvate.
   (*a*) Why are different products of glycogen breakdown observed in the two tissues ?
   (*b*) What is the advantage to the organism during a "fight or flight" condition of having these specific glycogen breakdown routes ?
6. What are the possible advantages in the synthesis of hormones as prohormones or preprohormones ?
7. Which of these hormones inhibits FSH production in a female mammal ?
   (*a*) estrogen
   (*b*) adrenalin
   (*c*) thyroxine
   (*d*) testosterone
   (*e*) luteinizing hormone
8. Does it make any difference when you take a thyroid hormone ?
9. During pregnancy, the chief source of progesterone is the :
   (*a*) ovary
   (*b*) placenta
   (*c*) pituitary gland
   (*d*) mammary gland
   (*e*) uterus
10. Name two animal hormones that are peptides (or proteins) and steroids.
11. Name two animal hormones that are antagonistic in their effects.
12. Why are diabetes patients more sensitive to skin diseases ?
13. On the basis of their physical properties, hormones fall into one of two categories: those that are very soluble in water but relatively insoluble in lipids (*e.g.*, epinephrine) and those that are relatively insoluble in water but highly soluble in lipids (*e.g.*, steroid hormones). In their role as regulators of cellular activity, most water-soluble hormones do not penetrate into the interior of their target cells. The lipid-soluble hormones, by contrast, do penetrate into their target cells and ultimately act in the nucleus. What is the correlation between solubility, the location of receptors, and the mode of action of the two classes of hormones ?

14. Which is the most widely-used drug in the world ?
(*a*) Aspirin (*b*) Morphine (*c*) Paracetamol (*d*) Progesterone

15. Acetylcholine is a :
(*a*) Neurotransmitter (*b*) B-vitamin (*c*) Enzyme (*d*) Protein

16. Name the gland which acts both as an 'endocrine' as well as an 'exocrine' gland.
(*a*) Pancreas (*b*) Pituitary (*c*) Thyroid (*d*) Adrenal

17. Can obesity affect life expectancy in humans ?

18. Why is edible salt iodized ?

19. The gland that degenerates during adolescence is :
(*a*) Thymus (*b*) Liver (*c*) Pituitary (*d*) Pineal

20. Excess production of thyroxine leads to :
(*a*) Diabetes (*b*) Cretinism (*c*) Gigantism (*d*) Exophthalmic goitre

21. Which is the hormone that lowers the concentration of calcium in the blood ?
(*a*) Somatostatin (*b*) Ocytocin (*c*) Calcitonin (*d*) Androgens

22. Which of the following contains iodine ?
(*a*) Insulin (*b*) Thyroxine (*c*) Ocytocin (*d*) Adrenalin

23. How do one benefits from taking a daily aspirin ?

24. Can bananas help fight depression ?

25. What foods have anti-inflammatory qualities ?

ANSWERS TO PROBLEMS

1. A high blood-glucose level would trigger the secretion of insulin, which would stimulate the synthesis of glycogen and triacylglycerols. A high insulin level would impede the mobilization of fuel reserves during the marathon.

2. Lipid mobilization can occur so rapidly that it exceeds the ability of the liver to oxidize the lipids or convert them into ketone bodies. The excess is reesterified and released into the blood as VLDL.

3. The phosphodiesterase in photoreceptor cells is homologous to PDE5 and is inhibited to some extent by sildenafil, leading to visual side effects.

4. It is presumed that starch increases blood glucose levels less than simple sugars do. Thus, there is less stimulation of insulin secretion. Insulin would tend to retard energy mobilization from intracellular stores– something not desirable during a marathon.

5. (*a*) Heart and skeletal muscle lack the enzyme glucose-6-phosphate phosphatase. Any glucose-6-phosphate that is produced enters the glycolytic pathway, and under O_2-deficient conditions is converted into lactate *via* pyruvate.

(*b*) Phosphorylated intermediates cannot escape from the cell, because the membrane is not permeable to charged species. In a "fight or flight" situation, the concentration of glycolytic precursors needs to be high in preparation for muscular activity. The liver, on the other hand, must release the glucose necessary to maintain the blood glucose level. Glucose is formed from glucose-6-phosphate and passes from the liver cells to the bloodstream.

6. Because prophormones and preprohormones are inactive, they can be stored in quantity in secretory granules. Rapid activation is achieved by enzymatic cleavage in response to an appropriate signal.

7. (*a*) estrogen.

8. Yes. Take it on an empty stomach and do not eat immediately afterward since food alters its absorption.

9. (*b*) placenta.
10. Peptide (or protein) hormones : Insulin and Growth hormone
Steroid hormones : Testosterone and Estradiol
11. Insulin and glucagon.
12. In diabetes mellitus, which is a disorder of the carbohydrate metabolism, glucose cannot enter the cells but remains concentrated in the blood instead. Thus, the cells starve in the midst of plenty. The increased amount of sugar in the blood also brings about adverse changes in the membrane transport system of the cells. The blood circulation is also impaired. A net result of all these changes is that opportunistic bacteria etc. flourish and skin infections especially at the extremities (toes, etc) are common. Increased incidence of skin infection is only one of a group of complications that may set in due to diabetes. Skin is easily affected as it is the most easily accessible part of the body.
13. Because of their low solubility in lipid, water-soluble hormones cannot penetrate the cell membrane; they bind to receptors on the outer surface of the cell. In the case of epinephrine, this receptor is an enzyme that catalyzes the formation of a second messenger (cAMP) *inside* the cell. In contrast, lipid-soluble hormones can readily penetrate the hydrophobic core of the cell membrane. Once inside the cell, they can act on their target molecules or receptors directly.
14. (*a*) Aspirin
15. (*a*) Neurotransmitter
16. (*a*) Pancreas
17. A new study suggests that being obese during adulthood decreases lifespan by 3 years. Obesity is defined as having a body mass index (BMI) of 30 or more. Obesity risks are similar to smoking risks
18. Iodine is an essential nutrient which is required in minute amounts for normal growth of animals and humans. Iodine deficiency results in impairment of the functioning of the thyroid gland, which produces 'thyroxine' hormone. *Since thyroxine controls the rate of body metabolism and growth, iodine deficiency leads to a variety of disorders like goitre in children or adults and impaired mental functioning, poor intellectual performance and muscular disorders in children.* Women with iodine deficiency give birth to stillborn babies. Iodization of salt offers the most convenient way of combating iodine deficiency. Firstly, because salt is an essential ingredient of our diet and secondly, one can never take too much of it. So by iodization of common salt, a minimum intake of iodine can be ensured for the entire population, thereby minimizing the incidence of iodine deficiency.
19. (*a*) Thymus
20. (*d*) Exophthalmic goitre
21. (*c*) Calcitonin
22. (*b*) Thyroxine
23. Small daily aspirin dose from 75 to 325 mg is widely accepted as a strategy for cutting the risk of heart attack by up to 44%. It has also been shown to reduce the risk of stroke, possibly colon cancer and dementia.
24. Bananas are a mood-enhancing food because they contain tryptophan, a type of amino acid that the body converts into serotonin. Serotonin is a hormone that helps you relax and improves mental well-being.
25. Grapes, broccoli, chilli, peppers and strawberries all contain salicylic acid, the same inflammation-fighting compound found in aspirin. Many studies have shown that these foods protect the heart, and they taste a lot better than aspirin.

CHAPTER

20 Plant Hormones

INSIDE

- Definition
- Auxins
- Gibberellins
- Cytokinins (= Kinins)
- Other Natural Growth Hormones In Plants
  - Ethylene
  - Traumatic acid
  - Calines
  - Vitamins
- Growth Inhibitors
  - Abscisic acid
  - Morphactins
- Plant Hormones Versus Animal Hormones

Effect of gibberellins on plant growth

The gibberellins cause stems to lengthen dramatically. The plant's stem "bolts" when levels of gibberellic acid rise.

(Courtesy : B.O. Phinney, University of California)

DEFINITION

Thimann (1948) designated the plant hormones by the term '*phytohormones*' in order to distinguish them from animal hormones. He defined a phytohormone as "*an organic compound produced naturally in higher plants, controlling growth or other physiological functions at a site remote from its place of production and active in minute amounts*."

This definition includes a variety of compounds, besides those responsible for growth curvatures in organs like *Avena* coleoptile. For example, it embraces the hormones which induce flowering, wound-healing and also those vitamins which act as growth factors.

A definition of plant hormones with still wider scope has been given by **Johannes van Overbeek** (1950). According to him, the plant hormones are defined as "*organic compounds which regulate plant physiological process— regardless of whether these compounds are naturally occurring and/or synthetic ; stimulating and/or inhibitory ; local activators or substances which act at a distance from the place where they are formed.*"

The migratory nature of hormones has been specifically emphasized by **Meirion Thomas** (1956) who stated that "all hormones are migratory correlating substances or correlators which play an essential part in the integration of plant behaviour."

Three types of plant hormones are usually recognized. These are *auxins, gibberellins* and *cytokinins*. These were discovered in the early decades of the twentieth century, in 1930's and in 1960's respectively. Naturally, the knowledge accumulated on auxins and gibberellins is far greater than that gathered for cytokinins.

AUXINS

Definition

Kögl and **Haagen-Smit** (1931) introduced the term 'auxin' (*auxein*[G] = to grow or to increase) for designating those plant hormones which are specially concerned with cell enlargement or the growth of the the shoots. An auxin may, thus, be defined as "an organic substance which promotes growth (*i.e.*, irreversible increase in growth) along the longitudinal axis when applied in low concentrations to shoots of the plants freed as far as practicable from their own inherit growth-promoting substances. Auxins may, and generally do, have other properties but this one is critical" (**Thimann**, 1948). This definition is nowadays widely accepted among plant physiologists.

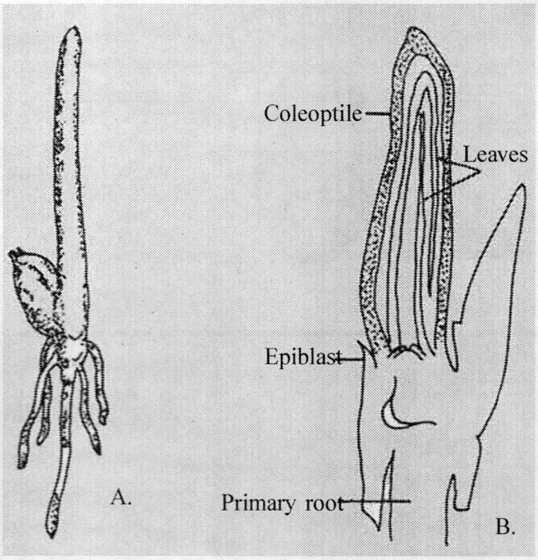

Fig. 20–1. Structure of oat coleoptile

Oat Coleoptile and the Auxins

Since long the action of auxin has been clearly demonstrated in the leaf sheath or coleoptile of oat plant (*Avena sativa*). The coleoptile (Fig. 20–1) is a tubular structure with a conical top and encloses the first-formed leaf in it. A transection of the coleoptile reveals that it consists of an epidermal layer and a few parenchyma cells with two vascular bundles running longitudinally.

The experimental utility of oat coleoptile lies in the fact that the process of cell elongation can be easily studied over here. After four days, the primary leaf breaks through the coleoptile at its tip and the organ stops growing. Thus, cell divisions cease relatively early in the development of oat coleoptile and further growth in length takes place as a result of cell elongation only. Elongation is confined to a region about 1 cm below the tip of the coleoptile. The rate of elongation is most rapid (about 1 mm per hour) when the seedling is about 3 days old. At this time, the coleoptile may measure about 3 cm in length. It is mostly at this stage that it is experimentally used for conducting growth experiments.

Hormone Concept

The presence of growth-regulating hormones in plants was first suggested by **Julius von Sachs** in 1980. He proposed that there were certain '*organ-forming substances*' in plants which were produced in the leaves and translocated downward in plant body.

Also in 1880, **Charles Darwin**, an evolutionist, studied the effect of unilateral light on plant movements. While conducting his experiments on canary grass (*Phanaris canariensis*), he found that if the coleoptile tip is provided light from one side only (*i.e.*, unilateral illumination), the tip would bend towards light. In the absence of illumination, however, no curvature could be induced (refer Fig. 20–2 for this and other subsequent experiments).

The material nature of the hormones was first conclusively demonstrated by a Dane, **Peter Boysen-Jensen** (1910). He first cut off (or decapitated) the coleoptile tip a few millimeters from the apex, then put a block of gelatin on the decapitated stump and ultimately replaced the cut tip on the gelatin block. Upon unilateral illumination, the coleoptile showed curvature towards light.

This he attributed to the fact that the 'stimulatory effect' can diffuse through gelatin and is a type of soluble material.

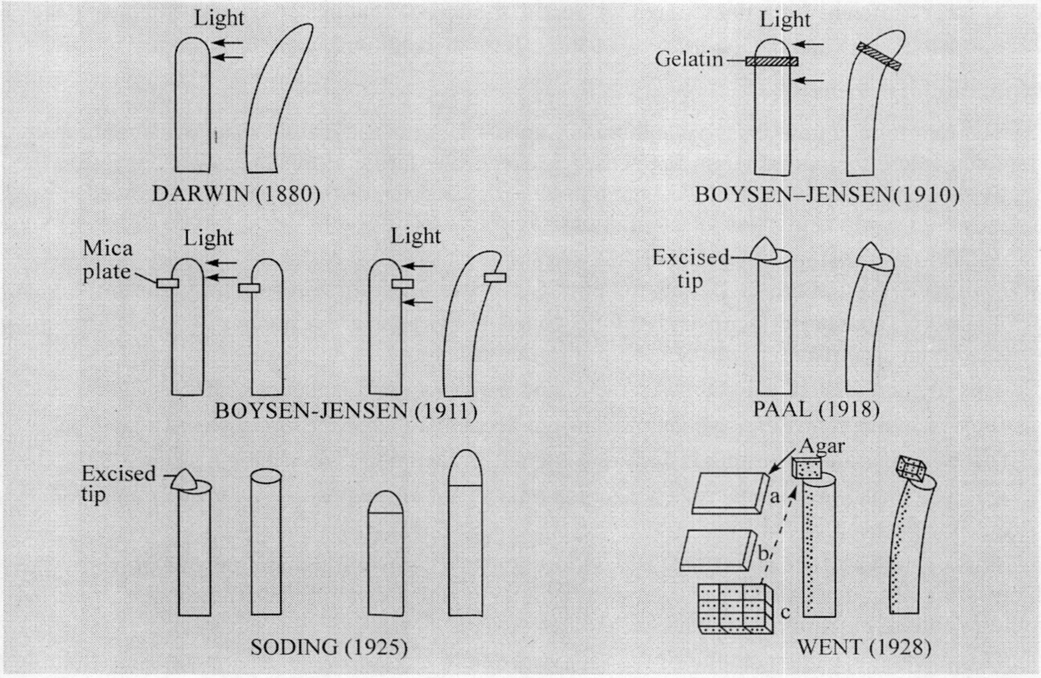

Fig. 20–2. Major experiments in the discovery of auxins

A year later (*i.e.*, in 1911) in another experiment, he made a transverse slit halfway through the coleoptile below the tip on one side and inserted a piece of mica in the slit. Upon unilateral illumination from the side opposite the slit, no phototropic curvature was observed. If, however, the operated coleoptile is illuminated on the side of the slit, the phototropic bending towards the source of light is seen. From this was inferred that the stimulus for bending passes down the dark side of the coleoptile.

In 1918, the Hungarian plant physiologist, **Arpad Paál** of the University of Budapest in an experiment decapitated coleoptile and replaced the tip eccentrically. He found that the coleoptile showed a negative curvature, *i.e.*, bent away from the side with the tip even in the dark. Paal, therefore, concluded that the material substance (called as '*correlation carrier*' by him) diffuses from the tip downwards and stimulates growth of the cells below the tip.

Soding (1925) found that decapitation of a coleoptile markedly retards the rate of cell division. But reheading of the tip on the cut stump resulted in resumption of vertical growth due to cell elongation.

The task of isolation, extraction and bioassay of these growth-promoting substances was admirably done by a Dutch botanist, **Frits W. Went**, in 1928. He worked in his father's laboratory at the University of Utrecht in the Netherlands. He placed numerous freshly-cut coleoptile tips on an agar block for some time. Later, this block was divided into small rectangular blocks (*a*, *b* and *c*). These agar blocks were, then, placed eccentrically on decapitated coleoptiles for 2 hours in dark. In all cases, the coleoptiles showed curvature towards the side opposite to the agar blocks (*i.e.*, negative curvature). **Went** also devised a method for the bioassay (to be described later) of these substances.

Cholodny in Russia also worked on similar lines independently and arrived at the same conclusions.

The availability of a technique of extracting the hormone opened up new vistas in this field and within a short span of about 8 years (from 1928 to 1936) extensive work was done. With the result, three different hormones (auxin *a*, auxin *b* and heteroauxin) were identified and their characteristics noted mainly on account of the efforts of **Kögl** and his colleagues in Holland.

Extraction of Auxins

The two forms of auxins (free and bound) appear to be in a dynamic state as there are many examples where the bound auxin is released in free state during extraction. With the result, the strictly separate measurement of free and bound auxins is often difficult. There are, however, two methods commonly employed for auxin extraction.

A. Diffusion method. It was devised by **Went** (1928) at Utrecht. In this method the growing tip (or other organ to be tested), under conditions of low transpiration, is severed (or cut) and is then placed on an agar block (usually of 1.5 concentration) for about an hour or so. During this period, the auxin diffuses from the cut tip into the agar block.

This method, though simple, has some major *drawbacks* and is, henceforth, not widely used.

(*a*) Excessive transpiration may prevent the accumulation of the auxin in the agar block.

(*b*) Severing the tip results in lowering the amount of auxin from the cut surface.

(*c*) The method cannot be widely adopted on account of the presence of growth inhibitors in many green plants.

B. Solvent extraction method. Here, the tissues are grinded in some organic solvents like chloroform, ether, ethyl alcohol or even water and the liquid is then filtered. The auxin is separated from the filtrate by chromatographic technique.

This method is widely employed for the extraction of auxins, *esp*., the bound auxins. But this one also suffers from certain *drawbacks*.

(*a*) Use of chloroform as a solvent causes slow accumulation of chlorine which is a toxic substance and probably an auxin inactivator too.

(*b*) Diethyl ether, if used as a solvent, brings about oxidation of the auxin in the presence of a spontaneously formed peroxide. This can, however, be avoided if the solvent is distilled with ferrous sulfate and calcium oxide before use.

(c) During auxin extraction, a new auxin may be produced which may thus contaminate the auxin to be extracted. The difficulty can be overcome by employing Gustafson's technique which involves the boiling of plant material for about a minute prior to extraction.

Bioassay of Auxins

The term *bioassay* refers to determining the amount of active substance present in the plant tissues. In the various methods employed for bioassay, the activity of the auxin is in general determined by making it available in certain concentration to a seedling (or an ovary or a root) and the degree of either acceleration or inhibition of growth is recorded.

One of the most commonly employed methods of bioassay of auxins, as devised by **Went** (1928), is known as **Avena curvature test**. The test involves the following steps (Fig. 20–3).

Went found that the degree of angular curvature of coleoptile tip is proportional, within limits, to the concentration of auxin present in the agar block. This fact he made the basis of his *Avena test*.

Kögl and **Haagen-Smit** (1931) used *Avena test* as the unit of measurement. It is termed *Avena Einheit* and is abbreviated as A.E. One A.E. is defined as the amount of auxin present in an agar block (2 × 2 × 1 cm) which produces a curvature of 10° to a decapitated *Avena* coleoptile when placed eccentrically on it for 90 minutes.

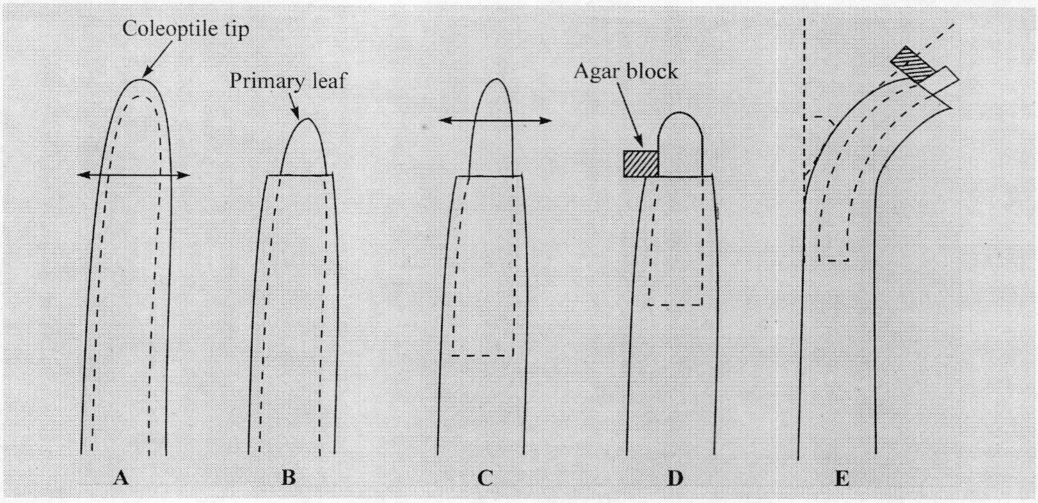

Fig. 20–3. Steps involved in Avena curature test

A. Decapitation of the coleoptile tip containing the primary leaf.

B. Pulling out of the exposed primary leaf so that its elongation may not dislocate the agar block.

C. Cutting the tip of primary leaf.

D. Affixing the agar block containing the auxin unilaterally to the cut tip.

E. Reading the angle resulting from movement of auxin into the side of coleoptile on which agar block is placed.

Biochemistry of Auxins

Application of *Avena* test to a wide variety of substances led to the discovery of auxins in human urine. **Kögl** and **Haagen-Smit** (1931) isolated 40 mg of auxin (which they named as *auxin a*) from as much as 33 gallons of human urine. Three years later, Kögl and his associates isolated two more compounds with auxin activity. These were named as *auxin b* and *heteroauxin* and were obtained from corn germ oil and human urine respectively. Their structure is shown in Fig. 20–4.

1 gallon	=	3.78533 litres (U.S. units)
	=	4.54596 litres (British units)

1. **Auxin a, auxentriolic acid.** It occurs at the meristematic apices (buds and growing leaves) both in the free state and bound to plasma proteins. It is a weak acid and is soluble in water, alcohol, ether and chloroform. It is stable in acid solutions but decomposes in alkaline soultions (*i.e.*, acid-stable and alkali-labile).

2. **Auxin b, auxenolonic acid.** It is present in corn germ oil, other vegetable oils, malt and a fungus called *Rhizopus*. It is also a weak acid and is soluble in water, alcohol, ether and chloroform. It is both acid-labile and alkali-labile. This as well as auxin *a* both are derivatives of cyclopentene.

3. **Heteroauxin, indole-3-acetic acid.** It is of universal occurrence in plants and is also synthesized by microorganisms including certain bacteria, yeasts and fungi like *Rhizopus*. It is resistant to alkalies whereas destroyed by acids and undergoes rapid decomposition on heating. Unlike the first two, it can be easily synthesized in the laboratory. Chemically, it is a monobasic acid of a relatively simple structure.

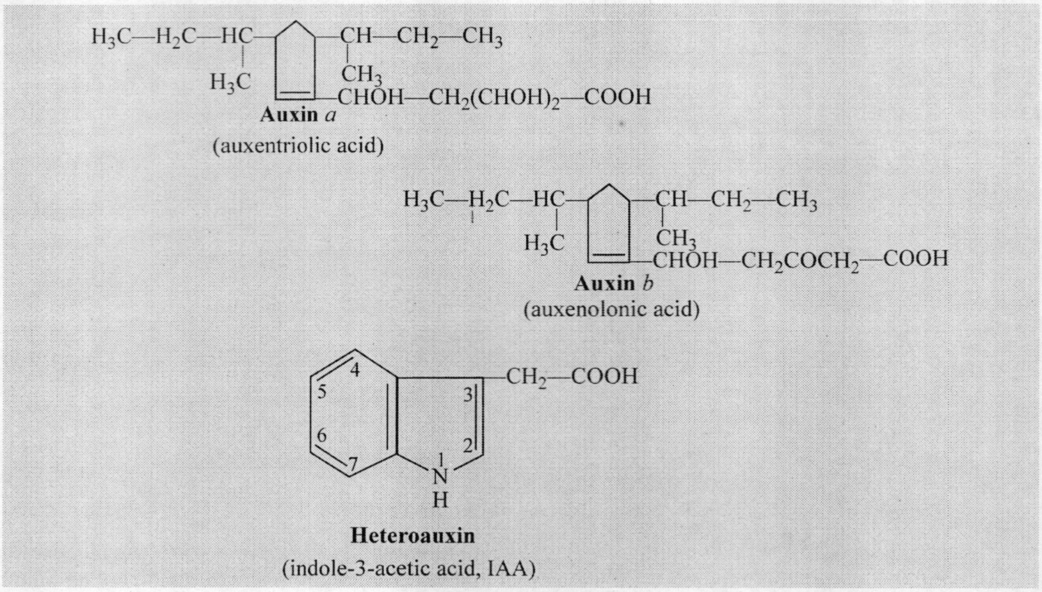

Fig. 20–4. Natural hormones

Biogenesis (= Synthesis) of Auxins

Folke Skoog (1937) of the University of Wisconsin, for the first time, experimentally proved that tryptophan is an auxin precursor in higher plants. Since then a wide variety of plant tissues (like leaf, stem, buds, coleoptile, ovary, pollen, embryo, endosperm and callus tissue) have been shown to convert tryptophan to indoleacetic acid (**Larsen**, 1951). In fact, it is quite probable that all living plant tissues may have the capability of bringing about this conversion. Although tryptophan is the primary precursor of IAA, about half a dozen indole compounds have been found to serve as potential precursors of IAA.

For the enzymic production of IAA, a plausible hypothesis has been suggested by **Wildman, Ferri** and **Bonner** in 1946. According to this hypothesis (Fig. 20–5), tryptophan, liberated by hydrolysis of proteins, undergoes either oxidative amination first and then decarboxylation or vice versa to yield indoleacetaldehyde (IAc). IAc is then oxidized to yield the free auxin. Indoleacetaldehyde, thus, acts as an intermediate metabolite as well as an immediate precursor of IAA.

The capability of living plant tissues to form auxin from tryptophan has been demonstrated for spinach (**Wildman** *et al*, 1947), pineapple (**Gordon** and **Nieva**, 1949) and many others. The three essential conditions for IAA synthesis are the presence of light, zinc and an enzyme system.

Distribution of Auxins

The greatest concentration of auxins is usually found in the growing apices of the plant, *i.e.*, in the coleoptile tip, in buds and in the growing tips of leaves and roots. However, auxin is found widely distributed throughout the plant body. In general, it may be stated that where there is active growth, there is auxin production. The formation of auxin by a mature organ like leaf, however, suggests that growth may not be the pre-requisite to auxin production.

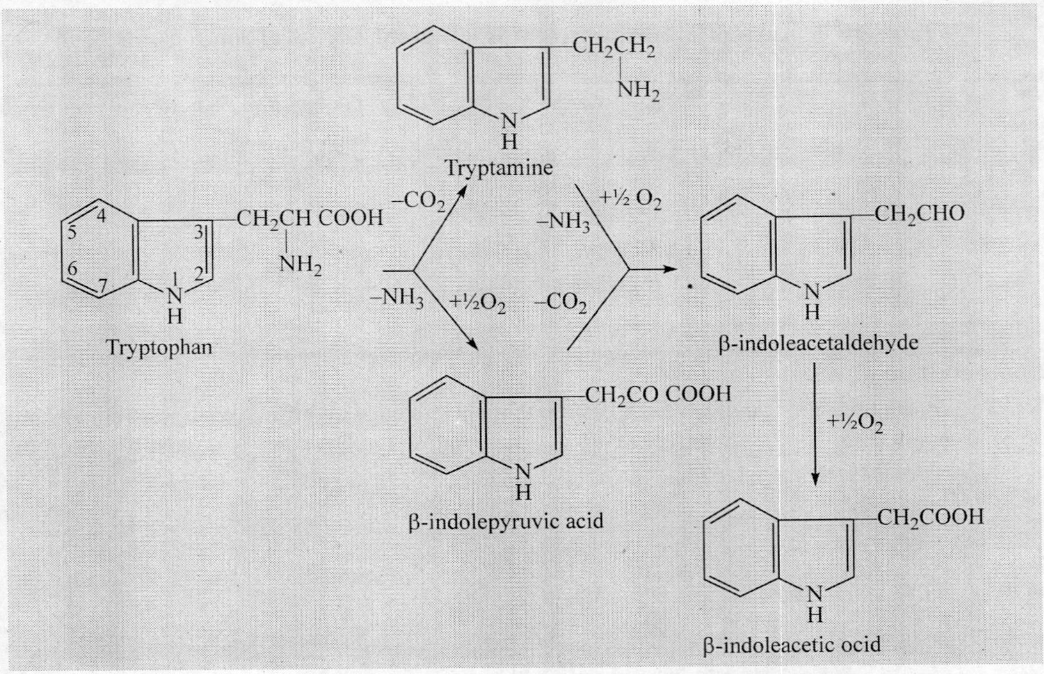

Fig. 20–5. Possible possible pathways of auxin synthesis from tryptophan

Thimann (1934) studied the distribution of auxin, in detail, in etiolated *Avena* coleoptile (Fig. 20–6). He found that the concentration of auxin drops as one progresses from the coleoptile tip to its base; the highest concentration being at the tip and the lowest at the base. If one progresses further from the base of the coleoptile along the root, there is a steady increase in auxin content till a maximum is reached in the root tip. Of the two maximal values, that for the stem tip is much higher than that for the root tip.

Thimann and **Skoog** (1934), while working on *Vicia faba* seedlings grown in light, found the concentration of auxins in various organs in the following descending order :

Apical buds > Young leaves > Mature leaves

The amount of diffusible auxin per hour for these organs was found to be approximately in the ratio of 12 : 2 : 1.

Van Overbeek (1947) studied the distribution of both free and bound auxins in pineapple. He found that large quantities of free auxin occurred in apical buds and lowest amount in mature leaves. For bound auxin, however, the condition was found to be reverse.

A few other examples of auxin concentration studies are given in Table 20–1.

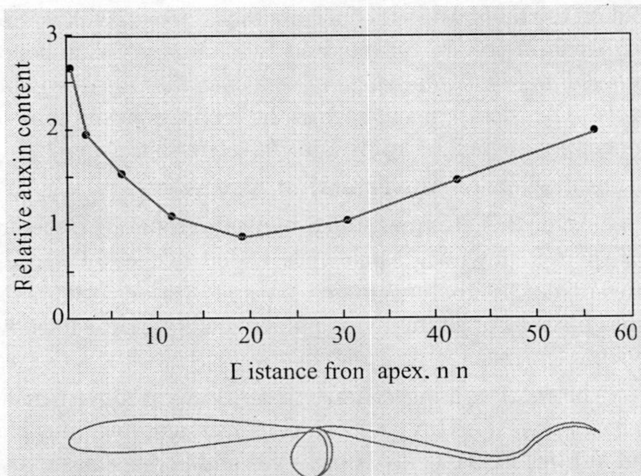

Fig. 20–6. Distribution of auxin in an etiolated Avena seedling

(*After Thimann KV, 1934*)

Table 20–1. Amount of auxin present in different organs of some plants

S.N.	*Plant*	*Organ*	*Maximum Concentration Found (in μg IAA equivalent)*
1.	Corn	Endosperm	105, 000
2.	Lily	Stem tip	83,900
3.	Oat	Grain	1,000
4.	Rice	Endosperm	250
5.	Turimp	Seed	250

Concentration of Auxins

The concentration of auxins, which has a profound influence on growth changes, varies from organ to organ (Fig. 20–7). The effect of auxin concentration on the shoots is quite different from

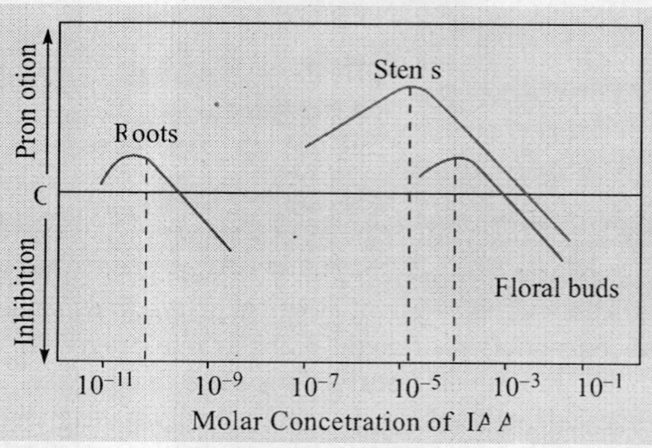

Fig. 20–7. Graph showing the interrelationship between growth promotion (or growth inhibition) and various plant organs such as roots, stems and floral buds

The optimum concentrations for growth promotion were found to be between 10–11 and 10–9 for roots, 10–8 and 10–3 for stems and 10–5 and 10–3 for floral buds.

(*After Leopold AC and Thimann KV, 1949*)

that on the roots. Higher concentration of auxin, which has growth-stimulatory effect on the shoots, is growth-inhibitory for the roots ; the latter growing better at much lower concentration. Thus, *in general, the optimum range of concentration for elongation in stems is much higher than for the roots*. The stems grow best at an auxin concentration of 1.0 mg/litre, whereas the optimum concentration for the roots is of 0.001 mg/litre.

Translocation (= Movement) of Auxins

The auxins are transported in plants from one organ to the other. The usual direction of auxin transport is downward but when added to the soil these are absorbed by the roots and carried upward along with transpiration stream to various plant organs. The prevailing downward movement takes place through the living phloem cells whereas the upward movement occurs through the dead xylem elements.

The most striking characteristic of auxin movement is its almost strict *basipetal* (from apex to base) *polarity*. This has been demonstrated in various plant organs such as oat coleoptiles (**Went** and **White**, 1939), petioles of leaves and herbaceous and woody strems (**Oserkovsky**, 1942). The actual proof of polar transport was furnished by **Went** (Fig. 20–8). He showed that if an agar block containing auxin be affixed to the morphologically upper end of a coleoptile segment and a block of pure agar to the lower end, auxin would move and collect in the agar block at the lower end.

But if the coleoptile segment is inverted, no translocation of auxin would occur. The translocation of auxins in oat takes place through the parenchyma tissue.

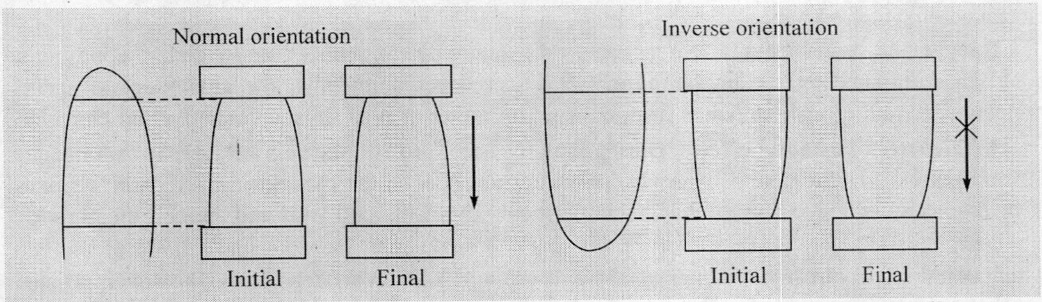

Fig. 20–8. Experiment demonstrating basipetal movement of auxin in oat coleoptile

(*After Went FW, 1935*)

In *Coleus*, however, **Leopold** and **Guernsey** (1953) have shown that the basipetal polarity becomes progressively weaker as the distance from the shoot apex increases. Further, in the flowering stems there is some acropetal (from base to apex) movement also even at the stem tip. **Jacobs** (1961) has also shown that in *Coleus* stem sections, the ratio of basipetal to acropetal transport of auxin is 3 : 1. In fact, there exists a polarity gradient in *Coleus* from a complete basipetal polarity in vegetative apex to a complete acropetal polarity in root apex with a gradual transition in between.

The velocity of auxin transport varies from 26 mm per hour to 6.4 mm per hour (**Rajgopal,** 1967). These rates are higher than the rates of diffusion. The velocity of auxin transport is unaffected by temperature, although the amount of auxin transported is proportional to the temperature. Also, the distance over which the transport occurs does not influence the velocity.

Mechanism of Auxin Action

The action of auxins, like growth itself, seems to be a complex of many functions. Although the mechanism of auxin action may, in part, be attributed to each of these functions, none of these can account for *in toto* the multifarious effects of these growth substances. The various views put forward to explain auxin behaviour may be, for convenience, grouped under following five headings :

1. Molecular reaction theories. **Skoog** (1942) expressed the view that the auxin may act like a coenzyme and serves as a point of attachment for some substrate onto an enzyme regulating growth. The molecular configuration of the auxin affects the activity by altering the fit and functioning of this molecular union. The higher auxin concentrations would inhibit growth owing to separate molecules combining with the enzyme and the substrate.

Muir *et al* (1949) advanced the hypothesis that the auxins (*esp.*, phenoxy acids) may combine with some material (*e.g.*, protein) in the cell at two points, the ortho position of the ring and the acid group of the side chain

Foster and his associates (1952) put forward a theory of auxin action by *2-point attachment*. It is presumed that the enzyme is attached to a substrate to form an enzyme-substrate complex and that the complex may, then, dissociate to produce the end product of the reaction (*i.e.*, growth) and regenerate the enzyme. Considering the enzyme as the material with which auxin reacts, the theory may thus be expressed as :

$$E + S \rightleftharpoons ES \longrightarrow \text{Growth} + E$$

[where E = auxin receptor ; S = auxin ; ES = complex]

2. Theories of enzymatic effects. The fact that growing tissues, upon treatment with auxin, show an increased activity of a number of enzymes has proved to be the basis of these theories. **Northen** (1942) pointed out that the auxin causes decrease in cytoplasmic viscosity and also brings about dissociation of the cytoplasmic proteins. The latter effect would result in an increase in water permeability and also in the osmotic value of the cytoplasm. These effects ultimately lead to enhanced enzymic activity.

Burger and **Avery** (1942) demonstrated that some *dehydrogenases*, under certain conditions, could be stimulated by auxin. **Thimann** (1951), however, found that the auxins act as agents protecting certain growth enzymes from destruction rather than as substances activating enzymes.

3. Theories of osmotic effects. During the process of growth, the cell increases in volume due to water uptake. The uptake of water occurs due to changes in the cytoplasm itself (*esp.*, changes in the osmotic value) or due to changes in permeability of the cell wall and the cell membranes.

Czaja (1935), for the first time, stated that the auxin may increase the volume of the cell which could result directly in water uptake and growth. **Van Overbeek** (1944), however, pointed out that growth is not necessarily associated with an increase in osmotic value.

Commonor and his associates (1942–43) suggested that since water uptake in growth may be linked with respiration, the process of growth may be explained as due to the osmotic uptake of water which is, in its turn, activated by respiratory uptake of salts.

Two important *objections* put against this theory are as follows :

(*a*) In some instances (*e.g.*, potato slices), growth cannot be a function of salt accumulation causing an increase in osmotic value (**Van Overbeek**, 1944).

(*b*) **Reinders** (1942) points out that the auxin-induced growth can occur in pure water in the absence of salt uptake. She describes water uptake as dependent upon oxidative metabolism.

Thus, water uptake is not the cause but the consequence of growth (**Burström**, 1953).

4. Theories of cell wall effects. Hyen (1940) attributed growth to the dynamic function of the cell wall instead of the cytoplasm. He observed that auxin application increases flexibility and extensibility of the cell wall. This results in lowering the wall pressure around the cell wall, thus permitting water uptake due to this simple drop in turgor pressure.

The elasticity of cells always increases at the start of cell stretching but decreases again before the cells have reached maturity. Thus, increasing elasticity cannot cause elongation but is connected with the elongation process. **Ruge** (1942) has suggested that cell elongation proceeds in two different phases :

(*a*) an increasing extensibility of the wall without synthesis of new wall material.

(*b*) a hardening of wall with a deposition of new wall material through either intussusception or apposition.

The theory envisages that the effect of auxin upon growth is attained through activating cell wall growth. The wall is believed to be made plastic enough for cell extension and the deposition of new cell wall material, then, causes cell enlargement. To keep up with the growing wall, the cytoplasm must be capable of taking up sufficient water through osmosis.

5. Theories of toxic metabolism. Besides promoting growth, the auxins can also inhibit elongation. In short this inhibition is brought about by high auxin concentration whereas in roots the inhibition is induced even by relatively low auxin concentrations. This inhibition is thought to be due to excess auxin molecules inactivating the sites of auxin action and, thus, checking maximum growth response.

Van Overbeek (1951) proposed that growth regulator toxicity may be a result of an alternation of metabolism in such a manner that unsaturated lactones are accumulated in plant tissues. These are toxic to plants when applied in higher concentrations. Such toxic compounds may accumulate in plant tissues following application of such hormones as 2,4-D.

Thus, it may be concluded that the mechanism of auxin action remains unsolved, although the various theories put forward provide some clues regarding various functions which the auxins perform.

AUXIN AS HERBICIDE

Auxin-like herbicides as efficient means of getting rid of weeds like dock

Some commercial weedkillers are chemically similar to auxin and have similar effects. They cause affected plant cells to elongate and the plant grows. The exact mechanism in not yet understood, but the auxin seems to interfere with DNA transcription and RNA translation. The amount of auxin applied in a weedkiller is far greater than the amount produced within the plant, and so the rate of growth produced is much greater than normal. The plant cannot sustain this rate of growth : it becomes weakened, unable to reproduce, and then it dies. The auxin-like herbicides have a much greater effect on dicots than on monocots, such as grasses. This is partly because dicot leaves have a larger surface area than monocot leaves and so absorb more herbicide (refer adjoining figure).

Physiological Roles of Auxins

It was previously thought that the sole function of auxins was to promote cell enlargement. But the work done in later years has proved them to be deeply associated with a variety of functions. In some cases they act as a stimulating agent, in others as an inhibitory agent and in still others as a necessary participant in the growth activity of other phytohormones such as gibberellins and cytokinins. The various growth processes in which the auxins (both natural and synthetic) play their role are discussed below :

1. Cell elongation. It is usually considered that cell elongation occurs only in the presence of auxins and also that the rate of elongation is directly proportional to the amount of auxin applied provided no other factors are limiting. But relatively high concentrations usually exert inhibitory effect on this phase of growth.

As already discussed, the various plant organs like roots, buds and stems all react in a comparable way to auxins : their growth being promoted by relatively low and inhibited by relatively high auxin concentrations. Elongation of roots is promoted only at very low concentrations ; at higher concentrations their growth is retarded. Stems and coleoptiles respond similarly except that optimum range of concentrations for elongation is much higher than for roots. Flowers require still higher concentration for growth.

Auxins also play a significant role in the elongation of petiole, mid rib and major lateral veins of the leaves. Thus, adenine favours enlargement in detached leaves of radish and pea. Similarly, coumarin has been shown to promote expansion of leaves in some plants.

2. Cambial activity. In the spring season, the trees exhibit growth by developing buds which later on open. This is then followed by elongation of the young stems. This resumption of growth by cambial cells is activated by the auxins which move basipetally in the stems from developing buds. **Snow** (1935) has shown that a steady supply of auxin a at 1/1,000,000 mg per hour (or of IAA at 1/500,000 mg per hour) from a gelatin block, upon affixing it to the cut end of a decapitated shoot of sunflower (*Helianthus annuus*) seedling, stimulated meristematic activity of the cambium.

The suggestion by **Jost** (1940) seems to imply that the major function of hormones that migrate from the developing apex of the epicotyl is to activate the differentiation of procambial strands.

3. Callus formation and galls. Besides acting as stimulants of cell elongation, the auxins may also activate cell division. This may be illustrated by applying 1% IAA in lanolin paste to a debladed petiole of a bean plant. This causes prolific division of parenchyma cells resulting in the formation of a swelling or callus tissue at a point where the auxin in applied. The amount of callus tissue formed is directly proportional to the concentration of IAA applied (**Ropp**, 1950).

Bezerinck, in as early as 1885, investigated the production of cecidomid galls in *Poa nemoralis*. He compared gall production to callus formation and thought of a substance coming from the larval body as a causative agent. He regarded this action of animal substance as analogous to that of the inner causes which lead to the formation of roots in normal plants.

4. Rooting of stem cuttings (= Formation of adventitious roots). It is a common observation that the presence of buds on a cutting favours development of roots when the lower end is dipped in a suitable rooting medium. Developing buds are effective in accelerating root formation. Young leaves also favour the initiation of roots on the cuttings. These observations led to the suggestion that the root formation is favoured by the auxins which are synthesized in the buds and young leaves and are later translocated to the basal part of the cutting.

The auxins most commonly employed for this purpose are IAA, NAA, 2,4-D, naphthalene acetamide (NAd) etc.

Auxin-induced rooting is not only of academic interest but also of enormous horticultural value as it helps propagation of certain plants by cuttings. Although some success has been achieved in bean and alike plants, yet certain other plants like apples do not respond to it (**Thimann** and **Behnke**, 1947).

5. Apical dominance. It has been generally observed that so long as the apical bud is intact on the plant, the growth of the lateral buds remains suppressed. Upon removal of the apical bud, the lateral bud nearest the apical bud establishes its dominance over the remaining buds, causing them to become inactive again. *This inhibitory effect of a terminal bud upon the development of the lateral buds is called apical dominance and produces a cone-shaped plant.* This is why a gardener keeps on trimming the hedge occasionally in order to obtain a denser growth. Plants that are tall and unbranched exhibit strong influence of apical dominance than those which are short and branched.

> The characteristic conical shape of most **Christmas trees** results from apical dominance. Almost all Christmas-tree-shaped trees are conifers. Ever since U.S. **President Franklin Pierce** (a politician, otherwise known for putting stickum on the backs of postage stamps) first had a Christmas tree in the White House in 1856; people have been choosy about these trees, preferring ones that are 2 metres high and 1.3 metres wide at the base. Scotch pine (*Pinus sylvestris*) is the most popular Christmas tree in India, the United States and elsewhere. It is a fast-grower and takes only 8 years to grow to a height of 2 metres. Although the tree has long needles, but they stay on the tree much longer than those of balsam fir (*Abies balsamia*) or spruce (*Picea* sp.), the previous best sellers.

The relation of apical dominance with the auxin supply was first reported by **Skoog** and **Thimann** (1934). They demonstrated that when agar block containing auxin *b* or IAA was kept on the decapitated shoot of broad bean (*Vicia faba*), the lateral buds, as might be expected, resulted in the usual suppression of growth as if the terminal buds were present. But when the same decapitated shoot was reheaded with an agar block containing no auxin, these lateral buds resumed growth. Similar results were also obtained with field-grown tobacco plants using NAA as the auxin.

Although the exact mechanism behind this growth correlation is not yet known, the best explanation for this has been furnished by **Snow** (1939, 40). According to him, under the influence of auxin, a growth inhibitor is formed that is responsible for the inhibition of growth of the lateral buds. The growth inhibitor is synthesized in some unknown manner when the auxin moves in its usual downward direction. The theory lends support from the fact that certain growth inhibitors have chemical structure not much different from that of the auxins. It may, thus, be visualized that such a synthesis of growth inhibitor from the auxin may take place in the tissues.

The knowledge of apical dominance has been utilized practically in solving the problem of storage of the potatoes. Potatoes, stored for some time, sprout and become sweet in taste, thus causing financial loss to the grower as the sweet taste is disliked by its consumers. But spraying the potatoes with auxins like indole butyric acid (IBA) and NAA would prevent sprouting (or in other words, would prolong dormancy) by inhibiting the development of buds or 'eyes' ; the effect persisting for as long a period as 3 years. Although such a treatment is in the interest of the breeder but certainly not in that of the consumer.

6. Delay (or inhibition) of abscission of leaves. The abscission of leaves can be delayed or inhibited by the application of auxins on the surface of the lamina or on the cut surface of a debladed petiole. The controlling behaviour of the auxins on the abscission was first noted by **Laibach** (1933) who showed that the extract of orchid pollinia is capable of preventing the leaf fall. Since then, enough work has been carried out in this direction. **Addicott** and **Lynch** (1955) have proved conclusively the delaying effect of IAA on the abscission of various plant organs.

As to the mechanism of abscission, it has been suggested that the leaf fall is retarded by the basipetal migration of a hormone from the blade to the base of the petiole. Removal of the leaf blade eliminates the supply of hormone to the abscission zone and thus induces leaf fall.

The correlation between the amount of diffusible auxin and the age of the organ concerned (Fig. 20–9) was demonstrated by **Shoji** *et al* in 1951. They found high auxin contents in the young leaf blades of bean plant as compared to their petioles. But on ageing, the auxin contents in the leaf blade and the petiole fall almost to the same level.

7. Flowering. A flowering hormone, *florigen*, is produced in the leaves under correct light and dark period. It moves first down the petiole and then up the stem to the growing apex where it causes the development of floral buds in place of vegetative buds (**Cajlachjan**, 1936). Auxins are useful in modifying flowering in one of the following ways :

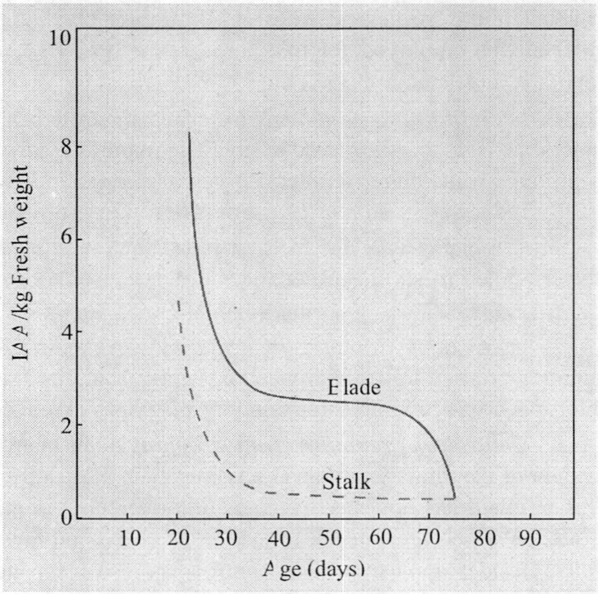

Fig. 32–9. Efect of ageing of the petiole and the leaf blades of bean plant on the auxin contents of these organs

(After Shoji et al, 1951)

A. *Altering earliness*— **Leopold** and **Guernsey** (1953) conducted experiments on oat, corn, barley, peas etc., and found that when their seeds were treated with auxins followed by low temperature treatment at 4°C for 5—15 days, they produced quantitative increase in earliness. **Chakravarti** (1955) also produced earliness and greater yield in mustard when auxin was applied before chilling. No gains were, however, obtained when the auxin was applied after chilling.

B. *Inducing flowering*— The flowering in pineapple has been successfully promoted in Hawaii islands by treating the fields with sodium naphthalene acetate at some time during growing period. In Caribbean areas, however, 2, 4-D is more commonly used.

Leopold and **Thimann** (1949) have, likewise, found low concentrations of auxins such as a-NAA or IAA as effective in inducing flowering in barley and Teosinte. Furthermore, 2, 4-D has been successfully used in control of flowering in sweet potato.

C. ***Preventing or delaying flowering***— Although natural flowering of many species of plants is inhibited by high endogenous auxin concentrations, attempts to attain this inhibitory effect by applying auxins exogenously (*i.e.*, from outside) usually fail as the plants are damaged before effective control is obtained. But experimentally in certain plants like cabbage and celery, the bolting is prevented by applying *p*-chlorophenoxypropionic acid during cold periods which would otherwise normally induce flowering.

8. Fruiting. Auxins play significant role in fruiting by modifying it in one of the following ways :

A. ***Fruit setting***—*Fruit set* refers to the changes in the ovary leading to the development of the fruit. These changes are usually induced after pollination and fertilization— the two processes which are in some way concerned with the release of some stimulus of hormone nature. But the development of fruit without fertilization, *i.e.*, parthenocarpy (*parthenos*^{G} = virgin; *carpos*^{G} = fruit), however, is also a common feature in the plant world and henceforth occurs frequently in nature.

Such a parthenocarpic development of fruits nowadays has also been induced artificially. For example, **Yasuda** (1934) demonstrated it by application of pollen extracts to cucumber flowers. An analysis of the extract showed that auxins were present in it. Later, **Gustafson** (1936, 39) also observed that ovaries of many plants (orange, lemon, grape, banana, tomato etc.) could be induced to develop into seedless fruits by application of IAA in lanolin paste to their stigmas. The various other auxins employed for this purpose are IPA, IBA, α-NAA, phenoxyacetic acid (POA), α-naphthoxyacetic acid (NOA) etc.

B. ***Fruit thinning***— In many instances, the trees bear extensively large number of fruits. This causes the trees to fail to produce average number of new flower buds. Such trees, therefore, have to produce fruits either at alternate years (*alternate bearing*) or if yearly, the number of fruits is greatly reduced (*infrequent bearing*). These trees, obviously, require thinning.

Fruit thinning was, for the first time, done in apple when naphthalene acetic acid applied to flowers failed to set the fruits and, in fact, caused a decrease in fruit set.

It is surprising to note that *naphthalene acetic acid appears to be the only successful auxin which brings about thinning of fruits*. However, other examples of auxins employed for fruit thinning are a-2,4,5-trichlorophenoxyacetic acid for thinning of pears (**Griggs**, 1951) and *p*-chlorophenoxyacetic acid for thinning of grapes (**Weaver** and **Winkle**, 1952).

C. ***Control of premature fruit dropping***— In many fruit trees, the unripe fruits fall off on account of the formation of an abscission layer, thus causing serious losses in yield to the gardeners. This problem has now been successfully overcome in many cases like apples by the application of auxins which prevent the formation of abscission layer and thus check preharvest drop of the fruits. Besides apples, such as control has also been induced in citrus fruits (like oranges and lemons) using 2,4-D and 2,4,5-trichlorophenoxyacetic acid as auxins.

D. ***Improving fruit quality***— The various processes like colouration, softening, sweetening and ripening are all involved in improving the quality of the fruit.

Auxin effects on fruit colouration are most pronounced in apples where use of 2,4, 5-trichlorophenoxyacetic acid has greatly increased red pigments. 2,4-D when applied to bananas hastened the process of ripening as the auxin facilitates the conversion of starch to sugars. By injecting 2,4-D, IBA or maleic hydrazide, the accumulation of sugars was reported in sugarcane.

9. Increase in respiration. James Bonner (1953), for the first time, recognized that auxins stimulate the process of respiration. And as such a direct relation between growth due to auxin treatment and the rate of respiration has been found. *Greater the growth, higher is the rate of respiration*.

Such a knowledge about the behaviour of auxins has been applied in controlling the development of weeds which grow obnoxiously in the crop fields. This has been successfully achieved by spraying 2,4-D which acts as a *weed killer*. In fact, this hormone, which operates only on broad-leaved herbaceous plants, increases the rate of respiration so much so that the plants die of over-oxidation and exhaustion. And fortunately most of the weeds are broad-leaved dicots and the crop plants are usually narrow-leaved monocots which, henceforth, escape destruction on spraying 2,4-D.

Another hormone 2,4-dichloropropionic acid (or dalapon), however, destroys graminaceous weeds.

10. Increased resistance to frost damage. In parsnip, the tops resist damage by frost on treatment with 2,4,5-T. Similarly, the application of 2,4,5-T in apricot fruits before the onset of frost resulted in less damage than the untreated fruits.

11. Great weapon of war. Auxins when applied in greater concentrations on enemy crop fields by air may cause devastation of land and thus form the basis of what is called *biological warfare*.

The synthetic auxins (such as 2,4-D, 2,4,5-T and NAA) have been widely used since 1940s. Their use has decreased production costs by reducing the amount of labour and mechanical weeding, needed to grow and effectively harvest a crop. Unfortunately, the effects of synthetic auxins have not all been positive. Most of the negative effects can be traced to a defoliant, once used in Vietnam War, by the code name Agent Orange. **Agent Orange** was a 1 : 1 mixture of 2,4-D and 2,4,5-T that was sprayed throughout the jungles of vietnam and followed by napalm bombs. This resulted in the destruction of hundreds of square kilometres of Vietnam forests.

> The defoliation in Vietnam forests was protested during the war by many botanists, but the problems have proved to be more serious than the destruction of the forests. Synthesis of 2,4,5-T, a component of Agent Orange, also produces **dioxin** (2,3,7,8-tetrachloro-dibenzo-para-dioxin). *Dioxin is one of the most toxic synthetic chemicals known*. It is also known to be carcinogenic as well as endocrine disruptor in nature and is highly persistent in the environment, besides being extremely resistant to chemical or physical breakdown. Thousands of U.S. pilots and Vietnamese citizens exposed to Agent Orange (which is often contaminated with dioxin) have had higher frequencies of miscarriages, birth defects, leukemia and other types of cancer. Although the **Environmental Protection Agency (EPA),** in 1979, banned the use of 2,4,5-T in the United States, its effects continue to plague many Vietnamese citizens, especially the veterans.

GIBBERELLINS

The gibberellins are another class of compounds whose minute quantities profoundly stimulate the growth of many plants.

Discovery

The gibberellins were discovered in an interesting and incidental way. In early part of the twentieth century, Japanese farmers noted that some plants in rice fields were taller, thinner and paler than the normal plants; had longer and narrower leaves markedly overgrowing their unaffected neighbours; and were sometimes devoid of fruits too. They named this disease as "*bakanae*", meaning foolish asseedlings. **Sawada** (1912) suggested that the disease is due to a 'substance' secreted by a parasitic as comycetous fungus, *Gibberella fujikuroi* (the perfect form, occurring only occasionally; the imperfect form is *Fusarium moniliforme*), in infecting the diseased plants. This suggestion was experimentally supported by **Ewiti Kurosawa** (1926) who demonstrated that sterile filtrates of the fungus could initiate symptoms of bakanae disease in healthy rice seedlings. Later in 1939, **Yabuta** and **Hayashi** isolated this growth promoting substance in crystalline form and named it as gibberellin A, which has now been shown as a mixture of many growth promoters, collectively known as **gibberellins**.

Since that time, gibberellins and allied substances have been found in higher plants also by **Mitchell** et al (1951), **West** and **Phinney** (1957) and **Sumiki** and **Kawarada** (1961).

Definition

A *gibberellin* (abbreviated as GA, for gibberellic acid) may be defined as a compound which is active in gibberellin bioassays and possesses a gibbane ring skeleton (refer page 428). There are, however, other compounds (like kaurene) which are active in some of the assays but do not possess a gibbane ring. Such compounds have been called *gibberellin-like* rather than gibberellins.

Isolation, Distribution and Biosynthesis

About 29 gibberellins were previously isolated and their chemical structures known. These have been named as gibberellin A_1 (GA_1), gibberellin A_2 (GA_2) and so on up to gibberellin A_{29} (GA_{29}). Of these, **Cross** *et al* (1961) have isolated 6 gibberellins from the fungus, *Fusarium moniliforme* and designated them as GA_1, GA_2, GA_3, GA_4, GA_7 and GA_9. The same year, **MacMillan**

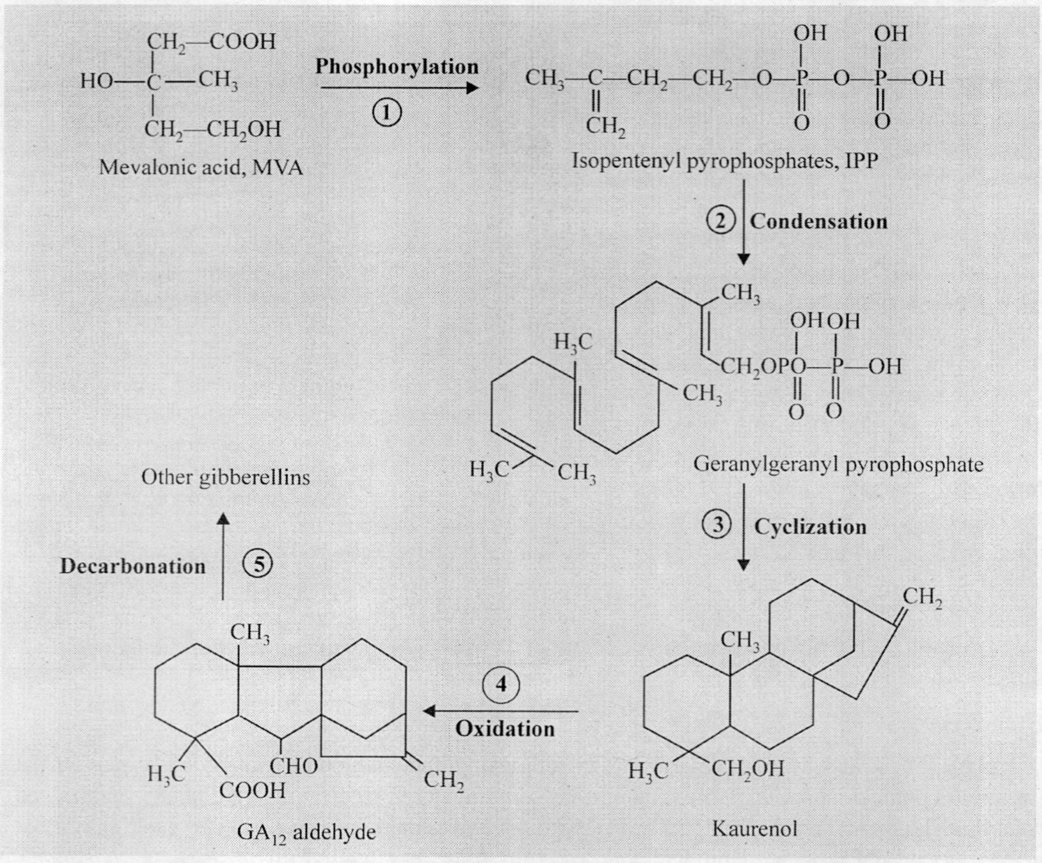

Fig. 20–10. The mevalonic acid (MVA) pathway for the synthesis of gibberellins

Although the biosynthesis of gibberellins from mevalonic acid occurs through some 18 or more steps, 5 key steps culminating in GA production have been outlined here. Mevalonate, in its turn, is produced from acetate. (*Adapted from Moore R, Clark WD and Stern KR, 1995*) *et al* isolated 3 gibberellins from bean seeds and named them as GA_5, GA_6 and GA_8. The GA_{10} and GA_{13} have been discovered by **Mulholland** (1963). All these compounds are sometimes referred to as constituting the **gibberellin A series**. Till date, more than 80 gibberellins have been isolated from various plant sources.

Although the gibberellins were originally isolated from a fungus, but now they have been shown to be present in almost all the groups of plant kingdom including angiosperms, gymnosperms, ferns, mosses and algae but are unknown in bacteria. For example, GA_1, and GA_5 have been isolated from immature seeds of *Phaseolus vulgaris* by **West** and **Phinney** (1959). Although all the organs of the flowering plants contain gibberellins, but the highest level has been detected in seeds. Young leaves and roots are also rich in them. It may, thus, be generalized that *rapidly growing and developing regions of the plant possess higher concentrations of gibberellins.*

Many angiospermous plants have now been used as bioassay for gibberellins and gibberellin-like substances. A few of them are *Avena sativa* (leaf section), *Pisum sativum* (intact seedling), *Triticum vulgare* (excised coleoptile) and *Rudbeckia bicola* (resetted plants).

Gibberellins are synthesized via the mevalonic acid (MVA) pathway. In fact, the biosynthesis of GA_3 from MVA proceeds via 18 or more steps or intermediates and about 15 related compounds. The biosynthetic studies were conducted using cell-free systems of *Gibberella fujikuroi* and systems from immature seeds of *Echinocystis macrocorpa* and *Cucurbita maxima*. In essence, the MVA pathway (Fig. 20–10) comprises of 4 major steps for the formation of C_{20} gibberellins, from which C_{19} gibberellins (such as GA_3) are produced by decarbonation, the 5th major step.

Many commercial compounds inhibit the synthesis of gibberellins. These inhibitors (Fig. 20–11), which are called **growth retardants**, include B-Nine, Cycocel (CCC), Phosphon D and Amo-1618. Growth retardants inhibit stem elongation whereby producing stunted plants. Growth retardants are frequently sprayed on growing chrysanthemums to produce flowers with thicker, sturdier stalks.

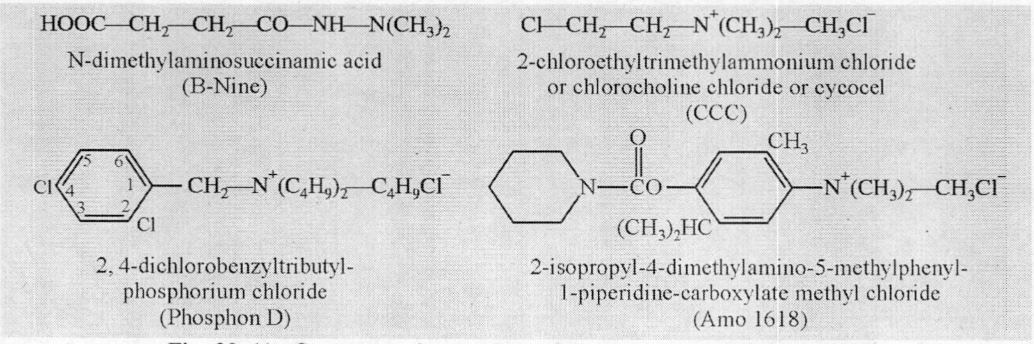

Fig. 20–11. Some growth retardants for the synthesis of gibberellins

Unlike the transport of IAA and other auxins, the transport of gibberellins is not polar: it moves in all directions (*i.e.*, nonpolarly) in the xylem and phloem.

Chemistry

The chemical structure of the gibberellins was established by **Cross** *et al* in 1961. They showed that the gibberellins (Fig. 20–12) are a group of closely related compounds with one or more carboxyl groups that impart acidic properties to the molecule and possess a common feature, the *gibbane ring skeleton*. The gibbane ring consists of a carbon skeleton with 4 interlocking rings, designated as A, B, C and D. They are, hence, described as tetracarbocyclic compounds. Of the 29 gibberellins, previously isolated, 19 are C_{19} compounds and the other 10 have 20 carbon atoms each. Eighteen gibberellins are monocarboxylic, 7 are dicarboxylic and 4 are tricarboxylic acids. The various gibberellins differ from each other in the number and position of the functional groups present in the molecule. In *gibberellin* A_1, the functional groups are a carboxyl, one ethylenic double bond, two alcoholic hydroxyl groups (one secondary and the other tertiary), a saturated lactone and one methyl group. *Gibberellin* A_3 differs from GA_1 in the presence of one more ethylenic double bond in the ring A. It is, thus, more unsaturated than GA_1.

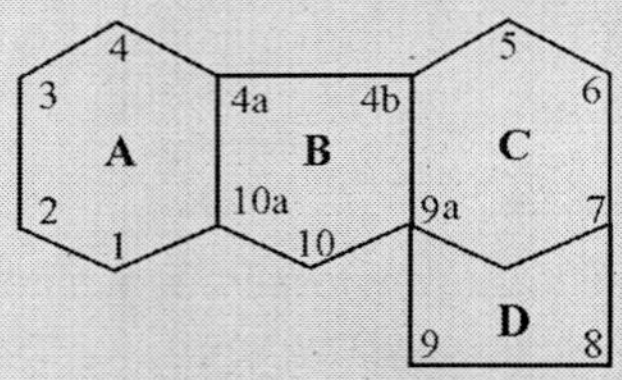

Gibbane ring

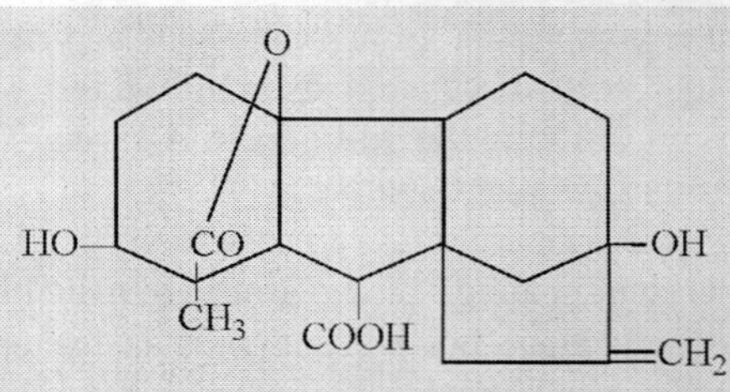

Gibberellin A_1, $C_{19}H_{24}O_6$

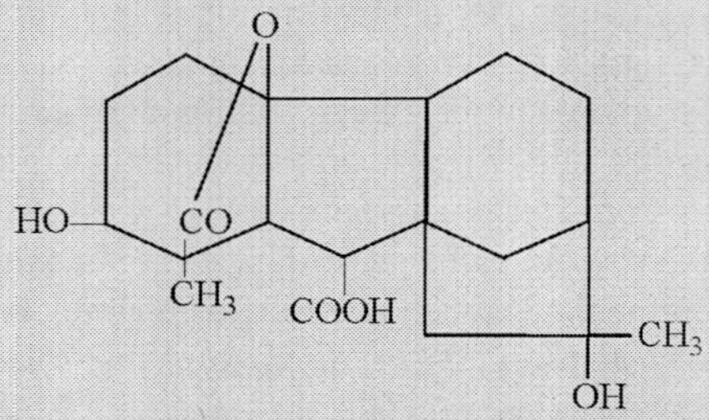

Gibberellin A_2, $C_{19}H_{26}O_6$

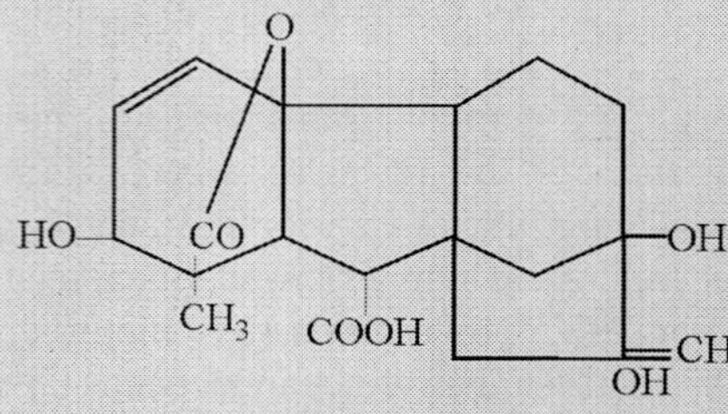

Gibberellin A_3, $C_{19}H_{22}O_6$

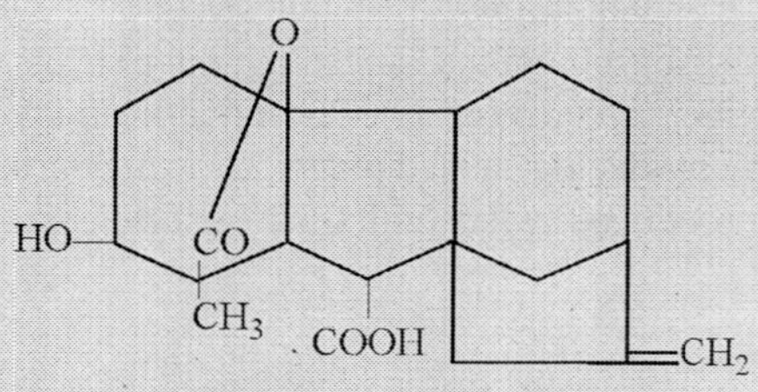

Gibberellin A_4, $C_{19}H_{24}O_5$

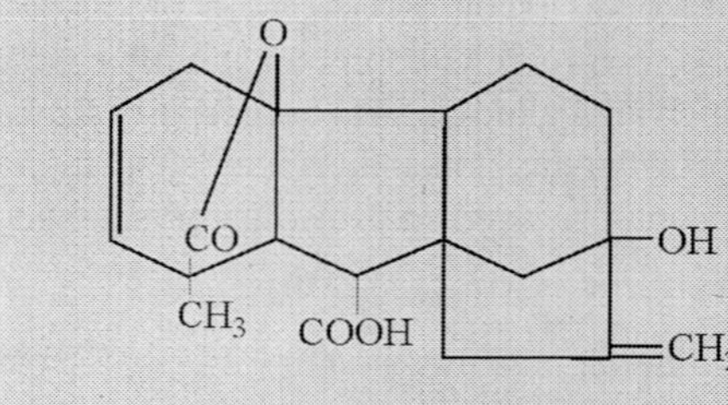

Gibberellin A_5, $C_{19}H_{22}O_5$

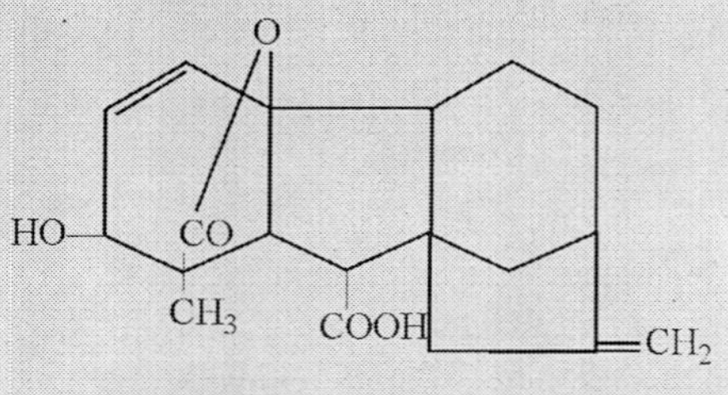

Gibberellin A_7

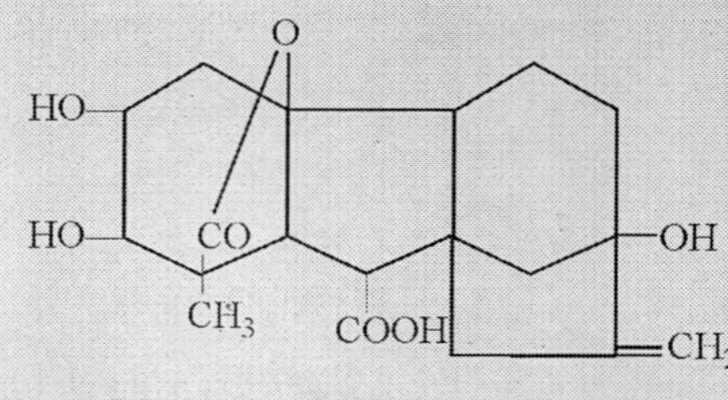

Gibberellin A_8

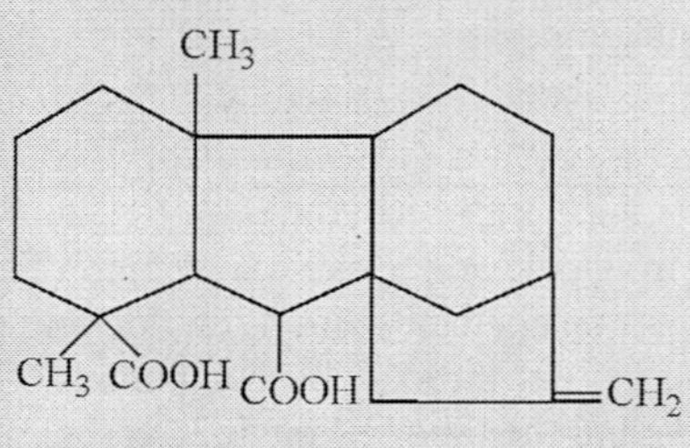

Gibberellin A_{12}

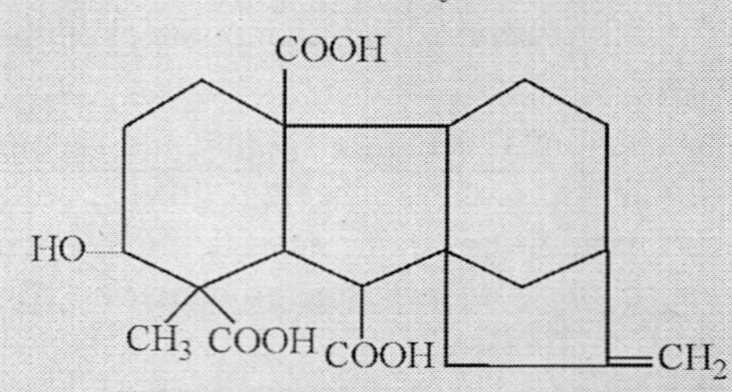

Gibberellin A_{13}

Fig. 20–12. Chemical structure of some gibberellins

Note the minor but significant differences between any two gibberellins. Differences in biologic potency are sometimes associated with presence or absence of double bonds, particular position of the OH groups(s), presence of the lactone ring and the number of carboxyl groups. Also note that while most of the gibberellins have only one carboxyl group, GA_{12} and GA_{13} have two and three carboxyl groups, respectively.

The *gibberellin A_2 and gibberellin A_4* both have structures similar to that of GA_1 except for the difference in position of the tertiary hydroxyl group and the absence of a double bond in GA_2 and in the absence of tertiary hydroxyl group in GA_4. The gibberellin A_5 is, in fact, a dehydrogibberellin A_1 where the secondary hydroxyl group is eliminated from ring A, making the compound more unsaturated.

Gibberellin A_3 has been usually shown to be biologically most active followed by GA_1, GA_4 and GA_2 in descending order of their activity.

$$GA_3 > GA_1 > GA_4 > GA_2$$

Physiological Roles

Gibberellins may be regarded as natural phytohormones on account of their wide range of distribution in plants and specificity of response of individual flowering plants to the exogenously applied gibberellins. The gibberellins, however, play important roles in the following processes:

1. Genetic dwarfism. In certain plants, dwarfism is caused by the mutation of a single gene. Such individuals are called '*single gene dwarfs*'. In these plants, dwarfism is due to shortening of internodes rather than a decrease in the number of internodes. Application of gibberellins on such dwarfs causes them to elongate so much as to become indistinguishable from the tall normal plants. Elongation of the stem, in fact, takes place due to an elongation in the internodes rather than an increase in the number of internodes. Thus, genetic dwarfism has been successfully overcome by gibberellin A_3 treatment in many single gene dwarf mutants like *Pisum sativum, Vicia faba and Phaseolus multiflorus* (**Brian** and **Hemming**, 1955). The gibberellins, thus, make most plants grow taller by causing the internodes to elongate considerably (Fig. 20–13).

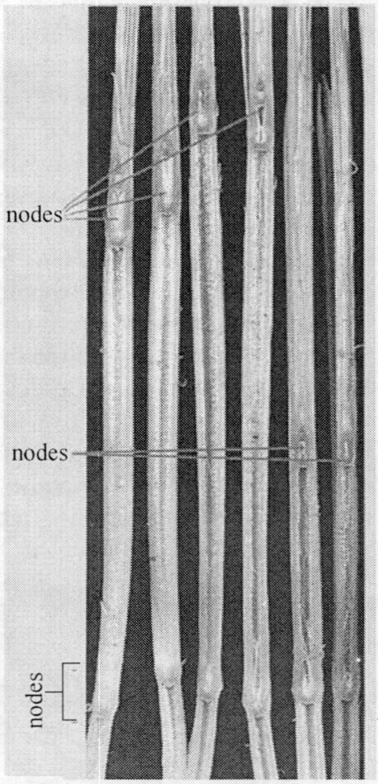

Fig. 20-13. Wheat stems, cut lengthwise, to show the nodes
The 4 plants on the left, treated with gibberellic acid, have internode lengths much greater than the internodes of the two untreated on the right.

2. Bolting and flowering. '*Rosette plants*' are characterized by their profuse leaf development and retarded internodal growth. But prior to the reproductive phase, there occurs striking elongation in the internode so that the plant attains 5 to 6 times the original height. Treatment of these `rosette' plants with gibberellins, under conditions that would normally maintain the rosette form, induces them to bolting (or shoot elongation) and flowering (**Lang**, 1957). By regulating the amount of gibberellin applied, it is also possible to separate shoot elongation from flowering ; *with low dosages of gibberellins, the plant will bolt but not flower* (**Phinney** and **West**, 1961).

As far as the use of gibberellins in agriculture is concerned, it may be possible to grow cold-requiring plants in warm countries and long-day plants in short-day conditions at lower altitudes.

3. Light-induced inhibition of stem growth. Light-grown plants reveal suppressed stem growth than the dark-grown (or etiolated) plants, indicating that light has an inhibitory effect on stem elongation. But this inhibitory effect of light on stem elongation can be reversed at least in some plants (*like Pisum sativum*) by the application of gibberellins on these plants. This clearly suggests that endogenous gibberellin is the limiting factor in stem elongation.

4. Parthenocarpy. Like auxins, the gibberellins are also capable of inducing parthenocarpic fruit-set. *Gibberellins are, in fact, more efficient than the auxins in inducing parthenocarpy.* For example, **Wittwer** and **Bukovac** (1957) have found gibberellin to be about 500 times more effective than IAA in inducing parthenocarpy in tomatoes. Moreover, there are cases where auxins have failed to induce parthenocarpy while gibberellins are effective, as shown experimentally for apples (**Davison**, 1960) and stone fruits (**Crane** *et al*, 1960).

Gibberellin-induced parthenocarpy has been reported in many plants such as *Cucumis sativus* (cucumber), *Solanum melongena* (brinjal) and *Zephyranthes* sp. Whether the production of parthenocarpic fruits is a direct action of gibberellins or an interaction with the natural auxins of the plant has not been conclusively proved.

5. Breaking dormancy of seeds. The light-sensitive seeds (lettuce, tobacco) show poor germination in dark and on exposure to light their germination starts vigorously. But when these seeds are treated with GA_3, the light requirement is alleviated and they germinate in dark.

6. Breaking dormancy of buds. In temperate areas, the buds produced in winter remain dormant until the next spring due to very low temperature. The dormancy in such cases is overcome by gibberellin treatment. Thus, GA_3 treatment to birch buds has replaced the light requirement for breaking dormancy (**Eagles** and **Wareing**, 1964). Gibberellins are also capable of breaking dormancy in potato tubers.

7. Role in abscission. GA_3 treatments have shown accelerated rate of abscission in explants of bean (**Chatterjee** and **Leopold**, 1964) and of *Coleus* (**Gupta** and **Kaushik**, 1969).

8. Stimulation of enzyme activity in cereal endosperm. Yomo (1960) and **Paleg** (1960) working independently showed that the gibberellins applied exogenously could stimulate *amylase* activity in isolated barley endosperm. It was then shown that it is the aleurone layer of the endosperm which is sensitive to the gibberellin. Subsequent researches by **Paleg** (1964) and **Varner** (1964) revealed that GA treatment of isolated aleurone can cause release of the enzymes, *amylase* and *proteinase*. Finally, **Jacobson** and **Varner** (1967) showed that the two enzymes (amylase and proteinase) induced by GA treatment arise through *de novo* synthesis. These enzymes

de novo is a Latin phrase, meaning anew.

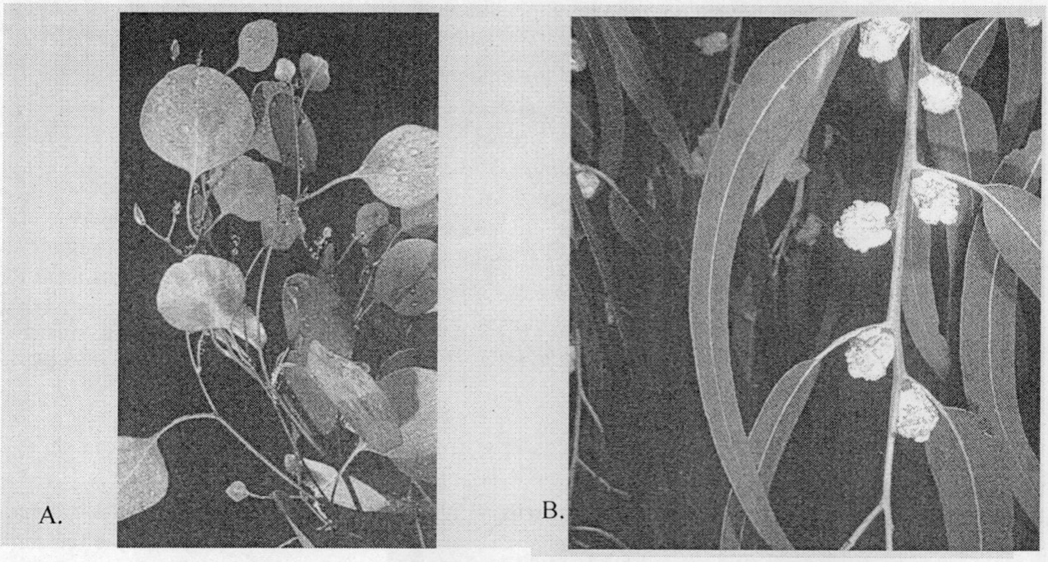

A. B.

Fig. 20–14. Photographs of two growth stages of *Eucalyptus globulus* leaves
A. Juvenile stage B. Adult stage

Juvenile and mature leaves are shaped differently. Juvenile leaves are softer and opposite each other. Adult leaves are hard and spirally arranged. The difference in the leaves of 2 stages is also exhibited in their anatomy. The juvenile leaves have palisade cells only on the upper surfaces, whereas the adult leaves contain palisade parenchyma on both sides.

participate in the breakdown of the stored starch to simple sugars. These sugars are then translocated to the growing embryo where they provide energy for growth.

9. Sex expression. Gibberellins are also capable of altering the sex of the flowers. **Galun** (1959) could induce maleness by foliar application of GA_3 to the female flowers of *Cucumis*. Also, the antheridia have been induced to develop in many fern gametophytes by GA_3 treatment.

10. Juvenility. Many plants in their life cycle exhibit two different stages of growth: a juvenile stage and an adult stage. For example, in a species of eucalyptus (*Euçalyptus globulus*), the juvenile leaves are oppositely-placed and are shorter, softer and almost obcordate with emarginate apex; while the leaves of the adult stages are spirally-arranged and are larger, harder and falcate with acute apex (Fig. 20–14). Gibberellins may help determine whether a particular part of a plant is juvenile or adult. For instance, the buds of adult branches usually develop only into adult branches, but treating them with gibberellin causes them to grow into juvenile branches.

Relationship between Auxins and Gibberellins

Auxins and gibberellins are similar to each other in that both promote cell elongation, flowering and parthenocarpy. These, however, differ form each other in many of the physiological activities. These differences are listed in Table 20–2.

Table 20–2. Differences between auxins and gibberellins

S. .N	*Physiological activity*	*Auxins*	*Gibberellins*
1.	Transport polar	Yes	No
2.	Promote root initiation	"	"
3.	Inhibit root elongation	"	"
4.	Delay leaf abscission	"	"
5.	Inhibit lateral buds	"	"
6.	Induce callus formation	"	"
7.	Promote epinastic responses	"	"
8.	Control dwarfism	No	Yes
9.	Promote seed germination and the breaking of dormancy	"	"
10.	Promote bolting and flowering in non-vernalized biennials and in long-day plants	"	"

(*Adapted from Galston AW and Purves WK, 1960*)

Accumulated evidences indicate that auxins and gibberellins act both independently and together, depending upon the type of plant and the conditions under which the plant grows. The fact whether the auxins and the gibberellins interact or not is not conclusively proved.

CYTOKININS (= KININS)

Discovery and Nomenclature

Auxins and gibberellins, besides inducing cell elongation, also do promote cell division under certain conditions. But this behaviour of them is an exception rather than a rule. However, there exist in plants many substances inducing cell division. For example, **Van Overbeek** *et al* (1941) found coconut milk as an active stimulant of cell division. Later, in 1955 **Carlos Miller** *et al* isolated a *"cell-division-stimulating factor"* from yeast DNA. It was named as kinetin because of its amazing power to stimulate cell division (cytokinesis) in the presence of an auxin. In subsequent years, many other compounds promoting cell division have been synthesized. **Miller** and his associates (1956) have grouped all such compounds including kinetin under a generic name kinin. **D.S. Letham** (1963) of New Zealand proposed the term cytokinins for such substances. This term is the most acceptable one.

Fairley and **Kilgour** (1966), however, prefer to use the term **'phytokinins'** for such substances in order to distinguish them from the peptide hormones of animal gastrointestinal tract.

Definition

Skoog, Strong and **Miller** (1965) have defined cytokinins as chemicals which, regardless of their activities, promote cytokinesis (cell division) in cells of various plant organs.

Fox (1969) has defined cytokinins as chemicals composed of one hydrophilic adenine group of high specificity and one lipophilic group without specificity.

Isolation, Distribution and Biosynthesis

Although kinetin does not occur in nature but other kinins are found occurring widely, if not universally, in plants. The naturally-occurring kinins do not occur free in nature but are normally bound to a pentose sugar, ribose and sometimes to an inorganic phosphate, the ribonucleotide.

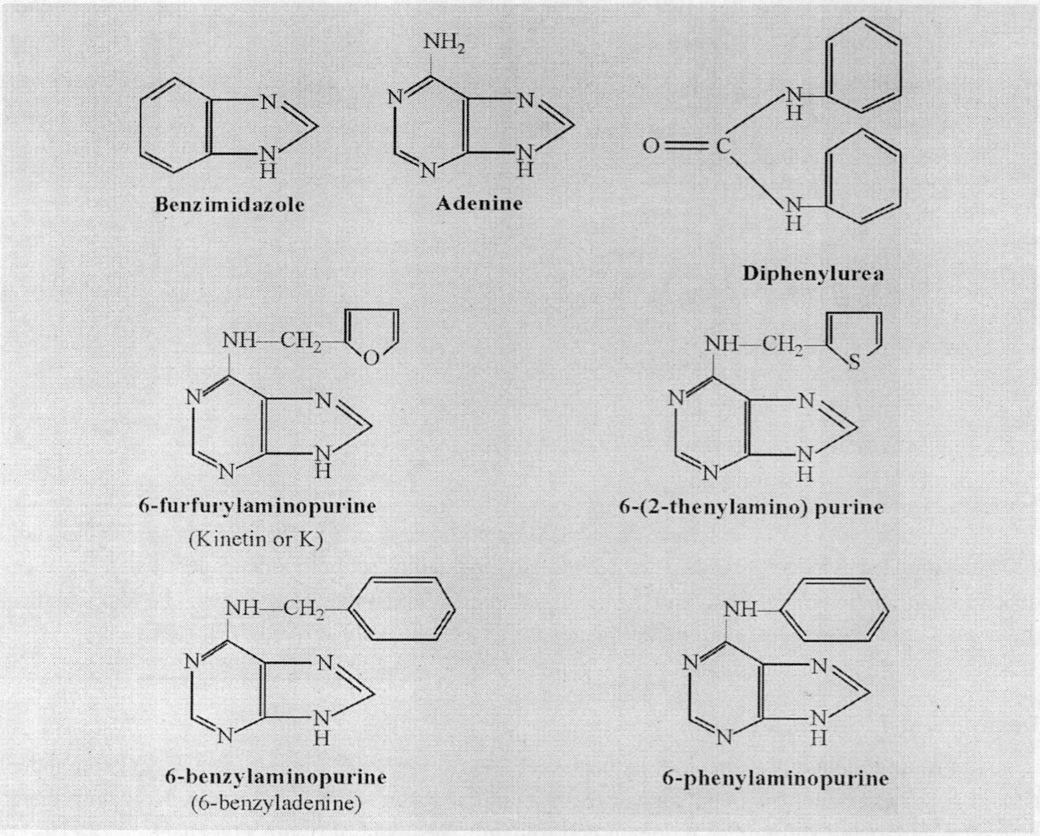

Fig. 20–15. Structural formulae of some kinins

Note that all cytokinins are structurally related to the purine derivative, adenine, *i.e.*, they have a side chains rich in carbon and hydrogen and attached to nitrogen protruding from the top of the adenine ring.

Fruits and endosperm are the richest sources of kinins. Coconut milk and corn endosperm possess the active substance. Substances with cytokinin activity have also been reported in tomato juice, in floral extracts of apples and pears and also in cambial tissues of certain plants. A *kinetin-like substance* is also present in peach embryo (**Powell** and **Pratt**, 1964) and sunflower root exudate (**Kende**, 1964).

In angiosperms, the cytokinins are synthesized mostly in roots and probably originate at the root tips. Whether the shoots also synthesize cytokinins or else receive their cytokinin requirement from the roots is not certain. Contrary to what was first believed, the cytokinins are not breakdown products of DNA. Rather, they are made *via* the mevalonate pathway, the same pathway used to make gibberellins. Like gibberellins, the cytokinins move nonpolarly in xylem, phloem, and parenchyma cells.

Chemistry

Chemically, kinetin ($C_{10}H_9ON_5$) is 6-furfurylaminopurine. It is formed from deoxyadenosine which is a degradation product of DNA (**Hall** and **de Ropp**, 1955). The structural formulae of kinetin and its 3 structural analogues are given in Fig. 20–15. All these substances promote cell division.

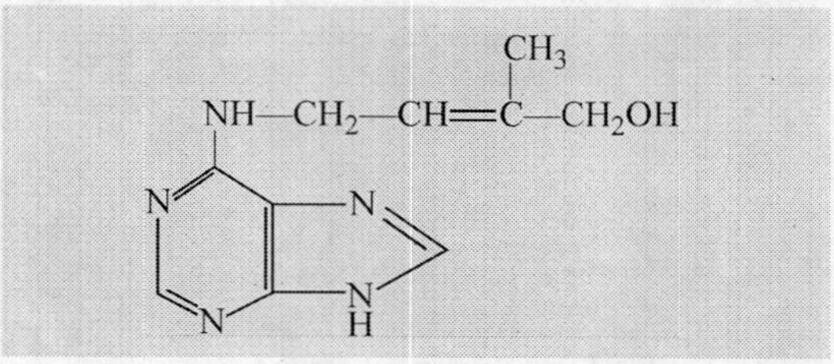

Fig. 20–16. Zeatin
[6-(4-hydroxy 3-methylbut-*trans-2*-enyl) aminopurine]

Apart from the above-mentioned kinins, **Letham** (1963) successfully isolated a cytokinin in pure crystalline form from immature maize seeds. It was named as zeatin (Fig. 20–16) and identified as 6-(4-hydroxy 3-methylbut-trans-2-enyl) aminopurine. *Zeatin is more powerful than any other known cytokinin probably because of the presence of a highly reactive allylic OH group in its side chain.* Zeatin riboside (**Letham**, 1966) and zeatin ribotide (**Letham**, 1966; **Miller**, 1967) also occur naturally in plants. A *cis*-ribosyl zeatin and a *ms*-ribosyl zeatin have also been extracted from plant tissue.

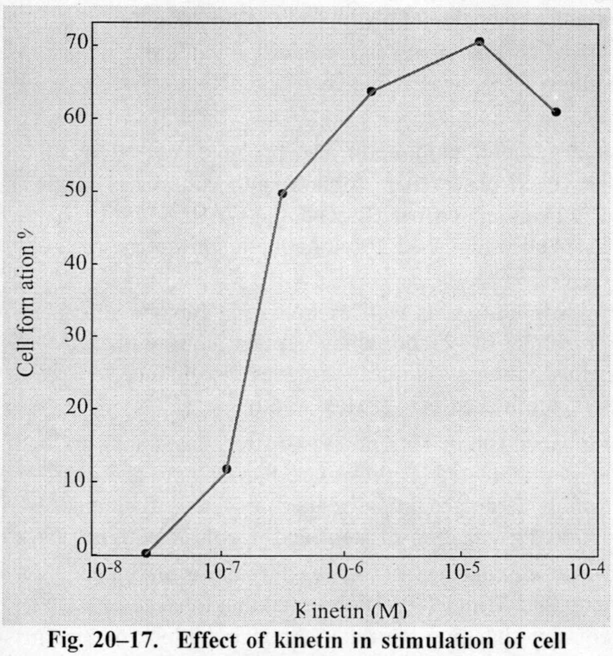

Fig. 20–17. Effect of kinetin in stimulation of cell division in tobacco pith cultures
[2mg/litre of IAA was also present in the medium]
(*After Skoog F and Miller CO, 1957*)

Physiological Roles

Certain physiological processes which are influenced by the cytokinins *esp.*, kinetin are given below :

1. Cell division. *Kinins are notable for their stimulatory effect on cell division.* Using tobacco pith cultures (Fig. 20–17), **Skoog** and **Miller** (1957) found that, in addition to IAA, kinetin is also needed for growth. The growth response is much more pronounced when both IAA and kinetin are used together in right ratio of concentrations. When either of them is used alone, a little response is produced which is due to the presence of small amounts of endogenous kinetin-like substances and IAA, already present in the tissues.

If a mixture of cytokinin and auxin is added to unspecialized cells (Fig. 20– 18), they will begin to differentiate. A high cytokinin to auxin ratio will lead to the formation of shoots, buds and leaves while a low cytokinin to auxin ratio will lead to root formation. The one treatment followed by the other provides a means of forming small plantlets. Such *in vitro* culture methods have become widely adopted for the rapid propagation of new plant varieties. The technique allows growers to produce very large numbers of plants, quite rapidly and in a small space. The alternative conventional propagation methods can take several years and cover a large area of land.

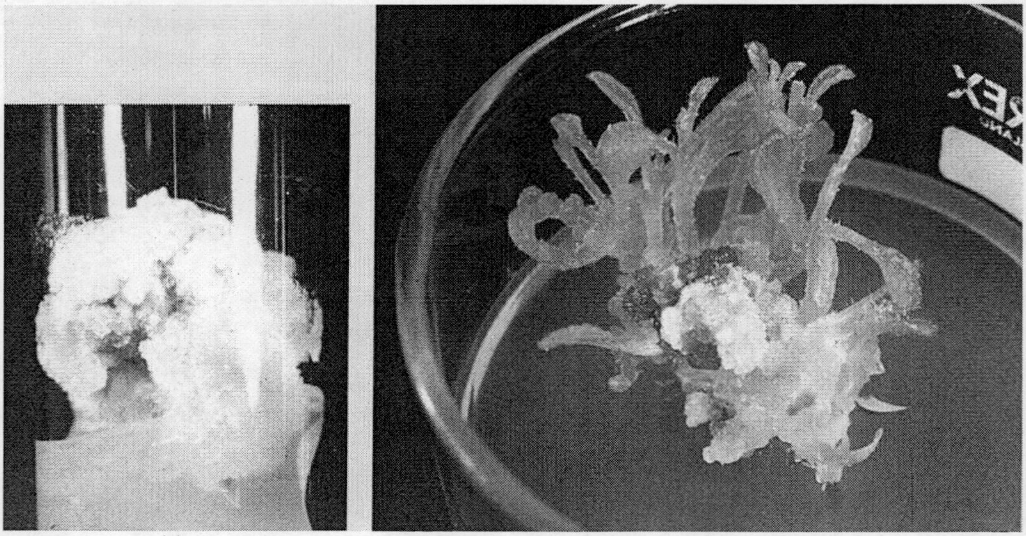

(a) (b)

Fig. 20–18. Callus formation and production of plants from callus cells

(a) If a small piece of pith from a shoot is placed on agar in aseptic conditions, it will grow into a mass of unspecialized cells called callus callus.

(b) Plants can be grown from these callus cells.

2. Cell elongation. Besides auxins and gibberellins, kinetin also promotes cell elongation. Such promotion after kinetin treatment has been observed in tobacco pith cultures (**Glasziou**, 1957), tobacco roots (**Arora** *et al*, 1959) and bean leaf tissues (**Powell** and **Griffith**, 1960). Since cell elongation induced by kinetin has been well established, the kinetin should not be regarded as exclusively a cell division factor.

3. Root growth. Kinetin is capable of stimulating as well as inhibiting root development. **Skoog** and **Miller** (1957) found stimulatory effect of kinetin, when applied along with IAA, on root initiation and development in stem callus cultures. Similarly, kinetins also induced increase in dry weight and elongation of the roots of lupin seedlings (**Fries**, 1960).

4. Shoot growth. The callus tissue of tobacco can be kept in an undifferentiated state so long as the proper balance of IAA and kinetin is maintained. If, however, the amount of kinetin is increased, leafy shoots are initiated to develop. Bean seedlings, soaked in kinetin solution, also showed an increase in dry weight and a marked elongation of stem and petioles (**Miller**, 1956).

5. Organogenesis. Cytokinins can cause organogenesis (*i.e.*, the formation of organs) in a variety of tissue cultures. For instance, **Skoog** and **Miller** (1957) observed that tobacco pith callus can be made to develop either buds or roots by changing the relative concentrations of kinetins and auxins. High kinetin and low auxin contents result in the production of buds. In reverse condition (high auxin and low kinetin), however, the roots appear on the pith.

The kinins also stimulate the production of buds in leaf segments of various plants such as

Saintpaulia ionantha, Bryophyllum sp and Begonia sp.

6. Counteraction of apical dominance. As discussed earlier (refer page 422), the auxins emanating from the apical bud inhibit the growth of lateral buds (apical dominance). **Wickson** and **Thimann** (1958) studied the antagonistic effect of auxin and kinetin in apical dominance using pea stem sections in culture solutions. They found, as might be normally expected, that the growth of lateral buds is inhibited when the culture medium contained IAA and is uninhibited when the culture medium does not contain IAA. They further noted that addition of kinetin, along with IAA, stimulates the growth of lateral buds.

The above workers also conducted experiments with entire shoots, *i.e.*, with the apical bud intact. As long as the apical bud is present, the lateral buds do not develop but removal of the apical bud leads to the stimulation of growth of the lateral buds. If, however, the intact shoot is soaked in kinetin solution, the inhibition of lateral buds is checked to a great extent or, in other words, the lateral buds tend to develop, although less vigorously, as if the apex of the shoot has been cut off. The above findings point out towards the possibility of controlling apical dominance by maintaining a proper balance of concentrations between IAA and the endogenous kinetin-like substances.

Studies conducted in subsequent years by **Sachs** and **Thimann** (1964, 67) and **Panigrahi** and **Audus** (1966) also indicated that the cytokinins are strong promoters of lateral bud growth.

7. Breaking dormancy of seeds. Cytokinins are also effective in breaking seed dormancy in lettuce, tobacco, white clover and carpet grass. **Thimann** (1963) suggested that the site of cytokinin action in such cases is the cotyledon. Furthermore, the inhibitory effect of infrared light on germination of lettuce seeds is also alleviated by kinetin treatment.

The seeds of parasites such as *Striga asiatica* require the presence of host plant for germination. But when treated with kinetin, the seeds germinate even in the absence of their host.

8. Delay of senescence (= Richmond-Lang effect). The term *senescence* refers to the ageing of the leaves which is associated with the loss of chlorophyll and the breakdown of proteins. **Richmond** and **Lang** (1957) showed that the senescence in the detached leaves of Xanthium could be postponed for many days by kinetin treatment. This effect of kinetin in retarding senescence (or ageing) is known as *Richmond-Lang effect.* According to **Mothes** and **Engelbracht** (1961), the cytokinins have the ability to attract certain substances including auxins and to prevent the movement of leaf components out of the treated area. However, the mobilizing effect of cytokinin may actually induce senescence in others parts of the plant. **Osborne** (1962) suggested that the high protein content in kinetin-treated areas is probably due to enhanced protein synthesis than their breakdown. The protein synthesis, in its turn, is dependent on RNA synthesis, a process governed by kinetins. It may, however, be emphasized that the cytokinin-induced delay in leaf senescence occurs only in detached leaves; cytokinins have little or no effect on senescence in attached organs. Leaf senescence is also delayed by the formation of adventitious roots. As the roots are rich in cytokinins, the transport of these cytokinins from roots to leaves could account for the delayed senescence.

A correlation between the age of the leaf and the kinetins has been established. Mature leaves of tobacco respond more vigorously to kinetin treatment in delaying senescence than the young leaves.

9. Role in abscission. Cytokinins can accelerate as well as retard the process of abscission in leaf petioles depending on the site of their application (**Osborne** and **Moss**, 1963). In explant petioles of *Coleus blumei*, **Gupta** and **Kaushik** (1969) reported accelerated abscission on kinetin application."

10. Effects on cotyledons. Cytokinins promote cellular division and expansion in cotyledons. Cellular expansion results from cytokinin-induced increases in wall plasticity that do not involve wall acidification. Cytokinins also increase the amount of sugars (especially glucose and fructose) in cells, which may account for the osmotic influx of water and the resulting expansion of cytokinin-treated cells in cotyledons.

OTHER NATURAL GROWTH HORMONES IN PLANTS

In addition to the 3 well-established categories of plant hormones described above, some other compounds with hormonal actions have been identified in plants which fall under 4 categories: ethylene, traumatic acid, calines and vitamins. Their account follows.

1. ETHYLENE

Discovery

During the 1800s, the city streets of Germany were illuminated by lamps that burned "illuminating gas". Soon after these lamps were installed, city residents made a curious observation: plants growing near the lamps had short thick stems and leaves falling from most of them. The mystery was solved in 1901 by a Soviet plant physiologist, Dimitry Neljubow who identified ethylene as the combustion product of "illuminating gas" that was responsible for defoliation and stunted growth of plants growing near the lamps. He also showed that only micro quantities are needed to bring about these effects, *i.e.*, only 0.06 ppm of ethylene.

ppm is part(s) per million;
1 ppm = 1 millilitre in 1,000 litres

Later in 1910, an annual report submitted to the Japanese Department of Agriculture recommended that *oranges be not stored with bananas*, because oranges released something that caused premature ripening of the bananas; this "something" was in 1934 identified by **R. Gane** as ethylene which is made by plants. Subsequent researches showed that ethylene has all the characteristics which warrant its inclusion under plant hormones, *i.e.*, it is made in one part of a plant and transported to another, where it induces a physiological response. Thus was discovered the *gaseous plant hormomes: ethylene*. Later, the presence of ethylene was shown in certain fungi (*Penicillium digitatum, Alternaria citri*) and in the leaves, flowers and fruits of many higher plants; its recognition as a natural plant hormone was confirmed by **Pratt** and **Goeschl** only in 1969.

Distribution and Biosynthesis

All parts of angiospermous plants produced ethylene but especially large amounts are released into the air by roots, the shoot apical meristem, nodes, senescing flowers and ripening fruits (for example, the dark flecks on a ripening banana peel are concentrated pockets of ethylene). Because most ethylene-induced effects result from ethylene in the air, *the effects of ethylene can be contagious:* ethylene made by one "bad" (*i.e.*, overripe) apple can "spoil" (*i.e.*, induce rapid ripening of) an entire bushel of apples. Ethylene also occurs in minute quantities in city gas and in tail gases of blast furnaces. It is a volatile gas of peculiar odour and is sparingly soluble in water but a little more in ethanol and ether. It is inflammable and hence the ignition of a mixture of ethylene with air leads to explosion.

$$H_2C = CH_2$$

Ethylene, ETH

Ethylene is made from methionine, a S-containing amino acid (Fig. 20-19). Its synthesis, which requires O_2, is inhibited by CO_2 When plants are placed in pure CO_2 (or O_2-free air), ethylene synthesis decreases dramatically.

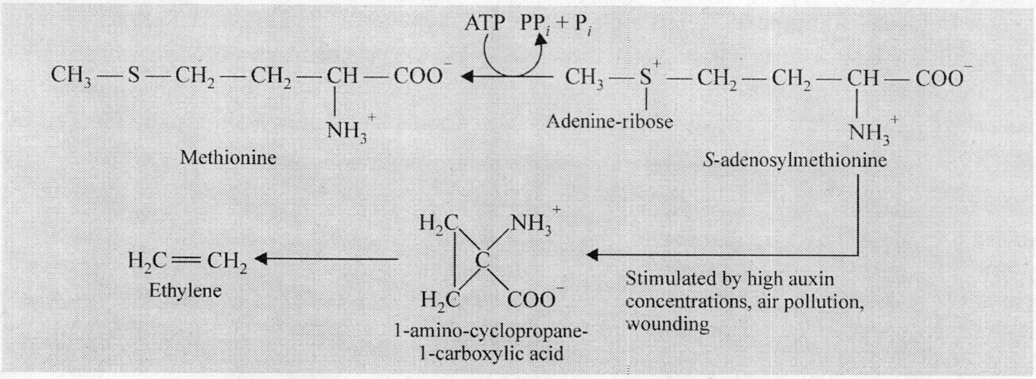

Fig. 20–19. Biosynthesis of ethylene from methionine

Note that a cyclic compound, 1-amino-cyclopropane-1-carboxylic acid is the immediate precursor of ethylene.

Physiological Roles

1. Stimulates fruit ripening. The ancient Chinese knew that fruits would ripen faster if placed in a room containing burning incense. The factor responsible for this hastened ripening was not heat, but ethylene released as the incense burned. The stimulation of fruit ripening by ethylene is a consequence of many ongoing processes such as:

(*a*) the breakdown of chlorophyll and synthesis of other pigments; for example, apples changing from green to red during ripening,

(*b*) fruit softening due to breakdown of cell walls by cellulase and pectinase, and

(*c*) conversion of starches and acids to sugars.

And ethylene stimulates each of these processes, leading ultimately to fruit ripening.

We often say that 'one rotten apple spoils the rest of the barrel' because pieces of fruit near to a rotten one start to go bad quickly. This is because a damaged on fungus-infected fruit starts to produce ethene. And in a closed barrel the concentration of ethene is high enough to quickly trigger the ripening process in neighbouring fruits making them more vulnerable to infection (Fig. 20–20).

Fig. 20–20. The oranges infected with the fungus *Penicillium digitatum*

The fungus itself produces large amounts of ethene, which accelerates the post-harvest maturation of oranges, making them more vulnerable to infection.

2. Promotes flowering. Although ethylene inhibits flowering in most species, but induces it in a few plants including mangoes, pineapples and some ornamentals. The Filipino mango growers and the Peurto Rican pineapple growers, who knew this effect long ago, set bonfires near their crops. The fires produced ethylene which initiated flowering of their plants. Nowadays, the pineapple growers in Hawaii produce pineapple fruits round the year by spraying plants with ethepon. Ethepon splits under neutral and alkaline conditions to release ethylene.

3. Hastens leaf abscission. Abscission zone in leaves causes the increased production of ethylene which triggers the breakdown of middle lamella, thus leading to the initiation of abscission (Fig. 20–21). This effect is also utilized by horticulturists to minimize the harvesting period of such fruits as cherries, grapes and blueberries. These fruits are sprayed with ethepon to coordinate abscission, thereby allowing growers to harvest their crops in shorter periods of time.

4. Induces leaf epinasty. When a plant's roots are kept submerged in water for long periods, water fills the intercellular spaces. Since these spaces are the primary routes of gas exchange with the atmosphere, the submerged roots become waterlogged and anaerobic. The symptoms of waterlogging (*e.g.*, leaf chlorosis, shorter and thicker shoots and wilting) can also be induced by

Fig. 20–21. Photograph of a plant showing leaf fall

Leaf abscission is strongly influenced by hormones, *esp.*, ethylene.

placing roots in O_2-free air. Because O_2 is required to produce ethylene, its synthesis is greatly checked in roots of waterlogged plants. The small amount of ethylene that is made in these roots is trapped, where it accumulates and eventually stimulates the activity of enzymes like cellulase and pectinase. These enzymes break down the cell walls. This leads to the formation of many intercellular spaces, characteristic of hydrophytes. Concomittantly, the ethylene precursors in the shoot are also converted to ethylene, which causes parenchyma cells on the upper side of the petiole to expand and point the leaf down, a physiologic response called epinasty.

5. Controls stem elongation. Mechanical disturbances such as shaking decrease stem elongation. This effect, which is called **thigmomorphogenesis**, is mediated by ethylene. Mechanical disturbances enhance ethylene production several times. Ethylene so produced causes cells to arrange their cellulose microfibrils *longitudinally.* This lengthwise reinforcement inhibits cellular elongation, causing cells to elongate radially; this leads to the formation of short and thick stems. This effect is opposite to that of auxin, which causes cells to orient their microfibrils *transversely*, thereby accounting for cellular elongation.

6. Determines sex expression. Both ethylene and gibberellins determine the sex of flowers on monoecious plants, i.e., plants having male and female flowers on the same individual. As an instance, cucumber (*Cucumis sativus*) buds, on treatment with ethylene, become carpellate (♀) flowers, whereas those treated with gibberellins become staminate (♂) flowers. Correspondingly, buds that ultimately become flowers produce more ethylene than do buds that become male flowers.

2. TRAUMATIC ACID

Most plants form callus when they are injured. This observation of **Gottlieb Haberlandt** (1913) led to postulate that the injured cells secrete a '**wound hormone**' which induces the neighbouring uninjured cells to become meristematic and divide till the injured part is healed up. Later, the active substance was isolated and identified as traumatic acid by Bonner and English in 1938. Traumatic acid is an open-chain dicarboxylic acid with a single double bond.

$$COOH—CH = CH—(CH_2)_8—COOH$$

Traumatic acid or **Traumatin** or **Nekrohormone**

Traumatic acid is specifically effective in inducing cell division in bean pods. It, however, seems to be of confined activity as it has failed to develop meristematic activity in a number of plant tissues including tobacco pith. **Davies** (1949), henceforth, regards this as a specific wound hormone for bean pods.

3. CALINES

Some indirect evidences accumulated during the past few decades have established beyond doubt the presence of some hormones which are needed for initiating the action of auxins on roots, stems and leaves. For example, the presence of leaves and buds is necessary for a better rooting caused by auxin application. These hormones have been collectively called as **calines** or **formative hormones**. The various postulated calines are :

A. *Rhizocaline*— It is a root forming hormone. It is produced by the leaves and transported down the stem.

B. *Caulocaline*—It is a stem forming hormone. It is synthesized mostly in the roots and translocated up in the stem.

C. *Phyllocaline*— It stimulates the development of mesophyll in the leaves. It is synthesized probably in the cotyledons from where it moves to its site of action.

None of these calines has been isolated from plants as yet.

4. VITAMINS

Their account follows in the succeeding two chapters.

GROWTH INHIBITORS

Introduction. While all the hormones discussed so far possess the power of growth promotion, the plants also do possess a few substances which inhibit growth. Although auxins when present in higher concentrations also inhibit growth, the growth inhibitors do so *irrespective of their concentration.* They retard such processes as seed germination, root and stem elongation and bud opening. As a matter of fact, the growth inhibitors act as chemical check upon plants preventing the seeds from germinating and the buds from opening under unfavourable conditions. For example, the sprouting of lateral buds during cold weather is inhibited by the presence of growh inhibitors produced within the plant body during the previous growing season.

Characteristics. The growth inhibitors are, in general, characterized by some common features (**Addicott** and **Lyon**, 1969 ; **Milborrow**, 1969). These are :

1. The amount of growth inhibitors decreases during active growth period of plants and increases during the period of growth suppression.
2. They counteract the activities of growth promoters.

3. They inhibit the growth of various isolated organs and tissues.
4. They do not evoke the strong stimulatory effects specific for auxins, gibberellins and kinins.
5. They are actively synthesized in green tissues and are found associated with auxins, gibberellins and kinins. However, they alone are accumulated in the absence of these growth substances in senescent and resting organs.
6. They are accumulated in woody plants during the period of dormancy.

Types. There are two types of growth inhibitors :

A. *Phenolic inhibitors*— These are better known and are widespread in plant organs. These are derivatives of either benzoic acid (*e.g.*, *p*-hydroxybenzoic, salicylic and gallic acids) or cinnamic acid (e.g., p-coumaric acid) and as well as coumarin.

B. *Abscisic acid and other substances*— Besides abscisic acid, these also include pyridine derivatives (*e.g.*, fusarinic and picolinic acids), quinones and flavonoids (*e.g.*, naringenin, phloridzin).

The structure of some common growth inhibitors is given in Fig. 20–22.

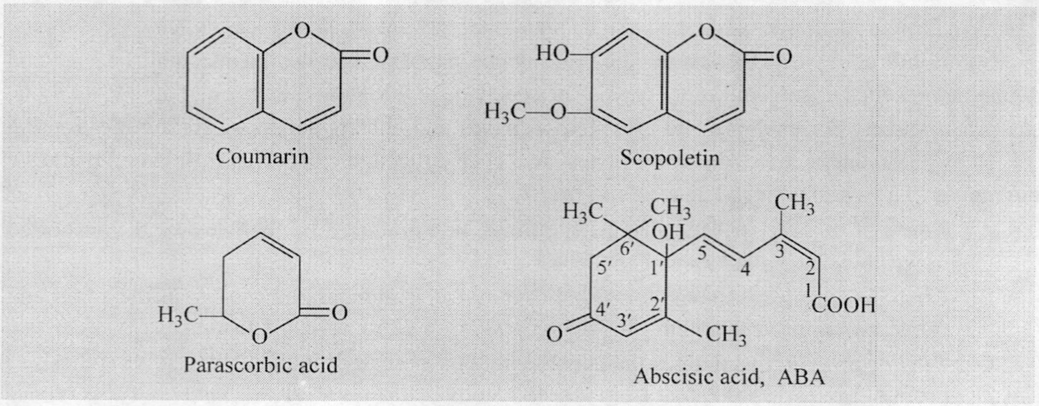

Fig. 20–22. Some common growth inhibitors

ABSCISIC ACID or ABSCISIN II or DORMIN

Discovery

Most of the effects first discovered for plant hormones were stimulatory. For example, IAA stimulates cellular elongation and cytokinins stimulate cell division. But near the end of the decade 1940s, **Torsten Hemberg** of Sweden reported that dormant buds of ash and potato contained inhibitors (rather than stimulators) that blocked the effects of IAA. When the buds germinated, the amount of these inhibitors decreased. Later, **Eagles** and **Wareing** (1963) isolated an inhibitor from the birch (*Betula pubescens*) leaves held under short day conditions. When this substance was reapplied to the leaves of birch seedlings, apical growth was completely arrested. As this substance induced dormancy, they named it as **dormin**. Later in 1965, **Ohkuma** *et al* isolated an inhibitor from cotton fruits and named it **abscisin II**. The same year, Cornforth and his associates isolated a growth inhibitor from sycamore and pointed out that both dormin and abscisin II are identical. Abscisin II is peculiar in that it is effective in much lower concentration than phenolic inhibitors and is accumulated under short day conditions. This compound was later named as abscisic acid (abbreviated as ABA)— an unfortunate name, because subsequent research has shown that *ethylene rather than abscisic acid controls abscission*. Abscisic acid (from *abscissus*[L] = cut off) is sometimes called *stress hormone*.

Distribution and Biosynthesis

Abscisic acid occurs in angiosperms and gymnosperms but apparently not in liverworts. ABA in plants is made from carotenoids. Once synthesized, ABA moves throughout a plant in xylem, phloem and parenchyma. *Like gibberellins and cytokinins, abscisic acid moves nonpolarly. There are no synthetic abscisic acids.*

Chemistry

Abscisic acid (Fig. 20–22) is a sesquiterpene with molecular formula $C_{15}H_{20}O_4$. Its molecule contains a corboxyl groups, 4 double bonds and an identical number of methyl groups.

Physiological Roles

1. Closure of stomata. It is a known fact that during drought, leaves synthesize large amounts of ABA which causes stomata to close. Thus, ABA acts as a messenger and enables plants to conserve water during drought. Because ABA-induced closure of stomata occurs within 1 to 2 minutes, this effect probably occurs independently of protein synthesis. As to its mechanism, ABA probably produces its effect by binding to proteins on the outer surface of the plasmalemma of guard cells. This renders the plasmalemma more positively-charged, thereby stimulating transport of ions (especially K^+) from guard cells to epidermal cells. The loss of these ions causes water to leave guard cells (via osmosis) which then collapse, thus closing the stomatal aperture.

2. Delays seed dormancy. In many species, applying ABA delays seed germination. Similarly, in many other plants, the amount of ABA in their seeds decreases, when seeds germinate. Thus, it may be inferred that ABA controls seed dormancy in some cases. However, this conclusion may not be generalized, since germination of many seeds occurs without any changes in the amount of abscisic acid.

3. Controls bud dormancy. Bud dormancy was previously thought to be controlled solely by ABA. But when leaves are treated with radioactive ABA, no radioactivity has been detected in buds. This suggests that, besides ABA, bud dormancy is probably also influenced by cytokinins and IAA-induced synthesis of ethylene.

4. Counteracts the effects of other hormones. ABA counteracts the stimulatory / inhibitory effects of other hormones. For example,

(*a*) ABA inhibits cell growth promoted by IAA.

(*b*) ABA inhibits amylase produced by seeds treated with gibberellin.

(*c*) ABA promotes chlorosis that is inhibited by cytokinins.

This may be due to the fact that *ABA is a* Ca^{2+} *antagonist* and its inhibition of the stimulatory effects of IAA and cytokinin may be due to its interference with Ca^{2+} metabolism. Although ABA usually inhibits growth, it is not toxic to plants as are inhibitors of RNA / protein synthesis. ABA often decreases gene acctivity, but there are instances of ABA stimulating genes. For example, ABA stimulates the synthesis of mRNAs for storage proteins in developing wheat grains.

MORPHACTINS

Recently, a group of growth substances called morphactins (meaning morphologically active substances) has diverted the attention of physiologists on account of its regulatory effect on the growth and development of plants (**Merck**, 1970). *The role of morphactins is usually of inhibitory nature.*

Chemically, the morphactins are derivatives of fluorene-9-carboxylic (Fig. 20-23) acid and are absorbed through the seeds, roots, and leaves. Their action is systematic but slow and can be transported both acropetally and basipetally. Considering from phylogenetic viewpoint, the morphogenic action of morphactins starts with the complex brown algae, the mosses and ferns and is exhibited well in higher plant groups. *Uniçellular plants and simple filaments are, however, not affected by morphactins.*

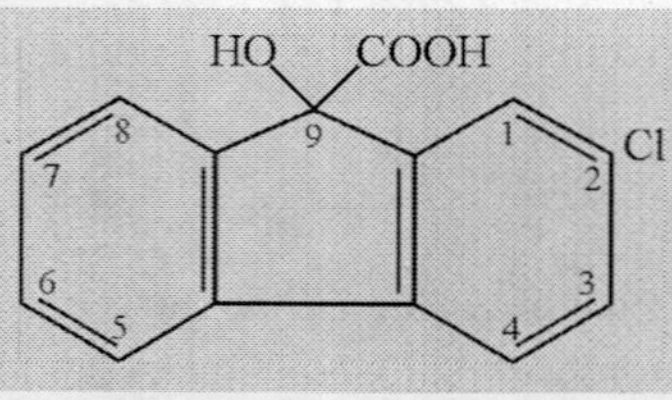

Fig. 20–23. Chlorofluorenol or morphactin

Although morphactins are much alike to abscisic acid in action, yet some of their effects are highly specific to them. Some such effects are :

1. inhibit sprouting of rosette plants
2. complete abolition of polarity in plants.
3. inhibition of mitosis in apical meristems.
4. stimulation of cell division in cambium of intact plants and cuttings.
5. reinforcement of the apical dominance in tap roots so that the formation of lateral roots is strongly inhibited.
6. abolition of geotropic and phototropic responses.
7. reduction of apical dominance in the main shoot so that the formation of branches is promoted, resulting in a broomlike growth of the plant.

PLANT HORMONES VERSUS ANIMAL HORMONES

Since plant hormones were discovered later than animal hormones, early research was based on the assumption that plant and animal hormones had basically similar physiological effects. However, there are important differences in plant and animal hormones (Table 20–3)

1. There is no evidence to indicate that the fundamental actions of plant and animal hormones are the same.
2. Plant hormones are not made in tissues specialized for hormone production, whereas animal hormones are produced in specialized organs, mostly in endocrine glands.
3. Plant hormones do not have definite target areas, whereas animal hormones do act at target .areas.
4. Animal hormones are usually very specific in their actions, while plant hormones seldom have such specific effects.

Table 20–3. Comparison of animal and plant hormones

Characteristic	*Animal hormones*	*Plant hormones*
Site of production	Specific endocrine glands specialized for homone production.	Produced by actively metabolizing tissues that have other functions.
Target tissues	Each hormone acts on a specific target tissue or organ.	Each hormone acts on a variety of tissues.
Number of hormones	Many, each with a specific function.	Relatively few, each with a variety of functions.
Primary function	Affect homeostasis and are regulatory in action; effects are reversible.	Affect growth and development; effects are permanent.

PROBLEMS

1. Describe two general chemical features that distinguish plant homones from animal hormones.
2. Name two plant hormones that are antagonistic in their effects.
3. The primary effect of gibberelllins is on :
   (*a*) mitosis (*b*) meiosis (*c*) cell enlargement
   (*d*) flowering (*e*) root growth
4. Which part of a plant produces auxins ?
   (*a*) Root hair
   (*b*) Meristematic regions of stem and root
   (*c*) Fruits
   (*d*) Leaves
5. Which of the following accelerates cell division in plants ?
   (*a*) Auxins
   (*b*) Cytokinins
   (*c*) Gibberellins
   (*d*) Florigens
6. The plant part which produces cytokinin is :
   (*a*) Roots
   (*b*) Stem
   (*c*) Leaves
   (*d*) Flowers
7. How do fruits ripen ?
8. Why do most leaves fall off in autumn ?

ANSWERS TO PROBLEMS

1. (*a*) Most plant hormones are terpenoid compounds.
   (*b*) There are no known peptide hormones in higher plants. Steroid-like compounds have been described in plants, but it is not clear that they act like steroid hormones.
2. Gibberellin and abscissic acid
3. (*c*) cell enlargement
4. (*b*) Meristematic regions of stem and root
5. (*b*) Cytokinins
6. (*a*) Roots
7. Unripe fruits are typically hard, green and have a sour or astringent taste due to the presence of organic acids such as malic, citric, and tartaric. They have high amounts of polysaccharides and low amounts of proteins. Ethylene gas that triggers the ripening process, emanates from fruits and causes rapid and dramatic changes. The green colour of the fruits disappears and red or yellow coloured pigments such as anthocyanins and carotenoids appear giving the fruits distinct colours. The polysaccharides are broken down to smaller sugars, which give the fruits the sweet taste. The fruits also begin to soften. Volatile substances are produced during ripening which give many fruits their distinct aroma.
8. A tree's roots, branches and twigs can endure freezing temperatures, but most leaves are not so tough. As sunlight decreases in autumn, the veins that carry sap into and out of a leaf gradually close. A layer of cells called the separation layer (or *abscission zone*) forms at the base of the petiole. When this layer is complete, the leaf is separated from the tissue that connected it to the branch, and it falls. The only exception is oak (*Quercus* sp.). The separation layer never fully detaches the dead oak leves, and they remain on the tree through winter.

CHAPTER

21 Fat-soluble Vitamins

INSIDE

- Historical Resume
- Definition
- General characteristics
- Classification
- Storage of Vitamins in the Body
- Daily Human Requirements of Vitamins
- Avitaminoses (Deficiencies of Vitamins)
- Vitamin A
- Vitamin D
- Vitamin E
- Vitamin K
- Coenzyme Q
- Stigmasterol

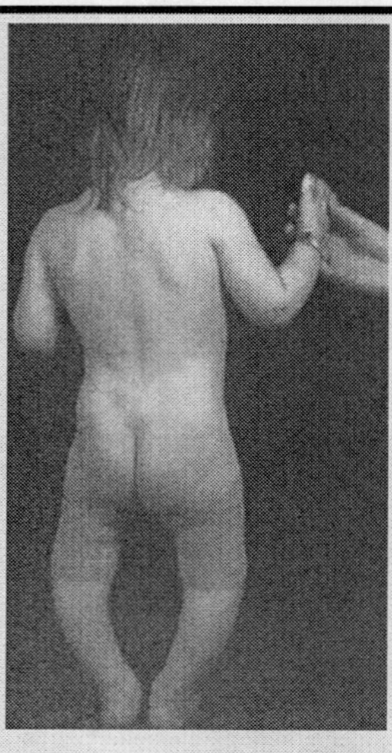

The badly-bowed legs of a rachitic child

The disease rickets develops due to the deficiency of vitamin D.

HISTORICAL RESUME

Vitamins have a back-dated history. The dreaded disease **scurvy** was one of the prevalent diseases in Europe during 15th and 16th centuries. The disease is said to have afflicted the crusadors. Scurvy was reported by **Vasco de Gama** during his sea voyages and **Jacques Cartier** in 1535 had reported death of about 25% of his sailing crew due to scurvy. On the recommendation of **Sir James Lancaster**, an English privateer, the ships of East India Company in 1601 carried oranges and lemons to prevent scurvy. In 1757, **James Lind**, a British naval surgeon, stated that fresh fruits and vegetables alone are effective to protect the body from various maladies and urged the inclusion of lemon juice in the diet of sailors to prevent scurvy. And some 40 years later, the Admiralty took his advice. After limes were substituted for lemons in 1865, British sailors began to be known as "limeys." Similarly, **rickets** was also attributed to faulty diets and **Guérin** (1838) produced it experimentally in puppies to prove the dietary connection.

In 1887, **Admiral Takaki**, Director-General of the Medical Services in Japan, demonstrated that another scourge **beriberi** could be prevented by enriching the diet with meat, vegetables and milk and at the same time decreasing the amount of milled rice in the case of Japanese sailors.

FREDERICK GOWLAND HOPKINS
(LT, 1861-1947)

Hopkins, a British biochemist, was born in Eastbourne, East Sussex, England on June 20, 1861. He was one of the founders of the science of Biochemistry, which he developed at Cambridge. His greatest work was in the identification and isolation of vitamins. He shared the **1929 Nobel Prize in Medicine or Physiology** for his discovery of essential nutrient factors now known as vitamins– needed in animal diets to maintain health, along with **Christiaan Eijkman** of The Netherlands for his discovery of vitamin B_1 (thiamine). Frederick Gowland Hopkins worked out the importance of vitamins in the diet in the early 1900s. His classic experiment, described in the adjoining graph, helped people understand the need for vitamins as additional food factors in a healthy diet.

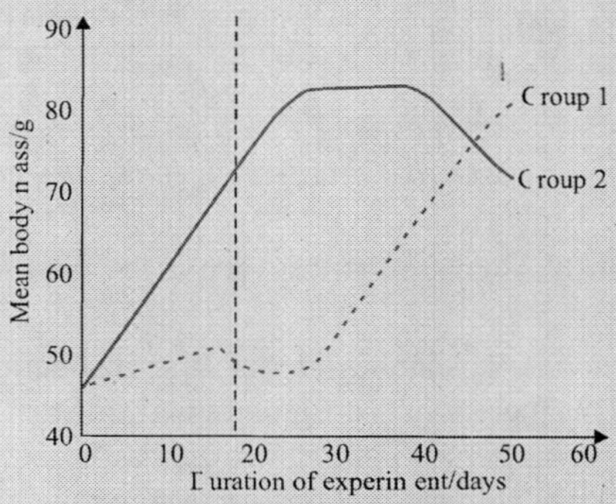

Basically, he showed that the milk-deficient rats did not grow properly. He suggested that the milk contained an 'accessory food factor', or 'vital amine'– hence the name *vitamin,* which the rats need to stay healthy. Later work by American researchers showed this compound to be fat-soluble. We now know this compound to be vitamin A.

Hopkins died on May 16, 1947 at Cambridge.

Between 1906 and 1912, Hopkins and his team at Cambridge University studied accessory food factors. In this experiment two groups of rats are used. Group 1 rats were fed on a diet of purified casein (milk protein), starch, cane sugar, lard and salts. Group 2 rats were fed the same diet but with a milk supplement.

Those on the artificial diet (1) stopped growing and lost body mass; those with the milk supplement (2) grew normally, even though the milk contributed only 4 per cent of the total food eaten.

Later, **Eijkman** (1897), a Dutch physician, found that experimental beriberi could be induced in hens when fed with polished rice without bran. Such hens could be cured by giving them the rice polishings. **Eijkmann**, for a time, believed that the rice polishings contained something that neutralized the beriberi toxin in the polished rice. In 1906, however, **Frederick Gowland Hopkins** ascribed the diseases such as scurvy and rickets to the lack of some '*dietary factors*'. Six years later (*i.e.*, in 1912), **Hopkins** in collaboration with **Casimer Funk** of Poland, who was working at the Lister Institute of London, suggested the **vitamin theory** which postulates that *'specific diseases such as beriberi, scurvy and rickets are each caused by the absence from the diet of a particular nutritional factor.'*

Funk, for the first time, also isolated the dietary factor from rice polishings which acted as antiberiberi substance. Since this factor was an amine and necessary to life, Funk at the suggestion of Dr. Max Nierenstein introduced the term 'vitamine' (*vita*[L] = life) to denote it. Since subsequent studies showed that not all these substances are amines, the terminal letter 'e' was dropped from its spelling at the suggestion of **Sir. J.C. Drummond** (1919), who also proposed their **alphabetical nomenclature**. In fact, the various vitamins have no structural resemblance to each other, but because of a similar general function in metabolism, they are studied together. Although these molecules serve nearly the same roles in all forms of life, but higher animals have lost the capacity to synthesize them.

CASIMER FUNK
(LT, 1884–1967)

Casimer, who was born in Poland, received his medical degree from the University of Bern, and became an U.S. citizen in 1920. In 1923, he returned to Poland to direct the State Institute of Health but returned to the United States permanently when World War II broke out. Funk, in collaboration with Hopkins, suggested the **vitamin theory.**

DEFINITION

The vitamin concept has undergone extensive revisions during the history of biochemistry. However, **Franz Holfmeister's** (LT, 1850-1922) definition— 'vitamins are substances which are indispensable for the growth and maintenance of the animal organism, which occur both in animals and plants and are present in only small amounts in food'— still holds good, although it has been interpreted in various ways.

> **Sine quo non** is a Latin phrase, which means an indispensable condition.

The term '**vitamin**', in its modern sense, usually refers to *the substances distinct from major components of food, required in minute quantities (i.e., oligodynamic in nature) and whose absence causes specific deficiency diseases*. As the living organisms cannot synthesize most of these compounds, a steady supply of them is *sine quo non* for life. Their ultimate source is the plant or bacterial world.

Differences with hormones— Indeed, for some time, it was believed that the distinction between vitamins and hormones was no longer tenable. But there exists a fundamental difference between these two classes of active substances : *the hormones are regulatory substances whereas the vitamins are accessory nutrients*. The vitamins also differ from the hormones in that they are supplied to the body from some external source (*i.e.,* chiefly from the food ingested) whereas the hormones are synthesized within some part of the body of an organism. Moreover, most vitamins and some hormones are involved either directly or indirectly in enzyme systems in order to carry out biochemical functions. Some vitamins are also known to be coenzymes.

A comparison of enzymes, hormones and vitamins is presented in Table 21–1.

Table 21–1. Differences among enzymes, hormones and vitamins

	Enzymes	*Hormones*	*Vitamins*
1.	These are specific catalytic substances.	These are regulatory substances.	These are accessory nutrients.
2.	All enzymes are proteinaceous in nature.	Hormones are steroids (estradiol, testosterone) or peptides (insulin, ocytocin) or amino acid derivatives (thyroxine, adrenalin).	Vitamins belong to diverse groups chemically.
3.	These are synthesized by the animal cells.	These are synthesized within some part of the body.	These are not synthesized by most animal cells but supplied to the body from outside source (mainly as food). Plants can, however, synthesize them.
4.	These can initiate and continue reactions.	These cannot initiate reactions but can influence the rate at which they proceed.	—
5.	These are not consumed during growth.	They are consumed during growth.	—
6.	These are involved in coenzyme syntems.	These are not involved in coenzyme systems.	These are involved in coenzyme systems.
7.	These operate in biological systems. *Examples* : Pepsin, Trypsin, Urease, Diastase, Lipase, Catalase, Maltase, Alcohol dehydrogenase, Cytochrome oxidase.	Only some hormones are in volved in enzyme systems. *Examples:* Estradiol, Testosterone, Relaxin, Progesterone, Insulin, Thyrotropin, Ocytocin, Secretin, Intermedin, Thyroxine, Adrenalin.	Most vitamins are involved in enzyme systems. *Examples* : Fat-soluble vitamins A, D, E, K ; Water-soluble vitamins B, C.

GENERAL CHARACTERISTICS

The vitamins are characterized for some general facts, which are listed below :

1. Vitamins are of widespread occurrence in nature, both in plant and animal worlds.
2. All common foodstuffs contain more than one vitamin.
3. The plants can synthesize all the vitamins whereas only a few vitamins are synthesized in the animals.
4. Human body can synthesize some vitamins, *e.g.*, vitamin A is synthesized from its precursor carotene and vitamin D from ultraviolet irradiation of ergosterol and 7-dehydrocholesterol. Some members of the vitamin B complex are synthesized by microorganisms present in the intestinal tract. Vitamin C is also synthesized in some animals such as rat.
5. Most of vitamins have been artificially synthesized.
6. All the cells of the body store vitamins to some extent.
7. Vitamins are partly destroyed and are partly excreted.
8. Vitamins are nonantigenic.
9. Vitamins carry out functions in very low concentrations. Hence, the total daily requirement is very small.
10. Vitamins are effective when taken orally.
11. Synthetically-made vitamins are just as nutritionally good as natural vitamins.
12. Old people need about the same amounts of vitamins as young people.

CLASSIFICATION

In 1913, **McCollum** and **Davis** described a lipid-soluble essential food factor in butter fat and egg yolk. In 1915, a water-soluble factor in wheat germ necessary for the growth of young rats was also described. Since then, two categories of vitamins are usually recognized : fat-soluble and water-soluble. These two groups discharge rather different functions.

A. Fat-soluble vitamins. These are oily substances, not readily soluble in water and their biochemical functions are not well understood. They contain only carbon, hydrogen and oxygen. Their examples are vitamins A, D, E and K. They, however, play more specialized roles in certain group of animals and in particular type of activities. For instance, they function in the formation of a visual pigment (vitamin A), in the absorption of calcium and phosphorus from the vertebrate intestine (vitamin D), in protecting mitochondrial system from inactivation (vitamin E) or in the formation of a blood clotting factor in vertebrates (vitamin K). The individual fat-soluble vitamins bear a closer resemblance to each other chemically. In fact, the 4 fat-soluble vitamins can be regarded as lipids. Vitamins A, E and K are terpenoids, and vitamin D is a steroid. All four are isoprenoid compounds, since they are synthesized biologically from units of isoprene (for structure, refer page), a building block of many naturally-occurring oily, greasy or rubbery substances of plant origin. Unlike the water- soluble vitamins (B and C), fat-soluble vitamins can be strored in the body, *e.g.*, an adult's liver can store enough vitamin A to last several months or longer. However, because *fat-soluble vitamins are storable*, their excessive intakes can result in toxic conditions (hypervitaminoses).

B. Water-soluble vitamins. Most of these are universally vitamins since they perform the same general functions wherever they occur. Besides C, H and O, they also contain nitrogen. They are catalytic factors and as such form vital links in the chains of biochemical reactions characteristic of all living objects. For instance, thiamine is required whenever sugars are oxidized aerobically to release energy. The individual water-soluble vitamins bear no closer resemblance to each other chemically. The biochemical or coenzyme function of nearly all of these is known. The common water-soluble vitamins are vitamins of B complex such as B_1 through B_{12} (vitamins B_4, B_8, B_{10} and B_{11}, however, do not exist) and the vitamin C. Choline, inositol, *p*-aminobenzoic acid, bioflavonoids and α-lipoic acid are frequently included in this category. Many nutritionists, however, do not consider them as true vitamins, although their dietary deficiencies in animals lead to the development of characteristic symptoms. Moreover, none of them except α-lipoic acid is a part of the coenzyme system. The B-series of vitamins, being water-soluble and excretable, are *required daily in meagre amounts* (in milligrams or even less) for the normal growth and good health of humans and many other organisms. It is virtually impossible to 'overdose' on them.

STORAGE OF VITAMINS IN THE BODY

The vitamins can be stored in the body to a slight extent. The liver cells are, however, rich in certain **fat-soluble vitamins**. For instance, the amount of *vitamin A* contained in the liver is sufficient enough to meet its requirement without any additional intake for about 6 months. Similarly, the quantity of *vitamin D* stored ordinarily in the liver is sufficient to maintain a person without any additional intake of vitamin D for about 2 months. The storage of *vitamin K* is, however, relatively slight.

The **water-soluble vitamins** are stored even in lesser amounts in the cells. Evidently, in cases of deficiency of *vitamin B compounds*, clinical symptoms appear rather early, that is within a few days. Similarly, absence of *vitamin C* can induce deficiency symptoms within a few weeks. Vitamin C is stored in the adrenal cortex.

DAILY HUMAN REQUIREMENTS OF VITAMINS

The requirement of vitamins varies considerably depending on the nature of the individual consuming them. Some general facts regarding vitamin requirement may be listed below :

1. Greater the size of the individual, higher are his vitamin needs.
2. Younger ones require higher quantities of vitamins than do the elders.
3. The vitamin requirements increase when a person performs exercise.
4. During ailments, the vitamin requirements are ordinarily enhanced.
5. Under certain specific conditions of metabolic disorders when the vitamins cannot be properly utilized, the requirement for one or more specific vitamins is at extreme.
6. Growing children require comparatively high quantities of vitamin D.
7. During pregnancy and lactation, the vitamin D requirement by the mother is greatly enhanced.
8. The requirements of vitamin B complex (*esp.*, that of vitamin B_1) are increased under conditions of greater utilization of carbohydrates.

The daily requirement of any vitamin (refer Table 21–2) for any individual is not a fixed quantity and varies according to the rate of metabolism. *In general, in all cases of high metabolic activity (such as heavy muscular work, during pregnancy and lactation and in growing children), the vitamin requirement is proportionately high*. Normally, a man doing ordinary work can obtain enough vitamin from his balanced diet.

AVITAMINOSES
(Deficiencies of Vitamins)

A lack of one or more vitamins leads to characteristic deficiency symptoms in man. Multiple deficiencies caused by the lack of more than one vitamin are, however, more common in human beings. Vitamin deficiencies occur rather frequently in certain parts of the world for socioeconomic reasons. Avitaminosis may be of following 2 types :

Table 21–2. Recommended daily dietary allowances*

	Age†	Weight	Height	Energy	Protein	Fat-soluble vitamins			Water-soluble vitamins						
	years	kg	cm	kcal	g	A	D_3	E	B_1	B_2	B_5	B_6	B_9	B_{12}	C
						μg‡	μg	mg	mg	mg	mg	mg	μg	μg	mg
INFANTS	0.0–0.5	6	60	650	kg × 2.2	420	10	3	0.3	0.4	6	0.3	30	0.5	35
	0.5–1.0	9	71	970	kg × 2.0	400	10	4	0.5	0.6	8	0.6	45	1.5	35
CHILDREN	1–3	13	90	1300	23	400	10	5	0.7	0.8	9	0.9	100	2.0	45
	4–6	20	112	1700	30	500	10	6	0.9	1.0	11	1.3	200	2.5	45
	7–10	28	132	2400	34	700	10	7	1.2	1.4	16	1.6	300	3.0	45
MEN	11–14	45	157	2700	45	1000	10	8	1.4	1.6	18	1.8	400	3.0	45
	15–18	66	176	2800	56	1000	10	10	1.4	1.7	18	2.0	400	3.0	60
	19–22	70	177	2900	56	1000	7.5	10	1.5	1.7	19	2.2	400	3.0	60
	23–50	70	178	2700	56	1000	5	10	1.4	1.6	18	2.2	400	3.0	60
	51+	70	178	2400	56	1000	5	10	1.2	1.4	16	2.2	400	3.0	60
WOMEN	11–14	46	157	2200	46	800	10	8	1.1	1.3	15	1.8	400	3.0	50
	15–18	55	153	2100	46	800	10	8	1.1	1.3	14	2.0	400	3.0	60
	19-22	55	163	2100	44	800	7.5	8	1.1	1.3	14	2.0	400	3.0	60
	23–50	55	163	2000	44	800	5	8	1.0	1.2	13	2.0	400	3.0	60
	51+	55	163	1800	44	800	5	8	1.0	1.2	13	2.0	400	3.0	60
Pregnant				+300	+30	+200	+5	+2	+0.4	+0.3	+2	+0.6	+400	+1.0	+20
Lactating				+500	+20	+400	+5	+3	+0.5	+0.5	+5	+0.5	+100	+1.0	+40

* The recommended allowances can be attained with a variety of common foods, including those nutrients also for which human requirements have been less well defined.

† All the entries for the age range represent allowances for the midpoint of the specified age range. However, in the case of age range 23–50 years, the reference is to the man and woman at age 25 years.

‡ Retinol equivalents : 1 retinol equivalent = 1 μg retinol or 6 μg β-carotene.

Note that there is no recommended dietary allowances (RDAs) for vitamin K, vitamin B_3 (pantothenic acid), and vitamin B_7 (biotin) as intestinal bacteria synthesize the body's requirement.

(Adapted from Food and Nutrition Board, National Academy of Sciences— National Research Council, Revised 1980)

A. *primary or direct*— This arises due to inadequate intake of vitamins resulting from chronic alcoholism, dietary fads etc.

B. *secondary* or *'conditioned deficiency'*— This arises due to other factors such as malabsorption, increased excretion, allergies, anorexia, gastrointestinal disorders etc.

Vitamin deficiency, whether primary or secondary, leads to:

(*a*) a gradual decrease in tissue levels of the vitamin(s) deficient,

(*b*) a biochemical lesion,

(*c*) an anatomic lesion, and

(*d*) finally cellular pathology and disease.

This sequence is schematically represented in Fig. 21–1.

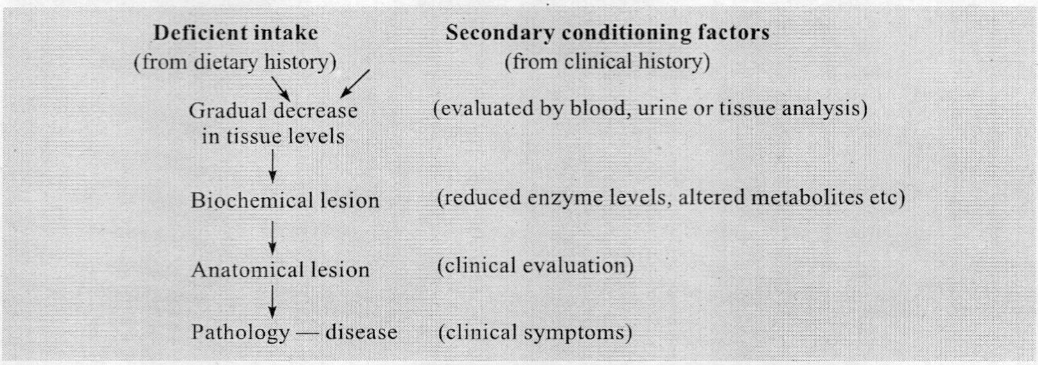

Fig. 21–1. Sequence of events occurring in a tyical avitaminosis

An account of **fat-soluble vitamins** is given below.

VITAMIN A

A. History. It was first recognized as an essential nutritional factor by **Elmer McCollum** in 1915 and then isolated from fish-liver oil by **Holmes** in 1917. On account of its established role in the visual process, it is often called as **antixerophthalmic factor** or the **"bright eyes" vitamin**. It was first synthesized in 1946 by **Milas**.

B. Occurrence. *Liver oils of various fishes are the richest natural sources of vitamin A.* Shark and halibut contain maximum amount whereas the cod-liver has lowest amount. Depending upon the species of fish and the time of year of catch, the fish livers contain 2,000 to 100,000 I.U. per gram of vitamin A. The amount present in human liver is much less (500 to 1,000 I.U.per gram). However, polar bear liver is an extremely concentrated source of vitamin A. Other noteworthy sources are butter, milk and eggs and, to a lesser extent, kidney. In its provitamin form (*i.e.*, as carotenes) it is supplied by all pigmented (particularly yellow) vegetables and fruits such as carrots, pumpkins, cantaloupes, turnips, peppers, peas, sweet potatoes, papayas, tomatoes, apricots, peaches, plums, cherries, mangoes etc. Yellow corn is the only cereal containing significant amounts of carotene. New-born infants have low quantities of vitamin A content that is rapidly augmented because colostrum and breast milk furnish large amounts of the vitamin. The milk of well-fed mothers, however, contains sufficient amounts of vitamin A (in ester form) for the infant's need. However, vitamin A is absent from vegetable fats and oils (olive oil, linseed oil, groundnut oil). It is added to margarine during its manufacture from these oils.

C. Structure. Vitamin A is found in two forms A_1 and A_2. The carotenoids that give rise to vitamin A in animal body is named as provitamin A. These inculde α-, β- and γ-carotenes and cryptoxanthin. β-*carotene is most potent of all these forms*. A molecule of β-carotene is made of eight 5-carbon isoprenoid units, linked to form a long chain of 40 carbon atoms with an ionone ring at each end. It is an orange-red hydrocarbon and upon hydrolysis yields 2 moles of vitamin A_1 (Fig. 21–2). During hydrolysis, a cleavage occurs at the mid point of the carotene in the polyene chain connecting the 2 β-ionone rings. This conversion usually takes place in livers of fishes and mammals. Vitamin A_1 is a complex primary alcohol called retinol, having the empirical formula, $C_{20}H_{29}OH$. The terminal hydroxyl group is ordinarily esterified. It contains β-ionone ring.

> **α- and γ-carotenes** as well as cryptoxanthin, however, yield only 1 molecule of vitamin A since in them one of the two rings present in their molecule differs in structure from that of vitamin A.

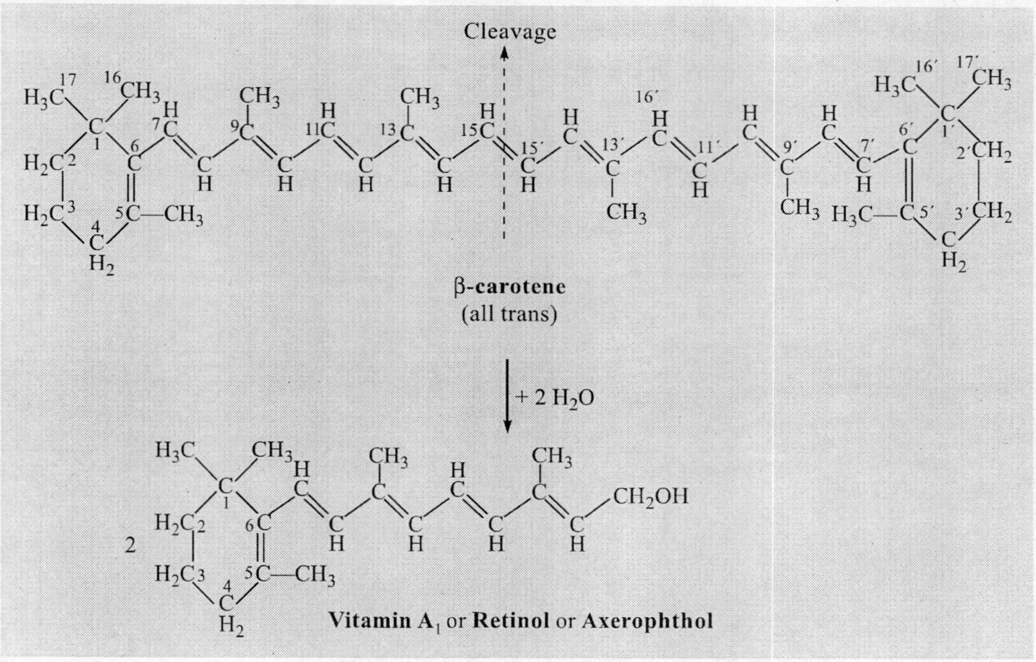

Fig. 21–2. Hydrolysis of β-carotene

Another form of vitamin A, present in fresh-water fishes, is known as vitamin A_2 (Fig. 21–3). It differs from vitamin A_1, which is found in salt-water fishes, in possessing an additional conjugate double bond between carbon atoms 3 and 4 of the β-ionone ring. Its potency is 40% that of vitamin A_1.

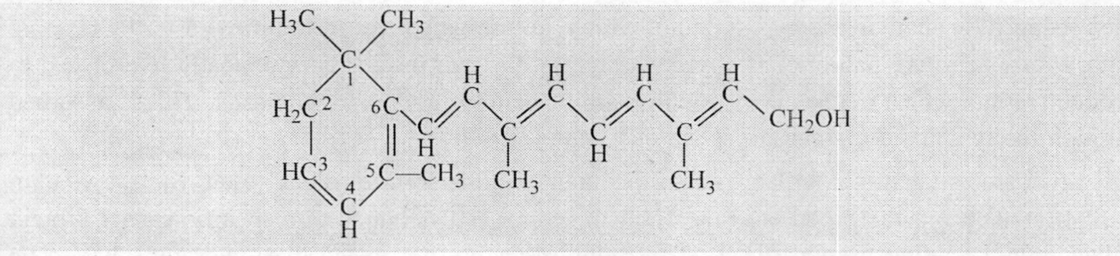

Fig. 21–3. Vitamin A_2 or 3-dehydroretinol

(all *trans*)

D. Properties. Ordinarily retinol is a viscid, colourless oil but by careful fractionation it has also been isolated as pale yellowish needles. It gives a characteristic absorption band in ultraviolet (UV) spectrum at 328 *m*μ. It is soluble in fat and fat solvents but insoluble in water. Loss of vitamin A in cooking, canning and freezing of foodstuffs is small; oxidizing agents, however, destroy it. It is destroyed on exposure to UV light. Vitamin A is relatively unstable in air unless protected by antioxidants including vitamin E.

E. Metabolism. In the tissues, the metabolic transformation of retinol is carried out by enzymes. The dietary β-carotene is split into retinal (Fig. 21–4) by an enzyme of the intestine. Retinal is

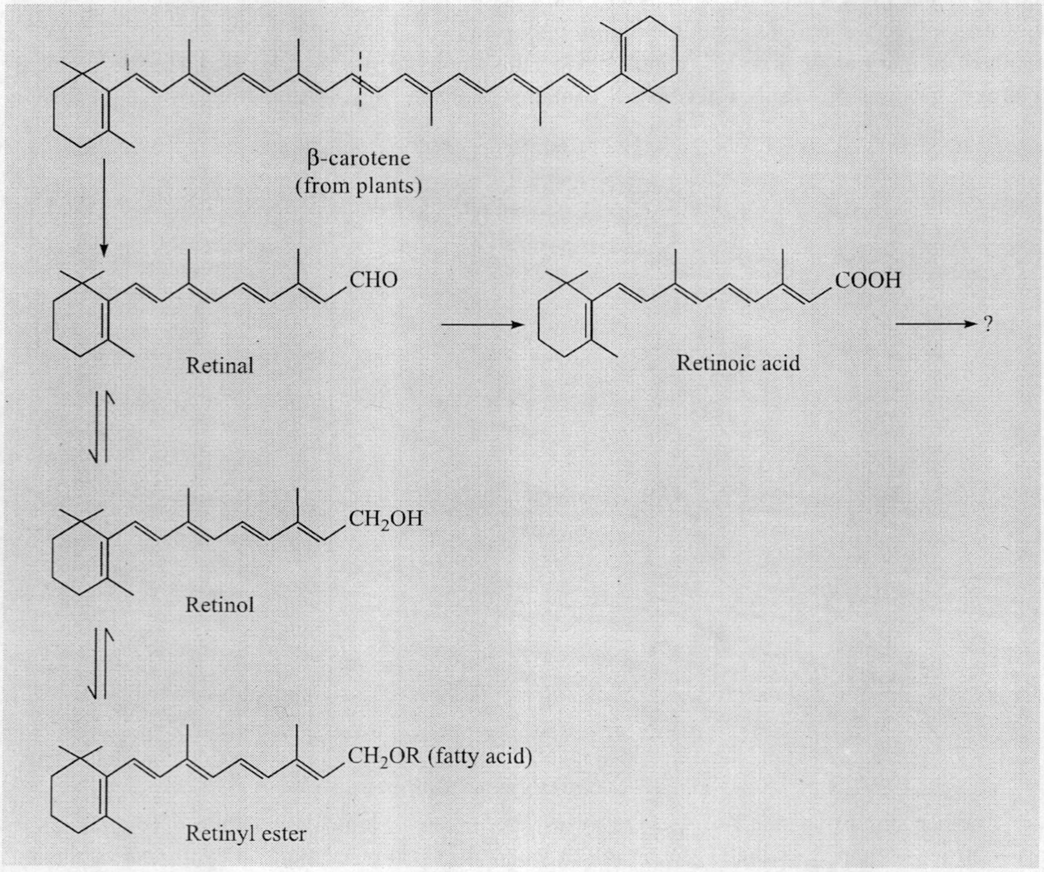

Fig. 21–4. Formation of retinoic acid

(Adapted from Ganguli J, 1973)

then reduced by another enzyme to retinol which, in turn, is converted to retinyl ester by reacting with a fatty acid like palmitic. The retinyl ester, on the contrary, is enzymatically hydrolyzed to produce retinol which is re-esterified with palmitic acid to produce retinyl ester. This is absorbed through the lymphatic system and is stored in the liver.

Liver stores vitamin A in large quantities mainly in the form of retinyl ester. But the vitamin A that circulates in the blood is in the form of retinol and is bound to a specific carrier protein called retinol binding protein (RBP). The retinyl ester present in the liver, therefore, has to be converted to retinol before it mixes with the blood. Liver can also successively convert retinol to retinal and retinal to retinoic acid. Retinoic acid is quickly absorbed from the intestine through the portal system and is rapidly excreted back into the intestine through the bile.

F. Deficiency. *Vitamin A is perhaps the most important as it affects the various metabolic processes in the body. It has profound effect on epithelial structures, in general.* Vitamin A deficiency leads to the onset of many diseases like nyctalopia or night blindness (inability to see in night), xerophthalmia (scaly condition of the delicate membrane covering the eyes), keratomalacia (softening of the cornea), phrynoderma or "toad skin" (hard and horny skin) and stunted growth.

Xerophthalmia or **xerosis** is a major cause of blindness in childhood and is still a major health problem in Far East such as Hong Kong, Jakarta, Manila and Dhaka. The disease affects large number of children but few adults. Xerophthalmia is characterized by drying of the eyes and hence so named (*xeros*^{G} = dry ; *ophthalmos*^{G} = eyes). The lacrymal glands become stratified and keratinized and cease to produce tears. This makes the external surface dry and dull. The ulcers develop. The bacteria are not washed away. The eyelids swell and become sticky. This results in frequent exudation of blood, causing severe infection to the eye. If left untreated, blindness results. Incredible, yet true, about 1400 cases of xerophthalmia were reported in 1904 among Japanese children. Besides, during World War I, many xerophthalmic cases were reported in Denmark because of the fact that butterfat was shipped out of that country in huge quantities and people had to live on substitutes with no retinol.

Keratomalacia (*kerato*^{G} = hair ; *malakia*^{G} = softness) is a corneal disease, occurring maximally in pre-school children of 3–4 years. This usually happens suddenly in young children with kwashiorkor *esp.*, after an episode of diarrhea or infection. At first the cornea loses its lustre, undergoes a necrosis and develops a few pin-point ulcers, which later coalesce to form a large, white ulcerative area. The perforation of the cornea may follow sometime, causing the release of aqueous humour and prolapse of the iris. Occasionally, instead of extensive ulceration and perforation of the cornea, the whole eyeball may shrink. Keratomalacia is unfortunately still prevalent on a wide scale in many parts of India (South and Eastern) and Indonesia and some other countries of Asia and Africa.

Phrynoderma is a skin lesion and is characterized by follicular hyperkeratosis. In it, the skin *esp.*, on the outer aspects of forearms in the regions of the elbows and of the thighs and buttocks becomes rough and spiky. In severe cases, the trunk is also affected.

Also, the damaged epithelial structures in diverse organs such as the eyes, the kidneys or the respiratory tract often become infected. It is for this reason that vitamin A has been called an '*anti-infection*' *vitamin.* The vaginal epithelium may become cornified, and epithelial metaplasia of the urinary tract may contribute to pyuria and hematuria. Increased intracranial pressure with wide separation of cranial bones at the sutures may occur.

In the alimentary tract, the deficiency of vitamin A causes damage to the intestinal mucosa, resulting in diarrhea.

Vitamin A is an important factor in tooth formation. In its deficiency, there is a defective formation of enamel so that the dentin is exposed. Evidently, sound tooth formation does not occur. Vitamin A deficiency may result in retardation of mental and physical growth and in apathy. Anemia with or without hepatosplenomegally is usually present.

G. Hypervitaminosis A. Vitamin A is less toxic to man and other animals when it is taken in large doses over a long period of time. The children receiving overdosages of 500,000 units of vitamin A per day exhibit tender swellings over the bones, limited motion and definite hyperosteoses.

Human adults consuming 500,000 units or more show nosebleed, weakness, headache, anorexia and nausea. Continuous excessive intake of the vitamin is dangerous because it results in fragile bones and abnormal fetal development. In the case of plant carotenoids, a dietary oversupply (for example, eating too many carrots on a daily basis) results in **carotenosis**, a condition readily diagnosed by yellowing of the skin. These toxic effects develop presumably due to the fact that viatmin A, like vitamin D, is not readily excreted and consequently tissue levels may build up dangerous concentrations.

H. Human requirements. The recommended dietary allowance (RDA) of vitamin A is about 5,000 International Unit (I.U.). Growing children, adults and pregnant and lactating mothers require high doses of up to 8,000 I.U. It is also possible that some individuals require more than the minimal requirement due to either faulty absorption or some other reason. For vitamin A, the World Health Organization (WHO) has chosen as one International Unit the biologic activity of 0.000344 mg (0.344 μg) of synthetic vitamin A-acetate, which is equivalent to 0.30 μg of retinol.

VITAMIN D

A. History. The first demonstration of the existence of vitamin D was shown by **Elmer McCollum** in 1922 who found that cod liver oil was effective in preventing rickets, a disease induced in rats by providing low calcium diet. On account of its preventive action on rickets, vitamin D is often called as **antirachitic factor**. It is also known as '**sunshine vitamin**' as its provitamin form present in human skin is easily converted to the active form by irradiation with ultraviolet light. At least 10 different compounds are known to have antirachitic properties and are designated as D_2, D_3 etc, but the two, namely, vitamin D_2 (ergocalciferol) and vitamin D_3 (cholecalciferol) are more important. Vitamin D_3 was, however, first isolated by **Brockmann** and others.

B. Occurrence. The best natural sources of vitamin D are the liver oils of many fishes such as cod and halibut. The flesh of oily fishes (*e.g.*, sardine, salmon, herring) is also excellent source. Egg yolks are fairly good but milk, butter and mushrooms are poor. The diets of infants may contain only small amounts of vitamin D; cow's milk contains only 0.1 to 1 μg/quart (1 μg = 40 IU). Cereals, vegetables and fruits contain only negligible amounts. Eggs and yolks contain 3 to 10 μg/g. Most marketed cow's milk is fortified with 10 μg of vitamin D per quart and most commercially-prepared milks for infant formulae are also fortified. Vitamin D_2 is of plant origin and is produced commercially by irradiation with ultraviolet light of a provitamin known as ergosterol which is found in plants, especially in ergot (hence so named) and yeast. Vitamin D_3, on the contrary, is of animal origin and can be produced from 7-dehydrocholesterol also by irradiating with ultraviolet light. The 7-dehydrocholesterol is also a provitamin found naturally occurring in animals. Vitamins D_2 and D_3 both have about the same degree of activity in the human beings. In nature, these vitamins occur as esters. *Like vitamin A, vitamin D is absent from vegetable fats and oils* and is added to margarine during its manufacture.

C. Structure. The transformation of ergosterol ($C_{28}H_{44}O$) to the active form D_2 takes place through a series of intermediate steps illustrated as below :

Ergosterol → Lumisterol → Protachysterol → Tachysterol → Precalciferol

→ Calciferol ↗ Toxisterol, ↘ Suprasterols

Similarly, cholecalciferol ($C_{27}H_{44}O$) is produced from 7-dehydrocholesterol through a series of intermediaries as follows :

7-dehydrocholesterol → Lumisterol_3 → Tachysterol_3 → Precalciferol_3
→ Cholecalciferol

During the activation of the provitamins, the ring B is cleaved between carbon atom 9 and 10 to produce vitamins D_2 and D_3 (Fig. 21–5).

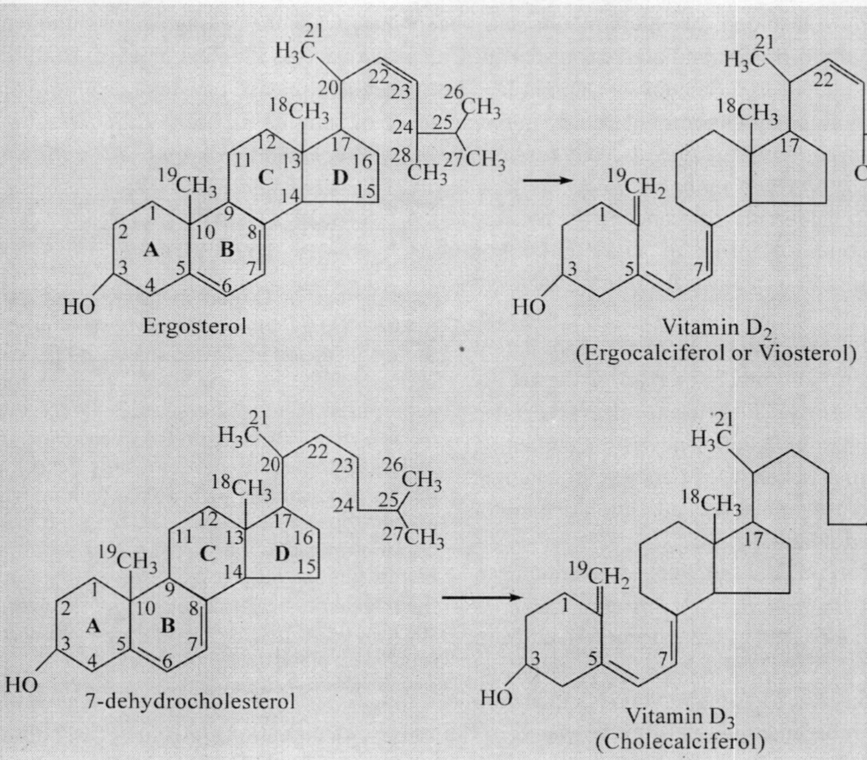

Fig. 21–5. Activation of provitamins of the vitamin D group

Note that in both the cases, the effect of irradiation is the opening of ring B.

It may, however, be noted that the two vitamins, D_2 and D_3 and also D_4 and D_5 (Fig. 21–6), differ only in their side chains attached to C_{17}.

Fig. 21–6. Side chains of vitamins D4 and D5

D. Properties. Vitamin D is a white and almost odourless crystalline substance, soluble in fat and fat solvents. It is fairly heat resistant and also relatively resistant to oxidation. It is not affected by acids and alkalies.

E. Metabolism. The provitamin D_3 can be synthesized within the human body so that it may, in fact, not be required in the diet. This may, henceforth, not be treated as a vitamin. In the past when man lived mainly outdoors and with minimum clothing, there was no hindrance for the penetration of ultraviolet light from the sun to convert it into the active form. In the far northern areas, however, the amount of light is not adequate for conversion and as such fish liver oils serve as excellent source of vitamin D in these areas. The increased need of this vitamin is usually felt in growth and in pregnancy to provide for the needs of the foetus.

Vitamin D plays an important role in calcification of bones and teeth. It encourages the absorption, into the blood, of calcium salts and phosphates. Calcium passage across duodenum occurs mainly by diffusion and active transport of Ca^{2+} occurs across the ileal mucosa. Both these processes are related in dificiency of vitamin D. The subsequent release of bound calcium is also markedly stimulated by vitamin D but only in the presence of parathyroid hormone. *On the whole, the function of vitamin D is to cause increased absorption, longer retention and better utilization of calcium and phosphorus in the body.*

There is considerable difference in the potency of these 2 common forms of vitamin D. For example, vitamin D_2 is a powerful antirachitic agent for man and for the rat but not for the chicken. Vitamin D_3, on the contrary, is much more potent for the chicken than either for man or the rat.

F. Deficiency. The most characteristic symptom of vitamin D deficiency is the childhood disease known as rickets. Deficiency of it in human adults leads to osteomalacia, a condition that might also be termed "adult rickets".

Rickets (origin doubtful. False derivation from *rachitis*[G], from *rhachis*[G] = spine + *-itis*[G] = inflammation; also said to have been derived from an old English word, *wrickken* = to twist) is primarily a disease of growing bones. In it, the deposition of inorganic materials on the matrix of bones (*i.e.*, calcification) fails to occur, although matrix formation continues. Rickets is unusual below the age of 3 months (mo). It may occur in older children with malabsorption. Clinical manifestations of rickets in children usually manifest in the first year or in the second year. One of the early signs of rickets is **craniotabes**, which is due to thickening of the outer table of the skull and is detected by pressing firmly over the occiput or posterior parietal bones. A ping-pong ball like sensation will be felt. Craniotabes near the suture line may, sometimes, be present in normal premature infants. Costochondral junctions become prominent to give appearance of a beaded ribs, the **rachitic rosary**. Thickening of the wrists and ankles are other early evidences of osseous changes. Increased sweating, especially around the head, may also be present.

CLINICAL IMPLICATIONS
RICKETS

As described by **Daniel Webster** in 1645, in rickets, " .... the whole bony structure is as flexible as softened wax, so that the flaccid and enervated legs can hardly support the superposed weight of the body ; hence the tibia, giving way beneath the overpowering weight of the frame, bend inwards ; and for the same reason the legs are drawn together at their tops ; and the back, by reason of the bending of the spine, sticks out in a hump in the lumbar region .... the patients in their weakness cannot (in the most severe stages of the disease) bear to sit upright, much less stand .....".

Signs of advanced rickets are easily identified. These are listed below :

1. Head. Craniotabes may obliterate before the end of the 1st year, although the rachitic process continues. The softness of the skull may result in flattening and, at times, permanent asymmetry of the head. The anterior fontenel is larger than normal; its closure may be delayed until after the 2nd year of life. The central parts of the parietal and frontal bones are usually thickened, forming prominences or bosses, which give the head a box-like appearance (*catput quadratum*).

2. Thorax . The sides of the thorax become flattened, and the longitudinal grooves develop posterior to the rosary. The sternum with its adjacent cartilage projects forward leading to protruding chest (*pigeonbreast*). Along the lower border of the chest develops a horizontal depression (*Harrison groove*), which corresponds with the costal insertions of the diaphragm.

3. Spinal column. Moderate degree of lateral curvature (*scoliosis*) is common and a *kyphosis* (=increased convexity in the region of thoracic spine) may appear in the dorsolumbar region when sitting. *Lordosis* (=forward curvature of the lumbar spine) may be seen in the erect position.

4. Pelvis. The pelvic entrance is narrowed by a forward projection of the promontory; the exit, by a forward displacment of the caudal part of the sacrum and coccyx. In the female, these changes, if they become permanent, add to the hazards of childbirth and necessitate caesarean section.

5. **Extremities.** The epiphyseal enlargement at the wrists and ankles becomes more noticeable. Bending of the softened shafts of the femur, tibia and fibula results in bowlegs (*knock-knees*) ; the femur and the tibia may also acquire an anterior convexity. *Coxa vara* is sometimes the result of rickets.

Deformities of the spine, pelvis and legs result in reduced stature, *rachitic dwarfism.*

6. **Ligaments.** Relaxation of ligaments helps to produce *deformities* and partly accounts for knock-knees, weak ankles, kyphosis and scoliosis.

7. **Muscles.** The muscles are poorly developed and lack tone. As a result, the rachitic children are late in standing and walking. Abdomen becomes protuberant (*potbelly*) because of marked hypotonia of abdominal wall muscles, visceroptosis and lumbar lordosis.

8. **Sense organ.** Avitaminosis D in early infancy results in *bilateral lamellar cataracts.*

9. **Dentition.** Eruption of temporary teeth may be delayed in rachitic children. The first tooth in such babies appears between 6th and 9th month, at which time it has appeared in half of the normal babies. In deficiency of vitamin D, the formation of teeth becomes defective and leads to the development of *dental caries.*

Rickets in itself is not a fatal disease but complications and intercurrent infections such as pneumonia, tuberculosis, and enteritis are more likely to cause death in rachitic children than in normal children.

Rickets is most prevalent where climate or custom prevents individuals from exposure to sun, whereby checking vitamin D production by irradiation of the skin. In the seventeenth century. this disease was so common in England that it used to be known as "*English disease*". A study conducted by the Indian Social Institute, New Delhi (1981) shows that about 168 children per 1,000 of live births die of rickets in India.

In **osteomalacia** (*osteon*^{G} = bone ; *malakia*^{G} = softness), the action of bones is essentially like that in rickets. However, the bones become softer than the rachitic bones and the C/P ratio does not remain constant. The loss of calcium is greater than that of phosphorus and there is a relative gain in magnesium content. The disease is prevalent in India, China and Arab, particularly among women because of the custom that keeps them indoor and also prevents them from exposure to sun. This is particularly true of Bedouin Arab women who are clothed so that only their eyes are exposed to sunlight. The serum calcium is reduced, sometimes to such an extent that tetany develops.

G. Hypervitaminosis D. Overdosing of calciferol to the children and adults as well produces demineralization of bone. Serum concentrations of both calcium and phosphate are greatly raised, resulting in metabolic calcification of many soft tissues and the formation of renal calculi. The latter disorder may block the renal tubules causing hydronephrosis.

The use of very high doses of vitamin D is not danger-free, however. The toxic effects caused by excess dosage include anorexia, nausea, polyuria, weakness, headache etc. The toxicity is due to the diminished excretion of this vitamin, rather than its storage in the liver. *The water-soluble vitamins, on the contrary, if given in excess pass out immediately in the urine and are henceforth nontoxic.*

Much of the whole milk available in urban areas and evaporated milk are fortified with vitamin D concentrate so that 1 quart of fresh, whole milk or a cane of evaporated milk contains the required amount (*i.e.*, 10 μg).

H. Human requirements. Vitamin D requirement is greatly influenced by the amount of ultraviolet light to which the individual is exposed. Half an hour of direct sunlight on the cheeks of a baby each day is sufficient to generate the minimal daily requirement of vitamin D. For adults also, exposure to sunlight for 30 minutes a day is believed to satisfy the daily requirement (about 10 μ*g* or 400 IU) of vitamin D. As effective UV rays do not penetrate glass windows, exposure to sun through window glass is of little importance. Smoke also hinders the progress of these rays and as such city sunshine is not much beneficial. It is for these and some other reasons that vitamin D should be included in the diet. This is particularly true for older people. The recommended daily allowance of vitamin D is 400 I.U. for infants, pregnant women and lactating mothers. For adults, 400 units are adequate. One International Unit is defined as the biologic activity of 0.025 μ*g* of pure crystalline vitamin D_3.

VITAMIN E

A. History. The presence of this active principle was first demonstrated in vegetable oils by **Herbert Evans** and **Mattill** independently in 1920. This was designated as vitamin E or **antisterility factor** on account of the development of sterility in animals in its absence. In 1936, two compounds with vitamin E activity were isolated from wheat germ oil by **Evans** and his associates and given the name, α- and β-tocopherol (*tokos*[G] = childbirth ; *pheros*[G] = to bear; *ol* = an alcohol). Subsequently, five other tocopherols were obtained from various cereal grains like wheat germ, corn oil, rice etc.

B. Occurrence. The tocopherols are of widespread occurrence in many plant oils such as wheat germ, rice, corn, cottonseed, soybean, coconut and peanut but not olive oil. Whole grain cereals (fortified breakfast cereals) and margarine butter are a good source of vitamin E. They are also present in small amounts in meat, milk, eggs, leafy vegetables (such as cabbage and lettuce) and some fruits. *Fish liver oils, so abundant in vitamin A and D, are devoid of vitamin E.* Of all the tocopherols discovered so far, the α-form has the widest distribution and greatest biologic activity. The relative *biologic potencies* of various tocopherols are :

α-tocopherol—100
β-tocopherol—25
γ-tocopherol—19

C. Structure. Vitamin E is the collective name for a group of closely related lipids called tocopherols. The tocopherols are derivatives of 6-hydroxychroman (also known as tocol) bearing an isoprenoid side chain at carbon 2. The structure of α-tocopherol ($C_{29}H_{50}O_2$) is given in Fig. 22–7.

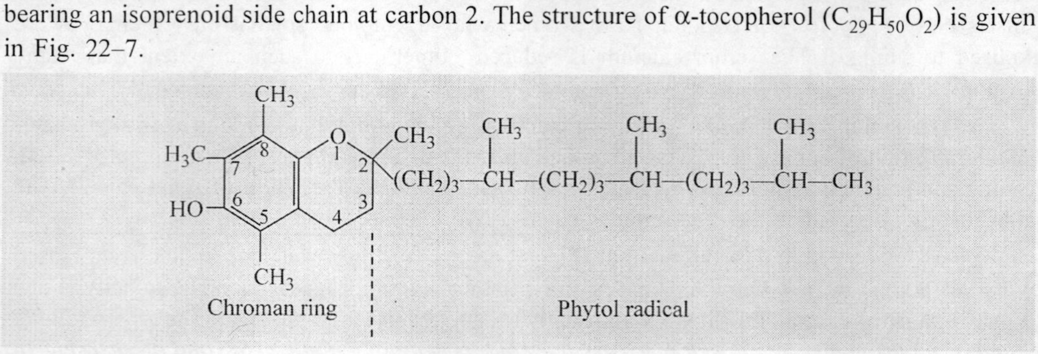

Fig. 21–7. α-tocopherol
(5,7,8-trimethyl tocol)

D. Properties. Vitamin E is a light yellow oil. It is resistant to heat (up to 200°C) and acids but acted upon by alkalies. It is easily but slowly oxidized and is destroyed by UV rays. The tocopherols are excellent antioxidants. They prevent other vitamins presents in food (*e.g.*, vitamin

A) from oxidative destruction. It is found in the nonsaponifiable fraction of the vegetable oils.

E. Metabolism. *Tocopherols act as antioxidants, i.e.,* they can prevent the oxidation of various other easily oxidized substances such as fats and vitamin A. It is for this reason that they are commercially added to foods to retard their spoilage. It may be recalled that vitamin A is essential for reproduction. *Whereas the beneficial action of vitamin A is mainly on the ectoderm and endoderm, that of vitamin E is on the mesodermal tissue*. But, very likely, vitamin E influences all the 3 germinal layers of the embryo by preventing the too rapid destruction of vitamin A. Certain substances such as phenols and vitamin C (ascorbic acid) stimulate the antioxidant property of vitamin E.

In fact, the biochemical activity of tocopherol lies in its capacity to protect mitochondrial system from inactivation by fat peroxides. Thus, in mitochondria obtained from vitamin E-deficient animals, a marked deterioration in activity is found due to peroxidation of unsaturated fatty acids which are usually present in these particles. Addition of vitamin E prevents this deterioration by acting as antioxidant for peroxidation.

F. Deficiency. The characteristic symptoms of experimentally-induced vitamin E deficiency vary from animal to animal. In mature female rats, **sterility** develops because of reabsorption of fetus after conception while in males, the germinal epithelium of the testes degenerates and the spermatozoa become nonmotile.

Avitaminosis E in herbivorous animals like rabbits and guinea pigs leads to acute **muscular dystrophy** (atrophy of muscle fibres), which ultimately results in creatinuria ; young chicks exhibit capillary damage and **encephalomalacia** ; hen eggs show **low hatchability** and monkeys reveal **hemolytic anemia**. There is, however, little evidence that man is ever short of vitamin E.

Finally, *as is true for almost all the vitamins, avitaminosis E prevents normal growth*. It also sometimes causes degenatation of the renal tubular cells.

G. Human requirements. Vitamin E is not a problem in human nutrition because it is ubiquitous in foods. However, the minimum daily requirement of vitamin E for adults is 30 I.U. for men and 25 I.U. for women. The pregnant and lactating mothers, however, require 30 I.U. daily. For infants and children,. the vitamin E requirement is at the rate of 1 to 1.25 I.U. per kilogram of body weight. One International Unit of dl-α-tocopherol is equivalent to the biologic activity of 1.1 mg of pure compound or 0.67 mg of d-α-tocopherol.

VITAMIN K

A. History. Henrik Dam (1929), a Danish investigator, found that newly-hatched chicks, fed on artificial diets, develop hemorrhage, a fatal disease characterized by prolonged blood-clotting period. The term vitamin K (*K* for Danish koagulations) was then proposed by **Dam** (1934) himself to designate the active factor which cured or prevented this disease. On account of its blood-clotting power, it is also called as **antihemorrhagic factor** or **coagulation vitamin**.

> **Henrik Carl Peter Dam** (LT, 1895-1976) was on a lecture tour of the United States and Canada when the Nazis occupied his homeland, Denmark, in 1940. He spent the war years at the University of Rochester, and it was during this period that he received his **Nobel Prize (1943)**.

Of the 2 naturally-occurring forms of this vitamin, vitamin K_1 was first isolated by **Dam** *et al* from alfalfa in 1939 and the other form, vitamin K_2 from fish meal by Doisy *et al*, also in 1939.

B. Occurrence. Vitamin K_1 occurs in green vegetables like spinach, alfalfa, cabbage etc. Fruits and cereals are poor sources. Vitamin K_2 is found in some intestinal bacteria. A rich source of K_2 is putrefied fish meal. Their relative *biologic potencies* are :

vitamin K_1—100
vitamin K_2—80

C. Structure. Chemically, the two forms of vitamin K (Fig. 21–8) are derivatives of quinones

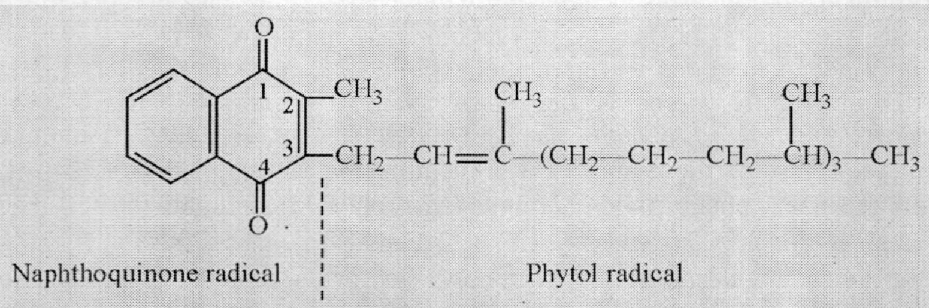

Vitamin K_1 or **phylloquinone** or **phytonadione**
(2-methyl-3-phytyl-1, 4-naphthoquinone)

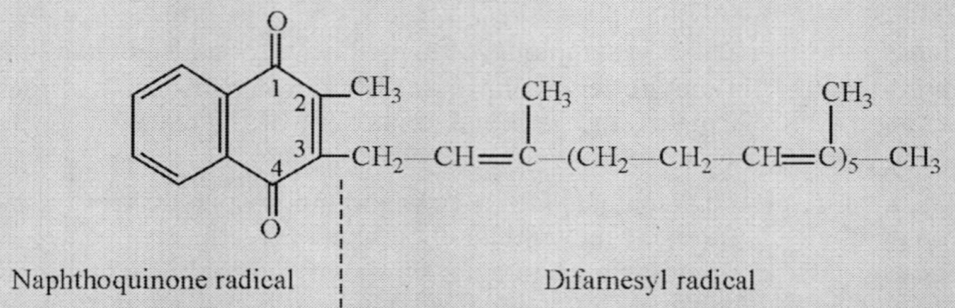

Vitamin K_2 or **flavinoquinone** or **farnoquinone** or **menaquinone**
(2-methyl-3-difarnesyl-1,4-naphthoquinone)

Fig. 21–8. Two vitamins of K group

and differ from each other in the composition of their side chain present at carbon 3 of the naphthoquinone ring. It is a phytol radical in vitamin K_1 ($C_{31}H_{46}O_2$) and a difarnesyl radical in vitamin K_2 ($C_{41}H_{56}O_2$). Vitamin K_1, found in plants, has 4 isoprene units in its side chain whereas vitamin K_2, found in animals, contains in its side 6 isoprene units, each with a double bond.

Analogues— The various analogues of naphthoquinone, however, have also been shown to possess vitamin K activity for animals. This is due to their structural resemblance. The common examples are menadione and phthiocol (Fig. 21–9).

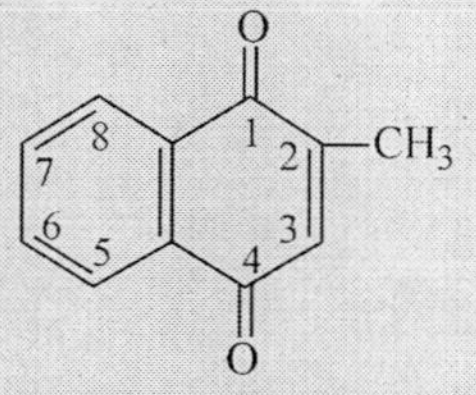

Menadione or **Vitamin K_3**
(2-methyl-1, 4-naphthoquinone)

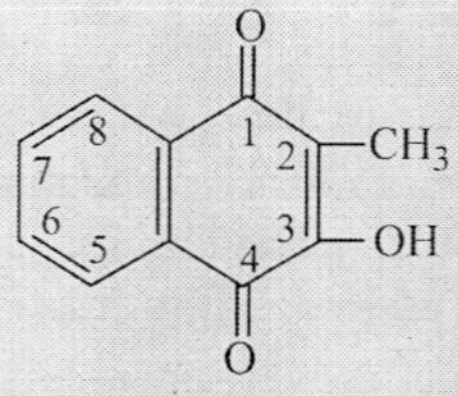

Phthiocol
(2-methyl- 3-hydroxy-1, 4-naphthoquinone)

Fig. 21–9. Two analogues of vitamin K

Menadione, which is sometimes referred to as vitamin K_3, is twice as potent as vitamin K_1. It is soluble in oil, sparingly so in water and is not oxidized in air when protected from light. Its diphosphate ester is water-soluble and is widely used clinically.

D. Properties. Vitamin K_1 is a yellow viscid oil but vitamin K_2 is a yellowish crystalline solid. It is sensitive to light and is, therefore, kept in dark bottles. It is destroyed by irradiation, strong acids, alkalies and oxidizing agents.

E. Metabolism. Vitamin K plays an essential role in the biosynthesis of prothrombin— a blood plasma protein needed in the process of blood clotting and produced in liver. The process of blood coagulation may be summarized as below.

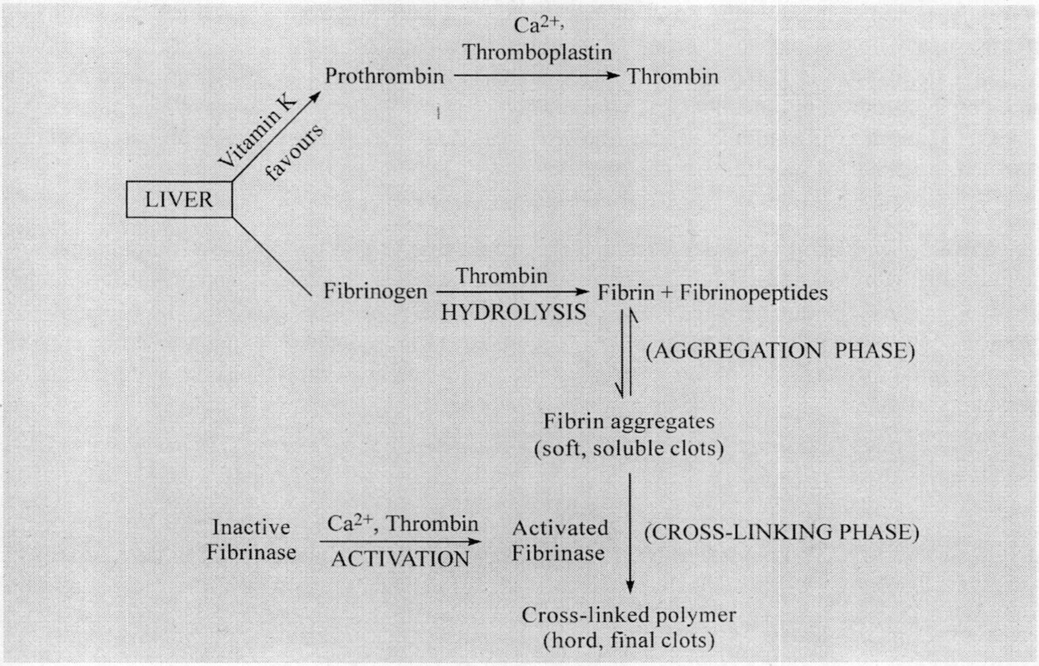

Fig. 21–10. Probable mechanism of blood coagulation

[Note the role of vitamin K in the synthesis of prothrombin which is a precursor of thrombin, the latter has dual function of (*a*) hydrolysing fibrinogen to fibrin and (*b*) activating fibrinase of FSF, which brings about clotting.]

The formation of the blood clot (refer Fig. 21–10) is caused by the enzymic hydrolysis of the soluble plasma protein, fibrinogen to the insoluble protein, fibrin and fibrinopeptides. This transformation is catalyzed by an enzyme, *thrombin*. Thrombin itself is not present in the blood but is produced from its precursor, prothrombin in the presence of Ca^{2+} ions and another protein called *thromboplastin* (= thrombokinase). In the next step, fibrin forms soft, fibrous networks (or soft clots). Then in the presence of Ca^{2+}, thrombin activates fibrinase, an enzyme precursor found in blood plasma. Fibrinase is also known as fibrin-stabilizing factor (FSF). Finally, fibrous networks of fibrin link with each other under the influence of activated FSF to produce cross-linked fibrin (or hard clots).

The vitamins K are fat-soluble and are absorbed only in the presence of bile. As a result, the absorption occurs in the upper portion of the small intestine where bile salts are present. The avitaminosis K may occur where bile is prevented from entering the intestinal tract. This is true for most of the fat-soluble vitamins but is, in particular, important in the case of vitamin K because of its blood clotting action.

Like vitamin E and coenzyme Q, the vitamin K has also been ascribed a role in electron transport system (ETS) and oxidative phosphorylation in mitochondria. The vitamin K_1 and K_2 both activate electron transport in the succinate oxidase of cardiac muscle preparations that have been made inactive by treating with isooctane. The bacterial extracts or liver mitochondria, when Antagonists irradiated, require vitamin K for oxidative phosphorylation. This suggests a possible role of vitamin K in oxidative phosphorylation. The specific site of action is believed to occur between NADH and cytochrome *b*. It has been suggested that a phosphate ester of vitamin K, upon oxidation, transfers phosphate to ADP to form ATP.

F. Deficiency. Deficiency of vitamin K causes loss of blood-clotting power. The infants may also show signs of vitamin K deficiency by developing **hemorrhage**. This disease persists by the time the bacteria develop in the intestine. Administration of this vitamin to pregnant mothers before parturition decreases the onset of this disease. In man, however, avitaminosis K results in **steatorrhea** with diminished intestinal absorption of lipids.

In general, vitamin K deficiency is rarely found in higher animals as this is provided by food and also synthesized by intestinal bacteria.

G. Human requirements. *There is seldom a lack of sufficient vitamin K in human beings.* As such, no standard requirement has been set.

COENZYME Q

Crane *et al* (1959) have demonstrated that coenzyme Q and certain other ubiquinones are components of mitochondrial lipids. These substances serve as electron transport agents and are involved in the formation of ATP at a cytochrome *a*. **Roles** (1967) has classified the coenzyme Q group as vitamins because of their ability to cure (or protect against) vitamin E deficiency in several animal species. Some of the coenzymes also participate in electron transport and/or oxidative phosphorylation.

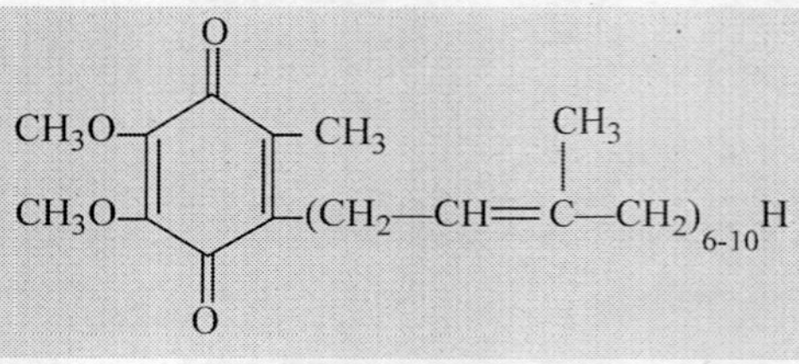

Fig. 21–11. Coenzyme Q

Various homologues of coenzyme Q, containing 6 to 10 isoprene units, have been isolated from various microbes, chloroplasts of green plants and mitochondria of animals. These have the same quinonoid nucleus (Fig. 21–11) but differ in the number of isoprenoid units in the side chain. For example, coenzyme Q from animal source has 10 isoprene units and is, henceforth, called coenzyme Q_{10} or *ubiquinone*$_{50}$ (50 carbon atoms, *i.e.,* 10 isoprene units, in the side chain), whereas the one from bacteria has less than 10 isoprene units. *Mycobacteria, however, contain no coenzyme Q.*

STIGMASTEROL

Another alleged fat-soluble vitamin is stigmasterol. It is a plant sterol and has been isolated from soybean and wheat germ oils. It in present in alfalfa and fresh cream. Chemically, it resembles ergosterol closely and contains only 2 double bonds at carbon postitons 5—6 and 22—23 (Fig. 21–12). The absence of stigmasterol causes stiffness of the wrists and elbows of the guinea pigs. For this reason, it is commonly called as '**antistiffness factor**'. The muscles atrophy and become streaked.

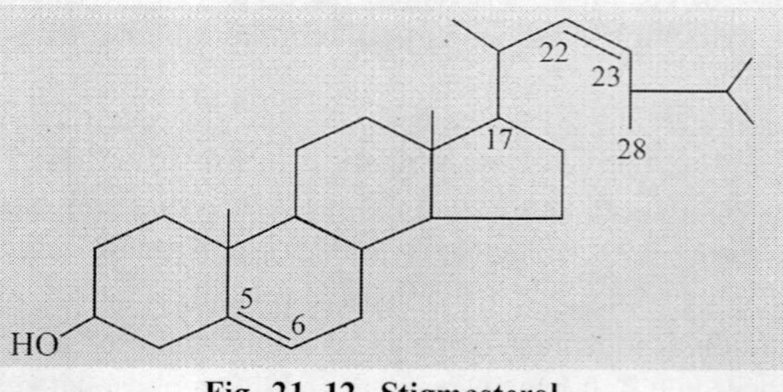

Fig. 21–12. Stigmasterol

A summary chart of the fat-soluble vitamins (Table 21-3) is given on the next page.

Table 21–3. Fat-soluble vitamins

Vitamin	Common name(s)	Chemical name	Sources	Metabolic functions	Deficiency diseases*	Daily requirement of man
A	Antixerophthalmic factor or 'Bright eyes' vitamin	A_1— Retinol $C_{20}H_{30}O$ A_2— 3-dehydroretinol $C_{20}H_{28}O$	Fish liver oils, butter, milk, eggs and kidneys; Pigmented (esp., yellow) vegetables and fruits	Vasual cycle; Membrane integrity	Demyelinization; Nyctalopia; Xerophthalmia; Keratomalacia; Phrynoderma	5,000 I.U.
D	Antirachitic factor or 'Sunshine' vitamin	D_2— Ergocalciferol $C_{28}H_{44}O$ D_3— Cholecalciferol $C_{27}H_{44}O$	Fish liver oils, egg yolks, milk, butter and mushrooms	Calcification of bones and teeth	Rickets (in children); Osteomalacia (in adults); Celiac disease	400 I.U.
E	Antisterility factor	α-tocopherol, $C_{29}H_{50}O_2$ β-tocopherol, $C_{28}H_{48}O_2$ γ-tocopherol, $C_{28}H_{48}O_2$	Plant oils like wheat germ, rice cottonseed, corn, soyabean etc; Also meat, milk and eggs	Act as antioxidants; Control O_2 consumption; Participate in nucleic acid metabolism	Sterility in rats; Muscular dystrophy in rabbits and guinea pigs; Encephalomalacia in young chicks	30 I.U.
K	Antihemorrhagic factor or Coagulation vitamin	K_1— Phylloquinone $C_{31}H_{46}O_2$ K_2— Farnoquinone $C_{41}H_{56}O_2$	Green vegetables like spinach, alfalfa, cabbage etc.; Also fruits and cereals	Biosynthesis of prothrombin; Oxidative phosphorylation; Electon transport system	Hemorrhage (in infants); Steatorrhea (in adults)	0.001 mg

* Deficiency diseases have been mentioned in relation to man, if not stated otherwise.

PROBLEMS

1. In contrast to water-soluble vitamins, which must be a part of our daily diet, fat-soluble vitamins can be stored in the body in amounts sufficient for many months. Suggest an explanation for this difference based on solubilities.
2. What is the chemical reaction in which vitamin K participates ? How is this reaction involved in blood coagulation and bone formation ?
3. If one takes vitamin E to protect his heart, how much should he take ?
4. Will taking vitamins give one extra energy ?
5. When taking vitamin E, is it okay to take other medicine ?
6. Should one take multivitamin pills containing vitamin D every day as one gets older ?
7. How much exposure to sunlight a person needs to supply his body with enough vitamin D ?
8. Vegetables oils are fortified with :
   (*a*) vitamin A
   (*b*) vitamin D
   (*c*) vitamin E
   (*d*) vitamin K
9. What needs to be stored in dark bottles ?
   (*a*) biotin
   (*b*) nicotinic acid
   (*c*) riboflavin
   (*d*) retinol
10. One claim put forth by purveyors of health foods is that vitamins obtained from natural sources are more healthful than those obtained by chemical synthesis. For example, it is claimed that pure L-ascorbic acid (vitamin C) obtained from rose hips is better for you than pure L-ascorbic acid manufactured in a chemical plant. Are the vitamins from the two sources different ? Can the body distinguish a vitamin's source ?
11. How much vitamin D a person needs daily to help prevent osteoporosis ?
12. Is it safe to take very high doses of vitamin D ?
13. Vitamin A is the essential dietary factor for the formation of :
    (*a*) rhodopsin
    (*b*) biliverdin
    (*c*) hemoglobin
    (*d*) biotin
14. Which of the following is both a vitamin and a hormone ?
    (*a*) ascorbic acid
    (*b*) calciferol
    (*c*) thiamine
    (*d*) riboflavin
15. When is a good time to take your multivitamin/mineral tablet ?
16. If a person takes a daily multivitamin and supplements, can he get away with less fruits,

vegetables and grains ?

17. Can one take vitamins with thyroid pills ?
18. The most nutritive substance on Earth is :
    (*a*) Spinach
    (*b*) *Spirulina*
    (*c*) *Laminaria*
    (*d*) Soybean
19. Is the white part of the orange peel edible ?
20. What is the rainbow diet ?
21. How one can decrease his chances of getting Alzheimer's disease ?
22. How helpful is adding more vitamin E to one's diet ?
23. How much antioxidant value does coffee have ?
24. What health benefits do vitamins offer ?

ANSWERS TO PROBLEMS

1. Water-soluble vitamins are more rapidly excreted in the urine and are not stored effectively. Fat-soluble vitamins have very low solubility in water and are excreted only very slowly by the kidney.
2. Vitamin K is necessary for carboxylation of specific glutamic acid residues in certain proteins to form γ-carboxyglutamic acid residues. In blood coagulation, this step is required for the conversion of preprothrombin to prothrombin. In bone formation, this is required to form the calcium-binding residues of the protein osteocalcin.
3. The usual recommended dose is 400 IU once a day.
4. Despite what many people believe, vitamins pills or vitamin shots do not give you extra energy.
5. No. If you take vitamin E while you are taking the anticlotting drug coumadin (warfarin), you could develop dangerous internal bleeding.
6. Yes, older people need extra vitamin D orally because they absorb less of it from their food and their skin makes less of it.
7. Ten to fifteen minutes of sunlight on the face and arms. Three times a week is usually enough.
8. (*b*) vitamin D
9. (*c*) riboflavin
10. The vitamins from the two sources are identical; the body cannot distinguish the source.
11. Ordinarily 400 I.U., but if one is over 50, 600 I.U. and if over 70, 800–1,000 I.U. a day
12. No. 5,000 I.U. a day can cause permanent kidney damage, but taking up to 2,000 I.U. a day is usually safe.
13. (*a*) rhodopsin
14. (*b*) calciferol (vitamin D)
15. After eating. This slows slows its movement through the intestine allowing more time for the pill to dissolve completely and be absorbed.
16. No. These foods contain thousands of substances that have not been identified yet and that they fight cancer and other health problems.

17. Not, if the vitamin pills contain iron. The iron can keep the thyroid pills from being adequately absorbed.

18. (*b*) *Spirulina*

19. Yes. The white membrane in the orange peel is rich in antioxidant compounds called limonoids. They have been shown to offer cancer protection.

20. The rainbow diet involves eating vegetables from each of the food colour groups to get a broader variety of important phytonutrients. For example, red tomatoes, purple grapes, orange sweet potatoes, yellow corn, green spinach or white broccoli.

21. Taking vitamins E and C in combination has been shown to have a synergistic effect in helping to prevent age-related dementia including Alzheimer's disease. Feed your brain by taking 100 – 400 IU of vitamin E and 100–500 mg of vitamin C daily. Get 400 mcg of folate (B_{12}) daily. Add omega-3 fats to your diet by eating fish 2-3 times per week and switch to canola or olive oil. Flax seed oil and fish oil are also good.

22. Studies have found that those who ate foods rich in vitamin E (like nuts, seeds, spinach and olive oil) cut their bladder cancer risk in half. Bladder cancer is 4 times more common in men than in women.

23. Coffee is truly a nutritional powerhouse. Green coffee beans contain more than 1,000 antioxidants. The brewing process adds 300 more. Even more healthful compounds are derived from the roasting process. In fact, coffee delivers 4 times the antioxidant benefits of its closest competitor, green tea. Coffee also beats cocoa, herbal teas and even red wine in antioxidant protection.

24. Multivitamin/mineral supplements have been shown to reduce the incidence of heart disease and stroke, protect against some cancers, halve the likelihood of some birth defects, delay the onset of cataracts and macular degeneration, and slow or prevent hip fractures by 20% or more.

CHAPTER

22 Water-soluble Vitamins

INSIDE

- Vitamin B Complex
- Vitamin B_1
- Vitamin B_2
- Vitamin B_3
- Vitamin B_5
- Vitamin B_6
- Vitamin B_7
- Vitamin B_9
- Vitamin B_{12}
- Vitamin C
- Bioflavonoids
- Vitamers (= Isotels)

Vitamine C in citrus fruits

The oranges, shown here, are rich in vitamin C, which acts as an antioxidant.

VITAMIN B COMPLEX

Originally, vitamin B referred to a vitamin whose deficiency causes beriberi in man and polyneuritis in birds. Later, **Goldberger's** researches on pellagra led to the view that vitamin B consisted of at least 2 factors : a heat-labile *antiberiberi factor* and a comparatively heat-stable *antipellagra factor*. Some called the former factor as vitamin B_1 and the latter as vitamin B_2. But the later researches conducted by **Richard Kuhn, Conrad Elvehjem** and others have established the fact that vitamin B complex, as represented by yeast, rice bran and liver extracts, contains still other factors. At present, the vitamin B complex is known to consist of a group of at least 13 components usually named as B1, B_2, B_3 etc. But to prevent confusion, their chemical names are now frequently used. The various members of the vitamin B complex are not related either chemically or physiologically, yet they have many features in common:

(*a*) All of them except lipoic acid are water-soluble.

(*b*) Most of them, if not all, are components of coenzymes that play vital roles in metabolism (refer Table 22–1).

(*c*) Most of these can be obtained from the same source, *i.e.,* liver and yeast.

(*d*) Most of them can be synthesized by the intestinal bacteria.

Table 22–1. Coenzyme derivatives of water-soluble vitamins

Vitamin	*Coenzyme form*
Vitamin B_1 (Thiamine)	Thiamine pyrophosphate (TPP)
Vitamin B_2 (Riboflavin)	Flavin mononucleotide (FMN)
	Flavin adenine dinucleotide (FAD)
Vitamin B_3 (Pantothenic acid)	Coenzyme A (CoA)
Vitamin B_5 (Niacin)	Nicotinamide adenine dinucleotide (NAD)
	Nicotinamide adenine dinucleotide phosphate (NADP)
Vitamin B_6 (Pyridoxine)	Pyridoxal phosphate (PALP), Pyridoxamine phosphate (PAMP)
Vitamin B_7 (Biotin)	Biocytin
Vitamin B_9 (Folic acid)	Tetrahydrofolic acid (THFA)
Vitamin B_{12} (Cyanocobalamin)	Deoxyadenosyl cobalamin
Vitamin C (Ascorbic acid)	Not known

VITAMIN B_1

A. History. Thiamine was the first member of the vitamin B group to be identified and hence given the name vitamin B_1. Thiamine was first isolated by **Jansen** (1949) in Holland and **Adolf Windaus** in Germany. On account of its curing action against beriberi, it is commonly known as **antiberiberi factor.** It is also known as **antineuritic factor** or heat-labile factor. In Europe, it is also designated **aneurin**.

Thiamine is also spelt as thiamin (**Frank B. Armstrong**, 1989 ; **Albert L. Lehninger, David L. Nelson** and **Michael M. Cox,** 1993)

B. Occurrence. *Thiamine is found practically in all plant and animal foods*. Cereals, heart, liver and kidney are excellent sources of it. In cereals, the outer layers of seeds are especially rich in thiamine (Fig. 22-1). In yeasts and animal tissues, however, it is present mainly as its coenzyme, thiamine pyrophosphate (TPP). Milk also contains thiamine, although in relatively low amounts. The milling of wheat flour lowers the thiamine content considerably, sometimes to the extent of even 80%. Consequently, wheat flour is usually enriched with thiamine at many places. Furthermore, improper cooking (*esp*., when the water in which foods are cooked is discarded) loses thiamine content. This is because of the solubility of thiamine in water. Therefore, it is desirable to use the "cook water" for soups and sauces. Thiamine is easily destroyed by heat in neutral or alkaline media. Because the covering of the grains of cereals contains most of the vitamin, polishing reduces its availability. Canning processes are, however, not particularly destructive.

Fig. 22–1. Crystals of thiamine (vitamin B_1)
Thiamine pyrophosphate (TPP) is a cofactor in a number of enzymes.

C. Structure. The chemical structure of thiamine (Fig. 22–2) was determined in 1935 by

Robert R. Williams and his associates in the United States and its chemical synthesis was achieved soon thereafter. Thiamine ($C_{12}H_{17}N_4OS$) is 2,5-dimethyl-6-aminopyrimidine bonded through a methylene linkage to 4-methyl-5-hydroxyethyl-thiazole. Thus, pyrimidine and thiazole are the two moieties present in its molecule. The pyrimidine is unique in that it is the only natural pyrimidine containing an alkyl group at C_2. Also, *with the possible exception of penicillin, thiamine is the only natural compound which contains a thiazole group*. It is interesting to note that plants can use a mixture of pyrimidine and thiazole compounds in place of thiamine itself. On the other hand, all the animals except pigeon require the complete vitamin.

> The monies that **Robert R. Williams** received from his patents on the chemical synthesis of thiamine were used to establish the **Williams-Waterman Fund** for the Combat of Dietary Disease, which supported nutritional research and field programmes to abolish malnutrition.

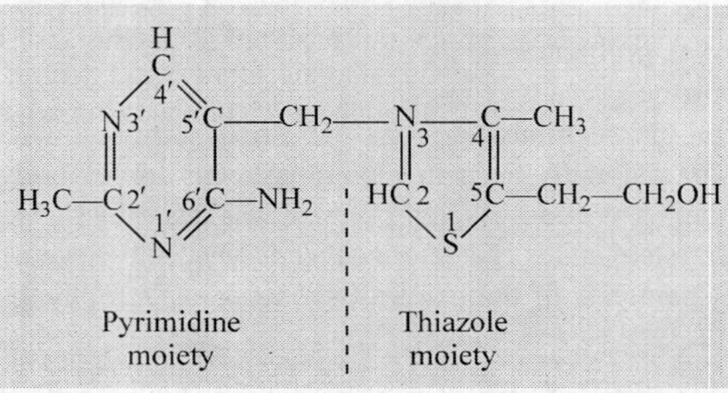

Fig. 22–2. Vitamin B_1 or thiamine or aneurin

D. Properties. Thiamine is a white crystalline substance, readily soluble in water, slightly so in ethyl alcohol but insoluble in ether and chloroform. Its odour resembles that of a yeast. The aqueous solution is optically inactive. Thiamine is destroyed at elevated temperature, unless the *p*H is low. It can stand short boiling up to 100°C. Hence, it is only partly lost in cooking or canning processes. Long boiling or boiling with alkali destroys it. But it is stable in acid medium. On oxidation, it produces *thiochrome*, which gives fluorescence.

E. Metabolism. The requirement of this vitamin is increased under high metabolic conditions such as fever, increased muscular activity, pregnancy and lactation and also under surgery and stress. A correlation also exists between the type of food taken and the vitamin B_1 requirement. Fats and proteins reduce while carbohydrates increase the amount of this vitamin required in the daily diet. Thiamine absorption decreases with gastrointestinal or liver disease.

Raw seafoods (*e.g.*, fishes and molluscs) contain an enzyme, *thiaminase* which destroys thiamine in the body. People consuming such foods may, therefore, reveal symptoms of thiamine deficiency. Thiaminase cleaves the thiamine molecule between the pyrimidine and thiazole rings.

If thiamine is administered in human body, a part of it is excreted or recovered in the urine and a part is converted to pyramin by the enzyme, thiaminase. Besides thiaminase, certain flavonoids of nonenzymic nature also work against thiamine. These have been shown to be present in ferns and certain higher plants.

Thiamine is phosphorylated with ATP to form thiamine pyrophosphate (TPP), which is also called diphosphothiamine (DPT).

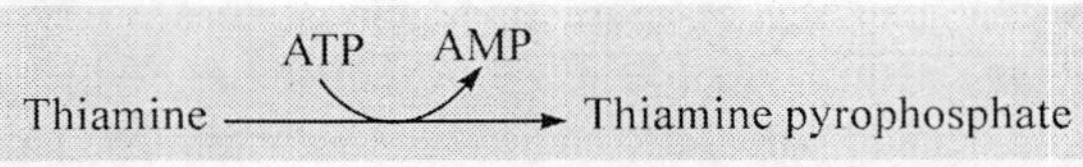

TPP, in association with lipoic acid, forms the prosthetic group, *cocarboxylase* for the enzyme *carboxylase*. TPP participates in many reactions, such as decarboxylation of α-keto acids, notably pyruvic and α-ketoglutaric and transketolation.

F. Deficiency. Vitamin B_1 deficiency leads to polyneuritis in animals and beriberi in human beings.

Polyneuritis in birds renders them unable to fly, walk or even stand. Rats develop, among other symptoms, a brachycardia (slowing of the heart rate).

Beriberi (*beri*^{singhalese} = weakness, which here means *I cannot*, symbolizing the incapacitated condition created by thiamine deficiency) has been and continues to be a serious health problem in Far East where polished or refined rice (rice from which husk has been removed) is eaten. The rice has a rather low content of thiamine. The problem is aggravated if the rice is polished because the husk contains nearly all the thiamine of rice. Beriberi is also occasionally seen in alcoholics who are severely malnourished. Even before the concept of vitamins was developed, beriberi was described as a deficiency disease. It is a disease of the nervous system and is characterized by polyneuritis (degeneration of the peripheral nerves) leading to partial paralysis of the extremities, muscular atrophy, cardiovascular changes and gastrointestinal disorders.

> **CLINICAL IMPLICATIONS**
>
> **BERIBERI**
>
> **Jacobus Bonitus**, a Dutch physician working in Java, had first described beriberi, in 1630, as follows : "A certain very troublesome affliction, which attacks men, is called by the inhabitants Beri-beri (which means sheep). I believe those, whom this same disease attacks, with their knees shaking and the legs raised up, walk like sheep. It is a kind of paralysis, or rather Tremor : for it penetrates the motion and sensation of the hands and feet indeed sometimes of the whole body ...."

At first, there is fatigue, apathy, irritability, depression, drowsiness, anorexia (loss of appetite), insomnia (sleeplessness), nausea and abdominal discomfort. This is followed by symptoms like peripheral neuritis with tingling, burning paresthesias of the toes and feet; decreased tendon reflexes; loss of vibration sense; tenderness and cramping of leg muscles; congestive heart failure and psychic disturbances. There may be ptosis of the eyelids and atrophy of the optic nerve. Hoarseness due to paralysis of the laryngeal nerve is a typical sign. Muscular atrophy and tenderness of the nerve trunks are followed by ataxia, loss of coordination, and loss of deep sensation. Paralytic symptoms are more common in adults than in children. Finally, the major symptoms may follow one of the following 3 courses (and accordingly beriberi is of 3 types) :

(*a*) Symptoms involving nervous system, causing *dry beriberi* : In it, the child may appear plump but is pale, flabby, listless and dyspneic; the heart beat is rapid and the liver enlarged.

(*b*) Symptoms associated with edema and effusions, leading to *wet beriberi* : In it, the child is undernourished, pale and edematous and has dyspnea, vomiting and tachycardia. The skin appears waxy. The urine may contain albumin and casts.

(*c*) Symptoms involving heart, resulting in *acute pernicious beriberi* : In it, the lesions may be found principally in the heart, peripheral nerves, subcutaneous tissue and serous cavities. The heart is enlarged, especially to the right and there is fatty degeneration of the myocardium. Generally edema or edema of the legs, serous effusions, and venous engorgement may be seen. Lesions in the brain include vascular dilatation and hemorrhage. Finally, death ensues due to heart failure.

Often the symptoms characteristic of more than one of these 3 types of beriberi appear simultaneously in individuals causing *mixed beriberi*. Although beriberi is caused due to thiamine avitaminosis, it is usually associated with deficiencies of other vitamins. This is true of all vitamin B complex-deficiency conditions in man.

G. Human requirements. The daily recommended dietary allowances are 1.2–1.4 mg for men and 1.0 mg for women. Pregnant and lactating mothers, however, require up to 1.5 mg daily. The thiamine requirement for infants is between 0.2 and 0.5 mg daily.

VITAMIN B_2

A. History. Riboflavin or vitamin B_2 was first isolated in 1879 from milk whey which is an essential dietary factor for rats. Since it was first isolated from milk, vitamin B_2 is also known as **lactoflavin.** Originally, it was also known as **ovoflavin** (*from eggs*) and **hepatoflavin** (*from liver*). Its synthesis was done by **Richard Kuhn** and **Paul Karrer.** It is popularly called as the "**yellow enzyme**" because of its colour.

B. Occurrence. In nature, it occurs almost exclusively as a constituent of one of the two flavin coenzymes, namely, flavin mononucleotide (FMN) and flavin adenine dinucleotide (FAD). Milk, cheese, eggs, liver, kidney, heart and brewer's yeast are excellent sources of this vitamin. Cow's milk contains about 5 times as much riboflavin as human milk. Leafy vegetables are good sources. They are usually richer in riboflavin than they are in thiamine. Fruits and most root vegetables contain moderate quantities. Whole grains, cereals and milled flour contain low riboflavin content. The riboflavin contents in cereals, however, increase strikingly during germination. The ordinary cooking processes do not affect the riboflavin content of the food. Roasted or boiled meat retains about 75% of the vitamin. *It is only very rarely that vitamin* B_2 *is present in free or uncombined state as in retina and spleen*. Fermentation residues from alcohol manufacture probably offer the richest large supplies.

RICHARD KUHN
(LT, 1900-1967)

Kuhn, a Austria-born German chemist, received his doctorate in 1922 on enzymes. He was awarded **Nobel Prize in Chemistry in 1938** for his work on carotenoids and vitamins but its acceptance was forbidden by the German government. After the war, he received the gold medal and diploma of the prize. His main contributions include: the separation of carotenoids (α and β) from carrots and thus isolated the third isomer (γ carotene), and the structure of lycopene.

C. Structure. Riboflavin ($C_{17}H_{20}N_4O_6$) belongs to a class of water-soluble pigments called *lyochromes.* A molecule of thiamine (Fig. 22–3) consists of a sugar alcohol, D-ribitol, attached to a chromogenic dimethyl isoalloxazine ring at position number 9.

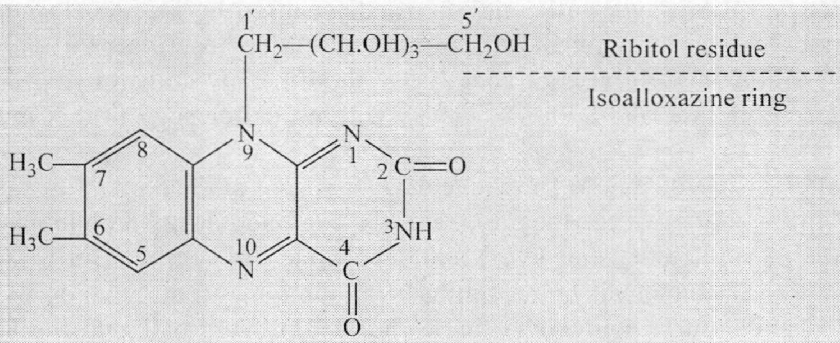

Fig. 22–3. Vitamin B_2 or riboflavin or lactoflavin
[6, 7-dimethyl-9-(1′-D-ribityl) isoalloxazine]

D. Properties. Riboflavin is a bright orange-yellow crystalline powder. It is soluble in water and ethanol but insoluble in ether and chloroform. It is stable to heat and acids but is easily decomposed by alkalies and exposure to light. The aqueous solution exhibits yellow-green fluorescence. It stands ordinary cooking and canning. On exposure to light, the ribityl residue splits off, forming a compound *lumiflavin* in alkaline solution and *lumichrome* in acidic or neutral solution.

E. Metabolism. Riboflavin is synthesized by all green plants, most bacteria, yeasts and moulds. *Ashbya gossypii*, an yeast, produces it in such large amounts that riboflavin crystals are formed in the culture medium. *Animals have, so far, not been shown to synthesize riboflavin.* In man, the ingested riboflavin is largely passed out as such or as its coenzyme, the FMN.

Experiments with plant tissues have suggested that riboflavin and flavoproteins may play a significant role in phototropic curvature of various plant organs (**Galston**, 1950).

Riboflavin is essential for growth and tissue respiration; it may have a role in light adaptation and is required for conversion of pyridoxine to pyridoxal phosphate.

When riboflavin is phosphorylated in the presence of an enzyme, *flavokinase*, it gets converted to FMN which is essential in the biosynthesis of fats.

$$\text{Riboflavin} + \text{ATP} \xrightleftharpoons{\text{Flavokinase}} \text{FMN} + \text{ADP}$$

FMN may undergo a further reaction with ATP, in the presence of an enzyme found in yeast and animal tissues, to produce FAD. It is a chief constituent of electron transport system (ETS). A decrease in the amount of FAD, therefore, would severely hamper the efficiency of ETS.

$$\text{FMN} + \text{AMP} \rightleftharpoons \text{FAD} + \text{PP}$$

The coenzymes undergo reversible oxidation-reduction in the presence of their enzymes and a suitable substrate.

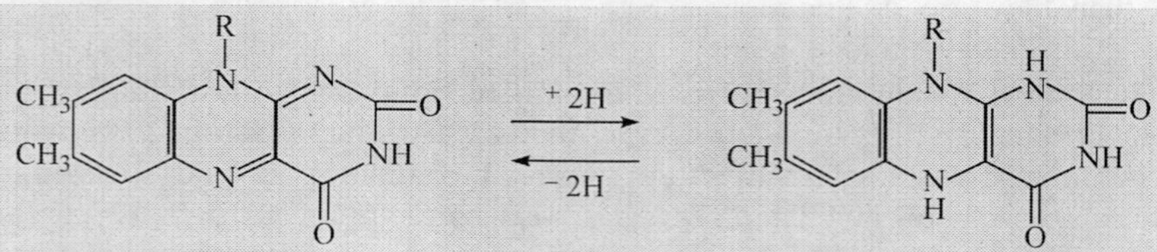

FMN or FAD FMNH$_2$ or FADH$_2$

The flavoenzymes play a key role in cell metabolism. They function in accepting hydrogen atoms from reduced pyridine nucleotides. They have been shown to participate in the enzymic oxidation of glucose, fatty acids, amino acids and purines.

F. Deficiency. Riboflavin deficiency is usually caused by inadequate intake. Faulty absorption may contribute in patients with biliary atresia or hepatitis or in those receiving probenecid, phenothiazine or oral contraceptives. Phototherapy destroys riboflavin content. It is interesting to note that riboflavin deficiency without deficiency of other member of the B complex is rare.

Persons deficient in vitamin B_2 show chelosis (fissuring at the corners of the mouth and lips), glossitis (inflammation of the tongue), keratitis, conjunctivitis photophobia, lacrimation, **corneal vascularization** (bloodshot eyes) and seborrheic dermatitis. But these symptoms are not specific to ariboflavinosis since similar symptoms may also develop in the absence of nicotinic acid and iron. **Cheilosis** (= perle′che) begins with pallor at the angles of the mouth, following by thinning and maceration of the epithelium. Superficial fissures often covered by yellow crusts develop in the angles of the mouth and extend radially into the skin for distances upto 2 cm. In **glossitis**, the tongue is smooth, and loss of papillary structure occurs. A normocytic and normochronic anemia with bone marrow hyperplasia is common.

However, patients suffering from pellagra and beriberi are usually also deficient in riboflavin content.

G. Human requirements. The minimum daily requirement of riboflavin varies from 0.6 to 1.7 mg for children and adults. During pregnancy and lactation, the women require up to 2.0 mg daily.

VITAMIN B_3

A. History. This was first isolated by **Roger J. Williams** in 1938 from yeast and liver concentrates. On account of its wide distribution, he named it as pantothenic acid (*pantos*^{G} = everywhere). The coenzyme form of this vitamin (coenzyme A or CoA-SH) was isolated and its structure determined by **Fritz A. Lipmann.** The chemical synthesis of this coenzyme was, however, described by **Khorana** in 1959. This vitamin is sometimes called as **filtrate factor** or the **yeast factor**.

B. Occurrence. Although widespread in nature, yeast, liver and eggs are the richest sources of it. The vegetables (potatoes, sweet potatoes, cabbage, cauliflower, broccoli) and fruits (tomatoes, peanuts) and also the skimmed milk, wheat bran, whole milk and canned salmon are some of the less important sources. In most animal tissues and microorganisms, it occurs as its coenzyme.

> **Roger J. Williams** was the brother of Robert R. Williams, who established the structure of vitamin B_1, **thiamine.** Although both shared similar interests in the B-vitamins and nutrition, their research careers were independent of each other and never involved collaboration.

C. Structure. Pantothenic acid ($C_9H_{17}NO_5$) is an amide of pantoic acid (α, γ-dihydroxy-β, β-dimethyl butyric acid) and β-alanine (refer Fig. 22–4).

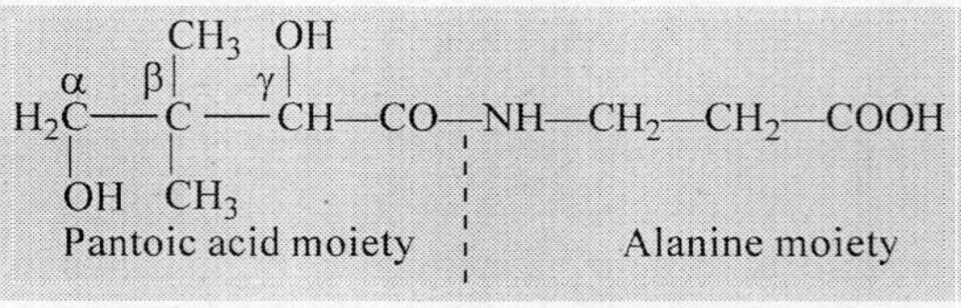

Fig. 22–4. Vitamin B_3 or pantothenic acid
(Pantoyl-β-alanine)

D. Properties. Pantothenic acid is a pale yellow viscous oil, soluble in water and ethyl acetate but insoluble in chloroform. It is stable to oxidizing and reducing agents but is destroyed by heating in an acidic and alkaline medium (*i.e.*, it is heat-labile).

E. Metabolism. Pantothenic acid can be synthesized by green plants and various microorganisms (*Neurospora, Escherichia coli, Bacteria linens*) but not by mammals. Hence, this must be present in the diet to serve as a starting point for coenzyme A (CoA). Coenzyme A is richly found in the liver and in poor quantities in the adrenals. There may be as much as 400 mg of CoA per kilo of liver. It functions in acetylation reactions. In order to be effective, CoA must be present in the form of acetyl-CoA. It may arise in many ways but the most common way of its production is that CoA, in the presence of ATP, acetate and a suitable enzyme, is converted into acetyl CoA. The overall reaction may be shown in 3 steps :

1. ATP + Enzyme $\rightleftharpoons$ Adenylic acid-Enzyme + Pyrophosphate
2. Adenylic acid-Enzyme + CoA $\rightleftharpoons$ CoA-Enzyme + Adenylic acid
3. CoA-Enzyme + Acetate $\rightleftharpoons$ Acetyl-CoA + Enzyme

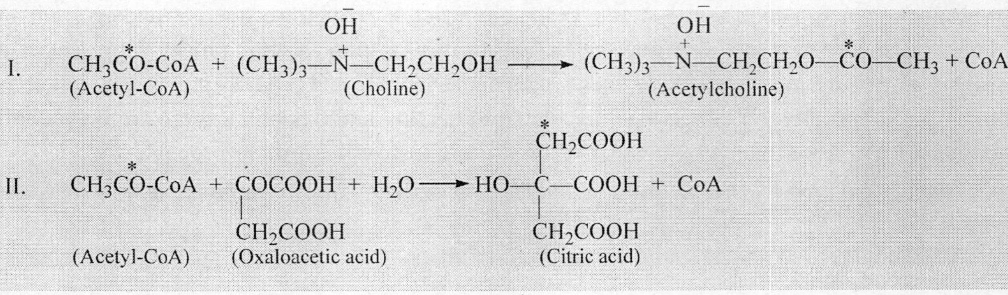

The acetyl groups may be transferred to an acetyl acceptor in the presence of a suitable acceptor. This may occur in two ways : either the acetyl group is attached to the accepting group at the carbonyl end (head reaction) or at the methyl end (tail reaction).

The only known metabolic fate of vitamin B_3 is its participation in the formation of the biologically important coenzyme A. It functions as a thioester of carboxylic acids.

F. Deficiency. A deficiency of pantothenic acid leads to depigmentation of the hair in rats, pigs and dogs and to depigmentation of feathers in chicks. Atrophy of the adrenal cortex with necrosis and hemorrhage may also occur in animals including rat. Corneal changes consisting of vascularization, thickening and opacity may be seen.

In human beings, no definite deficiency syndrome has been ascribed to pantothenic acid, probably because of the ubiquitous nature of this vitamin and because of the fact that a little amount of this vitamin can perhaps be synthesized in the body. Its correlation with **achromotrichia** (premature greying of the hair) has been described in the case of man, sometimes. But it seems too much to hope that grey hair can be averted by attention to diet ; rather it appears we must expect to go grey in spite of this vitamin.

G. Human requirements. The dietary allowance dose has not been officially worked out. Yet, 5—10 mg per day of vitamin B_3 has been suggested.

VITAMIN B_5

A. History. Vitamin B_5 refers to nicotinic acid and was named as **pellagra-preventive (PP) factor** by an Austrian-American physician of the U. S. Public Health Service, **Joseph Goldberger** (ca 1920) because of its curing action on pellagra (After Goldberger's death, vitamin B_5 was sometimes called **vitamin G** in his honour). The vitamin role on nicotinic acid was first recognized by **Conrad Elvehjem** and **D. Wayne Woolley** of Wisconsin University in 1937. As this vitamin has a curing action against blacktongue disease in dogs, it is also called as **antiblacktongue factor.** It was first isolated by **Funk** in 1911. Because the name `nicotinic acid' might mislead some people into thinking that tobacco is nutritious, nicotinic acid has been given the alternative official name *niacin* for public use.

> Although **Woolley** was blind since young adulthood, undaunted, he pursued a science career that established him as one of the most prominent biochemists in the United States.

B. Occurrence. Nicotinic acid is widely distributed in nature in plant and animal tissues mainly as its amide called nicotinamide (commercially called **niacinamide**, to avoid any misassociation with the alkaloid nicotine of tobacco). As dietary tryptophan can be converted, in restricted quantities, to niacin in the body, it can partially substitute for niacin, although other sources of vitamin B_5 are necessary. *Niacin is most abundantly found in yeast.* Liver, lean pork, salmon, poultry and red meat are also good sources, but most cereals contain only small amounts of it. Most vegetables and fruits are poor sources of it. Milk and eggs, which contain very little or practically no niacin, are good pellagra-preventive foods because of their high content of tryptophan. Since a number of stable vegetable articles of diet are not particularly rich in nicotinamide, the vegetarian's diet may be lacking in this vitamin. Nicotinamide, is present as a constituent in two pyridine nucleotide coenzymes namely NAD and NADP (previously called as DPN and TPN respectively). Since niacin is stable to heating and oxidation, there are only small losses in cooking. Like thiamine, most of vitamin B_5 is lost in the milling process.

C. Structure. *Niacin ($C_6H_5O_2N$) is simplest of all the known vitamins*. It is a pyridine derivative (Fig. 22–5).

Nicotinic acid or **niacin** (pyridine-3-carboxylic acid)

Nicotinamide or **niacinamide** (pyridine-3-carboxylic acid amide)

Fig. 22–5. Vitamin B_5 and its amide

D. Properties. Niacin is a white crystalline substance. It is soluble in ethyl alcohol but is less soluble in ether and benzene than nicotinamide. It is heat-stable. Nicotinamide, when pure, occurs as white needle like crystals. It is soluble in water and is stable in air and heat.

E. Metabolism. The conversion of niacin to niacinamide takes place in the kidney and brain slices and also in the liver slices, if glutathione is present. Nicotinamide is synthesized by amidation of nicotinic acid adenine dinucleotide and subsequent degradation of NAD thus formed.

The niacin in man and other animals is derived from the amino acid tryptophan, which also cures pellagra. The conversion of tryptophan to nicotinic acid in the body takes place through a series of intermediate steps, which are represented below :

Tryptophan ⟶ Kynurenine ⟶ 3-hydroxykynurenine

3-hydroxyanthranilic acid ⟶ ... ⟶ Quinolinic acid ⟶ Nicotinic acid

Nicotinamide undergoes methylation in mammalian liver to produce N′-methyl nicotinamide which is oxidized to give corresponding 6-pyridone. In many plant seeds nicotinic acid is, however, converted to trigonelline.

The two coenzyme forms of this vitamin, NAD and NADP, carry out 2 important functions in the tissues:

(*a*) Oxidation of alcohols, aldehydes, amino acids and hydroxy-carboxylic acids.

(*b*) Reduction of the flavin coenzymes.

F. Deficiency. A deficiency of niacin causes pellagra in man and blacktongue in dogs. **Pellagra** (of Italian origin, *pellis* = skin; *agra* = rough) is characterized by 3 "Ds", namely **d**ermatitis of the exposed parts, **d**iarrhea and **d**ementia. The early symptoms of pellagra are vague. Anorexia, lassitude, fatigue, burning sensations, numbness and dizziness may be prodromal symptoms. Their manifestation in children who have parasites or chronic disorders may be particularly severe. The most characteristic manifestations are the cutaneous ones, which may develop abruptly or insidiously and may be elicitated by irritants, esp., by intense light. They first appear as symmetric erythema of the exposed surfaces that may resemble sunburn. The lesions are usually sharply demarcated from the healthy skin around them, and their distribution may change very often. The lesions on the hands sometimes have the appearance of a glove (*pellagrous glove*), and similar demarcations are sometimes seen on the foot and leg (*pellagrous boot*) or around the neck (*Casal necklace*). The healed parts of the skin may remain pigmented. The cutaneous lesions are sometimes preceded by stomatitis, glossitis, vomiting or diarrhea. Swelling and redness of the tip of the tongue and its lateral margins may be followed by intense redness of the entire tongue and of the papillae and even ulceration. Nervous symptoms include depression, disorientation, insomnia and delirium. The histologic changes in the nervous system occur relatively late in the disease and consist of patchy areas of demyelinization and degeneration of ganglion cells.

The classic symptoms of pellagra are usually not pronounced in infants and children. Anorexia, irritability, anxiety and apathy are common in "*pellagra families*". They may also have sore tongues and lips and the skin is usually dry and scaly. Diarrhea and constipation may alternate and a moderate secondary anemia may occur. Pellagral children often have symptoms characteristic of other nutritional deficiency diseases. As coffee (*Coffea arabica*) is particularly rich in niacin, the heavy coffee drinkers usually do not develop pellagra. Other factors like thiamine-deficiency also seem to be responsible for this disease. Incredible as it may seem, over 600 deaths were attributed to pellagra in 1948. Pellagra is greatly aggravated in persons kept on a corn diet (as natives of Africa) because corn is very much deficient of tryptophan.

The canine **blacktongue** disease leads to complete loss of appetite. The inner surfaces of the lips and cheeks develop pustules ; the pustules may also develop on the thorax and abdomen. Intensive salivation and bloody diarrhea are other symptoms.

G. Human requirements. The recommended daily allowance of nicotinic acid is between 8 and 15 mg for children, between 15 and 20 mg for men and between 13 and 15 mg for women. Pregnant and lactating mothers may require up to 20 mg daily.

VITAMIN B_6

A. History. The name vitamin B_6 was suggested by **Albert Szent-Györgyi** (1934) to designate substances, other than thiamine and riboflavin, which cured a dermatitis (acrodynia) in rats. It was, henceforth, also named as **adermin** or **antidermatitis factor**. Vitamin B_6 group includes 3 compounds : pyridoxine, pyridoxal and pyridoxamine. Pyridoxine was first isolated, in 1938, from yeast and liver. Later, **Snell** (1942) discovered the other two compounds.

B. Occurrence. The B_6 vitamins are widely distributed in nature in plant and animal tissues. They are especially rich in cereals (wheat, rice), peas, turnip greens, brussels sprouts, carrots, potatoes, sweet potatoes, bananas, avocados, watermelons and yeasts. B_6 vitamins are also found in egg yolk, salmon, chicken, fish, beaf, pork and liver. Pyridoxine is adequately available in human and cow's milk. Pyridoxal (PAL) and pyridoxamine (PAM) also occur in nature as their coenzymes, namely, pyridoxal phosphate (PALP) and pyridoxamine phosphate (PAMP), respectively.

C. Structure. All the 3 forms of vitamin B_6 (Fig. 22–6) are derivatives of pyridine, C_5H_5N and differ from each other in the nature of substituent at position 4 of the ring. All the 3 forms are readily interconvertible biologically.

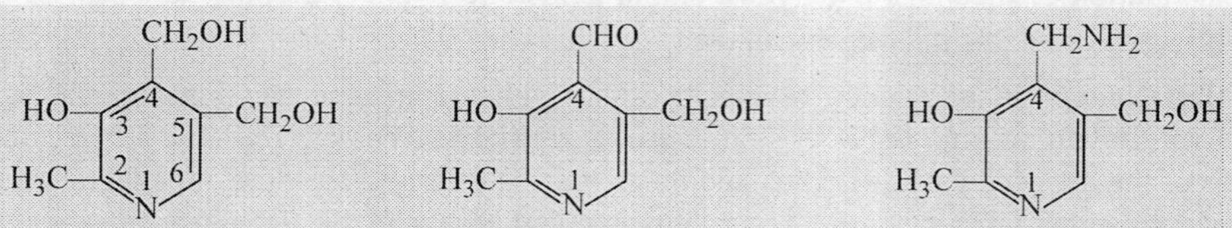

Pyridoxine **Pyridoxal** **Pyridoxamine**

[2-methyl-3-hydroxy-4, 5-di (hydroxymethyl)-pyridine]

Fig. 22–6. Vitamins of B_6 group

D. Properties. Pyridoxine is a white crystalline substance and is soluble in water and alcohol and slightly so in fat solvents. It is sensitive to light and ultraviolet irradition. It is resistant to heat (*i.e.*, heat-stable) in both acidic and alkaline solutions but its two allies pyridoxal and pyridoxamine are destroyed at high temperatures (*i.e.*, heat-labile).

E. Metabolism. The various forms of vitamin B_6 serve as growth factors to a number of bacteria. In addition, the 3 forms (pyridoxine, pyridoxal, pyridoxamine) are converted to pyridoxal-5-phosphate (Fig. 22-7), which acts as a coenzyme in various enzymic reactions involved in amino acid metabolism such as transamination, decarboxylation and racemization and in the metabolism of glycogen and fatty acids. It is also essential in the metabolism of hydroxy amino acids, sulfur-containing amino acids and also tryptophan.

Pyridoxal or its phosphate derivative also possibly acts as a carrier in the active transport of amino acids across cell membranes.

Pyridoxine can be converted to either pyridoxal or pyridoxamine (Fig. 22–7) but neither of them can be changed to pyridoxine. All these three can be detected in the urine after ingestion although 4-pyridoxic acid is the most important excretion product quantitatively. It is for this reason that when administered in the human body, about 90% of pyridoxine is oxidized to pyridoxic acid and excreted in human urine in this form.

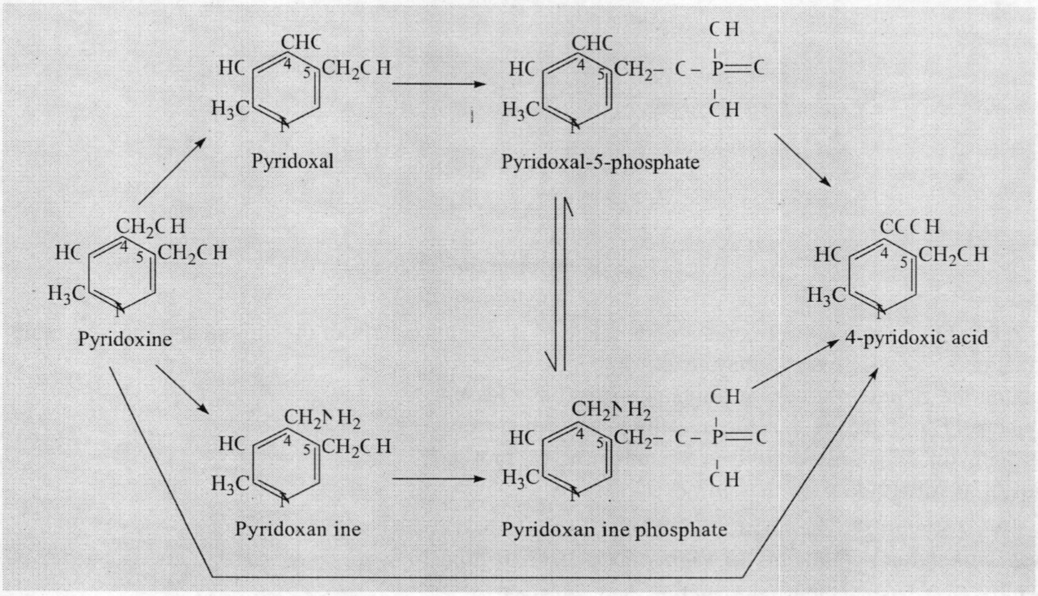

Fig. 22–7. Interrelationship between pyridoxine and its derivatives

F. Deficiency. Vitamin B_6 deficiency or apyridoxosis in rats leads to the development of acrodynia, a disease of dermatitis on ears, mouth and tail and accompanied by edema and scaliness of these structures. Dogs and chick develop anemia and nervous lesions in apyridoxosis.

In human infants, vitamin B_6 deficiency results in **convulsions, anemia**, dermatitis and gastrointestinal disorders such as nausea and vomiting. However, this deficiency is rare. Moreover, tryptophan metabolism is also disturbed. In adults, the vitamin B_6 deficiency is normally not found because the intestinal bacteria are capable of synthesizing vitamin B_6.

G. Human requirements. The minimum dietary allowance of vitamin B_6 is between 0.2 and 1.2 mg for infants and children and around 2.0 mg for men and women per day. During pregnancy and lactation, the recommended daily dose is 2.5 mg. Pyridoxine antagonists, such as isoniazid used in the treatment of tuberculosis, increase the requirements for pyridoxine as do pregnancy and drugs such as penicillamine, hydralazine and the oral progesterone-estrogen contraceptives.

VITAMIN B_7

A. History. In 1935, **Fritz Kögl**, a Dutch biochemist, isolated in crystalline form from 250 kg of dried egg yolks about 1 mg of a *'bios'* factor (growth promoting factor) necessary for yeast and named it as "biotin". Four years later, **Szent-Györgyi** *et al* conclusively proved that biotin is synonymous to the "**antiegg white injury factor**" which is responsible for the cure of egg white injury, induced in rats and other animals by feeding them with raw egg white. The raw egg white contains a biotin-antagonist protein, *avidin,* which combines with biotin in a firm linkage to form a compound that cannot be absorbed by the intestine and is therefore, excreted. It is also called as **coenzyme R** because it is a growth factor for the nitrogen-fixing bacterium, *Rhizobium.*

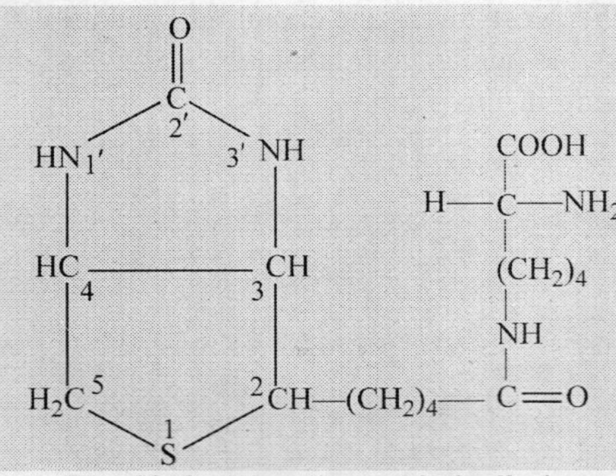

Fig. 22–8. Biocytin or biotinyllysine
(ε-N-biotinyl-L-lysine)

B. Occurrence. Biotin has a wide range of distribution both in the animal and the vegetable kingdoms. Yeast, liver, kidney, milk and molasses are among the richest sources ; peanuts and eggs have lesser amounts. Biotin occurs in nature usually in combined state as biocytin (Fig. 22–8). It is a bound form of biotin, linked as a peptide with the amino acid lysine.

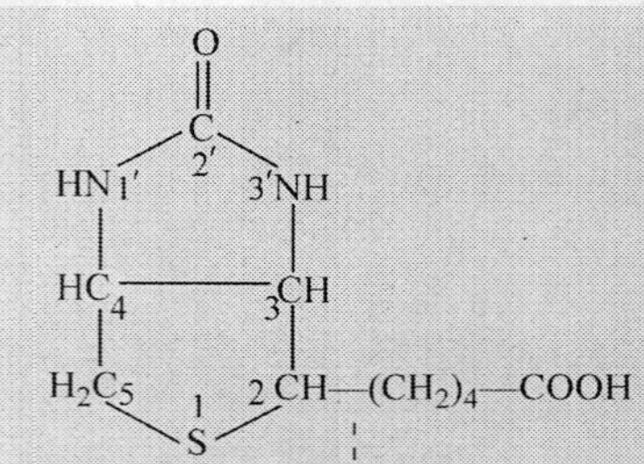

Fig. 22–9. Vitamin B_7 or biotin or vitamin H
(2′-keto-3, 4-imidazolido-2-tetrahydrothiophene-*n*-valeric acid)

C. Structure. The structure of biotin ($C_{10}H_{16}O_3N_2S$) was worked out by **Vincent du Vigneaud** in 1942. Biotin (Fig. 22–9) has an unusual structure and consists of a fused imidazole and thiophene ring with a fatty acid side chain. Two forms of biotin can exist, *allobiotin and epibiotin*. Biotin is optically active. Only the (+) biotin is active; the DL-biotin is half as active as the naturally occurring biotin. The oxybiotin, in which S atom of biotin is replaced with an O atom, has some activity. *Biotin and thiamine are the only sulfur-containing vitamins isolated to date*.

D. Properties. Biotin crystallizes as long needles. It is soluble in water and ethyl alcohol but is insoluble in chloroform and ether. It is heat-stable and is resistant to both acids and alkalies. It has a melting point of 230°C.

E. Metabolism. This vitamin serves as a prosthetic group for many enzymes. These biotin-containing enzymes catalyze the fixation of CO_2 into organic molecules, thus bringing about carboxylation. The carbon dioxide is carried as a carboxyl group attached to one of the ureidonitrogen atoms of biotin, forming *N-carboxybiotin complex* (Fig. 22–10). They also bring about synthesis of fatty acids such as oleic acid.

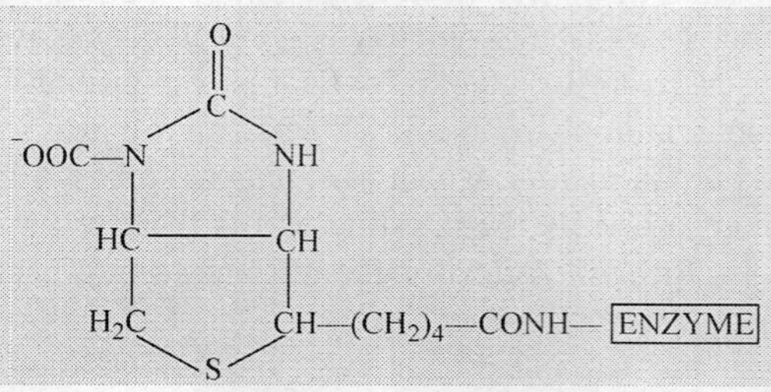

Fig. 22–10. N-carboxybiotin complex

F. Deficiency. In most animals including man, intestinal bacteria synthesize appreciable amounts of biotin. It is because of this reason that biotin-deficiency in human beings, fed on biotin-free diets, cannot be produced. However, biotin-deficiency may be induced by sterilization of intestine and by feeding with raw egg white. Avidin, the egg white protein, inactivates biotin by eliminating it from an otherwise complete diet. Such a deficiency in man leads to **dermatitis**, loss of hair, decrease in weight and edema. The lesions on skin appear with changes in posture and gait. These disorders may lead to death. Heating egg white destroys the avidin and prevents the so-called egg white injury.

G. Human requirements. The intestinal bacteria synthesize biotin in such appreciable amounts that the amount excreted in urine exceeds the intake. That is why the RDA for this vitamin has not been established. However, about 10 mg per day of biotin is sufficient for an adult.

VITAMIN B_9

A. History. Day, for the first time, showed the existence of this nutritional factor by demonstrating that yeast extract could cure cytopenia, a disease experimentally induced in monkeys. The potent factor was obtained from spinach leaf and this led to its nomenclature as folic acid, FA (*folium*^{L} = leaf). The official name of this vitamin is folacin. This is also known as **liver Lactobacillus casei factor** as it was isolated from liver and was shown as necessary for the growth of lactic acid bacteria. **Hogan** called this as vitamin Bc.

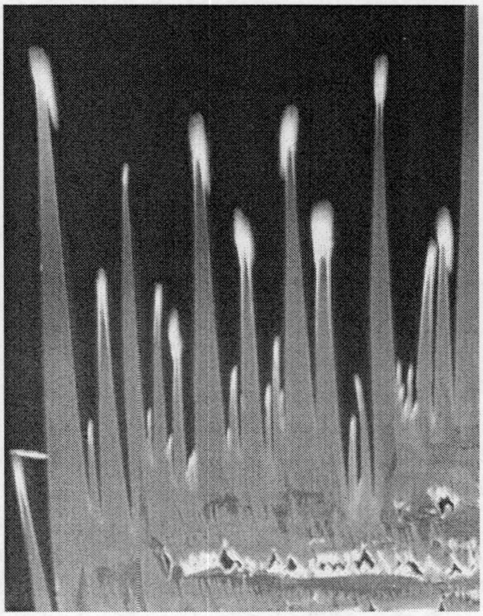

Fig. 22–11. Folic acid crystal

B. Occurrence. Folic acid (Fig. 22-11) and its derivatives (tri- and hepta-glutamyl peptides) are widely distributed in biological world. A few important sources are liver, kidney, tuna fish, salmon, yeast, wheat, dates and spinach. Root vegetables, sweet potatoes, rice, corn, tomatoes, bananas, pork and lamb contain little folid acid. With improper cooking, folacin contents are destroyed, like thiamine.

C. Structure. A molecule of folic acid (Fig. 22–12) consists of 3 units : glutamic acid, *p*-aminobenzoic acid and a derivative of the heterocyclic fused-ring compound pterin. Its molecular formula is $C_{19}H_{19}O_6N_7$.

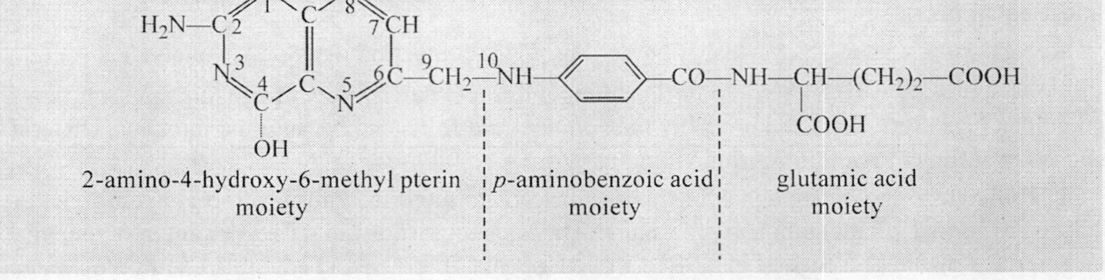

Fig. 22–12. Vitamin B_9 or **folic acid** or **folacin** or **vitamin M**
(pteroyl-L-glutamic acid, PGA)

D. Properties. Folic acid is a yellow crystalline substance, slightly soluble in water but insoluble in fat solvents. It is stable to heat in alkaline or neutral solutions only. It is inactivated by sunlight.

E. Metabolism. The reduction products of folic acid act as coenzymes. An enzyme, *folic reductase*, reduces folic acid to dihydrofolic acid (DHFA or FH_2), the latter compound is further

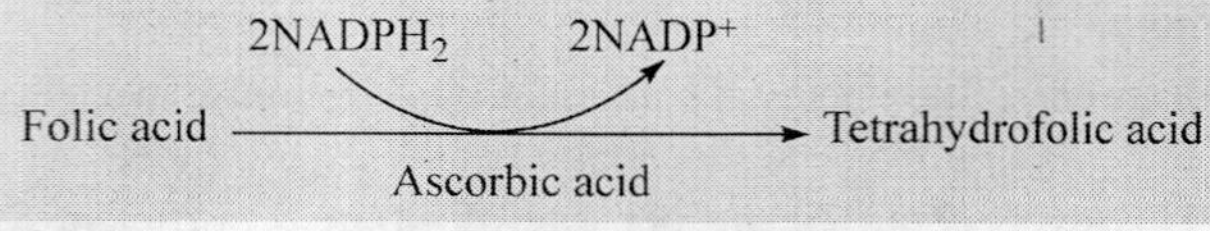

reduced by *dihydrofolic reductase* to 5,6,7,8-tetrahydrofolic acid (THFA or FH_4). The formation of FH_4 from FA is associated with the oxidation of NADPH or NADH and requires the presence of ascorbic acid.

The structure of FH_4 appears in Fig. 22–13.

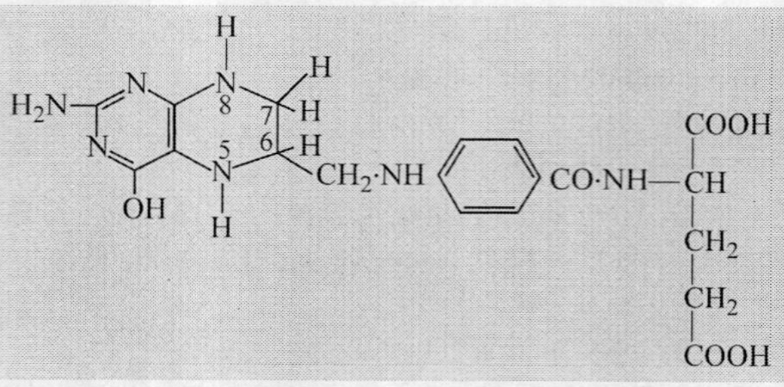

Fig. 22–13. Tetrahydrofolic acid, THFA or **FH_4**

(5, 6, 7,8-tetrahydropteroylglutamic acid)

The vitamins of B_9 group are involved in *one-carbon* metabolism in a way similar to the *two-carbon* metabolism in which CoA is involved. THFA acts as an acceptor of a one-carbon unit either from formate (in which case 5,6,7,8-tetrahydrofolic acid is formed) or from formaldehyde (in which case 5-hydroxymethyl-5,6,7,8-tetrahydrofolic acid is formed). THFA is also involved in the transfer of the methyl group and in the utilization of single carbons (formate) in the synthesis of serine, methionine, thymine, purines, choline and inosinic acid.

Folic acid, in conjunction with ascorbic acid, also appears to be related to tyrosine metabolism.

Citrivorum factor (CF) is 5 formyl derivative of tetrahydrofolic acid and is so named because it supports the growth of *Leuconostoc citrivorum.* CF is about one thousand times more potent biologically than folic acid. Its chemical name is folinic acid. During the conversion of FA to CF, vitamin B_{12} and ascorbic acid are also required. Citrivorum factor (and also folic acid to a lesser degree) are concerned in the production of an agent that stimulates the formation of normal RBCs.

It is interesting to note that the bacteria, which require PABA for growth, also utilize FA with almost equal ease.

Recent studies have shown that folic acid provides protection against Alzheimer's disease.

Folic acid is, however, essential for lactation in rats and hatchability of eggs in chicks, turkeys and guinea pigs. Rats, dogs and probably man do not need folic acid because the intestinal bacteria synthesize sufficient quantity of this vitamin.

F. Deficiency. In chicks, a lack of this factor leads to **anemia.** Rats develop **achromotrichia** (failure in normal pigmentation of the hair). The monkeys show **macrocytic anemia** (anemia characterized by the presence of giant RBCs), leukopenia, diarrhea and **edema** (retention of water by skin tissues).

On a worldwide basis, deficiency of folic acid is believed to be the most common form of vitamin undernutrition. In man, the folic acid deficiency leads to **megaloblastic anemia, glossitis** and gastrointestinal disorders. Pregnant women and infants are also particularly vulnerable. Folic acid deficiency is a major feature of **tropical sprue**, in which there is a general deficiency in absorption of many nutrients from the small intestine.

G. Human requirements. The daily dietary allowance of folic acid is 0.1 mg for infants, 0.2 mg for children and 0.4 mg for adult men and women. Pregnant mothers may, however, require up to 0.8 mg per day.

VITAMIN B_{12}

A. History. In 1926, two American physicians, **George Minot** and **George William Murphy** discovered that patients suffering from pernicious anemia could be cured by feeding them with about half a pound of liver a day. This landmark in medicine brought them Nobel Prize in 1934. In 1929, **Castle** suggested that gastric juice contained a factor (intrinsic factor) that, together with a factor present in the food (extrinsic factor), is responsible for the cure of pernicious anemia. This **anti-pernicious anemia factor (APA factor)** was, later, isolated in crystalline form in 1948 independently by **E. Lester Smith** in England and by **Edward Rickes** and Karl Folkers in the United States. It was then named as vitamin B_{12} or cyanocobalamin. It is the last B-vitamin to be isolated and is also known as *Factor X* or *L.L.D. factor*. The coenzyme form of this vitamin (deoxyadenosyl cobalamin or cobamide coenzyme) was first isolated by **Barker** of California. Coenzyme B_{12} has been called a "*biologic Grignard reagent*".

Cyanocobalamin is also spelt as cyanocobalamine (**Albert L. Lehninger,** 1984)

B. Occurrence. *Vitamin B_{12} has been found only in animals* ; the chief source is liver, although it is also present in milk, meat, eggs, fish, oysters and clams. Animal tissues contain it in varying amounts.

Under certain dietary conditions, vitamin B_{12} may be synthesized by the intestinal microorganisms. *In general, cyanocobalamin is not present in plant food*s except in *Spirulina,* a blue-green alga. However, it occurs in foods bound to proteins and is apparently split off by proteolytic enzymes.

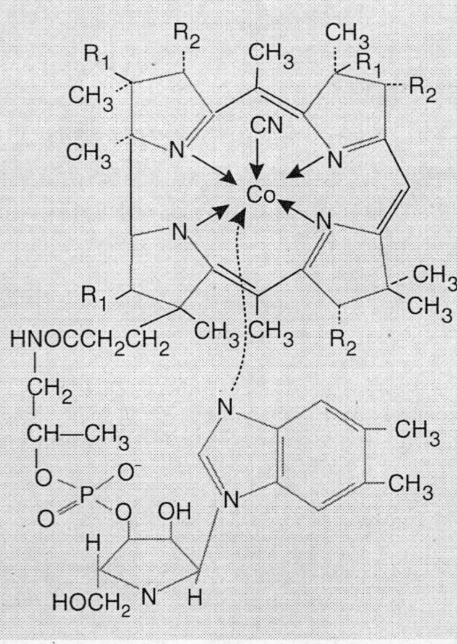

Fig. 22–14. Structure of vitamin B_{12} or cyanocobalamin

Animals and plants are unable to synthesize this vitamin. *Cyanocobalamin is unique in that it appears to be synthesized only by microorganisms especially anaerobic bacteria.* However, a process of producing vitamin B_{12} from waste products has been developed, in 1977, by the department of Chemical Engineering of the Indian Institute of Technology, Chennai, India.

C. Structure. The structure of vitamin B_{12} (Fig. 22-14), one of the most complex known, has been established, in 1957, by **Dorothy Crowfoot Hodgkin** (Nobel Laureate, 1964). Cyanocobalamin ($C_{63}H_{88}O_{14}N_{14}$ P Co) is a pigment alike to the tetrapyrrole ring structure of the porphyrins, *e.g.*, chlorophyll and haem. A unique feature of this vitamin (and other related compounds) is the presence, in its molecule, of an atom of a heavy metal cobalt in the trivalent state. *No other cobalt-containing organic compound has been found in nature*. The cobalt atom is centrally-situated and is surrounded by a macrocyclic structure of 4 reduced pyrrole rings (A, B, C and D) collectively called as corrin. It may be noted from the structural formula that the 6 coordinate valences of the cobalt atom (Co^{2+})are satisfied by the 4 nitrogens of the reduced tetrapyrrole, a nitrogen atom of 5, 6-dimethylbenzimidazole and a cyanide ion. Two of the pyrrole rings, namely A and D, are directly linked to each other and the corrin has lower degree of unsaturation with only 6 double bonds. The other two pyrrole rings, namely B and C, are joined through a single methene carbon. Another distinct feature of the vitamin B_{12} molecule is the presence of a loop of the isopropanol, phosphate, ribose and 5,6-dimethyl-benzimidazole in that order, the end of the loop being attached to the central cobalt atom.

The structure of *vitamin B_{12} coenzyme* (= 5′-deoxyadenosyl cobalamin) is similar to that of cobalamin except that here the CN group is replaced by adenosine and the linking with cobalt atom taking place at 5′ carbon atom of the ribose of adenosine (Fig. 22–15). *Vitamin B_{12} coenzyme is the only known example of a carbon-metal bond in a biomolecule.*

D. Properties. Vitamin B_{12} (molecular weight, ca 1,500) is a deep red crystalline substance. It is soluble in water, alcohol and acetone but not in chloroform. It is levorotatory. It is stable to heat in neutral solutions but is destroyed by heat in acidic or alkaline solutions.

E. Metabolism. Vitamin B_{12} is converted to coenzyme B_{12} by extracts from microorganisms supplemented with ATP.

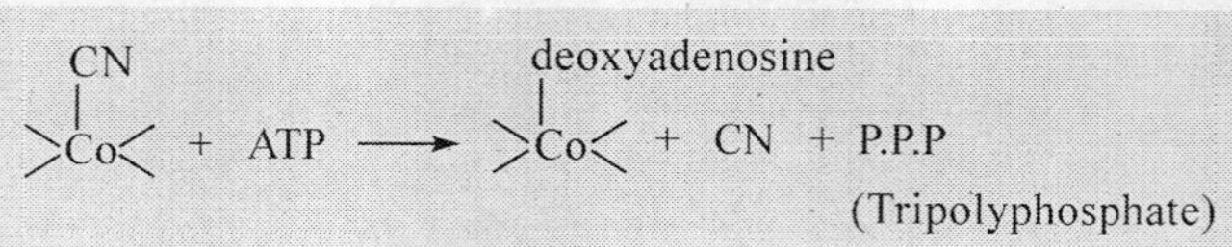

Coenzyme B_{12} is associated with many biochemical reactions :

(*a*) ***1,2 shift of a hydrogen atom*** : Coenzyme B_{12} catalyzes 1,2 shift of a hydrogen atom from one carbon atom of the substrate to the next with a concomitant 2,1 (reverse) shift of some other group, *e.g.*, hydroxyl, alkyl etc.

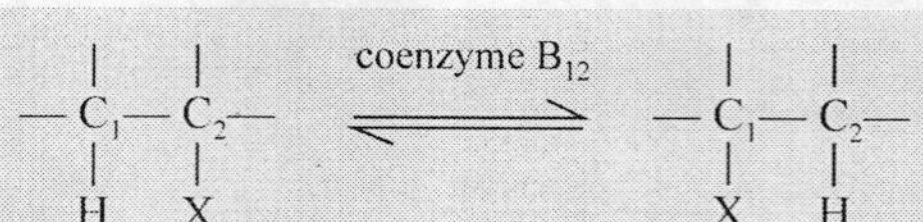

Conversion of methylmalonyl-CoA to succinyl-CoA is an example of 1,2-shift.

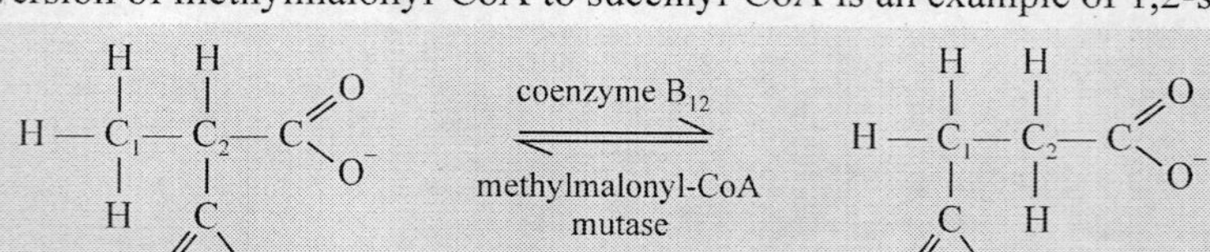

L-methylmalonyl-CoA Succinyl-CoA

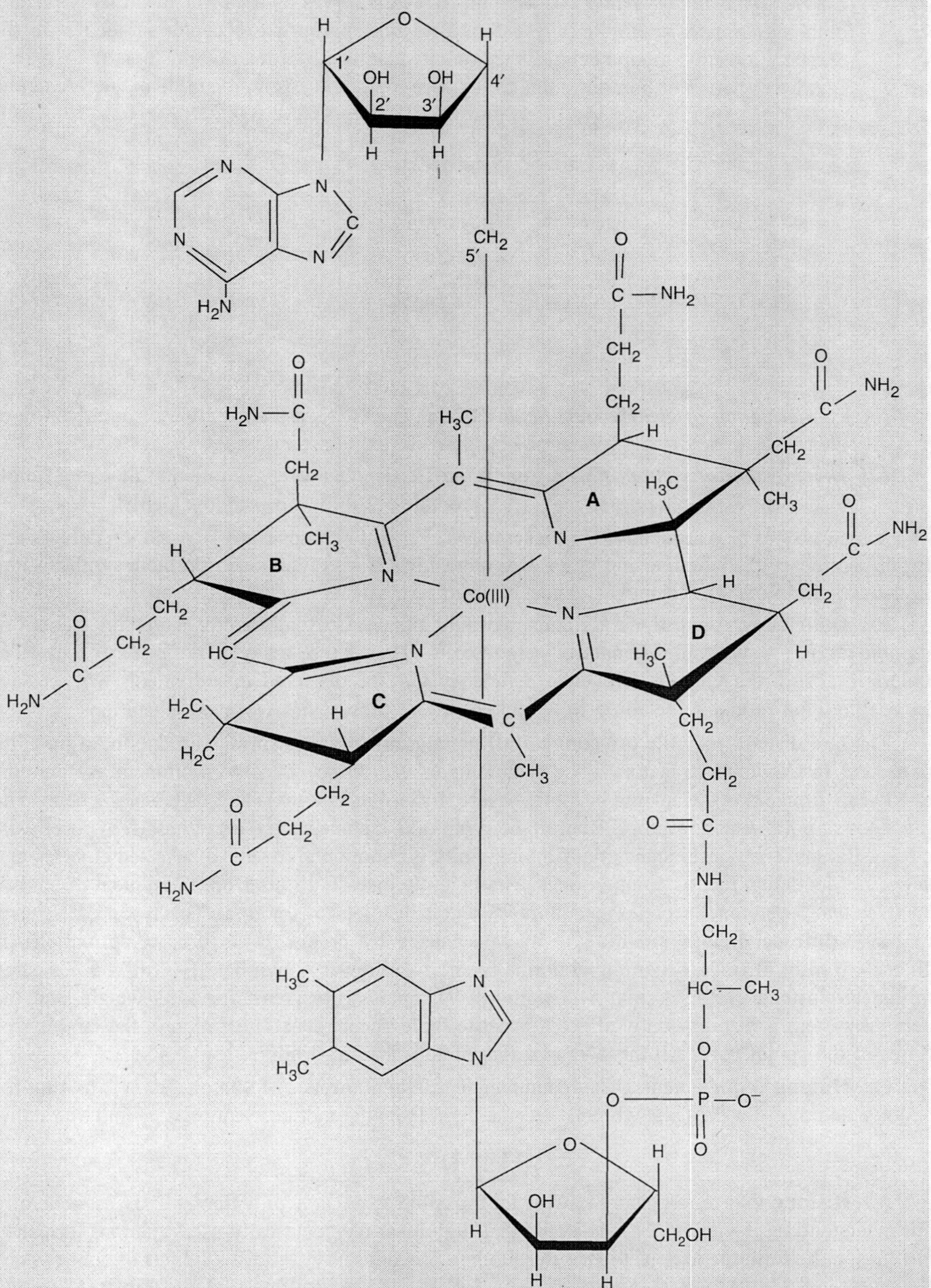

Fig. 22–15. Structure of 5′-deoxyadenosylcobalamin (coenzyme B_{12})

(*b*) ***Carrier of a methyl group :*** Coenzyme B_{12} also serves as a carrier of a methyl group, obtained from N^5-methyltetrahydrofolate, to the appropriate acceptor molecule. In the reaction, a methyl group occupies the 5-deoxyadenosyl coordination position of coenzyme B_{12}. Methylation of homocysteine to produce methionine is an example of such reaction.

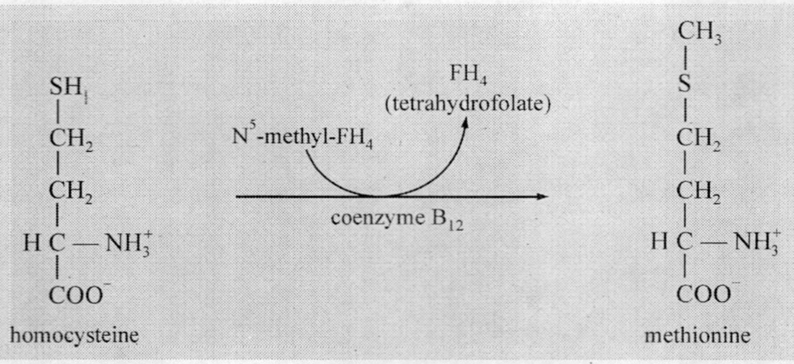

(*c*) ***Isomerization of dicarboxylic acids* :** Coenzyme B_{12} is associated with isomerization of dicarboxylic acids, *e.g.*, glutamic acid into β-methyl-aspartic acid.

(*d*) ***Dismutation of vicinal diols* :** Coenzyme 12 also catalyzes dismutation of vicinal diols to the corresponding aldehydes, *e.g.*, propane-1,2-diol into propionaldehyde.

Vitamin B_{12} is also needed for the biosynthesis of methyl groups from 1-carbon precursors and for the synthesis of thymidine and other deoxyribosides. It also functions in protein synthesis and in the activation of SH enzymes. Cyanocobalamin also affects myelin formation.

F. Deficiency. A nutritional deficiency of this vitamin is usually not observed on account of its ubiquitous (= widespread) nature in foodstuffs. Thus, most cases of deficiency stem from failure to absorb the vitamin. However, deficiency may be observed in individuals who abstain from all animal products including milk and eggs, *i.e.*, those who are strict vegetarians.

The rare disease **juvenile (or congenital) pernicious anemia** springs up due to an inability to secrete gastric intrinsic factors. The symptoms of this disease become prominent at 9 months to 10 years of age. As the anemia becomes severe, irritability, anorexia and listlessness occur. The tongue is smooth, red and painful. Neurologic symptoms include ataxia, paresthesias, hyporeflexia, clonus, Babinski responses and coma. Consanguinity is common in parents of affected children and suggests Mendelian recessive inheritance. The juvenile disease differs from the typical disease in adults in that the stomach secretes acid normally and is histologically normal. This typical deficiency disease, **adult pernicious anemia** (= anemia caused by failure of erythrocyte formation), is characterized by R.B.Cs. becoming abnormally large and fewer in number (1–3 million per cubic millimeter instead of the normal 4–5 million). The patient weakens, loses its weight and the nervous system is also gradually affected because there occurs demyelinization of the large nerve fibres of the spinal cord. All these changes ultimately lead to death.

G. Human requirements. The recommended daily allowance of vitamin B_{12} is 2 to 4 μg for children and 5 μg for men and women. Pregnant and lactating mothers require 8 μg and 6μg daily.

VITAMIN C

A. History. No other vitamin, with the possible exception of vitamin E, is as generally misunderstood as is vitamin C. It is ironic that the oldest therapeutically-used vitamin, furnished in 1750s in the form the lemons to British sailors to prevent scurvy, is still a subject of controversy. However, in 1928, **Albert G. Szent-Györgyi** isolated this crystalline vitamin from the paprika plant and named it **hexuronic acid.** Later in 1932, **C. Glen King** and **W.A. Waugh** in United States isolated this from lemon juice. It was synthesized by **Reichstein** in 1933. It is also called **cevitamin.** It has a curing action against scurvy and hence popularly called as **antiscorbutic factor.**

ALBERT VON NAGYRAPOCT SZENT-GYÖRGYI

(LT, 1893-1986)

Albert G. Szent-Gyorgi, a Hungarian biochemist, was one of the many distinguished biochemists, originally trained in medicine. His first research paper dealt with hemorrhoids. Later, he went into physiology, then biochemistry, and finally physical chemistry. He was awarded the prestigious **1937-Nobel Prize in Physiology or Medicine** for his researches on vitamin C. He was a prolific writer and wrote on many topics of Biology. His concept about research may be expressed in his own words, "The basic texture of research consists of dreams into which the threads of reasoning, measurement, and calculation are woven." Reflecting on his scientific career, he stated he has only one regret : "I started science on the wrong end." One of his famous quotes reads as follows :

"Discovery consists in seeing what everybody else has seen and thinking what nobody else has thought."

B. Occurrence. *In general, ascorbic acid is not as widely distributed as other vitamins.* Among plants, it is present in all fresh fruits and vegetables. *The richest source of vitamin C, known uptodate, is the acerola fruit (Malpighia punctifolia).* The fruit yeilds 1,000–4,000 mg of ascorbic acid per 100 g of edible matter. Citrus fruits (such as orange, lemon, lime), gooseberry, pineapple, guavas, tomatoes, melons, raw cabbage and green pepper are also rich sources of it. New potatoes contain relatively large amounts. Dried legumes and cereals contain very little vitamin C. Dry seeds, in general, are devoid of it but during their sprouting, the vitamin appears. Woody tissues also lack it. Vitamin C is synthesized by most mammals, but not by primates (such as apes, man) and guinea pigs which acquire it from their diets. In animals, the vitamin occurs in tissues and various glands or organs such as liver, adrenals, thymus, corpus luteum etc. Meat contains relatively low concentration. Human milk is 3 to 4 times richer in vitamin C contents than cow's milk. *Vitamin C is, however, absent from fish, fats and oils*. It is also not present in or required by microorganisms. Since this vitamin is a good reducing agent, it is lost under oxidizing conditions like aeration and heating. Thus, many cooked and canned foods contain little ascorbic acid. It is also found in the combined form as *ascorbigen*. The adrenal and lenses have particularly high contents of vitamin C.

C. Structure. The structure of ascorbic acid ($C_6H_8O_6$) was established mainly by **Haworth.** It is a derivative of a hexose called L-gulose. Chemically, it is 1-threo-2,4,5,6-pentoxyhexen-2-carboxylic acid lactone (refer Fig. 22–17). Although ascorbic acid is a small molecule when compared with DNA, RNA or proteins, its metabolic impact is no less considerable. The dienolic group consisting of hydroxyls on C2 and C3 with a double bond between them invests the ascorbic acid molecule with redox property.

D. Properties. Ascorbic acid is a colourless and odourless crystalline substance, slightly sour in taste and optically active. *Only the L-isomer has antiscorbutic properties*. It is soluble in water and alcohol but practically insoluble in chloroform, solvent ether and light petroleum. It is readily oxidized, particularly in the presence of copper and iron but not of aluminium. It is for this reason that the foods cooked in copper utensils lose ascorbic acid quickly. This vitamin is also rapidly destroyed by alkalies but is fairly stable in weak acid solutions. Therefore, baking soda has a deleterious effect but steam cooking destroys very little amount of ascorbic acid. Drying of vegetables and also their storage results in a loss of their ascorbic acid. However, freezing has no detrimental effect on this vitamin. Citrus fruit juices and tomato juice may be canned with but little loss of ascorbic acid. On account of its easily oxidizable nature, the ascorbic acid is a powerful reducing agent.

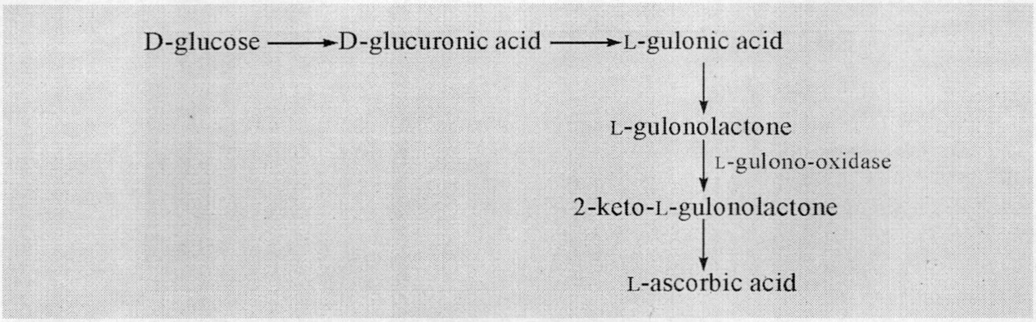

Fig. 22–16. Biosynthesis of vitamin C

E. Metabolism. Higher plants (like pea) and all known mammals except man, the primates and guinea pig can synthesize ascorbic acid from L-gulonolactone. The rat, for example, is resistant to scurvy. In animals, the liver and adrenals (cortical portion) are the main sites of synthesis. The biosynthesis of ascorbic acid in animals takes place according to the scheme as depicted in Fig. 22–16. The scheme also probably applies to the plants. *Man lacks the enzyme L-gulono-oxidase and as such is incapable of synthesizing ascorbic acid.*

Ascorbic acid can be readily oxidized (Fig. 22–17) to dehydroascorbic acid in the presence of metal ions. Dehydroascorbic acid is a much more powerful electron donor than even ascorbic acid by virtue of its unpaired electron. It is, in fact, the free radical form of ascorbic acid. Dehydroascorbic acid can be reduced, in the presence of H_2S or cysteine, back to ascorbic acid. The reduced form (*i.e.*, L-ascorbic acid) predominates in the plasma and also apparently in tissues at a ratio of about 15 : 1 of the oxidized form (*i.e.*, dehydro-L-ascorbic acid). Both of these are biologically active and are equally potent in carrying out their metabolic functions. When dehydro-L-ascorbic acid is hydrated, 2,3-diketo-L-gulonic acid is formed which is biologically inactive and cannot be converted back to either of the active forms in the body. Since the hydration reaction takes place automatically in the neutral medium, the oxidation of ascorbic acid, in other words, means its biologic inactivation. These reactions have been shown to occur *in vivo* in man and guinea pigs.

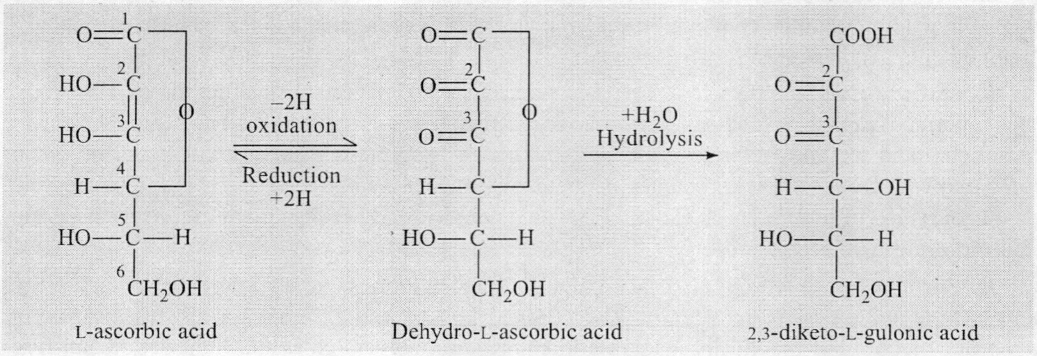

Fig. 22–17. Metabolism of vitamin C

Ascorbic acid functions in a number of enzymatic activities. A major function of ascorbic acid is the formation of tissue collagen or '*intracellular cement substance*'. In fact, ascorbic acid appears to be essential to the activity of the enzyme *collagen proline hydroxylase*, which catalyzes the conversion of proline to hydroxyproline. Hydroxyproline (Hyp) is found exclusively in collagen and is vital in maintaining the tertiary structure of this major vertebrate protein, *i.e.*, collagen.

Ascorbic acid is also involved in electron transport in the microsomal fraction. However, *in none of the biological oxidation systems, ascorbic acid has been shown to act as a specific coenzyme.*

The fact that in old age the deficiency of vitamin C is frequently observed point to a possible role of vitamin C as an anti-ageing agent. Free radicals are the major cause of ageing. These are oxygen molecules which lose an electron in the course of circulating through the blood-stream. These highly reactive molecules try to regain chemical stability by 'snatching' electrons from other molecules, a process which causes much damage. While normal metabolic processes produce some free radicals, their number increases by tissue injuries from infection, toxins, reduced blood flow, excessive exercise and environmental hazards like radiation, heat, and cold. And *the antioxidants like vitamins A, C and E neutralize these free radicals by 'donating' electrons.*

Vitamin C (and also vitamins A and E) have long been known to be beneficial for the skin. While vitamins A and E work by exfoliating the skin's surface cells, vitamin C works from the inside by boosting collagen production and repair. It also inhibits the excess production of melanin ($C_{17}H_{98}O_{33}N_{14}S$) which leads to a tan and hyperpigmentation. Some dermatologists encourage the use of vitamin C as an anti-inflammatory agent as its topical application speeds up the skin's healing process, reducing redness and irritation caused by sun exposure.

F. Deficiency : Avitaminosis C leads to **scurvy**, which may occur at any stage but is rare in the newborn infant. The majority of cases appear in infants 6-24 months (mo) of age. Breast-fed infants are protected as the breast milk contains adequate amounts of vitamin C. Clinical manifestations require time to develop. However, after a variable period of vitamin C depletion, contain vague symptoms of irritability, tachypnea (very rapid respiration), digestive disturbances and loss of appetite appear. The main symptoms which later develop are listed below :

1. *Tender bones :* There is evidence of general tenderness, esp., noticeable in the legs when the infant is picked up or when the diaper is changed. The legs assume the typical "*frog-like position*", in which the hips and knees are semiflexed with the feet rotated outward. This may be mistaken for paralysis and is hence aptly called **pseudo-paralysis.**
2. *Edematous swellings :* These develop along the shafts of the legs and in some cases a subperiosteal hemorrhage can be palpated at the end of the femur.
3. *Petechial hemorrhages :* The capillaries become brittle and burst, thus giving rise to red and purple spots (or *petechiae*) over the body. Petechiae may be seen in the skin and mucous membranes. Hematuria, melena, and orbital or

CLINICAL IMPLICATIONS
SCURVY

A vivid description of scurvy, a dietary deficiency disease, was given by **Jacques Cartier** in 1536, when it afflicted his men when they were exploring the Saint Lawrence River :

"Some did lose all their strength, and could not stand on their feet ... Others also had their skins spotted with spots of blood of a purple colour: then did it ascend up to their ankles, knees, thighs, shoulders, arms and necks. Their mouths became stinking, their gums so rotten, that all the flesh did fall off, even to the roots of the teeth, which did also almost all fall out".

The need for fruits like oranges, lemons and limes was first discovered in the seventeenth century when sailors on long sea voyages developed scurvy after many months of living on dried and stale meat.

But it was not until the 1940s that anyone showed that the substance in fruit which prevented these symptoms was vitamin C. In 1940, a young American surgeon, **John Crandon**, experimented on himself to investigate the effects of a diet deficient in ascorbic acid. After 3 months of deficiency, a skin cut healed normally, but after 6 months another cut would not heat at all. We now know that vitamin C deficiency leads eventually (but not immediately) to a severe reduction in the body's ability to produce collagen and intercellular 'cement'. Any damage to the skin or other tissues cannot therefore be repaired.

Today, some scientists think that some vitamins, including vitamin C may help protect the body against cancer. There is no hard evidence for this, but many people take vitamin supplements, just in case.

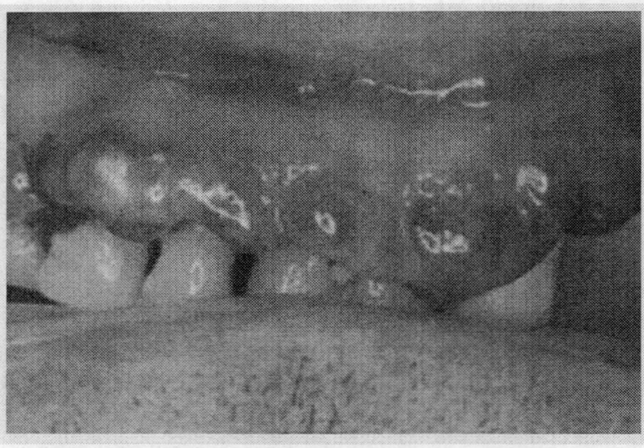

Fig. 22–18. Bleeding gums in a scorbutic patient

subdural hemorrhages may be found.

4. *Bleeding gums :* Changes in the gums, most noticeable when the teeth are erupted, are characterized by bluish-purple, spongy swellings of the mucous membrane, usually over the upper incisors (Fig. 22-18).
5. *Scorbutic rosary :* The costochondral junctions become prominent and appear sharp and angular, giving rise to a beaded structure, called scorbutic rosary. The angulation of the scorbutic beads is usually sharper than that of the rachitic rosary because it is produced by a separation of epiphyses of ribs and backward displacement of sternum rather than by widening of the softened epiphyses as occurs in rickets, where the prominence of the costochondral junction is dome-shaped and semicircular.
6. *Delayed wound healing :* The wound healing is delayed or, in many cases, even does not occur because of the failure of the cells to deposit collagen fibrils. The healed wounds may even break down.
7. *Cessation of bone growth :* The bones cease to grow. The cells of growing epiphyses continue to proliferate but no new matrix is laid down between the cells. Consequently, bones fracture easily at the point of growth because they fail to ossify. Moreover, when an already ossified bone fractures in a scorbutic individual, the osteoblasts cannot secrete a new matrix for the deposition of new bone. With the result, the fractured bone does not heal.
8. *"Sicca" syndrome :* Swollen joints and follicular hyperkeratosis may develop, as well as the "sicca" syndrome of Sjögren, which is usually associated with collagen disorders and includes xerostomia, keratoconjunctivitis sicca, and enlargement of the salivary glands.
9. *Anemia :* Anemia may reflect inability to utilize iron or impaired folic acid metabolism.
10. *Pyrexia :* Low-grade fever is usually present in scorbutic children.

Infants 6–12 months of age, who are fed on processed milk only, are very susceptible to this disease (*infantile scurvy*). Adult cases appear less frequently. Elderly bachelors and widowers who have to cook their own foods are especially prone to the development of vitamin C deficiency– a syndrome termed *'bachelor scurvy'*. Food faddists may also develop scurvy if their diet lacks fruits and vegetables

G. Human requirements. Since vitamin C is continuously oxidized in the body, the daily requirement of this vitamin is rather high. The recommended daily dose for children is 40 mg and for men and women, 50–60 mg. Formula-fed infants should, however, receive even lower doses, *i.e.,* 30 mg of ascorbic acid daily. Lactating mothers should take higher doses, *i.e.*, 100 mg daily.

According to a report published by the British Nutrition Foundation, the RDA for vitamin C on which most nutritionists base diets, is far too low. The 30 mg daily that is usually recommended in most countries is way below the United States at 60 mg, Germany at 75 mg and Russia at 100 mg. Ante diluvian though it sounds, the figure of 30 mg is based on the amount of vitamin C needed to prevent that scourge of seafarers, scurvy. The Nutrition Foundation believes that a more contemporary approach to vitamin C would be to consider at what level it actually promotes good health. Besides, vitamin C requirement in humans may vary with the time of day and time of year. Early autumn, for example, is when most poeple's vitamin C levels are at their lowest.

BIOFLAVONOIDS

A. History. Albert Szent-Györgyi and his associates, in 1936, reported the presence, in lemon peel, of a material which they named **citrin**. It consisted of a mixture of flavonoids and was shown to be associated with the maintenance of normal capillary permeability and fragility. The active principle in citrin was found to be **hesperidin**. It shows physiologic roles similar to those exhibited by other structurally-related compounds such as flavanones, flavones and flavonols. The term **vitamin P** (for permeability) was at first assigned to this group of compounds. They are now more commonly referred to as **bioflavonoids.**

B. Occurrence. The bioflavonoids are widely distributed in nature. *They are always of plant origin.* They are present in the juice, peel and pulp of citrus fruits, in tobacco leaves, in buckwheat (*Fagopyrum esculentum*), in grapes and in many other fruits and vegetables.

C. Structure. Hesperidin is 5,3′-dihydroxy-4′-methoxy-7-rhamnoglucosidoflavanone. Its aglycone hesperitin, rutin (5,7,3′,4′- tetrahydroxy-3-glucorhamnosidoflavone) and its aglycone, quercitin all have comparable physiologic roles.

D. Properties. The bioflavonoids are water-soluble.

E. Metabolism. The bioflavonoids act as antioxidants and thus protect ascorbic acid from oxidative destruction (**Clemetson** and **Andersen**, 1966). The effect is indirect due to the chelation of heavy metal ions (Cu^{2+} etc.,) that catalyze oxidative degradation of ascorbic acid. Bioflavonoids, thus, decrease oxidative losses of ascorbic acid from foods during storage or in intestinal tract.

F. Deficiency. Bioflavonoid deficiency in animals results in a syndrome characterized by increased capillary permeability and fragility. In man, however, the deficiency symptoms have not been observed.

G. Human requirements. The dietary allowances for man are not known.

VITAMERS (= ISOTELS)

The various forms of any vitamin are referred to as **vitamers**. Williams, however, prefers to call these as isotels or isotelic vitamins, since the name vitamer is misleading. Although the isotels, in general, are not the isomers but a few of them may be isomers. *All the fat-soluble vitamins and a few water-soluble vitamins (vitamins B_5 and B_6) have isotels.* The various isotelic forms of a vitamin may differ with respect to either the β-ionone ring (vitamin A), the side chain attached at carbon 17 of the steroid nucleus (Vitamin D), the substituents present at carbon atoms 6, 7 and 8 in the chroman ring (vitamin E) or the side chain attached at carbon 3 of naphthoquinone radical (vitamin K). The study of isotels helps in a better understanding of the various physiologic functions which the vitamins perform.

A summary chart of the water-soluble vitamins (Table 22-2) is given on the next page.

Table 22–2. Water-soluble vitamins

Vitamin	*Common name(s)*	*Chemical name*	*Sources*	*Metabolic functions*	*Deficiency diseases(s)**	*Daily requirement of man*
B_1	Antiberiberi factor or Antineuritic factor	Thiamine or Aneurin $C_{12}H_{17}ON_4S$	All plant and animal foods; Cereals, heart, liver, kidney and milk	Thiamine pyrophosphate (TPP) as coenzyme in many decarboxylation reactions	Beriberi (in man); Polyneuritis (in birds)	1.4–1.7 mg
B_2	'Yellow enzyme'	Riboflavin or Lactoflavin $C_{17}H_{20}O_6N_4$	Milk, liver, kidney, heart and green vegetables; Occur almost exclusively as a constituent of either FMN or FAD	Phototropic curvature of plant organs; FAD as a cofactor in respiration; Also in bioluminescence	Glossitis; Cheilosis; Corneal vascularization; Dermatitis	1.4–1.7 mg
B_3	Filtrate factor or Yeast factor	Pantothenic acid $C_9H_{17}O_5N$	Yeast, liver and eggs; Also fruits, vegetables and skimmed milk; In most animal tissues, it occurs as its coenzyme	Participates in the formation of coenzyme A	Adrenal cortical insufficiency in man; Depigmentation of hair in rats, pigs and dogs; Depigmentation of feathers in chicks	5.0–10.0 mg
B_5	Pellagra-preventive factor or Antiblacktongue factor	Niacin, $C_6H_5O_2N$ Niacinamide, $C_6H_6ON_2$	Yeast, liver, Meat, poultry, vegetables and fruits; Milk and eggs devoid of niacin	As a constituent in two pyrimidine nucleotide coenzymes, NAD and NADP	Pellagra (in man); Blacktongue (in dogs)	15–20 mg
B_6	Antidermatitis factor	Pyridoxine $C_8H_{11}O_3N$ Pyridoxal, $C_8H_9O_3N$ Pyridoxamine, $C_8H_{12}O_2N_2$	Cereal grams, yeast, egg yolk and meat ; Pyridoxal and pyridoxamine also occur as their phosphates	Serve as growth factor to a number of bacteria; Also act as a carrier in active transport of amino acids across cell membranes	Convulsions and anemia in human infants; Acrodynia in rats; Anemia and nervous lesions in dogs and chicks	2.0 mg

Table 22–2. (Cont'd)

Vitamin	*Common name(s)*	*Chemical name*	*Sources*	*Metabolic functions*	*Deficiency diseases(s)**	*Daily requirement of man*
B_7	Antiegg white injury factor or Coenzyme R	Biotin $C_{10}H_{16}O_3N_2S$	Yeast, liver, peanuts and eggs	As a prosthetic group for many enzymes which bring about carboxylation and synthesis of fatty acids	Very rare; Sterilization of intestine, however, may lead to dermatitis, loss of hair and edema	25 mg
B_9	Liver Lactobacillus casei factor	Folic acid or folacin $C_{19}H_{19}O_6N_7$	Liver, kidney, yeast and wheat	Enzymatic synthesis of purines, pyrimidines and amino acids	Megaloblastic anemia and Tropical sprue (in man); Macrocytic anemia (in monkey)	1.0—2.0 mg
B_{12}	Anti-pernicious anemia factor (APA factor)	Cyanocobalamin $C_{63}H_{88}O_{14}N_{14}PCo$	Animals and microorganisms only; Plants devoid of it	Nucleic acid metabolism	Pernicious anemia	1.0 mg
C	Antiscorbutic factor or Cevitamin	Ascorbic acid or Cevitamic acid $C_6H_8O_6$	Fresh fruits and vegetables-Acerola fruit, citrus fruit, tomatoes, green pepper and new potatos	Reducing agent; Biosynthesis of adrenal steriod hormines; Synthesis of colagen	Scurvy	75–100 mg
P	Bioflavonoids	Hesperidin	Of plant Origin–juice, peel and pulp of citrus fruits, in grapes etc.	As antioxidant; Maintenance of capillary permeability and fragility	Hemorrhage (in animals)	Not known

*Deficiency diseases have been mentioned in relation to man, if not stated otherwise.

PROBLEMS

1. Individuals with thiamine deficiency display a number of characteristic neurological signs: loss of reflexes, anxiety, and mental confusion. Suggest a reason why thiamine deficiency is manifested by changes in brain function.
2. What should a person eat to get all the vitamin C his body can use ?
3. If maize is the staple diet, the deficiency to develop is likely to be that of :
   (*a*) cholecalciferol
   (*b*) nicotinic acid
   (*c*) pyridoxine
   (*d*) thiamine
4. Macrocytic anemia can result from the deficiency of :
   (*a*) folic acid
   (*b*) vitamin B_2
   (*c*) inositol
   (*d*) vitamin B_{12}
5. If you are a complete vegetarian, do you miss any vitamins ?
6. Flour is often enriched with :
   (*a*) vitamin A
   (*b*) vitamin B_1
   (*c*) vitamin B_{12}
   (*d*) vitamin E
7. Can B vitamins cut the risk of developing Alzheimer's disease ?
8. Rain drops contain which of the following vitamins :
   (*a*) vitamin A
   (*b*) vitamin B_2
   (*c*) vitamin B_{12}
   (*d*) vitamin D
9. Can vitamins help in the treatment of depression ?
10. Why do doctors prescribe vitamin tablets along with antibiotics ?
11. Why humans need vitamin C supplementation in their diet ?
12. Can vitamin B_6 cure carpal tunnel syndrome in humans ?
13. Which microbe is highly sensitive to the presence of biotin (vitamin H) ?
    (a) Lactobacillus vugraicus
    (b) L. casei
    (c) Bacillus anthracis
    (d) Micrococcus orthobacter
14. Is vitamin C good for heart ?

ANSWERS TO PROBLEMS

1. Glucose is the primary fuel of the brain, and the brain is particularly sensitive to any change in the availability of glucose for energy production. A key reaction in glucose catabolism is the thiamine pyrophosphate-dependent oxidative decarboxylation of pyruvate to acetyl-CoA, and thus a thiamine deficiency reduces the rate of glucose catabolism.
2. Five servings of fruits and vegetables a day will give one plenty of vitamin C (200 mg) –you would not need many supplements.
3. (*b*) nicotinic acid
4. (*a*) folic acid
5. Yes, B_{12} is found only in animal products. To compensate, get B_{12} from a supplement or fortified products such as soy milk.

6. (*b*) vitamin B_1
7. A new study showed that people with high homocysteine levels were twice as likely to develop Alzheimer's as those with low levels. To cut the risk of Alzheimer's, take a multivitamin that has B_6, B_{12} and folic acid. Wheat germ is a good source of B vitamins and folic acid.
8. (*c*) vitamin B_{12}
9. Possibly. Vitamin B_{12} and folate deficiencies may cause symptoms of depression and may also cause a poor response to antidepressants.
10. Doctors prescribe vitamin tablets along with any antibiotic prescription to prevent our body from suffering vitamin deficiency. The antibiotics are powerful chemicals that can kill microorganisms, chiefly bacteria. When ingested these drugs kill not only the disease-causing bacteria but also the harmless ones normally residing in our intestines. Called the 'intestinal microflora', these friendly bacteria convert some of the food constituents into vitamin precursors which are absorbed into the body through intestine. But for this help some of the essential vitamins like those of vitamin-B complex will not be available for us, for our body lacks the machinery to convert food constituents into vitamin precursors. Use of powerful antibiotics kills these helpful bacteria and curtail the supply of vitamin precursors to the body. Therefore, in order to keep the body supplied with usual dose of vitamins, doctors prescribe readymade easily absorbable vitamin-B tablets along with antibiotics. Vitamins are also believed to help in early recovery from the ailments.
11. Higher plants and all known mammals except guinea pigs and primates (including man) can synthesize ascorbic acid from D-glucose (see page 486, Fig. 22-16). However, the enzyme L-gulono-oxidase that converts L-gulonolactone to 2-keto-L-gulonolactone is missing in guinea pigs and primates and hence need vitamin C supplementation in diet.
12. Some experts opine that more than 90% of carpal tunnel cases can be cured by vitamin B_6. Take 100 mg daily for 90 days to feel relief from symptoms. Taking more than 100 mg of B_6 daily for extended periods of time may cause nerve damage.
13. (*b*) *Lactobacillus casei*
14. Recent studies have shown that taking 515 mg of vitamin C daily for 2 months reduces C-reactive protein (CRP) in the bloodstream by 24%. CRP is a marker for chronic inflammation in the arteries, a risk factor in heart disease.

CHAPTER

23 Biochemical Techniques

INSIDE

- Observations on Tissues
  - Perfusion
  - Tissues Slices
- Homogenization
- Differential Centrifugation
- Chromatography
  - Paper Chromatography, PPC
  - Thin Layer Chromatography, TLC
  - Ion Exchange Chromatography
- Isotopic Tracer Technique
- Spectrophotometry
- Electrophoresis
- Ultracentrifugation

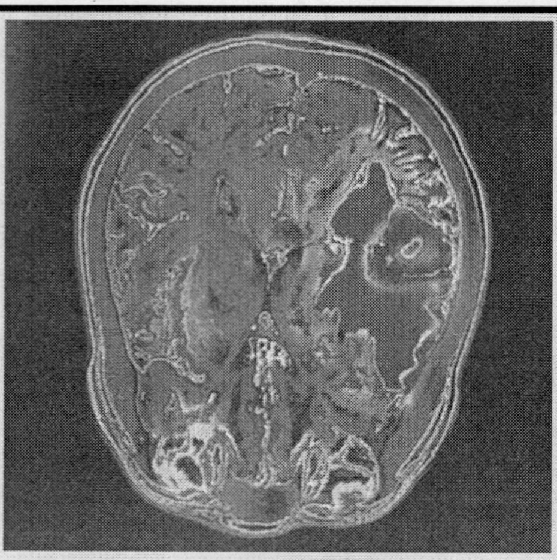

Magnetic resonance imaging

A brain tumour is shown by magnetic resonance imaging (MRI) as a *magenta* coloured area surrounded by damaged brain tissue in *scarlet*. Protons in tumours have longer relaxation times than protons in normal tissue, and this provides a way of locating them.

OBSERVATIONS ON TISSUES

It is often desirable to conduct biochemical studies on an individual organ or individual cells rather than on the body as a whole. But since the growing of single cells in multicelled organisms is not easy and also that after many generations cells from specialized tissues tend to revert to a more primitive type, the biochemist is forced to study either intact organs or preparations from these organs. He can also follow another technique of breaking up the tissues and studying the behaviour of a broken cell preparation.

PERFUSION

This is the simplest way to study intact cells organized in tissues. Here the whole organs is perfused *in situ* or *in vitro* with blood or an isotonic saline solution. The solution is kept recirculating and is also reoxygenated. Arrangements are also made to remove CO_2 from the system. The entire preparation is kept at 37°C by means of a water jacket. In such experiments the tissue itself can be analyzed at the end of perfusion and the changes noted and compared with reference to the composition of the organ before perfusion.

The method, however, suffers from certain *drawbacks* :

(*a*) During perfusion experiments, one is sure to lose certain of the regulatory mechanism such as hormonal and nervous control which operate upon the organ in its normal locus.

(*b*) Moreover, inferences drawn from such experiments are always those obtained under artificial conditions in which the tissue is operating.

TISSUES SLICES

This techinque has been much used in the past for organs such as liver, kidney and brain and also for certain plant tissues like roots and tubers. Here thin slices (approx. 50 μ thick) of tissues are cut. Many of the cells are, of course, damaged but some between 50–70% remain intact. Slicing also offers sufficient surface to the bathing fluid to permit adequate exchange of materials and waste products so that the tissues remain viable for several hours. These slices are then suspenced in buffered isotonic saline solutions in airtight vessel to prevent evaporation. Excess O_2 is given by filling the large airspace, present above the liquid, with pure O_2. The vessel is maintained at 37°C and is frequently shaken to promote diffusion. The changes in cellular composition are then studied.

The two serious *drawbacks* with this techniques are as follows :

(*a*) The cells are feared to 'die' early during such operations. Henceforth, it is best to conduct the experiments for the shortest possible time but consistent with accuracy in analysis of the various changes.

(*b*) Although a fraction of cell membranes is cut in slicing, most cellular constituents still remain contained within the cells.

HOMOGENIZATION

In this the cells are broken up completely so that a homogeneous mixture is obtained which can be adequately oxygenated. The suspension medium can also be changed at will here. The mincing of the material nowadays is done using either a Potter-Elvehjme homogenizer or a blender (see Fig. 23–1). **Potter-Elvehjem homogenizer** consists of a glass tube into which fits a pestle of glass or of polytetrafluoroethylene. The clearance between tube and pestle is between 0.1 and 0.15 mm. The pestle is driven by a motor at 2,000 rpm. A **blender**, on the other hand, consists of a glass vessel fitted with rotating knives. In both the instruments, the tissue is roughly chopped and suspended in a relatively large volume of suitable medium. In order to prevent local heating, the glass tube must be immersed in ice.

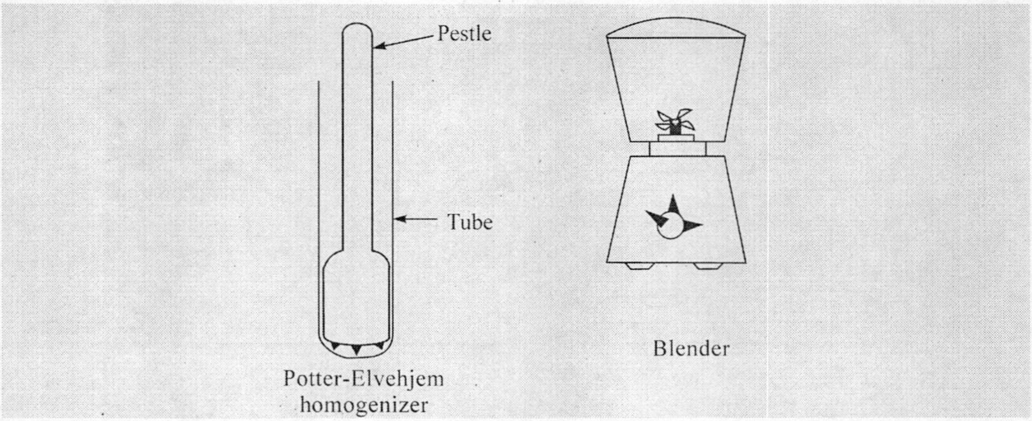

Fig. 23–1. Two common types of mincers

(*Redrawn from Datta SP and Ottaway JH, 1965*)

After homogenization, the various cell components–nuclei,mitochondria, microsomes etc., can be separated by differential centrifugation (see subsequent section). The various fractions so obtained can be identified by their biochemical composition and, to a limited extent, by histological staining. The intracellular organelles can, however, be observed under electron microscope.

This procedure has proved especially useful in studies concerned with determining the location of chemical processes within the cell.

DIFFERENTIAL CENTRIFUGATION

The centrifugal force exerted on a particle in the solution is expressed in multiples of the force exerted by gravity. *The centrifugal force is proportional to the radius of the centrifugal head and to the square of angular velocity*. Hence, it is more convenient to use relatively small heads rotating at high speeds. Thus, a head of approximately 10 cm diameter rotating at about 40,000 rpm will produce nearly 100,000 g. At such high speeds, the head is run in a vacuum to prevent the heat produced by air friction. The tube containing the homogenate is usually held at an angle to the axis of rotating to keep the path of particles through the solution as short as possible (Fig. 23–2).

> The broken cell preparation is loosely referred to as cell-free preparation or '**homogenate**'. Properly prepared, a homogenate is the sum of all the cell components obtained by rupturing the cell membrane.

The homogenate is first diluted and then centrifuged at a low speed to remove nuclei etc. The supernatant liquid is then poured off and centrifuged at a higher speed to remove next fraction. The

> **Supernatant** is that part of the cell material not sediemented by a centrifugal force known to sediment any subcellular structure.

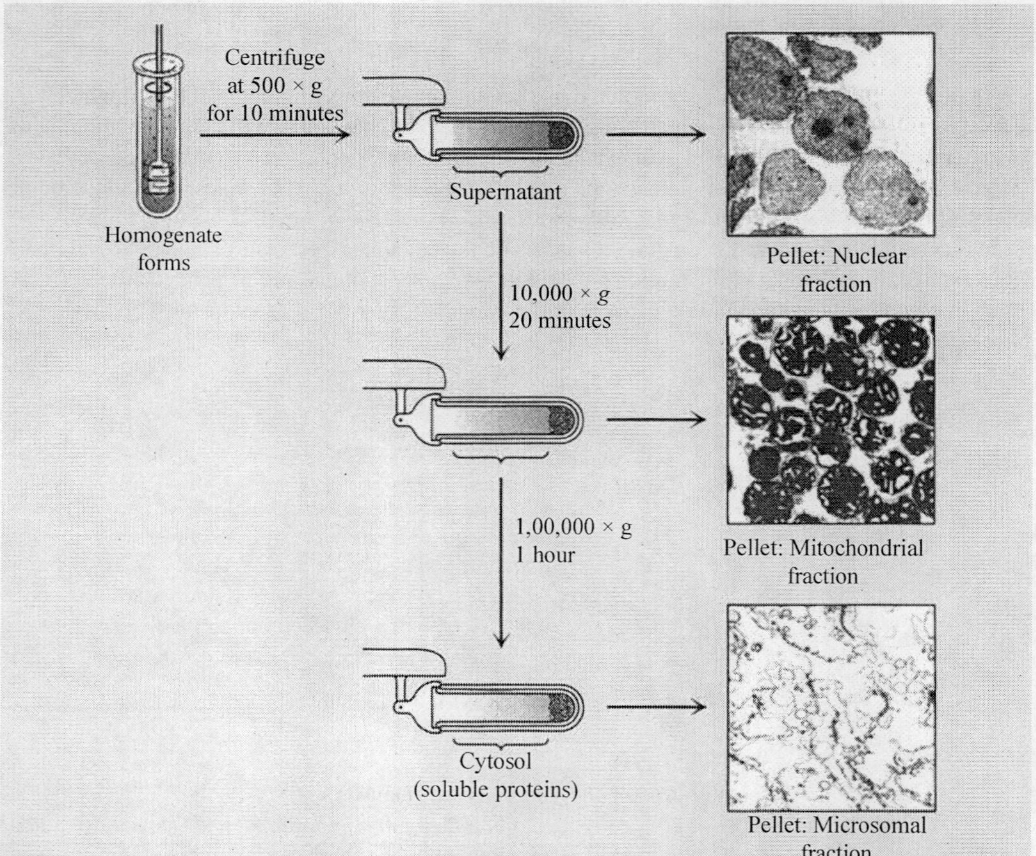

Fig. 23–2. Differential centrifugation

Cell are disrupted in a homogenizer and the resulting mixture, called the homogenate, is centrifuged in a step-by-step fashion of increasing centrifugal force. The denser material will form a pellet at lower centrifugal force than will the less-dense material. The isolated fractions can be used for further purification.

(*Photo courtesy : S Fleischer and B Fleischer*)

supernatant is then again contrifuged at still a higher speed and fraction separated. The removal of particles of a particular size depends on the magnitude of the field and also on the time for which the field is applied (refer Table 23–1).

Table 23–1. Separation of cell organelles by differential centrifugation

*Field**	*Time*	*Structure(s) separated*
700 g	10 min	Nuclei, cell membranes
5,000 g	10 min	Mitochondria, lysosomes
57,000 g	60 min	Microsomes, lysosomes
150,000 g	30 min	Ribosomes
Unsedimented		Cell sap, some ribosomes

* These centrifugal fields refer to homogenates in 0.25 M sucrose.

(*After Datta SP and Ottaway JH, 1969*)

In practice, a mixture of fractions is obtained on single centrifugation. In order to obtain pure fractions, the precipitates have to be recentrifuged many times.

CHROMATOGRAPHY

The term chromatography (*chroma*^{G} = a colour; *graphein*^{G} = to write) was orginally applied by a Russian chemist, **Michael Semonovich Tswett** (LT, 1872–1919), in 1906 to a procedure where a mixture of different coloured pigments (chlorophylls and xanthophylls) is separated from each other. He used a column of $CaCO_3$ to separate the various components of petroleum ether chlorophyll extract into green and yellow zones of pigments. He termed such a preparation as **chromatogram** and the procedure as **chromatography**.

Chromatography may be defined *as the technique of separation of substances according to their partition coefficients below (i.e., their relative solubilities in) two immiscible phases.* In this method, the separation of the components of a mixture is a function of their different affinities for a fixed or *stationary phase* (such as a solid or a liquid) and their differential solubility in a moving or *mobile phase* (such as a liquid or a gas). Separation starts to occur when one component is held more firmly by the stationary phase than the other which tends to move on faster in the mobile phase. Thus, the underlying principle of chromatography is first to adsorb the component of a mixture on an insoluble material and then to differentially remove (or elute) these components one by one with suitable liquid solvents. The adsorbent can be packed into a column (*column chromatography*) or can be in the form of a sheet (*paper chromatography*). A third form of chromatography is obtained with columns containing ion exchange resins (*ion exchange chromatography*).

The various chromatographic techniques fall principally under 2 categories : adsorption chromatography and partition chromatography. In **adsorption chromatography**, the stationary phase is a finely divided adsorbent such as alumina or silica gel and the mobile phase can be a gas or more commonly a liquid. **Partition chromatography** involves partition between two liquids rather than adsorption by a solid from a liquid. Here the stationary phase is a liquid which is held on an inert porous supporting liquid. The various forms of chromatography have been represented schematically in Fig. 23–3.

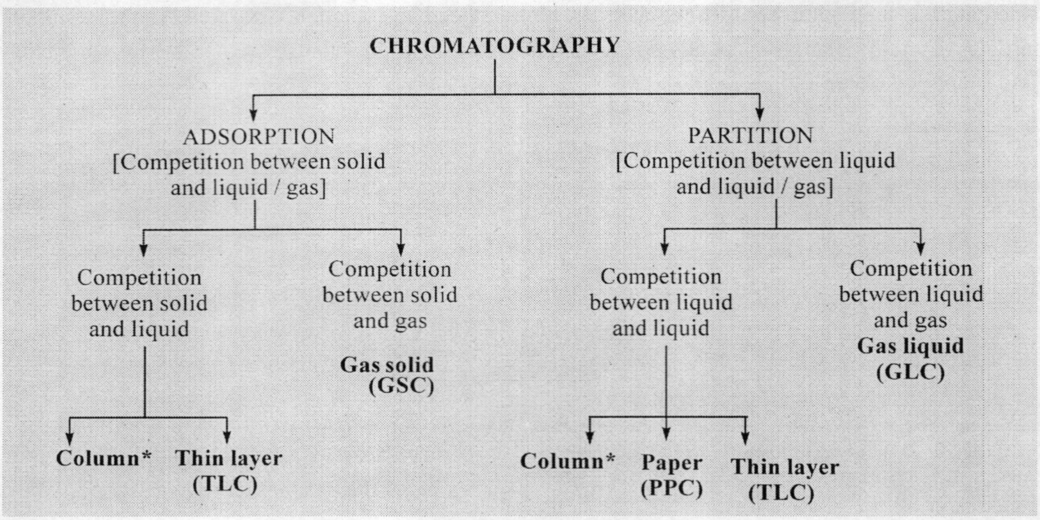

Fig. 23–3. Various forms of chromatography

* Other than gas chromatography.

A few chromatographic techniques are described below.

PAPER CHROMATOGRAPHY, PPC

Two Russian workers, **Izmailov** and **Schraiber** (1938) discovered this important technique. *This method is especially useful for the detection and separation of amino acids*. Here the filter paper strips are used to support a stationary water phase while a mobile organic phase moves down the suspended paper strip in a cylinder. Separation is based on a liquid-liquid partition of the compounds. Thus, this is essentially a form of partition chromatography between two liquid phases, although adsorption to the paper may also take place.

In this method, a drop of solution containing a mixture of amino acids (or other compounds) to be separated is applied at a marked point, about 3 cm from one end of a strip of filter paper. Whatman No. 1 paper is most frequently used for this purpose. The chemical composition of Whatman filter paper No. 1 is :

α-cellulose	98 – 99%
β-cellulose	0.3 – 1%
Pentosans	0.4 – 0.8%
Ash	0.07 – 0.1%
Ether soluble matter	0.015 – 0.1%

Whatman No. 3 MM is a thick paper and is best employed for separating large quantities of material; the resolution is, however, inferior to whatman No.1. For a rapid separation, **Whatman Nos. 4 and 5** are convenient, although the spots are less well-defined. In all cases, the flow rate is faster in the 'machine direction', which is normally noted on the box containing the papers. The paper may be impregnated with a buffer solution before use or chemically modified by acetylation. Ion exchange papers are also available commercially.

The filter paper is then dried and 'equilibrated' by putting it into an air-tight cylidrical jar which contains an aqueoues solution of a solvent. The most widely applicable solvent mixture is *n*-butanol : acetic acid : water (4 : 1 : 5), which is abbreviated as BAW. The end of the filter paper nearest the applied drop is inserted into the solvent mixture at the bottom of the jar, taking care that the marked point of application remains well above the level of the solvent in the jar. The paper is suspended in such a manner so that it hangs freely without touching the sides of the container. Thus, the solvent will *ascend* into the paper and this process is, therefore, termed **ascending chromatography** (Fig. 23–4).

Alternatively, the same end of the filter paper may be put into the solvent mixture contained in a narrow trough mounted near the top of the container. In this case, the solvent will *descend* into the paper and this process is then termed **descending chromatography** (Fig. 23–5).

Locating the compounds. Strip is removed when the solvent has migrated over most of the available space. The distance to which the solvent has run is marked. In most cases, the completed chromatogram is colourless with no indication of the presence of any compounds. Such a chromatogram is said as '*undeveloped*'. For locating the various compounds, the filter paper strip is first dried, then sprayed with 0.5% ninhydrin in acetone and at last heated for a few minutes at 80 – 100°C. The reaction occurs and the coloured spots appear at the sites of the amino acids. Such a chromatogram is now called '*developed*'.

However, if the solution chromatographed is one derived from a tracer experiment, the compounds will be radioactive and can be located either by using a Geiger counter or by placing the paper strip against a sheet of x-ray film. The β-rays from the radioactive compounds will expose the film. The negative, on developing, will show dark spots at the site of radioactive compounds. This method has been used most successfully in tracing the carbon pathway in photosynthesis.

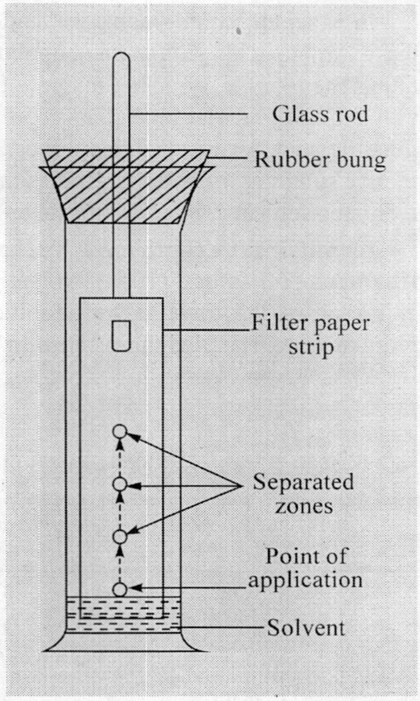

Fig. 23–4. One-dimensional paper chromatography
(Ascending type)

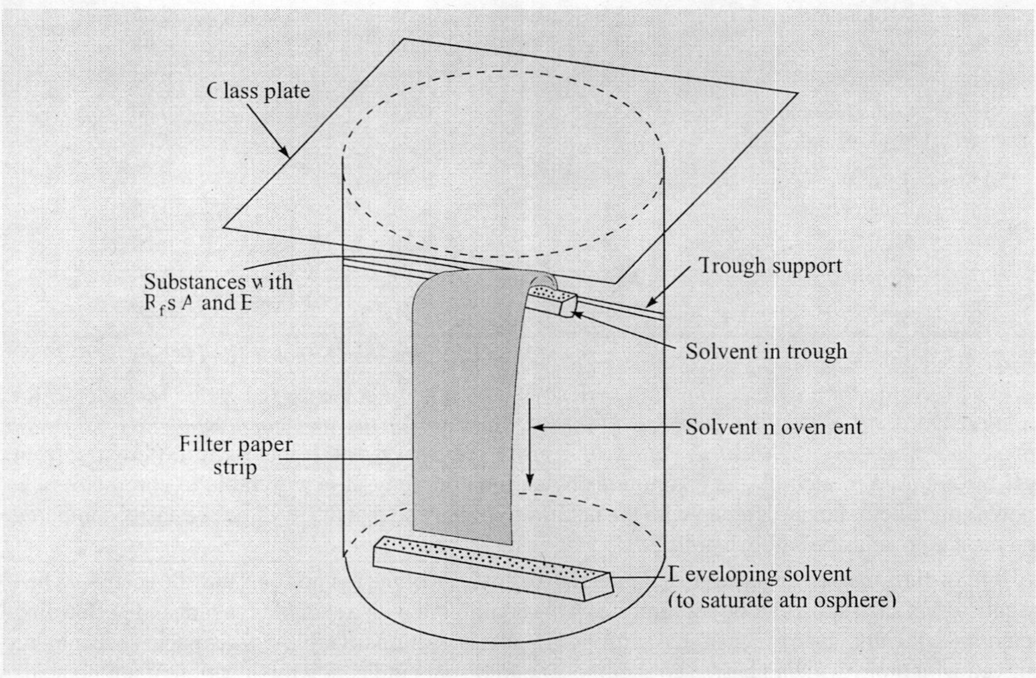

Fig. 23–5. One-dimensional paper chromatography
(Descending type)

In paper chromatography, the stationary cellulose phase is more polar than the mobile organic phase. Amino acids with large nonpolar side chains (leucine, isoleucine, phenylalanine, tryptophan, valine, methionine, tyrosine) magrate farther in *n*-butanol ; acetic acid : water (4 : 1 : 5) than those with shorter nonpolar side chains (proline, alanine, glycine) or with polar side chains (threonine, glutamic acid, serine, arginine, aspartic acid, histidine, lysine, cysteine). This reflects the greater relative solubility of polar molecules in the hydrophilic stationary phase and on nonpolar molecules in organic solvents.

Identifying the compounds. In chromatography, the distance travelled by a particular component of a mixture (or solute) is used to identify it. The ratio of the distance travelled by a component (*i.e.*, amino acid) to that travelled by the solvent front, both measured from the marked point of application of the mixture, is called the **resolution front** (R_f) value for that component. Thus,

$$R_f = \frac{\text{Distance from origin run by the compound}}{\text{Distance from origin run by the solvent}}$$

Table 23–2. R_f values of amino acids in various solvents

S. N.	*Amino Acid*	*Phenol : Water*	*Collidine : Water*	*Butanol : Acetic Acid : Water, BAW*
1.	Glycine	0.36	0.26	0.26
2.	Alanine	0.55	0.32	0.38
3.	Valine	0.72	0.43	0.60
4.	Leucine	0.80	0.55	0.73
5.	Isoleucine	0.83	0.53	0.72
6.	Serine	0.30	0.30	0.27
7.	Threonine	0.43	0.32	0.35
8.	Cystine	0.24	0.11	0.08
9.	Methionine	0.74	0.53	0.55
10.	Proline	0.88	0.34	0.43
11.	Aspartic acid	0.22	0.23	0.24
12.	Glutamic acid	0.23	0.27	0.30
13.	Phenylalanine	0.83	0.54	0.68
14.	Tyrosine	0.55	0.59	0.45
15.	Tryptophan	0.71	0.59	0.50
16.	Histidine	0.62	0.30	0.20
17.	Arginine	0.54	0.17	0.20
18.	Lysine	0.41	0.11	0.14

(*Adapted from Fruton JS and Simmonds S, 1958*)

The R_f value for a given compound varies with the type of solvent used (refer Table 23–2, for amino acids). *Similar R_f values, however, do not necessarily mean the identical compounds as different compounds may have identical R_f values in a given solvent system.* Henceforth, different solvent systems are used to identify the compounds by chromatography. It is preferable to chromatograph known amino acid standards along with the unknown mixture so that the R_f values be easily compared and the amino acids be safely identified.

Quantitation of the compounds. Quantitative analysis of the amino acids may be accomplished by cutting out each spot, eluting (or removing) the compound with a suitable reagent and performing a calorimetric (ninhydrin) or chemical (nitrogen) analysis. Alternatively, the filter paper strip may be sprayed with ninhydrin and heated so that the coloured spots indicating the location of amino acids may develop. The colour densities of these spots may be measured with a recording transmittance or reflectance photometer device.

The chromatographic technique described above allows *vertical* separation of amino acid and is termed as **one-dimensional paper chromatography**.

For obtaining better separation of components and to improve their quantitation, a **two-dimensional paper chromatography** is applied (Fig. 23–6). In this technique, a square sheet of filter paper, rather than a strip is taken. The test sample is applied to the upper left corner and chromatographed for some hours with one solvent mixture (*e.g.*, *n*-butanol : acetic acid : water). After drying to remove this solvent, the paper is tuned through 90° and chromatographed in a second solvent mixture (*e.g.*, collidine : water). This technique thus allows both *vertical* and *horizontal* separation of the amino acids.

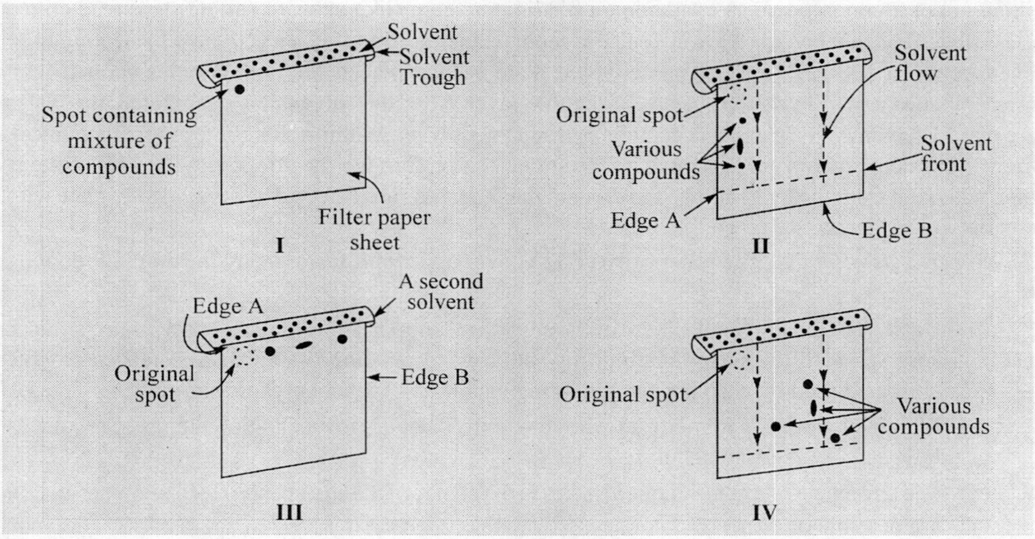

Fig. 23–6. Two dimensional (or '2-D') paper chromatography
(Descending type)

Forces in operation. The movement of the solute molecules on the chromatogram depends on the net result of a number of forces operating in the system. These forces are of 2 types : propelling and retarding.

(*a*) *Propelling forces* – These include the capillary force and the solubility force of the solvent. The Whatman paper is made up of numerous fibrils which are placed very close to each other, thus forming a network of capillaries. The solute rises through these capillaries as a result of **capillary force**. *The smaller the bore of the capillaries, the greater is the height to which the solute rises.*

The **solubility forces of the solvent** refers to the capacity of the solvent to dissolve the solute. The rise of solute also depends on its solubility in the solvent being used. *The greater the solubility of the solute in the solvent, the greater is the height to which it rises in the chromatogram.*

(*b*) *Retarding forces* – Concurrent with the forces of propulsion, certain retarding forces also operate in the system which try to drag the solute molecules from moving in either direction. These retarding forces include the gravitational force and the partition force. Under the ascending chromatography, the solute molecules have to move against the **gravitational force** which acts from below and tends to retard the movement of solute molecules in the upward direction. Under the descending chromatography, obviously, the gravitational force does not figure as a retarding force, rather it assists in propulsion and, henceforth, acts as a propelling force.

The **partition force** refers to the force between liquid and liquid molecules. The interstices between the fibrils on the Whatman paper are occupied by the solvent molecules and the solute molecules, for their movement, have to displace them. Should the spaces not been filled with the solvent molecules, the movement of the solute molecules through them would have been much more facilitated.

Thus, the movement of the solute molecules on the Whatman paper in either direction is the net result of the interaction between various forces of propulsion and retardation. Movement of the solute molecules is exhibited only when the propelling forces exceed the retarding forces in magnitude.

THIN LAYER CHROMATOGRAPHY, TLC

Thin layer chromatography is adsorption chromatography performed on open layers of adsorbent materials supported on glassplates. *This technique combines many of the advantages of paper chromatography with those of column chromatography*. Here a thin uniform film of adsorbent (like silica gel or alumina powder) containing a binding medium (like calcium sulfate) is spread onto a glass plate. The thin layer is allowed to dry at room temperature and is then activated by heating in an oven between 100° to 250°C. The activated plate is then placed flat and samples spotted with micropipettes carefully on the surface of the thin layer. After the solvent has evaporated, the plates are placed vertically in a glass tank containing a suitable solvent. Within a short time (5 to 30) minutes), the various components get separated by the solvent rising through the thin layer. The glass plate is then taken out from the tank, allowed to dry and then the spots are detected by spraying the plate with a variety of reagents.

Superiority of TLC – The superiority of TLC over paper chromatography lies in the following facts :

(*a*) Because of the inorganic nature of the adsorbent (supporting medium), concentrated sulfuric acid spray followed by heating may be used to develop (or locate substances on) the chromatogram by charring and rendering visible any spots of organic nature.

(*b*) Moreover, amino acid mixtures, which require 18 hours for separation on paper, require as little as 3 hours using cellulose TLC.

(*c*) The advantage of this technique also lies in the choice of the adsorbents which allow separation not possible on paper.

Thus, *the speed, efficiency and sensitivity of TLC has made this technique most valuable to the biochemists.* Lipids including sterols may be neatly separated by TLC on alumina. Also after localization of these substances on the glass plate with reagents, selected spots of the plate's surface may be scrapped off and the compounds isolated by extraction of the alumina powder with suitable solvents.

ION EXCHANGE CHROMATOGRAPHY

Ion exchange chromatography is in wide use for the separation of amino acids. It employs synthetic resins such as a strongly acidic cation exchanger, **Dowex-50** and a strongly basic anion exchanger, **Dowex-1**. The former is a polystyrene sulfonic acid and the latter a polystyrene quaternary ammonium salt.

Two common cellulose derivatives, carboxymethyl cellulose (CMC) and diethylaminoethyl cellulose (DEAE) are successfully used in protein purification. CMC is a cationic derivative and DEAE, an anionic derivative. An even more versatile substance for the separation of protein is a cross-linked dextran called **sephadex**. When packed in a column and equilibrated with buffer, the polysaccharide acts as a molecular sieve for protein molecules. Five grades of sephadex are available. Sephadex G-25 beads retain molecules with molecular weights up to 5,000. In other words, the 'exclusion limit' of G-25, is 5,000. The exclusion limits of the remaining four grades, G-50, G75, G-100 and G-200-- are 10,000, 50,000, 100,000 and 200,000 respectively.

Thus, electrostatic attraction of oppositely charged ions on a polyelectrolyte surface forms the basis of ion exchange chromatography. In principle, the ionized groups of the ion exchange materials form salts with ions of opposite charge and they exchange these ions for others when the relative salt concentrations in the solution are varied. Ion exchange chromatography is especially useful in separation and purification of nucleotides owing to their low molecular weights and the presence of ionizable groups.

In practice, the protein mixture is passed through a column of the modified cellulose contained in a glass tube and allowed to become attached to the column material. Increasing salt concentration and/or buffers of varying pH values are then passed through the column. Proteins will be washed off the column at different times depending on their molecular structure, net charge and side group. The essential process is depicted in Fig. 23–7.

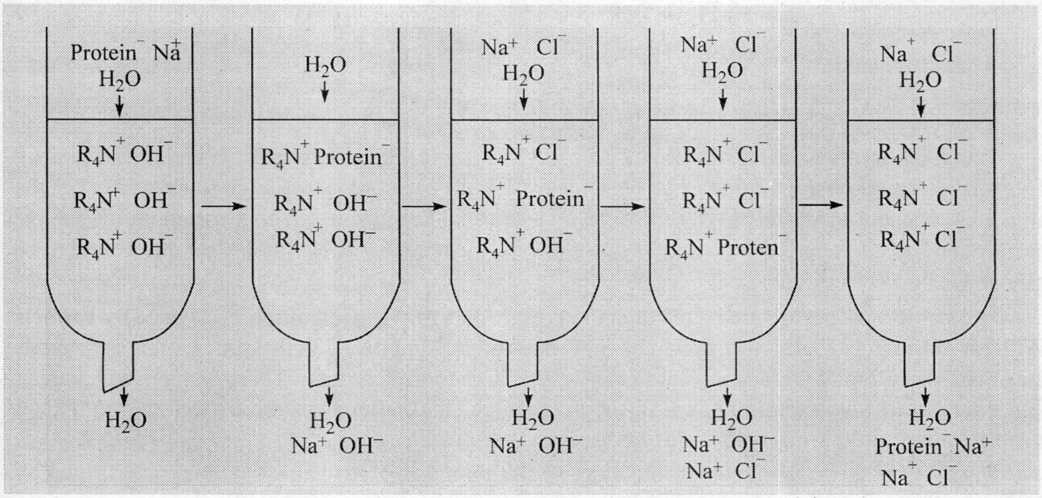

Fig. 23–7. Ion exchange chromatography

The various processes occurirng as protein is absorbed by and eluted from an ion exchange column are illustrated here. The ion exchange material is assumed to be an insoluble polymer having free substituted amine groups.

(*Adapted from Fariley JL and Kilgour GL, 1966*)

ISOTOPIC TRACER TECHNIQUE

In the studies of metabolic transformation of a particular substance, the primary objective of the biochemists remains to observe the fate of that substance *in vivo* under experimental conditions which cause minimum physiological disturbance to the test organism. Isotopic tracer technique has been most successfully used for such type of studies. Here one or more of the atoms in metabolite under study is "labelled" by means of one of the rare or artificially produced isotopes and its path followed while frequently testing the intermediary compounds at intervals.

The nucleus of each elementary species is characterized by an **atomic number**, which is equal to the number of protons and is also equal to the number of electrons around the nucleus when the atom exists in an electrically neutral state. All atoms of a particular element have the same atomic number. Each nucleus is also characterized by what is called as the **mass number** and which is equal to the total number of protons and neutrons contained therein. The difference between the two numbers is obviously the number of neutrons contained in the nucleus. *The various isotopes of an element have the same atomic number but different atomic mass*. Conventionally, the atomic number is written as subscript that precedes the elementry symbol and the mass number as superscript that follows the symbol. For example, nitrogen (atomic number 7) is found in nature both with a mass of 14 (designated as ${}_7N^{14}$) and with a mass of 15 (designated as ${}_7N^{15}$). Similarly, carbon with atomic number 6 has different mass number of 11, 12, 13 and 14 and the various isotopes are accordingly written as ${}_6C^{11}$, ${}_6C^{12}$, ${}_6C^{13}$ and ${}_6C^{14}$. Being implicit in the symbol, the atomic number is often not written. Thus, the two common isotopes of carbon are simply written as C^{12} and C^{14}.

Isotopes are of 2 types – stable and radioactive. The *stable isotopes* are those whose nuclei do not undergo spontaneous decomposition. N^{14}, N^{15}, O^{18}, C^{13}, S^{33} and S^{34} belong to this category. Atoms whose nuclei decompose spontaneously with the emission of radiations are termed *radioactive isotopes*. C^{14}, S^{35}, P^{32} and H^3 belong to this group.

Stable Isotopes. The stable isotopes of biological importance are available in enriched concentrations. The important ones are H^2, N^{15}, C^{13} and O^{18}. These atoms occur in nature in the following relative abundance :

$$H^1/H^2 \quad - \quad 99.98/0.02$$
$$N^{14}/N^{15} \quad - \quad 99.63/0.37$$
$$C^{12}/C^{13} \quad - \quad 98.09/1.10$$
$$O^{16}/O^{18} \quad - \quad 99.80/0.20$$

The concentration of these stable isotopes, 0.37% of N^{15} for example, is called the *normal abundance*. The concentration of a heavy isotope is usually measured as *atom per cent excess i.e.*, the amount in per cent by which the isotope exceeds its normal abundance. As an illustration, if a sample of nitrogen gas contains 3% N^{15} (and 97% N^{14}), the concentration of N^{15} in this sample is said to be (3.0–0.37) or 2.63 atom per cent excess.

The stable isotopes are measured quantitatively with a **mass spectrometer**. Devised by **Francis William Aston** in 1919, this instrument has now been greatly improved. In fact, this apparatus converts the uncharged atoms or molecules into positively charged ions (cations). These are then accelerated into the field of a powerful magnet. In the magnetic field, the ions will be deflected to an extent corresponding to their mass. *The path of the heaviest particles is bent least and that of the lightest particles is bent most*. The relative amounts of the ions of different mass is then determined by collecting them on a plate and measuring the current produced.

This technique for measuring stable isotopes is less sensitive and more tedious than that for measuring radioactive isotopes. However, since no useful radioactive isotopes of oxygen and nitrogen exist, the mass spectrometer remains an essential tool for studies involving these elements.

Table 23–3. Properties of useful radioactive isotopes

S.N.	*Isotope*	*Half-life**	*Type of Radiation*	*Energy of Radiation (in million electronvolts, mev)*
1.	H^3	12.5 years	β	0.018
2.	C^{14}	5760 years	β	0.155
3.	P^{32}	14.3 days	β	1.712
4.	S^{35}	87.1 days	β	0.167
5.	Cl^{36}	3.1×10^5 years	β	0.714
6.	K^{42}	12.5 hours	β	3.550
7.	Ca^{45}	152 days	β	0.255
8.	Fe^{55}	2.94 years	x-rays	–
9.	I^{125}	60 days	γ	0.035
10.	I^{131}	8.1 days	β	0.610
			γ	0.280

* **Half-life** is the time required for the loss of 50% of radioactivity. In other words, the time taken for half the atoms is a quantity of a radioactive substance to disintegrate is known as the half life.

Radioactive Isotopes. T*he radioactive isotopes are usually more useful as tracers than the stable isotopes since the analytical methods for their measurement are extremely sensitive*. The biochemically-important redioactive isotopes are H^3, C^{14}, P^{32}, S^{35} and Ca^{45}. All these emit β-rays and their nuclei, upon disintegration, produce electrons. The β-rays interact with the molecules through which they traverse causing ionization of the molecules. It is this ionization property which is used to measure quantitatively the amount of radioactive isotope present. Table 23–3 lists some properties of useful radioactive isotopes.

The most widely-used apparatus for the determination of radioactive isotopes is **Geiger-Müller counter or G-M tube** (Fig. 23–8). The functioning of this apparatus is based on the ability of the emitted radiation to ionize atoms. It consists of a large, round tube forming cathode with a fine wire stretched in the centre as anode. The fine wire is maintained at a high potential (1,000–2,500 volts) with respect of the outer cathode. The tube is filled with an easily ionized gas such as helium or argon and an organic quenching substance such as ethanol. The voltage as well as the gas filling is so adjusted that normally no current flows. The open end of the tube is covered with an extremely thin

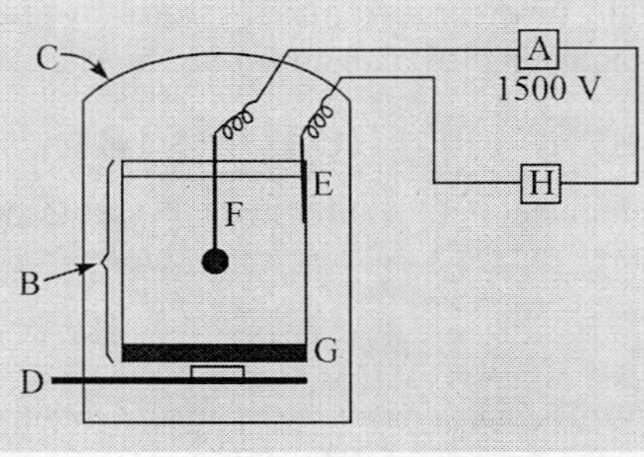

Fig. 23–8. Geiger-Müller tube and a scaler

[A = source of potential, B = Geiger-Müller tube, C = leadshield, D = sample pan holder, E = cathode shield, F = anode wire, G = mica window, H = scaler]

(*After Cowgill and Pardee, 1957*)

window of mica (1.5 to 2.0 mg/cm^2) or synthetic plastic. The radioactive material which is usually a solid is placed beneath this window. The radioactive particle enters the tube and ionizes the gas molecules with a subsequent release of a shower of electrons. These free electrons are then accelerated to the positive wire. As they progress through the gas, additional molecules are ionized. Thus, the tube becomes momentarily conductive. The resultant electrical pulse is collected in an electronic computing machine (called scaler) which records the number of such pulses in a predetermined time.

> The **radioactive sample** to be counted is, however, in some instances a gas and is actually included in the filling of the counter tube.

The radioactivity of a sample is given in terms of its *specific radioactivity, i.e.*, the number of counts per minute (*cpm*) per unit weight (milligram or micromole etc). The relationship of this quantity to the number of disintegrations per unit time depends on the counting efficiency of the system. The absolute unit of disintegration rate is the *curie*. A curie is defined as the amount of emitter which exhibits 3.7×10^{10} disintegrations per second (*dps*). More common units are a millicurie, mc (10^{-3} curie) and a microcurie, μc (10^{-6} curie).

The Geiger-Muller technique is being rapidly replaced nowadays by an alternative method called **scintillation counting**. It involves transformation of β or γ radiation into ultraviolet or visible light by the use of a *phosphor*. The phosphors are often highly aromatic substances such as *p*-terphenyl and have the property of emitting a light flash (or a scintillation) when they absorb radiation from a radioactive compound. The scintillations are detected by a sensitive photomultiplier and the resultant electrical pulses are collected in a scaler. *The scintillation technique has proved particularly useful for isotopes which emit low energy β particles such as* H^3, C^{14} etc.

Although extremely sensitive, this method also suffers from certain *drawbacks*. They are :

(*a*) quenching of light flashes by coloured samples, and

(*b*) limited solubility of highly polar compounds in nonpolar solvents.

Therapeutic applications of radioactive emissions. Radioactive emissions, particularly of γ-rays, are frequently used to kill unwanted tissues. Some common examples are the radioactive sterilization of surgical accessories, partial destruction of over-active thyroids by I^{131} and the irradiation of tumours by Co^{60} and by Au^{198} or Ra^{226}.

Neutron activation. The radioactive isotopes are made usually by bombarding a suitable target with neutrons in an atomic pile. But sometimes it becomes also possible to bombard a biological sample directly so that a little element is converted into a radioactive isotope. Thus, O^{18} has been converted into F^{18} by neutron activation. Of special interest was the dramatic example of bombardment of a single hair of the **Emperor Napoleon Bonaparte.** The hair was, in fact, taken at the end of his life and is preserved since then. The hair, after this treatment, was found to possess a radioactive isotope of arsenic which could only have come from arsenic already present in the hair. This finding was evidenced to the fact that Napoleon died of *arsenic poisoning*.

SPECTROPHOTOMETRY

Principle. When a beam of incident light of intensity, I_0 passes through a solution (Fig. 23–9), a part of it is reflected (I_r), a part absorbed (I_a) and rest transmitted (I_t), *i.e.*,

$$I_o = I_r + I_a + I_t$$

In colorimetric methods, I_r is eliminated because the measurement of I_o and I_t will be sufficient to determine I_a. For this purpose, the amount of light reflected (or I_r) is kept constant by using cells that have identical properties. I_o and I_t are then measured.

The mathematical relationship between the amount of light absorbed and the concentration of a substance can be shown by the following two fundamental laws, on which the spectrophotometry is based.

A. Lambert's law (= Bouguer's law). This law states that the amount of light absorbed is directly proportional to the length or *thickness* of the solution under analysis. Thus,

$$A = \log_{10} \frac{I_o}{I_t} = a_s b$$

[where A = absorbancy

a_s = absorbancy index characteristic for the solution

b = length or thickness of the medium.]

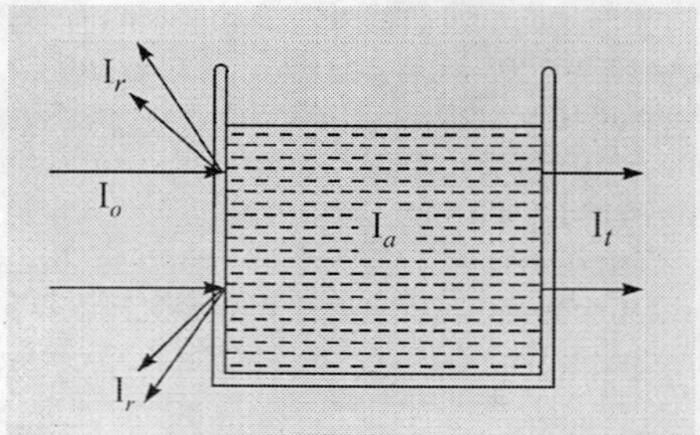

Fig. 23–9. Distribution of incident light passing through a solution contained in a cell

B. Beer's law. This law states that the amount of light absorbed is directly proportional to the *concentration* of the solute in solution. Thus,

$$\log_{10} \frac{I_o}{I_t} = a_s c$$

[where c = concentration of solute in solution.]

The combined **Beer-Lambert Law** then becomes :

$$\log_{10} \frac{I_o}{I_t} = a_s bc$$

If b is kept constant by employing a standard cell or cuvette, the above formula reduces to:

$$A = \log_{10} \frac{I_o}{I_t} = a_s c$$

The absorbancy index, a_s is defined as :

$$a_s = \frac{A}{CL}$$

[where C = concentration of absorbing material in gms/litre

L =distance in cms travelled by the light in solution.]

If one wishes to express the light absorption in terms of the molar concentration of the absorbing material, the molar absorbancy index, a_m will be equal to :

$$a_m = a_s M$$

[where M = molecular weight of the absorbing material.]

Spectrophotometer. A spectrophotometer has two fundamental parts : a source of radiant light and a monochromator. Fig. 23–10 outlines Beckman spectrophotometer. It consists of a prism. This disperses the radiant energy into a spectrum. A slit is also fixed which selects a narrow portion of the spectrum. The standard cell or cuvette is placed in a light-tight unit. The incident light strikes the standard cell and emergent light passes into a photocell. The photocell changes the emerging light energy into measureable electrical energy.

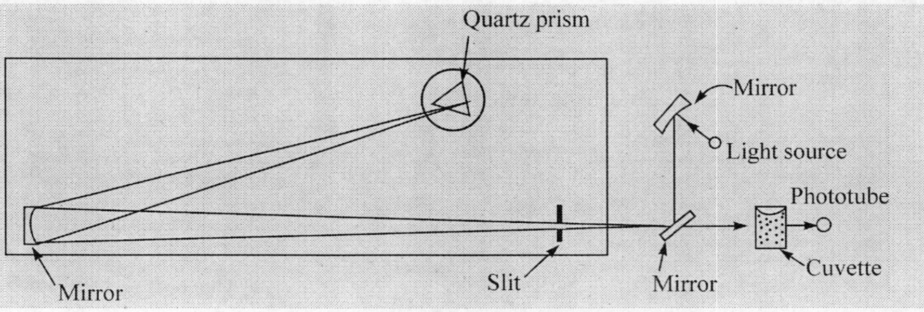

Fig. 23–10. Beckman spectrophotometer

(*Adapted from Conn EE and Stumpf PK, 1966*)

Applications. The technique finds many applications :

(*a*) It can be used in determining the concentration of a compound by measuring the optical density, provided the absorbancy index, a_s is known.

(*b*) The course of a reaction can be determined by measuring the rate of formation or disappearance of a light-absorbing compound.

(*c*) A compound can be identified by determining its absorption spectrum in the visible and ultraviolet region of the spectrum.

ELECTROPHORESIS

Principle. Migration of ions in an electric field at a definite *p*H is called **electrophoresis**. This method was developed by **Arne W.K. Tiselius** in 1937 and is based on the principle that the proteins migrate in an electric field except at the *p*H of their isoelectric point (refer page 184). And in a mixture of proteins, each protein with its characteristic electrical charge will respond differently to an applied electric potential. The rate of this electrophoretic migration (or 'mobility') depends on the pH of the medium, strength of the electric field, magnitude of the net charge on the molecule and the size of the molecule. A generalized diagram explaining the principle of gel electrophoresis for analyzing and sizing proteins is presented in Fig. 23–11.

ARNE W.K. TISELIUS

(LT, 1902-1971)

Tiselius, a Swedish physical biochemist, won the **1948 Nobel Prize in Chemistry** for the discovery of proteins in blood serum and for the development of electrophoresis as a technique for studying proteins.

Electrophoresis apparatus. Tiselius electrophoresis apparatus consists of compartmented cells forming a U tube, connected to an anode and a cathode compartment. The protein solution is placed at the bottom of the U tube. The U tube is kept immersed in a water bath at 4°C to minimize convection currents and the movement of the proteins is visualized by a Schlieren optical system.

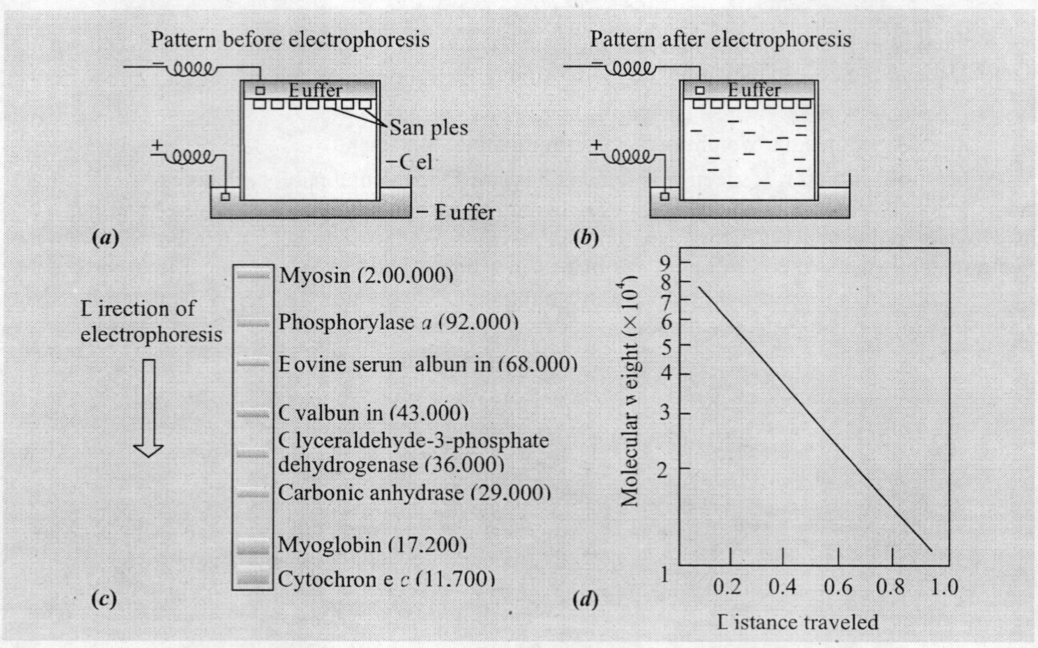

Fig. 23–11. Gel electrophoresis for analyzing and sizing proteins

(*a*) Apparatus for slab-gel electrophoresis. Samples are layered in the little slots cut in the top of the gel slab. Buffer is carefully layered over the samples, and a voltage is applied to the gel for a period of usually 1-4 h.

(*b*) After this time, the proteins have moved into the gel at a distance proportional to their electrophoretic mobility. The pattern shown indicates that different samples were layered in each slot.

(*c*) Results obtained when a mixture of proteins was layered at the top of the gel in phosphate buffer, pH 7.2, containing 0.2% SDS. After electrophoresis, the gel was removed from the apparatus and stained with Coomassie Blue. The protein and its molecular weight are indicated next to each of the stained bands.

(*d*) The logarithm of the molecular weight against the mobility (distance traveled) shows an approximately linear relationship.

(*Source: Data of K Weber and M Osborn*)

Consider that the protein solution is composed of 3 components, A, B and C and that these components carry charges of the same sign (either + or –) but of different magnitude. Before the electric field is applied, the solution will be homogeneous from ascending boundary to descending boundary. On the flow of current, the 3 components will separate. If the magnitude of the electric charge is in the order A > B > C, the components will separate as shown in Fig. 23–12. The faster moving component A will be present in pure form in the ascending boundary, as will C in the descending boundary. The component B will, however, always be mixed with A or with C. The rate of migration of the protein is measured by observing the movement of the boundary as a function of time. During electrophoretic migration, the concentration gradients will be set up at the boundaries. These gradients can be measured by optical systems since most of the proteins are colourless. Migration of coloured proteins like hemoglobin is, however, readily observed. The method described above is usually referred to as **free boundary electrophoresis** or simply **free electrophoresis** (*i.e.*, in free solution).

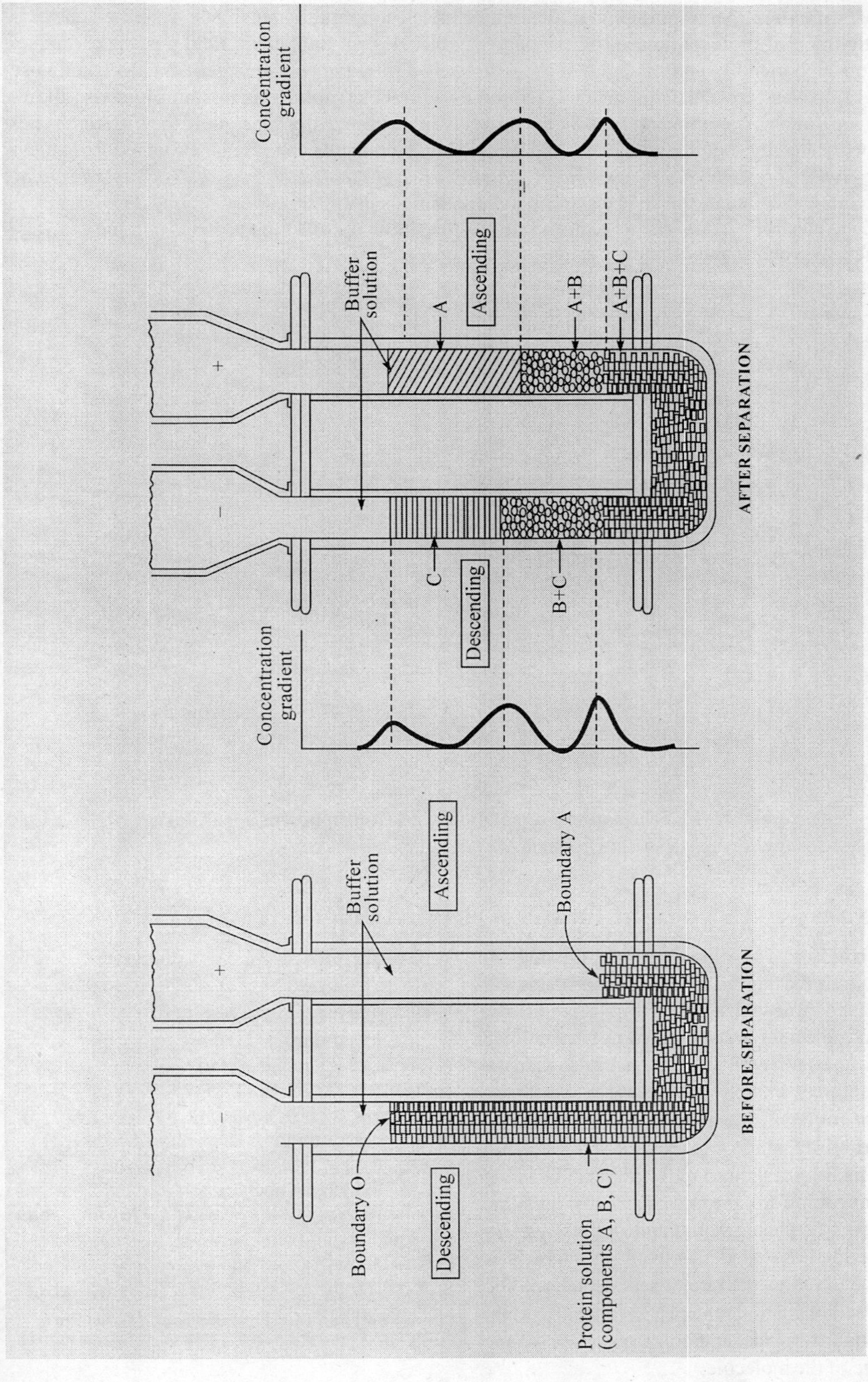

Fig.23–12. U tube of an electrophoresis cell before and after electrophoretic separation

Modification. An important modification of the electrophoretic technique described above is the migration of proteins (and of other charged molecules) in an electric field passing through a solution supported by inert materials such as moistened filter paper, starch gel, silica gel, cellulose sponges or glass powder. This method, termed as **zone electrophoresis** or **ionophoresis**, allows ready separation of components of different mobility into zones. After separation, the different zones may be located by staining with dyes. The individual components can also be extracted from these zones. *Ionophoresis technique is, thus, a combination of electrophoresis and chromatography.* A schematic diagram of the zone electrophoresis apparatus using gel appears in Fig. 23–13.

Zone electrophoresis has proved much useful for the study of serum proteins and the cleavage products of proteins and of nucleic acids.

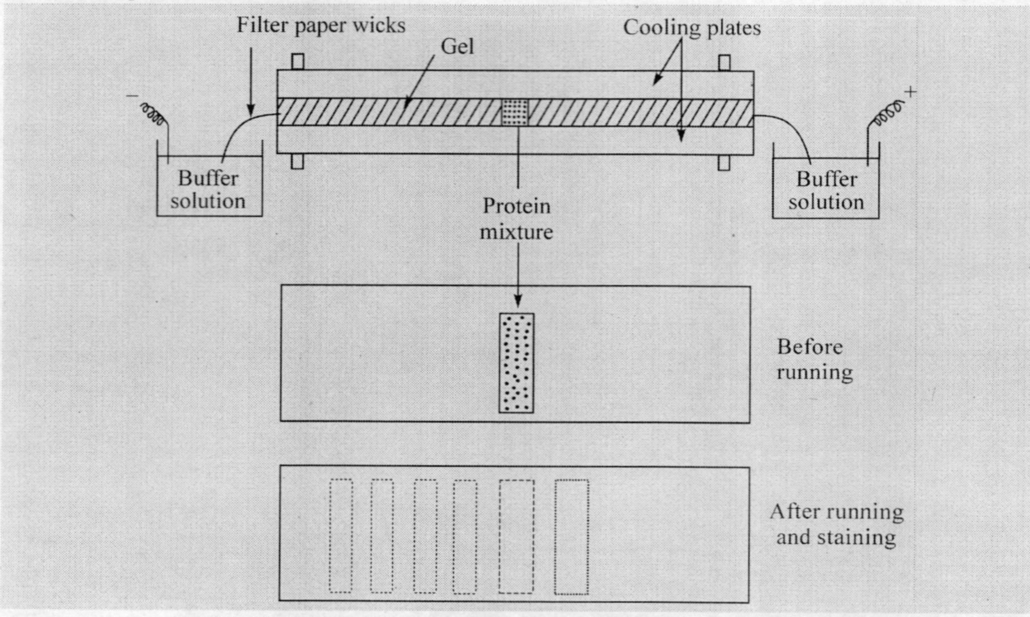

Fig. 23–13. Schematic representation of gel electrophoresis apparatus

[The two lower figures represent separation of protein mixtures.]

(*Redrawn from Fairley JL and Kilgour GL, 1966*)

ULTRACENTRIFUGATION

Principle. The ultracentrifuge method for determining the molecular weights of proteins was developed by **Svedberg.** In this method, the protein molecules are subjected to gravitational (centrifugal) forces greater in magnitude than the thermal forces whereby causing them to diffuses. Protein molecules are large-sized and hence described as macromolecules. Under high centrifugal forces, these molecules can be made to sediment toward one end of a centrifuge tube. By means of photographic and optical systems similar to those used in electrophoresis, it is possible to follow the rate of sedimentation. The rate of sedimentation for a protein under a certain centrifugal force depends on the density, shape and size of the molecule.

THEODORE SVEDBERG
(LT, 1884–1971)

Svedberg, a pioneer Swedish physical chemist, was responsible for much of the theoretical and practical development of centrifugation as a tool for studying large molecules. He was awarded the coveted **1926 Nobel Prize in Chemistry** for his work on colloids and macromolecular compounds. He also worked on nuclear chemistry, radiation biology, photographic processing and also worked out a method of making synthetic rubber during World War II.

The fundamental equation, devised by Svedberg, for the molecular weight (M) of a protein is :

$$M = \frac{RTs}{D(1 - Vp)}$$

where,
R = gas constant
T = absolute temperature
s = sedimentation constant
D = diffusion constant in cm_2/sec
V = partial specific volume of the protein
p = density of the solvent.

The rate of sedimentation is usually expressed in terms of sedimentation constant, s which has the dimensions of time per unit of gravitational field and which usually lies between 1×10^{-13} and 200×10^{-13} sec. For convenience, the sedimentation constant of 1×10^{-13} is referred to as 1 Svedberg unit (S) and all sedimentation constants are then expressed in Svedberg units. A *Svedberg unit* is, thus, defined as the velocity of sedimenting molecule per unit of gravitational field or 1×10^{-13} cm/sec/dyne/g. The S value of an organelle or macromolecule is related to its molecular weight and shape. Typical S values are 1.83 for cytochrome *c*, 4.4 for bovine serum albumin and 185 for TMV. For the sake of uniformity, the sedimentation coefficient is customarily corrected to the value that would be obtained at 20 °C in a solvent with the density and viscosity of pure water. This is symbolized $S_{20,w}$. Table 23-4 and Fig. 23-14 indicate the values of $S_{20,w}$ in Svedbergs for a variety of biological materials.

Table 23–4. Physical constants of some proteins

Protein	*Molecular mass (kD)*	*Partial specific volume, $V_{20,w}$ ($cm^3 . g^{-1}$)*	*Sedimentation coefficient, $S_{20,w}$ (S)*
Lipase (milk)	6.7	0.714	1.14
Ribonuclease A (bovine pancreas)	12.6	0.707	2.00
Cytochrome *c* (bovine heart)	13.4	0.728	1.71
Myoglobin (horse heart)	16.9	0.741	2.04
α-chymotrypsin (bovine pancreas)	21.6	0.736	2.40
Diphtheria toxin	70.4	0.736	4.60
Cytochrome oxidase (*P. aeruginosa*)	89.8	0.730	5.80
Lactate dehydrogenase H (chicken)	150	0.740	7.31
Catalase (horse liver)	222	0.715	11.20
Fibrinogen (human)	340	0.725	7.63
Glutamate dehydrogenase (bovine liver)	1015	0.750	26.60
Turnip yellow mosaic virus protein	3013	0.740	48.80

(*Adapted from Smith MH, 1970*)

The diffusion coefficient or constant D may be defined as the quantity of material that diffuses per second across a surface 1 cm^2 in area. The partial specific volume V is equal to the reciprocal of the density of the molecule. *For most proteins, however, the value of V is in the range of 0.70 to 0.75.*

Ultracentrifuge. Using the above principle, **Svedberg** in 1925 devised an instrument called ultracentrifuge (Figs. 23-15 and 23-16). By this instrument, it is possible to get information on the purity of a protein, its, molecular weight and approximate dimensions. Two types of ultracentrifuges are now known :

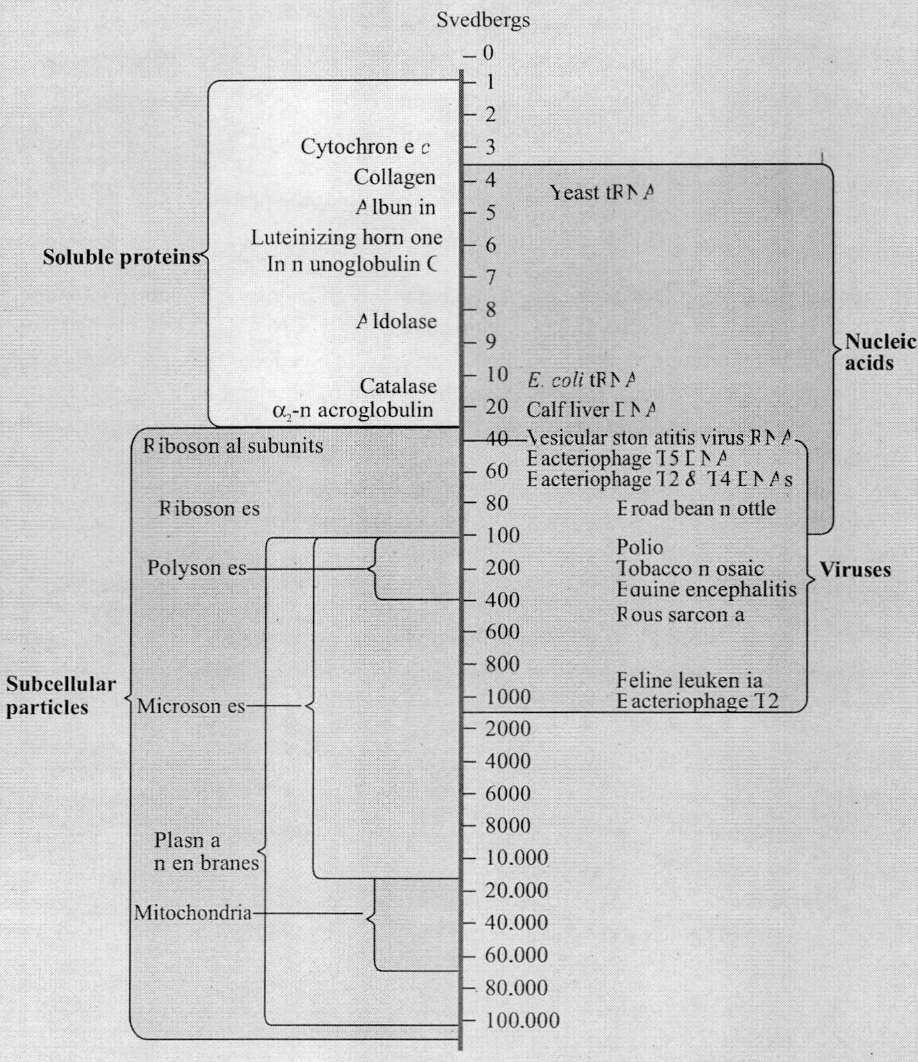

Fig. 23–14. The sedimentation coefficients in Svedbergs for some biological materials

(*Courtesy : Beckman Instruments, Inc.*)

(*a*) *Low speed type*, with centrifugal forces of the order of 100 to 8,000 times gravity and having the maximum speed of about 40,000 *rpm*.

(*b*) *Oil-turbine type*, with centrifugal forces of the order of 8,000 to 400,000 times gravity and having the maximum speed of about 80,000 *rpm*.

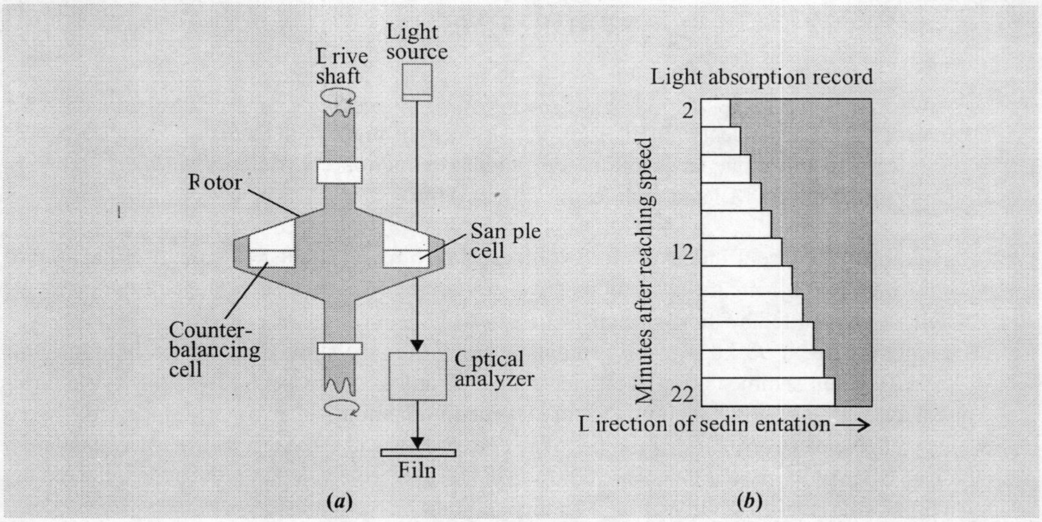

Fi.g 23–15. Apparatus for analytical ultracentrifugation

(*a*) The centrifuge rotor and method of making optical measurements.

(*b*) The optical recordings as a function of certifugation time. As the light-absorbing molecule sediments, the solution becomes transparent.

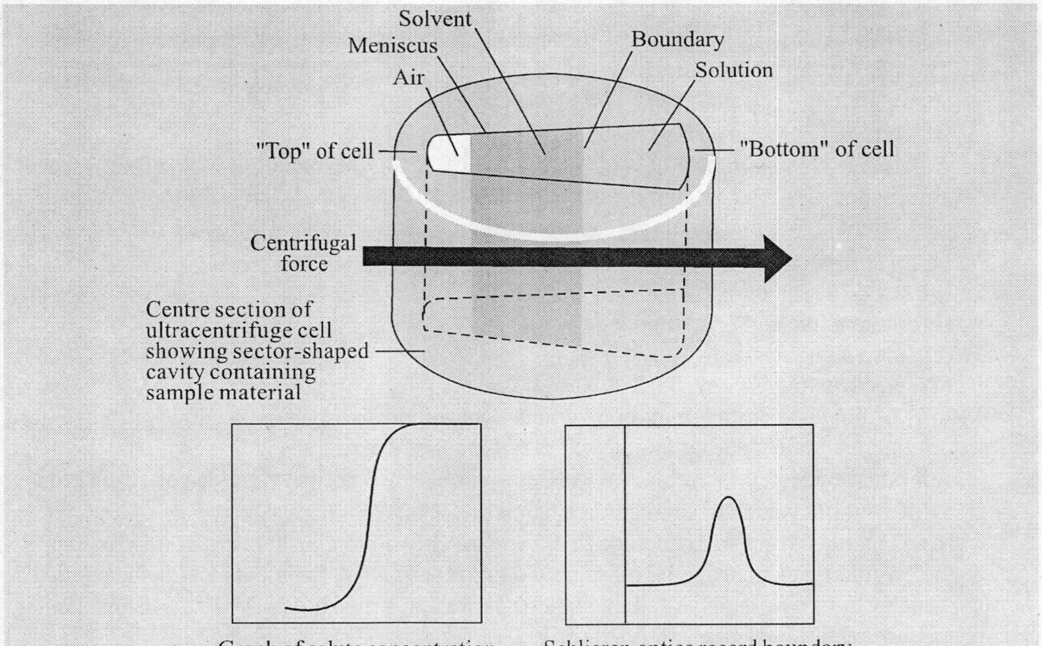

Fi.g 23–16. Centre section of ultracentrifuge cell

[The solute molecules which are initially evenly distributed are forced towards the botom of the cell by the centrifugal force. This migration creates at the top of the cell a region that is without solute molecules and contains only solvent molecules. A boundary is set up in the cell between solvent and solution in which concentration varies with the distance from the axis of rotation. The measurement of boundary movement which represents its movement of protein molecules, is the basis of analytical ultracentrifugation.]

(*Courtesy : Beckman Instruments, Inc.*)

PROBLEMS

1. The absorbance A of a solution is defined as

$$A = \log_{10} (I_0/I)$$

in which I_0 is the incident light intensity and I is the transmitted light intensity. The absorbance is related to the molar absorption coefficient (extinction coefficient) $\in$ (in M^{-1} cm^{-1}), concentration c (in M), and path length l (in cm) by

$$A = \in lc$$

The absorption coefficient of myoglobin at 580 nm is 15,000 M^{-1} cm^{-1}. What is the absorbance of a 1 mg ml^{-1} solution across a 1-cm path ? What percentage of the incident light is transmitted by this solution ?

2. Tropomyosin, a 93-kd muscle protein, sediments more slowly than does hemoglobin (65 kd). Their sedimentation coefficients are 2.6S and 4.31S, respectively. Which structural feature of tropomyosin accounts for its slow sedimentation ?

3. The relative electrophoretic mobilities of a 30-kd protein and a 92-kd protein used as standards on an SDS-polyacrylamide gel are 0.80 and 0.41, respectively. What is the apparent mass of a protein having a mobility of 0.62 on this gel ?

4. Suppose that an enzyme is dissociated into 4 identical subunits and that you want to test for the enzymatic activity of the individual subunits. However, you must be sure that there are no tetramers remaining in the sample.

(*a*) What chromatographic procedure would you choose to free the monomers from the tetramers ?

(*b*) How would you know where the tetramer would be if it were present ?

5. The density of DNA in CsCl is approximately 1.7 g/cm^3 and that of most proteins is approximately 1.3 g/cm^3. What would you expect to be the density of a typical bacteriophage which is 50% protein and 50% DNA ?

6. A solution of a substance at a concentration of 32 μg/ml, having a MW of 423, has an absorbance of 0.27 at 540 nm measured with a 1-cm light path. What is the molar extinction coefficient at 540 nm, assuming that Beer's law is obeyed at an absorbance of 0.27 ?

7. Details of the citric acid cycle were worked out by the use of :

(*a*) x-ray crystallography

(*b*) radioactive carbon compounds

(*c*) ultracentrifugation

(*d*) electron microscopy

8. In two-dimensional chromatography and electrophoresis, such as is employed in finger-printing technique, does it matter which is done first ? Explain.

9. A mixture of proteins is subjected to electrophoresis in 3 polyacrylamide gels, each having a different pH value. In each gel five bands are seen.

(*a*) Can one reasonably conclude that there are only 5 proteins in the mixture. Explain.

(*b*) Would the conclusion be different if a mixture of DNA fragments was being studied ?

10. Which would have a higher sedimentation coefficient, a rigid rod or flexible rod, both having the same radius and same mass ?

11. Name any one biochemical synthesized in the following :

(a) nucleus

(b) mitochondria

(c) rough endoplasmic reticulum

(d) Golgi bodies

(e) supernatant

ANSWERS TO PROBLEMS

1. A 1 mg/ml solution of myoglobin (17.8 kd) corresponds to 5.62×10^{-5} M. The absorbance of a 1-cm path length is 0.84, which corresponds to an I_0/I ratio of 6.96. Hence, 14.4% of the incident light is transmitted.
2. Tropomyosin is rod shaped, whereas hemoglobin is approximately spherical.
3. 50 kd.
4. (*a*) Gel chromatography or gel electrophoresis.

   (*b*) Add proteins of known molecular weight and obtain a standard curve in which either volume of the effluent (for chromatography) or the distance migrated (for electrophoresis) is plotted against molecular weight.
5. $(1.7 + 1.3)/2 = 1.5$ g/cm^3.
6. The molarity of the solution is (1/423) = 0.032. The molar extinction coefficient is 0.27/(0.032/423) = 3,569.
7. (*b*) radioactive carbon compounds
8. No. To move an object *x* units north and the *y* units east brings the object to the same place reached by moving it first *y* units east and then *x* units north.
9. (*a*) No, since differences in shape can affect mobility, *i.e.*, two proteins differing in both molecular weight and shape might have the same electrophoretic mobility.

   (*b*) Yes.
10. A flexible rod encounters less friction when moving through a fluid.
11. (*a*) DNA

    (*b*) ATP

    (*c*) Enzyme proteins

    (*d*) Insulin from proinsulin

    (*e*) NADPH

CHAPTER

24 Biochemistry Experiments

INSIDE

- Experiments on Carbohydrates (Expt. Nos. 1–5)
- Experiments on Nucleic Acids (Expt. Nos. 6–7)
- Experiment on Amino Acids (Expt. No. 8)
- Experiments on Proteins (Expt. Nos. 9–10)
- Experiments on Enzymes (Expt. Nos. 11–13)
- Experiments on Lipids (Expt. Nos. 14–18)
- Experiments on Vitamins (Expt. Nos. 19–21)
- Experiments on Milk (Expt. Nos. 22–23)

Application of computer molecular graphics for producing a three-dimensional model of a protein

CARBOHYDRATES

The carbohydrates, often termed as sugars, are the '*staff of life*' for most organisms. On the basis of mass, they are the most abundant class of biomolecules in nature. They are widely-distributed molecules in both plant and animal tissues. Plants are considerably richer in carbohydrates in comparison to the animals. They are indispensable for living organisms, serving as skeletal structures in plants and also in insects and crustaceans. They also occur as food reserves in the storage organs of plants and in the liver and muscles of animals. The term 'carbohydrate' was originally coined for this class of compounds as most of them were 'hydrates of carbon' or could be represented by the general formula, $C_x(H_2O)_y$. Later, it was found that some of them (such as deoxyribose, $C_5H_{10}O_4$) do not have the required ratio of hydrogen to water. In addition, certain other carbohydrates are now known to possess nitrogen (*e.g.*, glucosamine, $C_6H_{13}O_5$ N), phosphorus or sulfur and obviously do not coincide with above general formula. Furthermore, formaldehyde (H.CHO or CH_2O), acetic acid (CH_3.COOH or $C_2H_4O_2$) etc, which have C, H and O and the ratio of H : O is also the same as in water, but are not a carbohydrates. Hence, *the continued usage of the term 'carbohydrate' is for convenience rather than exactness.* To accommodate a wide variety of compounds, the carbohydrates are nowadays broadly defined as polyhydroxy aldehydes or ketones and their derivatives or as substances that yield one of these compounds on hydrolysis.

EXPERIMENT No. 1

A. Object. To perform colour test for carbohydrates, in general (Molisch test).

B. Requirements. Test tube, test tube holder, test solution, conc. H_2SO_4, α-naphthol solution.

C. Principle. Molisch test in based on the hydrolyzing and dehydrating action of conc. H_2SO_4 on carbohydrates. The acid hydrolyzes glycosidic bonds to give the monosaccharides, which are then dehydrated to 5-hydroxymethylfurfural (hexoses) or furfural (pentoses). These products then combine with sulphonated α-naphthol to give a purple complex. As the acid hydrolyzes acetal groups present, this fact makes this test the most general of all the tests for carbohydrates.

With *hexoses,* the reduction with strong mineral acid (*e.g.,* sulphuric acid) leads to the formation of 5-hydroxymethylfurfural, which on further heating is transformed to levulinic acid:

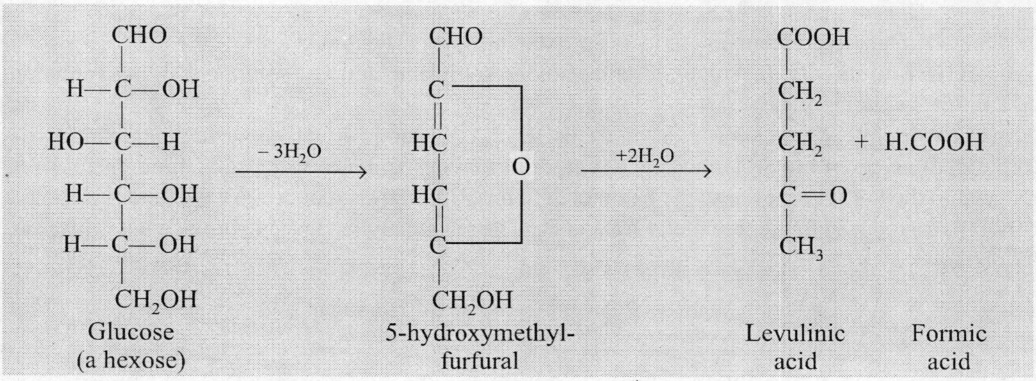

With *pentoses,* a product furfural is obtained on heating:

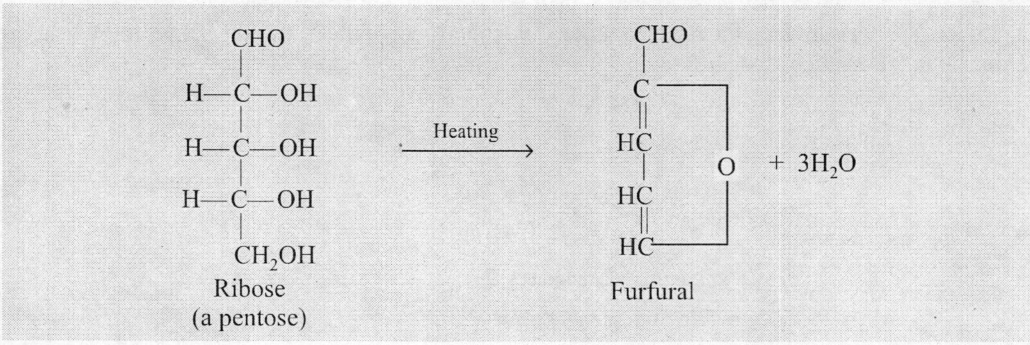

Tetroses and *trioses,* however, do not undergo this reaction as they do not possess the necessary minimum 5 carbon atoms for furfural formation.

D. Method. Take about 2 ml of test solution (or the extract of the material) and add 2–3 drops of freshly-prepared alcoholic α-naphthol solution (1% α-naphthol in 70% ethanol) to it. Then carefully add 1 ml concentrated H_2SO_4 solution down the inside of the test tube so as to form two layers (Do not allow the 2 liquids to mix). Observe any colour change at the junction of the two liquids. The formation of a purple colour zone at the junction of the two liquids indicates the presence of carbohydrates. Repeat the test, using water and 1% glucose in separate test tube instead of the carbohydrate solution.

E. Observations.

Name of test	*Experiment*	*Observations*	*Inference*
Molisch test	Add 2–3 drops of α-naphthol to 2 ml of glucose solution. Then pour gently 1 ml of conc. H_2SO_4 by the side of test tube.	A purple colour zone develops at the junction of the 2 liquids.	Presence of carbohydrate confirmed.

F. Result. The test solution is that of a carbohydrate.

EXPERIMENT No. 2

A. Object. To perform colour tests for reducing sugars (Fehling's test).

B. Requirements. Test tubes, test tube holder, spirit lamp, glucose solution, Fehling's solution A, Fehling's solution B.

C. Principle. Metal hydroxides like $Cu(OH)_2$, AgOH and $Bi(OH)_3$ oxidize the free aldehyde (or ketone) group of mutarotating mono- and disaccharides and are, at the same time, themselves reduced to the lower oxides or to the free metals. Thus, $Cu(OH)_2$ is reduced to Cu_2O and AgOH and $Bi(OH)_3$ are reduced to the free metal Ag and Bi. The reaction with cupric hydroxide takes place as follows:

$$\text{Reducing sugar} + \underset{\text{(Blue)}}{2Cu^{2+}} \longrightarrow \text{Oxidized sugar} + 2Cu^{+}$$

$$2Cu^{+} + 2OH^{-} \longrightarrow \underset{\text{(Yellow)}}{2Cu.OH} \longrightarrow \underset{\text{(Red)}}{Cu_2O} + H_2O$$

Cupric ion (Cu^{++}) is, in fact, the most common oxidizing agent of these. It is the active ingredient in Fehling's and Benedict's reagents.

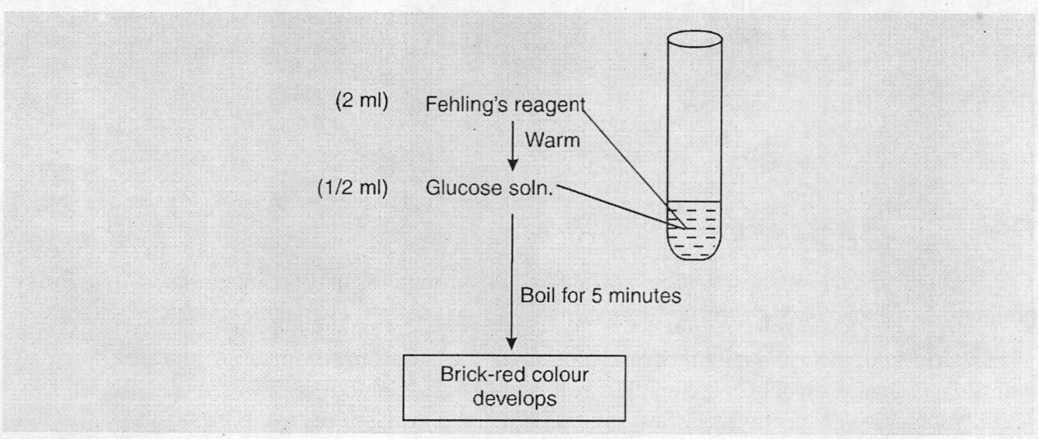

D. Method.

Preparation of Fehling's reagent. Fehling's reagent consists of 2 solutions: Fehling's A and Fehling's B.

Fehling's A solution. It is prepared by dissolving 35 g of Cu SO_4. $5H_2O$ in 500 ml of water.

Fehling's B solution. It is prepared by dissolving 50 g of KOH and 173 g of sodium potassium tartarate (also called Rochelle salt) in 500 ml of water. The 2 solutions are kept separate and mixed only just before use. This is because KOH consistently keeps the medium alkaline.

Take 2-3 ml of Fehling's solution in a test tube and warm on a spirit lamp. Add few drops of glucose solution. Boil it for few minutes. The result is the formation of a brownish-red precipitate.

E. Observations.

Name of test	*Experiment*	*Observations*	*Inference*
Fehling's test	Warm 2-3 ml of Fehling's solution and add glucose solution and boil it.	Solution turns brownish-red or *brick-red.*	Presence of reducing sugar confirmed.

F. Result. The given solution is that of a reducing sugar.

G. Precaution. If to the glucose solution, solution A and solution B (the two component solutions of Fehling's reagent) are added separately, usually a *green* colour develops after boiling, instead of a brick-red colour.

EXPERIMENT No. 3

A. Object. To perform colour tests for reducing sugars (Benedict's test).

B. Requirements. Test tubes, test tube holder, spirit lamp, glucose solution, Benedict's solution.

C. Principle. Metal hydroxides like $Cu(OH)_2$, AgOH and $Bi(OH)_3$ oxidize the free aldehyde (or ketone) group of mutarotating mono- and disaccharides and are, at the same time, themselves reduced to the lower oxides or to the free metals. Thus, $Cu(OH)_2$ is reduced to Cu_2O and AgOH and $Bi(OH)_3$ are reduced to the free metal Ag and Bi. The reaction with cupric hydroxide takes place as follows:

$$\text{Reducing sugar} + \underset{\text{(Blue)}}{2Cu^{2+}} \longrightarrow \text{Oxidized sugar} + 2Cu^{+}$$

$$2Cu^{+} + 2OH^{-} \longrightarrow \underset{\text{(Yellow)}}{2Cu.OH} \longrightarrow \underset{\text{(Red)}}{Cu_2O} + H_2O$$

Cupric ion (Cu^{++}) is, in fact, the most common oxidizing agent of these. It is the active ingredient in Fehling's and Benedict's reagents.

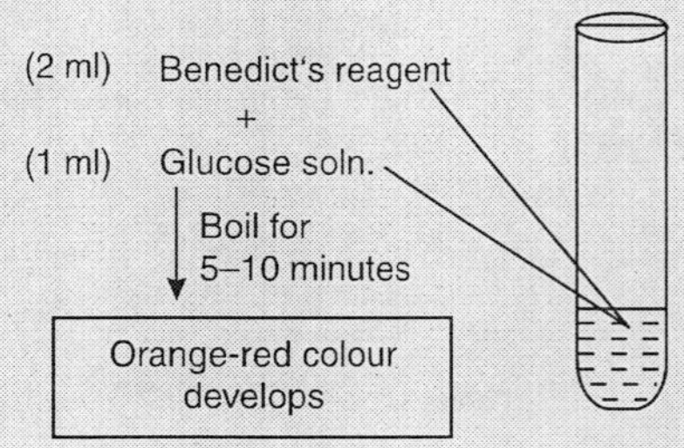

D. Method.

Preparation of Benedict's reagent. Dissolve 173 g of sodium citrate and 100 g of sodium carbonate in about 800 ml of warm (about 50°C) water. Filter the solution. In the mean time, dissolve 17.3 g of copper sulphate in about 100 ml of water. Add this solution to the first solution slowly while stirring continuously. Make up the total volume upto 1000 ml. This is Benedict's reagent and should be mixed well before use. Sodium carbonate reduces alkalinity to such an extent that, unlike the Fehling's reagent, the two solutions ($CuSO_4$ and sodium citrate) here are kept mixed indefinitely as one solution.

Take 5 ml of Benedict's solution in a test tube and add 2 ml. of test solution (glucose). Boil it (or better, place the test tube in a boiling water bath for about 10 minutes). A change in colour from blue to shades of green or orange-red indicates the presence of carbohydrate (simple sugars).

E. Observations.

Name of test	*Experiment*	*Observations*	*Inference*
Benedict's test	Add 2 ml of glucose solution in 5 ml. of Benedict's solution and boil it.	Solution turns green or *orange-red*	Presence of reducing sugar confirmed.

F. Result. The given solution is that of a reducing sugar.

EXPERIMENT No. 4

A. Object. To perform the colour test for nonreducing sugars.

B. Requirements. Test tube, test tube holder, glucose solution starch solution, Fehling's solution, Benedict's solution.

C. Principle. Nonreducing sugars like sucrose, starch, glycogen, inulin etc. upon hydrolysis, yield their constituent monosaccharide units. The monosaccharides are reducing sugars. Hence, the nonreducing sugars give a positive test with reagents like Fehling's, Benedict's and Barfoed's only when they are hydrolyzed to yield the reducing sugars. The hydrolysis reaction with sucrose and starch takes place as follows:

$$\underset{\text{Sucrose}}{C_{12}H_{22}O_{11}} + H_2O \xrightarrow[\text{with con. HCl}]{\text{Hydrolysis}} \underset{\text{Glucose}}{C_6H_{12}O_6} + \underset{\text{Fructose}}{C_6H_{12}O_6}$$

$$\underset{\text{Starch}}{(C_6H_{10}O_5)_x} + (x-1)\,H_2O \xrightarrow[\text{with con. HCl}]{\text{Hydrolysis}} \underset{\text{Glucose}}{x\,C_6H_{12}O_6}$$

D. Method. The unknown solution (starch) can be *hydrolyzed* by mixing the equal volume of this solution and con. HCl. It is boiled for five minutes and neutralized with sodium bicarbonate. Now it is tested for the presence of reducing sugars as in Experiments No. 2 and 3.

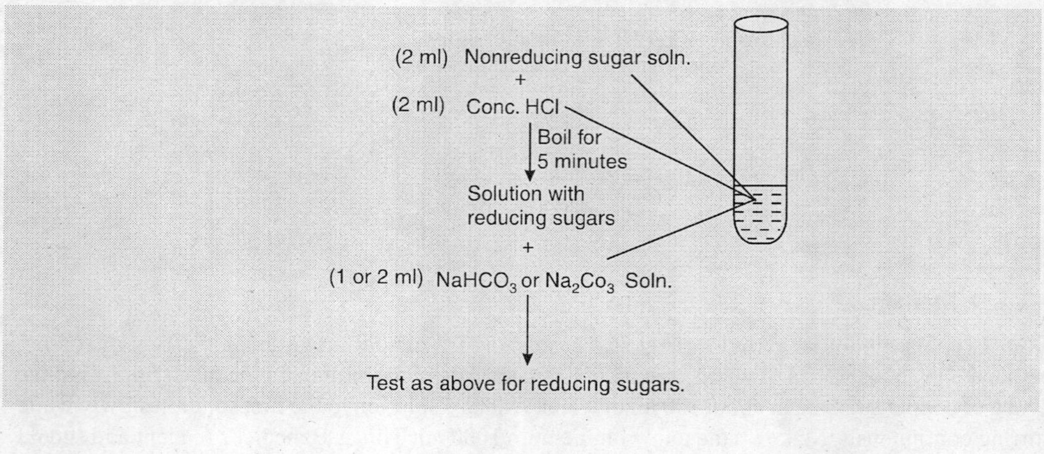

E. Observations.

Name of test	*Experiment*	*Observations*	*Inference*
Hydrolysis	Starch soln. + Con. HCl Boil for five minutes and then neutralize with $NaHCO_3$.	–	Starch will be hydrolyzed into glucose units.
Fehling's test	Warm 2-3 ml of Fehling's soln. and add hydrolyzed soln. and boil it.	Brownish-red precipitate.	Presence of reducing sugar confirmed.
Benedict's test	Add 2ml. of hydrolyzed soln. in 5 ml of Benedict's soln. and boil it.	Solution turns green or yellow-red.	Presence of reducing sugar confirmed.

F. Result. The given solution is a carbohydrate (nonreducing sugar) which is ensured by hydrolyzing it to reducing sugar and then testing.

EXPERIMENT No. 5

A. Object. To perform colour test for polysaccharides such as starch and glycogen (iodine test).

B. Requirements. Test tube, test tube holder, 1% starch solution, 1% glycogen solution, dil. HCl, iodine solution.

C. Principle. Most of the polysaccharides adsorb iodine to form coloured complexes. *Starch* is readily hydrolyzed by mineral acid with the final production of glucose. The course of hydrolysis may be followed by the gradual change in colour when treated with iodine (blue → purple → red → none) and also by the increase in concentration of reducing sugar.

Glycogen is also a nonreducing sugar and gives red colour with iodine. The red colour, however, disappears on boiling and reappears on cooling.

D. Method.

Preparation of Iodine solution. I_2 solution is prepared by dissolving about 8 g of solid iodine in 100 ml of 30 g/1 of potassium iodide solution.

Take about 2 ml of test solution (approximately 1% solution of starch) in a test tube. Add 1 or 2 drops of dil. HCl to this solution to acidify it. Then add 3-4 drops of iodine solution. A blue or purple colour appears in case of *starch* solution.

If the test solution is that of *glycogen,* a red colour appears. In both cases, the colour can be compared with the colour developed with water and iodine.

E. Observations.

Name of test	*Experiment*	*Observations*	*Inference*
Iodine test	2 ml of 1% starch soln. + 1 or 2 drops of HCl + 3 or 4 drops of I_2 soln.	A blue or purple colour develops.	Presence of starch confirmed.
Iodine test	2 ml of 1 % glycogen soln. + 1 or 2 drops of HCl + 3 or 4 drops of I_2 soln.	A red colour develops.	Presence of glycogen confirmed.

F. Result. The test solution is that of starch and glycogen.

NUCLEIC ACIDS

The nucleic acids (DNA and RNA) are the molecular repositories for genetic information and are jointly referred to as the '*molecules of heredity*'. Nucleic acids are macromolecules present in most living cells either in the free state or bound to proteins as nucleoproteins. Both DNA and RNA are composed of 3 component units: a phosphoric acid, a pentose sugar and 4 nitrogenous bases Nucleic acids can be tested on the basis of their pentose sugar which is either ribose (in RNA) or deoxyribose (in DNA).

EXPERIMENT No. 6.

A. Object. To perform colour test for both DNA and RNA (orcinol test).

B. Requirements. Test tube, test tube holder, ferric chloride, conc. H_2SO_4, orcinol, alcohol, nucleic acid saline solution, water bath.

C. Principle. Orcinol test is a general test for pentoses and hence both DNA and RNA give positive test with this method. The theory of orcinol test is as follows. When pentoses (*i.e.*, nucleic acids) are heated with conc. HCl, an aldehyde product furfural is produced. Furfural, so produced, reacts with orcinol in the presence of $FeCl_3$ (as a catalyst) to produce a green-coloured complex.

D. Method.

Preparation of orcinol solution. Orcinol solution is prepared by dissolving 250 mg of $FeCl_3$ in about 250 ml of conc. H_2SO_4 and adding to it 10 ml of 6% orcinol prepared in alcohol.

Take a commercial sample of nucleic acid and make a solution of the sample (approximately 1 mg / 5 ml) in saline water. Then take about 2 ml of nucleic acid solution in a test tube and add 3 ml of orcinol reagent to it. Heat on boiling water bath for about 15 minutes. Cool it. A green colour appears in the solution, indicating the presence of nucleic acids.

E. Observations.

Name of test	*Experiment*	*Observations*	*Inference*
Orcinol test	Nucleic acid saline solution (2 ml) + orcinol reagent (3 ml) ↓ Heat for 15 minutes. Then cool it.	A green colour appears in the solution.	Presence of nucleic acids confirmed.

F. Result. The test solution is that of nucleic acids.

EXPERIMENT No. 7

A. Object. To perform colour test for DNA (diphenylamine test).

B. Requirements. Test tube, test tube holder, diphenylamine, glacial acetic acid, conc. H_2SO_4, DNA saline solution, water bath.

C. Principle. This is a general test for all the compounds containing 2′-deoxypentose. In a DNA sample, the 2′-deoxyribose of the purine nucleotides gives this reaction. In this reaction, when an acidic solution of DNA is heated with diphenylamine, the 2′-deoxypentose of DNA is converted to β- hydroxylevulinaldehyde, which then reacts with the additional diphenylamine to produce a blue- or bark-coloured complex.

D. Method.

Preparation of diphenylamine solution. It is prepared by dissolving about 4 g of diphenylamine in 200 ml of glacial acetic acid and then adding about 5 ml of conc. H_2SO_4 to the mixture.

For testing, prepare a saline solution of DNA, as described in Experiment No. 6. Take about 2 ml of this solution and add about equal amount of freshly-prepared diphenylamine solution. Heat the mixture in a boiling water bath for about 15 minutes and then cool it. Appearance of a blue colour indicates the presence of DNA.

E. Observations.

Name of test	*Experiment*	*Observation*	*Inference*
Diphenylamine test	DNA saline solution (2 ml) + diphenylamine solution (2 ml) ↓ Heat for 15 minutes. Then cool it.	A blue colour appears in the solution.	Presence of DNA confirmed.

F. Result. The test solution is that of DNA.

AMINO ACIDS

Amino acids are nitrogenous compounds having both an acidic carboxyl (– COOH) and a basic amino (–NH_2) group. There are 20 different amino acids which are component of thousands of proteins present in living organisms. Amino acids are colourless crystalline substances and are soluble in polar solvents (water, ethanol) but insoluble in nonpolar solvents (benzene and ether).

EXPERIMENT No. 8

A. Object. To separate and identify a mixture of amino acids by paper chromatography.

B. Requirements. Chromatography paper, gas jar, amino acids, phenol, ninhydrin.

C. Principle. It is based on the principle that various substances have different degree of solubility in a particular solvent. Chromatography may be defined as the technique of separation according to their partition coefficients below (*i.e.*, their relative solubilities in) two immiscible phases. In this method, the separation of the components of a mixture is a function of their different affinities for a fixed or *stationary phase* (such as a solid or liquid) and their differential solubility in a moving or *mobile phase* (such as a liquid or a gas). Separation starts to occur when one component is held more firmly by the stationary phase than the other which tends to move on faster in mobile phase. Thus, the underlying principle of chromatography is first to adsorb the components of a mixture on an insoluble material and then to differentially remove (or *elute*) these components one by one with suitable liquid solvents.

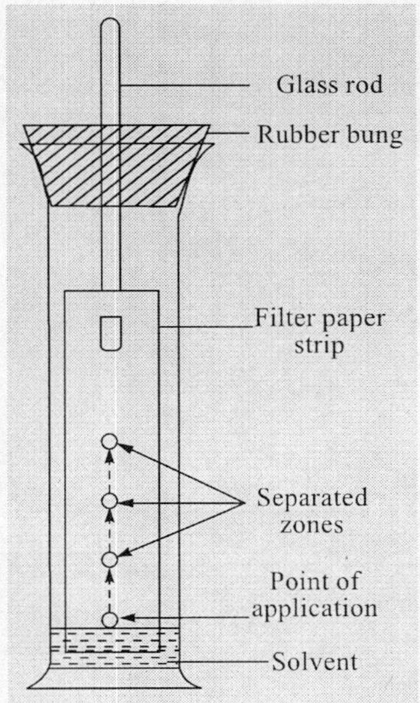

One-dimensional paper chromatography
(Ascending type)

D. Method. In this method, a drop of solution containing a mixture of amino acids (or other compounds to be separated) is applied at a marked point about 2 cm from one end of a strip of filter paper. Whatman No.1 paper* is widely used for this purpose.

* The chemical composition of **Whatman filter paper No. 1** is as follows:

α- cellulose	98 – 99%
β- cellulose	0.3 – 1%
Pentosans	0.4 – 0.8%
Ash	0.07 – 0.1%
Ether soluble matter	0.015 – 0.1%

Other grades of Whatman paper are also available. **Whatman No. 3 MM** is a thick paper and is best employed for separating large quantities of material; the resolution is, however, inferior to Whatman No. 1. For a rapid separation, **Whatman Nos. 4 and 5** are convenient, although the spots are less well-defined. In all cases, the flow rate is faster in the '*machine direction*', which is normally noted on the box containing the papers. The paper may be impregnated with a buffer solution before use or chemically modified by acetylation. Ion exchange papers are also available commercially.

The filter paper is then dried and '*equilibrated*' by putting it into an air-tight cylindrical jar which contains an aqueous solution of a solvent. 80% Phenol is used as a solvent (other solvents such as *n*-butanol: acetic acid: water (4 : 1 : 5) and collidine: water are also used). The end of the filter paper nearest the applied drop is inserted into the solvent mixture at the bottom of the jar. The paper is suspended in such a manner so that it hangs freely without touching the sides of the container. Thus, the solvent will ascend into the paper and this process is, therefore, termed *ascending chromatography*.

The strip is removed when the solvent has migrated over most of the available space. The distance to which the solvent has run is marked with a lead pencil. The completed chromatogram is colourless with no indication of the presence of any compounds. Such a chromatogram is said as '*undeveloped*'. For locating the various compounds, the filter paper strip is first dried, then sprayed with 0.5 % ninhydrin in acetone and at last heated for a few minutes at 80–100°C. The reaction occurs and the coloured spots (from pink to violet) appear at the sites of the amino acids. Such a chromatogram is now called as '*developed*'. The number of coloured spots corresponds to the number of components (*i.e.*, amino acids) present in the mixture.

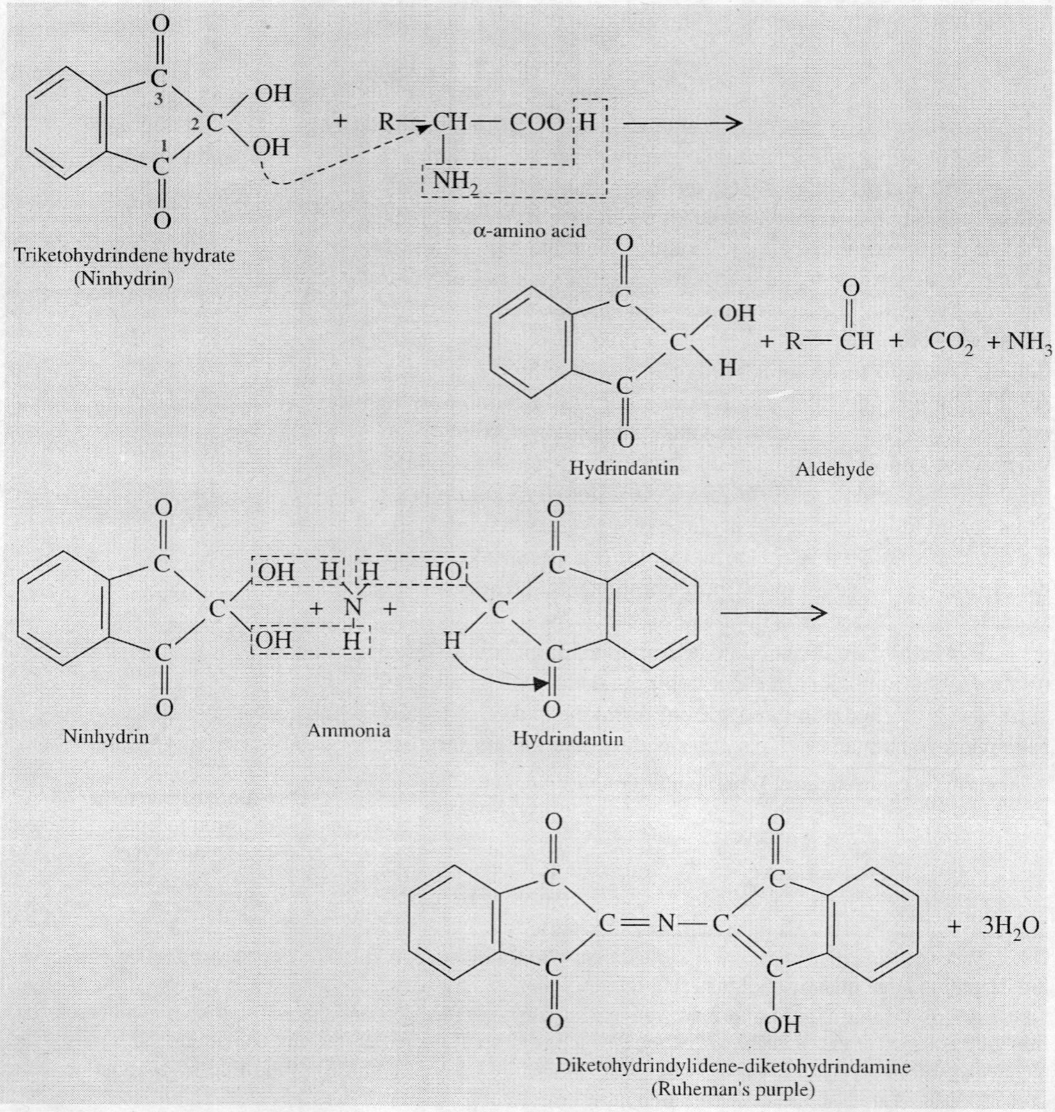

The net equation may, thus, be written as follows :

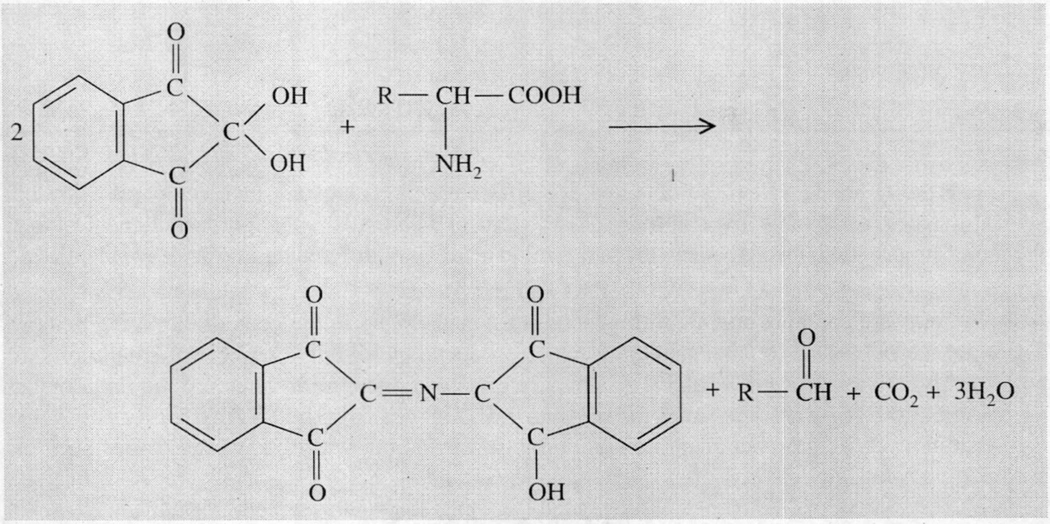

The distance travelled by a particular component of a mixture (or solute) is used to identify it. The ratio of the distance travelled by a component (*i.e.* amino acid) to that travelled by the solvent front, both measured from the marked point of application of the mixture, is called the *resolution front* (R_f) value for that component or amino acid. Thus,

$$R_f = \frac{\text{Distance (in cm) from origin run by the compound}}{\text{Distance (in cm) from origin run by the solvent}}$$

If the spots in the 'developed' chromatogram overlap and do not tally with reference /standard values, that shows that the resolution of the mixture is not complete. This can be done by following two-dimensional chromatography using a second solvent (phenol: water) and changing the direction of the paper.

Preparation of phenol : water solvent. Add 125 ml of water to 500 g of phenol; allow to stand overnight. Add few drops of ammonia before use and mix thoroughly.

The R_f values of a particular amino acid in different solvents usually differ from each other.

R_f values of amino acids in various solvents systems

S. N.	*Amino Acid*	*Phenol : Water*	*Collidine : Water*	*Butanol : Acetic Acid : Water, BAW*
1.	Glycine	0.36	0.26	0.26
2.	Alanine	0.55	0.32	0.38
3.	Valine	0.72	0.43	0.60
4.	Leucine	0.80	0.55	0.73
5.	Isoleucine	0.83	0.53	0.72
6.	Serine	0.30	0.30	0.27
7.	Threonine	0.43	0.32	0.35
8.	Cystine	0.24	0.11	0.08
9.	Methionine	0.74	0.53	0.55
10.	Proline	0.88	0.34	0.43
11.	Aspartic acid	0.22	0.23	0.24
12.	Glutamic acid	0.23	0.27	0.30
13.	Phenylalanine	0.83	0.54	0.68

14.	Tyrosine	0.55	0.59	0.45
15.	Tryptophan	0.71	0.59	0.50
16.	Histidine	0.62	0.30	0.20
17.	Arginine	0.54	0.17	0.20
18.	Lysine	0.41	0.11	0.14

(*Adapted from Fruton JS and Simmonds S, 1958*)

F. Result. The R_f values, so obtained, for different amino acids are then compared with the standard values (refer Table given above). The R_f value of a particular spot approaching nearest to a particular amino acid is taken to mean the same amino acid. Thus, different amino acids are identified.

PROTEINS

Proteins are complex organic nitrogenous substances found in the cells of the living beings. These are the *polymers*, *i.e.*, chain-like molecules produced by joining a number of small units of amino acids called *monomers*. Proteins occupy a central position in the architecture and functioning of living matter. Proteins are present in all tissues of the living body but the demonstration of the presence of protein is normally carried out with the storage tissues such as seeds.

EXPERIMENT No. 9

A. Object. To perform colour test for proteins, in general (Biuret test).

B. Requirements. Test tube, test tube holder, protein solution, $CuSO_4$ solution, NaOH solution.

C. Principle. Compounds containing peptide bonds produce a characteristic purple colour when treated with an alkaline 0.2% copper sulfate solution (or biuret reagent). This reaction is termed as 'biuret reaction' since it is also given by the substance biuret.

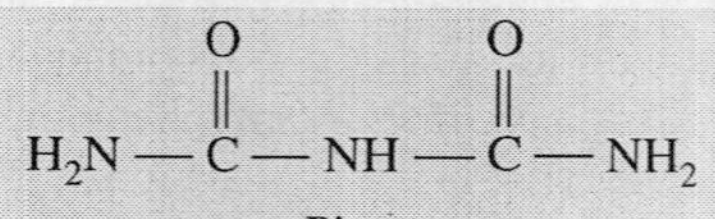

Biuret

The colour deepens as the number of peptide bonds is increased and the proteins produce a deep blue-violet colour due to the probable formation of a coordination complex whose structure is given below:

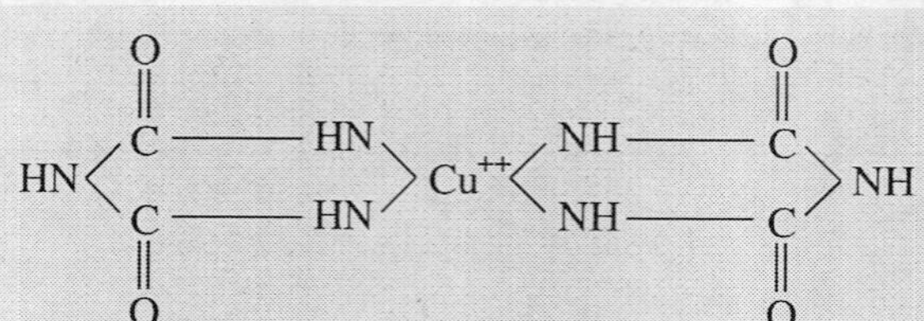

The test is, in fact, given by biuret as well as any similar structure having 2 amide or peptide bonds linked directly or through an intermediate carbon atom. The required unit is shown below between the two broken lines:

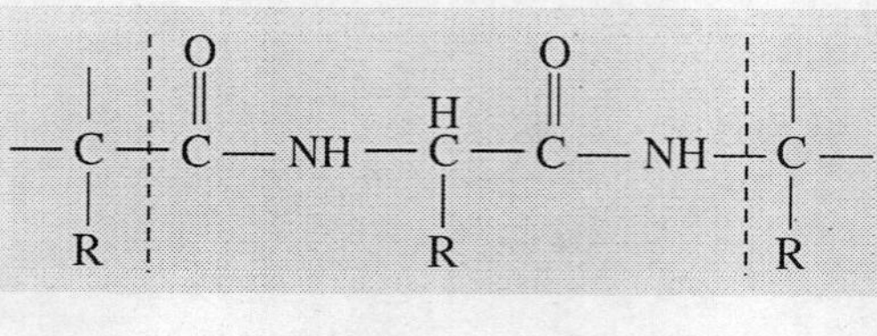

All proteins except dipeptides, therefore, respond to this reaction. This reaction is widely used both as a qualitative test for the detection of proteins and as a quantitative measure of protein concentration.

D. Method. Take about 1 ml of dilute suspension of proteins such as egg albumin, gelatin or casein in a test tube (If the protein is to be tested in a biological material, a similar amount of the tissue extract, normally in saline water, is taken). Then add 3 drops of $CuSO_4$ solution (10 g per litre) followed by about 1 ml of 10 N NaOH solution. Mix the ingredients thoroughly and note the change in colour. The appearance of purple to violet colour indicates the presence of protein.

E. Observations.

Name of test	*Experiment*	*Observations*	*Inference*
Biuret test	2 ml of protein soln. + 2 ml of Biuret reagent ↓ Wait for few minutes.	Pink to violet colouration develops.	Presence of protein confirmed.

F. Result. The test solution is that of a protein.

EXPERIMENT No. 10

A. Object. To perform colour test for proteins containing aromatic amino acids, *i.e.*, tryptophan, phenylalanine and tyrosine (xanthoproteic test).

B. Requirements. Test tube, test tube holder, protein solution, conc. HNO_3, conc. NaOH.

C. Principle. The proteins containing aromatic amino acids, on boiling with conc. nitric acid, form yellow-coloured derivatives. These are the nitro derivatives of these amino acids. In fact, there occurs nitration of the phenyl ring (of tryptophan, phenylalanine and tyrosine) to yield yellow substitution products. When this mixture is neutralized with an alkali (like NaOH), the nitro derivatives form their salts which are orange in colour.

$$\underset{\text{(with aromatic amino acids)}}{\text{Protein}} + \underset{\text{(Conc.)}}{HNO_3} \xrightarrow{\text{Xanthoproteic test}} \underset{\text{(=Picric acid, yellow)}}{\text{Trinitrophenol}} + NO\uparrow$$

$$\downarrow + \text{conc. NaOH}$$

$$\underset{\text{(orange)}}{\text{Salts of nitro derivatives}}$$

D. Method. Take about 1 ml of test solution in a test tube and add an equal volume of conc. HNO_3. Heat the mixture for about 5 minutes and then cool it. Appearance of yellow colour indicates the presence of proteins containing the aromatic amino acids. On addition of conc. NaOH, the colour of the solution changes to orange.

E. Observations.

Name of test	*Experiment*	*Observations*	*Inference*
Xanthoproteic test	1 ml of protein soln. + 1 ml of conc. HNO_3	A white precipitate develops.	
	↓ Heat and cool.	The white precipitate becomes yellow which	Presence of protein containing

		may partially dissolve to give a yellow solution.	aromatic amino acids confirmed.
	Add conc. NaOH to the above solution.	The solution becomes orange in colour.	Salts of nitro derivatives present.

F. Result. The test solution is that of a protein containing aromatic amino acids.

Note. Xanthoproteic test can also be employed for testing free aromatic amino acids in the sample. When both free amino acids and proteins are present in the same sample, proteins are separated first by precipitating out with equal volumes of 20% trichloroacetic acid (TCA) solution. The presence of amino acids can then be tested in the supernatant and that of protein in the precipitate.

ENZYMES

The chemical reactions normally occur very slowly at the low temperatures and the atmospheric pressures – the conditions under which living cells carry on their life processes. Yet in the living cells, these reactions proceed at extremely high rates. This is due to the presence of some catalysts produced and synthesized inside the body of the organisms. The term ' enzyme' was coined in 1878 by **Friedrich Wilhelm Kuhne** to designate these biological catalysts that had previously been called 'ferments'. As they quicken most of the chemical reactions occurring in the body, the enzymes have been designated as '*the manifestations of nature's impatience.*' The name 'enzyme' (*en*[G] = in ; *zyme*[G] = Yeast) literally means 'in yeast'. This was referred to denote one of the most noteworthy reactions wherein the production of ethyl alcohol and carbon dioxide through the agency of an enzyme, the *zymase,* present in yeast take place. This reaction is most popularly known as *alcoholic fermentation*. **Summer** and **Myrbä ck** (1950) have beautifully defined the enzymes as 'simple or combined proteins acting as specific catalysts.' They are soluble, colloidal molecules which are produced by living cells. All enzymes are globular proteins with a complex 3-'D' structure, capable of binding substrate molecules to a part of their surface, called the *active site*. They affect the life of an organism to such an extent that life has apply been called as "an orderly function of enzymes."

EXPERIMENT No. 11

A. Object. To test for the specificity of the enzyme diastase (= amylase).

B. Requirements. 2 test tubes, diastase, urease, iodine solution.

C. Method. Take starch solution in two test tubes, A and B. In test tube A, add diastase enzyme solution and in test tube B, add urease enzyme solution. Keep both the test tubes for 15 minutes and then test with iodine for the presence of starch (Figure on the next page).

D. Observation. The test tube A does not show the positive test for starch while the test tube B shows the positive test for starch.

E. Explanation. The enzymes are usually specific in their action, *i.e*., they act only on their own substrate; for example, the enzyme diastase acts only on its substrate starch and another enzyme urease acts only on its substrate urea. When diastase is added to starch solution (contained in test tube A) and the test tube is left for some time (for the reaction to occur), hydrolysis sets in and the enzyme splits starch into its component units, *i.e*., glucose. Thus, the test tube after hydrolysis will contain only the glucose molecules in solution,. Hence, a test for the presence of starch in the solution will be negative in this test tube. On the contrary, the starch present in another test tube B will not be hydrolyzed upon addition of urease because urease is specific only to urea and consequently starch will remain as such in the solution. Hence, test tube B will show a positive starch test.

F. Result. The enzyme diastase is substrate-specific.

$$\underset{\text{Starch}}{(C_6H_{10}O_5)_n} + \underset{\text{Water}}{n-1\,(H_2O)} \xrightarrow[]{\text{Diastase (= amylase)}} \underset{\text{Glucose}}{n\,C_6H_{12}O_6}$$

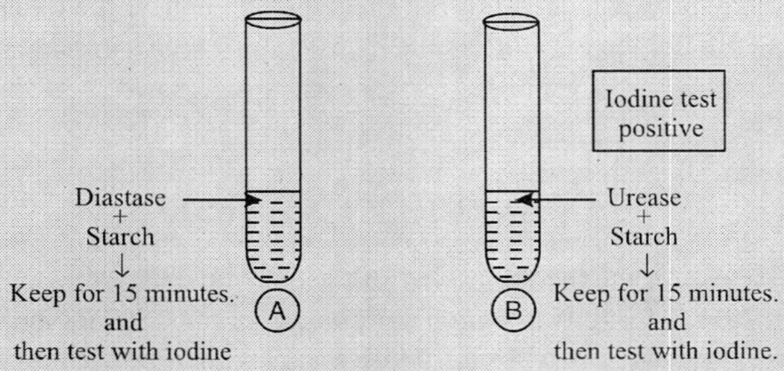

Test for the specificity of diastase

EXPERIMENT No. 12

A. Object. To test for the specificity of the enzyme, urease.

B. Requirements. 3 test tubes, urease, urea, phenolphthalein, water, spirit lamp.

C. Method. Take 3 test tubes, A, B, C. In test tube B, (Experimental), take 4 substances : urease (a pinch), phenolphthalein (5 drops), water (2 ml) and 10% urea (1 ml). In test tube A (Ist blank), take urease, phenolphthalein and water (but not urea) and in the test tube C (IInd blank), take phenolphthalein, water and urea (but not urease). Mix the contents of each test tube briskly and keep the 3 test tubes for about 10 minutes. Then heat gently each test tube and watch for the development of pink colour in them.

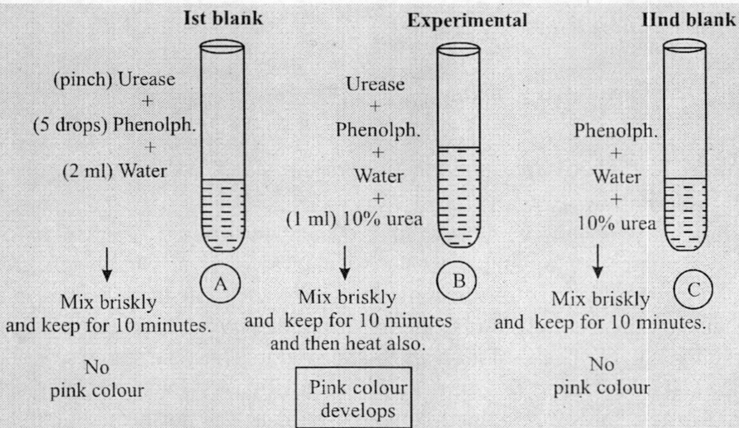

Test for the specificity of urease

D. Observations. Pink colour appears in test tube B only.

E. Explanation. The enzyme urease is specific in its action and hydrolyzes urea to carbon dioxide and ammonia. Ammonia, so formed, combines with water to form ammonium hydroxide.

$$\begin{matrix} \text{Urea} & H_2N-\overset{\overset{\large O}{\|}}{C}-NH_2 \\ + & + \\ \text{Water} & H-O-H \end{matrix} \xrightarrow{\text{Urease}} CO_2 + \underset{\text{(Ammonia)}}{2\,NH_3}$$

$$NH_3 + H_2O \longrightarrow \underset{\text{(Ammonium hydroxide)}}{NH_4OH}$$

Both ammonia and ammonium hydroxide make the medium of the test tube B alkaline. Hence, phenolphthalein already added to the test tube will give a pink colour in the alkaline medium. On the other hand, in the other two lest tubes (A and C), no pink colouration develops because in one case (test tube A), the enzyme was added but not the substrate and in the other case (test tube C), the substrate was added but not the enzyme, so the hydrolysis of urea by urease could not occur in either of the test tubes, A and C.

F. Result. The enzyme urease is substrate-specific.

EXPERIMENT No. 13

A. Object. To test for the thermolability of the enzyme diastase.

B. Requirements. 2 test tubes, starch solution, Taka diastase, iodine solution, spirit lamp.

C. Method. Take two test tubes (A and B), each containing 2 ml of Taka diastase enzyme solution. Heat test tube A and cool it. Then add 2–3 ml of starch solution in both the test tubes, A and B. Keep both the test tubes for 10 minutes and test with iodine for the presence of starch in both the test tubes.

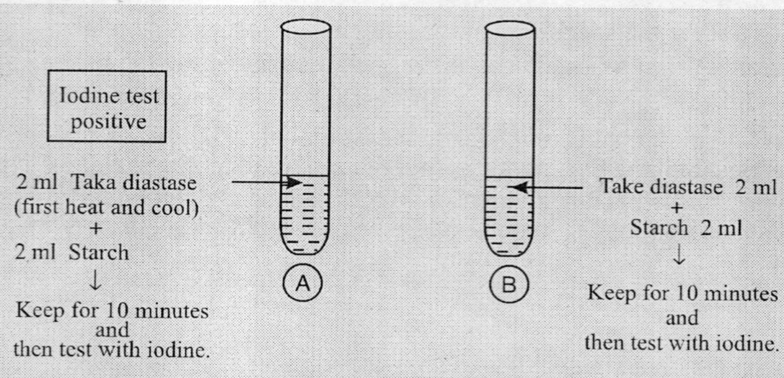

Test for the thermolability of diastase

D. Observation. Test tube A gives the positive test for starch while the test tube B does not give positive test for starch.

E. Explanation. The enzymes, being proteinaceous in nature, are very sensitive to heat. They work best in a specified range of optimum temperature of about 35 – 45°C. But at higher temperatures of 60 °C or above, they coagulate and thus become inactivated because there occurs an irreversible change in their structure. As the enzyme diastase solution in test tube A was first heated, it got inactivated and lost the power to hydrolyze starch solution when added to it. Thus, the starch solution remained present as such in the test tube. Hence, test tube A shows a positive test for starch. On the other hand, the diastase solution in test tube B remained active because it was not heated and consequently hydrolyzed starch solution when added to it, so that this test tube after hydrolysis contained glucose units. Hence, test tube B shows a negative test for starch .

F. Result. The enzyme diastase is thermolabile.

LIPIDS

The lipids are a heterogeneous group of compounds related to fatty acids and include fats, oils, waxes and other related substances such as cholesterol, lycopene, carotene, etc. These are oily or greasy organic substances, relatively insoluble in water and considerably soluble in organic solvents like ether, chloroform and benzene. They are, thus, *hydrophobic* in nature. In the body, the fats serve as efficient source of energy which is stored in the adipose tissues. They also serve as an insulating material in the subcutaneous tissues and around certain organs. Chemically, the *neutral fats* are defined as the esters of glycerol and fatty acids. Obviously, the biochemical tests of the neutral fats depend upon the tests of fatty acids and glycerol.

EXPERIMENT No. 14

A. Object. To perform colour test for lipids (Sudan III test).

B. Requirements. Test tube, castor oil, Sudan III.

C. Method. Take a small volume of castor oil in the test tube and add Sudan III (0.5% in 70% alcohol). Then shake it.

D. Observations.

Name of Test	*Experiment*	*Observation*	*Inference*
Sudan III test	Take few ml of castor oil and add few drops of Sudan III. Then shake it.	Red colour appears.	Presence of lipid confirmed.

E. RESULT. Lipid is present in the test solution.

EXPERIMENT No. 15

A. Object. To perform test for the presence of fatty acids in fats or oils (saponification).

B. Requirements. Test tube, test tube holder, unknown substance, alc. KOH, conc. H_2SO_4 or HCl, dilute NaOH (0.5%).

C. Principle. When fats or oils are heated (or hydrolyzed) with enough alkali (NaOH/KOH), free fatty acids and glycerol are liberated. The excess alkali present reacts with the liberated fatty acids to form the sodium or potassium salts which give a characteristic soapy appearance. This reaction is called *saponification*. Reaction with triolein, for example, takes place as follows:

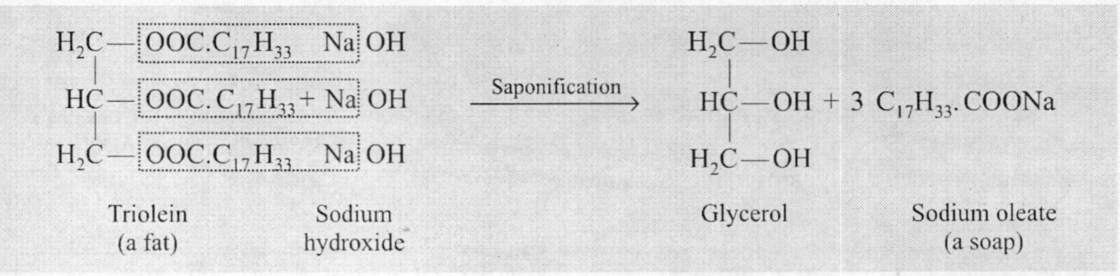

Soaps are soluble in water but are precipitated on addition of excess NaCl. Fatty acids in free state also give the same reaction.

D. Method.

Take about 1 ml of melted fat or oil in a boiling tube and add about 2 ml of 10% alcoholic KOH (about 10 g of KOH in 100 ml of alcohol) just to cover it. Boil the mixture for about 1 minute, then add about 5 ml of water and boil again for 5 minutes, then cool. Add about 0.5 ml of conc. H_2SO_4 or HCl slowly and carefully to make the mixture just acidic. The top layer of the fatty acid formed can be removed or taken out and tested for soap formation.

Take this layer of fatty acid in a separate test tube and heat with equal amount of dilute NaOH solution (about 0.5%). The solution becomes soapy, indicating the presence of fatty acids in the fat.

E. Observations.

Name of test	*Experiment*	*Observation*	*Inference*
Saponification	1 ml of melted fat/oil + 2 ml of alc. KOH. Boil for 1 minute. Add 5 ml of water and boil again for 5 minutes and cool. Add 0.5 ml of conc. H_2SO_4 or HCl. Remove top layer of fatty acid formed and add equal amount of dil. NaOH solution to it.	The solution becomes soapy.	Presence of fatty acid in the fat confirmed.

F. Result. The presence of fatty acid in the unknown substance indicates that it is a fat.
Note. Saponification test can also be employed for testing free fatty acids.

EXPERIMENT No. 16

A. Object. To perform test for the presence of glycerol in fats (acrolein test).

B. Requirements. Ignition tube, $NaHSO_4$ or $KHSO_4$, solid fat sample.

C. Principle. Fats when heated in the presence of a dehydrating agent, sodium or potassium bisulfate ($NaHSO_4$ or $KHSO_4$), produce an unsaturated aldehyde called *acrolein* from the glycerol moiety. Acrolein is easily recognized by its characteristic pungent odour and thus forms the basis of the test for the presence of glycerol in fat molecule.

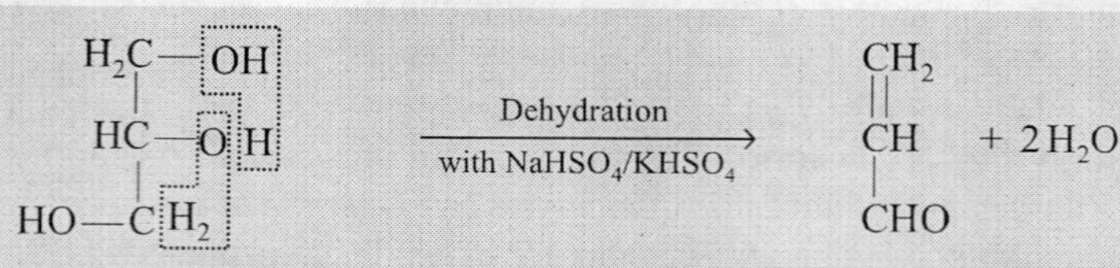

Glycerol Acrolein

Glycerol in free state also gives the same reaction.

D. Method.

Take about 1g of solid fat sample in an ignition tube. Cover the sample with an excess of $NaHSO_4$ or $KHSO_4$. The mixture is heated slowly. The pungent smell of acrolein, thus produced , indicates the presence of glycerol in the fat sample.

E. Observations.

Name of test	*Experiment*	*Observations*	*Inference*
Acrolein test	1 g of solid fat sample + excess of $NaHSO_4$/$KHSO_4$ ↓ Heat the mixture gently.	Pungent smell of acrolein is produced.	Presence of glycerol in the fat confirmed.

F. Result. The presence of glycerol in the unknown substance indicates that it is a fat.
Note. Acrolein test can also be employed for testing glycerol in free state.

EXPERIMENT No. 17

A. Object. To perform test for the presence of cholesterol in animal cells.

B. Requirements. Test tube, test tube holder, blood sample, alcohol, acetone, chloroform, acetic anhydride, conc. H_2SO_4.

C. Principle. Cholesterol is a derived lipid and is solely present in all animal cells, esp., in nerve tissues and in gallstones. It is undoubtedly the most publicized lipid in nature, because of the strong correlation between high levels of cholesterol and the incidence of diseases of the cardiovascular system in humans. The presence of cholesterol can be demonstrated easily in the blood samples. The sample is extracted with alcohol-acetone mixture (in the ratio of 1 : 1) which removes cholesterol and precipitates down the proteins. The cholesterol (which is present in the supernatant) can be taken out and reacted upon with a chloroform solution of acetic anhydride *in complete anhydrous conditions,* to produce a green-blue-coloured complex.

D. Method. Dissolve 5–10 mg of cholesterol sample in about 2 ml of chloroform. To this, add 2 ml of acetic anhydride–conc. H_2SO_4 mixture (in the ratio of 30 : 1) very carefully. Mix the ingredients well and keep the test tube in dark. A characteristic green-blue colour develops after a few minutes, indicating the presence of cholesterol.

E. Observations.

Name of test	*Experiment*	*Observations*	*Inference*
	5–10 mg of cholesterol + 2 ml of chloroform. To this, add 2 ml of acetic anhydride–conc. H_2SO_4 mixture (30 : 1 ratio). Mix the ingredients and leave the test tube in dark.	A characteristic green-blue colour develops.	Presence of cholesterol confirmed.

F. Result. The test sample contains cholesterol.

G. Precautions. All the glasswares must be completely dry and the reaction should be performed under anhydrous conditions, as far as possible.

EXPERIMENT No. 18

A. Object. To determine the saponification number of the given oil (castor oil).

B. Requirements. Erlenmeyer conical flask (250 ml), burette(50 ml), 2 pipettes (5 ml each), castor oil, alc. KOH (0.5N), H_2SO_4 (0.5 N), alcoholic phenolphthalein (0.1%).

C. Principle. Hydrolysis of fat/oil by means of alkali is called as *saponification.*

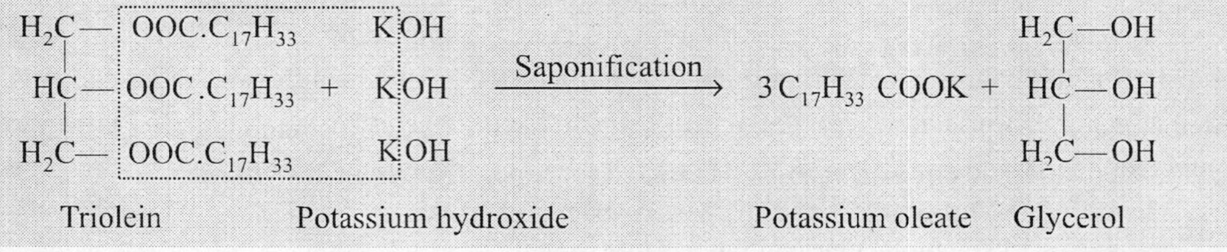

An idea about the length of the constituent fatty acid chain can be derived by determining the saponification number of fat/oil. The saponification number of a fat is "the number of milligrammes of KOH required to neutralize free fatty acids which are liberated from the complete hydrolysis of 1 g of fat/oil". *Shorter the fatty acid chain, higher is the saponification number and vice versa.*

After hydrolyzing the fat with the alkali, the residual amount of alkali can be calculated by titrating the alkali solution against acid in a control experiment. In the other actual or test experiment, the actual alkali hydrolysis is done and the titration is done like the control experiment. The difference between the two values of the acid used in titration denotes the equivalent amount of KOH used in hydrolysis.

D. Method. Take two flasks. In one flask, take 50 ml. of 0.5 N alcoholic KOH and in another flask, take 1 ml. of oil and add 50 ml. of 0.5 N alcoholic KOH. Boil both the flasks for about half an hour on a water bath. After removing the flasks from water bath, let them be cooled. To each flask, then add 1 ml of 0.1% phenolphthalein. The colour of the solution becomes pink. If pink colour does not retain in the solution, more phenolphthalein (*i.e.*, 0.5 ml) may be added.

Titrate the solution of both the flasks with 0.5 N H_2SO_4/HNO_3 and note the readings.

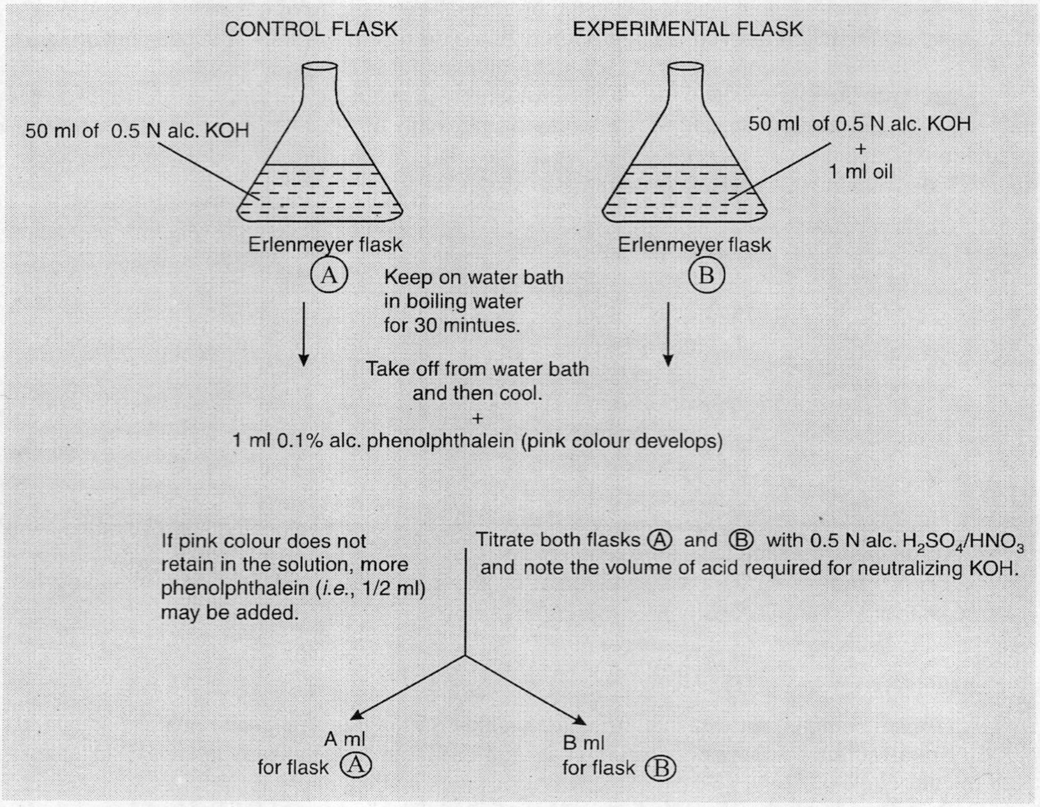

E. Calculation. Calculate the saponification number with the help of the following formula:

$$1 \text{ ml. of } 0.5 \text{ N acid} = 28 \text{ mg KOH}$$

If A is the volume of acid required for titration of KOH in the control flask (A), and B is the volume of acid required for neutralizing residual KOH in the flask (B) containing oil, (A-B) ml will be the volume of acid equivalent to KOH utilized for saponification of 1 ml of oil.

Hence, (A–B) ml of 0.5 N acid = 28 × (A–B) mg of KOH

[The molecular weight of KOH is 56.
0.1 N KOH contains 5.6 mg/ml.
0.5 N KOH will contain (5.6 × 5) = 28 mg/ml.]

The weight of oil can be obtained by multiplying the volume taken by specific gravity.

$$\text{Weight} = \text{Volume} \times \text{Specific gravity}$$
$$\text{Weight of 1cc oil} = 1 \times 0.86 = 0.86 \text{ gm}$$

Let the weight of oil be x grams. Thus, x gram oil requires 28 × (A–B) mg KOH.

Therefore, 1 gm oil will require $= \dfrac{28(A-B)}{x}$ mg KOH, or

the saponification number of the given oil $= \dfrac{28 \times (A-B)}{x}$

The difference between the two readings (A and B) should be around 5.5 ml.

$$\left.\begin{array}{l}\text{The value of A} = 11.5 \text{ ml} \\ \text{The value of B} = 6.1 \text{ ml}\end{array}\right\} 11.5 - 6.1 = 5.4 \text{ ml}$$

$$\text{The saponification number of the given oil} = \frac{28 \times (11.5 - 6.1)}{0.86}$$

$$= \frac{28 \times 5.40}{0.86} = 175.8$$

F. Result. The saponification number of the given castor oil, as determined by the experiment conducted within the experimental-error range, is 175.8, which value is within the standard-value range of 175-183 for castor oil.

VITAMINS

The term 'vitamin' usually refers to the substances distinct from major components of food, required in minute quantities (*i.e.*, *oligodynamic* in nature) and whose absence causes specific deficiency diseases. They are not confined to any particular class of organic substances and may be organic acids, amines, amino acids, esters, alcohols, steroids etc. As the living organisms cannot synthesize most of these compounds, a steady supply of them is *sine quo non* (an indispensable condition) for life. Their ultimate source is the plant or bacterial world.

EXPERIMENT No. 19

A. Object. To determine thiamine (vitamin B_1) in cereals.

B. Requirements. Cereal, mortar and pestle, H_2SO_4, centrifuge, stoppered tube, ferricyanide reagent, NaOH, *iso*-butanol, anhydrous Na_2SO_4, thiamine hydrochloride.

C. Principle. Thiamine is oxidized by ferricyanide. Thiochrome is fluorescent and hence this reaction is used for the detection and determination of vitamin B_1.

D. Method. Two grammes of cereal are ground in a mortar and extracted with 20 ml of 0.1 N H_2SO_4. Centrifuge the suspension. Discard the supernatant and extract the residue again with 10 ml of H_2SO_4. Take this extract in a stoppered tube and add 3 ml of ferricyanide reagent (0.03% potassium ferricyanide), followed by 3 ml of 15% NaOH. Make sure that the solution is alkaline and shake for 10 minutes. Add 10 ml of *iso*-butanol and shake again. The lower aqueous layer is discarded. Dry the top organic layer with anhydrous Na_2SO_4.

Prepare a stock thiamine solution in *iso*-butanol (5 mg thiamine hydrochloride per 100 ml). Dilute the stock solution with *iso*-butanol to give a series of values suitable for a standard curve and measure the fluorescence using a primary wavelength of 365 mμ and a secondary wavelength of 435 mμ. After calibrating the instrument, read the fluorescence of the test solution and determine the thiamine content of the cereal samples.

EXPERIMENT No. 20

A. Object. To determine the ascorbic acid (vitamin C) in fruit juice.

B. Requirements. Lemon juice, glacial acetic acid, 2, 6-dichlorophenolindophenol, ascorbic acid standard solution, chloroform.

C. Principle. Ascorbic acid ($C_6H_8O_6$) is a derivative of a hexose called L-glucose. Chemically, it is 1-*threo*-2,4,5,6-pentoxyhexen-2-carboxylic acid lactone. The dienolic group consisting of hydroxyls on C_2 and C_3 with a double bond between them invests the ascorbic acid molecule with redox property.

Ascorbic acid can be oxidized by the blue-coloured dye, 2, 6-dichlorophenolindophenol to Dehydro-L-ascorbic acid. The dye is also reduced to a colourless compound so that the end point of the reaction is easily determined.

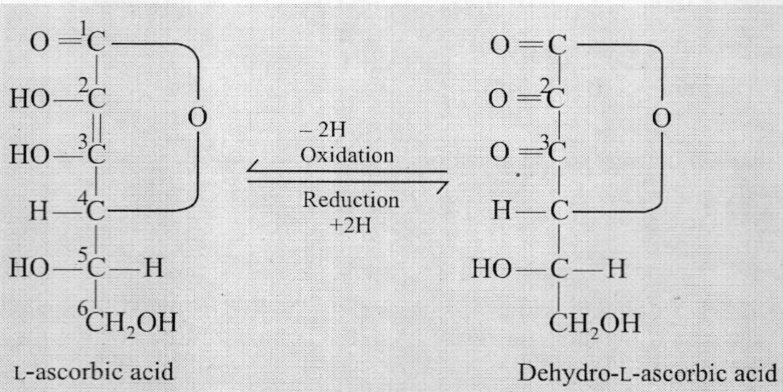

D. Method.

Preparation of 2, 6-dichlorophenolindophenol. Crush one tablet of 2, 6-dichlorophenolindophenol in 40 ml of hot water, cool and dilute to 50 ml.

Take 5 ml of diluted fruit juice (lemon juice, for example) into a boiling tube, add 1 ml of glacial acetic acid and titrate with the dye solution to a faint permanent pink colour. Record the titre (T). Repeat the titration with 5 ml of water for the blank (Bl) and 5 ml of ascorbic acid for the standard solution (St, 0.02 mg/ml) and calculate the vitamin C content of the test sample.

$$\text{Vitamin C of test sample (mg/100 ml)} = \frac{T - Bl}{St - Bl} \times 2 \times \text{Dilution}$$

If the test solution is highly coloured, add 1 ml of chloroform to the reaction mixture and the end point is obtained when a permanent pink colour is seen in the organic phase. The test and the blank are treated the same way.

EXPERIMENT No. 21

A. Object. To determine ascorbic acid (vitamin C) in germinating peas or beans.

B. Requirements. Germinating peas or beans, 5% trichloroacetic acid, glacial acetic acid, 2,6-dichlorophenolindophenol, ascorbic acid standard solution, chloroform.

C. Principle. Same as that described for Experiment No. 20.

D. Method. Remove the soil adhered to germinated seeds by washing in water. Then homogenize the whole seedlings with 5% trichloroacetic acid (TCA). Dilute to 20 ml with water, then leave to stand for about 15 minutes with occasional shaking. Centrifuge and measure the ascorbic acid in the supernatant, as in Experiment No. 20.

MILK

Milk, a source of nourishment for all mammals, is composed in part of a variety of proteins. It contains about 0.6 – 0.7% protein, which is not precipitated on acidification to pH 4.7. This represents about 20% of the protein contained in skimmed milk. These *whey proteins* are separated into 2 fractions: lactalbumin and lactoglobulin. The name *casein* is assigned to a fraction precipitated by acidifying milk to a pH of 4.7. It is present in cow's milk (3 – 3.5%) and human milk (0.3 – 0.6%). Casein may be further purified by redissolving and precipitating again. It is of 3 types : α, β and γ. These differ one another in their molecular weights, their rate of migration in an electric field and their phosphorus content.

EXPERIMENT No. 22

A. Object. To find out the specific gravity of milk.

B. Requirement. Glass lactometer, milk, glass jar.

C. Principle. Specific gravity represents the comparison of the density of a liquid with that of water which is taken as one. The lactometer is an instrument by which the quality and richness of the milk is ascertained. The richness of the milk depends principally on the quantity of fat or butter and curd or cheese, it contains. By measuring the specific gravity of the milk in different states, using a lactometer, it is possible to compare accurately the quality of the milk produced by different cows or other animals.

An unadulterated milk (= pure milk) has a specific gravity 1.032°. If milk is adulterated *i.e.*, if water is added to it, it will show a specific gravity lesser than the standard value. Greater the quantity of water in milk, lesser is the value of specific gravity. However, milk of thick consistency will have specific gravity greater than 1.032°.

D. Method. Lactometer consists of two portions: a lower broad '*bulb*' and an upper narrow '*stem*'. The bulb contains a heavy metal like mercury and the stem is graduated from the upper end. The graduations on the stem, 0° to 40° L are equivalent to 1.000° to 1.040° specific gravity. Due to the presence of heavy metal mercury, when the lactometer is kept in any liquid, it keeps on floating in an upright (= vertical) position.

Fill the glass jar 2/3rd with a freshly-drawn sample of the milk to be tested; immerse the lactometer in the glass jar, then read the division on the 'stem' which comes level with the surface of the milk. This represents the specific gravity of the milk, provided the temperature is the same as that for which the lactometer is usually adjusted, namely 60 °F or 84 °F, which is clearly marked on each instrument (see Temperature Correction Table, given on the next page).

Allow the milk to stand for 24 hours or until all the cream has risen. Remove the cream. Bring the skimmed milk to the temperature at which the lactometer has been adjusted (*i.e.* 60 °F or 84 °F) and take a reading of the specific gravity, as described above.

As the fat content, which is the lightest ingredient, has been removed in the form of cream, the specific gravity of the skimmed milk will be found greater than that for the full-cream (or unskimmed) milk. The larger the quantity of cream, the greater will be the difference in specific gravity. This difference, therefore, gives a direct indication of the relative quality of the milk.

Following two examples will make the above concept clear:

Example 1. If, on two samples of full-cream milk the reading is 29°L and the reading of skimmed milk of one is 35° L and that of the other 32°L, the difference in the first case is 6°L and in the second 3°L. The first sample of milk will, henceforth, yield approximately twice as much butter as that of the second sample.

Example 2. One sample of milk has a reading of 30° L which is increased to 34° L by the removal of the cream and the second sample has readings of 28° L and 32° L respectively, then as the differences are the same, the quantity of butter they will produce will be approximately equal.

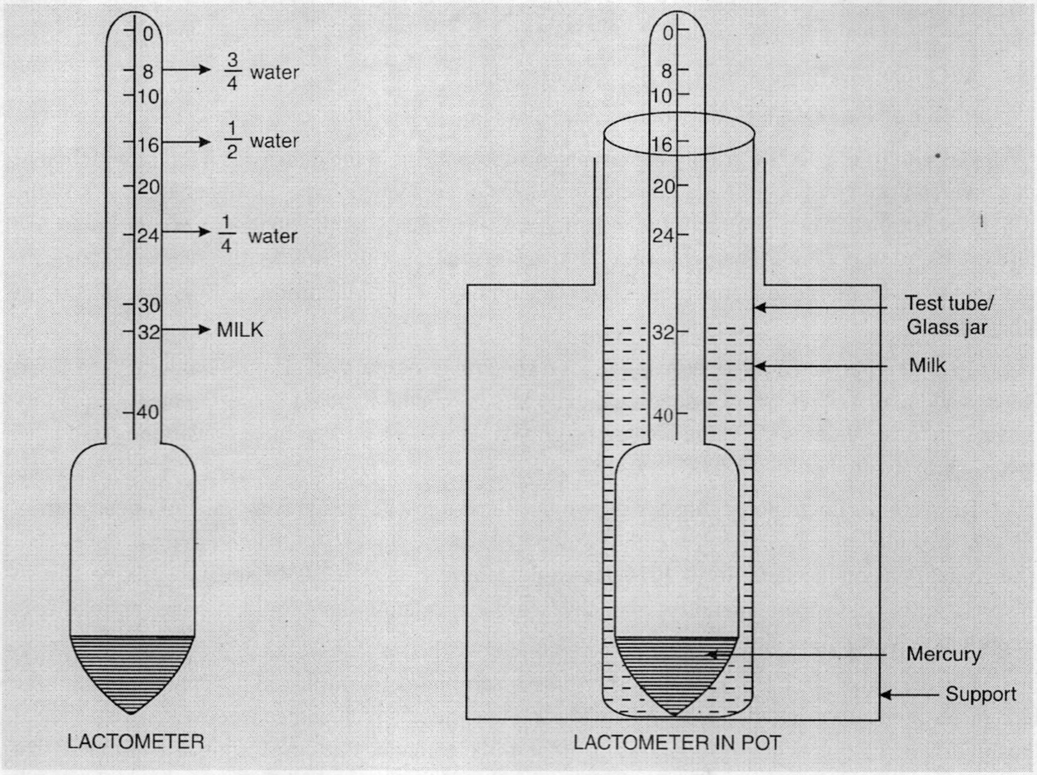

Temperature Correction Table

If due to any circumstances, it is not possible to take a reading of the milk at the temperature at which the lactometer has been adjusted, namely 60°F or 84°F, corrections should be made as follows:

For lactometers adjusted at 60°F		For lactometers adjusted at 84°F	
Temp. of milk (in °F)	Approximate correction	Temp. of milk (in °F)	Approximate correction
50	Subtract 2 °L	75	Subtract 20 °L
55	Subtract 1 °L	80	Subtract 1 °L
60	0 °L	84	0°F
65	Add 1°L	90	Add 1°F
70	Add 2 °L	95	Add 2 °F

Note. Should the temperature of the milk differ by more than 10 °F from the temperature at which the lactometer has been adjusted, this table should not be used. It is recommended that under these conditions (*i.e.*, when the temperature exceeds the adjustment by 10 °F), corrections should be extrapolated from this table with caution, if accurate results are required. Alternatively, the sample of the milk to be tested should be adjusted in temperature to within 10 °F of the standardized temperature of the instrument, in which case the above corrections can be used.

E. OBSERVATIONS.

Experiment	*Observations*	*Inference*
Immerse the lactometer in milk contained in a gas jar and take the reading.	Reading is 32 °L	The specific gravity of milk is 1.032°.

F. RESULT:- Since the specific gravity of milk is 1.032°, the milk is unadulterated *i.e.,* pure.

EXPERIMENT No. 23

A. OBJECT. To detect calcium and phosphorus in milk.

B. REQUIREMENTS. Milk, acetic acid, nitric acid, ammonium molybdate, ammonia solution, ammonium oxalate.

C. PRINCIPLE. On adding 1% acetic acid gradually to the milk, casein (a phosphoprotein) and fat are precipitated (pH = 4.6) and calcium and phosphorus remain in the filtrate. This white precipitate is discarded and the casein- and fat-free filtrate is then tested for the presence of calcium and phosphorus in it.

The white precipitate which appears upon addition of ammonium oxalate to the filtrate is that of calcium oxalate.

The yellow precipitate which appears upon addition of ammonium molybdate to the filtrate is that of ammonium phosphomolybdate.

CALCIUM

$$\underset{\text{(Acetic acid)}}{CaCO_3 + 2\,CH_3.COOH} \longrightarrow \begin{matrix} CH_3.COO \\ CH_3.COO \end{matrix}\!\!>\!Ca + \underset{\text{(Unstable)}}{\overset{(H_2O\ +\ CO_2\uparrow)\ \uparrow}{H_2CO_3}}$$

$$\begin{matrix} CH_3.COO \\ CH_3.COO \end{matrix}\!\!>\!Ca + \underset{\text{(Amm. oxalate)}}{\begin{matrix} COO.\ NH_4 \\ | \\ COO.\ NH_4 \end{matrix}} \longrightarrow \underset{\substack{\text{(Calcium oxalate)} \\ \text{White ppt.}}}{\begin{matrix} COO \\ | \\ COO \end{matrix}\!\!>\!Ca} + \underset{\text{(Amm. acetate)}}{2CH_3.COO.NH_4}$$

PHOSPHORUS

$$PO_4^{3-} + \underset{\text{(Nitric acid)}}{3\,HNO_3} \longrightarrow 3\,NO_3^- + H_3PO_4$$

$$H_3PO_4 + \underset{\text{(Amm. molybdate)}}{12\,(NH_4)_2\,MoO_4} + 21\,HNO_3 \longrightarrow \underset{\substack{\text{(Amm. phosphomolybdate)} \\ \text{Yellow ppt.}}}{(NH_4)_3\,PO_4.12\,MoO_3} + 21\,NH_4.NO_3 + 12\,H_2O$$

D. METHOD. Test for calcium. Take about 5 ml of casein- and fat-free filtrate and add little amount of ammonia solution to make it alkaline. Now add ammonium oxalate.

Test for phosphorus. Take about 5 ml of casein- and fat-free filtrate in a test tube and add about 1 ml of con. HNO_3 and 3 ml of ammonium molybdate.

E. OBSERVATIONS.

Experiments	*Observations*	*Inference*
1. In 10-15 ml of milk, add 1% acetic acid. Filter it and take this filtrate.	A white precipitate is formed.	Protein (casein) + fat are present in the precipitate.
2. Take 5 ml of filtrate and add few drops of ammonia to make the filtrate alkaline and add ammonium oxalate.	A white precipitate or turbidity of calcium oxalate appears.	The presence of calcium in milk confirmed.
3. In filtrate (5 ml), add 1 ml of HNO_3 and 3 ml of ammonium molybdate.	A yellow precipitate or colouration of ammonium phosphomolybdate appears.	The presence of phosphorus in milk confirmed.

F. RESULT. Milk contains both calcium and phosphorus.

APPENDICES

Only because water is so cohesive can trees raise it hundreds of feet off the ground. Individual California coast redwoods (*Sequoia sempervirens*) may grow to heights of 360 feet, taller than a 25-story building.

Appendix is "an addition to a document or book (generally at the end) which supplements or illustrates the text"

[*Source : (one of the meanings) : The New International* ***Webster's Dictionary*** *of the English Language, Trident Press International*]

APPENDIX I

SELECTED BIBLIOGRAPHY

1. **Abeles RH, Frey PA, Jenckes WP :** Biochemistry. *Jones and Bartlett, Boston. 1992.*
2. **Abraham C, Schepartz B :** Biochemistry. *4th edition. W. B. Saunders Co., Philadelphia and London. 1967.*
3. **Alberts B, Bray D, Lewis J, Raff M, Roberts K, Watson JD :** Molecular Biology of the Cell. *3rd ed. Garland Publishing, Inc., New York./London. 1994.*
4. **Armstrong FB :** Biochemistry. *3rd ed., Oxford University Press, New York. 1989.*
5. **Baldwin E :** Dynamic Aspects of Biochemistry. *4th edition. Cambridge Univ. Press, Cambridge, U.S.A. 1963.*
6. **Baldwin E :** The Nature of Biochemistry. *2nd edition. Cambridge Univ. Press, London. 1967.*
7. **Bennet TP, Frienden E :** Modern Topics in Biochemistry. *The McMillan Co., New York. 1969.*
8. **Berg JM, Tymoczko JL, Stryer L :** Biochemistry. *5th edition. W. H. Freeman and Company, New York. 2002.*
9. **Bonner JE :** Plant Biochemistry. *Academic Press, Inc., New York. 1966.*
10. **Burke JD :** Cell Biology. *The Williams & Wilkins Company, Baltimore (U.S.A.). 1970.*
11. **Conn EE, Stumpf PK, Bruening G, Dio RH :** Outlines of Biochemistry. *5th edition. John Wiley and Sons, Singapore. 1987.*
12. **Cunningham EB :** Biochemistry : Mechanism of Metabolism. *McGraw-Hill, New York. 1978.*
13. **Darnell JE, Lodish HF, Baltimore D :** Molecular Cell Biology. *2nd ed., Scientific American Books, New York. 1990.*
14. **Datta SP, Ottaway JH :** Biochemistry. *2nd edition. Bailliere Tindall and Cassell, London. 1969.*
15. **Devlin TM (Editor) :** Text Book of Biochemistry with Clinical Correlations. *5th edition. Wiley-Liss, New York .2002.*
16. **Doby G :** Plant Biochemistry. *Interscience Publishers, London and New York. 1965.*
17. **Edwards NA, Hassall KA :** Biochemistry and Physiology of the cell. *McGraw-Hill Book Company (UK) Ltd., London. 1980.*
18. **Fairley JL, Kilgour GL :** Essentials of Biological Chemistry. *2nd edition. Van Nostrand Reinhold Co., New York. 1966.*
19. **Florkin M, Stotz EH (eidtors) :** Comprehensive Biochemistry. *Amer. Elsevier Publishing Co., Inc., New York. 1962-67.*
20. **Freifelder DM :** Molecular Biology and Biochemistry : Problems and Applications. *W.H. Freeman and Company, San Francisco. 1978.*
21. **Fruton JS, Simmonds S :** General Biochemistry. *2nd edition, John Wiley and Sons, Inc., New York. 1958.*
22. **Green JH :** Basic Clinical Physiology. *3rd edition. ELBS and Oxford University Press, London. 1978.*
23. **Harper HA, Rodwell VW, Mayes PA :** Review of Physiological Chemistry. *17th edition.*

Lange Medical Pub., California. 1979.

24. **Hartman PE, Suskind SR :** Gene Action. *2nd edition. Prentice-Hall, Inc., Englewood Cliffs, N.J., U.S.A. 1969.*
25. **Jain JL, Jain S, Jain N :** Fundamentals of Biochemistry. *6th ed., S. Chand & Co. Ltd., New Delhi. 2005.*
26. **Karlson, P :** Introduction to Modern Biochemistry. *4th edition. Academic Press, Inc., New York. 1975.*
27. **Mahler HR, Cordes EH :** Basic Biological Chemistry. *2nd edition. Harper and Row, Publishers Inc., New York. 1968.*
28. **Mallette MF, Clagett CO, Phillips AT, McCarl RL :** Introductory Biochemistry. *1st edition. Robert E. Krieger Publishing Co., New York. 1979.*
29. **Mathews CK, van Holde KE, Ahern KG :** Biochemistry. *3rd Edition. Pearson Education, California.* 2000.
30. **Mazur A, Harrow B :** Textbook of Biochemistry. *10th editon. W.B. Saunders Co., Philadelphia/London. 1971.*
31. **McElroy WD :** Cellular Physiology and Biochemistry. *Englewood Cliffs, N.J., U.S.A. 1961.*
32. **McGlilvery RW, Goldstein :** Biochemistry : A Functional Approach. *3rd edition. W. B. Saunders Co., London/Toronto. 1983.*
33. **Mertz ET :** Elementary Biochemistry. *3rd edition. Burgess Pub. Co., Minneapolis. 1967.*
34. **Metzler DE :** Biochemistry. *Academic Press, Inc., New York. 1977.*
35. **Nelson DL, Cox MM :** Lehninger Principles of Biochemistry. *3rd edition. Worth Publishers, New York. 2000.*
36. **Orten JM, Neuhaus OW :** Biochemistry. *8th edition. The C.V. Mosby Co., Saint Louis. 1970.*
37. **Rawn JD :** Biochemistry. *Neil Patterson Publishers, Burlington, N.C. 1990.*
38. **Rethel FJ :** Concepts in Biochemistry. *McGraw-Hill Book Co., New York. 1967.*
39. **Smith EL, Hill RL, Lehman IR, Lefkowitz RJ, Handler P, White A :** Principles of Biochemistry : General Aspects. *7th edition. McGraw-Hill International Book Co., London/ Tokyo 1983.*
40. **Smith EL, Hill RL, Lehman IR, Lefkowitz RJ, Handler P, White A :** Principles of Biochemistry : Mammalian Biochemistry. *7th edition. McGraw-Hill International Book Co., London/Tokyo. 1983.*
41. **Stephenson WK :** Concept in Biochemistry. A programmed Text. Reprint. *Wiley Eastern Pvt. Ltd., New Delhi. 1973.*
42. **Voet D, Voet JG :** Biochemistry. *2nd edition. Wiley, New York. 1995.*
43. **Watson JD, Hopkins NH, Roberts JW, Steitz JA, Wiener AM :** Molecular Biology of the Gene. *4th ed., The Benjamin/Cummings Publishing Company Inc., California. 1987.*
44. **Wiel JH :** General Biochemistry. *6th ed., Wiley Eastern Limited, New Delhi. 1990.*
45. **West ES, Todd WR, Mason HS, van Bruggen JT :** Text-book of Biochemistry. *4th edition. Macmillan Co., 1966.*
46. **Wood WB, Wilson JH, Benbow RM, Hood LE :** Biochemistry. A Problems approach. *2nd ed., Benjamin/Cumings, Menlo Park, California. 1981.*
47. **Yudkin M, Offord R :** Comprehensible Biochemistry. *Longman. 1973.*
48. **Zubay GL, Parson WW, Vance DE :** Principles of Biochemisty, *1st edition, Wm. C. Brown Publishers, Dubuque. 1995.*

APPENDIX II

GREEK ALPHABET*

Greek letter		*Greek name*	*English equivalent*
Capital	*Small*		
Α	α	Alpha	a
Β	β	Beta	b
Γ	γ	Gamma	g
Δ	δ	Delta	d
Ε	ε	Epsilon	ĕ
Ζ	ζ	Zeta	z
Η	η	Eta	ē
Θ	θ	Theta	th
Ι	ι	Iota	i
Κ	κ	Kappa	k
Λ	λ	Lambda	I
Μ	μ	Mu	m
Ν	ϖ	Nu	n
Ξ	ξ	Xi	x
Ο	*0*	Omicron	ŏ
Π	π	Pi	p
Ρ	ρ	Rho	r
Σ	σ	Sigma	s
Τ	τ	Tau	t
ϒ	υ	Upsilon	u
Φ	ϕ	Phi	ph
Χ	χ	Chi	ch
Ψ	ψ	Psi	ps
Ω	ω	Omega	ō

* Greek letters are preferably used to indicate the positions of substituents in compounds described by trivial names, *e.g.,* α-hydroxypropionic acid (lactic acid). They are used to indicate the *class* of compound, *e.g.*, γ- and δ-lactones. The use of Greek letters to indicate the positions of substituents in a chain should, however, be avoided. Instead, the numerals should be used.

APPENDIX III

EXPONENTIAL NOTATION

Exponential notation is used to deal with very large and very small numbers. An **exponent** is a symbol written to the right and above a number or mathematical expression; it indicates number of times that figure is to be multiplied by itself. An exponent is also called the **power** of a number.

(*a*) *For numbers greater than one*, the exponents is positive and is equal to the number of zeroes following the one.

$$10 = 10^1 = \text{ten}$$
$$100 = 10^2 = \text{one hundred}$$
$$1000 = 10^3 = \text{one thousand}$$
$$10000 = 10^4 = \text{ten thousand}$$
$$1000000 = 10^6 = \text{one million}$$
$$1000000000 = 10^9 = \text{one billion}$$
$$1000000000000 = 10^{12} = \text{one trillion}$$
$$1000000000000000 = 10^{15} = \text{one quadrillion}$$

(*b*) *For numbers smaller than one*, the exponent is negative and is equal to the number of places the one is to the right of the decimal point.

$$.1 = 10^{-1} = 1 \text{ tenth}$$
$$.01 = 10^{-2} = 1 \text{ hundredth}$$
$$.001 = 10^{-3} = 1 \text{ thousandth}$$
$$.0001 = 10^{-4} = 1 \text{ ten-thousandth}$$
$$.000001 = 10^{-6} = 1 \text{ millionth}$$
$$.000000001 = 10^{-9} = 1 \text{ billionth}$$
$$.000000000001 = 10^{-12} = 1 \text{ trillionth}$$
$$.000000000000001 = 10^{-15} = 1 \text{ quadrillionth}$$

Any large or small number may be expressed as the product or a more convenient-sized number and a power of 10. Thus, 357 may be expressed as $3.57 \times 100 = 3.57 \times 10^2$. Similarly, 357,000,000 = 357×10^6, or 3.57×10^8.

In decimals smaller than one, the decimal point is moved to the right, and the resulting number is multiplied by a negative power of 10 equal to the number of places the decimal point was moved : $.0357 = 3.57 \times 10^{-2}$ or 35.7×10^{-3}; $.0000357 = 3.57 \times 10–5$ or 35.7×10^{-6}.

APPENDIX IV

THE INTERNATIONAL SYSTEM OF UNITS
(Generele de Poids et Mesures)
(Systéme International d'Unités)

The International System of Units is a metric system based on the *metre, kilogram and socond* (m.k.s. system) in place of the *centimetre, gram and second* (the old c.g.s system). These SI units (Syste′me International d'Unite′s) were approved in 1960 by the 11th General Conference of Weights and Measures and are being adopted by the scientific laboratories throughout the world. The SI units are a *coherent system of units* so that if two unit quantities are multiplied or divided, then the answer is the unit of the resultant quantity. In this way, the number of multiplies and submultiples of units now in use, will be reduced.

A. Basic Units

This system is now based on 8 base units in relation to 8 basic kinds of quantity (Table 1). Other units are based on these basic unit. The primary units of **length** is the *metre*. A metre is the distance between two lines, under standard conditions, engraved on a prototype preserved in the Bureau International des Poids et Measures, Se'vres, France.

The primary unit of **mass** is the *kilogram*, which is the mass of the prototype preserved in the same place.

Table 1. The Eight Base Units

S.N.	*Physical Quantity*	*Unit Symbol*	*Symbol*
1.	Length	Metre	m
2.	Mass	Kilogram	kg
3.	Time	Second	s
4.	Amount of substance	Mole	mol
5.	Thermodynamic temperature	Kelvin	K
6.	Electric current	Ampere	A
7.	Luminous intensity	Candela	cd
8.	Katalytic amount	Katal	kat

B. Derived SI Units

Besides the above basic units, there are also a number of derived SI units, obtained by appropriate combination of these basic units. For convenience, these derived units are given special names (Table 2).

Table 2. Some Derived SI Units*

S. N.	*Physical Quantity*	*Unit*	*Unit Symbol*	*Definition*
1.	Frequency	Hertz	Hz	S^{-1}
2.	Force	Newton	N	$kg\ m\ s^{-2}$
3.	Energy	Joule	J	$kg\ m^2\ s^{-2}$
4.	Pressure	pascal	Pa	$kg\ m^{-1}\ s^{-2}$
5.	Power	Watt	W	$kg\ m^2 s^{-3}$
6.	Electric charge	Coulomb	C	A s
7.	Electric potential	Volt	V	$kg\ m^2\ s^{-3}\ A^{-1}$
8.	Electric resistance	Ohm	Ω	$kg\ m^2\ s^{-3}\ A^{-2}$
9.	Electric capacitance	Farad	F	$A^2\ s^4\ kg^{-1}\ m^{-2}$
10.	Customary temperature	Degree Celsius	°C	°C = K – 273.15

* None of these units take the plural form, so that 2 volts is written as 2V, not 2 Vs and 3 metres is written as 3m, not 3ms.

C. Prefixes for the SI Units

Sometimes units may be too large or too small and in such cases in other to avoid writing too many zeros, a prefix is placed before the symbol of the unit. The recommended multiples or fractions of a unit change mostly by 1,000 each time (Table 3). Thus, 0.000025 mol is written 25 μmol and 12,500 m is written 12.5 km.

Table 3. Prefixes denoting decimal factors for the SI Units*

Magnitude	*Prefix*	*Symbol*
Multiples		
10^{24}	yotta	Y
10^{21}	zetta	Z
10^{18}	exa	E
10^{15}	peta	P
10^{12}	tera	T
10^{9}	giga	G
10^{6}	mega	M
10^{3}	kilo	k
10^{2}	hecto	h
10^{1}	deca	da
Fractions		
10^{-1}	deci	d
10^{-2}	centi	c
10^{-3}	milli	m
10^{-6}	micro†	μ
10^{-9}	nano	n
10^{-12}	pico	p
10^{-15}	femto	f
10^{-18}	atto	a
10^{-21}	zepto	z
10^{-24}	yocto	y

* These prefixes are applied to metric and other units. For example, a micrometer (μm) is 10^{-6}; a picoliter (pl) is 10^{-12}; and a kilogram (kg) is 10^{3} grams. Also applied to seconds, units, mols, hertz, volts, farads, ohms, curies, equivalents, osmols etc. Combinations of prefixes are no longer allowed, so that *n* is used instead of mμ and *p* instead of μμ.

† Formerly called *micron*, with the same abbreviation *i.e.*, μ. A micron is a unit of measurement of length or distance.

$$1\mu = 0.001 \text{ mm} = 0.000039 \text{ in.}$$

Millimicron (mμ) is a unit of measurement, commonly used in measuring wavelength of radiation.

$$1 \text{ m}\mu = 0.001\ \mu = 1/10{,}00{,}000 \text{ mm}$$

D. Non-SI Units

Although the SI system of units is coming into use the world over, there are a few recommended non-SI units which are still frequently used (Table 4). However, these will be abandoned in time but probably not until after the useful life of this text.

Table 4. Non-SI Units still in use

S.N.	*Unit*	*Symbol*	*Definition*
1.	Angstrom*	Å	10^{-10} m; 10–8 cm ; 10^{-7} mm; 10^{-4} μ
2.	Atmosphere	atm	760 mm Hg
3.	Calorie *a.* international	cal_{IT}	4.1868 J
	b. thermochemical	cal	4.184 J
4.	Dyne	dyn	10^{-5} N
5.	Erg	erg	10^{-7} J
6.	Gauss	G	10^{-4} T (testa) = Vsm^{-2} (volt)
7.	Litre	l	10^{-3} m^3 = dm^3
8.	Micron	μ	10^{-6} m = μm
9.	Millimicron	mμ	10^{-9} m = nm

* No longer used, nanometer used instead.

APPENDIX V

HINTS FOR USING THE METRIC SYSTEM

1. The metric system uses symbols rather than abbreviations. Hence, do not place a period (or full stop) after metric symbols (*e.g.*, 1 g, not 1 g.). Use a period after a symbol only at the end of a sentence, but always use a period with the symbol for inches (in.)
2. Metric symbols are always singular; they do not have a plural form (*e.g.*, 20 km, not 20 kms).
3. Capital initial letters are never used for units, except when named after famous scientists, such as N (Newton), J (Joule) and W (Watt).
4. Do not mix units or symbols (*e.g.*, 5.3 m, not 5 m 300 mm).
5. Symbols combined in a quotient can be written as, for example, metre per second or ms^{-1}. The use of the solidus (strope, /) is restricted to indicating the unit of a variable, such as temperature /°C.
6. Except for degree Celsius, always leave a space between a number and a metric symbol (*e.g.*, 70 mm, not 70mm; 10°C, not 10° C).
7. The comma is no longer used to separate groups of 3 digits, but a space is left instead (*e.g.*, 236 892 714 not as 236,892,714.
8. The raised decimal point is not correct. The internationally-accepted decimal sign is placed level with the feet of the numeral (*e.g.*, 3.182, not 3·182).
9. Express measurements in units that require only a few decimal places. For example, 0.4 m is more easily manipulated and understood than 400,000,000 nm.
10. Familiarize yourself with manipulations within the metric system. Work within one system, and do not convert back and forth between the metric and English (or the old c.g.s.) systems.
11. Use a zero before a decimal point when the number is less than one (*e.g.*, 0.25 m, not .25 m)
12. When measuring water, the metric system offers an easy and common conversion from volume measured in litres to volume measured in cubic metres to mass measured in grams : 1 ml = 1 cm^3 = 1 g.

APPENDIX VI

COMPARISON OF METRIC AND OTHER UNITS

Units of Length

1 centimetre	= 0.3937 inch	1 inch	= 2.5400 centimetres
1 mitre	= 3.2808 feet	1 foot	= 0.30480 metre
1 metre	= 1.09361 yards	1 yard	= 0.91440 metre
1 kilometre	= 0.62137 mile	1 mile	= 1.60934 kilometres

Units of Area

1 square cm	= 0.1550 sq. in.	1 square in.	= 6.4516 sq. cm.
1 square metre	= 10.7638 sq. ft.	1 square ft.	= 0.9290 sq. m.
1 square metre	= 1.1960 sq. yds.	1 square yd.	= 0.83613 sq. m.
1 square km.	= 0.38610 sq. mile	1 square mile	= 2.5900 sq. km.
1 hectare	= 2.47105 acres	1 acre	= 0.40469 hectare

Units of Volume

1 cubic cm. (c.c.)	= 0.061024 cubic in.	1 cubic in.	= 16.387 c.c.
1 cubic m.	= 35.3144 cubic ft.	1 cubic ft.	= 0.028317 cubic m.
1 cubic m.	= 1.3079 cubic yds.	1 cubic yd.	= 0.7645 cubic m.

Measures of Liquid Capacity

British Units (or Imperial Units)

pennyweight = 3.858 carats

1 gallon = 4 quarts = 8 pints = 32 gills

1 gallon = 4.54596 litres

1 litre = 0.2200 gallon = 1.7598 pints

U.S. Units

1 gallon = 4 quarts = 8 pints = 32 gills

1 gallon = 3.78533 litres

1 litre = 0.26418 gallon = 2.1134 pints

Apothecaries' Units (British)

1 gallon = 8 pints = 160 fluid ounces

1 fluid ounce = 8 fluid drachms = 24 scruples = 480 minims

1 gallon = 4.54596 litres

1 litre = 0.2200 gallon = 35.196 fluid ounces

Apothecaries' Units (U.S.)

1 gallon = 8 pints = 128 fluid ounces

1 fluid ounce = 8 fluid drachms = 480 minims

1 gallon = 3.78533 litres

1 litre = 0.26418 gallon = 33.81504 fluid ounces

Units of Mass

Avoirdupois Weight. This is a British and American system of weights, based on a pound of 16 ounces.

1 hundredweight = 4 quarters = 8 stones = 112 pounds
1 pound = 16 ounces = 256 drachms = 7,000 grains
1 grain = 0.064799 g
1 grain = 15.4323 grains
1 hundredweight = 50.80238 kg
1 kilogram = 0.0198461 hundredweight
1 millier = 1 metric ton = 10 quintals
1 quintal = 10 myriagram = 100 kilogram = 0.09842059 ton
1 kilogram = 2.204621 pounds
1 pounds = 0.4535926 kilogram

Troy Weight. This is a system of weights, used for gems and precious metals (gold, silver) and also for drugs.

1 pound = 12 ounces = 240 pennyweights
1 pennyweight (dwt.) = 6 carats = 24 grains
1 gram = 0.3215 ounce = 0.643 pennyweight = 3.858 carats
1 carats = 0.25920 gram

Carat : (1) A measure of weight of *diamonds* and other gems. (2) A measure of fineness of *gold*, expressed as parts of gold in 24 parts of the alloy. Thus, 24 carat gold is pure gold, 18 carat gold contains 18 parts in 24 or has a fineness of 750.

Apothecaries's Weight. This is a system of weights, used in pharmacy.

1 pound = 12 ounces = 96 drachms
1 drachm = 3 scruples = 60 grains
1 gram = 0.03215 ounce = 0.2572 drachm = 0.7716 scruple
1 scruple = 1.29598 gram

APPENDIX VII

Mathematical Signs and Symbols

$=$	equals
$\backsimeq$, $\cong$	equals approximately
$\neq$	is not equal to
$\equiv$	is identical to, is defined as
$>$	is greater than
$\gg$	is much greater than
$<$	is less than
$\ll$	is much less than
$\geq$, $\geqq$	is more than or equal to (or, is no less than)
$\leq$, $\leqq$	is less than or equal to (or, is no more than)
$\pm$	plus or mimus (*e.g.*, $\sqrt{4} = +2$)
$\propto$	is proportional to (*e.g.*, Hooke's law : $F \propto x$, or $F = -kx$
Σ	the sum of
$\bar{x}$	the average value of x

Mathematical Constants

$\pi = 3.14159$

$e = 2.71828$

$\log_e x = 2.303 \log_{10} x$

APPENDIX VIII

Relative Sizes of Structures, from Atom to Eggs

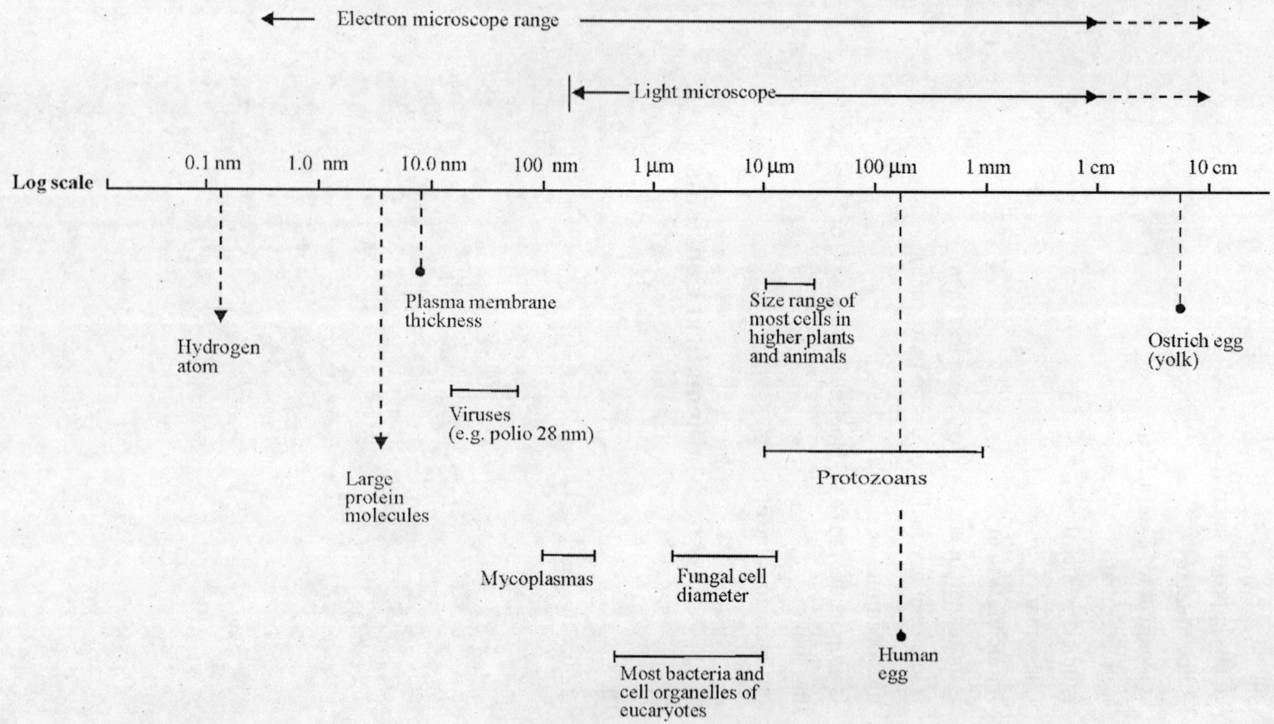

APPENDIX IX

THE NOBEL PRIZES

The Nobel Prizes are among the world's most venerated awards. Originally, they were a gesture of peace by a man whose discovery had unintentionally added to the destructive forces of warfare.

ALFRED BERNHARD NOBEL
(LT, 1833 – 1896)
The originator of the Nobel Prizes, the world's most venerated annual awards

A restless, melancholy man, **Alfred Bernhard Nobel** was one of the most imaginative of 19th century inventors. He was an engineer, a chemist, a philanthropist, a litterateur, an eccentric genius, and a linguist who could speak 6 languages. He was born on October 21, 1833 as 4th son to Immanuel and Caroline Nobel. His father Immanuel was a Swedish munitions expert, an inventor and engineer who had married Caroline Andrietta Ahlsell in 1827. The couple had 8 children, of whom only Alfred and 3 brothers reached adulthood. Alfred was prone to illness as a child, but he enjoyed a close relationship with his mother and displayed a lively intellectual curiosity from an early age. Nobel, who remained unmarried and had no childen, was closely associated with Countess Bertha Kinsky, who was a prominent Austrian pacifist. As a young engineer, he developed an interest in nitroglycerine, the oily substance that was 25 times more explosive than gunpowder. In 1863, Nobel obtained a patent for a detonator of mercury fulminate, and within 4 years he succeeded in solidifying nitroglycerine by mixing it with a type of sandy clay. The mixture was called **dynamite** (*dynamis*[G] = power).

Dynamite had a clear advantage over other explosives because it could be transported easily and handled without fear. It became an overnight success and was adapted to applications in mining, tunnel construction, and bridge and road building. Before long, it was being used in armanents on the battlefield. Nobel patented dynamite in 1867. He then turned his attention to military explosives, where the main requirement was for a smokeless powder, and achieved success with his combination of nitroglycerine with gun cotton known as **ballistite** (1888).

Nobel soon amassed a fortune through control of several European companies that produced dynamite. However, toward the end of his life, he became a pacifist and began speaking out against the use of dynamite in warfare. In 26 lines of his handwritten will, Nobel directed that the bulk of his estate should be used to award prizes that would promote peace, friendship, and service to humanity. It is certain that the awards he instituted reflect his lifelong interest in the fields of Physics, Chemistry, Physiology or Medicine, and Literature. However, there is also abundant evidence that his friendship with his one-time secretary Countess Bertha inspired him to establish the prize for peace. When Nobel first proposed a prize for world peace in 1883, it is interesting that he suggested it be given for the first 30 years only. He was optimistic that the world would realize the ravages of war and prize the peace dividend. Coincidentally, Hitler's putsch in Munich began just then. *Nobel prizes are not awarded posthumously.* The physics and chemistry prizes are awarded by committees chosen from the Swedish Academy of Sciences and the physiology prize by the Carolinian Institute, its medical equivalent.

After his childhood, Nobel only lived in Sweden for a few years (in the 1860s) to set up nitroglycerine manufacture, and again toward the end of his life when he took over the Bofors ironworks for conversion to a munitions factory. Otherwise he lived in France, in Italy and travelled constantly, "the wealthiest vagabond in Europe", he called himself. By 1895, Nobel had developed angina pectoris and he died of a cerebral hemorrhage at his village in San Remo, in Italy, on December 10, 1896.

After his death, the Governments of Sweden and Norway established **Nobel Memorial Prizes** in 5 categories: Chemistry, Physics, Physiology or Medicine, Literature, and Peace. A 6th category, Economics, was established by the Central Bank of Sweden in 1968 and first awarded in 1969 (Curiously enough, there is no provision for a Nobel prize in Mathematics). Each year, Nobel Laureates assemble at Oslo or Stockholm on December 10, the anniversary of Nobel's death. Each laureate receives a medallion, a scroll, and all or part of a cash award, currently valued at 7.8 million kronor (*i.e.*, about one million dollars) per category. Until 1980, the Nobel medal was made of 23-carat gold and weighed about 175 grams. Since then, the medallion is being made of 18-carat green-gold-plated with 24-carat gold.

The Nobel Medallion

In the background (on the left) is the sketch of Rabindranath Tagore, the 1913 Nobel Laureate for Literature from India.

Although Alfred Nobel made his fortune inventing and marketing dynamite, but is much more remembered the world over for bequeathing his legacy for the 6 annual prizes instituted in the 20th century. Only speculations can be made about the reasons for Nobel's establishment of the prizes that bear his name. He was reticent about himself, and he confided in no one about his decision in the months preceding his death. The most plausible assumption is that a bizarre incident in 1888 may have clicked the idea that culminated in his bequest for the Nobel Prizes. In fact, that his brother Ludwig Nobel, who himself made his millions in oil, had died while staying in Cannes, France. The French newspapers reported Ludwig's death but confused him with Alfred; one paper sported the headline "*Le marchand de la mort est mort*" ("The merchant of death is dead".). While other papers made critical comments on his explosive contribution for wartime destruction of man by man. Perhaps, Nobel established the prizes to avoid precisely the sort of posthumous reputation aired by the newspapers. Thus, the millionaire Alfred also provides the rare instance of a celebrity who read his own obituaries. In the end, Nobel came to dislike those very forces of military competition that had made him rich. Money never brought Nobel happiness, nor release him from loneliness and distrust. Such was a 'noble' Nobel !

Not surprisingly, then, a few countries have dominated in the Nobel stakes. Germany was first, and won the lion's share of science prizes until the 1930s, but the onset of Nazism and the flight of refugees to the U.S. brought about a clear American lead that has continued ever since. But even after World War II, many North American laureates have been immigrants. Up to 1986, United States citizens had won 47 physics prizes, 57 medicine and 29 chemistry prizes. Only in chemistry did Germany come anywhere near, with 24, while in physics and medicine/physiology Britain came a poor second with 20 in each, and Germany third with 15 and 11, respectively.

The first Nobel Prize were awarded in 1901. Among the recipients were :

1. **Wilhelm Konrad Roentgen** (LT, 1845-1923), a German physicist, for his discovery of x-rays which heralded the age of modern physics and revolutionized diagnostic medicine.
2. **Jacobus Henricus van't Hoff** (LT, 1852–1911), a Dutch physical chemist, for formulating the laws of chemical dynamics and of osmotic pressure.
3. **Emil (Adolf) von Behring** (LT, 1854-1917), a German biologist and the founder of the science of immunology, for his work on serum therapy, especially in developing a vaccine against tetanus and diptheria.
4. **Jean Henri Dunant** (LT, 1828-1910), a Swiss humanitarian, for peace. He was founder of the Red Cross, in 1864 (now Red Cross and Red Crescent).

Till 2005, six Indians have been honoured with the coveted Nobel Prize. They are :

1. **Rabindranath Tagore** (LT, 1861-1941), in 1913 for *Literature* for his classical work Gitanjali (1910). He was a Bengali poet, short-story writer, song composer, playwright, essayist, and painter (He published several books of poetry in the 1880s and completed Manasi (1890), a collection that marks the maturing of his genius), He was awarded a knighthood in 1915, but he repudiated it in 1919 as a protest against the Amritsar Massacre.
2. **Chandrashekhar Venkat Raman** (LT, 1888-1990), in 1930 in *Physics,* for his laws on light scattering, better known as 'Raman effect'.
3. **Har Gobind Khorana** (born, 1922), in 1960 in *Physiology or Medicine*, for deciphering genetic code . In 1970, he and his research team synthesized the first artificial copy of a yeast gene.
4. **Subrahmanyan Chandrasekhar** (LT, 1910-1995), in 1983 in *Physics* (along with William Alfred Fowler) for his theory on black holes. He is best known for what was later called as 'Chandrasekhar limit'. It is interesting to note that two of Subrahmanyan's former students T.D. Lee and C.N. Yang got 1957 Nobel Prize for discovering violations of the principle of parity. And the teacher, in a strange quirk of fate, had to wait for 26 years to join his students in the pantheon of Nobel Laureates.
5. **Mother Teresa** (LT, 1910-1997), in 1979 for *Peace,* in recognition of her missionary activities.
6. **Amartya Sen** (born, 1933), in 1998 in *Economic Science,* for his contributions to welfare economics and social choice and for his interest in the problems of society's poorest members. Sen was best known for his work on the causes of famine, which led to the development of practical solutions for preventing/limiting the effects of real/perceived shortages of food.

APPENDIX X

A SHORT CHRONOLOGY OF BIOCHEMISTRY

Certain key discoveries greatly influence the development of a particular discipline of study. This appendix presents some of the major landmarks in the progress of biology, biochemistry in particular and the individuals whose names are commonly associated with them. It is very difficult to appraise the historical development of any field of study. One investigator often gets the credit for an important discovery when many others should share in the honour. Many 'unlucky' others go in history unnoticed, while some of them get recognition posthumously. No one individual has a monopoly on ideas; advances in science are built on the work of many minds. Especially in recent years, important discoveries have resulted from the efforts of teams of investigators. Furthermore, because knowledge of a topic accrues over a period of time, it is often difficult to decide in which year the "discovery" *actually* occurred. In this brief outline, the student may be able to see some of the relationships that exist between one discovery and another. The discoveries are not completely isolated, as they may sometimes appear. One may also notice that fundamental discoveries in a particular branch of biology tend to be grouped fairly close together chronologically because that particular interest may have dominated the thought of the biological investigators at that time.

Biochemistry had its earliest origin in speculations of the role of air in the utilization of food and on the nature of fermentation. **Leonardo da Vinci** (1452-1519) was one of the first to compare animal nutrition to the burning of a candle. This sort of line of reasoning was further developed by **van Helmont** in 1648. However, the real history of biochemistry began during the late 18th century when the science of chemistry began to take form.

18th Century

1770-1774 **Joseph Priestley** –discovered oxygen and showed that it was consumed by animals and produced by plants.

1770-1786 **Karl Wilhelm Scheele** –isolated glycerol from olive oil, as well as citric, malic, lactic and urin acids from natural sources.

1780-1789 **Antoine Laurent Lavoisier** –demonstrated that animals require oxygen; recognized that respiration is oxidation; and first measured oxygen consumption by a human subject.

1783 **Lazaro Spallanzani**– deduced that protein digestion in the stomach is a chemical rather than mechnical process.

19th Century

1804 **John Dalton**–enunciated the atomic theory.

1828 **Friedrich Wöhler**–synthesized the first organic compound from inorganic components: urea from lead cyanate and ammonia.

1830–1840 **Justus von Liebig**–developed techniques of quantitative analysis and applied them to biological systems.

1833 **Payen** and **Persoz**–purified diastase (amylase) of wheat, showed it to be heat-labile, and postulated the central importance of enzymes in boilogy.

1837 **Jöns Jacob Berzelius**–postulated the catalytic nature of fermentation.

1838 **Gerardus Johannes Mulder**–carried out the first systematic studies of proteins; also proposed the name "proteine", later "protein," for the basic constituents of protoplasmic materials.

1842 **Herman von Helmholtz** and **J. R. von Mayer**–enunciated the First Law of thermodynamics and its applicability to living organisms.

1847 **Dubrunfaut**–degraded starch to maltose by diastase.

1850–1855 **Claude Bernard**–isolated glycogen from the liver and showed that it is converted into blood glucose.

1854–1864 **Louis Pasteur**–proved that fermentation is caused by microorganisms and demolished the spontaneous generation hypothesis. famous for his aphorism : *omne vivum e vivo* (every living thing from the living).

1864 **F. Hoppe-Seyler**–first crystallized a protein, hemoglobin which he also renamed.

1866 **Gregor Johann Mendel**–published his experiments leading to the principles of independent segregation and assortment of genes, thus formulated the first two laws of heredity.

1869 **Friedrich Miescher**–isolated DNA for the first time.

1877 **Friedrich Wilhelm Kuhne**–coined the term 'enzyme' and distinguished enzymes from bacteria.

1894 **Emil Fischer**–demonstrated the specificity of enzymes and the lock-and-key relationship between enzyme and substrate.

1895 **W. Roentgen**–observed that a new form of penetrating radiation, which he named x-rays, was produced when cathode rays (electrons) hit a metal target (Nobel Laureate, 1901).

1897 **Bertrand**–christened the term 'coenzyme'.

1897 **Eduard Buchner** and **Hans Buchner**–showed that cell-free extracts of yeast can ferment sugars to form CO_2 and $C_2H_5.OH$, laying the foundations of enzmology.

1897 **Christiaan Eijkman**–proved that beriberi is a dietary deficiency disease, and that a water-soluble component of rice polishings can cure it (Nobel Laureate, 1929).

20th Century

1902 **Emil fischer** and **Hofmeister**–demonstrated that proteins are polypeptides.

1903 **Carl Neuberg**–first used the term 'biochemistry'.

1904 **William M. Bayllis** and **Ernest H. Starling**–demonstrated the action of secretin, a hormone from the mucosa of the stomach; this marked the real birth of the science of endocrinology.

1905 **Knoop**–deduced the β-oxidation of falty acids.

1906 **Michael Tswett**–invented column chromatography, passing petroleum extracts of plant leaves through columns of powered chalk.

1906 **Frederick Gowland Hopkins** – explained dietary bya biochemical investigation of the lack of essential amino acids in the diet (Nobel Laureate, 1929).

1909 **Soren Sörensen**–showed the effect of pH on enzyme action.

1911 **Casimer Funk**–isolated crystals with vitamin B activity and coined the term 'vitamine'.

1913 **Leonor Michaelis** and **Maud Menten**–proposed a kinetic theory of enzyme action.

1914 **Edward C. Kendall**–isolated the hormone, thyroxine, which was later (1927) artificially synthesized by Harington.

1919 **Francis W. Aston** – discovered isotopes (Nobel Laureate, 1922).

1922 **Ruzicka**–recognized isoprene as the building block of many natural products, *i.e.*, propounded isoprene rule.

1922 **Elmer McCollum**–showed that lack of vitamin D causes rickets.

1925 **Walter Norman Howorth**–formulated sugars as pyranoses.

1925–1926 **George R. Minot, George William Murphy** and **G.H. Whipple**–discovered liver treatment of pernicious anemia (Nobel Laureates, 1934)

1925–1930 **Levene**–elucidated the structure of mononucleotides and showed that they are building blocks of nucleic acids.

1925–1930 **Theodore Svedberg**–developed the first analytical ultracentrifuge and used it to estimate the MW of hemoglobin as 68,000.

1926 **James B. Sumner**–first obtained crystals of an enzyme, uresase from extracts of jack beans (*Canavalia ensiformis*) proved it to be a protein (Nobel Laureate, 1946).

1926 **Jansen** and **Donath**–isolated vitamin B_1, thiamine from rice polishings.

1927 **Adolf Windaus**–showed ergosterol as a precursor of vitamin D.

1928 **Albert Szent-Györgyi**–isolated vitamin C (ascorbic acid) from paprika plant (Nobel laureate, 1937).

1928 **H. Wieland** and **Adolf Windaus** – established the structure of cholesterol molecule (Noble Laureates, 1927, 1928)

1929 **K. Lohamann, C. Fiske** and **Y. Subbarow**–isolated a labile phosphate (*i.e.*, ATP) and phosphocreatine from muscle extracts.

1929 **A. Butenandt** and **E.A. Doisy** – the first-ever isolation of a sex hormone (estrone) by these two investigators, independently (Doisy, Nobel laureate, 1943).

1930–1933 **John H. Northrop**–isolated crystalline pepsin and trypsin and proved their protein nature.

1931 **Linus Carl Pauling**–published his first essays on "The Nature of the Chemical Bond", detailing the rules of covalent bonding.

1932 **C. Glen King** and **W.A. Waugh**–isolated ascorbic acid from lemon juice.

1932 **Warburg** and **Christian**–discovered the 'yellow enzyme' (vitamin B_2 or riboflavin), which is a flavoprotein.

1933 **Hans Adolf Krebs** and **Kurt Hanseleit**–discovered the urea cycle.

1933 **Gustave Embden** and also **Otto Meyerhof**–demonstrated crucial intermediates in the chemical pathway of glycolysis and fermentation.

1933 **Arne W.K. Tiselius**–introduced electrophoresis for separating proteins in solution.

1933 **M. Goldblatt and U.S. von Euler** – discovered prostaglandins.

1933 **George Wald** – discovered the presence of vitamin A in the retina of the eye (Nobel lautrate, 1967).

1934 **Carl P. Henrik Dam** and **Edward A. Doisy** – isolated and synthesized vitamin K (Nobel Laureates, 1943).

1935 **Robert R. Williams et al**–deduced the structure of vitamin B_1 (thiamine).

1935 **Wendell M. Stanley**–first crystallized a virus, tobacco mosaic virus (Nobel laureate, 1946).

1935–1946 **Edward C. Kendall** and **P.S. Hench** – dicovered a hormone, cortisone (Nobel Laureates, 1950).

1935–1936 **Warburg** and **Van Euler**–isolated and determined the structure and action of pyridine nucleotides.

1937 **Hans Adolf Krebs**–postulated the citric acid cycle or Krebs cycle (Nobel laureate, 1953).

1937 **Carl F. Cori** and **Gerti T. Cori**–began their incisive studies of glycogen phosphorylase.

1937 **P. Köenig** and **Arne W. K. Tiselius** – developed electrophoresis.

1938	**Theodore Svedberg** – developed ultracentrifuge (Nobel laureate, 1926).
1939–1941	**Fritz A. Lipmann**–postulated the central role of ATP in the energy transfer cycle.
1939–1942	**Engelhardt** and **Ljubimova**–discovered ATPase activity of myosin.
1939–1946	**Albert Szent-Györgyi**–discovered actin and myosin and elucidated the role of ATP in muscular contraction.
1940	**George Beadle** and **Edward Tatum**–proposed the one gene-one enzyme hypothesis.
1941	**W.T. Astbury**–obtained the first x-ray diffraction pattern of DNA.
1941	**Carl F. Cori** and **Gerti T. Cori** – elucidated lactic acid metabolic cycle (Nobel Laureates, 1947).
1942	**A.J.P. Martin** and **R.L.M. Synge**–devolped partition chromatography, leading to paper chromatography two years later.
1943	**A. Claude –** isolated various cell constituents (such as mitochondria, microsomes and nuclei) by differential centrifugation (Nobel Laureate, 1974).
1943–1947	**Albert L. Lehninger**–showed the requirement of ATP and the stoichiometry of fatty acid oxidation.
1944	**Ostwald T. Avery, Colin M. Macleod** and **Maclyn McCarty**–demonstrated that bacterial transformation is caused by DNA, *i.e.*, DNA is genetic material.
1945	**Keith R. Porter** –analyzed the fine structure of endoplasimic reticulum.
1947–1950	**Fritz A. Lipmann** and **Kaplan**–isolated and characterized coenzyme A.
1948	**Alexander Ogston**–proposed a 3-point attachment hypothesis for citrate behaviour.
1948–1950	**Eugene P. Kennedy** and **Albert L. Lehninger**–discovered that the citric acid cycle, falty acid oxidation and oxidative phosphorylation take place in mitochondria.
1949	**Albert Szent-Györgyi**–showed that isolated myofibrils from skeletal muscle cells contract upon addition of ATP.
1949	**Ulf Svante von Euler** – isolated and identified norepinephrine as the neurotransmitter in the synpathetic nervous system (Nobel laureate, 1970).
1950–1953	**Erwin Chargaff**–discovered the base equivalences in DNA.
1951	**Fritz A. Lipmann** and **Lynen**–elucidated 'active' acetate *i.e.*, acetyl coenzyme A.
1951	**Albert L. Lehninger**–demonstrated that oxidative phosphorylation is coupled to electron transport in respiratory chain.
1951	**Linus Carl Pauling** and **Robert Corey**–proposed the structure of helical conformation of a chain of L-amino acids- the α-helix, and the structure of the β-sheet, both of which were later found in many proteins (Pauling, Twice Nobel Laureate, 1954 and 1962).
1952	**George E. Palade**–analyzed the fine structure of the mitochondrion (Nobel Laureate, 1974).
1952–1953	**Palade, Porter** and **Sjöstrand**–perfected thin sectioning and fixation methods for electron microscopy of intracellular structures.
1952–1954	**Paul Zamecnik et al**–discovered that ribonucleoprotein particles, later named ribosomes are the site of protein synthesis, also developed the first cell-free system to carry out protein synthesis.
1953	**James Dewey Watson** and **Francis Harry Compton Crick**–proposed the double-helix model of DNA, based on x-ray diffraction patterns obtain by Rosalind Franklin and Maurice Wilkins. (Watson, Crick and Wilkins, Nobel Laureates, 1962)
1953	**Vincent du Vigneaud**–carried out the first laboratory synthesis of the peptide hormones ocytocin and vasopressin.

1954 **Christian de Duve**–isolated lysosomes, and later peroxisomes by centrifugation (Nobel laureate, 1974).

1954 **Hugh Huxley**—proposed the sliding filament model for muscular contraction.

1955 **Frederick Sanger**–completed the analysis of amino acid sequence of bovine insulin, the first protein to the sequenced (Twice Nobel Laureate, 1958 and 1980).

1955–1957 **Arthur Kornberg** and **Severo Ochoa** – synthesized DNA (Kornberg) and RNA (Ochoa) artificially outside of cells, *i.e.*, in vitro (Nobel Laureates, 1959)

1956 **Dorothy Crowfoot Hodgkin**–determined the structure of coenzyme B_{12}.

1956 **Christian B. Anfinsen** and **White**–concluded that the three-dimensional conformation of proteins is specified by their amino acid sequence.

1957 **Earl Wilbur Sutherland** and **T.W. Rall**–discovered cyclic adenosine-3′5′ monophosphate (=cyclic AMP), an intracellular mediating agent, found in all living animal tissues (Sutherland, Nobel Laureate, 1971).

1957 **Hans Adolf Krebs** and **H.L. Kornberg**–discovered the glyoxylate cycle.

1957 **Mathew S. Meselson, Frank W. Stahl** and **Vinograd**–developed equilibrium density-gradient centrifugation in $CsCl_2$ solutions for separating nucleic acids.

1957 **Arthur Kornberg**–discovered DNA polymerase, the enzyme now used to produce labelled DNA probes.

1957 **Melvin Calvin** – elucidated chemical pathways in photosynthesis by using radioactive carbon 14 (Nobel Laureate, 1961)

1958 **A.F.J. Butenandt et al** – chemically analyzed the first pheromone-the sex attractant substance of silk moths (*Bombyx mori*) and named it as bombykol

1958 **Meselson** and **Stahl** – demonstrated that DNA is replicated by a semiconservative mechanism.

1959–1960 **Rosalyn Yalow** and **S.A. Berson** – developed the technique of radiommunoassay (Yalow, Nobel Laureate, 1977).

1960s–1970s **S. Bergstöm, B. Samuelsson** and **J. Vane** – characterized the prostaglandins, the chemical transmitters of intracellular and intercellular signals, Bergström is credited with isolating prostaglandins and determining their structures; Samuelsson with determining their biosynthesis and metabolism; and Vane for discovering prostacyclin (Nobel Laureates, 1982).

1960 **John C. Kendrew**–described the first detailed 3-'D' structure of a protein (sperm whale myoglobin) to a resolution of 0.2 nm.

Max F. Perutz–proposed a lower-resolution structure of larger protein hemoglobin..

1960 **William Moore** and **Stanford Stein**–determined the amino acid sequence of ribonuclease.

1961 **Marshall W. Nirenberg** and **J.H. Matthaei**–deciphered the genetic code (Nirenberg, Nobel Laureate, 1968).

1961 **Marshall W. Nirenberg** and **Severo Ochoa**–decoded the base code of nucleic acids.

1961 **Francǫis Jacob** and **Jacques Monod**–presented a model for the regulation of gene activity called operon model (Nobel Laureates, 1965).

1961 **Peter Mitchell**–proposed chemiosmotic coupling hypothesis of ATP formation.

1961 **Cross et al**–isolated gibberllins and elucidated their chemical structure.

1961 **Francǫis Jacob, Jacques Monod** and **Changeux**–proposed a theory of the function and mechanism of allosteric enzymes.

1962 **Efraim Racker et al**–isolated F_1 component of ATP synthetase from mitochondria and reconstituted oxidative phosphorylation in submitochondrial vesicles.

1963 **D.S. Letham**–coined the term 'cytokinins; also isolated zeatin.

1963 **C.F. Eagles** and **P.R. Wareing**–isolated a plant growth inhibitor, dormin.

1963 **John Cairns**–demonstrated the existence of circular DNA in a bacterium.

1965 **Robert W. Holley et al**–determined for the first time the base sequence of a nucleic acid (Nobel Laureate, 1968).

1966 **Marshall W. Nirenberg, Severo Ochoa** and **Arthur Kornberg**– elucidated the genetic code.

1966 **David C. Phillips et al**–described the 3-'D' structure of lysozyme, the first enzyme to be analyzed in detail.

1966 **Marshall W. Nirenberg** and **Har Gobind Khorana**–presented a complete dictionary of nucleotide triplets (Nobel Laureates, 1968).

1966 **Francis Harry Compton Crick**–proposed the wobble hypothesis.

1970 **David Baltimore, Howard M. Temin** and **Renatto Dulbecco**–studied the interaction between tomour viruses and the genetic material of the cell; discovered the reverse transcription (Baltimore and Temin, Nobel Laureates, 1975).

1970 **Gerald M. Edelman** and **Rodney R. Porter** – worked out the structure of gamma globulin (Nobel Laureates, 1972)

1971 **W.Y. Cheung** – discovered calmodulin.

1971–1973 **A.M. Cormack** and **G.N. Hounsfield** – invented the computer-assisted tomography (CAT) scanner (Nobel Laureates, 1979).

1972 **M. Brown** and **J. Goldstein** – discovered the LDL receptor on cell surfaces; also explained the intracellur regulation of cholesterol metabilism (Nobel Laureates, 1985).

1975 **Dobberstein** and **Günter Blobel**–demonstrated protein translocation across membranes in a cell-free system (Blobel, Nobel Laureate, 1999).

1976 **Walter Gilbert, A. Maxam,** and **Frederick Sanger** – devisd methods for determining base sequence in DNA (Gilbert and Sanger, Nobel Laureates, 1980).

1976 **J. Michael Bishop** and **Harold E. Varmus** – discovered oncogenes (Nobel Laureates, 1989)

1977 **Roger C.L. Guillemin** and **Andrew V. Schally**–synthesized peptide hormones.

Harrison and Rossman–determined the structure of 2 small spherical viruses.

1978 **Susumu Tonegawa**–demonstrated DNA splicing for an immunoglobulin gene.

1981 **Steitz**–determined the structure of CAP protein.

1982 **Stanley B. Prusiner** – discovered a new biological principle of infection, under the name 'prion theory' (Nobel Laureate, 1997).

1982–1984 **Sidney Altman** and **Thomas Robert Cech** – dicovered RNA catalysis (Nobel laureates, 1989).

1984 **Schwartz** and **Cantor**–developed pulsed field gel electrophoresis for the separation of very large DNA molecules.

1985 **Kary B. Mullis** – invented the polymerase chain reaction (PCR), a means of amplifying tiny amounts of DNA (Nobel Laureate, 1993).

1988 **Elion** and **Hitchings** – designed and synthesized therapeutic purines and pyrimidines.

1989 **Snyder et al** – purified and reconstituted the inositol-1,3,4-P_3 receptor.

1991 **Roger Beachy** – demonstrated that plants can acquire resistance to viral pathogens.

[illegible]

[illegible]

[illegible]

[illegible]

[illegible]

[illegible]

[illegible]

[illegible]

[illegible]

[illegible]

[illegible]

[illegible]

Walter Gilbert, [illegible] and Frederick Sanger [illegible]

[illegible]

[illegible]

[illegible]

[illegible]

[illegible]

[illegible]

[illegible]

[illegible]

[illegible]

INDEX-I

AUTHOR INDEX

Numbers in Roman normal type indicate the page number from the work of the author cited. Numbers followed by a D or T indicate the diagram or table, respectively, contributed by the author cited. Numbers followed by a C indicate the life history and contributions, presented in a box, of the author cited.

A

B

C

I

J

K

L

M

N

O

P

Q

R

S

T

U

V

W

X

Y

Z

INDEX–II

SUBJECT INDEX

Numbers in boldface type refer to pages on which the subject is illustrated or especially described.

A

B

C

F

G

H

I

J

K

L

M

N

O

P

Q

R

S

T

X

Y

Z